Lecture Notes in Networks and Systems

Volume 91

The series "Lecture Notes in Networks and Systems" publishes the latest developments in Networks and Systems—quickly, informally and with high quality. Original research reported in proceedings and post-proceedings represents the core of LNNS.

Volumes published in LNNS embrace all aspects and subfields of, as well as new challenges in, Networks and Systems.

The series contains proceedings and edited volumes in systems and networks, spanning the areas of Cyber-Physical Systems, Autonomous Systems, Sensor Networks, Control Systems, Energy Systems, Automotive Systems, Biological Systems, Vehicular Networking and Connected Vehicles, Aerospace Systems, Automation, Manufacturing, Smart Grids, Nonlinear Systems, Power Systems, Robotics, Social Systems, Economic Systems and other. Of particular value to both the contributors and the readership are the short publication timeframe and the world-wide distribution and exposure which enable both a wide and rapid dissemination of research output.

The series covers the theory, applications, and perspectives on the state of the art and future developments relevant to systems and networks, decision making, control, complex processes and related areas, as embedded in the fields of interdisciplinary and applied sciences, engineering, computer science, physics, economics, social, and life sciences, as well as the paradigms and methodologies behind them.

**** Indexing: The books of this series are submitted to ISI Proceedings, SCOPUS, Google Scholar and Springerlink ****

More information about this series at http://www.springer.com/series/15179

Elena G. Popkova · Bruno S. Sergi
Editors

The 21st Century from the Positions of Modern Science: Intellectual, Digital and Innovative Aspects

Springer

Editors
Elena G. Popkova
Plekhanov Russian University of Economics
Moscow, Russia

Bruno S. Sergi
Davis Center for Russian
and Eurasian Studies
Harvard University
Cambridge, MA, USA

ISSN 2367-3370 ISSN 2367-3389 (electronic)
Lecture Notes in Networks and Systems
ISBN 978-3-030-32014-0 ISBN 978-3-030-32015-7 (eBook)
https://doi.org/10.1007/978-3-030-32015-7

This Springer imprint is published by the registered company Springer Nature Switzerland AG
The registered company address is: Gewerbestrasse 11, 6330 Cham, Switzerland

Introduction

The 21st century opened a way into a new millennium, bringing deep transformation processes in almost all spheres of socioeconomic systems' activities. The most important task of the modern science is studying and explaining these processes, forecasting the tendencies of their development in the future and developing the methodological and practical recommendations for successful and highly effective management in the interests of humanity.

Under the influence of the technological progress in the 21st century, three tendencies of functioning and development of socioeconomic systems appeared. The tendency of intellectualization of socioeconomic systems is connected to the increase of significance of intellectual resources for production and distribution of goods and services in the economy. There are "knowledge-intensive" (usually, hi-tech) spheres of economy, in which intellectual resources are the key factor of production. Hi-tech products are an important direction of domestic production and export, as it determines the digital competitiveness of the economy.

The sphere of science and education has an important mission of creating intellectual resources for economy, including digital personnel, the demand for which grew recently due to the formation of the digital economy. This requires the development and application of new educational methodologies and technologies, including the development of remote (digital) education. The sphere of science deals with the creation of AI, and entrepreneurship uses intellectual technologies of decision support. The government faces high load on the protection of rights for the objects of intellectual property.

The tendency of digitization is manifested in all components of the noosphere. Under the influence of popular and widely used digital technologies, a modern human is subject to serious transformations: the type of thinking and algorithm of actions in various situations changes, new ideology forms, etc. This leads to the emergence of digital human as the main member of the digital society and the digital economy. Intensive innovative development of the economic systems takes place simultaneously, leading to a new technological revolution, which will result in emergence of intellectual technologies.

The tendency of innovative development of socioeconomic systems is connected to transition to a new quality of economic growth. Innovations are a driving force of the development of the modern economy. In the digital society, which is developing on the platform of "knowledge economy," innovations (new, unique knowledge, technologies and information) are very valuable, being a product (not only a production factor, as was the case in the twentieth century). At the same time, innovations have a contradictory influence on the modern economic systems.

On the one hand, they could cause the anti-crisis effect, increasing economy's sustainability against the negative influence of the internal and external factors and reducing its cyclic character. On the other hand, development and implementation of innovations are connected to high risk and could destabilize the economy. This happened in the early 21st century, leading to the global financial crisis of 2008. This contradiction requires attention to the issues of managing the innovative development of the modern socioeconomic systems.

This volume studies all the above tendencies and compiles the complex characteristics of the 21st century through the prism of the processes of intellectualization, digitization and innovative development of socioeconomic systems. The volume contains the best works that were presented at the 10th international conference that was organized by the autonomous non-profit organization "Institute of Scientific Communications"—"The 21st Century from the Positions of Modern Science: Intellectual, Digital and Innovative Aspects." The conference took place on May 23–24 in Nizhny Novgorod, Russia, with support from Minin Nizhny Novgorod State Pedagogical University.

The volume contains the materials on the following topics (formed according to the conference sessions):

1. Philosophical settings of socioeconomic development in the 21st century;
2. Digital society as the basis of development of economy in the 21st century;
3. Technological revolution in the 21st century: factors, conditions and perspectives;
4. Artificial intelligence: reality of the 21st century;
5. The role and value of intellectual resources in the 21st century and the scientific and educational platform for their training and development;
6. The prime of the innovative economy in the 21st century: expectations and reality.

One of the most interesting works in the volume is the paper by Elena G. Popkova and Kantoro Gulzat "Technological Revolution in the 21st Century: Digital Society vs. Artificial Intelligence", which is devoted to the issues of social adaptation to the conditions of the digital economy. The authors formulate and successfully solve the problem of opposition of the digital society and AI and present an economic view on the opposition of humans and machines in the conditions of the digital economy.

Another interesting paper is the work by Olga V. Konina "Scientific and Methodological Foundations of an Innovative Company Management." She offers a new approach to classification of the modern entrepreneurship according to the criterion of application of new (breakthrough, digital) technologies and separates the traditional and innovative entrepreneurship. She also substantiates the necessity for applying a new scientific and methodological approach to studying and managing innovative entrepreneurship and develops this approach.

One of the most progressive and creative works is a multi-disciplinary research by Elena B. Belik, Elena S. Petrenko, Georgy A. Pisarev and Anna A. Karpova "Influence of Technological Revolution in the Sphere of Digital Technologies on the Modern Entrepreneurship," which is performed at the joint of the information and communication technologies and the theory of entrepreneurship (economics). The authors prove that the influence of the technological revolution in the sphere of digital technologies, connected to transition to Industry 4.0, leads to transformation processes in the system of the modern entrepreneurship, which results in formation of its new type—entrepreneurship 4.0, which requires new organizational and managerial methods and approaches and requires new methods of its analysis, monitoring and regulation.

The leading studies in this volume include the work by Mansur F. Safargaliev, Yuliana A. Kitsay, Elena N. Egorova and Lilia V. Ermolina "Modernization of Modern Entrepreneurship on the Basis of Artificial Intelligence," in which the researchers outline the perspectives of using AI in the activities of modern companies and offer a new model of entrepreneurship that is based on AI. The authors perform a systemic analysis of the new form from the positions of its advantages and risks and offer the scientific and practical recommendations for its practical implementation.

Also, the editors could be proud of having such chapter as "Electronic-Digital Smart Contracts: Modernization of Legal Tools for Foreign Economic Activity" by Agnessa O. Inshakova, Alexander I. Goncharov and Daniil A. Salikov. The authors analyze the modern experience and outline the future perspectives of application of a new economic and legal tool of conducting foreign economic activities—electronic-digital smart contracts, as well as develop the normative and legal provision that is necessary for wide practical application of this tool in the interests of the development of the foreign economic activities.

All chapters in this volume are devoted to the current topics and are aimed at solving the problems of optimization of the influence of technological progress on the socioeconomic systems in the 21st century.

Contents

Philosophical Settings of Socio-Economic Development in the 21st Century

Electronic-Digital Smart Contracts: Modernization of Legal Tools for Foreign Economic Activity 3
Agnessa O. Inshakova, Alexander I. Goncharov, and Daniil A. Salikov

The Factor Model Study of the Development of Economic Systems 14
Tatyana N. Ivashchenko, Nadezhda V. Mironenko, Natalya E. Popovicheva, and Yuliia A. Zviagintceva

Digitalization as a Method to Enhance the Mechanism of Procedural Protection of Civil and Financial Rights 24
Olga V. Gritsai, Vera V. Kotlyarova, Ekaterina V. Mikhailova, Andrei V. Yudin, and Ramil' Z. Yusupov

Circadian Genes: Information and Legal Issues 30
Andrey A. Inyushkin, Alexey N. Inyushkin, Valentina D. Ruzanova, Vladimir I. Belyakov, and Elena S. Kryukova

Strategic Priorities of Digitalization of the Russian Economy 38
Elena Yu. Merkulova, Sergey P. Spiridonov, and Vera I. Menshchikova

Digitalization of Labor Relations in Agribusiness: Prospects of Legal Regulation and Labor Rights Protection 47
Apollinariya A. Sapfirova, Victoria V. Volkova, and Anna V. Petrushkina

Augmented Reality as Marketing Strategy in the Global Competition . . . 54
Marina V. Vilkina and Olga V. Klimovets

Computer Information as the Target and the Means of Commission of Crimes According to Criminal Legislation of the Russian Federation 61
Andrey V. Shulga and Vladimir A. Yakushin

The Algorithm of Creation of Territories of Rapid Socio-Economic Development in the Digital Economy 68
Tatiana A. Zabaznova, Elena V. Patsyuk, Natalia V. Shchukina, Svetlana E. Karpushova, and Olga A. Surkova

Appellations of Origin of Goods as a Tool of Socio-Economic Development and Preservation of Cultural Diversity of Russia in the Context of Globalization 77
Irina V. Samsonova, Elena V. Andreeva, Anna A. Neizvestnykh, and Vasilii V. Skriabin

Innovative Approaches to Corporate Financial Management in Business Systems 83
Liliya S. Valinurova, Tatyana B. Leibert, and Elvira A. Khalikova

The Transformation of Legal Personality as a Means of Achieving the Legal Bond Between the Participants of Tax Relations with the Use of Digital Technologies 96
Anastasia S. Alimbekova, Marina A. Katkova, Svetlana V. Rybakova, Elena N. Pastushenko, and Vasily V. Popov

Cognitive Modeling of the Mechanism of Partnership of Business Entities with Public Authorities 104
Veronica S. Epinina, Iakow I. Kayl, Roman M. Lamzin, Anzhelika N. Syrbu, and Yurij M. Kvintyuk

Digital Society as the Basis of Development of Economy in the 21st Century

Problems of Virtualization and Internetization of Social Space 119
Elena V. Gryaznova, Alexander A. Vladimirov, Svetlana M. Maltceva, Aleksey G. Goncharuk, and Nikolai V. Zanozin

Innovations and Motivation of Personnel as the Main Drivers of Development of Industrial Enterprises 125
Nikolay A. Zhdankin, Vladimir M. Suanov, and Bahrom K. Sharipov

Methodological Approach to Analysis of Management Systems Using the Graph Theory in the Digital Economy 134
Elena N. Letiagina, Yury V. Trifonov, Elena Y. Trifonova, Alexander N. Vizgunov, and Julia A. Grinevich

Expansion of Personal Development Space of a Student in the Sphere of Design 142
Elena E. Sherbakova, Maria V. Lvova, Evgenia K. Zimina, Irina S. Aboimova, and Alexandra A. Kulagina

The Impact of Digital Technologies on Various Activity Spheres and Social Development 149
Elena V. Yashkova, Nadezhda L. Sineva, Sergey V. Semenov, Olga I. Kuryleva, and Anastasia O. Egorova

Research of the Tools of Influence on the Behavior of Market Subjects 156
Alexander P. Garin, Irina A. Artashina, Svetlana A. Zaitseva, Zhanna V. Smirnova, and Yaroslav S. Potashnik

Organizational and Economic Foundations of the Management of the Investment Programs at the Stage of Their Implementation 163
Ekaterina P. Garina, Elena V. Romanovskaya, Natalia S. Andryashina, Victor P. Kuznetsov, and Elena V. Shpilevskaya

Formation of the Determinants of Socio-Economic Development of Territories 170
Ekaterina P. Garina, Natalia S. Andryashina, Victor P. Kuznetsov, Elena V. Romanovskaya, and Mahomed Z. Muzayev

Foresight Study of the Limits of Development and Application of the Modern Tools of GR Management in View of the Socio-Economic Specifics of Territories' Development 179
Iakow I. Kayl, Anna V. Shokhnekh, Roman M. Lamzin, Anzhelika N. Syrbu, and Yuliana Yu. Elsukova

Effective Technologies and Methods of GR Management that Stimulate Increase of the Level of Socio-Economic Development of Territories and Quality of Life 191
Iakow I. Kayl, Anna V. Shokhnekh, Veronica S. Epinina, Roman M. Lamzin, and Marina V. Samsonova

Special Economic Zone of the Tourist and Recreation Type as a Tool of Regional Policy 208
Natalia S. Beskorovaynaya, Elena V. Khokhlova, Ilya V. Ermakov, Zukhra S. Dotdueva, and Vitaly V. Lang

The Problems of Modernization of the Social and Economic Development in the Light of the Scientific and Technological Revolution 219
Irina V. Novikova, Nataliia N. Muraveva, Lilianna Y. Grazhdankina, Nina Y. Zhelnakova, and Larisa A. Tronina

Digitalization of Kyrgyz Society: Challenges and Prospects 229
Chinara R. Kulueva, Mirlanbek B. Ubaidullayev, Kurmanbek I. Ismanaliev, Victor P. Kuznetsov, and Elena V. Romanovskaya

Technological Revolution in the 21st Century: Factors, Conditions and Perspectives

Influence of Technological Revolution in the Sphere of Digital Technologies on the Modern Entrepreneurship 239
Elena B. Belik, Elena S. Petrenko, Georgy A. Pisarev, and Anna A. Karpova

The Transformation of Motor Vehicle Taxation in a Climate of Digitization of the Economy of the Russian Federation 247
Irina V. Gashenko, Yulia S. Zima, and Elena N. Makarenko

Teleology of Scientific and Technical Progress 254
Igor V. Astafyev and Dmitry P. Sokolov

Potential Horizons of Technological Development of Russia 264
Lyudmila V. Oveshnikova, Elena V. Sibirskaya, Evgeniya P. Tenetova, Natalya P. Kuznetsova, and Mariya O. Grigoryeva

Technology of Secure Remote Voting via the Internet 275
Marina L. Gruzdeva, Zhanna V. Smirnova, Elena A. Chelnokova, Olga T. Cherney, and Zhanna V. Chaikina

Transformation of the Industrial Potential in the Digital Economy 282
Elena V. Dudina, Ekaterina I. Mosina, Elena E. Semenova, Maria A. Stepanova, and Nataljya A. Baturina

Scientific Discussions of FES Russia on Digitalization of Economy of the 21st Century 291
Konstantin V. Khartanovich, Tatyana V. Starikova, Alexander V. Milenky, and Natalya E. Tikhonyuk

Risk Management System in the Engineering Design Practice 298
Ekaterina P. Garina, Natalia S. Andryashina, Bolotbek M. Seitov, Alexander P. Garin, and Gulzat T. Tashmurzaeva

Problems and Objectives of the Regional Socio-Demographic Potential Development 306
Yuri N. Lapygin, Denis Yu. Lapygin, and Pavel Yu. Makarov

Selection of Preferred Information Support of the Management Enterprise in the Sphere of Services Using the Method of Analysis of Hierarchies 316
Sergey N. Yashin, Sergey A. Borisov, Arthur E. Ambartsoumyan, Natalia A. Shcherbak, and Yana V. Batsyna

Motives Behind GCC Sovereign Wealth Funds' Investment in Foreign Securities and Their Role in Regional Capital Markets' Integration 325
Ibrahim A. Onour

Artificial Intelligence: Reality in the 21st Century

Technological Revolution in the 21st Century: Digital Society vs. Artificial Intelligence ... 339
Elena G. Popkova and Kantoro Gulzat

Modernization of Modern Entrepreneurship on the Basis of Artificial Intelligence ... 346
Mansur F. Safargaliev, Yuliana A. Kitsay, Elena N. Egorova, and Lilia V. Ermolina

Intellectual Production and Consumption: A New Reality of the 21st Century ... 353
Aleksei A. Shulus, Elena S. Akopova, Natalia V. Przhedetskaya, and Ksenia V. Borzenko

Imagination, Invention and Internet: From Aristotle to Artificial Intelligence and the 'Post-human' Development and Ethics ... 360
Qerim Qerimi

Application of Smart-Contracts When Using the Exclusive Rights to Results of Intellectual Activity ... 372
Agnessa O. Inshakova, Tatyana V. Deryugina, and Evgeny Y. Malikov

The Role and Value of Intellectual Resources in the 21st Century and the Scientific and Educational Platform for their Training and Development

The Flagship University's Model in Terms of Digitalization: The Case of Industrial University of Tyumen as a Center of Strategic Decisions in the Field of Smart-City, IoT/IIoT and Big Data ... 387
Viktoria A. Lezer, Liubov N. Shabatura, and Igor A. Karnaukhov

Investments in Human Capital as a Key Factor of Sustainable Economic Development ... 397
Angelika P. Buevich, Svetlana A. Varvus, and Galina A. Terskaya

Functional Model of the Head of the Third-Generation University Department ... 407
Irina V. Neprokina

Priority of Intellectual Resources for Development of Digital Entrepreneurship ... 415
Irina V. Mukhomorova, Aydarbek Giyazov, Gulzat K. Tashkulova, and Nurgul K. Atabekova

The Role of Managerial Competence of an Executive in Improving the Quality of Pre-school Educational Organization 422
Natalia V. Belinova, Irina B. Bicheva, Larisa V. Krasilnikova, Tatyana G. Khanova, and Anna V. Hizhnaya

Implementation of the Division Model of Pedagogical Labor in the Teacher Training System of a New Type 430
Tatyana K. Belyaeva, Evgeniy E. Egorov, Tatyana K. Potapova, Tatyana L. Shabanova, and Mikhail Y. Shlyakhov

Subjective Representation Study of University Teachers About the Significance of Changes in Higher Education 439
Elena G. Gutsu, Nadezhda N. Demeneva, Elena V. Kochetova, Oksana V. Kolesova, and Tatyana V. Mayasova

Development of Professional Creativity of Teachers in the System of Professional Safety Culture of Children in Transport 446
Galina S. Kamerilova, Marina A. Kartavykh, Elena L. Ageeva, Irina A. Gordeeva, and Marina A. Veryaskina

Forecasting the Development of Professional Education 452
Svetlana M. Markova, Svetlana A. Tsyplakova, Catherine P. Sedykh, Anna V. Khizhnaya, and Olga N. Filatova

Project Activities of University Students by Means of Digital Technologies 460
Elvira K. Samerkhanova, Lyudmila N. Bakhtiyarova, Aleksandr V. Ponachugin, Elena P. Krupoderova, and Klimentina R. Krupoderova

Information-Project Technology for the Formation of General Competencies of Students by Means of Electronic Information and Educational Environment 468
Alexandra A. Tolsteneva, Valeria K. Vinnik, Marina V. Lagunova, Anna A. Voronkova, and Natalia D. Zhilina

Intellectual Resource as a Factor of Ensuring National and Cultural Security in the Conditions of the Training Course "Teacher of Russian as a Foreign Language" 477
Elena M. Dzyuba, Victoria T. Zakharova, Natalia M. Ilchenko, Anna L. Latuhina, and Tatyana N. Sheveleva

The Role of Scientific and Educational Platform in Formation of the Innovative Economy of Kyrgyzstan: Foreign Experience, Realities, and Prospects 484
Chinara R. Kulueva, Inabarkan R. Myrzaibraimova, Gulbar B. Alimova, Victor P. Kuznetsov, and Elena V. Romanovskaya

Success of the Innovative Economy in the 21st Century: Expectations and Reality

Scientific and Methodological Foundations of an Innovative Company Management 495
Olga V. Konina

Transformation of Business Models in Terms of Digitalization 503
Olga B. Digilina and Irina B. Teslenko

Formation and Development of Regional Innovation Systems 510
Liubov V. Plakhova, Natalia V. Zakharkina, Ivan V. Ilin, Viktor P. Bardovskii, and Nikolay V. Pokrovskiy

Innovative Import Substitution in Russia and the World: A Comparative Analysis 521
Alla V. Litvinova, Natalya S. Talalaeva, and Marina V. Ledeneva

Innovation and Investment Potential of Regions as a Vector for Their Development in the 21st Century 534
Evgenia M. Kolmakova, Ekaterina M. Kolmakova, and Irina D. Kolmakova

Scenario of Hi-Tech Growth of Innovative Economy in Modern Russia 544
Anna I. Pakhomova, Rustam A. Yalmaev, Elena V. Belokurova, and Larisa V. Shabaltina

Innovative Economy in the 21st Century: Contradiction and Opposition of Developed and Developing Countries 552
Vladislav A. Shalaev, Elena A. Vechkinzova, Anna L. Shevyakova, and Oksana Y. Vatyukova

Structuring the Added Value of Biomedicine Products in the Innovative Economy 561
Natalia G. Varaksa, Sergey A. Alimov, Maria S. Alimova, and Victor A. Konstantinov

Organizational Innovation in Cost Management as a Factor of Increasing the Competitiveness of the Enterprise 570
Alexander N. Vizgunov, Yuri V. Trifonov, Anna A. Abrosimova, Tatyana E. Maslova, and Pavel S. Shalabaev

Industrial Policy: Peculiarities of Understanding and Dependence on the Level of Implementation 576
Oleg V. Trofimov, Andrey P. Kostyrev, Lyudmila V. Strelkova, Yuliya A. Makusheva, and Tatyana V. Trofimova

Methodological Aspects of Assessing the Creditworthiness of Municipalities as an Important Condition for Improving the Efficiency of Bank Lending Operations and Managing Municipal Borrowings 583
Nadezhda I. Yashina, Svetlana D. Makarova, Natalia N. Pronchatova-Rubtsova, Igor A. Makarov, and Oksana I. Kashina

Stress-Testing at the Bank of Albania: Methodology of Approaches and the Quality of Forecasting 590
Ela Golemi and Vasilika Kota

Macro Determinants of Real Exchange Rates: Albanian Case 608
Ermira Kalaj and Ela Golemi

Analysis of the Effects of Macroprudential Measures on GDP's Trend – Simulation Using a Macro Financial Model for Albania 615
Ela Golemi

Agricultural Lease as a Perspective Mechanism of Development of Infrastructure of Entrepreneurship in the Agricultural Machinery Market 624
Tatiana N. Litvinova

State Support for Digital Logistics 631
Vera V. Borisova, Tamila S. Tasueva, and Bella K. Rakhimova

Basic Approaches to the Understanding of Cooperation and Corporation: Historical and Philosophical Aspect 639
Olga Bezgina, Olga Evchenko, and Tatiana Ivanova

Analytical Procedures for Assessing the Risks of Introducing Innovative Technologies into the Organization's Activities 654
Maxim M. Kharlamov, Tatyana S. Kolmykova, Tatiana O. Tolstykh, Evgenia S. Nesenyuk, and Ekaterina P. Garina

Transformational Period of Russian Development in the Digital Economy 663
Svetlana N. Kuznetsova, Victor P. Kuznetsov, Elena P. Kozlova, Yaroslav S. Potashnik, and Sergey D. Tsymbalov

Methodological Bases of the Assessment of Sustainable Development of Industrial Enterprises (Technological Approach) 670
Elena P. Kozlova, Victor P. Kuznetsov, Ekaterina P. Garina, Elena V. Romanovskaya, and Natalia S. Andryashina

Understanding the Challenges in the Research and Innovation Ecosystem in India 680
Sandeep Goyal, Sumedha Chauhan, and Amit Kapoor

Author Index 691

Philosophical Settings of Socio-Economic Development in the 21st Century

Electronic-Digital Smart Contracts: Modernization of Legal Tools for Foreign Economic Activity

Agnessa O. Inshakova(✉), Alexander I. Goncharov, and Daniil A. Salikov

Volgograd State University, Volgograd, Russia
ainshakova@list.ru, goncharova.sofia@gmail.com, daniilsalikov@mail.ru, gimchp@volsu.ru

Abstract. The article explores the possibilities of improving the methods of managing foreign economic activity and increasing its efficiency with the help of smart contracts. The advantages of the blockchain technology and the legal contradictions associated with the use of smart contracts in the practice of economic entities are analyzed. The methodological basis consists of the authors' materialistic, positivistic worldview, general scientific methods of cognition, a number of special legal methods. The purpose of the study is to search for improvements in the legal support of electronic interaction of participants in foreign economic activity, registration of facts of property relations between them, which can be executed by smart contracts. As a result of the study, it was substantiated that a smart contract can neutralize many civil law problems, including questions about applicable law, judicial jurisdiction, verification of counterparty powers. Based on the analysis, it was substantiated the conclusion that neither at the domestic (national) level of legislation, nor at the international level within the framework of universal or regional agreements, a coherent systemic matter of legal norms regulating the conditions, procedure and limits of use of smart contracts was created.

Keywords: Smart contract · Blockchain · Distributed registry · Electronic registration · Foreign economic activity · International trade transactions

JEL code: K10 · K15 · K24

1 Materials

Justification of the need to modernize the legal instruments of foreign economic activity along the path of active use of smart contracts in modern economic activities of transnational organizations is based on a complex of sources. The authors studied the Federal Law Bill No. 419059-7 "On digital financial assets" dated January 25, 2018; in addition, the state program "Digital Economy of the Russian Federation", approved by the Russian Federation Government Order No. 1632-r dated July 28, 2017, was studied from regulatory documents. The scientific and theoretical basis was formed by studies

E. G. Popkova and B. S. Sergi (Eds.): ISC 2019, LNNS 91, pp. 3–13, 2020.
https://doi.org/10.1007/978-3-030-32015-7_1

of foreign (Giancaspro M., Geiregat S.) and Russian (Braginskij M.I., Vitrjanskij V.V., Inshakova A.O., Goncharov A.I., Savelyev A.I.) scientists.

The regulatory framework is based on the provisions of the positive law contained in national acts, namely, in the Civil Code of the Russian Federation, the Tax Code of the Russian Federation, the Federal Law “On Export Control”, the Federal Law “On Electronic Signature”, as well as in an international agreement - Vienna Convention 1980. The study also involved the codes of trade customs (Lexmercatoria), first of all, the rules of interpretation of the international terms INCOTERMS-2010 and the Principles of the International Commercial Contracts of UNIDROIT.

Theoretical approaches to the legal nature of a smart contract, built on its understanding, as a legal construction, as a way of fulfilling obligations and a computer program, were studied on the example of the works of Savelyev (2016), Kislyj (2017), Fedorov (2018). Solving the problem of digital marking of material objects with microchips in order to track and control their movements, including the use of Twin of thing technology, was studied on the example of the work of foreign scientists Reyna et al. (2018), Huckle et al. (2016).

2 Methods

The scientific development of the content of this article has been carried out on the basis of the universal scientific method of historical materialism from the standpoint of a positivistic worldview. General scientific methods of cognition were used: dialectical, hypothetical-deductive method, generalization, deduction and induction, analysis and synthesis, empirical description, classification. The study also used special methods: legal and dogmatic, comparative legal, structural and functional.

3 Introduction

Foreign economic activity in the context of accelerated growth in the volume of exchange of goods, services, growth of digital banking transactions and the number of investment projects between companies in different countries needs to simplify document management and modernize legal instruments for entering into non-economic transactions and fulfilling obligations for its. The achievement of these goals contributes to the introduction of electronic-digital smart contracts into the existing rules of legal regulation of foreign economic activity.

A uniform approach to determining the legal nature of smart contracts in Russian legal science has not yet been developed. Digital technology development immediately and inexorably requires the development of rules for regulating the use of cryptographic and blockchain technologies in business practice, and in accordance with these changes, the modernization of the legal consciousness of the legislator, legal practitioners and the scientific community is also needed.

The term “contract” suggests an agreement between persons, and the content of the term “smart” is reduced to information technology innovations, to something intellectually advanced and clever. Smart contracts are computerized transaction protocols which autonomously execute the terms of a contract. Smart contracts are disinter mediated and generally transparent in nature, offering the promise of increased commercial efficiency, lower transaction and legal costs, and anonymous transacting (Giancaspro 2017).

4 Legal Problems in Foreign Economic Activity and the Main Directions of Their Neutralization

International cooperation in the commercial, cultural, scientific and technical spheres is largely supported and developed due to the extensive transport communications between the states.

The main contractual structure in foreign economic activity is the international purchase and sale, which is the foundation of all possible transactions that mediate the movement of goods or complement the transfer of goods with specific conditions. Foreign trade operations are associated with the transportation of goods carried out by direct, indirect or several consecutive transportations, and also associated with the choice of the type of movement of goods: rail, road, air, sea transport or pipelines. Conflict rules and bindings are applied to international shipments: the shipment of goods governs the law of the state of departure, the issuance and receipt of goods by the final acquirer - the law of the state of destination, in other matters - the law of the carrier or the place of conflict's resolution.

The Vienna Convention of 1980, the norms of which apply only to international relations of commodity turnover, is a generally recognized and significant source of the law of international sale. In addition, there are a sufficient number of unified international rules and regulations, for example, the Principles of International Commercial Contracts approved by UNIDROIT or the INCOTERMS Code of International Trade Terms. However, despite the presence of international unifications, companies often do not include references or references to these acts when drawing up contracts, without giving them due weight. In turn, this practice, as a result, leads to disputes and disagreements regarding the applicable law, the identical understanding of the material and other conditions of contracts, etc. The problem of establishing a common understanding of the substance and conditions of the transaction exists everywhere in foreign economic activity in relations between counterparties.

The modern conjuncture in the world markets is characterized by the growing interdependence of exporters, importers and transit countries. At the forefront are questions about trusting not only counterparties to each other, but also to intermediaries (banking, transport organizations), to the legal instruments of transaction support, to the content of the document flow between them. Particularly acute is the problem of monitoring and tracking the route of movement of goods, checking the quality and quantity of the agreed range of the subject of the contract.

The party located at a distance of several thousand kilometers, when an improper performance was detected, expressed in the delivery of a damaged, distorted, possessing signs of a defect or defect, the counterfeit goods cannot promptly react to the fraudulent actions of the counterparty. As a consequence, the parties are forced to start disputes lasting for years only after the actual verification of the goods received. The most difficult is the verification procedure in the oil and gas sector of the world economy, where the transportation of raw materials is carried out through pipelines. A related problem of transportation of oil and gas products is safety on the continuation of the entire journey, since pipelines often transit through several neighboring states. There are cases of unauthorized connection to pipelines of gas, oil or derivative products, criminal selection of goods supplied by pipe. In this case, it affects not only the private interests of organizations, but also the public interests of states.

The solution of these problems can be facilitated by the modernization of foreign economic activity, namely, a combination of an electronic-digital form of a contract between the parties and a technology for digital tagging of goods with non-removable microchips. «Twin of Things is a solution for securing the ownership and provenance of everyday objects. The solution combines blockchain and cryptography to generate a hardware-based digital identity for all connected physical objects. Communication and transactions between devices is autonomously and securely performed thanks to blockchain. A highly secure crypto chip enables each device to become a blockchain node. The chip is produced in the form of an adhesive non-removable NFC tag and an Android application is used to carry out a blockchain transaction to register the unique, tamper-proof identity of the chip» (Reyna et al. 2018). Registration of goods transported across the border of Russia, in the mode of continuous electronic digital registration, will be possible due to the procedure of microchipping of goods. The specified mode of continuous attachment of microchips on goods provides for tracking the movement of these goods (microchips) through the corresponding electronic checkpoints and recording data in distributed registries, from which information on the movement, composition and quantity of goods is automatically sent to interested parties. An electronic-digital smart contract, as an encoded algorithm for performing predetermined actions, not only streamlines the private property relations of the parties, but also performs additional functions: a preventive and protective function of public law regulation, as well as a preventive and prophylactic function to reduce potential economic conflicts between business entities.

Electronic registration of legally significant facts in distributed registries will allow tracing all stages of goods movement, starting with the exit from the manufacturer's conveyor, then at each storage warehouse, at each loading and unloading, and at the final destination at the consignee's warehouse. It is also easy to trace the legally significant actions of the parties on the shipment-receipt of the goods, in the end, and calculations for the obligations. For example, the fact of sending goods is recorded, transmitted via electronic channels and entered into an array of the database, from which information to the buyer instantly about the date and time of shipment, quality and quantity, etc. The fact of crossing the customs border and subsequent movement is similarly displayed in the data chain, and specific information about the product is sent remotely to the servers of both the counterparty and the controlling state authorities. Finally, upon delivery of the goods to the destination, the microchip on each discrete unit of goods confirms the compliance of the received goods with the standards and stipulated conditions. The information about the receipt of the goods is transferred to the registry, then the counter-obligation is automatically fulfilled, for example, the obligation to pay for the goods by automatically transferring money from the buyer's account to the seller's account.

It should be emphasized that this method of tracking the movement of goods during its transportation from the supplier to the buyer, moreover, has a positive effect on the solution of a number of public law problems: tax, customs, currency regulation.

Article 1 of the Federal Law dated July 18, 1999 No. 183-FZ (as amended on July 13, 2015) "On Export Controls"[1], control is understood as a set of measures ensuring the implementation of the procedure for carrying out foreign economic activity in relation to goods information, works, services, results intellectual activity (rights to its). The purpose of control is to create conditions for the integration of the economy of the Russian Federation into the world economy. One of the methods of export control is "… customs control and customs clearance of controlled goods and technologies transported across the customs border of the Russian Federation, in accordance with the customs legislation of the Russian Federation" (Federal law 1999).

The execution of export transactions from the Russian Federation entails the appearance of the following objects of taxation: value added tax, tax on the profit of organizations, excises. On the one hand, strict accounting of transactions and their volumes will allow the tax service to timely and reliably receive the information necessary for tax purposes. On the other hand, the sale of many goods for export is taxed at zero or reduced interest rates. The basis for the application of such preferential regimes is the submission of documents, for example, provided for in Article 165 of the Tax Code of the Russian Federation: a taxpayer's contract with a foreign person for the delivery of goods outside the customs territory, a customs declaration and other customs documentation with a mark of the customs authority that issued the export goods.

Electronic-digital capabilities of the smart contract will allow to transfer the document flow completely into a paperless computer environment. Legal instruments for such a translation are actively being created; in addition, the regulatory framework is also developing. Federal Law of 06.04.2011 No. 63-FZ (as amended on 06.23.2016) "On Electronic Signature" (as amended and added, entered into force on 12.31.2017)[2] recognizes information in electronic form signed by a qualified electronic signature, an electronic document equivalent to a document on paper signed by a handwritten signature, which can be used in any legal relationship in accordance with the legislation of the Russian Federation (Federal law 2011).

Thus, the continuous registration of facts of property relations, carried out by electronic-digital fixation of microchips by minute location, which mark goods transported under foreign economic transactions, combined with the use of blockchain technology and smart contracts, will help solve the problems of modernization of legal regulation of foreign economic activity.

---

[1] Federal Law of 18.07.1999 No. 183-FZ (as amended on 07.13.2015) "On export control". // Rossiyskaya Gazeta, No. 146, 07.29.1999.

[2] Federal Law of 06.04.2011 No. 63-FZ (as amended on 06.23.2016) "On Electronic Signature" (as amended and added, entered into force on 12.31.2017). // Rossiyskaya Gazeta, No. 75, 04.08.2011.

5 Doctrinal Interpretations of the Legal Nature of Smart Contracts in Russia and Abroad

As British lawyers say, with the recent interest in the Internet of things and blockchains, there is an opportunity to create many exchange applications, for example. Peer-to-peer automatic payment mechanisms, currency platforms, digital rights management and cultural heritage, but these are just a few of them (Huckle et al. 2016). However, in the domestic doctrine of private law did not develop a common understanding and approach to the definition of a smart contract. A.I. Saveliev defines a "smart" contract as an agreement existing in the form of a program code implemented on the Blockchain platform, which ensures the autonomy and self-fulfillment of the conditions of such an agreement upon the occurrence of circumstances predetermined in it (Savelyev 2016). This position practically reflects the approach of the Ministry of Finance of the Russian Federation, formed in the draft law "On digital financial assets", which passed the first reading in the State Duma of the Russian Federation, but was openly criticized by the Bank of Russia (Inshakova and Goncharov 2018). Article 2 of this draft law contains the following definition: "a smart contract is an electronic contract, the rights and obligations of which are fulfilled by performing automatically digital transactions in a distributed register of digital transactions in a strictly defined sequence and upon the occurrence of certain circumstances" (Draft Federal Law 2018).

The Latin term "contractus", after the separation of obligations from contracts and delicts, is traditionally considered private law as an agreement enforced by a claim (conventio), that is, as an agreement (Braginskij and Vitrjanskij 2001). Such a view on a smart contract, in our opinion, will inevitably lead to confusion, since the phenomenon under investigation does not independently mediate public relations, but only merges into an integral part of a contract by fulfilling the task of fulfilling obligations. The smart contract does not have its own distinct subject matter from well-known contractual constructions. For example, if a supply is mediated by a social relation associated with the transfer of ownership from one person to another, then the subject of the contract is some property. Therefore, to take the term "contract" literally should not be.

The median position, which cannot be directly attributed to one of the two main points of view, is occupied by lawyers who believe that a smart contract is a way to implement agreements between the parties by executing a mortgaged algorithm that eliminates the human factor (Kislyj 2017). This approach is based on the functional purpose of the code, namely: automatic execution of commands, in this case, actions or omissions, the commission of which is stipulated by the obligations of the contract. The algorithm laid down in the code structure orders the sequence of conditions, and if one of them occurs, it sends a command to perform the corresponding action.

The diametrical point of view comes down to the fact that smart contracts are based on a computer protocol that, according to established rules, sends or receives certain information or modifies data. The basis is the program code laid down in the protocol (Fedorov 2018). Indeed, from a technical point of view, a smart contract is a computer program recorded in a distributed register and providing automatic fulfillment of contractual obligations or other legally significant actions. According to Article 1261 of the Civil Code of the Russian Federation, a computer program is an objectively presented set of data and commands intended for the operation of computers and other computer devices in order to obtain a specific result expressed in any language and in any form, including the source text and object code.

If a smart contract is an algorithm encrypted with a binary code, and its goal to achieve a result is the execution or enforcement of the transaction conditions (obligations) agreed by the parties, then the smart contract itself is an object of copyright whose purpose is to fulfill part of the obligations arising from the contract. The object of copyright extends the right of ownership, consisting of a triad of powers: possession, use and disposal. The value of a smart contract, as a computer program, is that its use ultimately allows the owner (the copyright holder of the exclusive right) to extract useful properties from it. Namely - the benefits expressed in reducing resource costs, time costs, verification of the counterparty. Various researchers of the continental legal family suggest that these agreements constitute a large multilateral treaty to which all parties are party. This multilateral agreement is automatically applied through intelligent contract technology (Geiregat 2018).

Thus, agreeing in general with this interpretation, we will make an interim conclusion that the smart contract is a computer program, the purpose of which is the timely and complete fulfillment of the obligations of counterparties upon the occurrence of certain conditions, without the possibility of unauthorized interference. This computer program arises due to the will of the parties to a particular property relationship settled by law, representing in the end an application to the contract - an electronic-digital algorithm for recording facts with legal and material consequences throughout the period of the parties' implementation of the contractual relationship.

6 Advantages of Using Smart Contracts in Foreign Trade Activities

Foreign trade transactions form the basis of international trade. It is in this area that the advantages of the blockchain technology and smart contracts can be most useful, both for the participants of cross-border property relations in the implementation of foreign economic activity, and for state control and regulation of exports and imports of goods. As noted above, the main type of transactions in international trade relations is the contract of sale, other transactions such as leasing or delivery are a type of sale, and insurance, transportation, storage are transactions that accompany the sale. In international commercial relations, there are always risks and difficulties in concluding transactions: physical distance and remote distance of counterparties, confirmation of the counterparty's eligibility to conclude a transaction, its export and import history and solvency, the proposed option of conducting cash calculations, their reliability.

In this regard, an important advantage of the smart contract is the elimination of the difficulties associated with the remote distance of counterparties. Digital technologies allow the exchange of information to counterparties while staying in the office in the country of their location. The parties remotely, using the information and telecommunications Internet, can adjust the terms of the main contract as their trade relations develop: regulate the volume and frequency of deliveries without the need to conclude additional agreements. It is necessary to enter into the distributed register data on the exchange of information on updating the terms of the transaction, this legal fact will be instantly registered, which will exclude further unilateral change of conditions or refusal to fulfill obligations.

The verification of the authority of the person making the transaction will turn from a complex task facing entrepreneurs to a standardized stage of registration on the blockchain platform. Electronic-digital registration of participants in property relations, especially those complicated by a foreign element, will make it possible to eliminate unlawful and unfair actions at the stage of entering the digital environment. The registration procedure is the primary screening stage filter, aimed, firstly, at increasing confidence in the subjects, as well as objects, about whose turnover in the future transactions will be concluded; secondly, to prevent possible fraudulent actions at the pre-contractual stages of interaction between potential counterparties. In addition to information about the executive body, location, other title information, information can be entered from the credit history, condition and availability of assets required for repayment of arrears in the process of trading. In general, this sector of the smart contracts functionality will initially eliminate a number of potential controversial issues. In addition, in the event of a misunderstanding that can escalate into a conflict, the parties without going to court through alternative means of resolving disputes, for example, through the mediation procedure, can resolve the situation on-line through the interface tools of the smart contract.

In practice, quite often the parties agree that their payment claims, which are mutually existing on both sides, must be redeemed by offset. If, when offsetting the value of supplies, an outstanding balance occurs, it can be settled by the supply of additional goods or by money payment. At present, the parties use such a mechanism for keeping records (valid account), which is usually maintained by a banking organization. (Inshakova 2018). The smart contract will do without intermediaries, all transactions will be reflected in the distributed registry and go through electronic registration, thus, all transactions and legally significant facts form a single chain of records, the change of which is subsequently impossible unilaterally.

A common form of payment for foreign trade transactions is a letter of credit. Taking into account that more than two persons take part in these relations (counterparties of the main contract, issuing bank, executing bank), and the level of trust between them often leaves much to be desired, the electronic-digital smart contract will level these problematic points.

In addition to improving the reliability and simplifying procedural relations, the digitalization of the economy aims to ensure the safety of participants in public relations involved in international economic turnover. In the program "Digital Economy of the Russian Federation", it is indicated that by 2024 the Russian Federation will be one of the world leaders in the field of information security (Order of the Government 2017). Such a striving, in our opinion, is intended to create conditions for the development and functioning of digital technologies that would limit interference in the domestic economy of Russia, thereby protecting Russian producers, investors and other participants in property relations. According to the results obtained by Swedish and Chinese researchers, the most frequently discussed problems and solutions in the literature are related to the security, confidentiality and scalability of the block chain and programmable intelligent contracts (Macrinici et al. 2018). Moreover, the technology of electronic-digital data capture in the interaction of the parties in the format of smart contracts will ensure high security of personal data and information of commercial value or legally protected secrets.

7 Conclusion

In conclusion, it should be concluded that the smart contract is an application to the contract in the form of a specific computer program arising from the will of the parties to the particular property relationship regulated by law. The smart contract generated by a completed property relationship is an individualized electronic-digital algorithm for fixing facts with legal and material consequences, throughout the period the parties implement this contractual relationship, has several advantages, including from legal positions.

By concluding a deal with an additional accessory in the form of a smart contract, participants of foreign economic activity reduce the time intervals required for pre-contractual procedures, warning local disputes regarding individual transaction conditions. In addition, a smart contract allows you to translate a number of routine actions (legal, accounting, financial) into an electronic-digital environment, eliminating the problem of geographical distances that divide counterparties. Reduction and optimization of time costs will lead to lower costs, which frees up significant resources from business entities.

Electronic-digital registration of facts of property turnover in foreign trade transactions is aimed at eliminating unfair actions at the stage of pre-contractual relations, compiling a single database of information about a potential counterparty - on the one hand, and recording the commission of legally significant actions of counterparties, exchanging information and coordinating the will regarding the terms of the transaction - with the other side. Electronic-digital registration also increases the level of mutual confidence of the parties in the tools by which commercial information, including confidential information, will be processed. Actions of the parties after the electronic registration are considered identically clear, proven and cannot be disputed. As a result, the number of potentially conflicting situations from the conclusion stage to the full execution of the transaction is significantly reduced.

The smart contract allows the parties to conduct settlements and make mutual settlements for the fulfillment of obligations without contacting intermediaries, transparently and openly, at the same time including every legal fact related to the movement of goods to the distributed register. It is impossible to supplement the database unilaterally, change or refuse to fulfill obligations in an unauthorized manner. To change the conditions under which the obligations are fulfilled, it is necessary to reach agreement of all parties to the transaction, by obtaining by all participants a cryptographic key. The blockchain technology, combined with the use of smart contracts, expands the possibilities for monetary settlement by the parties, in particular, simplifies and speeds up the execution of a cross-border letter of credit at a bank.

The procedure of electronic-digital registration of microchips, which mark the goods moved under foreign economic transactions, has a positive effect on improving the rule of law at the domestic level through more precise electronic customs, tax, and currency controls. In addition, in terms of the implementation of economic relations, electronic-digital registration performs a preventive-preventive function - reducing potential conflicts, since electronic registration of facts begins from the moment of conclusion of the contract and continues sequentially through the stages of fulfillment of obligations until the end of the deal.

At the same time, to realize the above advantages of smart contracts, at the moment it is necessary to develop and adopt a set of legal norms of legislative and subordinate level governing social relations in the digital economy, implementing business practices as blockchains, and smart contracts. In combination with the national modernization of the legal support of foreign economic activity, it is necessary to conclude universal conventions at the interstate level in the field of cross-border interaction of business entities using modern digital technologies. It is necessary to create digital foreign trade platforms that are accountable and controlled by the state - the jurisdiction of the party to the transaction, and at the international level - by intergovernmental organizations in which participants in foreign economic activity could register in distributed registries, make transactions using standard (standard) smart contracts, depending on product group sold for a particular foreign trade deal.

In the current period, applying smart contracts, it is necessary to rely on soft law norms and established business practice, combining them with the imperative method of state supervision and control over compliance with the requirements for the protection of confidential information, methods of protecting the legitimate rights and interests of persons involved in foreign economic relations.

Acknowledgments. The reported study was funded by RFBR according to the research project No. 18-29-16132.

References

Braginskij, M.I., Vitrjanskij, V.V.: Contract Law: General Provisions, 3rd edn., p. 848. Statut, Moscow (2001)

Draft Federal Law No. 419059-7 On Digital Financial Assets dated 25/01/2018. Minfin. https://www.minfin.ru/ru/document/?id_4=121810

Federal law No. 183-FZ On export controls. Dated 18.07.1999. Rossiyskaya Gazeta, 146 (1999)

Federal law No. 63-FZ On Electronic signature. Dated 06.04.2011. Rossiyskaya Gazeta, 75 (2011)

Fedorov, D.V.: Tokens, cryptocurrency and smart contracts in domestic bills from the perspective of foreign experience. Civil Law Bull. **2**, 30–74 (2018)

Geiregat, S.: Cryptocurrencies are (smart) contracts. Comput. Law Secur. Rev. **34**, 1144–1149 (2018)

Giancaspro, M.: Is a 'smart contract' really a smart idea? Insights from a legal perspective. Comput. Law Secur. Rev. **33**, 825–835 (2017)

Huckle, S., Bhattacharya, R., White, M., Beloff, N.: Internet of things, blockchain and shared economy applications. Procedia Comput. Sci. **98**, 461–466 (2016)

Inshakova, A.O.: International Private Law, p. 398. Jurait, Moscow (2018)

Inshakova, A.O., Goncharov, A.I.: The imperatives of financial policy in the sphere of the digital economy: impacts on increasing investment activity and tax potential of Russian regions. In: Proceedings of the International Scientific Conference «Competitive, Sustainable and Secure Development of the Regional Economy: Response to Global Challenges» (CSSDRE 2018), pp. 337–342. Atlantis Press, Amsterdam (2018)

Kislyj, V.A.: Legal aspects of the use of the blockchain and the use of cryptoactives. Law.ru (2017). https://goo.gl/tc2dZe

Macrinici, D., Cartofeanu, C., Gao, S.: Smart contract applications within blockchain technology: a systematic mapping study. Telemat. Inform. (2018). https://www.sciencedirect.com/science/article/abs/pii/S0736585318308013

Order of the Government of the Russian Federation No. 1632-P On approval of the program "Digital Economy of the Russian Federation" dated 28.07.2017. Government.ru. http://government.ru/docs/28653/

Reyna, A., Martín, C., Chen, J., Soler, E., Díaz, M.: On blockchain and its integration with IoT. Challenges and opportunities. Future Gener. Comput. Syst. **88**, 173–190 (2018)

Savelyev, A.I.: Contract Law 2.0: «Smart» Contracts as the Beginning of the End of Classic Contract Law. SSRN, 71 (2016). https://ssrn.com/abstract=2885241

The Factor Model Study of the Development of Economic Systems

Tatyana N. Ivashchenko[1(✉)], Nadezhda V. Mironenko[1], Natalya E. Popovicheva[1], and Yuliia A. Zviagintceva[2]

[1] Russian Presidential Academy of National Economy and Public Administration, 84 Vernadsky Ave., Moscow, Russia
my-orel-57@mail.ru, super-ya-57@mail.ru, 1278orel@mail.ru

[2] Orel State University of Economics and Trade, Orel, Russia
my-orel-57@mail.ru

Abstract. The scaling and multitasking of the modern economic processes testifies to the transformation of models of economic systems. The economic system is subject to change. The changes regulate the internal and external characteristics of the economic system. Preserving the sustainability of the development of the economic system remains a priority. Sustainable development of the economic system based on models with a theorized nature and based on historical aspects of the scientific knowledge. This factor indicates the need for new tools in the formation of a transformational model of sustainable development of economic systems. This model is factorial. The purpose of the study is to consider the key aspects of the factor model of sustainable development of economic systems. The objectives of the study are: to the formation of a differentiated representation of the concept of «the economic system»; to consideration of the agglomeration of the principles of development of the economic systems; to the allocation of a model of development of the economic systems; to formation of a heterogeneous factor model of sustainable development of economic systems. The methodological base of the research includes: the method of historical development, the method of theoretical premises, the method of abstraction, the method of formalization, the method of initial positions, the method of implication, the method of heterogeneity, the method of playing skills, the method of transparency. The scientific validity of the article includes the application of a differentiated approach to the concept of «the economic system». Practical significance is aimed at the application and further development of a heterogeneous factor model of sustainable development of the economic system.

Keywords: Model · Factor · Economic system · Instability · Principles · Differentiation

1 Introduction

Current processes in the modern economy provoke the emergence of new unstable systems. The formulated postulates fatalize the development of economic systems under the influence of the transformation of mechanisms and models of the modern

E. G. Popkova and B. S. Sergi (Eds.): ISC 2019, LNNS 91, pp. 14–23, 2020.
https://doi.org/10.1007/978-3-030-32015-7_2

economy. These aspects provoke the appearance of negative factors in the development of economic systems: singularity, invariance, recession.

The singularity of the development of economic systems based on the uniqueness of certain processes. Processes contribute to internal and external changes in the structure of systems. Internal changes form the main basis of the economic system. Mechanisms interact between elements of the economic system, principles of operation, methods for implementing the basic framework. Internal changes in economic systems make it possible to determine the strengths and weaknesses of the object under study. The singularity of development of economic systems denies this circumstance. External changes of economic systems are extensive factorial character. This process concerns the structuring of a model of the economic system (Perov 2007). Based on external changes, information flows are determined. Sets the functional requirements of the development of economic systems. This requirement determines the sustainability of the development of economic systems. The presence of a singularity of development of economic systems determines this model as a single, unchanged.

The invariance of economic systems determined by the singularity. The invariance of economic systems implies the existence of three characteristics. Characteristics are immutability, consistency, uniqueness. Highlighted characteristics deny changes. The complete absence of system changes is rather conditional. The immutability of the economic system implies constancy. The transformation of external environmental processes negates the aspect of constancy. Uniqueness of economic systems is an important factor of invariance. The uniqueness of the economic system argued at the initial stage of development and functioning of the economic system. The initial stage of development and functioning of the economic system based on the most stable production of all elements (Botterweg et al. 1998). In accordance with this fact, the uniqueness of the definition is present in the economic system.

The recessivity of the economic system determined by a non-critical structural change. This fact is associated with the transformation of economic systems. The economic system reflects the recessivity of its elements. The selected elements grouped into a single model of organization of the economic system. The emergence of new factors in the development model of the economic system brings it out of equilibrium. Subsequently, this system can be approximated in a recessive component.

The highlighted arguments determine the rather important role of the study of the sustainability of the development of economic systems. The formulated circumstances allow draw a conclusion about the development of economic systems. The economic system brings internal and external changes. Not a few important direction is the presence of factor characteristics. Factor characteristics are grouped in the context of a factor model. The factor model allows us to represent the process of development of economic systems. Based on this argument, we will conduct a study of the factor model of the sustainable development of economic systems.

The purpose of the study is to consider the key aspects of the factor model of sustainable development of economic systems. The objectives of the study are:

- to the formation of a differentiated representation of the concept of «the economic system»;

– to consideration of the agglomeration of the principles of development of the economic systems;
– to the allocation of a model of development of the economic systems;
– to formation of a heterogeneous factor model of sustainable development of economic systems.

Methods. The methodological base of the research determined by the methods of scientific knowledge. Methods of scientific knowledge reflect the conditions for writing a scientific article. The methods are theoretical, systemic and practical. Theoretical methods reflect the structured basis of the definitional prerequisites of scientific thought. Theoretical methods are:

– the method of historical development is the definition of the scientific basis of research based on time and evolutionary factors;
– the method of theoretical prerequisites is the selection of important elements of the essence of the problem being studied;
– the method of abstraction is the elimination of non-essential factors that don't affect the result of scientific research;
– the method of formalization is the selection of internal elements of the object under study with the aim of the most in-depth study of this component;
– the method of initial positions is a grouping of previously proven provisions that are the basis for a specific scientific research.

Systemic research methods aimed at specifying the research topic. The specification aimed at reviewing and forming a factor model. System methods are:

– the implication method is the formation of new assumptions in the framework of the scientific question under study;
– the method of heterogeneity is the grouping of different parts of one element.

The practical orientation of the study determines the effective methods for understanding the scientific issues. Practical orientation allows you to determine the strengths and weaknesses of the object. Practical methods are:

– the game skills method is the collaboration and diversification of the elements of the object under study for the modern realities of the development of the external environment;
– the transparency method is the formation of models as open systems for change.

Grouped research methods determine the framework nature of writing a scientific article. We form a differential representation of the concept of «economic system».

The Main Part of the Study. The concept «economic system» firstly appeared in the early 30s of the 20th century. The concept of the «economic system» identifying with the processes occurring in a single management environment (Smith and Ackere 2002). In general, the economic system throughout the 20th century viewed as a structure. The basic rule of the structure determined how much and what needs to produced at given rates of economic activity and the planned rate of profit (Sibirskaya 2010). The aspect of environmental variability and the installation of linking systems wasn't considered in the conceptual study of economic systems.

The modern concept of «economic system» is differentiated (Fig. 1).

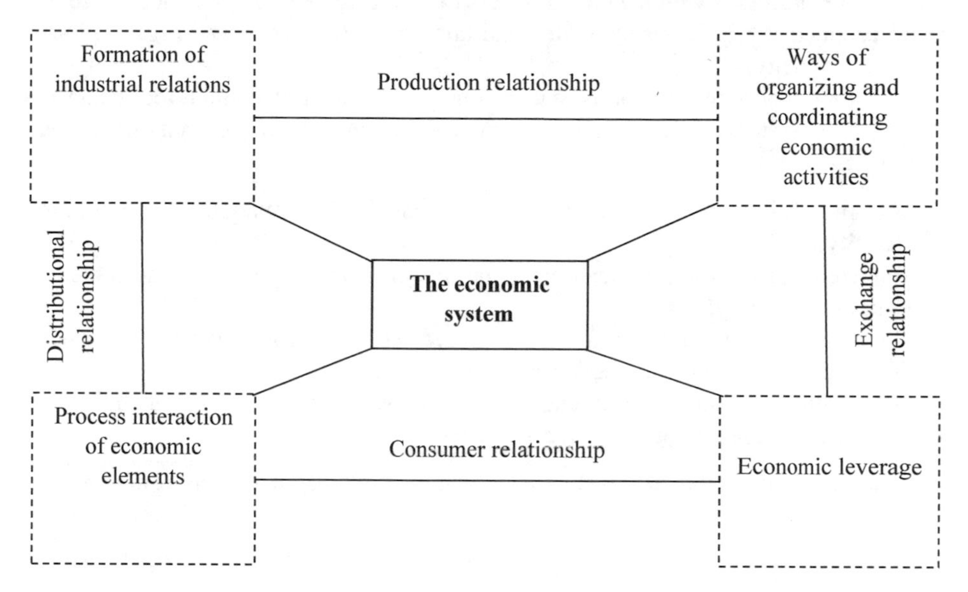

Fig. 1. Differential understanding of the concept of the «economic system»

Differentiation is manifold. Differentiation is the study of the essence of this concept. The concept of the «economic system» is presented in Fig. 1. The economic system reflects two important circumstances. Firstly, a structured rethinking of the definition of the «economic system». Secondly, the macroeconomic relationship builds up the economic system. Consider these theses in order.

There are four basic concepts of the «economic system»:

- the economic system is a set of economic elements functioning in the technological mode;
- the economic system is a place of concentration of resources and processes, distributed in predetermined proportions on the basis of levers of economic interaction (Gurieva 2018);
- the economic system is a set of economic mechanisms that serve for the formation of production relations (Karlis and Papadopoulos 2000);
- the economic system is a set of parameters of change and ways of organizing economic activity (Solodovnikov 2017).

The definitions reflect large parts of a single subject of the concept of the «economic system». The economic system is based on the interaction of economic elements. Economic elements are an essential definition. The relationship between these elements is formed by levers and tools. Levers and tools allow you to influence internal and external changes. On the basis of a clear design of the levers of security, economic interrelations are built between the elements of the economic system. In the framework of this relationship is the construction of production relations and the organization of economic activity. The economic system is a set of economic elements built on the basis of the processes with the help of levers to ensure economic interaction in order to build production relations and create conditions for coordination and organization of economic activity.

A rather important criterion is macroeconomic interrelations within the framework of economic systems. These relationships determine the role of the economic system. The interrelations of the economic system are:

– the production interconnection structures production relations and economic activity;
– the exchange relationship determines the role of economic levers in the context of economic activity;
– the distributive interrelation forms the process of interaction of economic elements in the construction of production relations;
– the consumer relationship regulates the relationship of economic elements in the application of levers of economic influence.

The highlighted relationship is based on the principles of development of economic systems (Fig. 2).

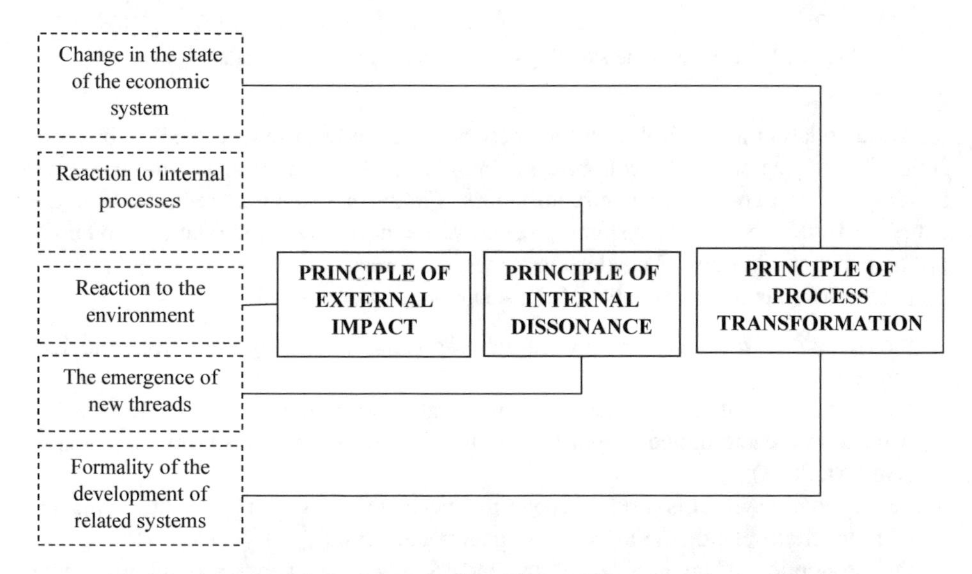

Fig. 2. Agglomeration of the principles of development of economic systems

Principles are identified with the main criteria. The criteria are peculiar to the organization of the development of economic systems. The study presents the agglomeration of the five main criteria. The criteria are formed in the context of the three principles of economic systems. Criteria for the development of economic systems are the main characteristics. These characteristics transform this phenomenon.

Principles of development of economic systems reflect the external prerequisite for the application of these criteria. Criteria for the development of economic systems are:

1. The change in the state of the economic system reflects the relative characteristics of transformation processes. The changes are related to the adjustment of the state of the economic system. This aspect indicates a stable, stable, recessive, imbalance state of the economic system.
2. The reaction to internal processes expresses the response of elements of the economic system to the adjustment of internal processes. The elements of the economic system imply this reaction. Internal processes correct the resulting reaction. The economic system transforms the resulting reaction into a framework formed by economic elements.
3. The reaction to the environment is the response of internal elements on the basis of external influence. The reaction to the external environment is direct. Her actions are due to the adaptive nature. The adaptive nature of the economic system to the external environment is determined by the flexible system of transformation processes. The absence of a flexible system of transformation processes does not allow us to speak of a stable state of the economic system.
4. The emergence of new flows is a combination of internal and external processes identified within the same flow direction. The stream component of the economic system is not permanent. The flow is determined by the constant movement of certain information within a given framework of the economic system. This criterion is the nature of the existence of an economic system.
5. The formality of development of adjacent systems divides the composition of economic systems into the main and adjacent systems. The main economic system reflects the cumulative development of related systems. Changing adjacent systems is reflected in the main system. The formality of the development of adjacent systems produces a negative dynamics of the formation economic system. The criterion of formality is to reflect the external features and the loss of the internal component of the economic system.

Differentiated agglomeration is formed in the context of three principles. The principle of external influence groups in itself the criterion of reaction to the external environment. The principle of external influence reflects changes in the external environment and have an impact on the economic system. The principle of internal dissonance classifies reactions to internal processes and the emergence of new flows. The principle of internal dissonance predetermines the structure of the economic system. The aspect of variability is characteristic of the principle of process transformation. The aspect of variability includes such criteria as changes in the state of the economic system and the formality of the development of adjacent systems. This principle is predetermined by the process component. This principle models in the framework of a given level of development of the economic system.

The economic system is classified on the basis of models. Models are divided into historical and formalized directions (Fig. 3).

The historical direction reflects the model of economic systems. Models have priority values at various stages of economic development. Models of economic systems in the historical direction:

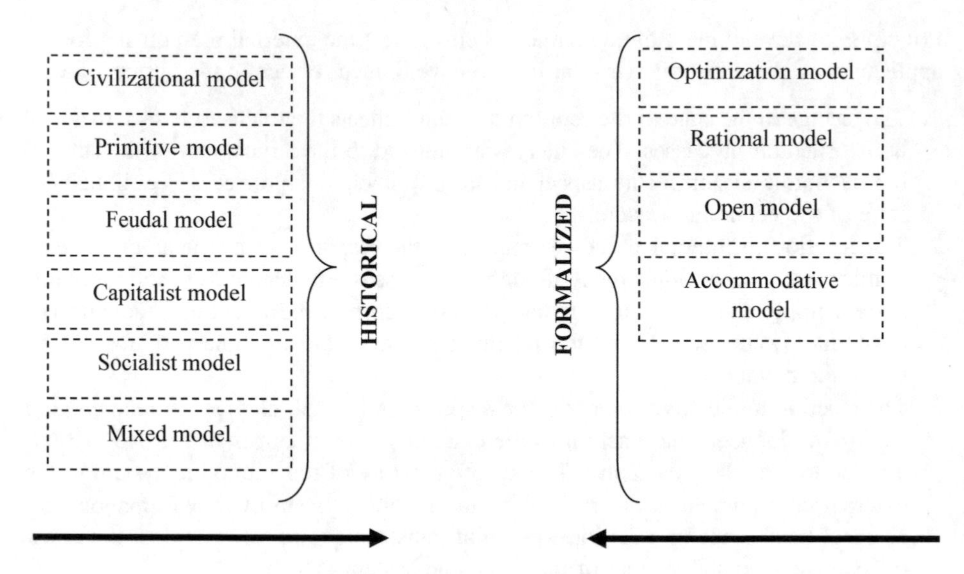

Fig. 3. Models of economic systems

- civilizational model reflects the evolutionary features of changes in production and distribution functions in the framework of economic parameters, forms of management;
- the primitive model reflects the traditional tenets of management and the functional social division of labor;
- the feudal model reflects the first appearance of private relations with the presence of rent as a source of income;
- the capitalist model reflects the presence of private property based on the entrepreneurial function;
- the socialist model reflects aspects of a socially-planned economy with the presence of a centralized distribution and concentration of resources for production purposes;
- the mixed model reflects the joint characteristics of the capitalist and socialist system with the presence of adjacent forms of ownership, hybrid forms of production, market formation with the presence of state influence.

Formalized direction reveals the internal structure, functional properties, factor characteristics. Models of economic systems in the formalized direction:

- the optimization model reflects the ability to minimize negative consequences due to the development of internal properties to maximize the factor advantages of this system;
- the rational model reflects the most regulated and rational use of economic resources on the basis of adequate and coordinated internal interconnections of the system;
- the open model reflects a transparent start in the process of production, distribution, consumption of economic resources;

– the accommodation model reflects the adaptive nature of the economic elements to the environmental conditions.

Selected models of economic systems determine theorized nature. These models consider some of the prerequisites of a functioning set of economic processes. The sustainability or dissatisfaction aspect in the development of the economic system is not taken into account. The author proposed a heterogeneous factor model of the sustainable development of the economic system (Fig. 4).

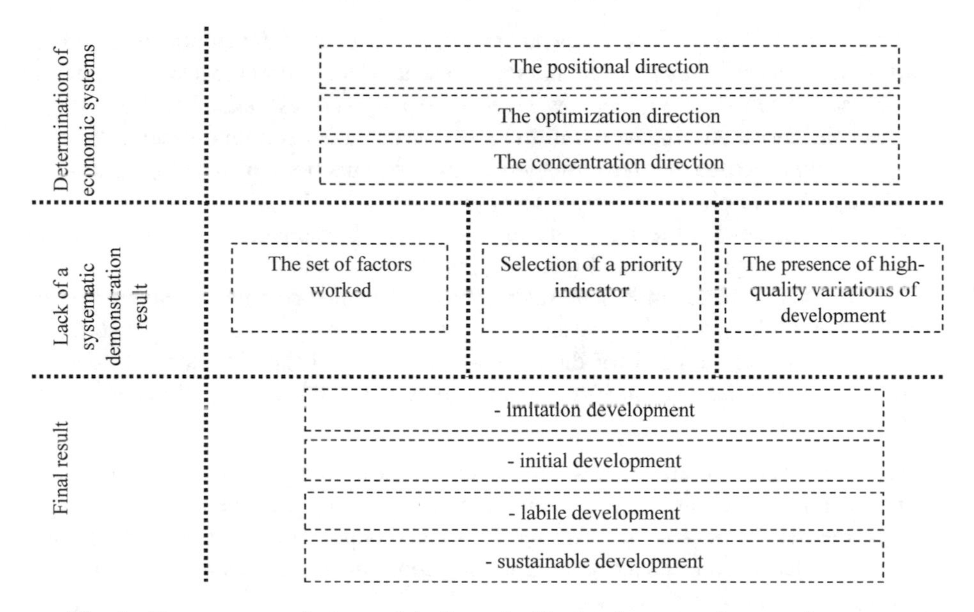

Fig. 4. Heterogeneous factor model of sustainable development of economic systems

Result. Heterogeneous factor model reveals the various properties of the economic system. The absence of clearly defined characteristics allows the model to be removed from the quantitative category. Heterogeneous factor model of sustainable development of economic systems forms the basis of a three-step mechanism. The three-step mechanism determines the state of development of the economic system. This state depends on the direction of development of the economic system. Consider this model in more detail. The direction of the economic system is determined within the framework of the priority, which serves as the basis for the development of this area.

In the framework of the selected model, the economic system includes: positional direction, optimization direction, concentration direction.

Selected directions allow you to create the initial level of a heterogeneous model. Directions determine the activities and industries requiring a certain positive impact. The second stage in this model implies the absence of a systematic demonstration result. This model does not adhere to a specific set of indicators and parameters that must be calculated in order to achieve a certain result. The development of these parameters depends on the directions formed above. The selection of parameters for a heterogeneous factor model is built through:

- the formation of a set of factors worked out is the selection of a certain number of factor parameters;
- the selection of a priority indicator is the definition of a single indicator of the development of the economic system;
- the presence of qualitative variations in development is the development of some proposals for improving the development of the economic system.

The final result of the heterogeneous model is the level of development. Levels of development determine the current and desired state of the economic system. Imitation development determines a rather low level of development associated with the duplication of directions and indicators in the most sustainable economic system. Primary development is focused on the formation of its own directions of development at the initial stage. The labile development shows the concentration of processes based on a slowdown in the development of the economic system. Sustainable development of this process is the most harmonious functioning of the economic system.

Heterogeneous factor model of sustainable development of the economic system forms the result. The result is a sustainable economic system. The lack of stability of this system leads to the need for the greatest structuring of the directions of development of the economic system. Conclusion as part of the study, the following conclusions were made.

1. The economic system reflects the basic properties and external characteristics inherent in this definition. The economic totality of economic elements functioning with the help of levers for ensuring economic cooperation in order to build production relations and create conditions for coordination and organization of economic activity.
2. The agglomeration of the principles of development of economic systems is based on five criteria. These criteria reflect the external prerequisite for the development of economic systems. Within the framework of these principles, the following criteria are highlighted: a change in the state of the economic system, a reaction to internal processes, a reaction to the external environment, the emergence of new flows, the formality of development of adjacent systems.
3. The dedicated models of the development of economic systems are formed in two areas: historical and formalized. The historical direction reflects the models of economic systems that have priority values at various stages of economic development. Formalized direction reveals the internal structure and functional properties of the model. The use of these models in the framework of this topic is not appropriate, which in turn is due to the lack of consideration of the sustainability aspect in them.
4. The heterogeneous factor model of sustainable development of economic systems concentrates the prerequisites for the implementation of quality characteristics. The quantitative parameter is differentiated and is selected in accordance with the directions of development of economic systems. The use of this model is associated

with the possibility of generating a result. The result determines the current level of development of the economic system.

References

Botterweg, P., Leek, R., Romstad, E., Vatn, A.: The eurosem–gridsem modeling system for erosion analyses under different natural and economic conditions. Ecol. Model. **108**(1–3), 115–129 (1998)

Gurieva, L.K.: The evolution of the socio-economic systems management system in the digital age. Humanit. Socio-Econ. Sci. **6**(103), 107–109 (2018)

Karlis, A.D., Papadopoulos, D.P.: A systematic assessment of the technical feasibility and economic viability of small hydroelectric system installations. Renew. Energy **20**(2), 253–262 (2000)

Perov, V.I.: Modern approaches to the selection of models of socio-economic systems management. Bull. North Caucasus State Tech. Univ. **1**, 114–116 (2007)

Sibirskaya, E.V.: Development of economic systems management systems. Reg. Syst. Econ. Manag. **2**(9), 131–135 (2010)

Smith, P.C., Ackere, A.: A note on the integration of system dynamics and economic models. J. Econ. Dyn. Control **26**(1), 1 (2002)

Solodovnikov, S.Y.: Economic system and institutional matrices: real – ontological hierarchy. Hours Econ. Reform **3**(27), 20–28 (2017)

Digitalization as a Method to Enhance the Mechanism of Procedural Protection of Civil and Financial Rights

Olga V. Gritsai(✉), Vera V. Kotlyarova, Ekaterina V. Mikhailova, Andrei V. Yudin, and Ramil' Z. Yusupov

Samara National Research University named after academician S. P. Korolyov, Samara, Russia
gritsayov@mail.ru, vera-er@mail.ru, evmihailova@yandex.ru, udin77@mail.ru, r.yusupov@bk.ru

Abstract. The article deals with the issues of ensuring the protection of rights and legitimate interests of legal entities in the field of civil jurisdiction with regard to the modern trend of civil procedure digitalization. The federal target program "Development of the Judicial System of Russia for 2013–2020" and some amendments to the Civil Procedure Code, the Federal Law "On Enforcement Proceedings" and other acts are intended to ensure the transition to a compulsory electronic enforcement proceeding. Therewith, the scientific community discusses the relevancy and expediency of introducing e-justice in Russia and using artificial intelligence technologies in adjudication. The international practice already uses particular elements of e-justice. In the course of the study, we analyzed the arguments of the discussing parties. In general, the judicial community and practitioners are wary of possible innovations, especially the idea of artificial intelligence technologies. A new finding of the research is the proposal on the algorithmization of writ proceedings as a digital service and the subsequent algorithmization of some elements of the enforcement proceedings.

Keywords: Digital Economy · Digitalization · Justice · Judicial authority · Civil law jurisdiction · Protection of civil rights · Writ proceedings

JEL code: K410

The main purpose of civil proceedings is to protect the rights and legitimate interests of legal entities. Furthermore, E.V. Vaskovsky wrote that the judicial system should be arranged in a way that a citizen claiming for the protection of his right could quickly and easily get it and, at the same time, that the court where the citizen filed a claim was able to conveniently allow it. The shorter and easier the procedure from filing a claim to a court judgment, the more perfect the process [1].

However, to date, there are a number of problems related to the quality of justice, the reasonable term of legal proceedings, the lack of citizen's awareness of the judicial activities, low performance of the courts, ineffective execution of judicial acts, unavailability of necessary conditions for justice administration, etc.

In accordance with the National Program "Digital Economy" in the Russian Federation, the Government should establish an ecosystem with digital data as a key factor

E. G. Popkova and B. S. Sergi (Eds.): ISC 2019, LNNS 91, pp. 24–29, 2020.
https://doi.org/10.1007/978-3-030-32015-7_3

in the development of all areas of social and economic activity, including cooperation between the state and society. However, this national program does not include measures on digitalization of the judicial authority.

There is an ongoing Federal Target Program "Development of the judicial system in Russia for 2013–2020" approved by the Resolution of the Government of the Russian Federation No. 1406 as of December 27, 2012 (last updated December 24, 2017). It provides for the fastest introduction of modern ICT into the judicial system, the system of compulsory execution of judicial acts and the forensic expert activity that allow designing an innovative approach to their development as well as achieving high-quality justice administration and reasonable term of legal proceedings, efficiency of expert evaluations conducted by forensic institutions, and ensuring the effective execution of court judgements.

Providing citizens with access to justice and ensuring the highest openness and transparency hereof as well as employing the principle of independence and fairness in adjudication are the main directions of further development of the judicial system [2].

The introduction of modern technologies in the execution system of acts of courts, other bodies, and officials, the creation of the essential conditions for the effective compulsory execution hereof will contribute to the establishment of an adequate system of enforcement proceedings and improvement of its transparency and accessibility.

The Program offers specific activities, such as:

- designing an electronic archive for storage of electronic documents in order to move towards compulsory electronic execution;
- developing a computing infrastructure of a data bank of enforcement proceedings at the regional level with a common data monitoring and replication environment to each territorial body of the Federal Bailiff Service (hereinafter referred to as FBS of Russia);
- building telephony and video conferencing systems based on a telecommunication system, which unites all 2.6 thousand entities of the territorial bodies of the FBS of Russia;
- ensuring information security of the telecommunications infrastructure of the FBS of Russia.

We should note that violated right or legitimate interest is in fact redressed mostly in enforcement proceedings, and one of the guarantees for the claim to be allowed is the fastest possible institution of enforcement proceedings and the application of enforcement measures. Given the heavy workload of bailiffs, providing this mechanism with currently-available digital technologies is useful and expedient.

In 2015, Part 1 of Art. 428 Code of Civil Procedure of the Russian Federation was supplemented by a rule allowing the issuance of electronic writs of execution with enhanced encrypted and certified digital signature of a judge in accordance with the statutory procedure of the Russian Federation. Currently, the system of internal e-workflow between the courts and the enforcement authorities is being implemented.

Digital technologies enforcement proceedings had been firstly introduced by the FBS of Russia more than 10 years ago [3]. So, a service "Databank of enforcement proceedings" appeared on the websites of the FBS of Russia and its territorial subdivisions [4]. There, everyone can get information about the validity of enforcement proceedings both against a natural entity and an organization in a simplified manner by the name of the citizen or name of the legal entity (Article 6.1 of the Federal Law “On Enforcement Proceedings”).

In accordance with the Resolution of the Government of the Russian Federation No. 606 as of June 29, 2016. “Concerning the Procedure for Sending E-notice Signed by the Bailiff with an Enhanced Encrypted and Certified Digital Signature via Information and Telecommunication Networks, and the Use of the Federal State information system “The Unified Portal of State and Municipal Services (Functions)” for Filing Applications, Explanations, Recusals and Complaints”, as well as Part 2.1 of Article 14 of the Federal Law “On Enforcement Proceedings” via the Portal of State Services to obtain information on the progress of enforcement proceedings, file petitions and pay debts. Bailiff officers can send notices to debtors via the personal account of the natural entity on the portal of state services. The notice is deemed delivered from the moment when the party in the enforcement proceedings entered the personal account using a single identification and authentication system. Notification of delivery is transmitted to the Federal Bailiff Service for the bailiff officer to adjudicate the enforcement proceedings.

The issues of the application of information technologies in the enforcement proceedings were discussed by experts at the 7th St. Petersburg International Legal Forum in 2017. The Forum noted that the Federal Bailiffs Service is currently one of the most technologically advanced Russian agencies. On the FBS website, there is an electronic bank of enforcement proceedings where you can learn about the debts of a natural or legal entity, the mobile application also has similar functions. Thus, 88 million people visited the “Bank of Enforcement Proceedings” on the FBS website in 2016.

Debtors are given the opportunity to pay the debt on the website without visiting the Bailiffs Department since ten widespread payment systems are connected to it.

The FBS of Russia has established electronic exchange of data on debtors with the territorial bodies of the Federal Tax Service of Russia, the Federal Migration Service, the State Road Traffic Safety Inspectorate of the Ministry of Internal Affairs of Russia and the Directorate of Internal Affairs for transport, branches of the Pension Fund of Russia, Sberbank of Russia, operational customs offices, Federal Agency for Intellectual Property, Patents and Trademarks (Rospatent), the State Inspection of Small Vessels, Federal Service for State Registration, Cadastre and Cartography (Rosreestr), Russian State Technical Authority, Federal Compulsory Medical Insurance Fund, mobile communication providers, committees on municipal property administration, FBS of Russia. E-workflow is as follows. A batch file-request is signed by an electronic signature of an official entitled to sign such documents and then delivered to the corresponding authority. The generated response is received within five-seven working days from the date of the request’s receipt. At the same time, the right of access to particular documents typical for the pre-digital era become out-of-date under pressure of new information technologies, which provide for on-line data exchange and the opportunity of continuous access and processing of data array. Therefore, the possibility of a quick exchange of particular documents or texts are no longer of interest and value, but the database and interrelated documents do. Today there such a uniform database on the body in Russia; the development hereof could assist the successful protection of violated rights.

The introduction of digital technologies in all spheres is the indispensable condition for the breakthrough development of the country. This was stressed by the President of the Russian Federation Vl. Putin in the Annual Address to the Federal Assembly in February 2019. Putin said that it is necessary to ensure the provision of all key public services by the end of 2020 "exactly in a proactive form when a person needs only to send a request for the required service, and the system should independently and automatically perform other operations" [5].

In this regard, it seems appropriate to address the issue of the possible introduction of e-justice in Russia and the use of artificial intelligence technologies in adjudication. Currently, elements of e-justice are being actively introduced in Russia. The Civil Procedure Code of the Russian Federation contains four groups of rules: the first group regulates the filing of documents to the court in electronic form, the second one concerns the audio recording of the course of legal proceedings and the introduction of video conferencing, another one regulates the delivery of acts, notices and other documents to participants of the legal proceedings, and the last one determines electronic interaction with the FBS.

Foreign countries have already held an online trial as an experiment using video conferencing, digital presentations and wireless networks [6]. On the one hand, it has some advantages: considerably accelerates the case hearing, ensures the safety of witnesses and other participants of the proceedings as well as saves public judicial costs to a great extent. A number of authors support the digitalization of the justice administration [7].

On the other hand, there are also risks common to the digitalization in any area of activity, for instance, cybersecurity, reliability of electronic data, the authenticity of data and information in the case papers. And if there is a technical disconnection of networks in the course of the trial, the continuity and other principles will be difficult to ensure.

We can give other arguments both confirming the necessity of digitalization and critical judgments. In K.L. Branovitskii' opinion, the introduction and extensive use of new technologies do not necessarily lead to higher access to justice in general, since a number of factors should be taken into account: available effective and sustainable system for identification of information system users, provision of an adequate system security and advanced training of public officials, and maintaining a balance between electronic and paper workflow (not all users can have access to modern technologies) [8]. In addition, the author rightly notes that there is no guarantee of a significant reduction in judicial costs from the budget since it is necessary to bear the costs of expensive information security systems, identification systems, data storage, etc.

The academicians are still debating on the possibility of admitting artificial intelligence programs to the justice administration of justice [9]. This issue is ambiguous and nowadays the majority of academic and research specialists are critical of "electronic robot judge". Indeed, one of the principles of civil proceedings is directness. In accordance with Part 1 of Article 196 of the Civil Procedure Code of the Russian Federation in the adjudication the very court evaluates the evidence, determines what facts are legally relevant to the proceedings, substantiated or not, what are the legal relations of the parties, what law should be applicable to this case and whether the claim should be allowed. On the other hand, a court decision can be executed in the form of an electronic document. When executing the decision in such a manner, a counterpart hereof is made on paper (Part 1, Article 197 of the Civil Procedure Code of the Russian Federation).

The judge interprets the evidence and reaches an opinion to adjudicate a comprehensive, complete, impartial and direct examination of the evidence of the case, which is tried (Part 1 of article 67 of the Civil Procedure Code of the Russian Federation). The judge should personally hear the parties to the conflict, witnesses, experts, specialists, see their psycho-emotional behavior and determine whether the information relevant to the case is reliable. In this situation, it is difficult to imagine the assignment of the judicial function on any robotic system at the current stage of scientific and technological development.

So, on April 4, 2019 a member of the Presidium of the Supreme Court of the Russian Federation V.V. Momotov at the 6th Moscow Legal Forum "The Russian Legal System in the Conditions of the Fourth Industrial Revolution" noted that "talking about replacing the judge with artificial intelligence is too early, and most likely – impossible," since the judge's inner conviction in evaluation of evidence is a more complex category than software algorithms [10].

At the same time, one of the types of civil legal proceedings, i.e. writ proceedings, provides for adjudication without the attendance of the participants in the case and a court hearing; only based on the documents submitted to the court that reflect the essence of the claim. The writ proceedings hear the non-contentious cases arising from the legal relations expressly specified in Art. 122 of the Civil Procedure Code of the Russian Federation. Nowadays, these are mainly cases of alimony recovery, collection of tax arrears, payments for public utilities, outstanding subscriptions of the property owners association. Cases of this category are adjudicated by justices of the peace and constitute a significant part of their workload.

It seems possible to develop and introduce service to facilitate hearing the cases of writ proceedings into the practice of the justices of the peace. On the website of the court or in a mobile application one can provide the application forms for issuing a writ. They are filled in by the claimants as answers to service questions that reflect information about the facts affecting acceptance, dismissal, discontinuance of the application for issuing a writ, the content hereof. Moreover, the person filling in the form and answering the proposed questions should be offered the options of answers that are considered for each type of writ proceedings.

The possibility to attach scanned documents that justify the claim should be also provided. Based on the answers chosen and filled by the claimant, the system automatically generates the text of the writ, which is verified by the justice of the peace and affixed with his digital signature. A debtor's counterpart of the writ should be also provided and delivered to the address specified in the case file for filing possible objections to the writ enforcement.

For this service to be developed it is necessary to collect the practice of justices of the peace in different regions of Russia and study the texts of writs for various claims relating to the list of cases established by Article 122 of the Civil Procedure Code of the Russian Federation.

Thus, issues of access to information and awareness hereof become crucially important in the economy, public administration, and private relations. However, in order to avoid mistakes and violations of the rights of citizens and companies, we should apply a low-keyed and delicate approach to the issues of justice digitization.

References

1. Vas'kovskii, E.V.: Civil Procedure Textbook. Statute (2016)
2. Sokolov, N.A.: Peculiarities of e-justice in the commercial (Arbitrazh) courts of Russia. Commer. (Arbitration) Civ. Proced. **4**, 3–6 (2018)
3. Letter of the FBS of the Russian Federation 2010 N 12/01-24016-АП as of October 26, 2010 "On the practical usage of information technologies in the enforcement proceedings. Bulletin of the Federal Bailiff Service, no. 1 (2011)
4. Federal Bailiff Service. Official site. http://FBSrus.ru/. Accessed 10 Feb 2019
5. Address of the President of the Russian Federation to the Federal Assembly as of February 20, 2019, Rossiyskaya Gazeta, no. 38 (2019)
6. Romanenkova, S.V.: The concept of e-justice, its genesis and introduction into the case law of foreign countries. Commer. (Arbitration) Civ. Proced. **4**, 26–31 (2013)
7. Nekrasov, S.Yu.: Information technologies in the ensuring access to justice in commercial (Arbitrazh) courts. Commer. Civ. Proced. **7**, 8–13 (2012)
8. Branovitskii, K.L.: Some issues related to the use of information technology in enforcement proceedings. Bull. Civ. Law Proced. **1**, 88 (2018)
9. Korolyova, A.N.: Judge with an artificial intelligence: myth or real future? Problems of harmonization of substantive and procedural protections of the law. In: Proceedings of the International Round-table Discussion, pp. 34–37. Publishing House of the Belarusian State University, Minsk (2018)
10. The 6th Moscow Legal Forum. Website of Moscow State Legal Academy. https://www.msal.ru/events/v-moskovskiy-yuridicheskiy-forum/. Accessed 4 Apr 2019

Circadian Genes: Information and Legal Issues

Andrey A. Inyushkin(✉), Alexey N. Inyushkin,
Valentina D. Ruzanova, Vladimir I. Belyakov, and Elena S. Kryukova

Federal State Autonomous Educational Institution of Higher Education
(FSAEI HE), Samara National Research University named after academician
S.P. Korolyov, Samara, Russia
inyushkin_a@mail.ru, ainyushkin, vd.ruz@mail.ru,
vladbelakov@mail.ru, kr-elena1203@mail.ru

Abstract. The article discusses the legal regulation of interrelated information on circadian genes and intellectual property items within the transition of the Russian Federation to the digital economy. We find out that the legal regime of copyright items will be the most appropriate for the introduction of circadian gene research findings into the economy of the Russian Federation. **The relevance of the issue under research** is connected to the essential improvement of the legislation on the regulation of genomic research set out in the national program "Digital Economy of the Russian Federation." **The purpose of the research** is to outline the main directions of legislation improvement that determine forms and methods of connecting the information on circadian genes and intellectual property items. **The main approaches of the research** are interdisciplinary, systemic dialectical. These approaches make it possible to investigate the relation between the legal regulation of information circulation and intellectual property in its entirety to solve the problems of the introduction of genomic research findings within the implementation of the national program "Digital Economy." **Findings:** the paper analyzed the legal regulation of genomic research and discovers the peculiarities of the relationship between circadian gene research findings and intellectual property. We assessed the applicability of some intellectual property items in the research information circulation and introduction of the genomic research findings into the economy. We find out that the legal regime of copyright items will be the most appropriate for the introduction of the genomic research findings into the economy. On the one hand, such a legal regime assists the introduction of information, and on the other hand, it allows classifying the information obtained as databases. **This article is of practical value** for multidisciplinary specialists, including geneticists, biologists, lawyers and economists who deal with the genomic research findings in their professional activity. Besides, the article is of interest to lawmakers.

Keywords: Circadian genes · Databases · Composite work · Legal regime · Genomic research findings · Intellectual property · Copyright law · Improvement of legislation

E. G. Popkova and B. S. Sergi (Eds.): ISC 2019, LNNS 91, pp. 30–37, 2020.
https://doi.org/10.1007/978-3-030-32015-7_4

1 Introduction

1.1 Establishing a Context

The development and scientific substantiation of the relationship between the genomic research findings and intellectual property seems the most forward-looking means for the introduction of genomic research findings into civil commerce. In particular, the approved mechanisms to use intellectual rights make it possible to lay the basis of the legal regime for the research findings in the field of the human and animal genome, including in the circadian genes. The most forward-looking legal regime for circadian gene research findings is the legal regime of copyright items. The works of science should be listed as a priority object for genomic research findings among the legally protected results of intellectual activity specified in article 1225 of the Civil Code of the Russian Federation (hereinafter referred to as the RF CC). However, despite the fact that the mentioned copyright object to the best possible way conveys the essence of the research conducted, the information obtained as a result hereof will not fit the corresponding legal regime.

However, it seems appropriate as a basis of model development for introduction of circadian gene information into civil commerce. In particular, the current criteria for the recognition of intellectual rights of a particular author-scientist generally reflect the activities typical for circadian gene research, however, the information obtained as a result hereof is so peculiar that it requires the combined use of a system of regulatory acts.

The Russian civil legislation provides a list of objects of civil rights, which include four groups: property (things and other property); results of work and provision of services; reserved results of intellectual activity (and equivalent visual identities); intangible benefits (Art. 128 of the Civil Code of the Russian Federation).

Therewith, the issue of the possible classification of information as an object of civil rights raises numerous discussions in Russian legal science (Kirichenko 2014; Malinin 2015). To determine the place of information on circadian gene research findings in the system of civil law objects, we should study the legal regulation of information. Foremost, it is expedient to address the Federal Law No. 149-ФЗ as of July 27, 2006 "On Information, Information Technologies and Protection of Information" (hereinafter referred to as the FL on Information), which is a special legal act that regulates the circulation of information in the Russian Federation. Article 1 hereof classify three groups of relations that are subject to regulation, namely: the exercise of the right to search, acquisition, transfer, production, and dissemination of information; application of information technology; ensuring the information security. Thus, the law contains an exhaustive list of areas of public relations covered by the FL on Information (Ryzhov 2011; Dolinskaya 2010). At the same time, there is a regulatory ban on the applicability of this law to relations arising in the legal protection of the of intellectual activity results and equivalent visual identities, except as provided in the FL on Information. The mentioned law does not contain direct references related to the legal protection of intellectual property, however, Federal Law No. 187-ФЗ as of July 02, 2013 "On Amendments to Certain Legislative Acts of the Russian Federation on the Protection of Intellectual Rights in Information and Telecommunication Networks" inserted the Article 15.2 "Procedure for Restriction of Access to Information Disseminated under Violation of Copyright and (or) Related Rights". It should be noted that the mentioned article regulates exactly relations connected with the circulation of information, but not with the protection of intellectual property. Meanwhile, specifying a violation of copyright and (or) related rights as a ground for restricting the access to information allows talking about the relationship between information and intellectual property.

The relationship between the circulation of information and the legal regime of intellectual property is obvious in the field of circadian genes. These genes are responsible for the biological rhythms of humans and animals.

Information about sleep and wakefulness has a potential commercial value because it allows for the selection of suitable staff to perform a particular job (a particular activity). Thus, it is reasonable to use the legal regime of trade secret (know-how), while the information on genomic research findings falls under the legal regime of databases (Bachilo 2009; Inyushkin 2016). The specific character of information circulation related to genomic research generates different approaches to the use of the intellectual property legal regime (Singh et al. 2007), also in the Russian legislation within the transition to a digital economy (Voynikanis 2013).

1.2 Literature Review

Current status of intellectual property and digital economy legislation is analyzed by Singh et al. (2007), Voikanis (2013), Inyushkin (2016), Mikhailov (2018), Ivardava (2019) Information as an object of property relations are investigated by Bachilo (2009), Chang and Zhu (2010), Inyushkin (2016). The general issues of the civil law system, including intellectual property law, are explored in the writings by Ruzanova (2018), Gavrilov (2019).

1.3 Establishing a Research Gap

Prior research of genomic legal regulation mainly considered the regulatory framework of intellectual rights, which did not allow revealing the peculiarities of the legal regime of information on genomic research findings. The research determines some intellectual property items acceptable to be used as carriers of information on genomic research findings. We use the information on circadian gene research as the subject-matter of the study.

1.4 Purpose of the Research

The purpose of the research is to determine the most appropriate legal regime taking into account the forms and methods of using the information on circadian gene research findings.

2 Methodological Framework

2.1 Research Methods

In the course of the study, we applied both general and private methods of cognition: systemic, comparative-law, historical, interdisciplinary, etc. Their combined use allows revealing the peculiar application of the legal regime of particular intellectual property items to the introduction of genomic research findings.

2.2 Research Background

The research background is proceedings of Russian and foreign academicians studying various aspects of legal regulation of information circulation, including the genomic research. The current information and civil law, also relating to intellectual property rights, serves as a research background.

2.3 Research Stages

The research takes two stages:

- the first stage is an analysis of the academic literature on the topic of the research as well as legislation relating to information, intellectual property, genomics;
- the second stage is making conclusions, also on the acceptable use of particular intellectual property items to the circulation of information on genomic research findings.

3 Results and Discussions

3.1 General Features of Legal Regulation of Information

Information means data (communications, facts) regardless of the presentation form (Zeinalov 2010; Fedoseeva 2008). Consequently, information can be expressed in any form: oral, written, graphic (Sebastian 2012). At the same time, the information of the human memory that cannot be materially expressed, as well as information obtained in the course of genomic research, will also be information in the sense stated in the Law on Information. In this regard, the first point of research is the analysis of genomic information as an object of legal relations. Article 5 of the FL on Information specifies that information may be an object of public, civil and other legal relations. It is necessary to pay attention to the fact that Article 1 hereof limited the scope of regulation of relations connected with information, including the genomic one. In particular, the issues of information storage without interrelation to information technologies are not regulated. In addition, the issues of its legal protection also placed beyond FL on Information. Other civil legal relations that are not related to the list provided in Article 1 hereof are also not covered by this special legal act. The second point of scientific discussions is the lack of information already described above in the list of civil law objects. In the theory of law, legal relations assume the subjects, object, and content as integral elements (Prokop'ev et al. 2015). This composition is also right for civil legal relations. Article 3 of the RF CC establishes that civil legislation consists of the Civil Code of the Russian Federation and other federal laws adopted pursuant hereto. Besides, civil law rules contained in other laws must conform to the Civil Code of the Russian Federation (Ruzanova 2015). Thus, as for civil law relations connected with the circulation of information, the priority in regulation should be given to the Civil Code of the Russian Federation, but not to the Federal Law on Information. However, since information is not object of civil rights, it cannot act in this capacity in civil law relations. It should be noted that in the original version of Article 128 of the Civil Code of the Russian Federation information was listed among civil law objects, however, it was later removed from this article.

Thus, civil relations on the genomic research findings require the application of the legal regime of intellectual property. The substantiation of the relationship between the genomic research findings and intellectual property seems the most forward-looking means for laying the foundation of legal regime. In particular, the approved mechanisms to use intellectual rights make it possible to establish the rules for the introduction of the research findings of human and animal genome into the economy.

3.2 The Application of the Trade Secret (Know-How) Legal Regime for Regulation of Genomic Research Findings

According to Art. 1465 of the Civil Code of the Russian Federation, trade secret (know-how) is information of any nature (production, technical, economic, organizational, and other) on the results of intellectual activity in the scientific and technical area and on the ways of conducting professional activity that have actual or potential commercial value due to non-public nature, if third parties do not have free legal access to such information and the holder hereof takes reasonable steps to keep the information secret, also by introducing a trade secret regime. Consequently, for the application of the appropriate legal regime in the regulation of genomic research findings, one should meet two criteria. The findings of genomic research should have actual or potential commercial value due to their non-public nature. Information about circadian genes, which particularly determines the sleep-wake schedule, exactly meets this criterion, since it allows selecting suitable for the performance of a particular job. The second criterion on confidentiality is provided by access control in educational and scientific institutions conducting genomic research.

Russian legislation requires that the information obtained as a result of the research should be data on the results of intellectual activity in the scientific and technical area. According to the Articles 1225 and 1259 of the Civil Code of the Russian Federation, academic proceedings belong to the works of science. Besides, the term "intellectual rights" specified in Clause 1 of Art. 1225 of the Civil Code of the Russian Federation is applicable to the results of intellectual activity, but have nothing in common with trade secrets (know-how) and visual identities (Eremenko 2014). Thus, the application of trade secret (know-how) regime with respect to the findings of circadian gene research requires formalization hereof. It seems suitable to use the regulatory framework of documented information, i.e. recorded on a tangible medium by documenting information with details, which allow identifying such information or, its tangible medium in the events established by the legislation of the Russian Federation.

3.3 The Application of Database Legal Regime for Regulation of Genomic Research Findings

Databases, as well as works of science, belong to copyright objects; they are protected as composite works. We should note that science already determines genetic code through the legal regime of copyright objects – computer programs. The definition of databases includes the systematization of information within the database and the relation with the computer (Inyushkin 2016; Mokhov and Yavorskii 2018). The legal regulation of circulation of information, which is the content of databases, as well as the rules that determine the database as a result of intellectual activity, are based on the RF CC and FL on Information. The scope of the law on information determines the relationship of information and content of databases. In particular, information technologies are understood as processes, methods of searching, collecting, storing, processing, providing, disseminating information and performing such processes. Systematization of information within the database should be based on certain methods, and processing thereof is a process that in total allows for the combined application of the FL on Information and the RF CC. Federal Law on Information stipulates that its provisions do not apply to relations arising from the legal protection of the results of intellectual activity and equivalent visual identities. Statutory ban on the legal protection of intellectual property by legal means enshrined in the FL on Information indirectly confirms that copyright protection of databases can be reduced to the protection of the algorithm of information location, that is, to the form of presentation, and but not to the information content hereof. This allows applying the rules on copyright objects to the legal protection of the circadian gene research findings. It should be mentioned that in the relations arising in connection with the database often occurs the term "information" or its derivatives. In particular, Art. 1274 of the Civil Code of the Russian Federation explicitly specifies possible free use of the work for information, scientific, educational or cultural purposes. In other words, we are talking about the legal regulation of the search, acquisition, and dissemination of information contained in the database.

Since academicians often publish scientific findings, this approach seems to be appropriate. We should note that the terms "data" and "information" are different. Data is the recorded signals from objects of the surrounding world. Information is data processed by appropriate methods that make a new product. Thus, information emerges and exists at the moment of the interaction of objective data and subjective methods. Information is influenced both by the properties of the data that make up its content and the properties of the methods that interact with the data during the information process. Upon completion of the process, the properties of the information are transferred to the properties of the new data, i.e., the properties of the methods can be transferred to the properties of the data. Concerning the findings of genomic research, there will be data processed during the research.

4 Conclusion

Determining the legal regime of circadian gene research findings contributes to the development of new sectors of the digital economy. The exploration of circadian genes necessitates the improvement of legislation and the choice of suitable approaches to the use of current legal regimes for the fast introduction of research findings into the economy. Current legal regimes of intellectual property are appropriate for efficient introduction of genomic research findings into commercial practice. Within the transition to a digital economy and required optimization of the legal regime for genomic research findings we recommend:

– to use a regulatory framework of documented information for implementation of the legal regime of the trade secret (know-how) in the relations on circadian gene research findings;
– to set the limits of application of information and civil laws to the relations on circadian genes;
– a legally enshrined admissibility of application of the legal regime of trade secret (know-how) for deposition of information collected in the course of circadian gene research;

Acknowledgments. The reported study was funded by RFBR according to the research project No. 18-29-14073.

References

Chang, J., Zhu, X.: Bioinformatics databases: intellectual property protection strategy. J. Intellect. Prop. Rights **15**(6), 447–454 (2010)

Singh, A., Das, S., Wilson, N.: Genomics and IP: an overview. J. Intellect. Prop. Rights **12**(1), 57–71 (2007)

Sebastian, T.: Copyright world and access to information: conjoined via the internet. J. Intellect. Prop. Rights **17**(3), 235–242 (2012)

Bachilo, I.L.: Information Law: Higher Education Textbook, 238 p. Higher Education, Yurait-Izdat (2009)

Voinikanis, E.A.: Intellectual Property Law in the Digital Era: A Paradigm of Balance and Flexibility, 208 p. Jurisprudence (2013)

Gavrilov, E.P.: The development of intellectual property law in Russia. Patents Licens. (1), 21–25 (2019)

Dolinskaya, V.V.: Information relations in civil commerce. Russ. Laws Exp. Anal. Pract. (4), 3–14 (2010)

Eremenko, V.I.: Some problems of codification of legislation on intellectual property. Legis. Econ. (2), 37–48 (2014)

Zeinalov, Z.Z.: The issues of defining information as an object of information relations. Inf. Law (1), 6–9 (2010)

Ivardava, L.I.: Changes in the scope and limits of legal regulation within digital economy. Bus. Secur. (1), 39–47 (2019)

Inyushkin, A.A.: Information in the system of objects of civil rights and its relation with intellectual property on the example of databases. Inf. Law (4), 4–7 (2016)
Kirichenko, O.V.: Information as an object of civil law relations. Mod. Law (9), 77–81 (2014)
Malinin, V.B.: Legal regulation of information. Leningr. J. Law (3), 120–129 (2015)
Mikhailov, A.V.: Issues of the digital economy establishment and the development of business law. Top. Issues Russ. Law (11), 68–73 (2018)
Mokhov, A.A., Yavorskii, A.N.: Genes and other gene-based formations as objects of intellectual property law. Civ. Law (4), 28–32 (2018)
Prokop'ev, A.Yu.: Absolute and relative civil law relations: general and particular points. Mod. Law (11), 34–39 (2015)
Ruzanova, V.D.: The significance of the civil code of the Russian federation for the further development of the civil-law system. The power of law. J. Res. Pract. **1**(21), 82–84 (2015)
Ruzanova, V.D.: Obligations related to business performed by the parties: issues of differentiation of legal regulation. Civ. Law (4), 18–20 (2018)
Ryzhov, R.S.: Legal regulation of relations on information technologies and information protection. Adm. Munic. Law (9), 64–68 (2011)
Fedoseeva, N.N.: The term "information" in modern science and legislation. Legal Educ. Sci. (1), 32–36 (2008)

Strategic Priorities of Digitalization of the Russian Economy

Elena Yu. Merkulova(✉), Sergey P. Spiridonov, and Vera I. Menshchikova

Tambov State Technical University, Tambov, Russian Federation
merkatmb@mail.ru, spiridonov_sp@bk.ru, menshikova2907@mail.ru

Abstract. The aim of the study is to analyze the conditions of economic growth and determine the strategic priorities of the digitalization of the Russian economy. The research toolkit includes methods to analyze static and dynamic information. The authors rely on the hypothesis that improving the quality of life of the population and ensuring Russia's national security depends on the pace of development of the "digital economy" based on the study of the global connectivity index. The study was conducted using the methods of economic and statistical analysis (comparison, calculation of absolute, relative and average values, tabular and graphic display of information), expert estimates. For the first time, the study made it possible to generalize part of the empirical data on the prospects for the implementation of an optimization model of the information economy in Russia. The authors define the digitization of the economy as the most important condition for economic growth in the modern world. The main strategic directions of digitalization of the Russian economy were identified: the main innovations in business models, and not in technologies; creating products and services on a project basis; common information space for the design, production, operation and organization of business; orientation on receiving income from servicing the sold products and services; selling sales through industrial ecosystems individually; customer retention through ecosystems through the implementation of standards, rules, methodologies; functioning of the economy on the basis of software-managed infrastructure.

Keywords: Digital economy · Strategic priorities · Information economy model · Digitalization · Digitalization index · Global Connectivity Index

JEL Code: O32 · O38

1 Introduction

Digital economy is a set of social relations taking shape through the use of electronic technologies, electronic infrastructure and services, data processing and forecasting technologies aimed at improving industrial production, distribution, exchange, consumption and the level of social and economic development of the country. Its main features include availability of information resources, which, along with traditional factors of production (labor, land and capital), ensures the process of reproduction of

E. G. Popkova and B. S. Sergi (Eds.): ISC 2019, LNNS 91, pp. 38–46, 2020.
https://doi.org/10.1007/978-3-030-32015-7_5

capital; an increase in the value of creative potential from the use of human capital; the reorientation of flows in the formation of the volume and range of products (if, during the industrial period, producers had an impact on demand, then production is currently based on customer needs); an increase in the speed of information flows (therefore, companies that follow the changing needs and demand of the population remain on the "wave"; the development of information technology to create different types of business on the Internet trading platforms, requiring smaller financial expenses, but quite a large amount of "knowledge"; capital virtualization, leading to an increase in the role of intangible assets and a decrease in wealth.

2 Methodology

The research toolkit includes methods to analyze static and dynamic information. The authors rely on the hypothesis that improving the quality of life of the population and ensuring Russia's national security depends on the pace of development of the "digital economy" based on the study of the global connectivity index (Global Connectivity Index 2017), the digitalization index (Digital Dividends: World Development Report 2016) and the global innovation index (Digital Russia: New Reality 2018).

The study was conducted using the methods of economic and statistical analysis (comparison, calculation of absolute, relative and average values, tabular and graphic display of information), expert estimates. For the first time, the study made it possible to generalize part of the empirical data on the prospects for the implementation of an optimization model of the information economy in Russia.

3 Results

The experts in the field of the digital economy (Bell 1999; Drucker 1993; Machhlup 1962; Efimushkin et al. 2017) believe that in the near future the labour market will undergo significant changes, since half of the workflows will be automated, which accordingly will cause the labour force disengagement. In the developed countries, this problem is widely discussed, and mechanisms for its solution are being developed. For example, Bill Gates proposed to introduce a tax on robots, and thus compensate for the social costs of the development of robotics.

According to the Rosstat forecast, the number of working-age population will significantly decrease in the coming years. This will compensate for labor force shortage, but the problem of disengagement of the least qualified labor resources remains. At the same time, we might expect the emergence of new highly paid jobs related to the implementation of digital technologies, as evidenced by recent studies (Dobrynin et al. 2016). As far as spatial-geographical factor is concerned, it is expected to explore the problem of equal opportunities for obtaining a digital resource in different regions of the country, in cities and villages, that is, how technological issues are resolved (Popov and Semyachkov 2017). Thus, there is a need to consider technological parameters, computer equipment, software, the Internet, mobile communications and other aspects of this problem (Sudarushkina et al. 2017; Petrosyan 2019).

The study showed that there are three approaches to the transfer to digital economy:

1. The process approach involves the introduction of relevant elements at each stage of the production chain (Fig. 1).

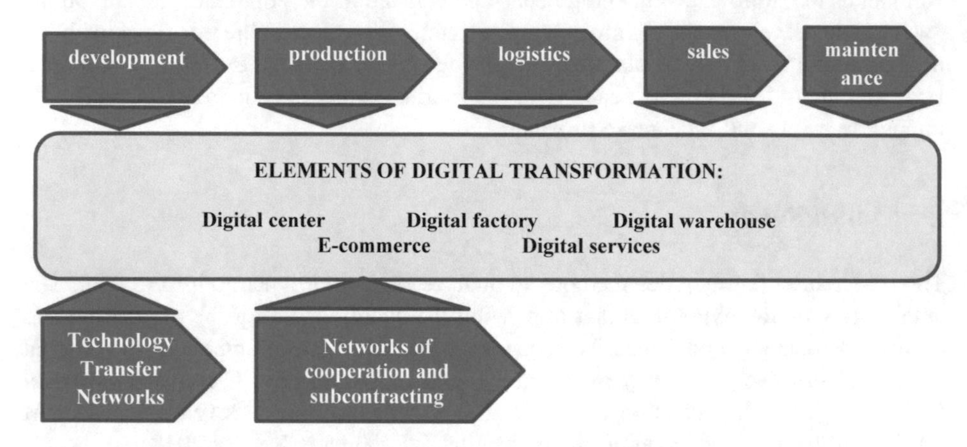

Fig. 1. Process approach to digital transformation (the authors' version)

2. The industry-based approach involves the introduction of appropriate digital technologies for various types of economic activity.

AeroNet is the direction which involves the development of unmanned technology, various aviation systems and the expansion of the range of services through the development of near-earth space systems. Today, the use of unmanned aerial vehicles in the country's economy is quite wide (observation of the fire situation, traffic arteries, search for missing objects, etc.). The objective of this global information market is to reduce the cost of services provided, to increase the reliability of their operation and to expand the possibilities for their use.

MariNet is a direction connected with the development of the resources of the seas and oceans, comprehensive solution of the problems in the field of navigation, innovative shipbuilding, building new underwater vehicles in order to conduct long dives and provide comfortable conditions for researchers, and in the future the development of tourist services is also possible.

AutoNet comprises technologies that allow for the use of automated control systems in transport. Increasing the environmental requirements of the vehicles used necessitates the use of new types of raw materials to reduce the amount of harmful emissions, and besides, researchers are returning to the idea of creating combined energy sources and developing unmanned vehicles.

HealthNet is a promising direction for the development of biotechnology and medical equipment to maintain human health, increase the duration and quality of life of the population of states.

NeuroNet is a neurocomputer interface technology that involves the development of neuromorph computers based on hybrid digital-analog architectures.

EnergyNet is a market for equipment, software, engineering and services for multi-scale integrated systems and services of intelligent energy.

FoodNet includes the technologies in the food sector of the information economy; they should be aimed at ensuring not only the appropriate level of food safety, but also at using "green technologies" in order to maximize the requirements of zero waste technology. Increasing competition in the global food market raises the problem of minimizing costs while improving the quality of agricultural products. To perform the tasks it is intended to maximize the use of biofuels, renewable energy sources, organic fertilizers, while it is desirable to bring the use of agrochemicals to the minimum acceptable level.

Another direction is expanding the scope of application of 3D printer technology (in architecture, medicine, industrial production).

"Smart City" comprises technologies used to automate the processes of managing security, transportation, lighting, parking, garbage collection, management of housing and public utilities and other aspects of the management of the urban landscape.

The implementation of the state-funded program "Electronic Government" contributes to reducing the time of the population to receive public services, which in turn reduces the corruption component, ensures equal access of the population to public services, automates the collection of information in the framework of management documents, which in turn has a positive effect on making management decisions.

3. The technological approach involves the use of appropriate digital technologies for the socio-economic development of the country, i.e. the Internet of things, the industrial Internet, quantum technologies, robotics, cloud technologies, etc.

The basic principles of building a digital economy are:

1. Provision of global access to resources without intermediaries. The main resources are computing and communication infrastructure, technologies, IP blocks, human resources, ready-made business models, intelligent online production, and finance.
2. Provision of resources for rent and the use of the volunteer model (Open Source); this form of resource provision, significantly affects the cost reduction, since it does not require the acquisition of expensive information resources, and it is possible to have them in the short-term use on lease terms.
3. Demand sales through global ecosystems; an ecosystem is a wide but finite circle of participants united by a single concept aimed at effective and successful competition with all other market participants. According to an analytical assessment of McKinsey experts, by 2025, ecosystems can account for about 30% of global GDP (about 60% of all non-production revenues of companies in the world) (McKinsey 2018).

Considering the structure of the "global innovation index" for 2018, it should be noted that Russia ranks 46th in the world. The main components of this index are human capital and science (ranked 22nd), the development of technology and knowledge economy Russia (ranked 47th), business development (ranked 33rd), import of services (ranked 28th), and scientific and engineering graduates (ranked 15th. Factors constraining the innovation index are infrastructure development (ranked 63rd), development of creative activity (ranked 72nd) and level of development of institutions (ranked 74th) (Digital Russia: New Reality 2018).

We consider the digitalization index of countries, which was calculated on the basis of 24 indicators to assess the level of use of digital technologies in the daily activities of consumers, companies and government agencies, as well as the provision of ICT infrastructure and the development of digital innovations. The analysis of the general level of digitalization (Table 1) of Russia shows that the country is among the leaders of the countries of "active followers"; the position is due to the active introduction of information technologies in government structures, the implementation of an active investment policy related to the expansion of ICT infrastructure. However, the lag of commercial enterprises in the application of digital technology, compared with the leading countries of the world is quite large.

Table 1. Comparison of the digitalization index of Russia with world economies

Group	Countries and regions	Overall ranking
Digital leaders	Singapore	High
	USA	
	Israel	
	Western Europe (United Kingdom, Germany, Italy, Norway, France, Sweden)	
Active followers	Russia	Average
	China	
	Central Europe (Poland, Czech Republic)	
	Brazil	
Lagging followers	Asia-Pacific region (Australia, Hong Kong, Indonesia, Malaysia, Thailand, Taiwan, Philippines, South Korea, Japan)	Below the average
	Kazakhstan	
	Middle East (Bahrain, Egypt, Jordan, Qatar, Kuwait, Lebanon, UAE, Oman, Saudi Arabia)	
	India	

Source: Compiled by authors based on the data from Digital Dividends: World Development Report (2016).

According to The Boston Consulting Group (BCG), the main reason for Russia's lagging behind the leaders of digitalization is the absence of coordinated actions on the part of all participants of the Russian economic system. Russia's lagging behind the leaders in digitalization, which currently accounts for about 5–8 years, will rapidly increase, and in five years, due to the high speed of global changes and innovations, it might reach 15–20 years - a gap that will be extremely difficult to narrow. The share of the digital economy in Russia is 2.1%, which is 1.3 times bigger than 5 years ago, but 3–4 times smaller than the leaders of digitalization (The BostonConsulting Group 2016).

It should be noted that in recent years Russia has noticeably advanced in many areas of digital development. The dynamics of the spread of broadband access and wireless networks of the Russian Federation is at the level of leading countries. According to the Rosstat statistics, from 2010 to 2016, the number of households with Internet access increased from 48.4% to 74.8%. At the same time, the average Internet speed in Russia in 2016 increased by 29%, which is comparable with France and Italy (Digital Economy of Russia Program 2017).

The current state and changes in traditional manufacturing and service industries, in organizing trade and procurement processes, related financial and logistics procedures, changing consumption patterns against the introduction of information technologies and digitizing economic processes create the basis for the formation of new markets and new conditions, determine their functioning, as well as new approaches to the analysis, forecast and management decisions. The leading asset of the state, business and civil society are "big data", which are formed as a result of the modernization of the economy. According to experts from the Higher School of Economics, the contribution of the ICT sector to the development of the economy in 2017 amounted to 2,211 billion rubles.

At present, the global connectivity index (GCI) is used to assess the level of digitalization of the economy in different countries of the world, the study and evaluation of which is made by the Chinese communications company Huawei Technologies Co Ltd. This integral index includes 40 indicators (by productivity and technological parameters of transformation into a digital economy) and is calculated for 50 countries of the world (Global Connectivity Index 2017). The TOP-30 of the leading countries in terms of GCl is presented in Fig. 2.

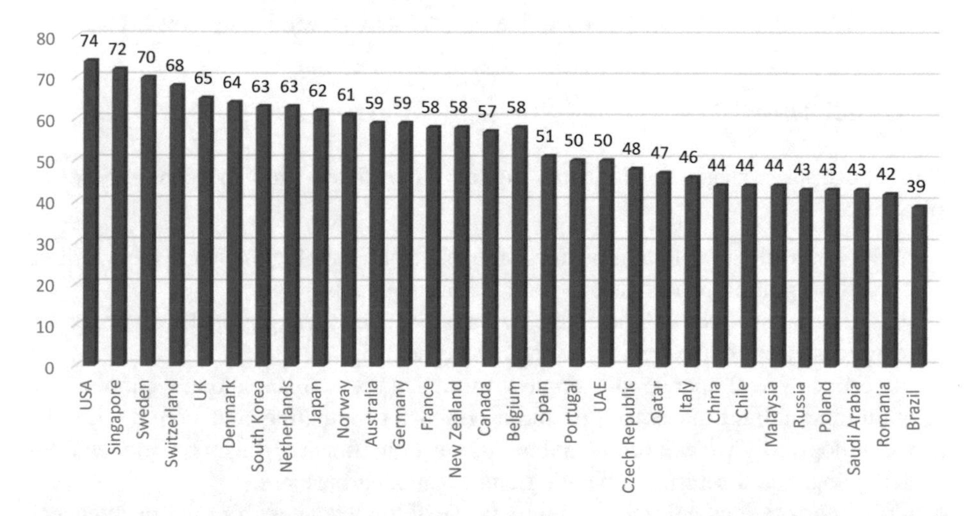

Fig. 2. TOP-30 countries-leaders in terms of GCl (global connectivity index) (Compiled by the authors from: Global Connectivity Index 2017)

According to the data obtained, all countries can be divided into three clusters: beginners, followers and leaders. Beginners are at the stage of development of ICT infrastructure; their main objective is to attract as many people as possible to the development of the digital industry. Followers, among which is Russia, currently have the most significant GDP growth rates due to the expansion of the information technology market. In particular, Russia is ranked 26th (GCl = 43) and has good prospects for the development of broadband mobile communications. There are prospects for the introduction of 5G networks, which will allow more active use of cloud technologies and ultimately change the business landscape. The first 16 countries in the ranking are leaders in the use of ICT; they have the most extensively developed e-commerce, and introduce technologies to improve the quality of life of the population.

It should be noted that the low ranking of the Russian Federation in terms of the global connection index is due to the following problems:

- the level of using information technologies in the system of education is significantly lower than that in the countries with developed information economies;
- the implementation of the developed innovations in the field of neurotechnologies, robotization, storage and distribution of energy to the world market remains the most problematic area;
- digital technologies are not actively introduced into modern business processes, since their impact on the performance of enterprises is underestimated.

The main problem is also an increase in thc gap between clusters of countries, which in turn allows developed countries to multiply their wealth, and outsider countries become even poorer compared to them. Thus, the lag in electronic technologies will inevitably affect the standard of living of the population. According to experts, the increase in the GCI index by 1 point allows for a rise in yields of 2.3%, an increase in innovation by 2.2% and a rise in national competitiveness by 2.1%.

4 Conclusions

Thus, the main strategic trends in the digitalization of the Russian economy are the following:

- major innovations will be in business models, not technologies; while there will be a deep integration of technology with business models;
- products and services will be created on a project basis using globally all necessary resources through leasing for the required time;
- the economy will operate in a single information space (using cloud technologies), combining all the necessary resources (human, computing and communication, technology, digital preforms, online production, finance, business models) for designing, manufacturing, operating and organizing business;
- global corporations will receive up to 100% of the revenues from the maintenance of sold products and services with 5–10% of the initial investment in the production of these goods and services (cloud volunteer model) and 0% in marketing;
- the operator of the global platform of industrial ecosystems will be the absolute beneficiary in the new conditions; in essence, this is a kind of production environment in new conditions (production of means of production);
- sales will be conducted through industrial ecosystems in a personalized manner (lack of warehouses and unsold goods);

- customer retention will occur globally through ecosystems through the introduction of standards, regulations, methodologies;
- the economy will operate on the basis of software-controlled infrastructure (computers, memory, networks);
- boundaries between industries will change significantly.

At the same time, the main effects of the implementation of digitization of the economy for Russia will include savings on the development of the innovation structure; acceleration of cooperation between the participants (economic agents) due to the common information space and integration possibilities on the digital economy platform; import substitution; growth of high-tech exports; global value added control by avoiding a purely raw model of the economy; transparency of economic management; providing customer "feedback"; economic growth and GDP by reducing redundant costs, accelerating production and services, eliminating intermediate links in the supply chain; availability of global resources for the Russian Federation; availability of products and services manufactured in Russia globally; lack of additional government investments.

At the regional level, the population will take advantage of the "social elevator" and professional mobility opportunities, which in turn will stimulate business activity and develop distance employment. The development of virtual infrastructure in the region will contribute to attracting investment.

The main effects of the digitalization of the economy for corporations and businesses are savings on resources, opportunities for effective cooperation with other companies, access to the end customer avoidance of long supply chains and a series of intermediaries, the lowered business entry threshold; the rapid development of new products and technologies, the possibility of using new business models, reduction in the cost of products and a multiple reduction in the time of their release; integration into networked intellectual production (the development of the industrial Internet), the creation of "smart" factories and network production.

References

Banke, B., Butenko, V.: Russia online? Catch up cannot be left behind. The Boston Consulting Group (BCG) (2016). http://gtmarket.ru/files/news/boston-consulting-group-russia-online. Accessed 12 Feb 2019

Bell, D.: The Coming of Post-Industrial Society: A Venture in Social Forecasting. Basic Books, New York (1999)

Digital Russia: a new reality. McKinsey (2018). http://apptractor.ru/info/analytics/otchyot-tsifrovaya-rossiyanovaya-realnost.html. Accessed 12 Feb 2019

Digital Dividends: World Development Report. World Bank (2016). http://www.worldbank.org/en/publication/wdr2016. Accessed 24 Feb 2019

Dobrynin, A.P., Chernyh, K.Yu., Kupriyanovskij, V.P., Kupriyanovskij, P.V., Sinyagov, S.A.: Digital economy - various ways to efficiently apply technology. Int. J. Open Inf. Technol. **4** (1), 4–11 (2016)

Drucker, P.: Post-Capitalist Society. HarperCollins, New York (1993)

Efimushkin, V.A., Ledovskih, T.V., Shcherbakova, E.N.: Infocommunication technological space of the digital economy. T-Comm. Telecommun. Transp. **11**(5), 15–20 (2017)

Global Connectivity Index (2017). http://www.huawei.com/minisite/gci/en/

Machhlup, F.: The Production and Distribution of Knowledge in the United States. Princeton University Press, Princeton (1962)

Petrosyan, A.: What you need to know about the digital economy and its perspectives. http://www.kommersant.ru/doc/3063024. Accessed 28 Feb 2019

Popov, E.V., Semyachkov, K.A.: Features of managing the development of the digital economy. Manag. Russ. Abroad **2**, 54–61 (2017)

The program "Digital Economy of Russia". Expert Council under the Government of the Russian Federation (2017). d-russia.ru/wpcontent/uploads/2017/05/programmaCE.pdf

Sudarushkina, I.V., Stefanova, N.A.: Digital economy. ANI Econ. Manag. **6**, no. 1(18), 182–184 (2017)

Digitalization of Labor Relations in Agribusiness: Prospects of Legal Regulation and Labor Rights Protection

Apollinariya A. Sapfirova[1(✉)], Victoria V. Volkova[2], and Anna V. Petrushkina[2]

[1] Kuban State Agrarian University, Krasnodar, Russia
pol499@yandex.ru
[2] North Caucasian Branch of Russian State University of Justice, Krasnodar, Russia

Abstract. The purpose of the article is to study the prospects of legal regulation of digitalization of labor relations in agribusiness, protection of labor rights in connection with the introduction of electronic personnel document management. The leading approaches to the study of this problem are the method of legal analysis, generalization, interpretation of law. As a result of the study, the authors made a number of conclusions. In particular, the necessity of digitalization of labor relations in agribusiness is obvious and not in doubt, but it is important not to allow the employer sought economic benefits in the replacement of employees on digital technology, since "digit" does not have to displace one of the parties of the employment relationship – the employee. Digitalization of labor relations in agribusiness will inevitably entail the need to protect personal data of employees, provide employees with enhanced qualified signatures, establish the maximum degree of protection of documents on labor activity of employees. Digitalization of labor relations in agribusiness is a costly but promising process for the employer in order to prevent violations of labor rights. However, it is not advisable to robotize (the highest degree of digitalization) labor relations, for example, in the field of plant and animal breeding, veterinary services, production of mineral fertilizers.

Keywords: Digitalization of labor relations · Agribusiness · Electronic personnel document management

1 Introduction

Human labor is now beginning to compete with machine work. Scientific and technological progress, of course, expands the horizons of the quality of life of every person, but it also creates "results" that need a legal shell. Otherwise, these results cannot be useful to society, and to use them legally no one will be possible. The gradual introduction of information technologies in labor relations (computers, printers, etc.) allowed employers to objectify and simplify the process of labor activity and, accordingly, reduce labor costs. In modern conditions, the update of information technologies allows rational use of not only the work of employees, but also the

E. G. Popkova and B. S. Sergi (Eds.): ISC 2019, LNNS 91, pp. 47–53, 2020.
https://doi.org/10.1007/978-3-030-32015-7_6

process of supervision by the authorities of compliance with labor legislation by employers. The appearance of the digit has opened up new opportunities for the employer to use labor. However, it remains to answer the many questions associated with the implementation of digital technologies in employment relationship. These questions reflect the challenges faced by employers, employees and supervisors in using the digit in labor relations. Especially sharply the question about the legal field of digitalization of the employment relationship is in agribusiness. The sectors covered by agribusiness are affected by informal employment and, in general, by the large number of labor relations unformed with employment contracts. The employer's motivation for the formalism of labor relations is practically absent.

At the same time, the increase in the number of flexible forms of employment in Russia and abroad contributes to the economic rebirth of these countries, as the labor force has always been the engine of the economy. With the introduction of digital technologies in labor relations, there have been changes in the sphere of registration, change and termination of labor relations, fixing the working time of employees, employers' responsibility for violation of labor rights of workers. There are also new categories of workers, including those using their labor in agribusiness, and new forms of their employment: remote workers, self-employed citizens, call-workers, etc. It is obvious that it is not possible to completely abandon traditional forms of employment in agribusiness, if only because the direct use of land and its wealth in the process of labor, and not digital technologies, is the basis for the development of agribusiness. The digitalization of labor relations will reduce the cost of agribusiness to the auxiliary resources, warn of a possible failure to comply with labor laws and increase the quality of protecting the rights of workers.

It is important to present a review of the literature in two parts. First, these are monographs by Russian scientists Tal (1913), Alexandrov (1948), Bugrov (2013) and a monograph ed. by prof. Yu. P. Orlovsky and prof. D. L. Kuznetsov "Features of regulation of labor relations in the digital economy" and foreign scientists Blanpain, Hendrickx (2016), which conducted a thorough study of the subjects and content of labor relations, as well as the employment contract, the order of its registration, change and termination of the terms, and also studied the forms of employment.

Secondly, it is the consideration of digital information as a subject of legal regulation in labor relations of agribusiness and protection of personal data. This is a novel that has just begun to be studied by Russian and foreign scientists in relation to labor relations: Otto (2016), Volkova (2014).

2 Methodological Framework

The study was conducted using general scientific (analysis, comparison, generalization) and private scientific methods (interpretation of the law).

The research is based on scientific views, ideas of Russian and foreign scientists, normative legal acts characterizing the introduction of digital technologies in the work of the employer.

The study of the problem was conducted in two stages. At the first stage, the selection of literature and its analysis, problem statement, definition of research goals and methods were carried out. The second stage is characterized by the conclusions we have received in the course of the study, the proposals formulated on their basis, and the preparation of the publication.

3 Results

Currently, information is the engine of economic processes. With the help of information technologies, industries are developing, their software is being improved, allowing to redistribute labor, eliminating the performance of unnecessary functions. The development of information technologies has led to the emergence of digital technologies, the introduction of which radically changes the approach to the use of labor.

The legislator of Russia, we must say, does not keep up with changes in digital technologies, and therefore there are no normative legal acts regulating the process of electronic personnel document circulation. At the same time, a number of draft federal laws have already been developed, which will be discussed below.

At one of the economic forums of 2017 in Sochi, the idea of digitalization of labor relations through electronic personnel document management was voiced, including the conclusion of electronic employment contracts, maintenance of electronic work books, electronic vacation schedules, keeping records of working hours, registration of business trips, etc. This idea found a response, and from March to October 2018, Rostrudof Russia conducted an experiment: it chose 8 employers, including JSC Russian Railways, JSC Gazprombank, JSC Rostelecom, etc., and "digitized" most of the documents confirming their employment relations with employees (personnel orders, employment contract, vacation schedules, etc.)[1]. Unfortunately, among the pilot organizations did not have even one of the largest agricultural holdings (for example, "Miratorg", "Promidex", "Agrocomplex" named after N. Tkachev and others), which, of course, is the omission of Rostrud.

The basis of the experiment was the local regulations establishing the procedure for the introduction of electronic personnel document management, and prepared corporate information systems. Financial costs for the introduction of electronic personnel document management, working time for its application, the impact on productivity, supervisory activities (conducting scheduled and unscheduled inspections) – these are the main parameters of the effectiveness of electronic personnel document management.

In part, the experiment met expectations and led to proposals for the introduction of electronic workbooks and employment contracts, as well as other documents on the employee's employment, as the cost of office equipment, consumables, time and process of interaction with supervisory authorities and employees decreased. At the same time, Rostrud of Russia conducted a sociological study on the consent of workers with the conduct of employment contracts in electronic form. According to the results

[1] The Order of the Ministry of labor of the Russian Federation from 26.03.2018 No. 194 "On conducting experiment on digitization of documents and information about the worker on labor relations" // https://rosmintrud.ru/docs/mintrud/orders/1290 (accessed 19 May 2019).

of the study, many employees supported this initiative, and insisted on the creation of a state database of such contracts, which will contribute, in their opinion, the highest degree of protection of information[2].

At the same time, the shortcomings of the introduction of electronic personnel document management became evident. In particular, the main drawback, according to employers, is the increased cost of electronic signatures to employees. The imperfection of the legislation on digital signatures shows the need to consolidate the possibility of providing employees with a universal qualified electronic signature for use in any legal relationship.

From these positions, it seems that it is unfair to impose only on the employer the costs of making a universal qualified electronic signature. Since the signature data must be updated annually, the financial burden for many employers would be unsustainable. And it is unclear why the costs should be borne by the employer, and the employee will use the signature not only in labor relations, but, for example, in civil law, family, etc.? In these cases, it seems appropriate to divide the financial burden between the employer and the employee in a certain proportion.

If the employee uses the electronic signature exclusively in labor relations, then the costs of its production, of course, should be borne only by the employer.

At the same time, the experiment showed that the maintenance of personnel documents exclusively in electronic form will make it difficult to protect the labor rights of employees, since there is a high risk of loss (theft) of information from the employer's corporate information systems, as well as software errors. In supervisory activities, these problems entail unjustified involvement of the employer to responsibility, as well as create problems with the objectivity of supervisory activities and the legality of the actions (inaction) of the employer.

The above analysis shows the primary need to change the legal support of the IT sphere, the creation of a legal framework for the effective and reliable activities of IT-specialists. Only after these measures have to be taken to update labor legislation and universally enter the electronic personnel document management. It seems to us that this is the right approach, since the management of personnel documents is not just the work of each employee, it is his subsequent social guarantees, and pensions, and benefits that depend on personnel documents.

That is why, on the basis of the experiment, Rostrud of Russia has developed a draft federal law[3] aimed primarily at improving the legislation in the IT-sphere, and then to change the labor legislation.

However, despite the negative result of the experiment in terms of imperfection of the legislation in the IT-sphere, the Ministry of labor of Russia simultaneously developed a draft federal law on electronic work books, which it plans to introduce from 2021 on the territory of Russia. Motivating by the fact that there are already

[2] Research on the transition to an employment contract in electronic form // https://www.rostrud.ru/openrostrud/ (accessed 19 May 2019).

[3] The draft Federal law "On amendments to the Labor code of the Russian Federation (in terms of the formation and maintenance of information about the employee's employment in electronic form)" (prepared by the Ministry of labor of Russia) // the text of the document is in accordance with the publication on the website http://regulation.gov.ru/ as of 06.12.2018.

electronic sick lists, electronic employment contracts with remote workers, officials responsible for the development of this project, consider it timely, important and quite capable of implementation by employers in modern conditions. It is possible that this is true, since the active use of electronic sick lists shows the positive dynamics of the first stage of digitalization of labor relations. At the same time, the effect of the introduction of the electronic work book will be the highest if the legislation in the IT-sphere is previously adjusted.

We emphasize that if there was at least one agricultural holding among the participants of the experiment, the problems of digitalization of labor relations would be identical. However, some of the issues reflecting the introduction of electronic personnel document management in agribusiness are still specific. In particular, all participants of the experiment are organizations that are quite numerous in terms of the number of employees, but carry out production activities in the industrial, banking or service sector, and agricultural holdings are the sphere of agriculture and auxiliary industries (processing, harvesting, fertilizer production, communications, construction, etc.), the specifics of which lies in their seasonality, temporary work, attracting foreign workers, the practical absence of remote labor.

Meanwhile, the digitalization of labor relations seems to us a key factor in the development of the labor market in agribusiness, since the seasonal, temporary nature of work involves non-permanent employment of citizens, employment documents, the dismissal of which must be periodically issued with the appropriate personal signatures of employees, which is difficult for employers of agribusiness.

It seems to us that the experiment also showed that, despite all the economic advantages of electronic personnel document management, it will not be possible to completely displace a person from production, and there should not be such a goal. "Digit" should serve man and facilitate his life, not replace it.

The dependence of the availability of documents on the labor activity of employees from the qualification of a programmer and an IT-technology specialist is certainly not acceptable, since it levels the protection of labor rights of employees and employers. Therefore, we emphasize once again that the conclusion about the primacy of changes in the legislation in the IT-sphere before the full-scale digitalization of labor relations in Russia is absolutely justified.

First of all, it is necessary to eliminate legal gaps in the IT-sphere and minimize the cost of a universal qualified electronic signature of an employee.

In the second place, to oblige employers to keep electronic work books, electronic employment contracts with all employees. It is important to allow the opportunity to have a paper version by the application of the employee. At the same stage, to provide Rostrud with periodic electronic scheduled inspections in accordance with the risk-oriented approach. Here it is necessary to explain separately that the checks in accordance with the risk-oriented approach will allow supervision without a long distraction of employers' specialists from the performance of their work functions, and with the use of electronic personnel documents – without distracting them from work. However, employers operating in hazardous and (or) dangerous working conditions should be aware that they will be under the constant supervision of Rostrud (every 2 years), because most often they are at high risk. On the contrary, low-risk employers are generally not subject to routine inspection until they violate the employee's labor rights (for example, the right to wages).

And only in the last, the third, turn it is possible to enter completely electronic personnel document management and respectively to digit all local regulations and other documents on labor activity as at this stage the employer will already have the protected corporate information systems.

4 Discussions

The issues raised in this article are only beginning to emerge as digital technologies are introduced into labor relations in agribusiness. However, the need for digitalization of labor relations is obvious and there is no doubt. Draft regulatory legal acts are being developed to allow the use of "digit" in the legal field, thereby preventing violations of labor rights; sociological surveys are conducted, the results of which are striking in scale and conclusions about the importance of electronic personnel document management and its applicability in personnel, accounting, economic departments.

However, it is also important that the introduction of digitalization of labor relations in agribusiness will entail the need to protect personal data of employees, provide employees with enhanced qualified signatures, establish the maximum degree of protection of documents on labor activity of employees. Digitalization of labor relations in agribusiness – a natural process, but costly and promising for the employer in order to prevent violations of labor rights. At the same time, an excellent degree of digitalization of labor relations - their robotization, for example, in the field of plant and animal breeding, veterinary care, it seems, is generally impractical.

5 Conclusion

This study has led to the conclusion that the digitalization of labor relations, including in agribusiness, is just beginning to "get a legal basis", that is, for the first time there were draft normative legal acts regulating the use of digital technologies in labor relations. This means the evidence of the prospects for the digitalization of the employment relationship. The challenge for society as a whole is to prevent digital technologies from completely replacing workers, because the "digit" should help the employer and employee in the process of labor relations, and not displace one of the parties of these relations. There are some labor functions that are impractical to replace with digital technologies (for example, the work of a nurse or agronomist, breeder, etc.). Meanwhile, the work of agricultural machine operators digital technologies will greatly simplify.

The positive conclusion is obvious when using digital technologies for the implementation of the supervisory authorities' functions in relation to employers. The digitalization of the employment relationship will allow to not distract employers during the period of audit by the State Labor Inspectorate. With a certain degree of confidence, it is possible to predict an increase in the degree of protection of workers' labor rights in proportion to the volume of inspections.

References

Alexandrov, N.G.: Employment relationship. Legal publishing house of the Ministry of justice of the USSR, 337 p. (1948)

Volkova, O.N.: Personal data in labor relations: on the history of the issue. Education. Science. Scientific personnel, no. 5, pp. 90–94 (2014)

Tal, L.S.: Employment contract. Civil law studies. Part 1, Yaroslavl, 150 p. (1913)

Bugrov, L.Yu.: Labor contract in Russia and abroad. Perm Stat. Nat. Research. Univ. – Perm, 641 p. (2013)

Orlovsky, Yu.P., Kuznetsov, D.L. (eds.): Features of regulation of labor relations in the digital economy. Law firm "CONTRACT", 99 p. (2018)

Blanpain, R., Hendrickx, F.: New Forms of Employment in Europe, 416 p. (2016)

Otto, M.: The Right to Privacy in Employment: A Comparative Analysis. Hart Publishing, Oxford and Portland, Oregon, 256 p. (2016)

The Labor code of the Russian Federation from 30.12.2001 No. 197-FZ ed. on 01.07.2017. Collected legislation of the Russian Federation. 07.01.2002. No. 1 (Part 1). Art. 3

The Order of the Ministry of labor of the Russian Federation from 26.03.2018 No. 194 "On conducting experiment on digitization of documents and information about the worker on labor relations". https://rosmintrud.ru/docs/mintrud/orders/1290. Accessed 19 May 2019

The draft Federal law. "On amendments to the Labor code of the Russian Federation (in terms of the formation and maintenance of information about the employee's employment in electronic form)" (prepared by the Ministry of labor of Russia). The text of the document is in accordance with the publication on the website http://regulation.gov.ru/. Accessed 06 Dec 2018

Research on the transition to an employment contract in electronic form. https://www.rostrud.ru/openrostrud/. Accessed 19 May 2019

Augmented Reality as Marketing Strategy in the Global Competition

Marina V. Vilkina[1(✉)] and Olga V. Klimovets[2]

[1] Russian Presidential Academy of National Economy and Public Administration, Moscow, Russia
m_klimovets@mail.ru
[2] Academy of Marketing and Social Information Technologies – IMSIT, Krasnodar, Russia
new_economics@mail.ru

Abstract. One of the promising developments in the field of mobile technologies is augmented reality technology. It is a technology of superimposing information in the form of text, graphics, audio and other virtual objects on real objects in real time. Interaction of computing devices with a picture of the real world distinguishes augmented reality from virtual reality. Augmented reality has significant potential to expand and support business efforts in promoting its products, as well as in increasing the competitiveness of manufactured goods and services. Marketing services companies can use augmented reality to provide contextual links between their offer to consumers, online resources and points of sale.

Keywords: International business · Marketing · Competitiveness · Outsourcing · Transnational corporations · Augmented reality · Technologies · World economy

JEL Code: E24 · F22 · J24 · P51

A key factor determining the firm's ability to compete successfully in the international market is the use of information technology. Today, thanks to IT-systems, familiar products become interactive. The software is present in large and small household appliances, medical equipment, toys, advertising and services (Klimovets 2018).

One of the promising areas of application of information technology to improve the competitiveness of the product are mobile technologies. Due to the fact that in recent years, mobile technologies are distributed at a very high speed, a large number of new developments in this area. These developments are used in various fields, such as marketing communications, sales, logistics, after-sales service, etc.

According to our estimates, in 2018 the Russian smartphone market totaled 25.3 million devices and exceeded the results of 2017 by 65% (Fig. 1).

Sales of smartphones continue to displace ordinary mobile phones. At the end of 2018, the share of smartphones in the total Russian market of mobile terminals reached 75%. Similar to the global average is higher, at 83%, but the backwardness of Russia at a unit share of smartphone sales is gradually declining.

The growth was the maximum for a year and a half, thanks to the expansion of the range of budget gadgets. Total revenue from smartphone sales in 2018 increased by

E. G. Popkova and B. S. Sergi (Eds.): ISC 2019, LNNS 91, pp. 54–60, 2020.
https://doi.org/10.1007/978-3-030-32015-7_7

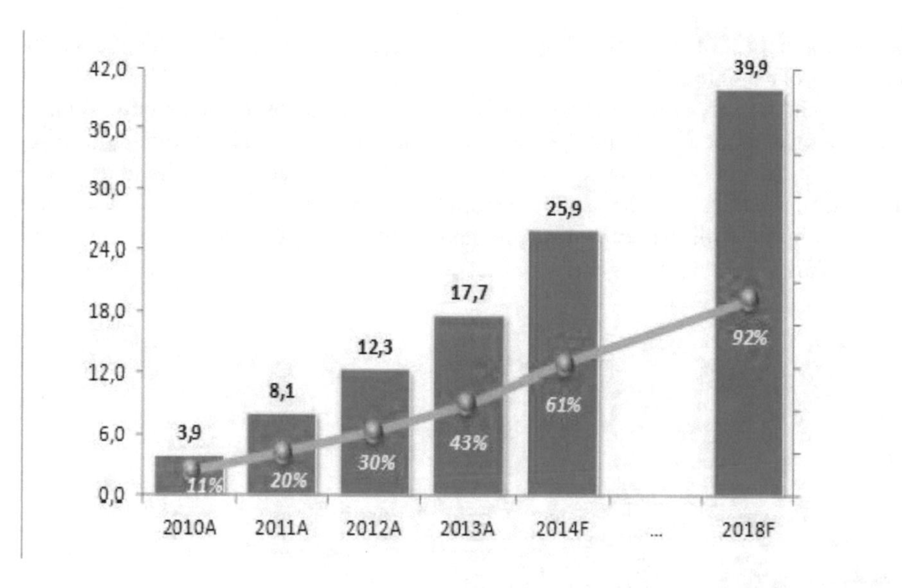

Fig. 1. Smartphone market volume in real terms in Russia, million pieces (Klimovets 2016)

28% compared to the same period in 2017. In General, the volume of the market of mobile gadgets in Russia in 2018 in monetary terms amounted to 120 billion rubles (an increase of 13.7%).

In 2018, the trend to abandon conventional phones has increased, and the speed of their replacement by smartphones has increased. The reason for the growing popularity of smartphones was the decrease in their average market price, which in 2018 reached a historic low of 5.69 thousand rubles, which is 26% less than in 2017 (Pozdnyakova et al. 2015). The share of budget smartphones worth up to 5 thousand rubles, in 2018 42.8% of the total number of devices sold (19.7% more than in 2017).

The smartphone market shows annual growth. The trend of replacing conventional phones with smartphones indicates a high degree of penetration of digital technologies to the masses, as well as an increase in the degree of understanding of users of the functionality of devices.

One of the promising developments in the field of mobile technologies is augmented reality technology. It is a technology of superimposing information in the form of text, graphics, audio and other virtual objects on real objects in real time. Interaction of computing devices with a picture of the real world distinguishes augmented reality from virtual reality. Augmented reality has significant potential to expand and support business efforts in promoting its products, as well as in increasing the competitiveness of manufactured goods and services. Marketing services companies can use augmented reality to provide contextual links between their offer to consumers, online resources and points of sale.

Until 2007, only specialists knew about the existence of augmented reality, and there were few attempts to use it in practice. However, already in 2009 dozens of companies were engaged in augmented reality. This technology is a new way to access data, but the impact of this technology on society can be comparable to the effect of the Internet, in our opinion.

The main objective of augmented reality: to expand the user's interaction with the environment, and not to separate it from reality and put in a virtual environment. Layers with content objects superimposed by means of a computer device on the image of the real environment are auxiliary and informative, thus, the information contextually related to objects with the help of augmented reality becomes available to the user in real time (Fig. 2).

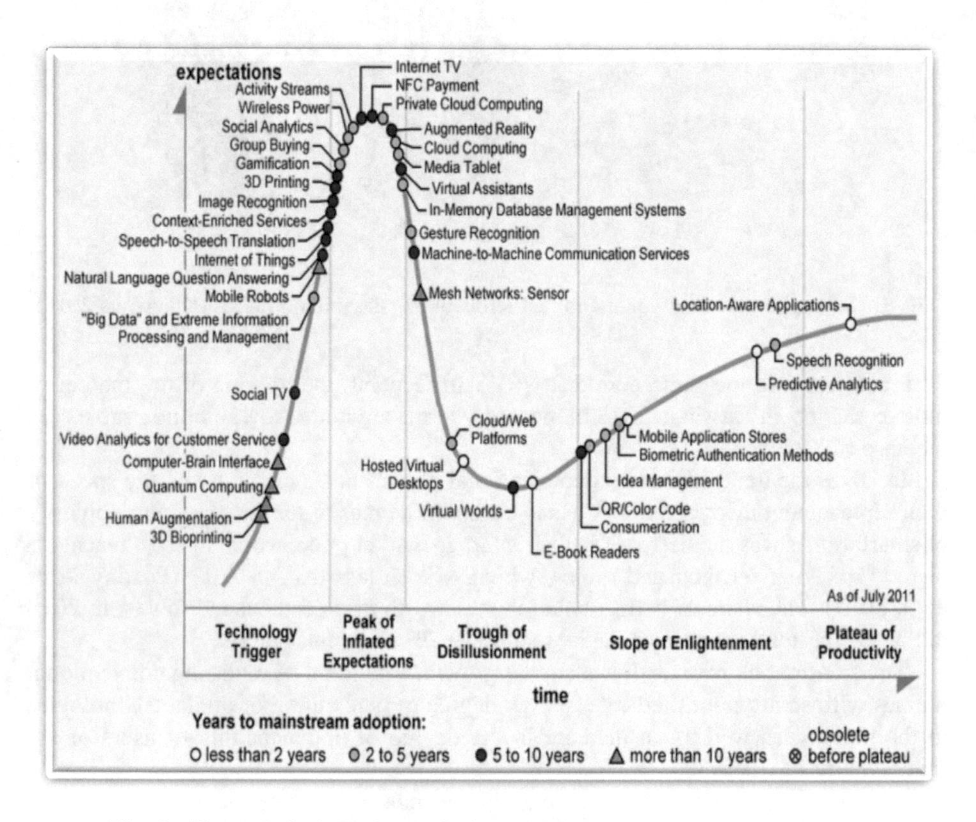

Fig. 2. The period of "Technological youth" technology "Augmented reality"

In December 2012, almost 2,000 applications using augmented reality can be found in the App Store (Klimovets 2015).

Augmented reality will move to a new level, users will be able to create a variety of content objects, animations and applications. In the near future, augmented reality will be available not only for developers but also for ordinary users, content for applications will be rapidly distributed over the Network.

Analysts believe, and we agree with them, that augmented reality has already entered the time of its technological "youth". This period is characterized by such problems as: lack of standards, lack of significant competition. But the technology will

Mature and will become dominant in the market on the horizon of 5–6 years (Klimovets 2017a, b and c). According to our calculations, in 2018 augmented reality approached the level of penetration of 50%, and by 2021 to double its achievement and reach the zone of maturity with almost 100% penetration.

Today, the augmented reality market is dominated by startups that promote this innovation. Industrial players are also actively developing in this area. Our research shows that first of all, the giants in the field of augmented reality will develop five areas: search, advertising, image recognition, mapping and 3D-visualization. The most attractive segments for augmented reality are: smartphone segment and tablet segment. Many startups from France, Germany, Austria, USA are already actively working in this market.

In the context of accelerated scientific and technological progress and increasing the number of mobile devices, augmented reality technology is expanding its presence in various areas of business. In our opinion, augmented reality is one of the promising and rapidly developing trends.

According to our forecasts, the value of the augmented and virtual reality market will continue to grow. Already today, augmented reality technologies are widely used in education. In the future, augmented and virtual reality technologies will be actively used in innovative projects due to their attractiveness and potential. Estimating separately, the market of specialized augmented solutions in 2018 will amount to 659.98 million dollars, and the market of virtual reality with immersion—407.61 million dollars (Klimovets 2017a, b and c). The main drivers of growth in these markets are advances in computer technology and Internet networks. Demand for augmented reality in the health sector is also projected to increase.

Global augmented reality market has grown from 60 million unique users in 2013 to 200 million in 2018, the Market has moved from the initial phase, in which he was in 2013, and dominated the niche gaming and navigation applications targeted at enthusiasts, to the Mature phase in 2018, when mobile augmented reality has become an integral part of the entire consumer ecosystem. We predict that augmented reality will be more actively used in social spheres. The geographical centers of attraction of augmented reality can already be called three key regions: North America, Western Europe and the far East, including China. These regions will benefit from having the most developed smartphone and tablet markets, on which augmented reality developers are more dependent. In the same regions before others began to use smart cards. The most valuable market in terms of one user in 2018 was Japan, and the US was only in third place, after Korea (Klimovets 2017a, b and c).

The growth of the global augmented reality market is driven by the following key factors: the growing need to improve user interaction; and the increasing number of augmented reality applications. In other words, augmented reality is able to meet the needs and interests of users, as well as to optimize user interaction in many areas of human activity. This increases competition in the market and leads to faster innovation.

The analysis of the global augmented reality market for 2014–2018 shows an increase in this market by an average of 130% per year. One of the key factors contributing to the growth of the augmented reality market is the growing number of applications using this technology.

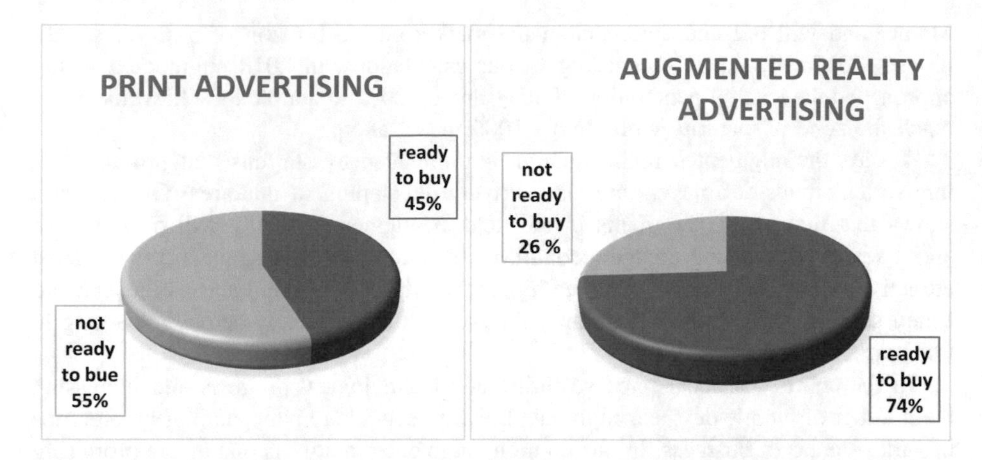

Fig. 3. The influence of augmented reality on the decision to purchase (compiled by the authors on the basis of Hidden Creative research) (Russian smartphone market 2012–2018 2019)

The growth of the augmented reality market testifies to its increasing introduction into business. There is also a trend of new devices that will increase the speed of technology implementation.

Augmented reality is a technology whose importance and usefulness is determined by the quality of content used for overlay in virtual layers. The capabilities of early applications with augmented reality technology will be in demand in four areas: marketing communications, sales, after-sales service and training, logistics (Klimovets 2017).

Augmented reality is not only a marketing tool to attract potential customers, but also an effective tool for direct sales. This technology allows you to show that for some reason it is difficult or not possible to demonstrate live.

Potential consumers of augmented reality can be: network retail, shopping centers, hotels, restaurants, services, transport, logistics, postal services.

In sales, augmented reality can improve the quality of sales, both in traditional sales channels and in e-Commerce. Additional data layers can include any useful information (product description, comparison with other alternatives, reviews, etc.) when viewing the product through the camera of a smartphone or other mobile device. For e-Commerce, augmented reality can become a tool for detailed study of the goods by a potential buyer in the context of its conditions (Global Augmented reality market 2014–2018 2019). One of the advantages of augmented reality is visibility. Augmented reality is more visual than physical models or videos, as it combines both physics and animation. Technology makes it possible to hold any product in your hands, and touch is an important sales tool. With the help of technology, it is possible to explain and show the device of complex equipment in a short time. Augmented reality transforms any traditional media: packaging, printed advertising, Souvenirs. Augmented reality provides:

- increasing the degree of involvement of the buyer in the advertising campaign.
- creation of WOW-effect for advertising media.

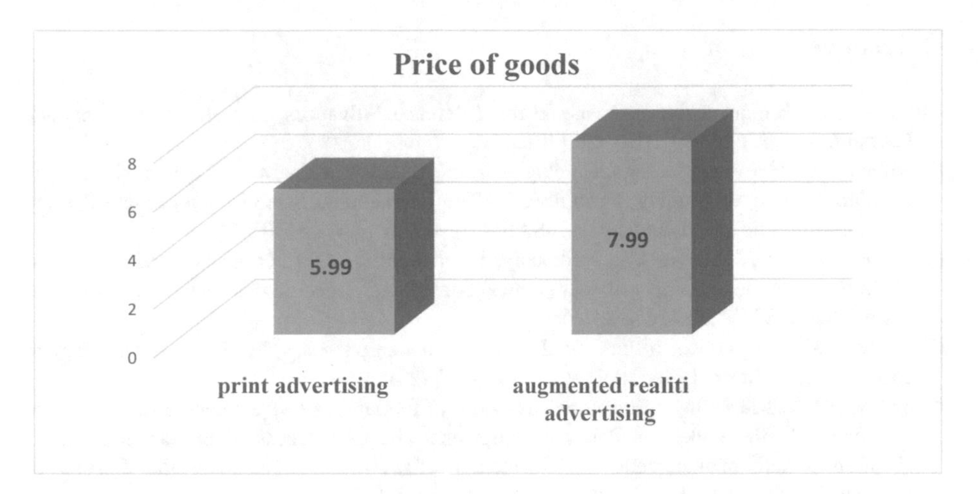

Fig. 4. The growth of consumer value of goods after the use of augmented reality (compiled by the authors on the basis of the study Hidden Creative)

– maintaining an integrated approach in advertising campaigns by placing virtual objects on any visual media.

Also, the technology allows you to show any number of products on one screen. In addition, augmented reality increases the consumer's desire to buy goods (Fig. 3).

Research company Hidden Creative conducted a small study and compared traditional print advertising and advertising with augmented reality. The essence of the study is that 100 parents were shown advertising children's toys in the traditional format, while the other 100 parents saw the same toy in the presentation with augmented reality. Participants were asked 2 questions:

1. Will you buy this toy for your child?
2. How much are you willing to pay for this toy?

After viewing a poster with a printed advertisement for a toy, 45% of parents out of a hundred said they would like to buy such a toy for their child. Of those parents who watched the presentation in augmented reality, 74% wanted to buy a toy (Fig. 4).

Those parents who saw the print ad were willing to pay $5.99 for the toy, which corresponds to the market price of this toy. Those who saw ads with augmented reality, on average, were willing to pay for this toy $7.99., this price is higher than the established market price of the product (Augmented Reality & Virtual Reality Market by Technology Types 2013).

This proves that augmented reality has a positive impact on the product selection process and increases the value of the product.

From the above it follows that augmented reality is an effective tool for direct sales and allows you to solve marketing problems. And information technologies have a significant impact on the speed of business development, creating additional competitive advantages, increasing its competitiveness at the international level.

References

Klimovets, O.: Human resources make all the difference. Advances in Intelligent Systems and Computing, vol. 622, pp. 315–320 (2018)

Klimovets, O.V.: New marketing technologies in international business. In the collection: living economics: yesterday, today, tomorrow. In: The International Scientific and Practical Web-Congress of Economists and Jurists. ISAE "Consilium", pp. 25–30 (2016)

Pozdnyakova, U.A., Dubova, Y.I., Nadtochiy, I.I., Klimovets, O.V., Rogachev, A.F., Golikov, V.V.: Scientific development of socio-ethical construction of ecological marketing. Mediterr. J. Soc. Sci. **6**(5S1), 278–281 (2015)

Klimovets, M.V.: Practice of outsourcing for strategic purposes by Russian and foreign companies. Mediterr. J. Soc. Sci. **6**(36), 193–200 (2015)

Klimovets, O.V.: Marketing of territories as a tool of formation of investment attractiveness of the region. In the collection: latest developments and the success of the development of Economics and management. In: Collection of Scientific Papers on the Results of International Scientific-Practical Conference, pp. 55–58 (2017a)

Klimovets, O.V.: Regional peculiarities of the implementation of the policy of import substitution. In the collection: modern scientific research: historical experience and innovations. In: Proceedings of the International Scientific-Practical Conference, pp. 6–12 (2017b)

Klimovets O.V.: Russian multinational corporations on the markets of Asia and Latin America. In the collection: cooperation between China and Russia in the framework of the initiative "One belt, one road". In: Collection of Materials of International Scientific-Practical Conference, pp. 150–156 (2017c)

Klimovets, M.V.: The Formation of international outsourcing in the new economy. In the collection: modern scientific research: historical experience and innovations. In: Proceedings of the International Scientific-Practical Conference, pp. 27–31 (2017)

Global Augmented reality market 2014–2018. http://www.technavio.com/report/global-augmented-reality-market-2014-2018. Accessed 25 Jan 2019

Russian smartphone market 2012–2018. http://www.advertology.ru/article123265.htm. Accessed 17 Jan 2019

Augmented Reality & Virtual Reality Market by Technology Types, Sensors (Accelerometer, Gyroscope, Haptics), Components (Camera, Controller, Gloves, HMD), Applications (Automotive, Education, Medical, Gaming, Military) & by Geography - Global Forecast and Analysis to 2013–2018. http://www.marketsandmarkets.com/Market-Reports/augmented-reality-virtual-reality-market-1185.html. Accessed 26 Jan 2019

Computer Information as the Target and the Means of Commission of Crimes According to Criminal Legislation of the Russian Federation

Andrey V. Shulga(✉) and Vladimir A. Yakushin

Kuban State Agrarian University named after I.T. Trubilin, Krasnodar, Russia
cshulga@rambler.ru

Abstract. The primary goal of this paper is to substantiate that while the information society is being built and computer technologies are being developed (information technologies based on digital or computed processing of information), computer information increasingly frequently constitutes not only the target of crimes, but also the means and the instrument of their commission. The method of achieving this goal consists in justifying the fact that the criminal offenses are increasingly frequently focused not only on computer information as such. In addition, when a person tampers with computer information, they do damage to other goods (targets) electronic payment facilities, property of others, state or bank secret, military objectives, critical infrastructure, as well as the interests of individuals (life and health), personal security, property, intellectual property, etc.).

Keywords: Crime · Target of crime · Instrument of crime · Means of commission of crime · Information · Computer information · Determining the nature of crimes · Penalty

JEL Code: K1 · K4 · O1 · O3

1 Introduction

Computer information is stipulated as the target of crimes embodied in Articles 272-2741 of Chapter 28 "Computer crimes" in Section IX of the Criminal Code of the Russian Federation. Computer information should be understood as information (messages, data) which are presented by means of electric signals, regardless of the ways of their processing, storage or transfer (Criminal Code 1996). This being the case, Chapter 28 of the Criminal Code of the Russian Federation exclusively deals with the protection of information located on a special medium – electronic. In other words, information that is stored on a different medium (for example, paper) becomes the target of crimes (for example, embodied in Articles 325, 326 or Article 327 of the Criminal Code of the Russian Federation, Chapter 32 of the Criminal Code of the Russian Federation "Crimes against administrative procedures", etc.).

E. G. Popkova and B. S. Sergi (Eds.): ISC 2019, LNNS 91, pp. 61–67, 2020.
https://doi.org/10.1007/978-3-030-32015-7_8

According to theorists, the Criminal Code of the Russian Federation does not specify any particular signs to assess the risk of crime depending on the significance of particular information. There is no difference in e-documents as the target of crime from paper or "common" documents. A question arises - what is the reason for such focused attention to electronic documents and their circulation? From the perspective of legal engineering, the placement of computer crimes in Section IX of the Criminal Code of the Russian Federation overloads the Criminal Code. In this regard, it has been argued that there is a need for the removal of Chapter 28 from the Criminal Code and the introduction of additions to the Criminal Code of the Russian Federation which might serve as a basis for correcting the deficiencies in legislation caused by the use of computer technologies during the commission of "common or conventional" crimes (Bytko, 2002).

However, the analysis of Articles 272-2741 of the Criminal Code of the Russian Federation shows that they have been introduced in the criminal law in a regular manner, with account of modern development of public relations. These legislative consolidations must not be placed on the same footing as the norms which fix responsibility for criminal acts with information fixed by means of common (paper) records.

As has already been pointed out, public danger of offenses against computer information as such is exactly associated with the formation and development of the information society in Russia. Information society is associated with processes of production, dissemination, and consumption of computer information, which thus becomes the most valuable common weal.

According to the Strategy for the Information Society Development in the Russian Federation, it is pointed out that information society is characterized by a significant level of adoption of information telecommunication technologies. Besides, it is characterized by their active use by public authorities, government authorities, citizens and entrepreneurs. The formation and development of the information society in Russia is based on such principle as national security protection in the field of circulation and use of information. Therefore, in order to achieve the abovementioned goals, the state should make provision both for the development of legislation and for the improvement of practice in the application of law in the field of the use of information telecommunication technologies (On the Strategy for the Information Society Development 2017).

The positive experience of foreign countries also bears record to the fact that cutting-edge technologies (information, telecommunication, etc.) are the driving force of both social and economic development today. Therefore, the overriding priority of these states is to guarantee free access of people to information.

Therefore, the need to improve the protection of information focused specifically on electronic media (computer information) served as a prerequisite for emphasizing an independent object of criminal and legal protection and fixing it in Chapter 28 of the Criminal Code of the Russian Federation "Computer crimes".

2 Methodology

In judicial practice, the infliction of damage in the commission of various crimes with the use of computer information (computer technologies) is treated as multiple crime (and these crimes are treated in the aggregate the crimes provided for under Chapter 28 of the Criminal Code of the Russian Federation "Computer crimes").

An alternative to this approach consists in the suggestion about the inclusion of the use of information technologies among descriptive attributes of offenses in the Criminal Code of the Russian Federation.

So, what is computer information? Is it a target of crime or a means of commission of it?

As is stated in legislation, the target of "computer crimes" is legally protected information, i.e. data for which the legislator stipulated a special legal status of protection (for example, commercial, official or state secret, personal details, etc.) (Method guidelines 2014).

It is hard to escape a conclusion if the information is protected, then it affects some area of public relations, regulates these relations (for example, relations in the field of state secret, trade secret, etc.). Therefore, when a crime is committed, an offender encroaches not on the information as such, but through this information – on the abovementioned public relations. The information only serves as the means of commission of crimes in particular public relations. For example, a "hacker" tampers with a computer program which controls the operation of a nuclear reactor at the nuclear power plant and commits a terrorist act by so doing.

Such practice should be used in cases when the purpose of the offender has been determined, when their criminal intent has been identified - what kind of public relations the offender intends to do damage to through the use of computer information. Therefore, corresponding Articles of the Criminal Code of the Russian Federation should be amended by adding a descriptive attribute or a particularly descriptive attribute: "the same action that was committed by a person with the use of information telecommunication technologies".

It would be incorrect to assess these crimes in the aggregate with computer crimes (Articles 272-2741 of the Criminal Code of the Russian Federation), since computer crimes imply that public relations in the area of public security and public order are the object of the offense. In the example above, the damage is primarily caused to relations in the field of economy, and not relations in the field of public security. In this case, illegal access to computer information should be considered as a method of stealing. Resolution of the plenum of the Supreme Court of the Russian Federation No. 29 of 27.12.2002 "On Judicial Practice in Cases of Theft, Robbery and Assault with Intent to Rob" (On Judicial Practice 2002) in Paragraph 20 resolves a similar qualification issue. In this paragraph, it is pointed out that if an offender illegally entered a room, dwelling, or storage by breaking doors, window grates, locks, etc. in case of stealing, robbery or assault with intent to rob, then the nature of such actions should be determined in line with Articles 158, 161, 162 of the Criminal Code of the Russian Federation without any additional qualification under Article 167 of the Criminal Code of the Russian

Federation, since the destruction of property during illegal infiltration is only a method of stealing (Shulga, 2018).

Thus, breaking locks, doors, and window grates while stealing tangible things constitutes an analogy to illegal access to computer information when stealing a virtual property.

If an offender's goal – what public relations he or she is attempting to inflict harm through a particular offense against computer information – is not defined, the criminal intent for the infliction of harm to certain relations is not established, then an action must be treated exclusively under Article 272-2741 of the Criminal Code of the Russian Federation. Such practice will provide an opportunity to prevent crimes that are committed with the use of computer information at a stage of attempt.

Accordingly, Part 2 of Article 272, Part 2 of Article 273, Part 2 of Article 274 and Part 5 of Article 2741 of the Criminal Code of the Russian Federation which says about causing major damage, causing severe consequences, must be supplemented with a statement that such consequences were caused by inadvertence.

3 Results

In connection with the adoption of Federal Law No. 187-FZ "On the Security of the Critical information Infrastructure of the Russian Federation" (On security 2017) of 26.07.2017, Article 2741 of the Criminal Code of the Russian Federation "illegal influence on the critical information infrastructure of the Russian Federation" was added to the Criminal Code of the Russian Federation pursuant to Federal Law No. 194-FZ of 26.07.2017.

In accordance with Article 1 of this law "The scope of application of this Federal Law", it is pointed out that a regulatory act regulates relations in the field of security of critical information infrastructure of the Russian Federation to provide for its proper functioning when it is exposed to computer attacks.

In addition, in accordance with Article 2 of the same regulatory act "Basic terms used in this Federal Law", the following has been established. Critical information infrastructure should be understood as facilities of this infrastructure, including electric communication networks that are used for the management of interaction of these facilities. Whereas facilities of critical information infrastructure are represented by automated control systems, information and telecommunication networks, or information systems of particular facilities of critical information infrastructure.

Public authorities, state institutions, Russian legal entities and private entrepreneurs who own these information systems, automated control systems, or information and telecommunication networks operating in healthcare, transportation, science, energy production, communication, banking and other financial sectors, fuel and energy sector, as well as in defense sphere, sphere of nuclear energy, mining, aerospace, chemicals and metallurgical sector, on the right of ownership, tenancy, or due to other right, as well as entities which make provision for the interaction of abovementioned networks or systems are qualified by this law as entities.

In view of the above, Article 2741 of the Criminal Code of the Russian Federation provides for criminal responsibility for illegal influence on the security of entities in public relations in such spheres as science, healthcare, communication, transportation, banking, energy production, and other sectors of the financial market, in the sphere of nuclear energy, fuel and energy sector, aerospace sector, defense sphere, metallurgic, mining and chemicals sector. And this criminal offense is committed through an action on information and telecommunication networks, information systems, automated control systems (i.e. on computer information) of these entities.

Hence, this confirms once again that computer information is a means or an instrument of violation of security of abovementioned entities. Whereas the targets of these crimes (injured persons) may be, for example, individuals, means of transport, electric power plants, communications, defensive and other buildings, money, other property, etc. Therefore, we should consider that the main direct target of crime is represented not by relations in the field of computer information, but by relations in the field of security of transportation, science, healthcare, etc.

4 Discussion

In relation to offense embodied in Articles 272-274 of the Criminal Code of the Russian Federation, offense provided for by Article 2741 of the Criminal Code of the Russian Federation is a special offense distinguished in law according to specific features of the target of crime - computer information contained in the critical information infrastructure of the Russian Federation (that is, computer information that is processed by a significant facility of critical information infrastructure of the Russian Federation), or critical information infrastructure of the Russian Federation directly.

In accordance with Paragraph 1 of Article 7 of abovementioned Federal Law No. 187-FZ of 26.07.2017, components which form critical information infrastructure must be significant. In other words, they must belong to one of certain categories of significance (either first, or second, or third); besides, they also be included on the register of significant facilities of critical information infrastructure.

Moreover, in accordance with Paragraph 12 of this Law, the category of significance to which a particular significant facility of critical information infrastructure belongs, may change (and the procedure and cases of such changes are specified). Moreover, since this statutory provision is not specifically stipulated, this category of significance can be either increased or reduced. And there is no special prohibition to change this category of significance any number of times.

This being the case, when a criminal action is treated under Article 2741 of the Criminal Code of the Russian Federation, a conflict might arise, when the target of designated crime lost the category of significance (that is, lost its features). Hence, there will be a need to treat such action under other Articles of Chapter 28 of the Criminal Code of the Russian Federation concerning computer crimes. And this may happen at any stage of existence and development of criminal relationship (starting from the prosecution and ending with the expungement of record). It is clear that such practice will by no means simplify the work of law enforcement authorities.

If we compare actions fixed in Articles 272-274 of the Criminal Code of the Russian Federation, then an offense provided for by Article 2741 of the Criminal Code of the Russian Federation fixes responsibility for actions that are more injurious to the public. Such a difference in the level of public danger should be primarily reflected in the sanctions for abovementioned offences.

However, if we compare the sanctions of above mentioned offenses provides for a penalty of imprisonment, we can see that the difference in public danger is not clearly defined in the law.

For example, the sanction that is fixed in Part 1 of Article 2741 of the Criminal Code of the Russian Federation (in special offence) fixes a penalty of two to five years of imprisonment. However, the sanction of Part 1 of Article 273 of the Criminal Code of the Russian Federation (in general offence) fixes a penalty of up to four years of imprisonment.

5 Conclusion

In our opinion, such amount of sanctions (which do not even distinguish between the categories of gravity and include the abovementioned crimes – therefore, both special offence and general offence belong to the same category – medium-gravity crimes) are not indicative of increased public danger fixed in Part 1 of Article 2741 of the Criminal Code of the Russian Federation. Not to mention the fact that the sanction of a new offense that was added to the Criminal Code of the Russian Federation entitles the courts to fix a penalty less than five years of imprisonment.

Furthermore, one should consider the absence of the lower limit of sanction provided for by Part 3 of Article 2741 of the Criminal Code of the Russian Federation (up to six years of imprisonment) as its drawback. Besides, the sanction in the standard fixed in Part 1 of Article 274 of the Criminal Code of the Russian Federation, also fixes a penalty of imprisonment, yet its term does not exceed two years. Reasoning from this fact, in case of commission of crime provided for by Part 3 of Article 2741 of the Criminal Code of the Russian Federation, the court may fix the same penalty – two years of imprisonment. In this case, when the nature of abovementioned actions (computer crimes) is determined, the difference in the assessment of their public danger is eliminated.

Such practice is conceivable (equal terms of imprisonment can also be appointed by the court) when the offenders are held criminally liable for the same crimes, but committed in privity (in accordance with Part 4 of Article 2741 of the Criminal Code of the Russian Federation and Part 3 of Article 272, Part 2 of Article 273 of the Criminal Code of the Russian Federation). For example, a sanction is envisaged for a crime in Part 4 of Article 2741 of the Criminal Code of the Russian Federation (special offense), which fixes a penalty of three to eight years of imprisonment. At the same time, Part 3 of Article 272 or Part 2 of Article 273 of the Criminal Code of the Russian Federation (general offense) fixes a penalty for a collective crime of up to five years of imprisonment. If this is the case, how can the increased level of public danger of a crime be reflected in the judgement of court with due consideration of specific features of its target, if for an act provided for by Part 4 of Article 2741 of the Criminal Code of the Russian Federation, the court will fix a penalty of three, four or five years of imprisonment?

The same conclusion can also be made while analyzing sanctions specified in Part 5 of Article 2741 of the Criminal Code of the Russian Federation and Part 4 of Article 272 of the Criminal Code of the Russian Federation, Part 3 of Article 273 of the Criminal Code of the Russian Federation which fix penalties for the commission of abovementioned computer crimes which resulted in severe consequences. Part 5 of Article 2741 of the Criminal Code of the Russian Federation fixes a penalty of five to ten years of imprisonment. Part 4 of Article 272 of the Criminal Code of the Russian Federation and Part 3 of Article 273 of the Criminal Code of the Russian Federation fixes a penalty of up to seven years of imprisonment. In this case, the abovementioned crimes are be classified among serious crimes (which also reduces the differentiation of their level of public danger). If such crimes were committed, the court may appoint an equivalent penalty – for example, five, six or seven years of imprisonment.

Hence, an understandable question arises - what is the purpose of emphasizing general and special offenses in the criminal law from among computer crimes in this situation, if they belong to the same category of severity and may carry the same punishments?

It appears that this problem can be resolved through a more distinct differentiation of public danger of abovementioned crimes by means of sanctions enshrined in law, stipulated for the commission of computer crimes.

References

Criminal Code of the Russian Federation of May 24, 1996. Official Gazette. 1996. No. 25, Article 2954

On the Security of the Critical information Infrastructure of the Russian Federation. Federal Law No. 187-FZ of 26.07.2017. Rossiyskaya Gazeta No. 167 of 31.07.2017

On the Strategy for the Information Society Development in the Russian Federation for 2017–2030. Decree of the President of the Russian Federation No. 203 of 09.05.2017

On Judicial Practice in Cases of Theft, Robbery and Assault with Intent to Rob. Resolution of the plenum of the Supreme Court of the Russian Federation No. 29 of 27.12.2002. Rossiyskaya Gazeta. No. 9 of 18.01.2003.

Method guidelines for the procuracy supervision over the law enforcement in the investigation of computer crimes (approved by the Office of the Prosecutor General of the Russian Federation), 15 April 2014. http://genproc.gov.ru

Bytko, S.Y.: Certain issues of criminal responsibility for the crimes that are committed with the use of computer technologies. A thesis research of the candidate of legal sciences. Saratov, pp. 23–24, 44, 143, 144 (2002)

Shulga, A.V., Galiakbarov, R.R.: Criminal responsibility for illegal influence on the critical information infrastructure of the Russian Federation (Article 274.1 of the Criminal Code of the Russian Federation). Gumanitarnyie, Sotsialno-Ekonomicheskie i Obschestvennyie Nauki. No. 5. pp. 238–242 (2018)

The Algorithm of Creation of Territories of Rapid Socio-Economic Development in the Digital Economy

Tatiana A. Zabaznova(✉), Elena V. Patsyuk, Natalia V. Shchukina, Svetlana E. Karpushova, and Olga A. Surkova

Sebryakov Branch of Volgograd State Technical University, Volgograd, Russia
tazabaznova@yandex.ru, elenapatsyuk@yandex.ru, schuka-amurskay@yandex.ru, sfkse@yandex.ru

Abstract. Purpose: The purpose of the paper is to develop an algorithm of creation of territories of rapid socio-economic development in the digital economy and to approbate it by the example of city of Mikhaylovka, Russia.

Design/Methodology/Approach: Based on the data of the Federal State Statistics Service for 2018, the authors analyze GDP per capita in Volgograd Oblast and compare it to the average value for the Southern Federal District and Russia. Dynamics of income and expenditures per capita in Volgograd Oblast in 2013–2017 are analyzed, and a forecast with all other conditions being equal for 2018–2019 is compiled.

Findings: An algorithm of creation of territories of rapid socio-economic development in the digital economy is created. A conceptual model of territory of rapid socio-economic development in the digital economy in Mikhaylovka is offered. Recommendations for the indicator value of the results of creation of a territory of rapid socio-economic development in Mikhaylovka in 2024 are provided.

Originality/Value: The offered hypothesis is proved; it is substantiated that the digital economy enables creation of the territories of rapid socio-economic development and preservation of their production specialization. It is shown by the example of Mikhaylovka that within the national tendency of digital modernization it is possible to attract the federal financing of implementation of digital technologies into top-priority spheres of the territorial economy and start hi-tech productions.

The territory will receive diversification of economy within the current sectorial specialization at production of construction materials. Creation of a territory of rapid socio-economic development in Mikhaylovka will allow for systemic modernization of the construction sphere and starting of digital construction (with the usage of construction robots and manipulators) and construction of "smart" buildings, thus increasing quality of life (including on the inclusive basis) in Volgograd Oblast on the whole and Mikhaylovka in particular.

Keywords: Territory of rapid development · Socio-economic development · Digital economy · Digital construction · "smart" construction

JEL Code: L74 · O18 · O31 · O32 · O33 · O38 · R12

E. G. Popkova and B. S. Sergi (Eds.): ISC 2019, LNNS 91, pp. 68–76, 2020.
https://doi.org/10.1007/978-3-030-32015-7_9

1 Introduction

One of the most important problems of the modern regional economy is vivid disproportions in development of regional centers and peripheral cities. In Russia, this problem is expressed in prioritized federal financial support for regional centers, while peripheral cities lack financing, which restrains their development. However, a lot of peripheral cities have a large potential of socio-economic development and play an important role in the region's economy.

An example of this problem is the city of Mikhaylovka (hereinafter - Mikhaylovka) in Volgograd Oblast (Russia, Southern Federal District), which is the region's 4th city as to population (87,100 as of January 1, 2019) (Volgogradstat 2019) – i.e., 2.31% of population of Volgograd Oblast, which total number is 2,521,276 (Federal State Statistics Service 2019). Mikhaylovka has a large concentration of construction materials, which predetermines its important role in the economy of Volgograd Oblast; the share of construction in its GRP constituted 7.71% in 2018 (Volgogradstat 2019).

Development of the residential and infrastructural (including housing and utility, transport and logistical, and industrial) construction is included into the list of the tasks of the Strategy of socio-economic development of Volgograd Oblast until 2025, adopted by the Law of Volgograd Oblast dated November 21, 2008, No. 1778-OD (Volgograd Oblast Duma 2019). Mikhaylovka is a monocity – most of the population work at two companies, which specialize in production of construction materials.

That's why increase of living standards and quality of life requires increase of the power of sellers in the labor market via diversification of entrepreneurship of the territory with preservation of its production specialization. Here we offer a hypothesis that the digital economy opens new opportunities for acceleration of socio-economic development of the territory. The purpose of the work is to develop the algorithm of creation of territories of rapid socio-economic development in the digital economy and to approbate it by the example of Mikhaylovka.

2 Materials and Method

The scientific and theoretical foundations and the existing practical experience of creation of territories of rapid socio-economic development are studied in the works Bannikova et al. (2019), Bogoviz et al. (2018), Peshkova (2017), Ragulina et al. (2019), and Vukovich et al. (2018). Specific features of socio-economic development of the territories in the digital economy are studied in the works Bezrukova et al. (2017), Boulesnane et al. (2018), Lozano and Gaona-Ramirez (2018), Popkova (2019), Pritvorova et al. (2018), Sergi et al. (2019), Sibirskaya et al. (2019), and Vanchukhina et al. (2018).

Sufficiency of the literature sources on the selected topic shows its through elaboration from the scientific point of view. However, the problem of creation of territories of rapid socio-economic development in the digital economy is new for the modern economic sciences and is thus insufficiently studied.

From the scientific and methodological point of view, the basis for creation of the territories of rapid socio-economic development in a region is low living standards.

According to the Federal State Statistics Service (2019), the 2018 GDP per capita (RUB 292,565.7) in Volgograd Oblast was below the average GDP per capita in the Southern Federal District (RUB 298,585.7) by 2.02% and below the average GDP per capita for Russia (RUB 472,161.9) by 38,04%. Dynamics of money revenues and expenditures per capita in Volgograd Oblast in 2013–2017 and the forecast with all other conditions being equal for 2018–2019 are given in Fig. 1.

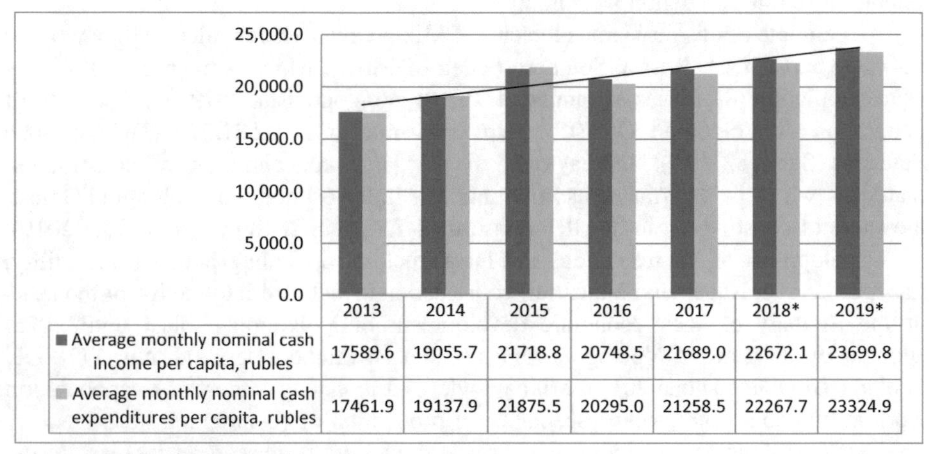

Fig. 1. Dynamics of money revenues and expenditures per capita of Volgograd Oblast in 2013–2019. Source: calculated and compiled by the authors based on the materials of Volgogradstat (2019).

Figure 1 shows that in 2013–2019 money revenues and expenditures of the population of Volgograd Oblast were at the same level, with an upward trend (dotted line). Together with low living standards, as compared to other regions of the Southern Federal District and Russia, this shows the topicality of creation of the territories of rapid development of in Volgograd Oblast.

Let us provide more detailed information on the research object – the city of Mikhaylovka. The city district includes 56 communities; according to the Law of Volgograd Oblast dated June 28, 2012, No. 65-OD "Regarding unification of rural communities in the Mikhaylovka municipal district of Volgograd Oblast with the city district of the city of Mikhaylovka of Volgograd Oblast, implementation of changes into the Law of Volgograd Oblast dated March 25, 2005, No. 1033-OD 'Regarding establishment of borders and assigning of status of city to Mikhaylovka of Volgograd Oblast', and loss of effect of certain legislative acts of Volgograd Oblast".

The main purpose of creation of a territory of rapid socio-economic development is diversification of the economy of a city district or overcoming of mono-dependence – reduction of the share of systemic companies in a city district's economy through development of various types of economic activities that are not connected to the activities of systemic companies. As of now, state initiatives on creation of a territory of rapid socio-economic development in Mikhaylovka are already in effect.

According to the Decree of the Government of the Russian Federation dated April 12, 2019, No. 428 "Regarding creation of a territory of rapid socio-economic development 'Mikhaylovka'", the following types of economic activities are envisaged:

- crop research and cattle breeding, hunting and provision of the corresponding services in these spheres;
- production of food products, production of soft drinks;
- production of furniture;
- maintenances and assembly of machines and equipment;
- activities on provision of food products and drinks, activities in sports, leisure, and entertainment;
- activities in the sphere of healthcare;
- development of computer software, consultation services in this spheres, and other corresponding services; activities in the IT sphere;
- production of chemical products, production of plastic items;
- transport processing of cargoes, activities on warehousing and storing.

Besides, the item (c) Requirements to investment projects that are implemented by the residents of territories of rapid socio-economic development, created on the territories of mono-profile municipal entities of the Russian Federation (monocities) set that "implementation of an investment project does not envisage execution of contracts with a systemic organization of the monocity or its subsidiary organizations and/or receipt of revenues from realization of goods and provision of services to the systemic organization of the monocity or its subsidiary organizations in the volume that exceeds 50% of all revenues from realization of goods (services) and performed works as a result of implementation of an investment project".

We think that the set list of the types of economic activities, which are to diversify the economy of Mikhaylovka, contradicts the natural market conditions on the territory of the city district. That's why creation of a territory of rapid socio-economic development will be either slow or too expensive for state budgets of Mikhaylovka, region (Volgograd Oblast), and the consolidated federal budget of the Russian Federation (under the condition of provision of federal financing). Thus, an important task is harmonization of the types of economic activities based on which diversification and acceleration of development of Mikhaylovka, with its market environment and existing sectorial specialization, will take place.

3 Results

For creation of territories of rapid socio-economic development in the digital economy we offer to use the following algorithm (Fig. 1).

Figure 2 shows that the offered algorithm is implemented in five consecutive stages. The first stage "Preparation" envisages analysis of the conditions – determining the territory's socio-economic position and its technological mode (level of digitization). The second stage "Selection of 'growth points'" envisages analysis of the territory's sectorial specialization and selection of its most perspective "growth points" in view of the capabilities of the digital economy.

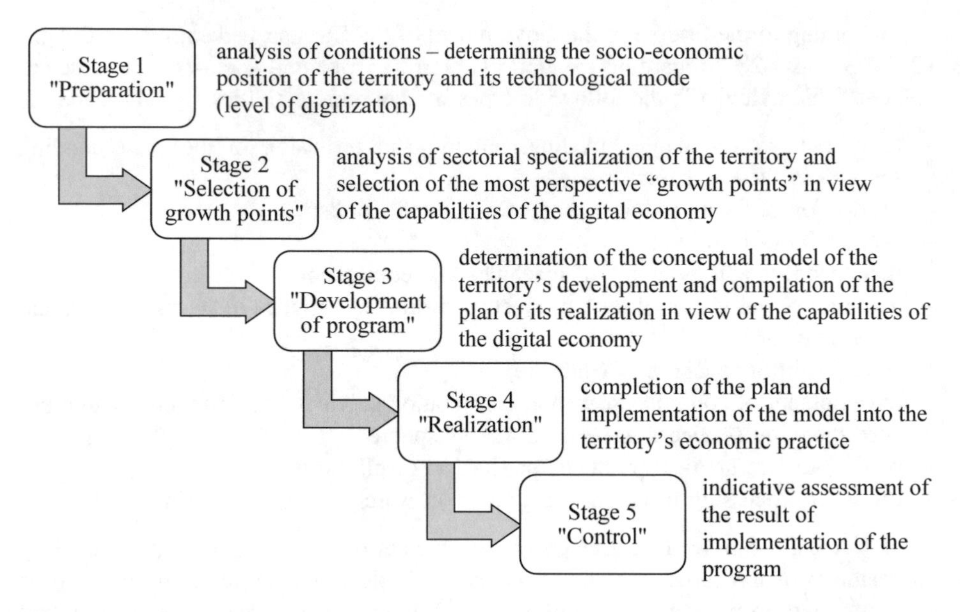

Fig. 2. The algorithm of creation of the territories of rapid socio-economic development in the digital economy. Source: compiled by the authors.

The third stage "Development of program" envisages determining the conceptual model of development of the territory and compiling the plan of its implementation in view of the capabilities of the digital economy. The fourth stage "Realization" envisages execution of the plan and implementation of the model into the territory's economic practice. The fifth stage is indicative evaluation of the results of the program's implementation (execution of the plan).

Let us perform approbation of the developed algorithm by the example of Mikhaylovka. Capabilities of the digital economy on the studied territory are moderate – same as for Volgograd Oblast, which was ranked 38th among 75 regions of Russia as to the level of digitization in 2018 (D-Russia 2019). Mikhaylovka specialized in production of construction materials. However, this production is narrow and limited by production of cement and asbestos-cement products.

The "growth point" of the selected territory is expansion of the list of production of construction materials. A perspective means of acceleration of the territorial development could be territorial cluster of the construction sphere, which unifies the whole complex of productions of construction materials (Fig. 3).

In the Decree of the President of the Russian Federation dated May 7, 2018, No. 204 "Regarding the national goals and strategic tasks of development of the Russian Federation until 2024", the top-priority goals include modernization of the construction sphere and increase of the quality of construction. This goal is to be achieved in the context of implementation of the "technologies of information modeling" and other digital technologies into the practice of construction (President of the Russian Federation 2019).

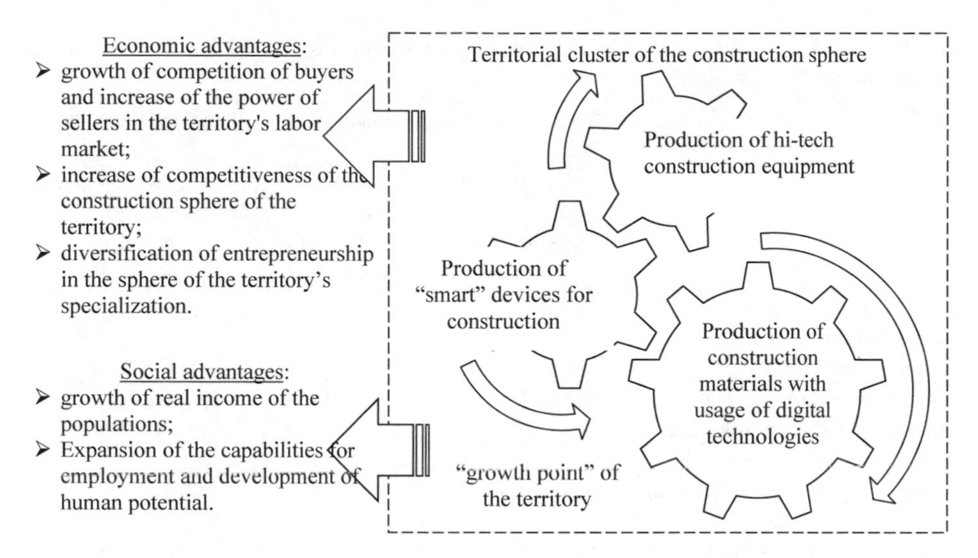

Fig. 3. A conceptual model of a territory of rapid socio-economic development in the digital economy in Mikhaylovka. Source: compiled by the authors.

Within this goal, it is possible to obtain federal financing of digital modernization of construction productions in Mikhaylovka. The territorial cluster of the construction sphere will be based on the existing production of construction materials, which will be optimized on the basis of digital technologies – e.g., automatized conveyor lines based on manipulators and robototronics. This will increase security (asbestos production is harmful for human health) and environmental friendliness of the products.

The cluster is to become a basis for production of hi-tech construction equipment and devices – e.g., robots, manipulators, sensors, etc. Also, it is recommended to start production of "smart"-devices for construction of "smart"-buildings (commercial and residential, which ensure high level of their authomatization of their usage and inclusive character). This will allow forming infrastructural provision for digital modernization of construction in Volgograd Oblast on the basis of construction materials and equipment that are manufactured in Mikhaylovka.

The economic advantages for Mikhaylovka include growth of buyers' competition and increase of sellers' power in the territory's labor market, increase of competitiveness in the territory's construction sphere, and diversification of entrepreneurship in the sphere of the territory's specialization. Social advantages for Mikhaylovka include growth of real disposable income of the population and expansion of the capabilities for employment and development of human potential.

For indicative evaluation of the result of creation of a territory of rapid socio-economic development in Mikhaylovka we shall use the following scale (Table 1).

Table 1 shows that the indicators of efficiency of creating a territory of rapid socio-economic development in Mikhaylovka, according to the offered models, are as follows:

Table 1. Indicators for evaluating the efficiency of creation of a territory of rapid socio-economic development in Mikhaylovka.

Indicator	Scale for qualitative treatment of the indicator's values in 2024		
	Values that show critical underrun from the plan	Target (planned) values of the indicators	Values that show significant excess of the plan
Share of hi-tech production construction materials in the structure of GDP, %	Below 5	10	Above 15
Share of production of "smart"-devices for construction, in the structure of GDP, %	Below 5	10	Above 15
Annual growth rate of GDP in constant prices, %	Below 2.5	5	Above 7.5
Index of tension in the labor market[a]	Above 1.5	1	Below 0.75
GDP per capita, RUB (in the 2018 prices)	Below 298,585.7	300,000.0	Above 320,000.0

[a]Ratio of unemployed to the number of vacancies
Source: compiled by the authors.

- share of hi-tech production of construction materials in the structure of GDP: it should constitute 10%;
- share of production of "smart"-devices for construction in the structure of GDP: it should constitute 10%;
- annual growth rate of GDP in constant prices: it should constitute 5%;
- index of tension in the labor market: it should constitute 1 – i.e., balance of demand and offer in the labor market should be established;
- GDP per capita, RUB (in the 2018 prices): it should constitute RUB 300,000 (it should increase the average level of the Southern Federal District).

Planned values of the indicators are established for 2024, as it is the period of termination of digital modernization of the Russian economy, based on the forecast of the start of practical implementation not before 2020. The values of the indicators are determined – they show critical underrun from the plan, and the values, which show significant excess of the plan.

4 Conclusion

The offered hypothesis is proved; it is substantiated that the digital economy enables creation of territories of rapid socio-economic development with preservation of their production specialization. It shown by the example of Mikhaylovka that within the national tendency of digital modernization it is possible to attract federal financing of implementation of digital technologies into the top-priority spheres of territorial economy and to start hi-tech productions.

The territory will receive diversification of economy within the current sectorial specialization at production of construction materials. Creation of a territory of rapid socio-economic development in Mikhaylovka will allow for systemic modernization of the construction sphere and starting of digital construction (with the usage of construction robots and manipulators) and construction of "smart" buildings, thus increasing quality of life (including on the inclusive basis) in Volgograd Oblast on the whole and Mikhaylovka in particular.

References

Bannikova, N.V., Astrakhantseva, E.Y., Morozova, I.A., Litvinova, T.N.: Mastering of the information and communication technologies by labor resources of rural territories as a new vector of their development in modern Russia. Adv. Intell. Syst. Comput. **726**, 701–708 (2019)

Bezrukova, T.L., Popova, E.V., Korda, N.I., Kuznetsova, T.E., Bezrukov, B.A.: Institutional traps of innovative and investment activities as an obstacle on the path to the well-balanced development of regions. Contributions to Economics, pp. 235–240 (2017)

Bogoviz, A.V., Alekseev, A.N., Kletskova, E.V., Kuznetsov, Y.V., Cherepukhin, T.Y.: Territories of advanced economic development as the most favorable environment for the development of a modern man. Qual. - Access Success **19**(S2), 161–165 (2018)

Boulesnane, S., Bouzidi, L., Varinard, C.: CT and territory development: Issues of digital mediations | [Tic et développement des territoires: Enjeux des médiations numériques]. Commun. et Manag. **15**(1), 75–86 (2018)

D-Russia: Rating of regions as to the level of development of digitization "Digital Russia" (2019). http://d-russia.ru/vyshla-polnaya-versiya-rejtinga-regionov-po-urovnyu-razvitiya-tsifrovizatsii-tsifrovaya-rossiya.html. Accessed 10 May 2019

Lozano, L.A., Gaona-Ramirez, G.A.: Digital observatory of social appropriation of knowledge of a territory. Commun. Comput. Inf. Sci. **942**, 3–15 (2018)

Peshkova, M.K.H.: Feasibility study of management of man-made mineral formations in the territories of advanced socio-economic development. Gornyi Zhurnal **6**, 39–43 (2017)

Popkova, E.G.: Preconditions of formation and development of industry 4.0 in the conditions of knowledge economy. Stud. Syst. Decis. Control **169**, 65–72 (2019)

Pritvorova, T., Tasbulatova, B., Petrenko, E.: Possibilities of blitz-psychograms as a tool for human resource management in the supporting system of hardiness of company. Entrepr. Sustain. Issues **6**(2), 840–853 (2018). https://doi.org/10.9770/jesi.2018.6.2(25)

Ragulina, Y.V., Semenova, E.I., Avkopashvili, P.T., Dmitrieva, E.A., Cherepukhin, T.Y.: Top-priority directions of implementing new internet technologies on the territories of rapid economic development. Adv. Intell. Syst. Comput. **726**, 182–188 (2019)

Sergi, B.S., Popkova, E.G., Bogoviz, A.V., Ragulina J.V.: Entrepreneurship and economic growth: the experience of developed and developing countries. In: Entrepreneurship and Development in the 21st Century, pp. 3–32. Emerald publishing limited (2019)

Sibirskaya, E., Popkova, E., Oveshnikova, L., Tarasova, I.: Remote education Vs traditional education based on effectiveness at the micro level and its connection to the level of development of macro-economic systems. Int. J. Educ. Manag. **33**(3), 533–543 (2019). https://doi.org/10.1108/IJEM-08-2018-0248

Vanchukhina, L.I., Leybert, T.B., Khalikova, E.A., Khalmetov, A.R.: New approaches to formation of innovational human capital as an element of institutional environment. Espacios **39**, 22–32 (2018)

Vukovich, G.G., Makushenko, L.V., Bateykin, D.V., Titova, O.V., Dobrosotskiy, V.I.: Support of the territories of advanced economic development on human capital: theory and practice. Qual. - Access Success **19**(S2), 157–160 (2018)

Volgograd Oblast Duma. The Strategy of socio-economic development of Volgograd Oblast until 2025, adopted by the Law of Volgograd Oblast dated November 21, 2008, No. 1778-OD (2019). http://docs.cntd.ru/document/819076044. Accessed 10 May 2019

Volgogradstat. Volgograd Oblast in numbers: a short statistical collection (2019). http://volgastat.gks.ru/wps/wcm/connect/rosstat_ts/volgastat/resources/bacce3804560f219910cdfc4d78fa45b/002_2017.pdf. Accessed 10 May 2019

President of the Russian Federation. Decree dated May 7, 2018, No. 204 "Regarding the national goals and strategic tasks of development of the Russian Federation until 2024" (2019). https://base.garant.ru/71937200/. Accessed 10 May 2019

Federal State Statistics Service. Regions of Russia. The main socio-economic indicators of cities – 2018 (2019). http://www.gks.ru/wps/wcm/connect/rosstat_main/rosstat/ru/statistics/publications/catalog/doc_1138631758656. Accessed 10 May 2019

Appellations of Origin of Goods as a Tool of Socio-Economic Development and Preservation of Cultural Diversity of Russia in the Context of Globalization

Irina V. Samsonova[1(✉)], Elena V. Andreeva[2], Anna A. Neizvestnykh[3], and Vasilii V. Skriabin[3]

[1] Academy of Sciences of the Sakha Republic, North-Eastern Federal University by M. Ammosov, Yakutsk, Russia
irsam@list.ru
[2] North-Eastern Federal University by M. Ammosov, Yakutsk, Russia
klio_e@mail.ru
[3] Academy of Sciences of the Sakha Republic, Yakutsk, Russia
anna08062009@rambler.ru

Abstract. One of the causes of tension and instability of the ethnic group as a social system, today are the processes of globalization. Issues of preservation and development of traditional knowledge of the ethnic group through folk art crafts, as a tool for preservation of cultural diversity and social basis of society in conditions of globalization. It has received great attention from both scientists and practitioners.

The purpose of this study is a comparative analysis of approaches to legal protection of products of traditional crafts as objects of intellectual property. Consideration from the point of view of evolutionary, systemic approaches the development of crafts in various countries of the world and Russian Federation.

The methodology for studying the content of discussions consists of: development of conceptual apparatus, opinions of experts, statistical (quantitative) methods, evaluation methods of state policy in the field of protection of intellectual rights of producers of folk handicrafts and the preservation of traditional knowledge.

This article presents results on the analysis of folk art crafts state as a sector of market economy and social stability of society in risks conditions of ethnic identity loss by globalization processes. The role of the industry in the economy of countries has been determined, as well as the most optimal mechanisms for protecting the legitimate interests of producers have been identified. Protection of traditional knowledge from distortion and favorable conditions for the development of creativity is identified also.

Recommendations on updating of international agreements in the field of preservation of traditional crafts, including protection of intellectual rights of their producers, taking into account positive experience regulation of countries practicing the use of appellations of origin.

E. G. Popkova and B. S. Sergi (Eds.): ISC 2019, LNNS 91, pp. 77–82, 2020.
https://doi.org/10.1007/978-3-030-32015-7_10

Keywords: Ethnos · Culture · Traditions · Folk art crafts · Goods · Appellation of origin · Geographical indication · Traditions · Objects of intellectual property · Globalization

JEL code: F69 · L79 · O34 · Z19

1 Introduction

Problems of preservation, revival, and development of folk art crafts are not a problem of the economy of the country, which have intersectoral, interdepartmental, interregional and government quality. This is, first of all, lack of cultural and social attention, both from government and regions. On this basis, solved the following tasks: define the role of traditional knowledge in the preservation of cultural diversity in the context of globalization; consider the sphere of arts and crafts as the social basis of society; legislation in various countries to determine the most effective system for protection of the intellectual rights of producers of handicrafts and preservation of traditional knowledge.

Sections following after introduction show the place of this work among studies on the preservation and development of traditional ethnos knowledge through folk arts and crafts. Describes the data and method used in this study. Below are results of the analysis in the context outcomes of previous studies. The last part presents conclusions, highlights limitations of this work, as well as the directions for further research.

2 Methodology

This study used an interdisciplinary approach based on the synthesis of such scientific disciplines as cultural studies, ethnography, economics, and law. Traditional knowledge, expressed in handicrafts is a unit of goods in domestic and foreign markets.

In order to achieve objective of study, solve the problems and ensure high reliability of results. We used a systematic approach to the object under study, methods of induction and deduction, synthesis, analysis, method of comparative law, statistical (quantitative) methods of cognition and comparative analysis.

3 Results and Discussion

One of scientific approaches to study the ethnos is the evolutionary and historical direction of primordialism according to which ethnos is considered as a social community of people, embodied in a historical form as a result of social and historical development and has special unique characteristics such as language, identity, and culture, whose content and features are preserved in ethnos and transmitted individually (Bromley 1978).

One way of self-preservation of ethnic groups in such circumstances is to strengthen their cultural identity. UNESCO adopted the International Convention for the Safeguarding of the Intangible Cultural Heritage in 2003, which called for the protection of forms of representation and expression, skills, manners, knowledge and related instruments, including artifacts and cultural spaces recognized by ethnic groups as their cultural heritage (UNESCO 2003).

For example, in Russia one of the main directions of national policy in Russia is development of national cultures and languages of peoples. In China guidelines were adopted for assessing the government intangible cultural heritage in order to enhance the protection of China's national intangible cultural heritage, including artifacts in 2005 (Hu et al. 2013). In France, all kinds of handicraft arts and crafts are recognized as the most important part of the artistic and cultural heritage.

Development of crafts in various countries of the world has always been based on ethno-cultural traditions, which helped to shape the image of territories and attention of investors. Nowadays craft, as an exclusive sphere of economy, exists in all countries of the world. In the countries of the European Union (EU), handicraft activity is the basis of social and economic stability of Europe, due to increase in the share of handicrafts in Europe has formed a powerful "middle class". In the European Union, artisans have a tax load on value added, which is at least 15%. The share of the gross domestic product (GDP) of handicrafts sector in a number of European countries is as follows: Greece—3.0%, Netherlands—3.5%, France—5.1%, Germany—9.6%, Italy—12.0%, Luxembourg—15.0%. Different definitions of handicrafts have been adopted in different European countries. For example, in Italy, artistic crafts, along with construction, transport services, mass services, are an opportunity to organize a small business. In France, artisans are persons who carry out work at their own expense, run their firm and usually take part in their activities personally, have a professional qualification (length of service confirmed by a certificate of the Chamber of Commerce or diploma), and the company employs only members of their families, students or no more than five partners. In Poland, craftsmanship fully coincides with the concept of small business in industrial production. The craft includes professional production activity of a natural person (or their society), with participation of personal skilled labor, on behalf of that person and at his expense (with employment up to 50 employees). In Germany, a person or group who is registered in the Chamber of Commerce belongs to one of 124 professional federations for handicrafts or has a certificate confirming professional skills.

The market of folk arts and crafts is influenced by raw material base of a country. It's market specialization within the world economy. Well-developed folk crafts in countries where the economy is heavily dependent on tourism - Egypt, India, Italy, Czech Republic.

In the Russian Federation legal status of folk, art trade is given to legal entities of any organizational and legal forms of ownership that sell products of recognized art virtues made in accordance with traditions of the place of distribution.

Folk art crafts in the Russian economy also play a big role. To date, according to the Ministry of Industry and Trade of the Russian Federation, manufacture of objects of folk art crafts, including 15 types of productions, is engaged in more than 250 organizations located in 64 constituent entities of the Russian Federation. The main part of Russian crafts refers to processing production - porcelain, pottery, glass products (35%), garment and textile production (26%), wood processing (23%), etc. About 74% of products, according to official statistics, are manufactured at enterprises of six subjects of Russia. These are Moscow region, Vologda, Vladimir, Tver regions, St. Petersburg and Republic of Dagestan.

Today in the Republic of Sakha (Yakutia) production of art products and souvenirs is carried out by 25 economic entities in 12 districts and cities. 89 Yakutian are engaged in individual entrepreneurship in this sphere. More than six thousand craftsmen from all over the country work in 108 associations of masters. Folk handicrafts in the total production of non-food consumer goods in the Republic of Sakha (Yakutia) account for about 10%. One of dynamically developing segments of a market is production and sale of mammoth tusk products, share of which is more than 18% of the total production of artistic and souvenirs.

In order to protect the legitimate interests of producers, in relation to products of folk art crafts, such legal institute of protection of objects of intellectual property may be applied as an appellation of origin.

In world practice, there is no single definition of the concept of "appellation of origin". The appellation of origin as a term is first mentioned on the Paris Convention for the Protection of Industrial Property of 1883 (Paris Convention 1883). The first paragraph of article 2 of the Agreement on the Protection of Appellations of Origin and their International Registration defines for the first time semantic meaning as geographical name of a country, area or locality used to indicate an article originating from a given country, area or locality, the quality and characteristics of which are explained solely or principally by geographical environment, including natural and human factors" (Lisbon Agreement 1958).

However, article 22/1 of TRIPS introduces the concept of "geographical indication" which is designation by which goods from a certain geographical area are identified, locality or region, quality, reputation and other characteristics of goods are largely determined by its geographical origin. This definition does not establish a direct link between the characteristics of a commodity with natural conditions and human factors.

There is no unity in the legal systems of different countries regarding protection of an appellation of origin and geographical indication due to differences in their concepts and status as objects of intellectual property. In some countries, protection of an appellation of origin and geographical indications are based on the legal institution of registration of collective or certification marks.

For example, the Republic of Korea has a system for protection of geographical indications, as interpreted by TRIPS, based on registration of a collective mark with the national patent office (O'Connor et al. 2017). In other countries, appellations of origin and geographical indications are protected on the basis of registration with the patent office or on separate regulations of governmental authority.

The World Intellectual Property Organization (WIPO), in an effort to harmonize existing national systems for the protection of geographical indications, recommends using intellectual property special approach (sui generis). It is necessary to identify shortcomings in the existing intellectual property law enforcement system that sui generis should work on (WIPO 2013).

The main difference between appellation of origin and geographical indication is that handicrafts should be produced only on the territory of geographical object which is traditional place of existence of folk arts and crafts. Special properties are determined by the geographical environment, including natural conditions and human factors.

Place of origin due to its conditionality and dependence on the human factor is part of the culture of a certain ethnic group living in the territory of a specific geographical object.

In Russia, there is currently a legal institute, whose concept corresponds to the provisions of the Lisbon Agreement for the Protection of Indications of Origin (Lisbon Agreement 1958). To register the appellation of origin of goods must be established dependencies of special properties on characteristic for geographical or natural factors. Also, traditions of folk art crafts and its artistic and stylistic features inherent in places of their traditional existence, technological processes of manufacturing products, performed on the basis of the creative work of the masters of folk arts and crafts. These actions are aimed at substantiating the involvement of presented products in an ethnic culture within which creative variation is made.

4 Conclusions

Although there are two approaches to the question of the territorial affiliation of handicrafts to a particular geographical object, institutions as a geographical indication and appellation of origin. In turn, a designation of a product on market as the appellation of origin allows manufacturer to attract to the largest range of consumers, as the presence appellation of origin implies manufacture of goods in strict compliance with traditions. In addition, a manufacturer of goods has more opportunities to confront counterfeit products and unfair competitors.

Since individualization of products as an appellation of origin is legitimate in the case of special registration procedures which are based on the establishment of history and traditions of manufacture of the product in a certain area, depending on specific properties of products on natural conditions of traditional place of production. It seems that this legal institution is ideal for the purpose of preserving and developing the ethnos and culture in the context of a world economy. At present, this object of intellectual property needs to be developed, for which it is necessary to revise international agreements affecting the study area with a positive experience regulation of countries practicing the application of this law institution.

References

Convention for the safeguarding of the intangible cultural heritage, Paris, 17 October 2003. UNESCO (2003). unesdoc.unesco.org/ark:/48223/pf0000132540

Paris Convention for the Protection of Industrial Property of March 20, 1883, as revised at Brussels on December 14, 1900, at Washington on June 2, 1911, at The Hague on November 6, 1925, at London on June 2, 1934, at Lisbon on October 31, 1958, and at Stockholm on July 14, 1967, and as amended on September 28, 1979. www.wipo.int/treaties/en/text.jsp?file_id=288514)

Lisbon Agreement for the Protection of Appellations of Origin and their International Registration of October 31, 1958, as revised at Stockholm on July 14, 1967, and as amended on September 28, 1979. www.wipo.int/lisbon/en/legal_texts/lisbon_agreement.html

Uruguay Round Agreement: TRIPS Trade-Related Aspects of Intellectual Property Rights. The TRIPS Agreement is Annex 1C of the Marrakesh Agreement Establishing the World Trade Organization, signed in Marrakesh, Morocco on 15 April 1994. wipolex.wipo.int/en/text/305907

WIPO. Intellectual property, traditional knowledge and traditional cultural expressions/folklore a guide for countries in transition, Version One (2013). www.wipo.int/edocs/pubdocs/en/wipo_pub_transition_9.pdf

Bromley, Iu. V.: Ethnos and ethnography, AN SSSR. The Institute of Ethnography. N.N. Miklukho-Maklai, M. Nauka (1978)

O'Connor, B., de Bosio, G.: Economies Ius Gentium: comparative perspectives on law and justice. In: van Caenegem, W., Cleary, J. (ed.) The Global Struggle Between Europe and United States Over Geographical Indications in South Korea and in the TPP. Springer, Cham. IUSGENT, vol. 58, pp. 47–79 (2017). https://link.springer.com/chapter/10.1007/978-3-319-53073-4_3. Accessed 14 Apr 2019

Hu, Z., Xiong, W., Sun, Z., Wang, S., Huang, L.: Intangible cultural heritage and geographical indication of specialty resources: a case study of Shiyan City. Agribusiness, no. 10, 77–87 (2013). https://ageconsearch.umn.edu/record/160723/files/18.PDF. Accessed 14 Apr 2019

Innovative Approaches to Corporate Financial Management in Business Systems

Liliya S. Valinurova[1](✉), Tatyana B. Leibert[2], and Elvira A. Khalikova[2]

[1] Bashkir State University, Ufa, Russia
valinurovalilia@mail.ru
[2] Ufa State Petroleum Technological University, Ufa, Russia
lejjbert@mail.ru, ydacha6@yandex.ru

Abstract. This article reveals the current trends of international and Russian practice of introducing financial innovations in a business system. We outline the prospective areas of innovative development of corporate management in business systems within the framework of the industrial revolution, including blocks of corporate development, financing sources and innovative technologies of financial corporate management.

Based on the analysis of the current situation and trends related to the development of innovative technologies in the financial market, we determined the promising lines of financial innovations in the corporate management of the business system.

The paper offers a method of corporate analysis of the priority lines for financial innovations on the back of hierarchical structure analysis. The specific feature of the method is to choose the best financing option from provided alternatives by specified selection criteria: corporate development blocks, financing sources, types of activities, types of innovative corporate management tools.

The suggested method was approved on the example of one of the largest Russian business systems, the concept of sustainable development hereof is to develop a digital transformation culture aimed at the effective management of the business model. Application of the proposed method by financial and economic service in companies will allow justifying the distribution of financing sources for the introduction of financial innovations.

Keywords: Business system · Financial innovations · Corporate analysis · Corporate management · Financial forecasting technologies · Method of hierarchical structure analysis

JEL Code: G30 · G32 · G39

1 Introduction

Over the past decade, business systems have undergone dramatic changes in the conditions of informational openness, which entailed the transformation of approaches to corporate financial management. This is primarily determined by the fact that modern business systems have become more complicated in organizational and functional structure and the number of managers involved. The business is highly-diversified and a

E. G. Popkova and B. S. Sergi (Eds.): ISC 2019, LNNS 91, pp. 83–95, 2020.
https://doi.org/10.1007/978-3-030-32015-7_11

number of transactions in sales markets are enormous. Also, in the context of informational openness and high interest of the stakeholders, business systems are obliged to publish the results of their activities in the IT environment according to the international standards. The above-mentioned prerequisites have a significant impact on the corporate financial management system, which should be flexible and adjustable to the uncertainties of the external environment. Therefore, business systems intensively use innovative approaches to financial management, including both accounting and analytical tools, as well as modern financing methods.

The authors of the article show the Russian practice of deploying financial innovations in the corporate management of a business system using the methods of corporate analysis.

2 Materials and Methods

2.1 Financial Innovations in the Corporate Management System of the Company

As a rule, a business system is a large industrial company composing of individual organizational units that are entitled to manage finance, assets and other resources and are responsible for the generation of net profit as the main source of equity capital. The performance of such a business system consists in ensuring the reasonable and efficient use of all available resources and enhancing the shareholder's value. The expected result can be achieved through adequate and expert financial management by the top management of the business system and the engagement of financial innovations.

In this research, we understand financial innovations as new financial products, new financial services, new financial technologies based on innovative digital technologies, which makes it possible to advance the term of financial transactions between participants of financial and economic relations.

The combination of innovative methods, tools, and approaches aimed at enhanced performance of financial transactions, the efficiency of financial resource distribution, additional profit-making from redistribution hereof and mitigated risks of financial transactions constitutes a system for management of financial innovations. This system is closely related to the external and internal environment of the business system. As the author of the article states (Bilalova 2012), information on corporate financing is generated from the external environment, and information on needs for financial resources is derived from the internal environment to carry out the operation and investment activities of the business system. The relationship between the financial innovation management system and the internal and external environment is presented in Fig. 1.

The main functions of financial innovations of a business system are the following:

(1) Application of modern methods to finance basic and supporting business processes for the implementation of the general strategy and objectives hereof in the area of innovative and financial development of the business system by stepping up the upgrade of the production, organization, management and infrastructure framework;

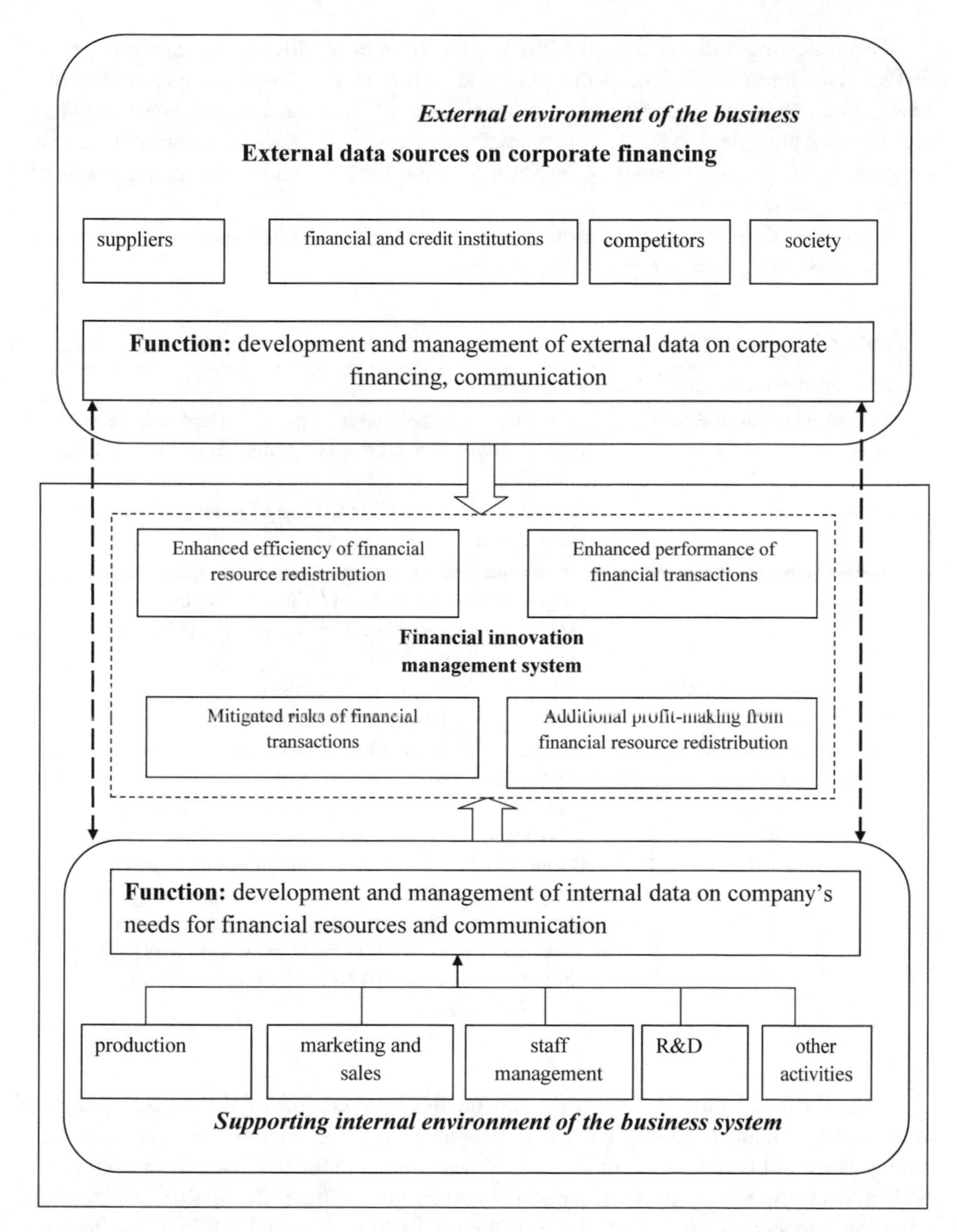

Fig. 1. The relationship between the financial innovation management system and the external and internal environment of the business system

(2) Deployment of self-financing and lending tools to attract funds in expanded production based on determining the amount and sources of financing, as well as an assessment of their availability.

The following authors (Krupka 2013), (Popkova et al. 2018), (Lugamanova et al. 2016), (Nikulin et al. 2016), (Krasnykh et al. 2017) in their research papers describe financial innovations wide-spread in modern international and Russian practice. They include new financial products, new financial services, new financial technologies, new organizational forms of financial institutions, new tools in corporate management of business system finance.

Examples of financial innovations in modern Russian and international practice for the specified categories are presented in Table 1.

Table 1. Examples of financial innovations in modern Russian and international practice

Type of financial innovation	Examples
Innovative financial products	Floating-rate mortgage lending, exchange and mutual funds, credit derivatives, swaptions, innovative leasing options
Innovative financial services	Internet banking, online trading of financial instruments, online loans
Innovative financial technologies	Electronic accounting and circulation of securities, digital money, distributed ledger technology (blockchain), blockchain protocols for verification of corporate financial transactions
Innovative organizational forms of financial institutions	Electronic banks, electronic insurance, exchange electronic systems, venture funds
Innovative tools of finance corporate management	Corporate treasury, blockchain in financial management of business systems (smart contracts, own counterparty check system, international non-bank payments, verification of contracts by user identification, smart contracts for value transactions, investment funding, financial consulting, XBRL - reporting), robotization in financial management (automatic payment processing, customer database), financial status forecasting (Deep Machine Learning and Reinforcement Learning technologies)

Thus, financial innovations are based on the introduction of digital technologies both in the financial sector (banking, insurance, stock exchange) and in corporate management of business system finance. Financial innovations provide the acceleration of big data processing and prompt implementation of both the major and service business processes related to the fulfillment of the financial obligations hereof. Moreover, financial innovations ensure the higher performance of corporate financial management of a multiple and complicated business system, which covers a large number of both business activities and organizational and supervising units with financial and property autonomy.

2.2 Methodological Approaches of Corporate Analysis to Justify the Priority Lines for Business System Process Funding

One of the lines in the financial innovation management system is to determine the scope of finance of innovative tools for corporate financial management in a business system, justify the distribution of financial resources between key areas and choose a source of financing.

The authors of the article propose a method of corporate analysis of the priority areas of innovation funding, which is based on hierarchical structure analysis. This method allows choosing one of the alternatives presented in accordance with the specified criteria.

The following authors (Mishin 2004), (Saati 1993) in their research papers thoroughly study the methodological approaches to the analysis of hierarchical structures, which served as a background for the development of analytical tools.

The idea of the method lay in drawing up some hierarchy based on a set of alternative options of criteria that determine the choice of the best solution. The criteria from the proposed set are compared in a pairwise way in accordance with a rating scale. Besides, the rating scale should have discrete values. The rating scale suggested by authors (Mishin 2004), (Saaty 1993) for use in this research is presented in Table 2.

Table 2. The suggested rating scale of criteria or comparable elements of the hierarchical structure

Degree of importance	Value	Explanation
Equal	1	Equally important contribution of the two criteria to achievement of the goal
Moderate superiority of one over another	3	Experience and judgments give a slight superiority of one criterion over another
Significant or strong superiority	5	Experience and judgments give strong superiority to one criterion over another
Significant superiority	7	One criterion is given such a strong superiority that it becomes practically significant
Huge superiority	9	The obvious superiority of one criterion over another has the strongest recognition
Intermediate decisions between two adjacent judgements	2, 4, 6, 8	Typical for a compromise

The author of the following scientific publication (Saati 1993) proposed an algorithm to select key lines for innovation financing based on the method of hierarchical structure analysis that consists of five stages:

(1) calculation of local priorities of the judgment matrix (a, b, c);
(2) calculation of the eigenvalue of the judgment matrix (P_1, P_2, P_3);
(3) determination of the consistency index of the judgment matrix (i);
(4) hierarchical synthesis of priorities;
(5) determination of the consistency index of the entire hierarchy.

The vectors of local priorities of the judgment matrix are determined by the formula:

$$a = \sqrt[3]{\frac{x_1}{x_1} \times \frac{x_1}{x_2} \times \frac{x_1}{x_3} \times \ldots \times \frac{x_1}{x_n}}, \tag{2}$$

$$b = \sqrt[3]{\frac{x_2}{x_1} \times \frac{x_2}{x_2} \times \frac{x_2}{x_3} \times \ldots \times \frac{x_2}{x_n}}, \tag{3}$$

$$c = \sqrt[3]{\frac{x_3}{x_1} \times \frac{x_3}{x_2} \times \frac{x_3}{x_3} \times \ldots \times \frac{x_3}{x_n}}, \tag{4}$$

where a, b, c are intermediate values;

x_1, x_2, x_3 are the weighting coefficients of pairwise comparisons of the relative importance of decision-making criteria.

At the second stage we determine eigenvectors of priorities by the formula:

$$P_1 = \frac{a}{\alpha + b + c}, \tag{5}$$

$$p_2 = \frac{b}{a + b + c}, \tag{6}$$

$$p_3 = \frac{c}{a + b + c}, \tag{7}$$

where p_1, p_2, p_3 are the eigenvectors of the priorities of judgments of the first, second, and third order respectively.

1. First, each column of the judgment matrix is added together, then the total of the first column is multiplied by the magnitude of the first component of the normalized vector of priorities, etc., the total of the second column is multiplied by the second component, etc., then the final numbers are added. Consequently, it is possible to obtain the largest eigenvalue of each judgment matrix.
2. Consistency can be assessed as follows.

Multiplying the comparison matrix on the right by the obtained assessment of the decision vector, we get a new vector U, which is calculated by the formula:

$$\begin{bmatrix} \frac{x_1}{x_2} & \frac{x_1}{x_2} & \cdots & \frac{x_1}{x_n} \\ \frac{x_2}{x_1} & \frac{x_2}{x_2} & \cdots & \frac{x_2}{x_n} \\ \cdots & \cdots & \cdots & \cdots \\ \frac{x_n}{x_1} & \frac{x_n}{x_2} & \cdots & \frac{x_n}{x_n} \end{bmatrix} \times \begin{bmatrix} P_1 \\ P_2 \\ \cdots \\ P_n \end{bmatrix} = \begin{bmatrix} U_1 \\ U_2 \\ \cdots \\ U_n \end{bmatrix}. \tag{8}$$

The algebraic problem in the case of consistency is to solve the equation: $A \times x = n \times x, A = \frac{X_i}{X_j}$, and the general problem with antisymmetric judgments are to solve the equation:

$$A' \times x' = a_{max} \times x', A' = a_{ij}, \tag{10}$$

where α_{max} is the largest eigenvalue of the judgment matrix A.

Dividing the total of the components of this vector by the number of components, we calculate approximate a_{max} to assess the consistency reflecting the proportionality of preferences. The closer the value of a_{max} to n (the number of compared elements), the more consistent the result.

Deviation from consistency can be expressed by the consistency index I_{cor} and determined by the formula:

$$I_{cons} = \frac{(a_{max} - n)}{(n-1)}. \tag{11}$$

For antisymmetric matrix $a_{max} \geq n$.

If I_{cons} is divided by random consistency of a matrix of the same order (R), then we obtain the value of consistency T with level no more than 10%. If the value of T will exceed this level, the re-comparison is advised.

3. Then priorities are summarized starting from the second level down as follows. Matrices are compiled in a way that private priorities are located in relation to each criterion located in the top row. Each column of vectors is multiplied by the priority of the corresponding criterion, and the result is added along each row. Thus, generalized priorities are calculated.
4. To find the consistency of the entire hierarchy, we should multiply each consistency index by the priority of the corresponding element at a higher level add the numbers obtained. Then the result is divided by the same expression, but with a random index of consistency corresponding to the size of each weighted matrix priority. The admissible value is no more than 0.1. Otherwise, the quality of judgment should be improved.

The proposed hierarchy structure for innovation financing in business systems is presented in Table 3.

A sequential algorithm for corporate analysis to determine the priority direction of innovation financing is presented in Fig. 2.

The basis of the pairwise comparison of the preferences by the directions of financing by the hierarchical levels is the method of expert evaluation. The experts should be the head of the finance and economics department, the heads of departments of business performance improvement programs, the head of the digital transformation directorate, the head of the investment development directorate and other experts.

The degree of superiority of one criterion over another by hierarchical levels is determined in accordance with the score from 1 to 10. Then weighting coefficients and a consistency index are calculated.

The pairwise comparison matrix of corporate development blocks is presented in Table 4.

The data in Table 4 show that financial innovations are of high superiority for corporate financial management at the executive management level. The matrix element located at the intersection of the executive and strategic levels has 7 scores.

Table 3. The proposed hierarchy structure for financing financial innovations in business systems

Hierarchy levels	Level denomination	Lines of financial resource application			
1	Corporate development blocks	Strategical		Tactical	Operating
2	Sources of funding	Own		Traditional	Innovative
3	Types of activity	Operating		Investment	Financial
4	Types of innovative tools for corporate management	Corporate treasury	Digitalization of management with counterparties (smart contracts)	BIG DATA	Financial forecasting technologies

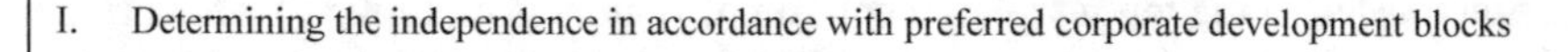

II. Determining the independence in accordance with preferred sources of financing, types of activity and types of corporate management innovative tools

III. Developing a matrix of corporate development block comparisons

IV. Developing the matrix of pairwise comparisons of financing sources by the degree of importance in relation to corporate development blocks

V. Developing the matrix of pairwise comparisons of company's types of activities by the degree of importance in relation to each source of financing

VI. Developing the matrix of pairwise comparisons of corporate development blocks by the degree of importance in relation to each type of business system's activity

VII. Calculating a vector of priorities for sources of financing, types of activities, types of corporate management innovative tools

VIII. Computing the weighting coefficients of types of business system's activity, sources of financing, types of activities, and corporate management tools by priority vectors

XI. Verifying the consistency of the entire hierarchy to reveal the key sources of financing sources by types of corporate management innovative tools

Fig. 2. A sequential algorithm for corporate analysis to determine the priority direction of innovation financing based on the hierarchical analysis

Table 4. Matrix of pairwise comparison of corporate development blocks

Corporate development block	Corporate development block			a, b, c values	Vector of weighting coefficients of corporate development blocks	Consistency ratio, percent
	Strategical	Tactical	Operating			
Strategical	1	1/3	1/7	0.3625	0.081	5.59
Tactical	3	1	1/5	0.8434	0.188	
Operating	7	5	1	3.2711	0.730	
Total	11.00	6.33	1.34	a_{max}	3	

Further, we develop a matrix of pairwise comparisons of financing sources by the degree of importance with respect to corporate development blocks (Table 5).

The data of Table 5 indicate that it is expedient to use all types of financing sources at the strategic level in different proportional ratios, but own sources have a great superiority. At the operating management level, it is advisable to use own financing sources and traditional ones, for example, loan funds from financial and credit institutions.

Table 5. Matrix of pairwise comparisons of financing sources by the degree of importance with respect to corporate development blocks

Corporate development block	Financing sources of the company's activity	Financing sources of the company's activity			Vector of priorities	Consistency ratio, percent	a, b, c values
		Own	Traditional	Innovative			
Strategical	Own	1	3	3	0.584	11.7%	2.080
	Traditional	1/3	1	1/3	0.135		0.481
	Innovative	1/3	3	1	0.281		1.000
	Total	1.67	7.00	4.33			
Tactical	Own	1	1/3	3	0.281	11.7%	1.000
	Traditional	3	1	3	0.584		2.080
	Innovative	1/3	1/3	1	0.135		0.481
	Total	4.33	1.67	7.00			
Operating	Own	1	1/3	1/5	0.105	3.3%	0.405
	Traditional	3	1	1/3	0.258		1.000
	Innovative	5	3	1	0.637		2.466
	Total	9.00	4.33	1.53			

Further, we develop a matrix of pairwise comparisons of types of activities with respect to financing sources and a matrix of pairwise comparisons of types of innovative corporate management tools with respect to types of activities (Tables 6 and 7).

Table 6. Matrix of pairwise comparisons of types of activities with respect to financing sources

Financing sources	Types of activity	Types of activity			Vector of priorities	Consistency ratio, percent	α, β, γ values
		Executive	Investment	Financial			
Own	Operating	1	3	1/5	0.202	11.7%	0.843
	Investment	1/3	1	1/5	0.097		0.405
	Financial	5	5	1	0.701		2.924
	Total	6.33	9.00	1.40			α_{max}
Traditional	Operating	1	3	5	0.618	11.7%	2.466
	Investment	1/3	1	5	0.297		1.186
	Financial	1/5	1/5	1	0.086		0.342
	Total	1.53	4.20	11.00			α_{max}
Innovative	Operating	1	1/3	5	0.279	5.6%	1.186
	Investment	3	1	7	0.649		2.759
	Financial	1/5	1/7	1	0.072		0.306
	Total	4.20	1.48	13.00			α_{max}

After matrix development follows the calculation of the weighting coefficients of the key lines of innovation financing for each hierarchical criterion. So, for example, the weighting coefficient of financing sources is equal to the sum of productions of the i-th weight of the corporate development block and the i-th weight of the financing source (importance of the i-th source in relation to the j-th block of corporate development).

Table 7. Matrix of pairwise comparisons of the company's development blocks by the degree of importance with respect to each type of activity

Types of activity	Types of innovative tools	Types of innovative tools				Vector of priorities	Consistency ratio, percent	α, β, γ values
		Corporate treasury	Smart contracts	BIG DATA	Financial forecasting technologies			
Operating	Corporate treasury	1	3	1/7	1/7	0.082	3.6%	0.497
	Smart contracts	1/3	1	1/7	1/7	0.047		0.287
	BIG DATA	7	7	1	1	0.435		2.646
	Financial forecasting technologies	7	7	1	1	0.435		2.646
	Total	15.33	18.00	2.29	2 2/7			
Investment	Corporate treasury	1	1	7	7	0.435	3.6%	2.646
	Smart contracts	1	1	7	7	0.435		2.646
	BIG DATA	1/7	1/7	1	3	0.082		0.497
	Financial forecasting technologies	1/7	1/7	1/3	1	0.047		0.287
	Total	2.29	2.29	15.3	18			
Financial	Corporate treasury	1	1	1/3	1/5	0.102	1.6%	0.508
	Smart contracts	1	1	1/3	1/5	0.102		0.508
	BIG DATA	3	3	1	1	0.347		1.732
	Financial forecasting technologies	5	5	1	1	0.449		2.236
	Total	10.00	10.00	2.67	2 2/5			

Table 8. Calculation of the weighting coefficients of business system financing sources

Financing sources	Corporate development blocks			Weighting coefficients of financing sources
	Strategical	Tactical	Operating	
	0,081	0,188	0,731	
Own	0,047	0,053	0,077	0,177
Traditional	0,011	0,110	0,189	0,310
Innovative	0,023	0,025	0,465	0,514

Table 9. Calculation of the weighting coefficients of the business system types of activity

Types of activity	Corporate development blocks			Weighting coefficients of financing sources
	Strategical	Tactical	Operating	
	0.177	0.310	0.514	
Operating	0.036	0.191	0.143	0.370
Investment	0.017	0.092	0.333	0.442
Financial	0.124	0.027	0.037	0.187

The calculated weighting coefficients for all hierarchical levels are presented in Tables 8, 9 and 10.

The calculated vectors of the weighting coefficients of financing sources show the priority of innovative ones for the business system (0.514) and the superiority of traditional ones (0.310) over own funds (0.177).

The calculated weighting vectors of the types of activity show the priority of the investment activity of the business system (0.442), and superiority of operating activity (0.370) over financial one (0.187).

3 Results

Thus, it is better to use both innovative sources (0.333) and traditional ones (0.092), but not own funds (0.017) for the investment activity of the business system As well as for operating activity, it is expedient to use both traditional sources (0.191) and innovative ones (0.143), but not own funds (0.036). As for financial activities, the priority is on own funds (0.124), rather than on traditional (0.027) and innovative (0.037) sources.

The calculated vectors of the weighting coefficients of corporate management innovative tools reflect that the priority direction of operating activities from the financial point is BIG DATA technologies (0.262) and financial forecasting (0.266). As for financial activities, BIG DATA technologies (0.065) and financial forecasting technologies (0.084) are also preferable.

Besides, the obtained weighting coefficients specified in the last column of Table 10 can be used in distributing the total amount of financing allocated for the development of financial innovations of corporate management.

Thus, the corporate analysis made it possible to determine the priority lines of financial innovations of the business system with regard to the sources of financing.

Table 10. Calculation of the weighting coefficients of innovative corporate management tools

Types of innovative tools of corporate management	Types of activity			Weighting coefficients of financing sources
	Operating	Investment	Financial	
	0.370	0.442	0.187	
Corporate treasury	0.030	0.192	0.019	0.242
Smart contracts	0.017	0.192	0.019	0.229
BIG DATA	0.161	0.036	0.065	0.262
Financial forecasting technologies	0.161	0.021	0.084	0.266

4 Discussion

A great deal of research papers is devoted to the developmental issues of corporate management of business systems in the context of the transition of the national and global economy to Industry 4.0. Groups of scholars in their works (Sergi et al. 2018), (Popkova et al. 2018) outline the conceptual provisions of business system management based on the principles to enhance innovation activity and ensure sustainable development. In the context of transition to Industry 4.0 arises a need for new and original organizational and management tools in corporate management, which allow making forecast scenarios for the development of business systems to assess the achievement of strategic goals and priorities.

Today, organization and management innovations in corporate management are based on the digitalization of business processes. Thus, the following scientific proceedings (Krupka 2013), (Krasnykh et al. 2017) describe in detail blockchain-based financial innovations of corporate management and state the main advantages of this technology.

In addition to the digitalization of business processes in the corporate financial management system, another field of research is the financing of organization and management innovations. A group of authors (Kyurdzhiev et al. 2020) in their scientific publication propose theoretical and methodological approaches to financing of innovation in terms of differentiation of financing sources and methods depending on the stages of innovation activity.

5 Conclusion

Thus, the proposed method of corporate analysis allows justifying the distribution of financial resources for the investment of the introduction of corporate management innovative tools. The idea of the approach lies in a multi-stage choice of alternative lines of financing in terms of their sources, availability and the prospects of business development blocks.

Thus, the introduction of a corporate analysis of the priority lines for innovation financing in the corporate management of a business system will allow for prompt decision-making at different levels of management.

References

Kyurdzhiev, S.P., Pashkova, E.P., Mambetova, A.A.: Financial provision of innovative activity in the Russian Economy. In: Smart Innovation, Systems, and Technologies, no. 138, pp. 444–454 (2020)

Sergi, B.S.: Exploring the Future of the Russian Economy and Markets: Towards Sustainable Economic Development. Emerald Publishing, Bingley (2018)

Bilalova, N.D.: Financial innovations in the system of corporate management. Current Trends in Economics and Management: New Vision, no. 14, pp. 82–87 (2012)

Krasnykh, S.S., Andreeva, E.L.: Development of financial innovations in the global economy and specific features of blockchain-based financial innovations. In: The 15th International Research and Practice Conference of Young Scientists "Development of Territorial Socio-Economic Systems: Theoretical and Practical Issues". Ekaterinburg: Publishing House of the Institute of Economics, Ural Branch of the Russian Academy of Sciences, pp. 336–338 (2017)

Krupka, I.: Financial Innovations in Ensuring the Development of the Financial Market and the National Economy. Bulletin of Kiev National University named after Taras Shevchenko. Economics, no. 145, pp. 35–37 (2013)

Mishin, S.P.: Optimal Management Hierarchies in Economic Systems, p. 190. PMSOFT, Moscow (2004)

Popkova, E.G., Sergi Bruno, S.: Will Industry 4.0 and other innovations affect the development of Russia? In: Sergi, B.S. (ed.) Exploring the Future of the Russian Economy and Markets: Towards Sustainable Economic Development, pp. 51–68. Emerald Publishing, Bingley (2018)

Saati, T.: Decision-Making. Method of hierarchy analysis. Radio and Communication 314 p. (1993)

Lugamanova, I.F., Yarullin, R.R.: Development of financial innovations in banks. Innovative Science, 3–1, 153–155 (2016)

Nikulina, V., Pechenin, K.K.: Development and introduction of financial innovations in banking (on the example of Sberbank PAO). The economy of Sustainable Development, 1(25), 283–293 (2016)

The Transformation of Legal Personality as a Means of Achieving the Legal Bond Between the Participants of Tax Relations with the Use of Digital Technologies

Anastasia S. Alimbekova[1(✉)], Marina A. Katkova[2], Svetlana V. Rybakova[1], Elena N. Pastushenko[1], and Vasily V. Popov[1]

[1] Saratov State Academy of Law, Saratov, Russia
pokina@yandex.ru, svrybakova@mail.ru, past_en@mail.ru, vpopov1970@rambler.ru
[2] Saratov State Technical University, Saratov, Russia
mkatkova@mail.ru

Abstract. The paper deals with the analysis of the content of legal personality of participants of tax relations which is subject to the transformation processes occurring in view of digitization of economy and law. Prerequisites for the formation and change of rights and obligations of participants of tax relations arising within the tax administration process have been identified. The content of such concepts as "tax administration" and "transformation of legal personality" has been developed. Directions of the transformation of legal personality have been identified for various groups of participants of tax relations, arising in connection with the fulfilment of tax obligations: taxpayers, tax agents, tax authorities, banks (lending institutions), implementing taxpaying transactions and having the opportunity to take part in the tax control due to banking transactions involving opening and management of bank accounts for individuals and entities, public authorities possessing the information necessary for the maintaining of records of taxpayers and taxable items. Through the example of systemic analysis of particular provisions of the Tax Code of the Russian Federation, regulatory basis of transformation of legal personality of the listed participants of tax relations is presented. The innovations of the transformation of tax legal norms and legal engineering in the field of legal regulation of electronic document flow have been identified. It has been proved that the process of the transformation of legal personality has its advantages and disadvantages. Their list is presented as a conclusion.

Keywords: Electronic document flow · Legal personality · Tax administration · Participants of tax relations · Transformation in law · Fulfilment of tax obligations

E. G. Popkova and B. S. Sergi (Eds.): ISC 2019, LNNS 91, pp. 96–103, 2020.
https://doi.org/10.1007/978-3-030-32015-7_12

1 Introduction

Digitization is one of the factors that stipulates global transformations in the most diverse fields of public life, including in law. It appears that legal aspects of digitization involve, at the very least, two fields: the first one is digitization as a process that mediates legal ties between the subjects of law; the second one is digitization as the object of legal regulation. In the first case, digitization acts as a mediating phenomenon, while in the second case it acts as a self-sufficient phenomenon. Certainly, both fields require adequate scientific conceptualization, legal projects and technologies, and, finally, formalization in legislation.

It is important to note that many segments of law are already covered by the digital transformation process, since the challenges of time demand that a legislator takes counter measures as soon as possible. However, not everything depends on it in this case. Digitization is also an object of regard of the experts in information technology. That is why, from the point of view of the law, today most legal norms are concentrated in the field of public law, since the resources of this platform allow developing and using digital technologies on a larger scale. Certain issues in the field of private law are also subject to regulation by the law, but, as a rule, if this concerns the protection of public interests.

This study is focused on the field of digitization of tax relations, to be more precise – fulfilment of tax obligations. In this regard, it is important to identify the problem of the legal status of participants of tax relations, or rather their legal personality.

In tax relations, apart from the very fact of payment of taxes by a taxpayer, there is a prevalent range of problems related to mediatory, service, delegated and other similar aspects which implies the possession of relevant rights and obligations by general public, someway or other involved in the process of fulfilment of tax obligations. In particular, it is referred to banks, implementing taxpaying transactions, public authorities possessing the information necessary for the maintaining of records of taxpayers and taxable items, etc.

A comprehensive set of rights and obligations of participants of social relations which they can fulfill by their own actions or entrust them to other persons, commonly referred to as legal personality.

Certain rights and obligations or their comprehensive set can be assigned to other persons under the power of attorney or the law. Thus, Chapter 4 of the Tax Code of the Russian Federation (hereinafter referred to as the Tax Code) regulates the institution of representation in relations, regulated by legislation on taxes and fees. An innovation of the Tax Code of the Russian Federation consists in the introduction of the possibility of fulfilment of tax obligations on behalf of the taxpayer by any other person (relevant law came into force on November 27, 2018). This law does not establish criteria for such an "other person", from which it may be deduced that they can be any other person, regardless of the form of ownership and some other signs: both an individual subject (individual) and a collective subject (entity).

It is important to note that the question of whether it is possible to transfer (delegate) powers to another person is important both for private law and for public law. Such a transfer is possible provided that the execution of the corresponding action is not directly related to the identity of the performer. As it appears from the above example related to the payment of taxes, it is referred to the possibility of self-delegating (assumption). One can transfer rights to others and assume obligations. Self-delegating is a unique legal remedy for the emergence of legal capacity.

The transformation and delegation processes are fairly closely related. This relationship is due to the base "-trans" (to transfer). A fundamental difference is of importance here. When delegating, there is a preservation of the essence of the right (obligation), whereas the practical implementation is carried out by another person, i.e. there is kind of translation of the law (transmission). In case of transformation, the essence of right (obligation) is transformed; they seem to be transformed, although implemented by their carriers. In case of delegation (transmission), transformed rights and obligations can be transferred. These concepts are not mutually exclusive.

In view of the above, it is important to point out that in this research the point at issue is about the transformation of legal personality, to be specific, about the transformation of legal personality of participants of tax relations.

2 Materials and Methods

There are two large groups of social relations in the field of taxation. One of them is associated with the establishment of tax obligations – the field of taxation, while another is associated with its enforcement – the field of tax administration. Finally, one may identify the field of implementation of the concept of tax liability as a kind of financial and legal liability, as the third (delictual) group. The transformation of legal personality in the aspect under study is largely caused by the field of tax administration.

In order to use the term "tax administration", it is necessary to narrow down its understanding, since it appeared in Russian legislation relatively recently, although it has been known for a fairly long time in foreign law: English - tax administration (Bird and de Jantsher 1992).

At first it was enshrined in bylaws (2004, Order No. SAE-3-30/290 of the Ministry of Taxes and Duties of the Russian Federation of April 16, 2004. "On arrangement of work on tax administration of major taxpayers and approval of criteria for the assignment of Russian legal entities to the major taxpayers that are subject to tax administration at the federal and regional levels". The term "tax administration" has been abundantly used in professional vocabulary (Mironova and Khanafeev 2005). It was formalized at the legislative level in 2006 without the wording of the definition of the concept in the Federal Law No. 137-FZ "Concerning the Introduction of Amendments to Part One and Part Two of the Tax Code of the Russian Federation and in certain legislative acts of the Russian Federation in connection with the accomplishment of measures to improve the tax administration system".

The term "tax administration" is suggested to mean a scope of measures aimed both at promotion of conscientious fulfillment of obligations assigned to taxpayers, and at their enforcement. It would seem that the state exercises exactly administrative functions in the field of taxation, when it enforces the implementation of legal norms assigning tax obligation as a complex obligation, according to their correct calculation, timely and full payment. The tax administration techniques include: maintaining of records of taxpayers and taxable items; tax return formalization; a set of enforcement measures (pledge of property, suretyship, etc.); calculation of amounts of taxes by tax authorities; the use of an institute of tax agents; the use of the banking system for the purposes of payment (collection) of taxes; credit or reimbursement of taxes; tax inspections, etc. (Rybakova 2008).

Since the scope of tax administration covers the basic process of interaction of participants of tax relations, having decided upon the terms, we can pass on to the analysis of legal personality in circumstances where digital technologies are used.

In the first place, we should specify prerequisites for the creation of legal personality in the field of taxation. They include: the law which consolidates tax obligation; possession of a taxable item by a person; tax administration mechanism.

The law which consolidates tax obligation – Tax Code, as well as the laws of the subjects of the Russian Federation and the acts of representative self-governing authorities, the complex of which enables us to state the fact of establishment and introduction of a particular tax in a certain territory, that is the establishment of tax obligations.

As Mitskevich, A.V. reasonably points out, legal personality is a "capability", "possibility", prerequisite of enjoyment of a right. In relation to specific subjective rights and obligations arising in legal relationship, legal personality is a general (or abstract) prerequisite (Mitskevich 1962).

Possession of a taxable item by a person – possession on the right of ownership, control or operative management, for example, of a vehicle, land plot, income, income and other items which are classified as such in accordance with legislation on taxes and fees.

Tax administration mechanism is a comprehensive set of statutory measures providing for the procedure of voluntary or forced fulfilment of tax obligations.

The distinguished items constitute the list of basic prerequisites for the creation of legal personality of participants of tax relations.

The participants of abovementioned relations are as follows: taxpayers, tax agents, tax authorities, banks, authorities and other entities possessing the information about taxpayers for the maintaining of their records (authorities of the Ministry of Internal Affairs of the Russian Federation, authorities of the Federal Service for Public Registration, Cadastre and Mapping, etc.). The interaction between the specified participants in the context of the use of digital technologies is mediated by electronic document flow, which actually causes the transformation of legal personality.

3 Obtained Results

The analysis of issues arising in connection with the digitization of the process of fulfilment of tax obligations, as well as in connection with the tax administration process in general, provides us an opportunity to draw the following conclusions.

There are two ways of fulfilment of tax obligations: voluntary and forced. They imply a legal regulation of the use of a system of interrelated tax obligations **rules** and **procedures** that acquire their own peculiarities in the context of the use of **digital technologies**.

It is possible to identify several rules that concern the use of the tax administration mechanism through the use of digital technologies:

- providing a legislative framework to create an obligation of taxpayer or other person to calculate and pay taxes;
- resolving the issue of recourse payments (tax reimbursement from the budget);
- implementation of procedures of registration with the tax authorities;
- exchange of information that is significant for payment of taxes;
- system of monitoring over opening and closing of accounts in lending institutions, performance of account transactions;
- protection of tax secret, etc.

Electronic document flow in regard to fulfilment of tax obligations is regulated by several articles of the Tax Code: 80, 85, 85.1, 86, etc.

We can analyze the content of Article 85 as an example: "The obligations of authorities, institutions, organizations and officials to submit information associated with the maintaining of records of entities and individuals, to tax authorities".

Thus, authorities that carry out the registration (migration registration) of individuals at place of residence, the registration of act of civil status of individuals, guardianship and custodianship agencies, authorities which issue work permits or patents to foreign nationals or persons without citizenship, diplomatic missions and consular offices of the Russian Federation, authorities that carry out state cadastral registration and state registration of real property titles, authorities that carry out vehicle registration, authorities that issue and reissue documents which certify the identity of citizens of the Russian Federation in the territory of the Russian Federation, regulatory bodies of the Pension Fund of the Russian Federation and its regional offices shall submit relevant information to the tax authorities in electronic form. Currently, the obligation to submit information in electronic form has not been formalized yet for other authorities. Maybe this is not an obligation but a right. This issue is quite tricky: to determine whether the question is about an obligation or a right in this particular case.

The legislative technique that is used in Article 85 of the Tax Code prevents from clear differentiation between the powers of subjects. It is thought that one should be guided by the following principle: if the right to choose the form remains with a taxpayer – it is a subjective right, if it remains with a legislator – it is a subjective obligation. If a particular form is preferable for a taxpayer – it is a subjective right; if it is burdensome for them – it is a subjective obligation.

The analyzed Article 85 contains Paragraph 11 of the following content: "11. Authorities referred to in Paragraphs 3, 4, 8, 9.4 of this Article, shall submit relevant information to the tax authorities in electronic form. The procedure for the submission of information to the tax authorities in electronic form is stipulated in the agreement of the interacting parties". Under that logic, authorities listed in other Paragraphs shall submit information by way of common (non-electronic) document flow.

It would seem that in Paragraph 11, the question is most likely about an obligation to submit information in electronic form, while in other paragraphs the question is most likely about the submission of information in other form. Perhaps the answer to the question consists in time and court costs that are necessary for the transition to electronic document flow.

The analysis of the regulations of the Federal Tax Service regulating the rules of electronic document flow, testifies that all requirements that apply to them come down to the three types: **form, format, and procedure for completing a form**. It appears that in view of the introduction of electronic document flow, participants of tax relations accrued additional obligations associated with the fulfilment of requirements with regard to form, format and procedure for filling out a form, non-fulfilment of which, in point of fact, is indicative of the commission of an offence, since one of potential adverse effects is a statement of the fact of non-submission of a document or late submission of it.

The form of the document assumes the presentation of its content as a form; the format describes the requirements for XML data transfer files in electronic form (hereinafter referred to as the exchange file), procedure for filling out a form – a detailed description of the rules for filling out a form.

As has been already noted, the next step in the implementation of electronic document flow consists in the fact of conclusion of agreement of the interacting parties. The relevant agreements are public agreements (quasi-agreements) by nature. They define: the subject of the agreement, the principles and forms of interaction of the parties, the dates of entry into force and other issues, mainly organizational. For example, the following forms of interaction are formalized: creation of joint working groups, development and approval of joint action plans, holding of general consultations, seminars and meetings. After conclusion of agreements, the parties and government agency supervising the relevant aspect of interaction, adopt the procedure for the exchange of information in the form of an order in electronic form.

This is the rulemaking procedure in terms of the definition of the rules of electronic document flow concerning the submission of information that is necessary for the maintaining of records of taxpayers and taxable items, to the tax authorities. Many legal norms that are used during this process are technical legal norms.

Judging from the above analysis, it can be seen that not only legal personality associated with the submission of information, but also legal personality associated with the implementation of powers for establishing the interaction procedure, is subject to the transformation. Both cases draw on innovations caused by the issues of digitization.

Document flow which arises in connection with the fulfillment of obligations to exercise tax control by the banks (Article 86 of the Tax Code).

In particular, the bank is obliged to submit the information about opening or closing of account, deposit, about the change of account details, deposit of an entity, a private entrepreneur, an individual who is not a private entrepreneur, about the granting or termination of the right of an entity or private entrepreneur to use corporate electronic payment facilities for electronic money transfers, as well as about the change of account details of corporate electronic payment facilities, to the tax authority at the place of its location. Such information must be submitted in electronic form within three days of the corresponding event.

The forms and formats of submission of information by the bank to the tax authority are stipulated by the Federal Tax Service of Russia, and the procedure for the submission of information in electronic form by the bank - Central Bank of the Russian Federation subject to agreement with the mentioned Service.

Electronic declaration is yet another form of electronic document flow. Paragraph 3 of Article 80 of the Tax Code stipulates that tax declaration (tax return) shall be submitted to the tax authority at place of registration of the taxpayer (tax agent) presented in the prescribed form in hard copy or in accordance with the prescribed formats in electronic form.

4 Discussion

Researchers who have been studying the issues of digitization of the field of tax control, place special emphasis on its risk-oriented nature that, on the one hand, allows to provide an opportunity of the timely submission of information about the potential violation of legislation on taxes and fees for a taxpayer (for example, in the case of creation of the automated system of typification of tax avoidance schemes), on the other hand, presume the misbehavior of a taxpayer (for example, in the context of automated creation of tax risk groups, a conscientious payer of value added tax may become the victim due to actions of their unconscientious contracting parties, whom the program assigned to the higher risk group) (Migacheva 2019).

5 Conclusion

Consequently, the analysis of legal regulation of the electronic document flow system in the field of fulfilment of tax obligations and associated interaction of participants of tax administration process enables us to suggest essential advantages and grave disadvantages of the digitization process.

The advantages of electronic document flow include, in particular:

- information transmission rate;
- transmission process operability;
- capability to ensure comparableness of information;
- a wider range of entities participating in electronic document flow;
- reduction of financial, time and court costs.

  That said, it is expedient to specify the following disadvantages:
  - the hazard of disclosure of information classified as tax secret;
  - the risk of unjustified assignment of a taxpayer to the qualification group with low indices;
  - the hazard of loss of electronic money.

Nevertheless, the current process of the transformation of legal personality of participants of tax relations interacting with each other in regard to fulfilment of tax obligations and utilizing electronic document flow, is indicative of the following:

- new obligations emerge;
- the field of tax delicts becomes wider;

- there arises a need for the legislative consolidation of additional rights of participants of electronic document flow, related to the protection of information classified as secret as well as personal data;
- quasilegal personality of taxpayers is generated, caused by the automatic inclusion in a certain tax risk group both in connection with the quality of own fulfilment of tax obligations and in connection with the settlement of transactions with contracting parties.

Acknowledgments. The research was performed with financial support from the Russian Fund for Fundamental Research within the scientific project No. 18-29-16102 "Transformation of legal personality of the participant of tax, budget, and public banking legal relations in the conditions of development of the digital economy".

References

Bird, R.M., de Jantsher, M.C.: Improving Tax Administration in Developing Countries, p. 403. International Monetary Fund, Washington (1992)

Migacheva, E.V.: Tax control in the context of development of digital economy. Financial law in the context of development of digital economy: a monograph, under the editorship of Tsindeliani, I.A., pp. 192–207. Publishing House, Moscow (2019)

Mironova, O.A., Khanafeev, F.F.: Tax administration: a learning guide for students with a specialization in "Taxes and taxation". M (2005)

Mitskevich, A.V.: Subjects of Soviet Law, 212 p. State Publishing House of legal literature, M. (1962)

Rodygina, V.E.: Digital economy and tax administration: conflicts and ways of their solution. Financial law in the context of development of digital economy: a monograph, under the editorship of Tsindeliani, I.A., pp. 243–260. Prospekt Publishing House, Moscow (2019)

Rybakova, S.V.: Involvement of lending institutions in the tax administration process: theoretic issues. Bankovskoe Pravo, No. 2, pp. 44–46 (2008)

Satarova, N.A.: Coercion in financial law, under the editorship of Kucherov, I.I., 392 p. Yurlitinform Publishing House, M. (2006)

Cognitive Modeling of the Mechanism of Partnership of Business Entities with Public Authorities

Veronica S. Epinina[1(✉)], Iakow I. Kayl[2], Roman M. Lamzin[2], Anzhelika N. Syrbu[2], and Yurij M. Kvintyuk[2]

[1] Volgograd State University, Volgograd, Russia
econmanag@volsu.ru
[2] Volgograd State Socio-Pedagogical University, Volgograd, Russia
kailjakow@mail.ru, surbyan@mail.ru, rom.lamzin@yandex.ru

Abstract. The paper deals with special aspects of management and effective functioning of the mechanism of partnership of business entities with public authorities. It has been substantiated that at the current stage of development of economic and social relations, the goal of their regulation which is common for business entities and public authorities consists in the fixation of the limits of government (municipal) intervention in entrepreneurial activities, as well as priorities of participation of the state in certain market segments. It has been proved that the interaction between business entities and public authorities makes provision for mutual interest. The formation of such a stable and reliable partnership ensures, among other things, the integrity of the economic space and the improvement of the business climate. It has been suggested that the cognitive approach can be used as a tool for the study of partnership of business entities with public authorities and modeling of the mechanism. Based on the research findings, the authors have put forward their guidelines for achieving the target values of development of the mechanism of partnership of business entities with public authorities in the Russian Federation.

Keywords: Cognitive modeling · GR management · Partnership · Mechanism · Efficiency · Interaction between business entities and public authorities

1 Introduction

In the context of implementation of the strategic objective of efficient and constructive development of the economy of Russia, formation of the efficient management concept is becoming increasingly relevant. This being said, the basic element of the management of such administration is the establishment of partnership of business entities with public authorities and, as a consequence, development of the institution of GR management.

The core issue of partnership of business entities with public authorities consists in the identification of the level of responsibility of each party for solving the existing social and economic problems. One of the principal tasks of public authorities consists

E. G. Popkova and B. S. Sergi (Eds.): ISC 2019, LNNS 91, pp. 104–116, 2020.
https://doi.org/10.1007/978-3-030-32015-7_13

in determining the available advantages with due consideration of all factors influencing the regulation of social and economic processes. This is why the efficient interaction of public authorities and business entities is based on the due consideration of mutual interests and forms the basis of stable development of any modern state, which eventually determines the conditions for solving the problem of employment and reduction of social tension.

The overriding priority of business entities consists in making a profit which is impacted by varied circumstances, primarily the attitude of public authorities to them, which forces business entities to search for means and direction of impact on government.

It is suggested to use cognitive modeling for the analysis of the mutual interaction in the mechanism of partnership of business entities with public authorities. The cognitive map as a final product of cognitive structuring may be treated as a proper tool of research and optimal representation of weakly structured problems which can serve as the basis for the determination of the tasks of formation of coordinated managerial decisions, including those in the area of GR management.

2 Materials and Method

Academic papers dealing with the problems of establishing partnership relations between business entities and public authorities, as well as cognitive modeling, were used as scientific matter. The authors of the paper have put to good use the methods of cognitive modeling, comparative and systemic analysis, as well as the logical approach, which provide the necessary degree of elaboration of the scientific challenge.

3 Discussion

The categorical framework of research is based on GR (Government Relations) management which shall be understood to mean the management of partnership relations between business entities and public authorities, and the goals of research are to promote and defend the interests of business development at all levels of public administration.

In the pursuance of the research, the authors draw on fundamental and applied research of the Russian authors in the area of study of the interaction between business entities and public authorities, which include the works by Dibie et al. (2017), Gabsa (2017), Gadzekpo (2017), Hussien and Dibie (2017), Igbokwe-Ibeto (2017), Kawewe and Dibie (2017), Kayl et al. (2016), Klochko and Prokhorova (2015), Lapina and Chirikova (1999), Nur (2017), Perepelitsa (2006), Plotnikova (2014), Turovsky (2010), Yasin (2002).

In addition, the authors of research use the information from papers dealing with cognitive modeling of economic and social systems, which include the publications by Axelrod (1976), Busemeyer and Diederich (2010), Gorelova et al. (2014), Holt and Osman (2017), Kostikova et al. (2016), Laskowskaya et al. (2008), Lewandowsky and Farrell (2010), Makarenko and Khrustalev (2007), Moseyko et al. (2015), Pachur and

Scheibehenne (2017), Prezenski and Brechmann (2017), Prokhorova (2011), Ragulina et al. (2015).

That said, in consequence of the comprehensive content analysis of the available academic literature on the subject of the research, the authors failed to find any publications dealing with the formation of the cognitive map of mechanism of partnership of business entities with public authorities, which makes it possible to treat this topic as understudied in contemporary economic science and requiring further academic pursuits.

4 Results

The essence of interaction between business entities and public authorities is manifested in the establishment of relevant partnership relations which imply mandatory coordination of interests of these actors. The starting point for this is the recognition that each partnership actor has their personal goals and a desire to solve their personal problems.

In the process of such interaction, business entities anticipate satisfying their interests by means of established contacts with public authorities. Public authorities, in turn, wait for such proposals. The social significance of such a partnership is expressed, on the one hand, in improving the quality of products and services provided to the society, and, on the other hand, the focus of public authorities on the performance of control and administrative functions.

Undoubtedly, business is promoting its influence on the managerial decision-making process by public authorities by means of involvement of the clubbish set in all stages of formation of government agencies. Big business can initiate transformations (reforms) to level administrative barriers that limit economic freedom. In general, business aims to maximize the limitation of participation of the state in the economy and support of functioning of the adequate economic system of the country, where every party exercises its own functions: the government creates comfortable conditions for the entrepreneurs (business-enabling environment), and the entrepreneurs, in turn, implement manufacturing processes and pay taxes.

The impact of business entities on managerial decision-making process by public authorities is a prerequisite for the implementation of their own economic interests. Therefore, the efficient impact of business entities on making state (municipal) management decisions results from the combination of the objective needs and subjective capabilities of these participants. That said, the actors of GR interaction stick to the following basic conditions of interaction (Fig. 1).

Business entities interact with public authorities at federal, regional and municipal levels, which give reason to identify such relations as multitiered partnerships. Thus, the clubbish set (members of the Council for Entrepreneurship under the government of the Russian Federation, members of the Bureau of the Management Board of the Russian Union of Industrialists and Entrepreneurs and some other entrepreneurs having economic clout and autonomous political resources) resolve issues of tax and administrative reform, monetary and customs regulation in cooperation with representatives of the top echelons of power. Businessmen of the second and lower levels, representing

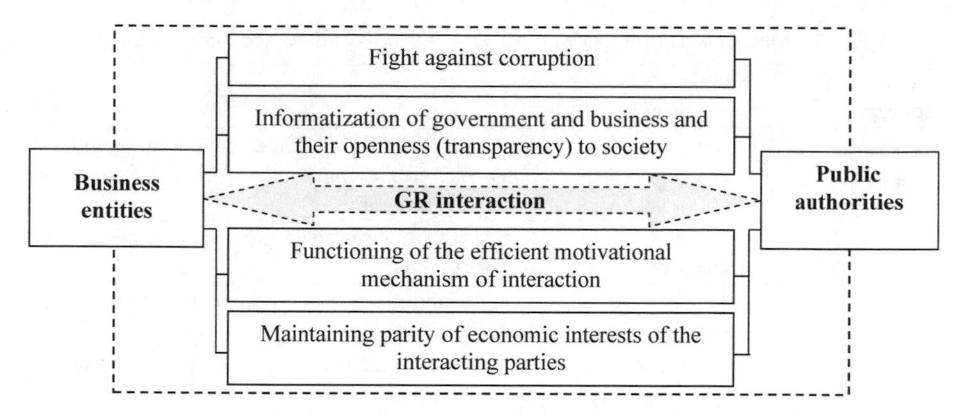

Fig. 1. Principal conditions of GR interaction between business entities and public authorities

the industry (company), practice submission letters to public authorities, published in open mass media and begging for assistance in solving a particular problem.

Significancy and relevance of the interaction between business entities and public authorities are increased in the periods of revolutionary social and economic changes. Therefore, the partnership relations between business entities and public authorities in contemporary Russia can only be rationalize through the joint development and implementation of the efficient mechanism of interaction respecting the interests and requirements of social and economic modernization of the country.

Mechanism of interaction between business entities and public authorities describes the cooperation between these actors for achieving the sustainable development of the economic and social system of the territory as a single entity. The efficiency of functioning of the mechanism of interaction between business entities and public authorities determines the achievement of socioeconomic indices which, in turn, determine the level of development of territories, the speed and the quality of the implementation of state-run programs with the involvement of business entities, social standards and quality of living, as well as the rate of innovations in economic activity.

Various aspects of functioning of the mechanism of interaction between business entities and public authorities are discussed in papers of many researchers (Table 1).

Without doubt, in real-life situation there is often a combination of several mechanisms of interaction between business entities and public authorities; however, one prevailing mechanism can be distinguished in the majority of cases.

The entrepreneurs can put pressure on government agencies with a view to forcing them to act definitely and effectively. For that purpose, indirect methods of impact are used with the involvement of intermediate parties (lobbying), delegating the representatives to government agencies (to senators, deputies, governors, ministers etc.), informal personal relations with state and municipal officers. At the same time, the mechanism of interaction between business entities and public authorities within the scope of GR management implies active use of the two priority forms of interaction of these actors (Fig. 2).

Table 1. Mechanisms of interaction between business entities and public authorities

Authors of the concept	Distinguished mechanisms	Summary
Turovsky, R.F.	Functional mechanism	Mutual distancing of business entities and public authorities, which implies that they will solve their tasks independently
	Partnership mechanism	Public authorities establish partnership relations with business entities
	State patronage mechanism	Public authorities are aimed at controlling over the activities of business entities
	Conflict mechanism	Lack of stable relations between business entities and public authorities
	Symbiotic mechanism	Splicing of government and business, generally where there is a dominant actor
Lapina, N.Y., Chirikova, A.E.	"Patronage" mechanism	It implies administrative-executive attitude of public authorities towards business entities
	"Partnership" mechanism	Dialogue and mutual aid of business entities and public authorities
	Mechanism of "privatization of power"	Business entities established control over public authorities
	Mechanism of "suppression"	Weakness of public authorities which failed to create a well consolidated team, to put forward an efficient development program and recommend a competent leader
Yasin, E.G.	Mechanism of "white area"	Forming the level playing field for all business entities and their indiscriminate coercion to comply with these rules on the part of public authorities
	Mechanism of "black area"	Covers informal criminal practices, corruption in the first place
	Mechanism of "grey area"	Covers informal practices of informal fees collected from business entities and practices of informal bargaining between entrepreneurs and public authorities for the business environment

Selection of the optimal form of establishment of GR relations is determined by a number of factors, namely:

- specific features of business activities and economic features of a relevant business entity;
- the degree of impact of the business entity in a relevant economic sector and territory;
- the amount of financial resources at the disposal of the business entity;
- activities of specially trained GR experts and top managers with the necessary communication skills;
- the scope and the degree of economic value of tasks solved by business entities;
- a range of areas of GR activity implemented in the business community.

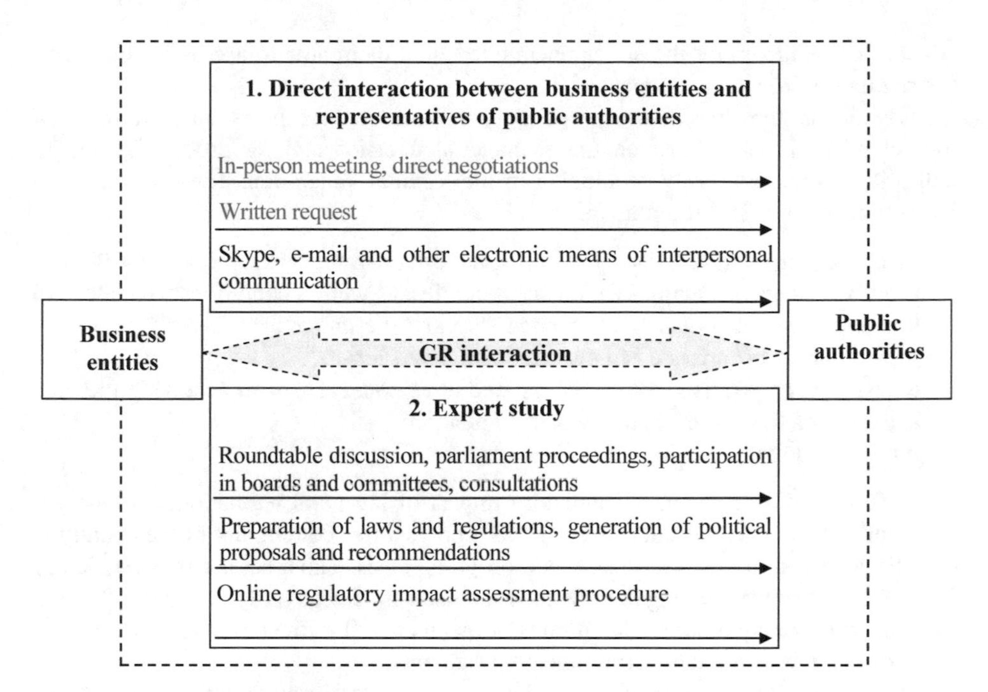

Fig. 2. Priority forms of interaction between business entities and public authorities within the scope of GR management

Business entities come forward with their initiatives, substantiated proposals and projects in the public administration system for the implementation of business interests. Direct interaction between business entities and public authorities within the scope of GR management is carried out by means of in-person meetings, negotiations, written requests, as well as by means of the use of electronic communications which do not imply the involvement of third parties. In turn, the form of expert study falls short of the evident feedback which is typical for the direct interaction between business entities and public authorities. Indeed, when we talk about feedback, we mean efficiency, communication performance; whereas any expert activity implies discourse, but does not guarantee any feedback.

A rather efficient line of action for business entities consists in the interaction with public authorities through the involvement of representatives of business entities in coordination and expert boards, as well as in joint committees under the auspices of federal and regional public authorities, which enables business entities to identify their issues and settle the deadlines and develop the mechanism for solving them jointly with public authorities.

The use of the documentary type of channels is the most typical method for direct expert study: the reference entity is engaged in the preparation of laws and regulations that are submitted to government officials. Although the electronic channel is also developing today, for example, online regulatory impact assessment procedure: the administrations of certain territorial entities of the Russian Federation even hold educational seminars for the entrepreneurs, teaching them how to use online platforms for regulatory impact assessment.

Overall, at the present time regulatory impact assessment as an instrument of impact of business entities on the managerial decision-making process by public authorities is quite actively developing in the Russian Federation. The tasks of regulatory impact assessment are as follows:

- identifying in draft legal acts the terms that establish unreasonable responsibilities, prohibitions and limitations for business entities, as well as conditions which lead to unwanted expenses of the entrepreneurship and the budget;
- prevention of occurrence of new administrative barriers;
- assessment of positive results and expenditures connected with achieving them;
- selection of the better alternatives of regulation;
- public consultations with the business community.

A format of assessment of regulatory impact of laws and regulations on business environment has been introduced in Russia. The business community of the country is directly involved in this assessment on a on-going basis. The assessment of regulatory impact of draft laws and regulations must be used by the government of the Russian Federation to timely identify superfluous norms that lead to the unfounded increase in costs of entrepreneurs and restrain their investments.

Regulatory impact assessment in the context under study is an element of the "smart regulation" policy; it is a tool for getting the standpoint of stakeholder groups across to public authorities regarding the business activities and is treated by the government of the Russian Federation as one of the key elements in the improvement in lawmaking quality. Thus, regulatory impact assessment procedure allows to:

- mitigate risks associated with the introduction of new regulation;
- improve the business climate and increase the investment attractiveness of business entities;
- increase the level of trust of business entities to managerial decisions made by public authorities;
- ensure the selection of the most efficient decision-making options of public authorities;
- cut down expenditures of the regional and municipal budgets in the introduction of new regulation or amendment of current regulation;
- make the decisions made by public authorities realizable;
- reduce the tension in relations between public authorities and representatives of the business community;
- arrange for participation of representatives of business communities in the adoption of draft laws and regulations and assessment of real effect of regulation;
- provide state and municipal services of good quality.

The identified priority forms of interaction between business entities and public authorities within the scope of GR management make it possible to produce the relevant mechanism of partnership of these actors with the use of cognitive modeling.

The concept "cognitive" implies the capability to decompose, analyze and synthesize. The concept of cognitive modeling which is inherently interdisciplinary, was introduced in 1976 by R. Axelrod. In the late 1980s his idea were embodied in the form of fuzzy cognitive maps, proposed by Bart Kosko as a consequence of fusion of fuzzy logic and system dynamics. Currently, these cognitive maps constitute the basis for modern systems of dynamic modeling in the economic domain.

The tools of cognitive analysis enable us to use the both qualitative and quantitative data in the modeling process; what is more, the degree of utilization of the latter may increase depending on the possibility of quantitative assessment of cooperating factors, analyzed during the modeling process. This being said, one of the most important factors of successful implementation of cognitive modeling methods is the determination of elements (in our situation – priority forms of interaction), determining the stability of GR interaction between business entities and public authorities and used in this context as peaks of a cognitive map.

Starting out from the priority forms of interaction between business entities and public authorities within the scope of GR management which we have identified and which will serve as peaks of a cognitive map in a more detailed description for achieving the efficient functioning of the mechanism of partnership of the mentioned actors, this map for the Russian Federation can be presented in the form shown in Fig. 3.

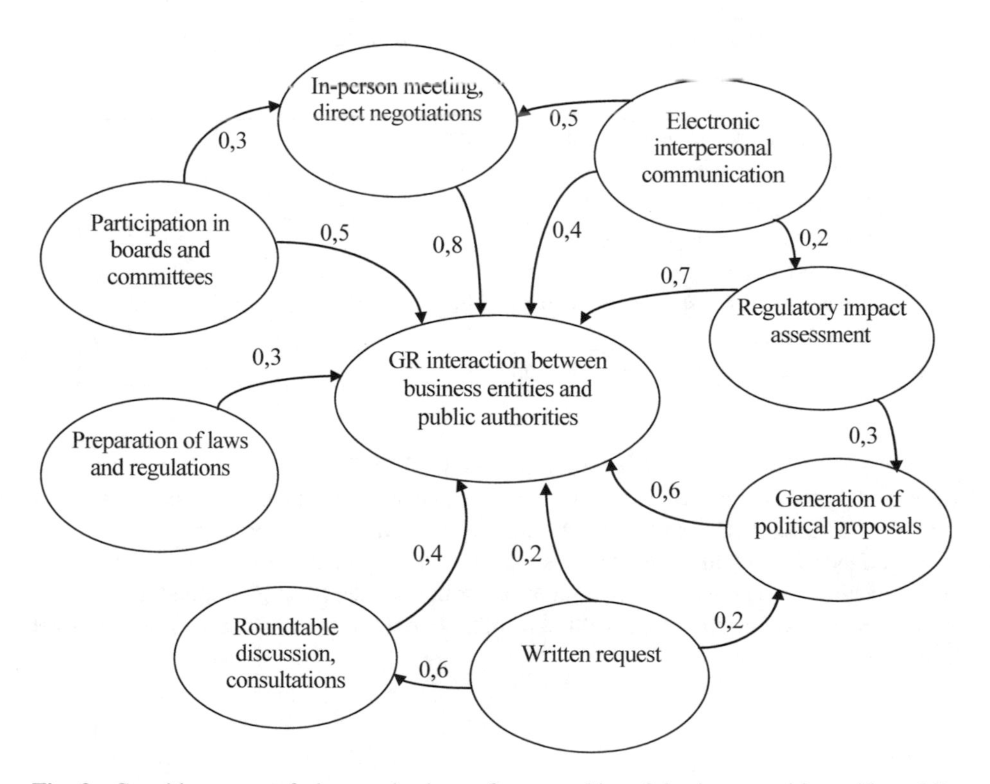

Fig. 3. Cognitive map of the mechanism of partnership of business entities with public authorities in the Russian Federation

The researchers have analyzed the impact of priority forms of interaction in the chains of cause-effect relationships within the scope of the suggested cognitive map ensuring the efficiency of GR interaction between business entities and public authorities. In order to carry out structured analysis and dynamic modeling of development of the situation within the scope of the cognitive map, vectors of initial trends, target vectors and management vectors have been identified (Table 2). The value of initial trend determines the direction and the rate of the change of value of each form of interaction. The values are a reference point in modeling, i.e. the input information for the map. The target value is only determined for the target forms. This is the value that we would like to observe in the efficient and properly developing mechanism of partnership of business entities with public authorities. The management vectors (Nos. 1–3) constitute combinations of measures (or "managing factors" from the modeling perspective), for each of which the efficiency should be checked, i.e. how much each combination will bring the researcher closer to the intended result (or "target values" in terms of model).

Table 2. Priority forms of cognitive map of the mechanism of partnership of business entities with public authorities in the Russian Federation and management vectors

No.	Name of interaction form	Initial trend	Target value	Management vector		
				№1	№2	№3
1	In-person meeting, direct negotiations	0.8	–	–	–	–
2	Electronic interpersonal communication	0.4	0.6	0.3	0.5	0.4
3	Regulatory impact assessment	0.7	–	0.5	0.4	
4	Generation of political proposals	0.6	–	0.4	0.5	0.5
5	Written request	0.2	0.3	0.2	0.1	0.3
6	Roundtable discussion, consultations	0.4	0.5	0.3	0.5	0.4
7	Preparation of laws and regulations	0.3	0.4	0.3	0.5	0.2
8	Participation in boards and committees	0.5	–	0.6		0.5

We shall consider in detail the target vector (Table 3). It characterizes the most desired dynamic pattern and the line of development of mechanism of partnership of business entities with public authorities in the Russian Federation. The authors assigned the assessment of dynamic pattern, i.e. desired motion direction, to each form of interaction. Then they identified the most critical goals, having assigned a particular rank to each of them; they assessed the initial, i.e. current trends, and the desired target values.

Table 3. Vector of development goals of the mechanism of partnership of business entities with public authorities in the Russian Federation

No.	Name of interaction form	Assessment of dynamic pattern	Importance	Initial trend	Target value
1	Electronic interpersonal communication	+1	1	0.4	0.6
2	Written request	+1	4	0.2	0.3
3	Roundtable discussion, consultations	+1	2	0.4	0.5
4	Preparation of laws and regulations	+1	3	0.3	0,4

Note: +1 – positive change in the form value is preferable; – 1 – negative change in the form value is preferable.

In our opinion, in order to achieve the target values of development of mechanism of partnership of business entities with public authorities in the Russian Federation, it is necessary to:

1. Approving measures that block the use of the significant market power of specific organizations by influencing the development and adoption of state (municipal) management decisions:
   - development and approval of legal foundation for the interaction of business entities with public authorities;
   - adoption of the law on lobbying with the obligation of public announcement of the presence of private interests;
   - alleviation of taxes burden, the use of the tax holiday tool for small enterprises established in important social areas of economic activity;
   - legislative support for the development of various forms of mutual assistance of small businesses in the loan, insurance and other areas that contribute to the consolidation of small businesses in financial and economic networks and increase its social influence;
   - adoption of the law on mandatory diversification of ownership packages of persons starting performing functions of executive, legislative and judicial authorities, as well as members of their families with a view to counteracting the economic tools of domination of these persons in economic entities;
   - development of the institute of regulatory impact assessment, mandatory involvement of major business entities in the process of coordination of decisions made by public authorities at all levels of public administration.
2. Improving the interaction of business and government in the modern economic paradigm within the scope of government support for business:
   - formation of the efficient system of information communications;
   - development of labor productivity conditioned by retooling, digitization of economic processes, use of innovative technologies, as well as improvement of

the efficiency of business processes (an important role here will be played by the further adjustment of the educational system);
– development and implementation of adequate educational programs in the area of GR management in the leading universities of the country.

3. Development of feedback and public monitoring mechanisms of state (municipal) management decisions in the area of business development:
   – development of various formats of interaction between public authorities and representatives of the business community;
   – regular consideration and discussion of business development initiatives at the "open government" site, as well as at the sites of leading business associations;
   – development of regulatory impact assessment as a tool for supporting the teamwork of entrepreneurs for the improvement of business environment.

5 Conclusion

The approach of cognitive modeling was suggested and used for solving the issues in the research of the modern mechanism of partnership of business entities with public authorities; it is focused on the qualitative analysis of complex situations that are characterized by the lack of accurate numerical information about the processes occurring in them, and include "qualitative" variables. The existing approach to the assessment of the development of this mechanism in the context of GR management makes it possible to keep qualitative and quantitative accounts of long-term effects of state (municipal) management decisions, it affords an opportunity to search for the solution in a climate of the sharply-changing environment, as well as the lack of information about the factors of the model and the degree of their mutual influence.

The cognitive analysis presented above is an evidence of significant dependence of efficient functioning of the mechanism of partnership of business entities with public authorities on the expediency and qualitative use in-person meetings, direct negotiations, regulatory impact assessment, development of political proposals, as well as participation of representatives of business entities in boards and committees among priority forms of such interaction in the first place.

Specific lines of development of the mechanism of partnership of business entities with public authorities in the Russian Federation have been put forward based on the research findings for achieving the target values: approving measures that block the use of the significant market power of specific organizations by influencing the development and adoption of state (municipal) management decisions; improving the interaction of business and government in the modern economic paradigm within the scope of government support for business; developing the mechanisms of feedback and public monitoring of state (municipal) management decisions in the area of business development.

Acknowledgments. The reported study was funded by the Russian Foundation for Basic Research, grant No. 19-010-00103 "Cognitive modeling of innovative mechanism of partnership of business entities and public authorities (GR-management) in the conditions of digitalization".

References

Axelrod, R.: The analysis of cognitive maps. In: Axelrod, R. (ed.) Structure of Decision: The Cognitive Maps of Political Elites, p. 395. Princeton, Princeton University Press (1976)

Busemeyer, J.R., Diederich, A.: Cognitive Modeling, p. 225. Sage, Thousand Oaks (2010)

Dibie, R.A., Dibie, J., Quadri, O.M.: Business and Government Relations in Kenya. Business and Government Relations in Africa, pp. 340–370. Taylor and Francis (2017)

Gabsa, W.: Business and Government Relations in Cameroon. Business and Government Relations in Africa, pp. 291–313. Taylor and Francis (2017)

Gadzekpo, L.: Government and Business Relations in Ghana. Business and Government Relations in Africa, pp. 132–146. Taylor and Francis (2017)

Gorelova, G.V., Zakharova, E.N., Martyshina, T.V., Pankratova, N.D.: Cognitive process modeling of sustainable regional development. Bull. ASU **2**, 166–174 (2014)

Holt, D.V., Osman, M.: Approaches to cognitive modeling in dynamic systems control. Front. Psychol **8**, 2032 (2017)

Hussien, M., Dibie, R.A.: Business and Government Relations in Sudan. Business and Government Relations in Africa, pp. 162–195. Taylor and Francis (2017)

Igbokwe-Ibeto, C.J.: Business and Government Relations in Tanzania. Business and Government Relations in Africa, pp. 314–339. Taylor and Francis (2017)

Kawewe, S., Dibie, R.A.: Business and Government Relations in Zimbabwe. Business and Government Relations in Africa, pp. 196–219. Taylor and Francis (2017)

Kayl, I.I., Lamzin, R.M., Epinina, V.S.: GR-management: realities and prospects of development. Fundam. Res. **9–3**, 597–600 (2016)

Klochko, E.N., Prokhorova, V.V.: GR-management as a tool for business community interaction with government authorities. Econ. Entrepreneurship **6–2**, 831–834 (2015)

Kostikova, A.V., Tereliansky, P.V., Shuvaev, A.V., Parakhina, V.N., Timoshenko, P.N.: Expert fuzzy modeling of dynamic properties of complex systems. ARPN J. Eng. Appl. Sci. **11**(17), 10601–10608 (2016)

Lapina, N., Chirikova, A.: Regional Elites in the Russian Federation: Behaviors and Political Orientations, p. 129 (1999)

Laskowskaya, T.A., Saveliev, E.P., Lindina, T.A.: Theoretical bases of modeling of interactions in the construction industry based on the cognitive approach. Bull. Chelyabinsk State University **7**, 24 (2008)

Lewandowsky, S., Farrell, S.: Computational Modeling in Cognition: Principles and Practice, p. 365. Sage, Thousand Oaks (2010)

Makarenko, D.I., Khrustalev, E.U.: Cognitive technologies in the theory and practice of management. Prob. Theory Pract. Manag. **4**, 25–33 (2007)

Moseyko, V.O., Korobov, S.A., Tarasov, A.V.: Cognitive modeling in the formation of management decisions: the potential of resource-factor analysis. Creative Econ. **5**, 629–644 (2015)

Nur, Y.A.: Business and Government Relations in Somalia. Business and Government Relations in Africa, pp. 147–161. Taylor and Francis (2017)

Pachur, T., Scheibehenne, B.: Unpacking buyer-seller differences in valuation from experience: a cognitive modeling approach. Psychon. Bull. Rev. **24**, 1742–1773 (2017)

Perepelitsa, G.V.: Improving the interaction of business and government in the modern economy. Russ. Entrepreneurship **8**, 28–32 (2006)
Plotnikova, N.A.: Features of the interaction of business and government in the regulation of socio-economic problems of the territory. Concept **S28**, 56–60 (2014)
Prezenski, S., Brechmann, A.: A cognitive modeling approach to strategy formation in dynamic decision making. Front. Psychol. **8**, 1335 (2017)
Prokhorova, V.V.: Cognitive modeling of sustainable economic development of an enterprise. Econ. Manag. **1**, 24–25 (2011)
Ragulina, Y.V., Stroiteleva, E.V., Miller, A.I.: Modeling of integration processes in the business structures. Modern Appl. Sci. **9**(3), 145–158 (2015)
Turovsky, R.F.: Regional models of interaction between business and government elites: modern processes and their socio-political consequences. Final analytical report, p. 42 (2010)
Yasin, E.: Burden of the state and economic policy. Questions Econ. **11**, 7–10 (2002)

Digital Society as the Basis of Development of Economy in the 21st Century

Problems of Virtualization and Internetization of Social Space

Elena V. Gryaznova[1(✉)], Alexander A. Vladimirov[2], Svetlana M. Maltceva[1], Aleksey G. Goncharuk[1], and Nikolai V. Zanozin[3]

[1] Minin Nizhny Novgorod State Pedagogical University, Nizhny Novgorod, Russia
egik37@yandex.ru, maltsewasvetlana@yandex.ru, aleksgon75@yandex.ru
[2] Volga State University of Water Transport, Nizhny Novgorod, Russia
[3] Branch of the SBOU in Nizhny Novgorod State Engineering and Economic University, Nizhny Novgorod, Russia
nzanozin@mail.ru

Abstract. Relevance. In conditions of modern development of information in society, the problem of studying processes of informatization of social space becomes relevant for the methodology of philosophical, humanitarian and social knowledge. However, as shown by analysis of scientific literature on this issue, the categorical apparatus that determines the essence of the information society remains not yet fully developed. In particular, both in foreign and domestic scientific literature introduced the concepts of "virtualization", "virtual reality" and "internetization". However, introduced concepts do not have their own clear meaning, which leads to the identification of concepts "social" and "virtual". As a result, when analyzing the processes of Internet social space, researchers can not clearly dissociate such concepts as "social reality" and "virtual reality".

The Purpose of the Study: analyze problems of defining the concepts of "virtualization" and "internetization" of social space.

Authors identify problems associated with the development of concepts of "virtualization" and "internetization" of social space; substantiate the need to clarify these categories for the study of social space consequences of information development phenomena of modern society: transformation of social interaction between subjects, replacement of the real social subject with information quasi-subject in the process of socialization of personality, transformation of social space on level of structure, erosion of cultural and ethnic boundaries, globalization and unification of the economic processes of various subjects of civilization. The category "virtualization" in the meaning used in modern social and humanitarian knowledge, duplicates the concept of "Internet" and "social reality" and requires a deeper philosophical comprehension.

Keywords: Information society · Internet · Virtual reality · Social reality · Information socialization · Cyberspace · Internetization · Virtualization

JEL Code: O100

E. G. Popkova and B. S. Sergi (Eds.): ISC 2019, LNNS 91, pp. 119–124, 2020.
https://doi.org/10.1007/978-3-030-32015-7_14

1 Introduction

As you know, concepts of the information society are beginning to be developed around the middle of the twentieth century. In the XXI century, the direction of theoretical thought from the analysis of the conceptual content of the term "information society" passed to analysis of the concept of "virtual society" (Kuznetsov et al. 2018).

We believe that such a reorientation is related not so much to the inadequacy of the concept of the information society, but to the narrow boundaries of forecasts in understanding the possibilities of a technology of the future. It is known that modern society is in a phase of transition, which is called the "Epoch of the Internet", which is at the stage of rapid development of the mobile Internet, in contrast to the stage of "home based", which E. Toffler wrote about (Toffler 1980).

This period of civilization development of Internetization indicates that electronization, computerization, informatization, and mediatization by society are almost overcome. Modern researchers designate the process of internetization as the next stage of development of the information society, and virtualization as the next stage after (Kuznetsov et al. 2017). However, the content of these terms practically does not differ, moreover, both terms coincide with the notion of "social reality".

2 Methodology

As the main methods of research were used: analysis, generalization, comparison, analytical review of literature, and method of dialectics. The study was conducted according to the following stages: analysis of the content of concepts "virtual reality", "virtualization" and "internetization" in domestic and foreign literature. Identification of problems and their causes when using this terminology in research of social problems of informatization of society.

3 Results

Analyzing works devoted to the study of the concept of "virtual reality", we found that it is interpreted in a narrow and broad sense. In a narrow technical sense, it means information reality, which is created on the basis of information technology. As you know, the phrase "virtual reality" was put into circulation at the end of the 20th century and meant a space created by a computer into which a person could use with the help of special equipment. Today, in a narrow, technical sense, virtual reality means an informational interactive simulation of the real environment. The basic principle of its interactivity is the ability to influence the senses of a person and react to his reactions (Saler 2012; Rucker 1993).

In a broad sense, the concept of "virtual reality" means all phenomena that were previously defined as "ideal", "mental", "subjective", "social", etc., for example, such as mental states, social reality or its individual layers. This approach is widely used in modern domestic works, for example, (Kovalevskaya 1998; Bychenkov 2001; Bondarenko 2001).

Expansion of the meaning of the concept of virtual reality originates from post-modern traditions, among the main concepts of which can be distinguished pluralism of reality and theory of simulacrum. So, Adolphe Buhl made an attempt to create the theory of "virtual society". He proved that virtual reality technologies create in social space "parallel" worlds, in which there are analogues of real system elements of society: economic and political structures on the Internet, interaction with artificial characters of computer games, etc. By virtualization he meant substitution based on computer technology virtual real social space (Buhl 1997). Virtualization of the social space of M. Paetau understood as the emergence in the structure of the social system of new elements, such as analogs of real communications (Becker 1997). In turn, A. Kroker and M. Weinstein mean virtualization not the process of replacing one reality with another, but the process of alienating a person from the body by means of computer technology (Kroker and Weinstein 1994).

You can see that the concept of "virtualization" by foreign scientists is considered in two ways: as a new technological process embedded in the social system and as a new social, computerized mechanism for social interaction. It is the second option that creates in the future a broad interpretation of virtual reality as a social phenomenon, leading to the identification of the concepts of "social" and "virtual" reality.

Extensive understanding of virtual reality, which substitutes many social phenomena considered earlier by social philosophy and sociology received and in the domestic literature. Researchers write: "Virtualization, in this case, is any replacement of reality by its simulation, in a way - not necessarily with the help of computer technology, but necessarily with the use of virtual reality logic".

In the works of domestic researchers there are concepts in which all social reality is recognized as virtual. The main feature of this reality is: "… phantom, fictitious, covering subjective elements." Thus, the following are considered as virtual: social institutions, a set of signs for expression of thoughts, monetary system, professional customs, social norms, etc. The question arises: "What, then, remains a social reality?" It turns out that the whole sphere of social relations is virtual reality. In the social reality, the American sociologist M. Castells concludes, nothing social remains - everything is virtual, but virtual because it is symbolic (Castells 2002).

Continuing the conversation about the introduction of the concept of "virtual reality" into the philosophical and sociological categorical apparatus, we find that here, there is a reorientation of the previously established concepts derived from the category "social" to "virtual". Examination of works on the problem of virtualization of society showed that when speaking about virtualization, researchers reduce its analysis, as a rule, to consider problems associated with implementation in all spheres society of new information technologies. The proof of the last statement can be a rapid surge of works on sociology of the Internet, which analyzes problems of creating a network community, which is an alternative to the traditional society and social relations in it (Bondarenko 2001).

4 Results

No matter how widely interpreted virtual reality, new information society is a society of information quasi-subjects, not virtual. Only they are able to make global changes in society. This is due to the fact that information quasi-subjects become a new type of subjectivity, with which the real subject enters into information interaction (Gryaznova 2018).

As we have shown, the concept of "virtual reality" does not reflect the essence of the new reality. It would be more correct to speak of the social reality created by means of information technology. This reality, created by technology, is organized in such a way that, being in it, a person interacts with information quasi-subjects and objects as it would interact in the real reality with real people and objects. Even at the present stage of development, the Internet is independent self-organizing information social rather than a virtual system that can act as an informational imitation of the real society, its alternative for example, educational space or complete replacement (Petrova et al. 2018; Grishina and Volkova 2018a, b). There is a situation where the way to the information society lies through the passage of the social system of following stages:

(a) Creation of an autonomous information environment for interaction to improve the effectiveness of all spheres of human activity;
(b) Integration of this artificial environment into the infrastructure of society and its parallel functioning with the traditional spheres of activity;
(c) Formation of the information environment as the basis of the public system;
(d) Displacement of traditional ways of interaction of elements of the social structure of information;
(e) Full replacement of traditional ways of the interaction of elements of the social structure by information means of interaction.

Most advanced countries of the world community have already reached stages (b) and (d), but no country has completely moved to stage (e), which makes it impossible to give an objective assessment of the reality of information society. That is why today the forecasts of positive and negative consequences of functioning of the information society are adjusted taking into account the Internet and introduction of virtual reality technologies in all sphere of human activity.

The greatest fears among philosophers are expressed about transformation of human consciousness in the era of the Internet. The problems of socialization of the individual become more urgent, the main mechanisms of which are not so much traditional education, upbringing and training in the teacher-student system, as informational simulation (Semarkhanova et al. 2018; Petrova et al. 2018; Grishina and Volkova 2018a, b). Two systems of socialization still exist in parallel, and it would seem that the Internet and other computer technologies are adapting the evolving personality to an era of rapid technological change, compacting social time. But a significant part of personal time is spent on the development and implementation of a new type of activity—information communication, which changes both the type of thinking and all psychology, and human physiology. In the most general form, you can safely discuss the transformation of a person feeling (direct communication) into a person reading and writing (indirect communication), and then into a piece of person information (network communication).

5 Conclusion

Without using the term "virtualization" it is possible to define that the Internet today is an information social institution of society, existing on the basis of information technologies. It gives the opportunity to realize the functions of communication of people freed from ethnic, social, religious barriers and state borders, which leads to globalization - the next and most significant stage of the formation of the information society, which is characterized by the formation of a global economy with a global division of labor on the basis of information resources, without which can no longer exist and resources material, energy and spiritual. The processes of organization and self-organization of such a social system differ from the traditional one. The introduction of the term "virtualization", which does not have an independent meaning, does not give rise to scientific knowledge. The terms "virtual reality" and "virtualization" require rethinking and refinement of socio-humanitarian knowledge and clearer limits of use. This is the subject of our next study.

References

Becker V., Paetau M.: Virtualisierung des Sozialen. Die Informationsgesellschaft zwischen Fragmentierung und Globalisierung, Frankfurt a. M. (1997)

Buhl, A.: Die virtuelle Gesellschaft. ukonomie, Politik und Kultur im Zeichen des Cyberspace, Opladen (1997)

Castells, M.: The Internet Galaxy: Reflections on the Internet, Business, and Society, 292 p. OUP Oxford (2002)

Grishina, A.V., Volkova, E.N.: Psychological factors for computer game addiction of young adolescents. Int. J. Eng. Technol. (UAE), **7**(3.14 Special Issue 14), 327–330 (2018a)

Gryaznova, E., Kozlova, T., Sulima, I.: Forming and realizing a pedagogue's philosophical culture «The Turkish Online Journal of Design, Art and Communication-TOJDAC», September 2018, Special Edition, pp. 2136–2142 (2018)

Gryaznova, E., Maltseva, S.M., Zanozin, N.V., Goncharuk, A.G., Kozlova, T.A.: Information culture person: problems and perspectives. In: 5th International Multidisciplinary Scientific Conference on Social Sciences and Arts SGEM, Vienna ART Conference Proceedings, 19–21 March 2018, vol. 5, no. 2.1, pp. 241–248 (2018)

Kroker A., Weinstein, M.: Data trash. The theory of the virtual class, Montreal (1994)

Kuznetsov, V.P., Garina, E.P., Andryashina, N.S., Romanovskaya, E.V.: Models of modern information economy conceptual contradictions and practical examples, 361 p. Emerald Publishing Limited (2018)

Kuznetsov, V., Kornilov, D., Kolmykova, T., Garina, E., Garin, A.: A creative model of modern company management on the basis of semantic technologies. In: Communications in Computer and Information Science (2017)

Petrova, N.E., Ilchenko, N.M., Patsyukova, O.A., Samoylova, G.S., Moreva, A.N.: Man as the subject of possible/impossible in the Russian nominations of the feature of the subject. In: Advances in Intelligent Systems and Computing, vol. 622, pp. 163–169 (2018)

Saler, M.: As If: Modern Enchantment and the Literary Prehistory of Virtual Reality. Oxford University Press, Oxford (2012)

Semarkhanova, E.K., Bakhtiyarova, L.N., Krupoderova, E.P., Krupoderova, K.R., Ponachugin, A.V.: Information technologies as a factor in the formation of the educational environment of a university. In: Advances in Intelligent Systems and Computing, vol. 622, pp. 179–186 (2018)

Toffler, A.: The Third Wave, p. 544. Morrow, New York (1980)

Bondarenko, S.V.: Virtual network communities: specifics of formation and functioning: author's thesis of PHD Sciences: 22.00.04, Rostov-on-Don, 26 p. (2001)

Grishina, A.V., Volkova, E.V.: Structure of subjectivity of teenagers with different levels of gaming computer dependence. J. Bull. Minin Univ. **6**(1) (2018b). https://doi.org/10.26795/2307-1281-2018-6-1-14

Gryaznova, E.V.: Model of information culture management of education of the municipality: prevention of risks of inefficient use of achievements of informatization. J. Bull. Minin Univ. **6**(2) (2018). https://doi.org/10.26795/2307-1281-2018-6-2-18

Innovations and Motivation of Personnel as the Main Drivers of Development of Industrial Enterprises

Nikolay A. Zhdankin(✉), Vladimir M. Suanov, and Bahrom K. Sharipov

National Research Technological University "MISIS", Moscow, Russia
regul-consult@mail.ru, suanovv@mail.ru, sharipov.misis@mail.ru

Abstract. This article is devoted to the problems of development of industrial enterprises, which are connected with the lack of processing plants, high level of wear of fixed assets, small capacity of the domestic market, low productivity, outdated technologies, etc. All this is a consequence of the raw economy of the country, which is the lack of innovation and weak motivation of the staff. Purpose of the work: (1) understand what is happening, why innovation is not adopted in Russia, (2) identify the reasons for low staff motivation, and (3) propose solutions to create innovation and increase motivation.

To solve the problems of industrial enterprises an innovative approach is applied, which with the help of known methods of innovative management (analysis of a problem, star-shaped, fishbone and ladder charts) allow us to analyze problems, identify root causes and develop measures for their elimination. For analyzing motivation the approach of measuring and analyzing the level of motivation by enneagram method of nine factors is used.

Ranking of the developed activities showed that the main one for creating innovations is the development of an effective strategy taking into account introduction of innovative technologies and development of small and medium business, which gives an explosive effect. It is necessary to create a system of monitoring staff motivation at an enterprise and use it to analyze motivation and develop recommendations for its growth.

Effective solutions for development of industrial enterprises through creation and introduction of innovations, which is launched through an effective system of personal motivation, developed on the basis of measurements, are found.

Keywords: Industrial enterprises · Development problems · Analysis of problems · Innovations · Motivation of personnel

JEL Code: O30

1 Introduction

The main problems of industry in Russia (and CIS countries), especially basic industries, are associated with a high level of wear and tear of fixed assets, a high percentage of defects, small capacity of the domestic market, low productivity, old

E. G. Popkova and B. S. Sergi (Eds.): ISC 2019, LNNS 91, pp. 125–133, 2020.
https://doi.org/10.1007/978-3-030-32015-7_15

technologies, etc. (Zhdankin and Romanycheva 2016). All this is combined by almost total lack of innovation, low motivation of staff and low production efficiency, weak participation of small and medium-sized businesses in the industry, a small number of new jobs, and high unemployment (Zhdankin and Gurin 2013).

In these circumstances, enterprises in all branches of industry need to focus on the tasks of increasing labor productivity, reducing costs, improving production efficiency, and product quality. To do this, they need to develop and implement effective strategies to improve competitiveness and strengthen their position in the domestic market. At the same time, the main directions of these strategies should be related to innovation, new technologies and staff motivation (Zhdankin 2013; Yashin et al. 2018).

2 Methodology

2.1 Innovation

Innovations determine the main vector of the modern development of all enterprises in different branches without exception. Innovations in our difficult time are absolutely necessary. Despite these obvious moments, Russia does without its own innovations, losing the pace of development and not receiving additional income. Today, most of the innovations that appear in our market have come and continue to come from abroad. With huge intellectual potential, we can not realize it, and take advantage of the results of someone else's innovation activity. It is necessary to reverse this trend, and to start mass-creating innovations and producing modern innovative products.

The study shows that the most important reasons that create problems for the development of an industrial enterprise are the following:

- There is no clear strategy for the development of the country in general and industry in particular
- Innovative technologies and programs to develop and implement them are small or virtually non-existent
- Lack of financing of innovative technologies
- Improper stimulation of personnel in enterprises, as well as lack of highly qualified workers
- Low level of innovation infrastructure development
- Lack of willingness of investors to invest in enterprises with large risks
- Management misunderstanding of the importance of developing innovative technologies in the industry, etc.

Figure 1 shows a **ladder diagram**, which allows establishing a causal relationship between the identified main causes of the problem, to arrange all the causes in the order of this connection (cause-consequence) and identify the root causes of the problem.

2.2 Staff Motivation

The issue of development and implementation of innovations, as the most important direction of development of industrial enterprises, based on low motivation of staff in

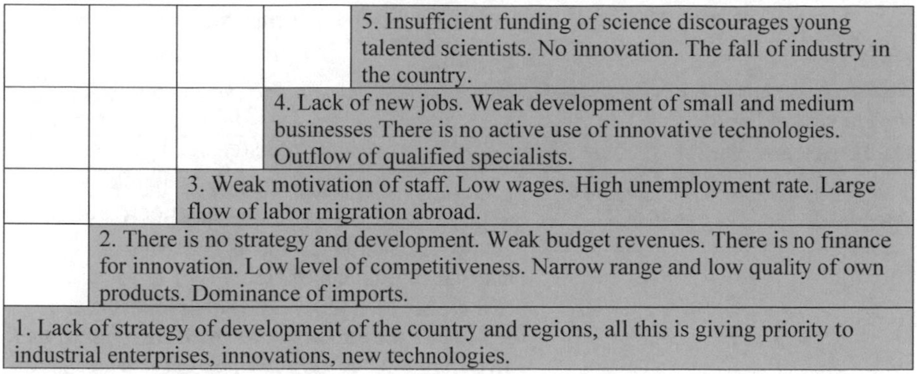

				5. Insufficient funding of science discourages young talented scientists. No innovation. The fall of industry in the country.
			4. Lack of new jobs. Weak development of small and medium businesses There is no active use of innovative technologies. Outflow of qualified specialists.	
		3. Weak motivation of staff. Low wages. High unemployment rate. Large flow of labor migration abroad.		
	2. There is no strategy and development. Weak budget revenues. There is no finance for innovation. Low level of competitiveness. Narrow range and low quality of own products. Dominance of imports.			
1. Lack of strategy of development of the country and regions, all this is giving priority to industrial enterprises, innovations, new technologies.				

Fig. 1. Ladder diagram

general and motivation to create innovations in particular, that should be the main strategic objective of development.

Well-known approaches of strategic analysis (Ansoff 2008; Boumen 2007; Vikhansky 2008), used in strategy development, the main focus of study on the competitiveness which is depending on the financial performance: volume of investments, profitability, indicators of capital return, financial leverage, fixed and working capital, capital turnover, etc. However, all of these parameters are known *to be secondary* because they are the result of certain actions of employees that can be effective or cause losses and lead to the bankruptcy of an enterprise. The level of enterprise development depends on the *root* causal factors—professionalism and motivation of people who make certain management decisions and perform work on their implementation. At present, there is an urgent need to include in the strategic analysis motivation of a company's staff and their professionalism, as previously suggested by the authors (Zhdankin and Suanov 2017).

3 Results

3.1 Innovation

To improve efficiency of the innovative development of industrial enterprises, it is necessary to remove the main obstacles to creation and implementation of innovations. Finding solutions to the problem was done using method of *stimulating the process of achievement of goals* (Table 1). At the same time, we take the ladder chart (see Fig. 1), because to solve this problem, it is necessary to eliminate root causes.

With a certain set of ideas to solve this problem, we can proceed to their assessment. The assessment of the main ideas was carried out using the method of expert assessments on four criteria. The results of the expert group on assessments of proposed ideas are given in Table 2. A 10-point evaluation scale was used.

Table 1. Method of stimulating the process of achieving goals.

Goal *Improving the level of development of the country and regions*	**Problem** *Industrial development and innovation*
Goal characterization	Solutions
1. Development of the country and regional development strategy	1. Develop an effective strategy for the development of the country and regions, giving priority to industry, industrial enterprises, innovations, innovative technologies, etc. 2. Increase the level of innovation infrastructure 3. Increase the number of industrial enterprises by diversifying production and increasing the depth of processing of raw materials 4. Create a system of continuous improvements in the country, etc.
2. Develop businesses in the industry	1. Analysis of foreign practice, where SMEs gave positive results (examples of China, USA, Spain) 2. Privileges and development of SME support programs in the initial stages 3. To develop competition among SMEs 4. Targeting them to develop innovations to improve competitiveness, etc.
3. Increase staff motivation	1. Develop a program of staff motivation with identification of motivation and anti-motivation factors 2. In order to identify trends in personal preferences, conduct constant monitoring of the level of motivation 3. Creation of a system of bonuses and benefits for employees of industrial enterprises, etc.
4. Increase funding for education and science	1. Increase in faculty salaries to attract motivated professionals with experience in the specialty 2. Increase of scholarships for students 3. Allocation of funds for advanced training courses and development for faculty members 4. Targeted funding of faculty members, which has projects, grants and other proposals related to innovations, etc.
5. Development of innovative technologies	1. To develop innovative technologies taking into account the peculiarities of development of the country and industry 2. Financing of developments, which are being implemented by leading technical and technological institutes 3. Hire highly qualified specialists for the development of innovative technologies 4. To modernize enterprises taking into account innovative development of technologies, etc.

As you can see, the results were very interesting. On the *1st place* was *"Development and implementation of the strategy of innovative development of the country and regions"*. And this is obvious. Without a strategy, no development is possible. On the *2nd place*—the idea *"Improve technologies of production of innovative products of industrial enterprises"*. Here you need to do everything possible to develop or search for such technologies (Kuznetsov et al. 2017). These can be already known technologies, but not yet used in production, or completely new ones developed specifically for our conditions (Potashnik et al. 2018). On the *3rd place*, the idea *"Create new jobs through the development of SMEs in the industry"*. This is the root of future development. Small and medium business is the driver of production, job and GDP growth. Worldwide, SMEs provide up to 70–90% of the population's employment and up to 60–80% of GDP. It is a gift for the rapid development of the economy and growth of its efficiency. At the same time, SMEs through greater mobility and work in a competitive environment create innovations, innovations, and other techniques in order to bypass competitors and win, and thus get at the same time maximum income and profit. The development of SMEs will give an influx of money to the country's budget due to increased tax revenues.

Table 2. Results of the ranking of idea-projects.

Solutions	Cost	Income	effect/outcome	time	Amount	Place
Specific weight	0.25	0.33	0.33	0.09	1	
1. Development and implementation of the strategy of innovative development of the country and regions	6	9	10	4	8.13	1
2. Improvement of technology of production of innovative products	5	9	9	4	7.55	2
3. Development of the program and system of benefits for SME at the initial stages	2	8	8	3	6.05	9
4. Create new jobs through the development of SMEs in industry	6	8	8	5	7.23	3
5. Increase of expenditure on innovation development in the country	4	2	7	6	4.51	11

(*continued*)

Table 2. (*continued*)

6. Allocate sufficient funding for science	6	6	7	7	6.42	5
7. Increase student exchange and international academic mobility programs	6	2	5	5	4.26	12
8. Forums with the participation of international experts	7	6	6	4	6.07	8
9. Development of competition in the country, strengthening the fight against corruption	6	6	7	4	6.15	6
10. Attracting new foreign investors, in addition to bank loans	4	5	7	3	5.23	10
11. Create a system of continuous improvement in the country	7	6	7	5	6.49	4
12. Carrying out activities to stimulate and increase motivation of staff	6	6	6	7	6.09	7

On *4th place* is the idea of *"Creating a system of continuous improvement in the country"*, which is very important from the point of view of involving a wide range of people in the development process. It is necessary to create conditions for public participation in this process, allowing the authors of ideas and proposals to try to implement them with support of management.

5th place - is science. Implementation of the idea *"Allocate sufficient funding for science and education"* will give a powerful boost to innovation for SMEs, which will contribute to its active expansion and development, as well as to the development of science. On the *6th place—"Development of competition in the country, strengthening the fight against corruption"*, which is the most important government task for strengthening transparency of the economy and relations in society between government and people.

Further on the importance of activities is motivation of staff. Note that these ideas received quite high marks of more than 6 points, which indicates their importance in solving the problem.

3.2 Staff Motivation

As an example of motivation measurement in Fig. 2, *an enneagram* of dynamic change in the level of motivation in the sales division of a large metallurgical enterprise by years.

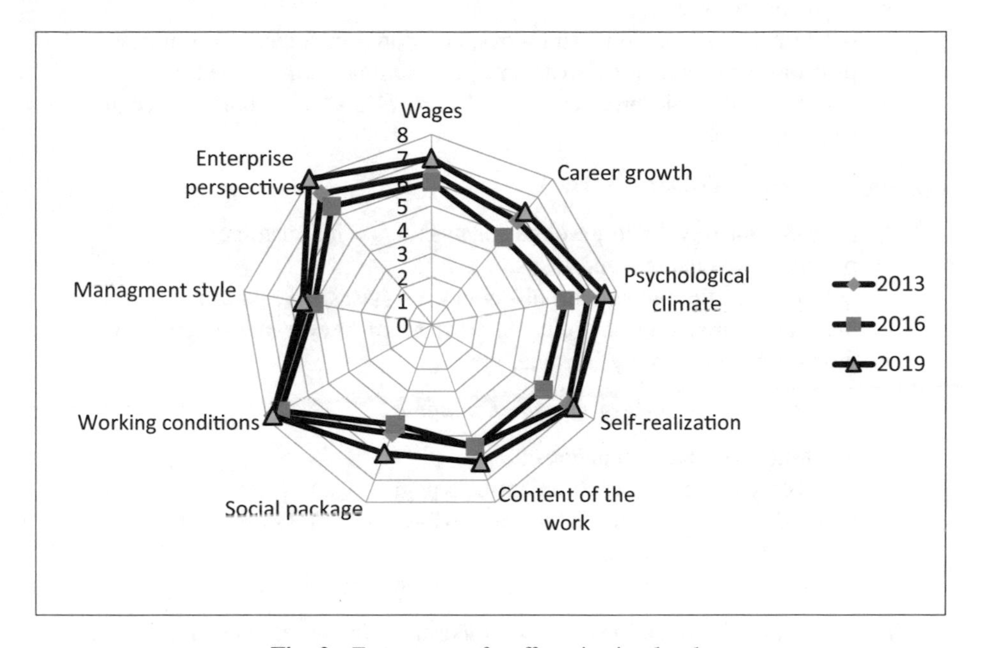

Fig. 2. Enneagram of staff motivation level.

As we can see, in the pre-crisis of 2013 the highest marks were received by working conditions, prospects of the enterprise, psychological climate, and self-realization. However, the crisis made adjustments, and already in 2016 there was a reduction in the level of motivation of staff and a decrease in its efficiency, which was reflected, by the way, in a decrease in sales volumes. All the grades (perhaps, except the content of the work) went down, but fell especially strongly and fell into the problematic zone of career development, leadership style, social package, self-realization, and psychological climate, although wages have remained at an acceptable level, although they have also declined.

This clearly demonstrates that the current staff motivation and incentive system was not adapted to changing circumstances. This adjustment is especially necessary in difficult crisis conditions.

In what direction is it necessary to develop motivation? You just need to turn to those for whom this system is made—to people. That is why the development of an effective motivation system should begin by inviting people to express their problems, ideas and "dreams" to improve the system, bringing the will receive a package of valuable offers from the employees themselves. As it was made at the enterprise under consideration.

The result was the collection of proposals from employees and their use as wishes to improve the motivation system, where:

(a) *group of methods of financial stimulation:*

- bonuses to wages;
- payment for training;
- additional payments for extra work, additional payments for seniority;
- promotion of innovative ideas and projects that optimize work;
- reward for the sold products not only for sales volume but also for the profit from sales, etc.

(b) *on the social package:*

- payment for travel and gasoline for meetings with customers;
- preferential vocations and loans;
- assistance in the preparation of documents (visas, passports);
- medical insurance in paid clinics, payment of lawyer and psychologist;
- promotion of a healthy lifestyle, etc.

(c) *on the group of methods of moral stimulation:*

- gratitude from the management;
- valuable gifts for over fulfillment of a plan;
- promotion of time, introduction of flexible working hours;
- training tours on production;
- reward board of honor, as well as boards of shame, etc.

The result of the introduction of these proposals in the work of an enterprise was an increase in the level of motivation of the staff. Very small at first, and then more, which was finally formed in 2019 (see Fig. 2), when it even exceeded the level of motivation of 2013, which indicates that the system of motivation is successful.

Separately it should be noted that an excellent result was obtained with *extra days off* when an employee who fulfilled a plan for more than 100% received additional days for vocation, which served as a good motivation for staff.

4 Conclusion and Recommendations

In general, we can say that all ideas on creation of innovations are extremely important for development, and their ranking shows certain *priorities* in solving problems of industrial enterprise. This is *an algorithm* for their effective solution so that you can solve not everything at once, but in a certain sequence, highlighting the main directions for the influence. At the same time, we emphasize that the development of an effective strategy for the development of the country and regions introduction of innovative technologies and development of SMEs is the main thing.

All this should be based on an effective system of motivation developed on the basis of feedback in the form of measurements of staff motivation level. This contributes to the management's understanding of how people evaluate their work and what they want the development of the enterprise to be effective and irreversible.

References

Ansoff, I.: New Corporate Strategy, 501 p. Peter, SPb (2008)

Bowman, K.: Fundamentals of Strategic Management, 389 p. UNITI, M. (2007)

Vikhansky, O.S.: Strategic Management, 601 p. Gardarika, M. (2008)

Zhdankin, N.A.: Choice of innovative strategy of enterprise development. Gornyi J. (10), 61–66 (2013)

Zhdankin, N.A.: Innovation Management: Textbook (for bachelors), 315 p. KnoRus, M. (2017)

Zhdankin, N.A.: Staff Motivation. Measurement and Analysis, 272 p. Finpress, M. (2010)

Zhdankin, N.A., Gurin, D.V.: How to create new jobs by innovative approaches. Innov. Manag. (5), 42–53 (2013)

Zhdankin, N.A., Romanycheva, S.K.: Modern trends in real labor productivity change in metallurgy. Metallurg (3), 11–17 (2016)

Zhdankin, N.A., Suanov, V.M.: Motivation of personnel as a key factor of strategic development of industrial enterprises of Russia. Manag. Today (4), 256–272 (2017)

Zhdankin, N.A., Sharipov, B.K.: Socio-economic problems of industrial region development and innovative approach to their solution. Creative Econ. **13**(1), 151–168 (2019)

Kuznetsov, V., Kornilov, D., Kolmykova, T., Garina, E., Garin, A.: A creative model of modern company management on the basis of semantic technologies. In: Communications in Computer and Information Science (2017)

Potashnik, Y.S., Garina, E.P., Romanovskaya, E.V., Garin, A.P., Tsymbalov, S.D.: Determining the value of own investment capital of industrial enterprises. In: Advances in Intelligent Systems and Computing, vol. 622, pp. 170–178 (2018)

Yashin, S.N., Trifonov, Y.V., Koshelev, E.V., Garina, E.P., Kuznetsov, V.P.: Evaluation of the effect from organizational innovations of a company with the use of differential cash flow. In: Advances in Intelligent Systems and Computing, vol. 622, pp. 208–216 (2018). https://doi.org/10.1007/978-3-319-75383-6_27

Methodological Approach to Analysis of Management Systems Using the Graph Theory in the Digital Economy

Elena N. Letiagina[1(✉)], Yury V. Trifonov[1], Elena Y. Trifonova[1], Alexander N. Vizgunov[2], and Julia A. Grinevich[1]

[1] National Research Nizhny Novgorod State University name after N.I. Lobachevsky, Nizhny Novgorod, Russia
helenlet@yandex.ru, itime@iee.unn.ru, trifonova.elen@gmail.com, julia-grinevich@mail.ru

[2] National Research University Higher School of Economics, Moscow, Russia
vizgunovhse@yandex.ru

Abstract. Relevance. In the context of formation of the digital economy and fierce competition. It is very important to introduce new effective approaches to improve enterprise management systems for business analysis of huge data flows. The aim of this research is to develop a methodological approach to the study of management systems of organizations with use of graph theory. Theoretical methods are used: description, analysis and synthesis, additive method, method of functional analysis, as well as combinatorial methods of theoretical and multiple analysis.

Results. On the basis of formalization of main characteristics and interrelations of organizational structure the system of indicators of balance, manageability, and stability of systems and integral criterion is proposed evaluation of management system effectiveness.

Discussion. Article offers ways and methods of using graphs in order to obtain new knowledge, generalization and deepening understanding of facts and theories in the field of development of organizational structures. The top of the graph is proposed to designate heads of organization and its subdivisions. In this case, edges of the graph will be management links of a management system.

Conclusion. Research of enterprise management systems with the use of developed methodical approach and estimation tools of graph theory will allow defining strategic directions of adjustment management decisions on formation of the organizational structure.

Keywords: Management system · Organizational structure · Management · Graph theory · Management relations · Delegation of authority

JEL Code: L250 · M1

E. G. Popkova and B. S. Sergi (Eds.): ISC 2019, LNNS 91, pp. 134–141, 2020.
https://doi.org/10.1007/978-3-030-32015-7_16

1 Introduction

Huge data flows have created a new paradigm for business analysis processes, increasing the potential of advanced analytical and cognitive tools of the digital economy. Large data sets and new technologies have created conditions for organizational change and formation of new types of management systems, which require a strategic approach to implementation of business decisions processes. (Bratasanu 2018; Zastupov 2019).

Relevance of development of a new methodical approach is due to the need to study problems of enterprise management in conditions of digitalization of the economy. Development and analysis of management systems in conditions of digitalization of the economy are of practical importance in the field of improvement of management activity of enterprises (Kuznetsov et al. 2018).

Organizational structure is one of elements that form the basis of the management system. Therefore, the right choice of an organizational structure and its compatibility with other elements affect performance of the entire system (Surie and Singh 2013). Issues of building management systems of organizations are considered in writings of many scientists and practitioners (Arjaliès and Mundy 2013; Böckers and Heimeshoff 2014; Johnson et al. 2018; Novikov 2013; Zhu and Jiao 2013). At the present stage of development of organizations in the digital economy, there may be a problem of building new management systems (Kuznetsov et al. 2017). For the effective implementation of this process, it is very important to develop a scientifically based approach to the analysis of management systems (Grochla and Szyperski 2018; Halk 2018).

2 Methodology

The article considers possibilities of application of graph theory to solve organizational problems and design of management systems of organizations.

Ancestor of the theory of graphs is Euler, who published an article on the Koeningsberg bridges in 1736. With development of research of electrical networks, crystallography and organic chemistry, next impulse to graph theory was almost 100 years later. The Hungarian mathematician König et al. (1936) published in 1936 the book "Theory of finite and infinite graphs" - first textbook in the field of graph theory. The introduction of probabilistic methods in graph theory, especially in studies by Paul Erdős and Alfréd Rényi (1959) on the asymptotic probability of graph connectivity, gave rise to another branch known as random graph theory. Graphic representation of graph (Barabási et al. 1999) is used to demonstrate the topology of exponential and smaller-scale networks.

Using graph theory, researchers solve a large number of applied optimization problems, simulates behavior of network elements and their interaction, including combining elements in clusters or system.

The formation of a set of all possible organizational structures of enterprises is proposed to be carried out using combinatorial modeling methods (Glamazdin and Zinchenkov 2011; Thulasiraman and Swamy 2011). For this purpose, organization is divided into subsystems, for example, responsibility centers. This approach allows an expert analyst to compare individual variants of control systems according to different criteria in order to select the most suitable ones.

Investigated organizational structure can be represented in the form of graph G1:

$$G1 = \{a_0, a_1, a_2, \ldots a_9, b_1, b_2, \ldots b_{15}, c_1, c_2, \ldots c_{14}\}$$

Sources of formation and recipients of management influences are peaks of the graph. So the top of a_0 can characterize the head of the organization, a_1 - deputy head, a_2 - financial director, a_3 - development director, a_4 - technical director, etc.

Management relations in management system—(a_0 a_1)—from head to deputy head, (a_0 a_2)—from head finance director, and further, edges of the graph (a_0 a_1), (a_0 a_2)…(a_0 a_9), (a_1 b_1), (a_1 b_2) and etc.

G1 graph is a tree structure containing cycles. Effectiveness of the graph G1 can be determined by calculating indicators (Salov 2008):

1. The balance of the structure;
2. Integrity;
3. Manageability;
4. Stability.

The following criteria should be applied to assess the balance of the structure:

1. Indicator of information load at the highest level of hierarchy - the first head - head of the organization. This indicator is determined by the number of management relations at the head and is associated with the rule of manageability, which was developed by V. Greikunas, L.F. Urvik, L. Gulyk and F. Nichols (Graicunas 1937; Nickols 2000) defined this indicator as "the rate (volume) of attention." For construct graph G1 is the number of vertex edges a_0 - degree of homogeneity deg a_0. In research and practice, the optimal number of information links among managers of different levels is seven. The number of information channels more than seven leads to information overload of persons making managerial decisions, and, consequently, the quality of managerial influences falls.
   Therefore, quantitative assessment of the information load of a manager, making managerial decisions, can be calculated by the formula:

$$\lambda_0 = \lambda_{a0} = \mathrm{deg} a_0 / 7$$

2. As the next characteristic of balance of the organizational structure, it is necessary to calculate the maximum information load max degG1 managers of other links. In order to improve systemic management efficiency, efforts should be made to ensure that information load is distributed as evenly as possible among all participants of management process.

3. The next indicator of analysis of management structure is the degree of uneven information load in a management system. The index is determined by the ratio of maximum and minimum degrees of homogeneity of vertices of the graph (μ):

$$\mu = \text{maxdeg}G/\text{mindeg}G$$

3 Results

Study of the effectiveness of a control system includes a system of indicators characterizing integrity, manageability, and stability of the system:

1. The number of centers of the graph—n.
   Center of the graph is vertex a_i, relative to which the minimax condition is met, i.e.

$$n(G) = \max_{b \in V(G)} d(a_i, b) = \min_{a \in V(G)} \max_{b \in V(G)} d(a,b)$$

   Center of the graph is the most important vertex with the worst-case optimization property, i.e. minimizing the path length (distance) to furthest vertex of the graph (Salov 2008). In graph centers it is necessary to develop management decisions that will have maximum efficiency and intensity of management influences.
2. The diameter of the graph d (G) - the maximum distance between its two vertexes a and b

$$d(G) = \max_{a,b \in V(G)} d(a,b)$$

   where a and b are arbitrary vertexes of a graph,
   V (G) is a set of all vertexes,
   d (a, b)—distance between d (a, b) is the smallest chain length connecting vertices a and b.
   The length determines a size, compactness of control system, length of the shortest simple chain connecting the two vertexes with minimum distance.
3. Radius of the graph r (G) characterizes minimum of maximum distance from a fixed vertex to all vertices of graph V (G)

$$r(G) = \min_{a \in V(G)} \max_{b \in V(G)} d(a,b)$$

Radius of the graph determines minimum length of a control chain to the most remote elements of the control system.

Manageability of the management system can be assessed by following indicators:

1. The number of information channels on which management impacts are transmitted is determined by the number of edges of the graph (N) connecting different vertices of the management structure. More channels of managerial influence in a structure, more opportunities for managing the appropriate structure have.

2. Number of feedback loops in control system is a number of closed control loops (cycles) in an organizational structure. Than more closed circular cycles of information transmission in a control system, the higher its manageability. Number of closed loops in graph G1 corresponds to a cyclical rank of graph—ν (G1).

For connected planar graphs, which include enterprise management systems, cyclical rankν is determined using the Eulerian formula based on its vertex number n and number of edges N:

$$n = N - n + 2$$

Structure with the largest number of control loops is preferred by a number of control loops.

Stability of the management system is determined by following indicators:

1. External stability index β (G).
   A set of vertices of a graph (V) has external stability if each vertex not part of the set V is adjacent to some vertex of V. Number of vertices in the smallest externally stable set is a number of external stability of the graph—β (G).
   The power of the smallest externally stable set determines minimum number of persons making managerial decisions directly affecting all participants of the organizational structure, i.e. identifies the number of key decision makers who directly manage all events and processes. The larger β, i.e. the more such managers, as well as stable and reliable the structure becomes.
   In the opposite case, with stability β, completeness of management is provided by a small number of managers (at the limit of one with absolute, unlimited power) and then the loss of operability or failure of one or a small number of these managers leads to a loss of manageability.
2. Number of internal stability of the graph is α (G).
   A set of vertices of graph V is internally stable, or independent, if no two vertices of that set are adjacent, i.e. do not belong to the same edge. Number of vertices in the greatest internally stable set is the number of internal stability of the graph - α. It determines the maximum number of directly unrelated participants in the organizational structure. The greater α, the more the number of persons who have independence in making and implementing management decisions, more stable and reliable organizational structure has in case of loss operability or failure of individual elements of a structure.

The graph of the optimal management system of an organization must meet the following criteria:

$$\begin{cases} F_b(\mu, \lambda_i, d) & \rightarrow \quad \min \\ F_y(\beta, N, \nu, \alpha) & \rightarrow \quad \max \end{cases}$$

where F_b is the balance function of a control system;

F_y - function of controllability and stability of a control system;
β is number of external stability structure;
N - number of information channels through which management impacts are transmitted;
ν - number of feedback loops;
α is number of internal stability;
μ is a degree of unevenness of information load;
λ_i - information load of the manager of i level;
d is the diameter of the system;
P is the number of control centers.

Developed approach was tested in the study of six territorial generating companies in Russia. The studies are presented in Table 1.

Research shows that at present none of the territorial generating companies in Russia has an optimal organizational structure and its development is an important direction improvement of the process of management in Russian electric power industry.

It is necessary to develop as much as possible sub-systems from the standpoint of energy efficiency management of generating companies and on this basis to form

Table 1. Quality indicators of management structures of territorial generating companies of Russia

Indicators	Territorial generating companies					
	JSC "TGC-1"	JSC "TGK-6"	JSC "Volga TGK" (TGC-7)	JSC "TGK-9"	JSC "TGK-11"	JSC "Yenisei TKG" (TGK-13)
Number of level 1 management relationships	15	21	8	10	18	20
Information load at 1 level manager	2.14	3.00	1.14	1.43	2.57	2.86
Maximum number of management links at other levels	16	8	15	8	11	18
Maximum information load for managers at other levels	2.29	1.14	2.14	1.14	1.57	2.57
Number of managers with critical information load	2	2	3	2	2	7
The degree of unevenness of information load	8	10.5	7.5	5	9	10
Number of the graph centers	2	5	2	3	2	3
Diameter of the graph	4	4	4	4	6	6
Radius of the graph	3	3	3	3	4	4
Number of information channels through which management influences are transmitted (number of edges of graph)	75	80	45	41	56	116
Number of feedback loops in organizational structure	3	2	1	1	12	17
Number of external stability	14	20	7	8	10	15
Number of internal stability	61	60	39	34	36	86

common to all selected strategies directions of adjustment of proposed solutions for development of the country's energy sector, from standpoint of ensuring requirements of energy efficiency.

As a result of combination of management systems of different generating companies in each of territorial generating zones, a graph of a structure of generation management of the country as a whole can be obtained.

4 Conclusions

Conducted research allows to integrate tools of graph theory into processes of research and construction of effective management systems to achieve business goals of organizations. Development of systems management of enterprises with the use of proposed methodical approach and evaluation tools of graph theory allows to define strategic directions of adjustment of management decisions on formation of the organizational structure. Designing optimal management system in the digital economy is proposed to be carried out by developing a set of sub-institutional from position of efficiency of organizational structures and on the basis of received options to determine common direction for all selected strategies of decisions adjustment on development.

References

Barabási, A.L., Albert, R., Jeong, H.: Mean-field theory for scale-free random networks. Physica A **272**, 173–187 (1999)

Rényi, A.: On Random Graphs, p. 1959. Publ. Math, Debrecen (1959)

Arjaliès, D-L.a., Mundy, J.b.: The use of management control systems to manage CSR strategy: a levers of control perspective.Manag. Account. Res. **24**(4), 284–300 (2013)

Böckers, V., Heimeshoff, U.: The extent of European power markets. Energy Econ. **46**, 102–111 (2014)

Bratasanu, V.: Leadership decision-making processes in the context of data driven tools. Qual. Access Success **19**(3), 77–87 (2018)

Grochla, E., Szyperski, N., de Gruyter, W.: GmbH & Co KG, 3 December 2018

Glamazdin, E.S., Zinchenkov, V.I.: Management structure the management company. J. Manag. Large Syst. **5**, 152–157 (2011)

Graicunas, V.: Relationship in organization. In: Gulick, L., Urwick, L. (eds.) Papers on the Science of Administration, pp. 181–189. Columbia University's Institute of Public Administration, New York (1937)

Halk, L.: Independent by structure: a study of the structural nesting of organizational diversity. Organization **25**(2), 242–259 (2018)

Johnson, M.P., Midgley, G., Wright, J., Cicero, G.: Operational research community: innovation, internationalization and development programs. Eur. J. Oper. Res. **268** (2018)

König, D.: Theorie der endlichen und unendlichen Graphen. Akademische Verlagsgesellschaft, Leipzig (1936). First Textbook. Translated from German Richard McCoart, Theory Finite and Infinite Graphs, Birkhäuser (1990)

Kuznetsov, V.P., Garina, E.P., Andryashina, N.S., Romanovskaya, E.V.: Models of Modern Information Economy Conceptual Contradictions and Practical Examples, 361 p. Emerald Publishing Limited (2018)

Kuznetsov, V., Kornilov, D., Kolmykova, T., Garina, E., Garin, A.: A creative model of modern company management on the basis of semantic technologies. In: Communications in Computer and Information Science (2017)

Nickols, F.W.: The span of control and the formulas of V.A. Graicunas. Paper presented at the NSPI Journal (2000). http://www.nickols.us/graicunas.htm. Accessed 12 Jan 2019

Novikov, D.A.: Theory of control of organizational systems, p. 211. Moscow Psychology-Social Institute (2013)

Surie, G.Ab., Singh, H.B.: A theory of the emergence of organizational form. The dynamics of cross-border knowledge production by Indian firms. Emergence Complex. Organ. **15**, 37–75 (2013)

Salov, A.G.: Analysis of the effectiveness of structures and management of energy enterprises. Working paper, Izvestiya vuzov. North Caucasus region. Technical Sciences, No. 1, pp. 32–37 (2008)

Thulasiraman, K., Swamy, M.N.: Graphs: Theory and Algorithms, p. 480. Wiley, New York (2011)

Zastupov, A.V.: Innovation activities of enterprises of the industrial sector in the conditions of economy digitalization. In: Advances in Intelligent Systems and Computing, vol. 908, pp. 559–569 (2019)

Zhu, S., Jiao, H.: Organizational structure and corporate performance: insights from 6065 listed corporations. Chin. Manag. Stud. (SSCI) **7**(4), 535–556 (2013)

Expansion of Personal Development Space of a Student in the Sphere of Design

Elena E. Sherbakova[1(✉)], Maria V. Lvova[2], Evgenia K. Zimina[2], Irina S. Aboimova[2], and Alexandra A. Kulagina[2]

[1] National Research Nizhny Novgorod State University N.I. Lobachevsky, Nizhny Novgorod, Russia
e.e.1806@yandex.ru

[2] Minin Nizhny Novgorod State Pedagogical University, Nizhny Novgorod, Russia
mashasherb89@mail.ru, zimina-evg@bk.ru, i.aboimova@mail.ru, aleksa.culagina2011@yandex.ru

Abstract. In the article, the basics of design education were studied and actual essential features of design formation were revealed. Possibilities of educational influence of design education on students are considered. The aim of the research is to develop theoretical and program-methodical substantiation of students' mastering the basics of design approach to subject-transformative activity and obtaining a scientific -reasonable idea of its effectiveness. Theoretical methods were used: theoretical and methodological analysis of philosophical, psychological and pedagogical, methodical literature, normative documents on the problem of research, modeling, systematization, generalization, comparison; empirical - observational, diagnostic, experimental. The results are the analysis of research, which confirms the formed system of continuous art education, giving certain aspects of design education, in particular, aesthetic foundations, and proves the established scientific school. Performed analysis allows to summarize: taking into account the preparation of the general education system, it is necessary to create conditions for additional education for a program that creates a certain mindset for teenagers, which is characterized by realization of design as a creative process aimed at the transformation of the environment, realization of the sense of style and aesthetic attitude to the subject-transformation work. The scientific novelty of the research contains a substantial characteristic, development of the theoretical basis and definition of a complex of organizational and pedagogical conditions of the design approach to subject-transformative activities.

Keywords: Pedagogical conditions · Applicants · Personal development · Technology of project training · Design activity

JEL Code: I26

E. G. Popkova and B. S. Sergi (Eds.): ISC 2019, LNNS 91, pp. 142–148, 2020.
https://doi.org/10.1007/978-3-030-32015-7_17

1 Introduction

Design education is a special type and quality of personality education, giving significant aesthetic axioms in the world perception of a person, contributing to the development of project thinking, which is necessary in any sphere of social activity and is the starting point for further effective self-realization of personality in professional interaction with the subject world, as the basis of the design activity creating a design product not only corresponding to the present reality, but more meeting the requirements of the future.

The main reference point of the national doctrine of education in the Russian Federation– is the formation of the basis for stable socio-economic development of the government and ensuring a high quality of life. In turn, the strategic objective of the design development concept of our country is to create a harmonious subject-spatial environment, information, visual and communication, social, contributing to the development of the economy and aimed at improving the quality of life of citizens (Kuznetsov, V., Kornilov, D., 2017). Thus, the potential of design, as an activity, is the most important tool in the socio-economic sphere of creation of the subject world.

Objective of the research: to develop theoretical and program-methodical justification of learning the basics of design approach to subject-transformation activities and get science-based understanding of its effectiveness.

Theoretical Basis of Research. At the present stage of development of society, design activity becomes relevant, socially and economically demanded, and, as a result, it is necessary to expand the development of the student in the scope of this activity, for faster formation of professional competitiveness (Kuznetsov [13]).

The established principles of artistic and technological training do not meet the progressive claims made by the society to the personality capable of self-realization in the field of subject-transformative activities. More suitable conditions for initial experimentation and testing of own forces in the direction of design, for obtaining propaedeutic knowledge are formed in additional education. We study the features of additional design education.

The problem of ensuring multilevel and continuity of design training is relevant and promising at the present stage of design education.

Design activity is based on mutual exchange of information and technologies of related professional fields (engineering, architecture, art, etc.), it includes aesthetic and technological components that are inseparably linked and involve the development of spatial thinking, automation of motor actions, the skill of three-dimensional vision of the represented object, visual memory, emotional and sensual, imaginative perception of reality, etc.

In a number of dissertation researches, the development of certain elements of design activity is considered. Thus, the analysis of research and publications confirms the established system of continuous art education, which gives certain aspects of design education, in particular, aesthetic foundations, and proves the established scientific school [T.Ya. Shpikalova, L.V. Ershova, N.P. Makarov et al.].

In general education school, classes of arts are aimed at the general development of students, but the task of development and acquisition of artistic skills is not set. From experimental observations of N. D. Kalina, it is also determined that in the classes of arts little attention is paid to the study and comprehension of spatial images, which does not contribute to the comprehensiveness of the volumetric spatial images in the minds of students.

The level of the generated artistic skills in the period of pre-university training, which are the basis of development of design activity, was analyzed by Y.Y. Artemyeva. The author made a conclusion about the artistic skills that are acquired in secondary school, specialized classes and art school. Comprehensive school provides primary spatial analysis skills based on geometry and drawing, which does not contribute to the development of the eye; visual skills are acquired on the basis of sensual reflection of reality. In specialized classes, where more profound study of visual literacy takes place, individual simple skills are developed in building the form, shading, the ability to determine the size of the depicted object "by eye"; but understanding of the form design and volumetric-spatial solution is given little time.

Education in an art school contributes to the formation of elementary art skills, but the task of in-depth study and understanding of design activities is not set. In the opinion of V.Yu. Medvedev, the design activity, in relation to artistic and technical, is a design-creative one and is characterized by the inclusion and identification of artistic essence in the design ideology of design.

Currently, the concept of practice-oriented technological education gives way to the concept of advanced technological education, which is convincingly proved by the analysis of domestic and foreign experience research conducted by V.M. Zhuchkov. Technological education requires mastering technological culture. That is the ability to transform the world around us not only to adapt to it in accordance with its interests and goals but also to change it using traditional and modern technologies.

One of the personally-oriented technologies is the technology of project training, which allows preparing students for information and creative work of the modern world.

A special type of project activity is project-design, based on the scientific foundations of object modeling and combining scientific and artistic principles in the design of an object that meets not only the requirements of today but also allows one to foresee the aesthetic demands of the future, going beyond the boundaries of ordinary world objects. Certain problems of the organization of design education in the system of general education, its conceptual foundations are considered by E.V. Tkachenko, S.M. Kozhukhovskaya. Study of V.F. Sidorenko considers the introduction of a program of design training and appropriate methods of knowledge in specialized schools.

Yu.S. Porshnevaya, and N.V. Nedoumova note in their research that studying and observing students of real-world objects, with a subsequent process of processing and creating an artistic image, contributes to the formation of an aesthetic attitude to the objective world and to gain experience in evaluating their activities. The problems of determining the content and teaching methods in the field of design, the use of patterns of design in the subject-transformational activities of students were covered by E.V. Ilyasheva, E.N. Koveshnikova, V.D. Krakinovskaya, I.A. Spichak, L.M. Kholmyansky and others.

These studies help to understand that the acquired artistic and technological skills provide benefits for the development of design skills, but clearly structured pedagogical technologies, taking into account individual experience and subjectivity of the student, to master the basics of design activity.

Methodology of Research. The study of the methodological basis of additional education, described in the works of A.G. Asmolov, V.A. Gorsky (2011), etc., allows to note the proper function of additional education: continuity of education, ensuring the satisfaction of the need for the creative perception of the world and oneself in the given world and creating a strong motivation of knowledge [16–18]. In the study of I.E. Nikitina (2006) design is considered as an integral system, aggregating different aspects of human activity: cognitive, subject-transformative, axiological [8, 11, 14].

The above researchers confirm that the current system of additional education in the field of art and technology, differentiated by the standards of educational intentions, does not contribute to the formation of a teenager's need for creative self-expression and regulated self-determination [7]. Acquired artistic and technological skills provide advantages for the development of design activities, but clearly structured pedagogical technologies are required, which allow individual experience and subjectivity of the student to master the basics of design activity [9].

One of the personal-oriented technologies is the technology of project training, which allows preparing applicants for information and creative work of the modern world [4–6]. A special type of project activity is project design, based on scientific bases of object modeling and combining scientific and artistic principles in object design [10, 12, 15].

Research methods: theoretical and methodological analysis of philosophical, psychological and pedagogical, methodical literature, normative documents on the problem of research, modeling, systematization, generalization, comparison; empirical: observational, diagnostic, and experimental.

2 Analysis of the Results of the Study

Software and methodical support include a program of gradual development of the design approach to the subject and transformation activities, methodical support of the process, including methodical recommendations, system of creative tasks, evaluative and criteria-based tools.

Organizational and pedagogical conditions that ensure the effectiveness of the process of mastering the fundamentals of the design approach by students in the context of additional education of children are: implementing the capacity of institutions to create an atmosphere of tolerance, respect for students, and orientation of schoolchildren to individual and group co-creation, taking into account educational needs, age-related features.

Scientific Novelty of the Research

- refined substantial characteristics of development students the basics of design approach as an educational process, implementing the student self-strategy in domain-reform activities;

- developed theoretical foundations for students to master the basics of a design approach to subject-transformational activities, based on the axiological, personality-activity, social and pedagogical approaches, including principles, structural and functional model of the process;
- a set of organizational and pedagogical conditions for students to master the fundamentals of the design approach to subject-transforming activities, ensuring its integrity and effectiveness, was determined.

Theoretical Significance of the Study is:

- social and pedagogical messages, outstanding quality of teenagers mastering the basics of design layout for subject-transformative work in additional education of children are opened;
- the main approaches are analyzed: axiological, personal-activity, socio-pedagogical; the structural and functional model defining scientific requests to the organization of the process of mastering the design layout in additional education, complementing the concept of pedagogy in the section "Methodology of additional education";
- revealed the basics and organizational and pedagogical conditions that determine the requests to the organization of the process of mastering the basics of design approaches of subject-transformation work, which expand the ideas about methods of organization of additional education of students;
- describes the boundaries of teenagers mastering the basics of the design approach, expanding abstract ideas about the content of study and the process of self-realization in the subject-transformative design work.

3 Discussion

The test allows us to state that taking into account training in the education system, it is necessary to assess the possibility of building an educational program that makes young people have a specific way of thinking, which is an understanding of design, as a creative process aimed at reconstruction around the environment, a sense of manner, aesthetic matter to the subject-transformative work.

Practical Significance of the Results of the Research:

- theoretical model is presented in the educational process of V.P. Chkalov design workshop "Outline" and has proved its own effectiveness. The opportunity to work for the search of fresh pedagogical conclusions in educational practice of educational institutions of different types, in teacher training institutes and in the system of increasing the qualifications of educational staff;
- scientific and methodological support has been created to enable adolescents to design and implement a step-by-step process of mastering the basics of design to the subject - transformative work.

– methodological recommendations for teachers on the organization of subject-transformation activities of schoolchildren, presented and replicated at the interregional training seminars and master classes, within the framework of the experimental site of the "University of the Russian Academy of Education" for participants of volunteer events.

4 Conclusion

There is objectively impartial dialectical dependence of additional education on the basic, being a means of humanization of the education system as a whole, together with the basic, additional education provides personality development, expands, deepens and complements basic knowledge, identifies and develops probabilistic possibilities, and this happens in a comfortable situation for the applicant.

References

1. Aboimova, I.S., Sherbakova, E.E., et al.: Principles of building the subject-spatial environment in the educational organization. Int. Electron. J. Math. Educ. **11**(10) (2016). http://iejme.com/makale/1614
2. Zimina, E.K.: Design approach to the subject world in additional education of schoolchildren: monograph. Zimina, E.K., Lagunova, M.V. NGPU named after. K.Minin, N. Novgorod, 111 p. (2013)
3. Zimina, E.K.: Additional education of children - a platform of professional formation and self-determination of the future designer. Zimina, E.K. Actual problems of modern science and education: materials of All-Russian scientific conference with international people. participation. T.VII, pp. 137–139. RIC Bash GU, Ufa (2010)
4. Lvova, M.V., Pleshkov, A.V., Sherbakova, E.E.: Development of complex model of formation of creative approach in conditions of continuous education. Lvova, M.V., Sherbakova, E.E., Pleshkov, A.V. Problems of modern pedagogical education. Collection of scientific works, Issue 61-Part 2, pp. 261–263. RIO GPA, Yalta (2018)
5. Markova, S.M.: Professional values in conditions of continuous multi-level education. Bulletin of Minin University, no. 3 (2013). http://www.mininuniver.ru/scientific/scientific_activities/vestnik/archive/no3. Accessed 13 Nov 2018
6. Medvedeva, O.P.: Pedagogical conditions of creative self-development of the student's personality in the institution of additional education: on the material of the design studio activity: 13.00.01. Medvedeva, O.P. Rostov-on-Don, 200 p. (2003)
7. Mukhina, T.G., Sherbakova, E.E., Sherbakova, M.V.: Design education: features of development of creativity of future designers: Monograph. Sherbakova, E.E. et al. Minin University, N.Novgorod, 90 p. (2015)
8. Mukhina, T.G., Sherbakova, E.E., Sherbakova, M.V.: Pedagogical support of creativity of students in conditions of continuous education in the field of design: Textbook/Sherbakova, E.E., et al. 152 p. NGASU, N.Novgorod (2016)
9. Mukhina, T.G., Pleshkov, A.V., Sherbakova, E.E.: Frame-technology as a condition for the development of creativity of students. In: Modern Problems of Science and Education, no. 6 (2016). http://www.science-education.ru/ru/article/view?Id=25591

10. Norenkov, S.V., Sherbakova, E.E., Sherbakova, M.V.: State and trends of architectural environment development for children in preschool education. Volga Sci. J. no. 4, IF -0.145, June 2016
11. Sherbakova, E.E., Sherbakova, M.V.: Dynamics of formation of creativity in students in conditions of continuous education in the sphere of design. In: Sorokoumova, G.V. (ed.) Conditions of Development of Professional Personality, 98 p. Collective monograph. NF URAO, N. Novgorod (2016)
12. Sherbakova, E.E., Sherbakova, M.V.: Pedagogical support of formation of creativity of future designers, no. 4(41). World of science, culture, education, Gorno-Altaysk (2013)
13. Kuznetsov, V.P., Garina, E.P., Andryashina, N.S., Romanovskaya, E.V.: Models of Modern Information Economy Conceptual Contradictions and Practical Examples, 361 p. Emerald publishing limited (2018)
14. Kuznetsov, V., Kornilov, D., Kolmykova, T., Garina, E., Garin, A.: A creative model of modern company management on the basis of semantic technologies. In: Communications in Computer and Information Science (2017)
15. Young, S., Shaw, D.: Profiles of effective college and university teachers. J. High. Educ. **70**, 670–686 (2009)

The Impact of Digital Technologies on Various Activity Spheres and Social Development

Elena V. Yashkova[1(✉)], Nadezhda L. Sineva[1], Sergey V. Semenov[2], Olga I. Kuryleva[1], and Anastasia O. Egorova[1]

[1] Minin Nizhny Novgorod State Pedagogical University, Nizhny Novgorod, Russia
elenay2@yandex.ru, sineva-nl@rambler.ru, kurylev-nnov@mail.ru, nesti88@mail.ru

[2] Nizhny Novgorod State University of Engineering and Economics, Knyaginino, Russia
svsemenov@gmail.com

Abstract. The purpose of this research was to study the specifics of digitalization processes in various sectors of the national economy. Authors consider invasion tendencies of digital technologies into public life.

In particular, article deals with the integration of virtual economy in Russia and countries of the Eurasian Economic Union. It is noted that the implementation of digital technologies directly affects the growth of national welfare and material profit of traditional forms of management.

Authors used scientific, research and interdisciplinary approaches to achieve this goal. The main methods of research were analysis and synthesis, methods of statistical analysis of information.

Revealed difference between integration of digital technologies in the business and the actual state of human resource management. In addition, the real need to implement digital technologies in the future for future generations is verified. Emphasized the fact that companies using the digital economy in their activities are more successful and competitive, in particular in the media, banking, education spheres, industrial and extractive industries. Issues of transition of enterprises from traditional model to technological one through integration of Digital and HR-strategies and their incorporation into the mechanism of functioning of the organization are considered separately.

It is concluded that process of digitization of human resources services will allow to implement a competence-based approach in providing organizations with qualified personnel as a long-term investment.

The ubiquity of digitalization in various sectors of the economy and public life is noted as a potential for increasing labor productivity, capitalization, resource efficiency, competitiveness and quality of life.

Innovation of conducted research lies in the substantiation of the need to implement a competence-based approach to digitalization in various spheres of scientific knowledge, mainly in management of human resources.

Keywords: Digitalization · Digital economy · Information society

JEL Code: J24 · F63

E. G. Popkova and B. S. Sergi (Eds.): ISC 2019, LNNS 91, pp. 149–155, 2020.
https://doi.org/10.1007/978-3-030-32015-7_18

1 Introduction

In the age of scientific and technological progress there was an invasion of technology in all spheres of human existence. Where information systems are successfully used and various processes are automated, which makes them more efficient. In confirmation of this research carried out foreign and domestic researches in various scientific fields [9–15].

Digan M., McCarthy W. (2012), Thompson Klein J., (2014); theorized the digital cultural heritage of Cohen D.J., Rosenzweig R. (2005), Cameron F., Kenderdine S. (2007), Brown J., Brian J., Brown T. (2005), Simmons A. (2015), Keller J. (2015) Morra S. (2014); the problems of digital economy were addressed by Brignolfsson E. and Kahin B. (2000), Singh, N. (2003) and Tapscott, D. (2014); issues of digital transformation to create new business models, strategy and leadership were studied by Bryson J., Crosby B., Bloomberg L. (2014), Berman S.J. (2012), Kuznetsov, V. (2017); research on informatization of the sphere of management and public service employees was conducted by Stolterman E., Krunn Force A. (2004), Bannister F., Connolly R. (2014), Danlevy P., Margetts H., Bastow S., Tinkler J. (2016), Canberra Times (2016); simulation of digital transformation, software and systems was addressed by Gray J., Rumpe B. (2017) and Janssen M., Esteves E. (2013), transformation of the digital government was considered by Lindgren I., Van Weenstra A. F. Margetts H., Danlivi P. (2018), et al.

In Russian research practice in the field of digitalization, including network intelligence in electronic digital society was carried out by Stepanov V.K. (2001); digital economy Kozyrev A.N. (2011) and Koshelava A.V. (2017); digital transformation and communications in computer science Dobrolyubov E., Efremov A., Aleksandrov O. (2017), et al.

In Russia on July 6, 2017 digital economy development program was adopted, which will allow to integrate the Russian virtual economy with the same sphere of the Eurasian Economic Union. In accordance with this program, it is planned to create technical and financial conditions for the rapid development of the new industry. Russian President Putin V.V. compared this global large-scale program in importance with the country's general electrification, which is unprecedented in its impact on economic progress. Implementation of such a government project becomes possible due to the huge accumulated intellectual potential. Altogether, the purpose of this study is to define trajectories of development of public spheres of activity in the conditions of digitalization of the Russian economy.

2 Theoretical Basis of Research

The digital revolution that has taken over the world economy is impressive not only in scale and speed but also in geography. Digital innovations actively spread around the world and replaced each other waves coming from scientific epicenters of the USA, Europe, and the USSR [11] In his time the leader of the USSR L.I.Brezhnev said the following: "The economy should be economical." In modern conditions of digitalization, this slogan would sound like this: "The economy must be electronic." This is because the digital revolution, which we also call the Third Industrial Revolution, brought with it universal computing, development of software, personal computers and the connected world of computing through the emergence of global digital infrastructure and the Internet. But most computing technologies we are using today have developed through one classical paradigm of the calculation process, which was created in the 1940s. Researchers and entrepreneurs are now working on other computing capabilities that can enrich our capabilities and increase expectations regarding information storage, processing, and transmission (Kuznetsov, V.P., 2018.).

In 1991, Mark Weiser wrote: "The most fundamental technologies are those that disappear. They are interwoven into the matter of everyday life, becoming indistinguishable from it." As a result of the victorious democratic march of Moore's law, digital computers lost their value of discrete objects: computers today are more than just an important part of new cars, consumer electronics, and most household appliances. They are integrated into fabric and clothing, as well as into the infrastructure surrounding us — roads, traffic lights, bridges, and buildings. We live in a world built by computers [7].

Methodology of Research. Methodological basis of research:

- Theoretical and methodological analysis and synthesis of existing special domestic and foreign scientific and methodical literature, conceptual analysis of scientific articles and publications on the topic;
- Studying and summarizing both domestic and foreign developments, on introduction of digitization in all spheres of social development;
- Application of methods of generalization, comparison, and forecasting.

3 Analysis of Results

The digital economy is the result of the transformation of various effects of new technologies of general purpose in the sphere of information and communication, affecting all sectors of the economy and social activity (Garina, E.P., 2017). Attention to the digital economy is due to the fact that information technologies are becoming increasingly important in the economic development of all countries of the world. Accordingly, the digital economy is seen as an economic activity, where the key factor is data in digital form, processing of large volumes the use of results traditional forms of management, efficiency of various types of production, technologies, equipment, storage, sale, delivery of goods and services.

Formation of the digital economy of Russia is characterized by the increasing importance of information and digital means in public and financial life. Practice shows that there is a real need for such technologies designed in the future for the so-called "Facebook generation" - young people living literally in the network, which causes and business in the future become digital (Tables 1 and 2).

Today we can safely say that digitalization is already a widespread reality and examples of digital technologies are different. Companies using the "digital economy" in media, retail and banking sphere become the most advanced [9]. At the same time,

Table 1. Digital competences in Russia and world [6]

%	World average	Western Europe	Eastern Europe	Russia
Internet access, % of population	53%	90%	74%	74%
Use of social media, % of population	42%	54%	45%	47%
Mobile phones (connections), % of population	112%	119%	157%	176%
Use of mobile social media, % of population	39%	44%	37%	39%
Mobile Internet (connections), % of population	63%	98%	92%	98%

Table 2. Dynamics of the digital literacy index, 2015–2017 [6]

Indices	2015	2016	2017
Digital literacy index	4.79	5.42	5.99
Digital consumption subindex	5.17	5.49	5.35
Digital competence subindex	4.48	5.27	6.48
Digital security subindex	4.86	5.57	5.43

digitalization penetrates into such traditional industries as oil and gas production, in educational sphere is manifested in development and implementation of electronic educational innovative product in order to increase the potential of a future graduate [8]. Application of a digital strategy integrated with business and HR strategies in the modern world is necessary to ensure the future competitiveness of a company through its transition from traditional to technological. Such technologies in the management of human resources contribute to improvement of labor productivity, continuous development, and improvement of quality of work life of employees. As research proves, it is necessary to create infrastructure, institutional environment, the development of digital culture, human resources [9], the commission of the so-called "Digital Disruption" [1]. It is obvious that active digitalization, requirements will inevitably increase every year, removing part of the routine work from them and transferring it to IT systems.

Human resources management system successfully uses information systems and automates the processes of attracting, using, developing and releasing personnel, making them more effective [13]. Moreover, modernization of Russian education provides for search of innovative ways and approaches in the professional training of graduates in demand on a labor market [14]. As researchers predict in the coming 2019 there will be a strengthening of trends of digital HR-transformation and their introduction into the everyday life of a huge number of business companies: HR-automation, HR-analytics, HR-marketing, Smart-recruitment, e-learning [8].

The use of services and applications for automation by HR is gaining momentum: internal document management and recruitment: search and selection of candidates for vacancies; assessment and certification of campaign personnel; training of personnel;

evaluation of loyalty and involvement of personnel; monitoring of the company's HR brand; hr-analytics (Fig. 1).

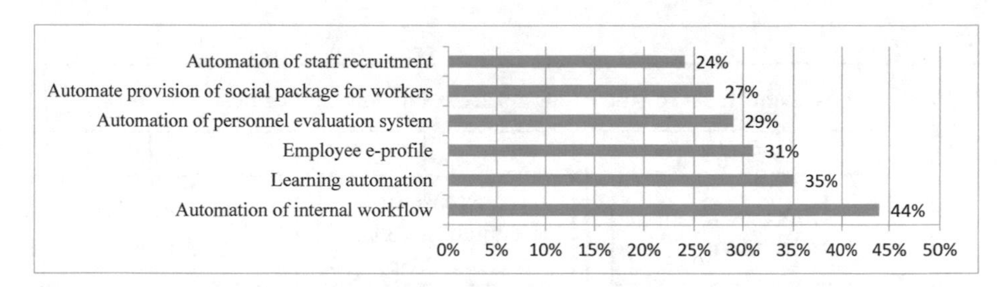

Fig. 1. Application by HR managers of services and applications for automation [8]

At the same time there is a requirement to significantly increase the speed of response of a specialist: it is given the opportunity to quickly calculate and predict the situation, provided the program with the necessary algorithm, but it requires from a living person a greater speed of response to a request of the counterparty [4]. Figure 2 describes the reasons for the introduction of HRM systems in Russian companies.

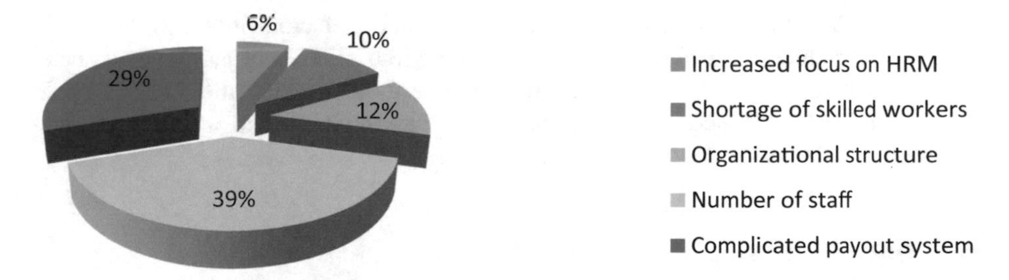

Fig. 2. Reasons for implementation of HRM systems in Russian companies

Today, an example of successful digitalization was the Russian Digital Pyramid 2018 Award for Achievements in the Field of Digital Transformation of Companies. The winners of the Prize were announced during the VI International Forum of Technologies of Management of Companies and People "HR TECH 2018". Grand Prix and honorary diplomas for achievements in the field of digitalization of business and the sphere of human capital management were presented in four categories: Digital-transformation of the year, HR Digital-solution of the year, Digital-project of the year, Digital-startup of the year. In nomination "Digital-transformation of the year — 2018" the winner at ceremony of awarding HR TECH AWARD 2018 was the company "Sberbank Technologies" (JSC "SberTech") for implementation of the project "System "MUSS — Sberdgale structure management module". DeltaCredit Bank and TELE 2 LLC mobile operator took in this nomination second and third places [6].

Consequently, introduction of new technologies (production, financial, managerial, social and any other) can lead to a huge number of positive effects and consequences for the economy (Fig. 3):

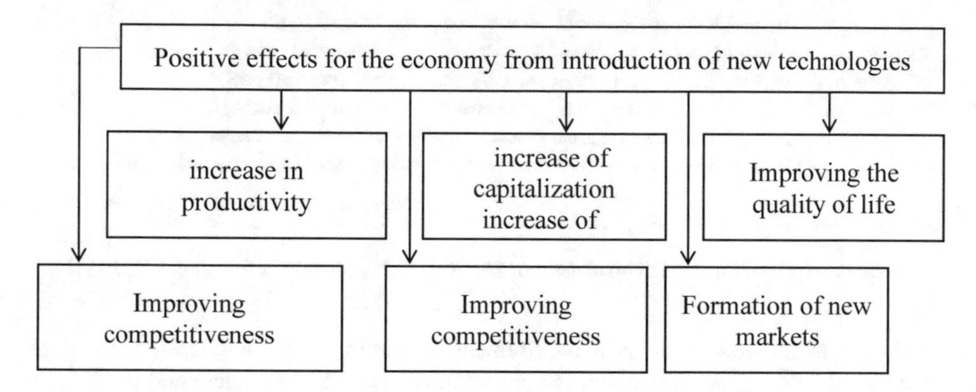

Fig. 3. Positive effects for the economy from introduction of new technologies

4 Conclusion

This research made it possible to identify a number of factors that will especially influence introduction of digital technologies in various spheres of activity and social development in the next decade. Digitalization is a modern trend in the social life in Russia, actively influences consumer behavior, manifests itself in the mobility and desire of companies to continuous improvement.

References

1. Belaichuk, A.: Digital disruption: without the internet, your business will not survive. Executive.ru. 23.09.2015. [Electronic resource]. http://www.e-xecutive.ru/management/marketing/1982777-tsifrovoi-perevorot-bez-interneta-vashemu-biznesu-ne-vyzhit. Accessed 03 Feb 2018
2. Boroday, V.A.: Problems of digitization of HR processes in the service. In: High Technologies and Modernization of Economy: Achievements and New Vectors of Development: Collection of Scientific Works on the Materials of the I International Scientific-Practical Conference, pp. 78–85
3. Keshelava, A.V., Budanov, V.G., Rumyantsev, V.Yu., et al.: Introduction to the "Digital" economy under the general editon of Keshelava, A.V., Zimnenko, I.A. VNII of Geosystems, 28 p. (2017) (On the threshold of the "digital future". Book one)
4. Zarina, I.: Digital transformation in HR [Electronic resource]. http://hr-portal.ru/blog/cifrovaya-transformaciya-v-hr. Accessed 08 Feb 19
5. Mekshun, N.: HR-branding in Digital format. http://key-solutions.ru/files/seminar/HR/Mekshun_N.A.pdf

6. Yashkova, E.V., Sineva, N.L.: Experience of development of career guidance course in e-educational environment Moodle in pedagogical university. Probl. Mod. Pedagogical Educ. **51–3**, 306–313 (2016)
7. Garina, E.P., Kuznetsova, S.N., Garin, A.P., Romanovskaya, E.V., Andryashina, N.S., Suchodoeva, L.F.: Increasing productivity of complex product of mechanic engineering using modern quality management methods. Acad. Strategic Manage. J. **16**(4), 8 (2017)
8. Tolstykh, T.O., Shkarupeta, E.V., Shishkin, I.A., Dudareva, O.V., Golub, N.N.: Evaluation of the digitalization potential of region's economy. Adv. Intell. Syst. Comput. **622**, 736–743 (2018)
9. Hurwitz, J., Nugent, A., Halper, F., Kaufman, M.: Big Data for Dummies, pp. 336. Wiley (2013)
10. Weller, T. (ed.): History in the Digital Age. Routledge, London, New York (2013)
11. Keller, J.: The art of telling digital stories. http://creativeeducator.tech4learning.com/v04/articles/The_Art_of_Digital_Storytelling. Accessed 03 Feb 2019
12. Kuznetsov, V.P., Garina, E.P., Andryashina, N.S., Romanovskaya, E.V.: Models of modern information economy conceptual contradictions and practical examples, 361 p. Emerald publishing limited (2018)
13. Kuznetsov, V., Kornilov, D., Kolmykova, T., Garina, E., Garin, A.: A creative model of modern company management on the basis of semantic technologies. In: Communications in Computer and Information Science (2017)
14. Thaller, M.: Controversies around the digital humanities: an agenda. Hist. Soc. Res. **37**(3), 7–23 (2012)
15. Thompson Klein, J.: Interdisciplining Digital Humanities: Boundary Work in an Emerging Field. University of Michigan Press, Ann Arbor (2014)
16. Shkunova, A.A., Yashkova, E.V., Sineva, N.L., Egorova, A.O., Kuznetsova, S.N.: General trends in the development of the organizational culture of Russian companies. J. Appl. Econ. Sci. **12**(8), 2472–2480 (2018)

Research of the Tools of Influence on the Behavior of Market Subjects

Alexander P. Garin[1(✉)], Irina A. Artashina[2], Svetlana A. Zaitseva[1], Zhanna V. Smirnova[1], and Yaroslav S. Potashnik[1]

[1] Minin Nizhny Novgorod State Pedagogical University, Nizhny Novgorod, Russia
rp_nn@mail.ru, szl0473@yandex.ru, z.v.smirnova@mininuniver.ru, yaroslav.sandy@mail.ru
[2] Nizhny Novgorod Institute of Management—branch of RANEPA, Nizhny Novgorod, Russia
artashina@mail.ru

Abstract. Relevance of the research topic is due to the importance of market mechanisms development that largely determines the economic development of the country. An experimental case study method was used as a methodological basis and heuristic basis of current research process. In the article, it is determined that the current efficiency of market functioning is determined by a sufficiency of resource supply, reaction to changes and uncertainty, degree of adaptability to requests of consumer. It is proved that a branch structure of the market is determined by the degree of concentration of producers on it, institutional development, and differentiation of products presented on it. The object of research is a market of the domestic automotive industry, where the main factor of economic growth – competition of producers of goods and services – burdened with opposing trends: dominance of individual companies, attempts to preserve monopolism even in advanced, high – tech industries, and lack of developed infrastructure. A solution to this problem lies in studying motives and features of the economic behavior of enterprises, factors of their competitiveness, and instruments of government influence. Analysis of the market according to the paradigm "Structure – Behavior – Result" allowed to determine that there are initial basic conditions for the behavior of market subjects: both on demand and supply side; that the structure of the market is determined by the number of sellers and buyers; market concentration, barriers to entry for new firms; product differentiation; integration and product differentiation. Based on what the further strategy of development of the industry should be built on.

Keywords: Market · Market subjects · Instruments of influence · Effectiveness · Development

1 Introduction

The automotive industry is the leading industry of domestic engineering, largely determining the economic development of the country. In addition, in developed countries, the industry creates about 1 million jobs in dealer companies and generates

E. G. Popkova and B. S. Sergi (Eds.): ISC 2019, LNNS 91, pp. 156–162, 2020.
https://doi.org/10.1007/978-3-030-32015-7_19

more than 10% of the national gross national product (Schätz 2016). In parallel with the production of vehicles, related industries are developing: construction of roads, extraction of raw materials, fuel production, a wide range of components, and banking sphere. However, the current performance of market is determined by the insufficiency of resource provision, uncertainty of consumer requests, and restructuring of institutions. All this predetermines the necessity of formation of market assessment according to the paradigm "Structure—Behavior—Result" followed by formation of instruments of influence on behavior of market subjects.

2 Methodology

Methodological basis and heuristic basis of the research process were a project research methodology, case study method of natural processes, and method of the pilot case study.

3 Results

In Russian Federation, the development of this sector is interconnected with many social, political and economic processes and is determined (Romanovskaya 2011):

1. Sectoral structure of the market, where the degree of concentration of producers on it is of particular importance. Concentration Index (CRm)—characterizes the ratio of amount of market shares of several largest firms in the industry to the total sales volume of goods by all firms operating in this industry market. Assessing the data for 2015–2018 on the example of "Lada", Kia, and Renault we determine that the market in terms of concentration is practically the same. CR32015 = 17,4 + 7,8 + 6,8 = 32%; CR32016 = 15,2 + 7,7 + 7,6 = 30,5%; CR32017 = 18,7 + 11,2 + 10,2 = 40,1%. CR deficiency is overcome with the help of the Lind index for studied companies: in 2015: $l_1 = 2{,}3836$; $l_2 = 1{,}8530$; $L_m = 2{,}1183$. In 2016: $l_1 = 1{,}9870$; $l_2 = 1{,}5066$; $L_m = 1{,}7468$. B 2017 г. $l_1 = 1{,}7477$; $l_2 = 1{,}4657$; $L_m = 1{,}6067$.

Resulting Lind Index values for years: in 2015–13; in 2016–8; in 2017 > 16. A defined boundary can characterize the market for a rigid (2–3 firms in the market) or blurred (6–8 firms) oligopoly, thus allowing to define the intended circle of subjects that can collusion together and take concerted action for limiting of competition. As can be seen from calculations, Russian automotive market for the analyzed period was not characterized by rigid oligopoly. In 2016, on car market was the possibility of blurred oligopoly, but in 2017 the probability of occurrence of such event became much less.

In determining the sectoral structure of the market, calculation of the Gerfindahl-Girschmann index is also important: $HHI_{2015} = 657(\%)^2$; $HHI_{2016} = 619(\%)^2$; $HHI_{2015} = 737(\%)^2$. According to estimates, the car market by brand is a weakly concentrated industry. In 2017, the level of concentration of firms in the market increased compared to previous years. This was due to changes in the share of firms in the market.

An additional measure of concentration is the entropy coefficient. It shows the degree of disordered and chaotic market (Mizikovsky 2011). The greater the degree of uncertainty, the higher the level of competition. Calculation of the absolute entropy coefficient (E) for the period under study allowed to determine the absolute entropy coefficient by years: E = 3.126; E = 3.142; E = 2.923. If we compare the absolute entropy coefficient, we can conclude that in 2017 the concentration level is higher than in previous years.

Next, we calculate the index of relative entropy. Accordingly, by year it is: $Eotn_{2015}$ = 3, 126/ln (47) = 0,8119; $Eotn_{2016}$ = 3, 142/ln (45) = 0,8254; $Eotn_{2017}$ = 2, 923/ln (44) = 0,7724. The relative entropy index in 2017 decreased compared to 2015 and 2016, indicating a decrease in uncertainty in the market. For the state of a homogeneous market, this coefficient tends to 1. Let us summarize these and other indicators in Table 1.

Table 1. Market concentration indicators by brand

Indicator	2015	2016	2017
CR concentration factor for three firms	32%	30,5%	40,1%
	Lm = 2,1183	Lm = 1,7468	Lm = 1,6067
Linda Index (L)	13	8	> 16
HHI	$657(\%)^2$	$619(\%)^2$	$737(\%)^2$
Absolute entropy coefficient (E)	3.126	3.142	2.923
Relative entropy index (Eotn)	0.8119	0.8254	0.7724
Rank concentration index (HT)	0.0576	0.0571	0.0712
Index of maximum share (I)	0.7821	0.7449	0.7833
Inverse share index (K)	0.15	0.18	0.14
Variation coefficient (V)	1.445	1.3348	1.6082
Relative Concentration Ratio (Kotn)	0.22	0.23	0.18

Next, we will analyze the industry market according to the paradigm "Structure—Behavior—Result". The basic conditions on the demand side are: 1. High elasticity of demand for the price; 2. Availability of substitute products (rail, air, water (sea and river); 3. Seasonal demand (more in summer, less in winter); 4. Extent of demand dynamics; 5. Consumer dislocation - Central Russia. In turn, on the side of the offer: 1. Technology meets market and environmental requirements; 2. Considerable investments are made in R & D, 3. Continuous growth in the scope of activities, 4. Effectively organized the system of products sales; 5. Wide links with suppliers of necessary resources.

Structure of the market is characterized by: 1. The number of sellers (44 companies), with a lot of buyers; 2. Market concentration - low; 3. Barriers to entry for new firms - significant; 4. Product differentiation—present; 5. Vertical integration—present; 6. Diversification—present.

Public policy includes: State regulation; 2. Antimonopoly policy; 3. Barriers to market entry (licensing system, a system of state control over prices, above the level of profitability); 4. Macroeconomic policy ("Strategy of development of the automotive industry of the Russian Federation for the period up to 2020", 5. "Plan of priority measures to ensure sustainable economic development and social stability in 2015").

Market performance: 1. Sufficiency of organizational, human and social resources; 2. Quite fast response to the changes taking place, flexible adaptation to the emerging conditions; 3. Tendencies to improve the quality of products.

As the analysis shows, the main non-strategic barriers to entry of firms into the industry market include:

- Positive impact on scale and minimally effective output;
- Vertical integration;
- Foreign competition;
- Licensing system of activity of firms, a system of state control over prices, above the level of profitability.

The main strategic barriers of the market in turn include:

- Diversification of the company's activities (diversification reflects the distribution of the company's output between different target markets). Diversification of activities allows the firm to reduce the risk of business associated with a particular market. A diversified firm is more sustainable due to its ability to compensate for the possible losses that the company incurs on the other market with profit. In the automotive market, diversification of the company's activities can mean production of cars of different classes and sizes by one manufacturer. Diversified domestic producer is AvtoVAZ OJSC. In the last few years began to actively develop car assembly enterprises on the territory of Russia. This option provides foreign producers with significant tax incentives. The high risk associated with the complexity of exit from a market is a disincentive for potential competitors to enter the industry.

Government is trying in all possible ways to support the domestic car industry, in a difficult international situation, facing sanctions by Western countries against Russia. The basic documents of strategic regulation in the industry are the "Strategy of development of the automotive industry of the Russian Federation for the period up to 2020", approved by the order of the Ministry of industry and trade of the Russian Federation of April 23, 2010 No. 319 taking into account changes in 2013 and "Plan of priority actions to ensure sustainable development of economy and social stability in 2015". In order to stabilize a situation in the automotive market, 30 billion rubles were allocated from the federal budget. Since 2013, government has also carried out targeted actions to stabilize car market: concessional lending and recycling program. Unfortunately, these types of support for significant results during the crisis period do not bring and there is a sharp decline in sales in the market. At present, we are studying the issue of the possibility of spending the funds of the parent capital for the purchase of vehicles, since in our time the car has become the basic need for life.

The highest competition in prices and quality of products, the need for significant investments, the existence of customs and other barriers, the tightening of environmental requirements, relatively low the rate of return, rising prices for raw materials—

all these and other factors make the car market difficult to reach for new participants. Despite this, young Asian automotive companies (primarily Chinese and Indian) are actively seeking to obtain traditional markets. According to some analysts, these companies formed relatively recently and not without the participation of the world's leading manufacturers of automotive equipment, in the future are quite able to reach more than high level. Chinese and Indian car manufacturers, which have received modern technologies not yet subject to globalization trends, are now in a relatively strong position. Chinese and Indian companies offer small cars with low fuel consumption. Appropriate effective solutions in modern realities can begin global expansion.

Market power of suppliers. Suppliers are able to exercise market power over market participants by the threat of higher prices or lower quality of goods and services supplied. Passenger car manufacturers are closely linked to such areas as metallurgical, petrochemical, electrical, textile and other products, and auto components. All of these industries should be regarded as suppliers to car manufacturers, and their market power varies. The degree of market power of steel suppliers, which is an important starting point in the automotive industry, is comparatively higher. In Russia, the level of prices for cars is a very important factor for consumers, manufacturers of automotive equipment are very limited in the ability to raise their own prices in response to price increases for metallurgical products consumed in the process of automobile production. In addition, the automotive industry, though important, is not the only consumer of metallurgical products, whose suppliers serve a number of other areas of activity. The degree of market power of suppliers of components for cars is negligible. First, the concentration of automotive components is much lower than that of automobiles. Second, manufacturers of automotive equipment are key consumers of automotive components. Therefore, the success of automotive components manufacturers is closely related to the success of car manufacturers, so they strive to respect interests of car manufacturers not only in pricing but also in support of such areas as R & D.

Market power of buyers. The market power of buyers is manifested in their ability to force prices down, demand higher quality products or more services on the market, and push competitors against each other. The production of passenger cars in Russia is primarily focused on domestic consumer: current share of exports in production is only 7%. The All-Russian Center for Public Opinion Research conducted a survey of consumers in Russian car market in March 2017. It is noteworthy that 39% of respondents stated that they were unwilling to purchase a car because of high cost and their inability to make such a purchase without serious implications for their financial security.

Threat from substitute products (substitutes). Definition of substitutes means finding products capable of performing the same function as those of participants in a given market. The alternative to road transport is rail, air, water (sea and river) modes of transport, each of which has both advantages and disadvantages. Currently, cars are the most common form of transport. It is younger than railway or water transport. Its advantages are maneuverability, flexibility, speed. However, despite the obvious advantages, road transport has many disadvantages. Passenger cars are one of the most wasteful forms of transport in terms of the cost of transferring one passenger. Most of the environmental damage to the planet is related to road transport. However, the level of motorization in the country is growing rapidly, which means that more and more people buy cars. In general, the answer to a question of how high the threat posed by substitutes to manufacturers of motor vehicles is uncertain. The choice between cars and other modes of transport is affected by: level of consumer expenses associated with buying and owning a car, importance of the factor of time, comfort and convenience when traveling, frequency of trips, and personal preferences of a consumer. Thus, the analysis of competitive conditions showed that combined impact of these five forces on the car market participants in Russia can be estimated as very intense.

4 Conclusion

The solution of industry development by means of adaptive influence on behavior of market subjects lies in the sphere of motives and features of the economic behavior of enterprises, factors of their competitiveness, and instruments of government influence. Analysis of the market according to the paradigm "Structure—Behavior—Result" allowed to determine that there are initial basic conditions for the behavior of market subjects: both on demand and supply side; that the structure of the market is determined by the number of sellers and buyers; market concentration, barriers to entry for new firms; product differentiation; integration and product differentiation. Based on what the further strategy of development of the industry should be built on.

References

Andrashina, N.S.: Analysis of best practices of development of domestic machine-building enterprises. Bull. Saratov State Socio-Econ. Univ. **1**(50), 24–27 (2014)

Garina, E.P., Garin, A.P., Efremova, A.D.: Research and generalization of design practices of product development in the theory of sustainable development of production. Human. Socio-Econ. Sci. **1**(86), 111–114 (2016)

Garina, E.P., Kuznetsova, S.N., Garin, A.P., Romanovskaya, E.V., Andryashina, N.S., Suchodoeva, L.F.: Increasing productivity of complex product of mechanic engineering using modern quality management methods. Acad. Strateg. Manage. J. **16**(4), 8 (2017)

Gruzdeva, M.L., Smirnova, Zh.V., Tukenova, N.I.: Application of internet services in technology training. Vestnik Minin Univ. vol. 6, no. 1(22), p. 8 (2018)

Druzhilovskaya, T.Yu.: Problems of practical application of fair value for estimation of accounting objects. Int. Account. vol. 21, no. 9(447), pp. 1086–1099 (2018)

Kuznetsov, V.P., Andrashina, N.S.: Innovative activity of industrial enterprises: problems and prospects. Izvestiya Penza State Pedagogical University named after V.G. Belinsky, no. 28, pp. 408–410 (2012)

Markova, S.M., Narkoziev, A.K.: Production training as a component of professional training of future workers. Vestnik Minin Univ. vol. 6, no. 1 (2018). https://doi.org/10.26795/2307-1281-2018-6-1-4. Accessed 12 Dec 2018

Mizikovsky, I.E.: Harmonization of indicators of internal control. Audit statements, no. 12, pp. 62–66 (2011)

Mizikovsky, I.E., Druzhilovskaya, T.Yu., Druzhilovskaya, E.S.: Accounting and reporting of non-credit financial organizations: Textbook. Nizhny Novgorod, 511 p. (2018)

Romanovskaya, E.V.: Formation of organizational and economic mechanism of restructuring of engineering enterprises: dissertation for the degree of candidate of economic sciences. Nizhny Novgorod State University. N.I. Lobachevsky. Nizhny Novgorod, 162 p. (2011)

Potashnik, Y.S., Garina, E.P., Romanovskaya, E.V., Garin, A.P., Tsymbalov, S.D.: Determining the value of own investment capital of industrial enterprises. Adv. Intell. Syst. Comput. **622**, 170–178 (2018)

Schätz, C.: A methodology for production development: doctoral thesis. Norwegian University of Science and Technology, 126 p. (2016)

Yashin, S.N., Trifonov, Y.V., Koshelev, E.V., Garina, E.P., Kuznetsov, V.P.: Evaluation of the effect from organizational innovations of a company with the use of differential cash flow. Adv. Intell. Syst. Comput. **622**, 208–216 (2018). https://doi.org/10.1007/978-3-319-75383-6_27

Organizational and Economic Foundations of the Management of the Investment Programs at the Stage of Their Implementation

Ekaterina P. Garina[1(✉)], Elena V. Romanovskaya[1], Natalia S. Andryashina[1], Victor P. Kuznetsov[1], and Elena V. Shpilevskaya[2]

[1] Minin Nizhny Novgorod State Pedagogical University, Nizhny Novgorod, Russia
e.p.garina@mail.ru, alenarom@list.ru, natali_andr@bk.ru, kuzneczov-vp@mail.ru

[2] Rostov Institute of the All-Russian State University of Justice (RPA of the Ministry of Justice of Russia), Rostov-on-Don, Russia

Abstract. The main principle of formation of the investment budget for implementation of projects is project financing, i.e. long-term targeted financing within which inefficient management is possible, which determines the relevance of research topic. Subject of research is the problem of management of investment projects in terms of planning of operational budgets, monitoring of expenditure of own and attracted financial resources, ensuring the safety of project implementation at the stage of attracting partners through outsourcing and outstaffing. Methodology of project studies is used as a methodological and heuristic basis of the research process. Formation of proposals for organization of work is based on the objectives of minimizing risks of causing damage and achieving KPI of projects. The goal is to regulate the process of management of investment projects of business units; formation of an investment program within "controlled area". Study proposes: timely forecasting and identification of risks during execution of a project, not defined during the planning phase; control over the achievement of process objectives and regulation of business processes; development of a documented process of project management; actions to minimize risks of infringement of rights of a patent holder, etc.

Keywords: Project management · Investment planning · Monitoring · Damage prevention

1 Introduction

Concept of a project is an interdisciplinary category, and in the context of its practical application it is usually implemented in the form of activities with characteristics "project", rather than projects in the strict sense of this words. Projects differ from operations that represent routine, ongoing activities and from programs that represent a group of projects managed in coordination to obtain benefits that are impossible in case

E. G. Popkova and B. S. Sergi (Eds.): ISC 2019, LNNS 91, pp. 163–169, 2020.
https://doi.org/10.1007/978-3-030-32015-7_20

of implementation of projects separately. Correct interpretations of categories of project, program or operation are important from a practical point of view and need a certain structural expression, used procedures, methods of tools, resource provision, which determines the relevance of the research topic.

In more detail on the categories, we define that the investment activity of the manufacturer is documented through programs and projects. An investment project is characterized by the attribute "development" and implies methodical structuring of reality in the future, focuses on the amplitude of actions, means and methods achieving goals based on alternative solutions (Yashin et al. 2018). Programs are less accurately identified and, in most cases, one or more projects need to be detailed in order to implement them. On the basis of which:

– An investment program represents the planned or budgeted costs of capital investments of an enterprise, usually in the form of a hierarchical structure. Investment program is used to manage projects that belong to a single "controlled area", valid in several balance sheets. Traditionally, within the hierarchy of an investment program, capital expenditures are planned from the bottom up, and in the budget - from the top down (Mizikovsky et al. 2016);
– Investment planning is the process of formation of programs and projects (identification, preparation, selection, planning taking into account limited resources), implementation of which is carried out within the a period of time in order to improve the process of generating assets and accumulating capital in order to obtain some desired future benefits for a producer.

Implementation of the investment project is to carry out activities to achieve KPI and monitor the process, which can be defined as project control for its compliance with a plan, according to achieving of final results (Potashnik et al. 2018). Experience in the implementation of projects proves that through "feedback" achievement of project results is stimulated, in the context of proportionality of cost and quality of the different elements; potential problems and simplifies the decision-making process, especially if it is corrected.

2 Methodology

Methodological basis and heuristic basis of the research process were a project research methodology, case study method of natural processes, and method of the pilot case study.

3 Results

The longer a project is implemented, the more important is systematic tracking of events to avoid too much deviation from the original plan and beyond a project. Specialized literature also traditionally highlights the importance of spending own and borrowed financial resources, ensuring security at the stage of attracting partners through outsourcing and outstaffing (Gruzdeva et al. 2018).

Let's dwell on the last question and use the factual material of the object of research —business units of LLC "Management Company" GAZ Group" on the issue of control of spending own and borrowed financial resources. As the open source study (Kuznetsov and Garina 2013) shows, the Company's investment program is formed on the basis of the Consolidated Investment Portfolio and is part of the long-term financial model, which is a tool for the implementation of the adopted Development Strategy of the Company and serves as the basis for the development of the annual investment budget. Sources of investment financing program are:

- Own funds (retained profit, depreciation deductions as part of proceeds from sales of products, works, services);
- Shareholder's funds (contributions to the capital of the company);
- Borrowed funds (bank loans)
- Government budget funds;
- Merchant finance (commodity loans of equipment suppliers);

The main principle of formation of an investment budget for the implementation of projects is project financing, i.e. long-term targeted financing for the implementation of a certain project (Hurjui 2018).

Monitoring and control of the progress of execution of investment and other projects in GAZ Group are carried out according to the projects included in the approved investment program. Monitoring is carried out until reaching the return on investment in the project for investment projects (with direct economic effect), or until the project is completed for other projects (without direct economic effect). In the case of closure of an investment or other projects, monitoring is stopped.

The most significant of projects within an investment program during the reporting year are the projects (Garina et al. 2017; Andrashina and Garin 2016):

"Creation and production of NEXT Gazelle with all-metal van"—the amount of financing is 44.9 million rubles.;

"Creation and production of the LDT Next family of medium-tonnage cars"—the amount of funding is 257,8 million rubles.;

"Creation and production of an increased capacity shuttle bus" - the amount of financing is 11.3 million rubles.;

"Organization of production for welding and painting of frames for Mitsubishi"—the amount of funding is 6.2 million rubles.

In the framework of individual investment projects, due to the shortage of production personnel, outstaffing is used with the involvement of specialized recruitment agencies (Markova and Narkoziev 2018). For example, as part of the implementation of the investment project "VW/Skoda—contract assembly" in JSC "AZ" GAZ", qualified workers from enterprises of NCE "GAZ Group" were sent to work on production facilities of JSC "AZ "GAZ". In the business unit for outstaffing actually incurred expenses for the reporting period amounted to 5.6 million rubles.

Specialists systematically monitor execution of the investment program of a business unit, as a result of which a number of violations were revealed. During the period under review, unjustified and undocumented expenses of GAZ OJSC in the total amount of RUB 3.7 million were revealed during an audit. For example, in a separate area, 38 heating metering stations were actually put into commercial accounting, of which 11 nodes were received under a guarantee letter with subsequent revision to the new heating season. The remaining accounting units are either modernized or write-off in the course of refinement. As a result of liquidation or conclusion of the project perimeter of the remaining accounting units, losses in the amount of 33.1 million rubles.

Attention should also be paid to the ineffective management of investment projects of the reporting period for the creation of joint ventures of Bozal-GAZ LLC, and Bulten-GAZ LLC by the Director for Development due to poor-quality development of certain aspects of strategic plans and the lack of proper analysis of possible risks and threats during implementation of projects. In more detail on projects, we define the following.

Project Bozal-GAZ LLC. In April of the reporting year, a decision was made to establish a joint venture Bozal-GAZ LLC with a global manufacturer of exhaust gas exhaust system components Bosal Nederland B.V. for the production of high quality product elements for AZ GAZ LLC and third-party consumers. Creation of a joint venture was due to the need to extract additional profits for the GAZ Group, as well as to minimize the cost of producing the product components by avoiding purchases with an intermediary markup of neutralizers (catalysts) from a third-party supplier - MobilGazService LLC. LLC Bozal-GAZ is a supplier of components for LLC Avtozavod GAZ, which include the following products: Receiving and exhaust pipes, silencers, neutralizers. The authorized capital of this project was adopted in the amount of 160,000 thousand rubles, of which accounted for: Bosal - 50.1% - 80 160 thousand rubles and JSC "GAZ" - 49.9% - 79 840 thousand rubles. The production was located on areas of a workshop of pipes for the production of fittings, wheels and steering departments of JSC "GAZ" (production area—1650 m^2, area of household premises—71.34 m^2).

Control over implementation of the joint venture arrangement was entrusted to a director of business unit development. However, due to the low competence of the manager the joint venture was under the threat of liquidation, and JSC "GAZ" under the threat of significant financial losses, due to the use of 11 patented components of the product without the appropriate permission of a patent holder. Such consequences could be avoided in the event of the following actions:

- Timely determination of the patent cleanliness of components used in the manufacture, even at the stage of investment design, in negotiations on the creation of the JV "Bozal-GAZ" LLC: since the rights to use patents owned by third parties are protected by the legislation of the Russian Federation, the issue of patent purity of products intended to be created, is one of the most important ones to be discussed at the stage of negotiations with a potential partner, i.e. with representatives of Bozal's parent company;
- Carrying out the necessary actions provided by the legislation to minimize risks of infringement of the rights of a patent holder when using his patents on the part of a project executor. In fact, prior to commencement of production of products, a director of business unit development was not convinced of a patent purity of elements.

In addition to risks associated with the lack of patent frequency verification, a section related to marketing was weak and undeveloped: while reducing the demand for the sale of an expensive component. At that time, the joint venture did not provide the planned revenue for the project and its payback. Unearned income for the reporting year amounted to 629.7 thousand rubles (Table 1).

Table 1. Results of monitoring of investment project implementation

Indicator	Plan according to approved BP	Fact	Deviation
Revenue	960,506	330,840	–629 666
Gross profit	115,776	69111	–46 665
Net profit	95 226	56827	–38399

Project of LLC "Bulten-Rus". Goal of the project is: 1. Organization of production of fasteners at JV "Bulten-Rus" by cold disembarkation method for JSC "GAZ" and local consumers Bulten high quality; 2. Attraction of new customers - AvtoVAZ, Ford, Nissan, Renault, VW, contact assembly (VW, GM, Daimler) for business expansion. Share of GAS Group - 37%, share Bulten - 63%.

According to an analysis of the project, there are a number of negative trends in the reporting period:

- Start of production has shifted by 11 months, which will lead to a shortfall in revenue of the joint venture in the next period in the amount of RUB 64.6 million, and then—in the amount of RUB 574.8 million. It should also be noted that there was a refinement of nomenclature, and an adjustment of volumes according to data led to a reduction in the volume of the study year from 399 million units to 253 million units.);
- Volume of investments increased by 123.7 million rubles due to the growth of an exchange rate, which necessitated the performance of additional work (additional fire prevention measures, preparatory work, repair of the roof);
- EBITDA during the project period decreased from 647 million rubles to 370 million compared to the approved one;
- Increasing volume of investments in the project has determined the deterioration of economic efficiency indicators (NPV, IRR, PP).

In addition, financial model of the project did not take into account the market value of the equipment transferred to the joint venture in the amount of 74 million rubles. And now there is a high risk of "non-return" investments. Therefore, in order to achieve previous performance indicators in the next period, it is necessary to consider the issue of excluding GAZ Group investments in the amount of 52 million rubles.

4 Conclusion

Based on a performed initial analysis, in the future, in our opinion, it is necessary:

1. To monitor investment projects in the following areas:
   - Formation and maintenance of a directory of investment projects;
   - Formation and maintenance of payment planning documents in the context of investment projects;
   - Formation and maintenance of documents on adjustment of payment planning in the context of investment projects (Druzhilovskaya 2013);
   - Formation of the fact of payments in the context of investment projects;
   - Analysis of plan and fact discrepancies in terms of payments of investment projects;
   - Formation of the fact of costs in the context of investment projects according to the primary documents of a legal entity.
2. To carry out continuous monitoring by DSB over the implementation of the regulatory framework when choosing counterparties and fulfilling contractual obligations on their part (Shpilevskaya 2016);
3. To carry out continuous monitoring of the conclusion and execution of contracts in close cooperation with lawyers, financiers, and responsible persons.
4. Revise payment plans for investment projects (for key projects to consider the possibility of transferring payments to the 2nd half of the year), do not take new obligations;

In order to regulate the process of management of investment projects of a business unit, it is necessary to introduce a documented management procedure to establish common management requirements (Kuznetsov and Romanovskaya 2011). As key indicators of the efficiency of the process of management of investment projects to adopt:

- % of projects successfully passed the procedure of initiation and included in the investment portfolio;
- % of agreed and accepted for implementation of business plans from the total number of initiated projects;
- Timely forecasting and identification of risks during the execution of the project, not defined at the initiation/planning phase, and elaboration of proposals to minimize the predicted or caused damage;
- Achievement of approved goals and indicators of economic efficiency of a project.
- The ratio of planned costs for the completion/closing phase of the project in accordance with the approved budget, to the actual costs, incurred;
- The ratio of the planned completion/closing date of a project in accordance with the approved calendar schedule of work, in the actual completion/closing date of a project.

To control the achievement of objectives of the process and to regulate business processes, establish the following procedures and rules:

- Operational project management control by a project manager at all stages of the project life cycle;
- Control of compliance of an investment project strategy of the business unit by a Development Director.

All information used in the development of a documentation package, including the timing of activities in accordance with an organizational plan of a project, must be confirmed by managers of the relevant functional units of a business unit.

References

Andrashina, N.S., Garin, A.P.: Evaluation of complex development of the product on the basis of modern methods of quality management (on the example of separate production). Econ. Humanit. no. 4(291), pp. 74–88 (2016)

Gruzdeva, M.L., Smirnova, Zh.V., Tukenova, N.I.: Application of internet services in technology training. Vestnik Minin Univ. vol. 6, no. 1(22), p. 8 (2018)

Druzhilovskaya, E.S.: Fair value: problems and prospects of use in accounting. Syst. Manage. no. 4(21), p. 14 (2013)

Kuznetsov, V.P., Romanovskaya, E.V.: Analysis of methods of restructuring of industrial enterprise in modern conditions. Bull. Cherepovets State Univ. vol. 1, no. 2(29), pp. 59–62 (2011)

Kuznetsov, V.P., Garina, E.P.: Studying solutions for product development in industry. Bulletin of Nizhny Novgorod University named after. N.I.Lobachevsky, no. 3-3, pp. 134–141 (2013)

Markova, S.M., Narkoziev, A.K.: Production training as a component of professional training of future workers. Vestnik of the Minin University, vol. 6, no. 1 (2018). https://doi.org/10.26795/2307-1281-2018-6-1-4. Accessed 12 Dec 2018

Shpilevskaya, E.V.: Economic security of the country: threats and ways of its provision. Int. Sci. Res. J. no. 5-1(47), pp. 188–193 (2016)

Garina, E.P., Kuznetsova, S.N., Romanovskaya, E.V., Garin, A.P., Kozlova, E.P., Suchodoev, D. V.: Forming of conditions for development of innovative activity of enterprises in high-tech industries of economy: a case of industrial parks. Int. J. Entrepreneurship **21**(3), 6 (2017)

Hurjui, I., Hurjui, M.: Investment projects: general presentation, definition, classification, characteristics the stages. The Annals of The "Ştefancel Mare" University Suceava. Fascicle of The Faculty of Economics and Public Administration, no. 8 (2018). http://seap.usv.ro/annals/ojs/index.php/annals/article/viewFile/33/32

Mizikovsky, I.E., Bazhenov, A.A., Garin, A.P., Kuznetsova, S.N., Artemeva, M.V.: Basic accounting and planning aspects of the calculation of intra-factory turnover of returnable waste. Int. J. Econ. Perspect. **10**(4), 340–345 (2016)

Potashnik, Y.S., Garina, E.P., Romanovskaya, E.V., Garin, A.P., Tsymbalov, S.D.: Determining the value of own investment capital of industrial enterprises. Adv. Intell. Syst. Comput. **622**, 170–178 (2018)

Yashin, S.N., Trifonov, Y.V., Koshelev, E.V., Garina, E.P., Kuznetsov, V.P.: Evaluation of the effect from organizational innovations of a company with the use of differential cash flow. Adv. Intell. Syst. Comput. **622**, 208–216 (2018). https://doi.org/10.1007/978-3-319-75383-6_27

Formation of the Determinants of Socio-Economic Development of Territories

Ekaterina P. Garina[1(✉)], Natalia S. Andryashina[1], Victor P. Kuznetsov[1], Elena V. Romanovskaya[1], and Mahomed Z. Muzayev[2]

[1] Minin Nizhny Novgorod State Pedagogical University, Nizhny Novgorod, Russia
e.p.garina@mail.ru, natali_andr@bk.ru, kuzneczov-vp@mail.ru, alenarom@list.ru
[2] Pension Fund of Russia in Achkhoy-Martanovsky District of Chechen Republic, Chechen Republic, Russia

Abstract. Relevance of the study is due to the need to determine the potential of social and economic development of systems, followed by the formation of an effective mechanism for the management of processes, both in economic and social spheres. The object of research is a separate region of the Russian Federation. Purpose of the study – is to assess the level of socio-economic development of the territory as a system, to identify the "points of inhibition" of development with the subsequent formation of a determinant of success. Estimated data allowed to determine as the initial prerequisites of decrease in purchasing power of population, decrease in investment and infrastructure potential, reduction of productive potential of the region, a large share of unemployment against sufficient resource potential of the territory. In more detail on the last statement, modern management thinking is more focused on the use of the potential of the territory, taking the form of mobilization of underutilized resources, increasing the capacity and value of existing resources, and attracting new resources to the region. Authors proceed from this position that the potential growth is possible due to the development of infrastructure that allows to ensure an economic performance of the region. And taking into account the postulate that a territorial organization of society and spatial organization of the economy is a function of the government. Growth vector of a territory may also include providing a favorable environment for the development of territories using the terms of "non-trade" goods (culture as a basis for tourism, etc.), using an emerging multiplier effect. Modernizing a regional social system by combining institutions in the field of social services for the population and entry into the market of social services by NPOs and commercial organizations.

Keywords: Efficiency · Management · Potential of development of the territory · Determinants of success

E. G. Popkova and B. S. Sergi (Eds.): ISC 2019, LNNS 91, pp. 170–178, 2020.
https://doi.org/10.1007/978-3-030-32015-7_21

1 Introduction

In works of researchers, the "economic development" of territories is often considered as an increase in the "equilibrium" income per capita and is determined by the determinants: (1) development is an intermittent process, is less a matter of implementation than initiation and management of the process of change [3, p. 548]; (2) spatial dimension when development, implies a certain result, occurs on a certain territory (for example, in local development programs in urban and rural areas), but is always present in regional policy; (3) the potential of the territory in the time horizon [9]; (4) quantifying the results of management in digital terms (incomes, jobs, etc.), as well as the quality of work, environment, educational opportunities, etc. [8]

In the context of assessing socio-economic development plans of the region, it is important that the specific objectives of the program are relevant to socio-economic problems it intends to address, and also meet the needs of residents. Where relevant are issues related to choosing the best strategy or improving its quality, objectives, effectiveness of planned measures, utility, and sustainability of achievements.

2 Methodology

Methodological basis and heuristic basis of the research process were a project research methodology, case study method of natural processes, and method of the pilot case study.

3 Results

In a historical context of the issue of ensuring the economic development of territories (until the end of the 1970s), the main emphasis was placed on management of consumer demand through regional subsidies and subventions (for example, payments to unemployed). Modern thinking is more focused on the use of potential of a territory, taking various forms, such as the mobilization of underutilized resources, increasing the capacity and value of existing resources, and attraction of new resources to the region [1, p. 250].

Going into detail on the subject of research, we determine its available potential. An overview of the region is presented in Table 1.

Table 1. Main indicators characterizing the Nizhny Novgorod region, as of 2017 [12]

Population of Nizhny Novgorod region	3,234,676 people
Distance to Moscow	439 km
Percentage of urban population	79,53%
Population density	42.23 persons/km^2
Area of territory	66 thousand km^2

(*continued*)

Table 1. (*continued*)

Population of Nizhny Novgorod region	3,234,676 people
The proportion of dilapidated and emergency housing	1,90%
Number of employees per pensioner	1.63
Average per capita income	30 543.17 rubles per month
GRP per capita	363 328 rubles
Gini coefficient (income concentration index)	0.40

Economic indicators of the region are shown in Table 2.

Table 2. Economic indicators of Nizhny Novgorod region for 2017 [7]

Indicator	Amount, billion rubles.
Gross regional product	1069.2
Shipment volume	1208.8
Retail turnover	652.8
Foreign trade turnover	4.755
Capital investments	219, 7

According to estimates of data from open sources of the Russian Federation and Nizhny Novgorod region, for 9 months of 2018 economic growth in the region reached 102–103%.

General dynamics of the calculated indicators of socio-economic development of Nizhny Novgorod region in the reporting period in a context of plant and fact is presented in Table 3.

Table 3. General dynamics of social and economic development of Nizhny Novgorod region in 2016–2017, thousand rubles. (period—quarter) [13]

Indicator	2017	2016	Change
Private consumption	13026	12143	883
Nominal fixed investment (gross fixed capital formation)	5047	4614	433
Real fixed investment (gross fixed capital formation)	4223	2726	1 497
Government consumption	4499	4514	–15
Real government consumption	2686	2658	28
Investments	6833	5549	1 284
Real investment	3663	3529	134
Nominal gross domestic product	27007	24846	216
Real gross domestic product	16668	15927	741
Real private consumption	7866	7538	328

According to calculation dynamics of indicators of socio-economic development of Nizhny Novgorod region in 2016–2017 is ambiguous. There is a relative growth, but no development, and taking into account the planned decisions against the Russian Federation in 2019 by the world community, with possible deterioration in the forecast period.

In order to determine "breaking points" of development, we will focus on the assessment in more detail, using calculations of private potentials values:

1. Production (an aggregate result of economic activity of the population in the region) is presented in Table 4.

Table 4. Production (total result of economic activity of the population in the Nizhny Novgorod region in 2016–2017, tens of thousands of rubles [8]

Indicator	Period	2017	2016	Change
Industrial production	Monthly	124.2	116.1	8.1
Change in inventory	Quarter	1786	935.1	851

2. Consumption potential (total purchasing power of the population of the region) is shown in Table 5;

Table 5. Consumer potential (total purchasing power of the population) in Nizhny Novgorod region in 2016–2017, tens of thousands of rubles [13]

Indicator	Period	2017	2016	Change
Consumer price index	Monthly	176.8	175.3	1.5
Producer Price Index (PPI)	Monthly	100.6	100.7	0.1
Consumer trust	Quarter	–17	–14	3
Real Retail Sales	Monthly	327.26	275.94	51.32
Retail Sales	Monthly	3306	2763	543
Personal income	Quarter	31986	32357	371

According to the estimated data, the total purchasing power of population in the Nizhny Novgorod region in 2016-2017 has decreased relatively, which coincides with the decrease in production activity of economic activity of subjects on territory.

3. Investment potential (includes a set of indicators, including market indicators) is shown in Table 6;

Table 6. Investment potential of Nizhny Novgorod region, % [12]

Indicator	Period	2017	2016	Change
Key Rate	Daily	7.75	7.75	–
Loan rate	Daily	7.75	7.75	–

(continued)

Table 6. (*continued*)

Indicator	Period	2017	2016	Change
Treasury bills (more than 31 days)	Annual	7.79	7.65	0.14
Average long-term government bond	Annual	8.88	8.77	0.11
Stock market index	Daily	1190	1215	25
Money market course	Annual	8.74	6.8	1.94
Gross external debt of territory, RUB	Quarterly	49655655706	62041272706	12385617000
Unpaid public debt of territory, RUB	Annual	200188877962	181869350008	18319527954

During the reporting period, there has been a significant decrease in the investment potential of the Nizhny Novgorod region. This is evidenced by indicators of the possibility of attraction of additional sources of financing on a territory.

4. The labor force (taking into account the number and level of education of the active population) is shown in Table 7;

Table 7. Labor potential of Nizhny Novgorod region in 2016–2017 [13]

Indicator	Period	2017	2016	Change
Labor force	Monthly	76.3	76.2	0.1
Employment	Monthly	72.6	72.55	0.05
Unemployment rate	Monthly	4.8	4.8	–
Unemployment	Monthly	3.7	3.65	0.05
Employment in tertiary industries	Annual	50239981	50550127	–310146
Secondary industrial employment	Annual	20330926	20532631	–201705
Employment in agriculture	Annual	5067792	5119251	–51459
Total employment	Annual	71746	68430	3316
Wages (average)	Monthly	30543	27839	2704

Significant decrease in a number of indicators of labor potential for the period under study is explained by a number of reasons: reduction of production potential of the region, a large share of "unregistered" unemployment, the proximity of industrial-developed territories such as Moscow, Kazan, Kaluga, etc. Also, a decrease in quality labor happened because potential of a territory was not used.

5. Financial capacity is reflected in Table 8;

Table 8. Dynamics of indicators of financial potential of Nizhny Novgorod region in 2016–2017 (period—quarter)

Indicator	2017	2016	Change
Current account balance	27663	18436	9227
Export of goods	110646	108721	1925
Import of goods	62676	63356	–680
Balance of goods	47969	45365	2604
Exports of goods and services	8399	7771	628
Pure export	2577	2344	233
Imports of goods and services	5822	5426	396
Real exports of goods and services	5159	4658	501
Real import of goods and services	2481	2537	56

6. Infrastructure potential of the territory (in terms of housing construction) is shown in Table 9. Data on other infrastructure projects cannot be systematized due to limited access to source information.

Table 9. Infrastructure potential of the territory (in terms of housing construction) in the Nizhny Novgorod region in 2016–2017

Indicator	Period	2017	2016	Change
Housing price index for existing houses	Quarter	101.81	100.7	1.11
Cost of new housing	Quarter	61831	60952	879
The cost of housing on the secondary market, rubles per m^2	Quarter	54923	53948	975
Price index for new housing	Quarter	101.44	101.64	–0,20
Housing stocks in the region, m^2	Annual	24.9	24.4	0.5

7. Resource potential of the territory development in terms of the possibility of monitoring the level, quality of life of the population (Table 10).

Table 10. Resource potential of the territory development (in terms of level, quality of life of the population) in Nizhny Novgorod region in 2016–2017.

Indicator	Period	2017	2016	Change
Number of births	Monthly	120133	126668	–6535
Number of deaths	Monthly	145440	138749	6691
Population	Monthly	146.8	146.8	–
Network migration	Annual	799998	n.d.	–
Mortality	Annual	12.9	n.d.	–
Birth rate	Annual	12.9	n.d.	–

Data on population change indirectly indicate the low level of living as well as the quality of life of the population. In more detail on the last point, authors proceed from the position that improving the quality of life of the population as possible at the expense of the development of social infrastructure. Moreover, the social sphere of territory development in modern conditions, which is separate types of infrastructures, should be a priority project—a favorable environment for the development, allowing through its formation to provide economic indicators of the region, including through the use of the emerging multiplier effect [6].

These and previous assessments require a number of solutions to improve them, both from authorities and from business community of the territory. We take into account the postulate that the development of the regional economic system cannot be carried out only on the basis of market self-regulation and care for a territorial organization of the society [2]. It is proposed to carry out government regulation of regional development by scientists through the implementation of theories:

- Comparative advantage, involving the use of the potential of growth of the territory taking various forms: comparative advantage in trade (goods and services), in terms of "non-trade position" goods (landscape or culture as a basis for tourism), in terms of agricultural production, etc. [10];
- Growth point, assuming the concentration of efforts of authorities on the "growth points" of territory, taking into account that in the short term initially, these decisions stimulate differences and contradictions [14].

As the study showed, the priority directions of development of Nizhny Novgorod region can be:

Promotion of investment potential of Nizhny Novgorod region; support of investors at all stages of investment project implementation;

Help with projection of growth vectors [3, p. 551];

Assistance in finding partners and markets; attraction of additional financial resources;

Infrastructure development (social, investment) [11];

Management of social and economic development of Nizhny Novgorod region can also include state support at the level of the region, including:

- Provision of tax benefits (for payers of income tax and property tax),
- Provision of free land;
- Partial reimbursement of loan interest,

Taking into account the volume of decisions, let's focus on one of them—"Modernization of the regional social system, including restructuring, unification of institutions in the sphere of social services and entering the social services market of NGOs and commercial organizations." Social services for citizens in the region provide social services. While the legislation allows social services to be provided not only to public organizations but also to the commercial sector. According to the authors, entering a market of social services of the population for commercial organizations will allow:

Expand the range of competitive services provided to citizens;

Save budgetary funds on the part of public authorities. Materiality of such payments is evidenced by the reporting data.

In Nizhny Novgorod region there is a register of providers of these services.

Introduction of this proposal will significantly reduce the waiting time of citizens in the queue, improve the quality of service and establish the process of communication between specialists and clients at a friendly and trusting mood.

Proposed stages of modernization of the social services system in this case include:

Stage I: Restructuring (organizational effect)

Stage II: Improving the quality of service to citizens when applying to social services for the registration of an individual program using the tool “Single Window” (organizational and social effects)

Stage III: Reorganization of social service institutions by combining them (organizational, social and economic effects (16854 thousand rubles))

Stage IV: Involvement of business structures and NGOs in the sphere of social services of the population (socio-economic effect)

Taking into account a client orientation and the need to save budget funds, a new structure of the institution is formed, in which it is proposed to apply the tool of “single window” in the department of social services of the population, which was not previously used. In order to modernize, a structure and staffing structure can be changed without increasing wage fund, as well as without increasing the staff size of the management. A positive aspect is a fact that the calendar period of time for the implementation of the new structure has no effect on a result.

Restructuring of the organization will achieve the following positive results: economic effect; social effect, organizational effect (possibilities of elimination of duplicative functions and avoidance of “double” control).

4 Conclusion

Therefore, the modern theory of regional management pays considerable attention to the search for determinants of relative economic growth indicators at the level of individual territories. In these approaches, “economic development” is often seen as an indicator of the “growth point”, while the social component acts as an “accompaniment”. The study revealed the dependence of economic development and social development of the region. Assessing the level of development of the region should not underestimate social indicators, for example, such as quality of life of the population, the comfort of living and customer orientation of the region. Also, the definition of such an economic concept as a region should be formulated taking into account the aggregate regional interests. It is proved that the modernization of the regional social system should be carried out in several stages by restructuring and consolidating institutions in the sphere of social services for the population on the principles of uniformity and accessibility of social services. Also, organization of reception on social issues can be solved by reducing reception spots and implementing reception to provide social support measures in multifunctional centers. A positive aspect is the stage of entering the market of social services of NGOs and commercial organizations

References

1. Ayplatova, I.I., Garin, A.P.: Evaluation of the efficiency of entrepreneurial activity at the level of the regions of the Russian Federation. Social and technical services: challenges and ways of development: collection of articles of the IV All-Russian scientific-practical conference. NGPU named after K.Minin, pp. 248–252 (2018)
2. Garin, A.P., Ayplatova, I.I., Yarov, R.B., Kartasheva, I.A.: Assessment of investment policy of Russia at the present stage. Economics and Entrepreneurship, no. 9-2 (86), pp. 546–554 (2017)
3. Kozlova, E.P., Romanovskaya, E.V.: Content of the mechanism of sustainable development of industrial enterprise. Bulletin of Nizhny Novgorod University named after N.I. Lobachevsky. Series: Social sciences, no. 2(50), pp. 25–30 (2018)
4. Kuznetsov, V.P., Sudaeva, Zh.A.: Influence of implementation of the road map on development of socio-economic system (region). Scientific Review, no. 4, pp. 162–168 (2016)
5. Parshina, A.A., Shpilevskaya, E.V., Garina, E.P.: Improving the competitiveness of domestic production through the implementation of the concepts of management of complex product development. Humanitarian and socio-economic sciences, no. 3(100), pp. 113–118 (2018)
6. Romanovskaya, E.V.: Development of industries and technologies: educational-methodical textbook. Minin University, Nizhny Novgorod, 84 p. (2018)
7. Redko, M.D.: On the assessment of socio-economic conditions for groups of territories in the formation of regional development strategy (on the example of urban districts of Chelyabinsk region). News of Higher Education institutions, no. 2, pp. 52–56 (2013)
8. Sarchenko, V.I.: The concept of rational use of urban territories taking into account their hidden potential. Housing construction, no. 11, pp. 9–13 (2015)
9. Website of the Legislative Assembly of the Nizhny Novgorod region. http://www.zsno.ru. Accessed 10 Apr 2019
10. CEIC. Russia. Population of the Nizhny Novgorod region. https://www.ceicdata.com/en/russia/population-by-region/population-vr-nizhny-novgorod-region. Accessed 02 Mar 2019
11. Gruzdeva, M.L., Smirnova, Zh.V., Tukenova, N.I.: Application of Internet services in technology training. Vestnik of the Minin University, vol. 6, no. 1(22), p. 8 (2018)
12. Markova, S.M., Narkoziev, A.K.: Production training as a component of professional training of future workers. Vestnik of the Minin University, vol, 6, no. 1 (2018). https://doi.org/10.26795/2307-1281-2018-6-1-4. Accessed 12 Dec 2018
13. Garina, E.P., Kuznetsova, S.N., Romanovskaya, E.V., Garin, A.P., Kozlova, E.P., Suchodoev, D.V.: Forming of conditions for development of innovative activity of enterprises in high-tech industries of economy: a case of industrial parks. Int. J. Entrepreneurship **21**(3), 6 (2017)
14. Yashin, S.N., Trifonov, Y.V., Koshelev, E.V., Garina, E.P., Kuznetsov, V.P.: Evaluation of the effect from organizational innovations of a company with the use of differential cash flow. Adv. Intell. Syst. Comput. **622**, 208–216 (2018). https://doi.org/10.1007/978-3-319-75383-6_27

Foresight Study of the Limits of Development and Application of the Modern Tools of GR Management in View of the Socio-Economic Specifics of Territories' Development

Iakow I. Kayl(✉), Anna V. Shokhnekh, Roman M. Lamzin, Anzhelika N. Syrbu, and Yuliana Yu. Elsukova

Volgograd State Socio-Pedagogical University, Volgograd, Russia
kailjakow@mail.ru, shokhnekh@yandex.ru, rom.lamzin@yandex.ru, surbyan@mail.ru

Abstract. The article presents the main conceptual and methodological directions of foresight study of the limits of development and application of the modern tools of GR management in view of the socio-economic specifics of territories' development. The influence of the conditions of globalization, growth of uncertainty, and informatization of society's life activities on effectiveness and flexibility of the policy of managing the interaction of all levels of authorities (municipal, regional, and federal) with business is reflected. Flexible policy of management in the system of interaction between public authorities and business synthesizes forecasting, planning, and controlling over the socio-economic specifics of development of territories, including analysis of the change of the situation of external environment and evaluation of dynamics of internal organization. Application of new modern tools of interaction of authorities and business as a form of GR management will allow determining the specifics of socio-economic development of territories. Foresight study of the limits of development and application of the modern tools of GR management in view of the socio-economic specifics of territories' development will allow determining the risk of crises and evaluating, identifying, and eliminating the threats, as well as recognizing and using the opportunities in external environment and in internal environment of the system of interaction of public authorities and business.

Keywords: Foresight · Foresight study · GR management · Tools of GR management · Development of territories

1 Introduction

Government Relations – GR management – appeared as a type of professional activities in the 2nd half of the 20th century for increasing the effectiveness of interaction with public authorities in the process of growth of the role of government in regulation of the economy. In 1960's – 1980's, first GR departments formed in large business structures of the USA and countries of the Western Europe. Active development of international transnational corporations initiative the appearance of Government

E. G. Popkova and B. S. Sergi (Eds.): ISC 2019, LNNS 91, pp. 179–190, 2020.
https://doi.org/10.1007/978-3-030-32015-7_22

Relations in the process of globalization of the world economy. For conducting business, transnational corporations had to build effective interrelations with national governments and take into account the national character of doing business in certain countries, understanding legal and ethnic peculiarities of entrepreneurship in various states. GR departments appeared first in transnational corporations and then in other organizations.

At present, the progressive tool of forecasting of the future based on the result of interaction of public authorities and business in the form of GR management is a foresight study. Foresight study as totality of methodological means and tools, which include the procedures of forecasting of socio-economic and innovative development (state, sector, sphere, region, municipalities, companies, and husbandries) allows building a strategy of vision of the future and influencing it by determining the opportunities and risks (Sidunova et al. 2017; Rogachev et al. 2017; Shokhnekh et al. 2016; Sidunova et al. 2017).

According to the experts of CORDIS, foresight study presents a plan of actions that is oriented at consideration, discussion, and outlining of the future. Analysis and forecasting of the future envisages in the system of foresight research the developments and application of technologies for designing long-term trends and coordinating the decisions. Experience of application of foresight studies has been recently manifested in Europe, which is peculiar for distinguishing the priorities of the modern research on the basis of the basic scenarios of development of science, technology, society, and economy.

The purpose of foresight study is to build potential future; to create a desired image of the future; to determine the strategy and paths of achievement. The results of foresight study are manifested in public decision making for socio-economic development of territories. Foresight study is a method of forecasting of long-term future of technologies, economy, and society – which will allow performing identification of the stages of strategic direction, based on which is possible to develop foresight technologies of formation and evaluation of alternative scenarios of development. B.D. Moguev (Moguev 2013) states that foresight development influenced the formation of technological policy of the UK and Germany and the EU on the whole; in Ireland, foresight development allowed distinguishing strategic top-priority directions in the scientific and educational policy, where the main directions are IT and biotechnologies. Almost all countries of the EU and East Asia use the concept and methodology of foresight development. Russia has also chosen the course of national development with application of foresight modeling of the future.

The above definitions were generalized in the notion that was formulated by the UN – foresight study is seen as a systemic formation of the variants of possible long-term future of science, methodology, technology, economy, and society for the purpose of identification of parameters and strategic research for creating innovative technologies that could being large socio-economic benefits" (Clayton 2005).

N.D. Emirov states that foresight is a completely new approach to determining the scenarios of the future. The essence of foresight study consists in achievement of the tasks for future decades during mobilization of labor resources and people from various groups (from ordinary consumers and entrepreneurs to managers of research centers and political persons) (Emirov 2012).

According to S. Pereslegin, the "methodology of foresight study" was offered not for forecasting of the future but as a means of coordinating the positions of parties that make decisions. As a matter of fact, this methodology is a process of exchange of opinions at the round table, regarding the perspectives of development in various spheres (Pereslegin 2009).

2 Materials and Method

The scientific material includes scientific publications on the issues of forecasting of the possibility of formation of partnership interactions of public authorities of all levels with business for socio-economic development of territories on the platform of foresight study. The authors use the methods of deduction, induction, comparative analysis, logical conclusions, and graphic transcription of the results of the author's concept, which allowed ensuring the necessary elaboration of the scientific problem.

3 Discussion

The modern conditions of economy show that the strategy of development of the tools of GR management in view of the socio-economic specifics of territories' development is to be often formed in the conditions of uncertainty and minimum of information. In the strategic management, it is recommended to use the following methods:

- method of strategic management for weak signals;
- method of management in the conditions of strategic surprises;
- method of management via ranking of strategic tasks (Ansoff 1989).

Such system of planning is built on the minimum volume of information that is constantly changed and supplemented. That's why it is necessary to have a method that would allow forming he variants of development and application of the tools of GR management for achieving the desired future in the conditions of insufficiency of information in view of usage of the levels of awareness on the state of development of business (Miropova 2013; Sokolov 2007).

According to A. Sokolov, topicality of application of the technology of foresight development will be growing due to a range of important reasons:

(1) scientific studies become more expensive and multidisciplinary, and even rich countries have to cut their research budgets;
(2) network interactions that appear in the process of foresight development between multiple experts are a driver of development in the new economy;
(3) active stimulation of foresight development by the EU and other international organizations (Miropova 2013; Sokolov 2007).

Functional features of the methods of foresight study of the limits of development and application of the modern tools of GR management in view of the socio-economic specifics of territories' development are manifested in leveling the risks of strategic drift with the method of expert evaluation.

4 Results

Foresight study of the limits of development and application of the modern tools of GR management envisages literary forecasting, expert panels, and creation of scenarios. The tools of GR management on the basis of foresight study include workshops, extrapolation of trends, brainstorm, determining critical technologies of interview, surveys, Delphi method, analysis of mega-trends, and SWOT analysis. The most modern tools of GR management on the basis of foresight-research envisages creation of technological road maps, modeling, simulation, essays, and scanning of the environment. The following tools are rarely used: development of citizens panels, creation of stakeholders maps, structural analysis, analysis of mutual influence, usage of relevance tree, and morphological analysis.

It should be noted that any changes that take place in the external and internal environments of the system of interaction of business and public authorities could be classified as a challenge. These challenges, which eventually are certain stimuli for development, acquire the form of risks and future opportunities for the subjects of the system of interaction between public authorities and business (Fig. 1). If management of business sees the risks and opportunities, timely preventive measures could increase the competitiveness of a local or territorial entity and of the whole country. However, the opportunities that are not determined or implemented could transform into real risks.

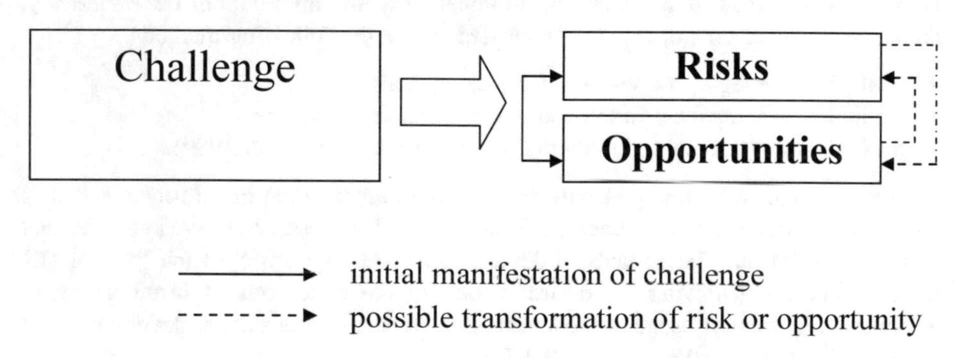

Fig. 1. Variants of development of GR management in the system of foresight study. Source: developed by the authors.

During determination and identification of risks that threaten the process of development and application of the modern tools of GR management, which create the conditions for opposing the movement along the vector of strategy, it is necessary to develop the measures for their minimization and full elimination in case of a negative situation. Evaluation of the influence of possible and current risks to GR management envisages modification of undesirable processes into drivers of effectiveness. Foresight study helps to prepare to probable surprises in the process of development and application of the modern tools of GR management, from the point of view of the past and present – which will allow developing the strategy in view of events in the future.

Considering foresight study of the limits of development and application of the modern tools of GR management as the key basis of the entrepreneurial resource, it is necessary to focus on large complexity of socio-economic specifics of development of territories where the drift (termination of development, “retreat”) is manifested, which is caused by uncertainty, changeability of the political environment and international relations. Complexity of overcoming the situation of stagnation is manifested in absence of “loyalty of approaches” to managing the mechanisms of interaction of business and authorities. In any case, the tools of GR management, which are used during interaction of business and public authorities, are aimed at overcoming the strategic drift of socio-economic development of territories. It is necessary to take into account that results of GR management are used by the state, which is either developing or stopping in development, or stagnating, being in the conditions of strategic drift. Studies on strategic drift of economic systems were introduces by Charles Handy, who studied the development of business and taught in the London Business School. Charles Handy worked as an executive manager in an oil company and then economist in Shell. According to him, strategic drift is small untraceable changes of the strategy in the course of its implementation, which are not visible for managers at first but which then accumulate and become visible at the stage when it’s too late to change anything (Sidunova 2017).

It should be note that simplicity and transparency are a principle of the functional practical management. Forecasting models of strategic management allow formulating new ideas for solving complex problems. However, the manager’s actions could lead the business system to a “strategic drift”, which will not allow for development but will lead to stagnation, bankruptcy, and downfall.

Therefore, foresight studies are aimed at determining the risk of strategic drift of GR management as negative phenomena, situations, and processes in the system of interaction of business and public authorities, which form in the period of strategic management of changes, where all efforts and means of managers, employees, and strategic partners are aimed at overcoming the external and internal barriers of development, corruption component, business corruption, etc.

It should be acknowledges that the conditions in which unexpected difficulties for business (risks of strategic drift) appear become widespread in the statistical discrete range, and, therefore, they could form any time during organization and promotion of development of strategic changes in management of economic systems. Strategic drift could be defines as a current complicated position in which business systems deviated from the purpose and achieve the results that are expected from GR management.

Also, the situation in which GR management loses focus from the factual plan and gives priority to “another course of actions” created the conditions for traps of “strategic drift”. “Another course of actions”, which goes beyond the limits of ethics and culture of communication in the system of interaction of public authorities and business is seen as corruption aspects, extortions, blackmail, building barriers for business by government workers, hiding favorable government tendencies for business, and other unethical actions. The studies show that the main cause of a strategic drift of GR management is inability of business to adapt to the changing conditions in the sphere, sector, and macro-economic processes which formation is influenced by the normative and legal acts of public authorities of all levels (local, regional, and federal) (Dibie et al. 2017; Gabsa 2017; Gadzekpo 2017; Kayl et al. 2016).

Therefore, foresight studies of the limits of development and application of the modern tools of GR management are aimed at determining the indicators of effective and cultural interaction of business and public authorities at all levels of management, including:

(1) indicators of evaluation of a flexible environment of business as to participation of lower and middle management in the process of formation of cultural (ethical) tools of GR management;
(2) indicators of evaluation of communication settings, where public authorities and business are ready to negotiate in the process of formation of cultural and ethical tools of GR management and to stimulate feedback;
(3) indicators of evaluation of skills and intuition, which allow determining the character of new tasks and their priorities as to their importance in the system of interrelations of public authorities and business;
(4) indicators of evaluation of planning and forecasting of complex sets of the strategies of development and application of the tools of GR management;
(5) indicators of controlling of development and implementation of the tools of GR management at each stage of implementation of the system of interrelations of public authorities and business;
(6) indicators of evaluation of the mechanism of formation of reactive corrections of the limits of development and application of the tools of GR management;
(7) systems of feedback as a result of controlling of the level of culture of development and application of the tools of GR management:
(8) indicators of growth of the level of socio-economic development of territories based on application of the tools of GR management.

In the process of interaction of public authorities and business the decisions are made which are logical continuation of each other, where history of existence of business influences its development. Accordingly, the managing driving forces make based not on "ideal conclusion logic" but on turbulence of the conditions.

In the turbulent conditions, which influence the system of interaction of business and public authorities, it is possible to enter the "Strategic drift". Definition "turbulence", as a transcription of the Latin word "turbulentus", is treated as "without an order" (Makarov 2012) and as complex "physical phenomenon, which is peculiar for irregular interconnected movements of the parts of the volumes of environment (liquid or gas) and their mixture, which is accompanied by chaotic change of gas and dynamic variables in time and space" (Sidunova 2017). In the modern science, the definition "turbulence" is used not only in mechanics of liquid and gas but also as to various systems that are peculiar for a change from laminar flow to chaotic flow (appearance of non-linear processes).

Turbulent state is manifested in business, government structures, and the system of interaction of public authorities and business, which could be influenced by the risks of reduction of activity of the national economy, which influence the dynamics of growth of socio-economic development of territories.

A strategic direction is provision of economic security in the system of interaction of public authorities and business, where the mechanism of elimination of risks of a strategic drift in the culture of development and application of the tools of GR management is built.

In the state standard, the notion of security is defined as "absence of inadmissible" risk and is noted that "security is achieved with an alternative way of reducing the level of risk to the allowable level – therefore, allowable risk is considered" (Aspects of security. Rules of inclusion in the standards 2002).

It is possible to consider such process as leveling the risks of strategic drift in the culture of development and application of the tools of GR management (bringing the level of risk down to the minimum) for provision of security by organized procedures (processes) for minimization of the volume of risk to a certain limit.

Implementation of the facts of economic life in the government sector and business (financial and non-financial sectors) is accompanied by the risks that are to be determined, eliminated, or prevented. The risks could be classified of determining the directions of interaction of pubic authorities and business for the following types:

- entrepreneurial;
- strategic risk (risk of strategic drift);
- resource depletion;
- financial depletion;
- market crisis of demand;
- credit provision;
- investment insolvency;
- risk of management of economic systems (Sidunova 2017).

The tool of GR management, which is aimed at leveling the entrepreneurial risk, envisages discussion of the directions of the government policy of business and authorities at various levels for determining the priorities of socio-economic development of territories in view of their specifics for understanding: (1) possible support in the form of state support for strategically important territories for the business activities; (2) availability of state programs with distinguished directions of public-private partnership; (3) possibility of private initiative of business for the specifics of socio-economic development of territories at the local, regional, and federal levels.

Entrepreneurial risk grows in the conditions of isolation of business in its own strategies without considering the needs of public authorities for development of a certain territory. Public authorities at the local, regional, and federal levels use the programs that are implemented through public-private partnership, government support, and investment support for business.

The tool of GR management, aimed at leveling the risks of strategic drift (strategic risk) envisages minimization of the possibilities by a situation or process that emerge in the period of strategic management of changes, where all efforts and means of business and strategic partners are aimed at overcoming the government barriers for development. Also, the external environment transforms under the initiative manifestations of new demand of consumers of products, goods, works, and services in the conditions of update of the innovative technological capabilities. So it is important to adapt business with the tools of GR management in the system of interaction. Organization of anti-crisis negotiations of business and public authorities should be accelerated in order to prevent business's losing competitiveness, profitability, and sustainability. However, the problem is manifested when the managing structure of business neglect the adaptation cultural tools of GR management in the situation of a strategic drift and start preferring the corruption interactions with public authorities.

The tool of GR management, aimed at leveling the risks of financial depletion is applicable at all stages of business activity and is an inseparable condition of economy. Risk of financial depletion is a threat of potential financial losses, which provoke strategic drift in the structure of public authorities, business, and, as a result, socio-economic development of territories and population's well-being – and, thus, leads to a "strategic drift" in management. In such conditions, negotiations should lead to a ban on government inspections of business (by controlling bodies), tax vacations for business, and delay for tax and social payments. The government should issue grants for development of business projects, which support strategic projects for a corresponding region of Russia.

The tool of GR management, aimed at leveling of the risks of market crisis of demand, envisages minimization of dependence of business's profitability on the change of market prices. In the conditions of the digital economy, the state conducts active work on provision of information on prices and dynamics of consumption in various segments of the market. Federal State Statistics Service of Russia publishes annual reports and analytical information that is presented by all economic subjects of the market. Success of a specific business is thus predetermined by its flexibility and maneuverability in the market, individual business skills, and experience of forecasting, as well as timely request for support from public authorities at all levels.

The tool of GR management, aimed at leveling the risks of credit provision, envisages minimization of possible non-execution of the obligations of returning the debts by buyers and customers. In order to prevent this risk, it is possible to start an initiative on financial support for business by crediting organization of the non-financial sector. The tool of GR management is the possibility of interaction between public authorities and business regarding government credits with low interest rate.

The tool of GR management, aimed at leveling the risks of investment insolvency, envisages minimization of threats of depreciation of capital that is invested into production and securities and usage of the investment for incorrect purposes. In the conditions of leveling the risk of investment failure, business may request public authorities of various levels to ensure support for the investment projects that are aimed at socio-economic development of territories. Government's support in elimination of barriers will allow increasing the low investment activity in a lot of spheres of society's activities.

The tool of GR management that is aimed at leveling the risks in management of effectiveness of the system of interaction between public authorities and business envisages timely reaction of the managing structures in specific situations. In the process of leveling of this risk, public authorities and business could negotiate state quotas for training of personnel for application of digital information & technical equipment. It should be taken into account that business envisages entrepreneurial activities which are a specific type of social behavior in the system of market uncertainty and risks. These factors predetermine the necessity for leveling the risks of management through development and application of the tools of GR management, where, on the one hand, it is possible to obtain support from public authorities for business in the process of usage of economic resources (land, labor, and capital), and, on the other hand, ensure state's expectation of socio-economic development of territories in the conditions of business's forming jobs, paying wages and taxes, and manufacturing national product within import substitution. Also, business environment is formed under the influence of external factors, which include the following:

- political threats;
- socio-economic threats;
- ecological influences;
- scientific and technical inventions;
- international conflicts.

These factors lead to emergence of external risk, which are difficult to forecast and control; therefore, they stimulate the development of the risks of strategic drift, where all efforts and assets and government structures and business and results of the system of their interaction are aimed at overcoming the barriers.

Risks of internal environment are predetermined by the results of doing business, where determining the source of risks with the tool of GR management could reduce their influence, thus developing favorable interrelations of public authorities and business and developing the country's territories in view of their socio-economic specifics. Here it is possible to speak of emergence of synergetic effect, where the route of information has a large horizon of foresight study of the future result from application of the tools of GR management for socio-economic development of territories (Fig. 2).

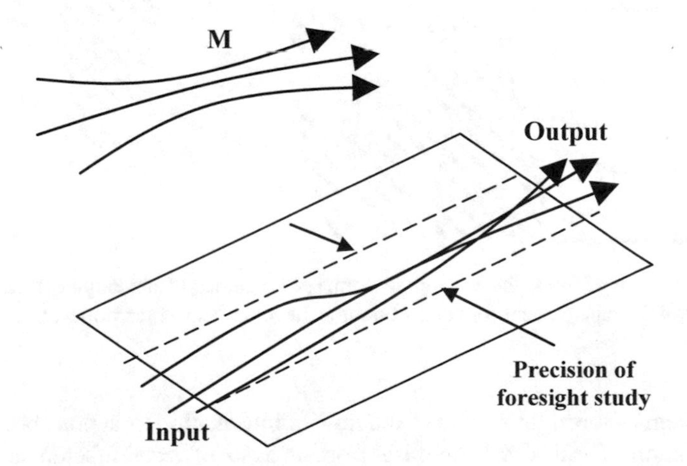

Fig. 2. Scheme of development and application of tools of GR management for socio-economic development of territories.

Foresight study of the limits of development and application of the tools of GR management for socio-economic development of territories envisages studying the concept of variation of the routes and behavior of flows, which determine that a large number of real objects in the phase space have the regions of the flows of information or actions that could change the system from one stage of the phase space into another. The function of the information flows or actions consists in intermittent transition of the system from one route to another.

Figure 3 shows two routes (M_1 and M_2) and 3 three flows of information or action (F_1 F_2, F_3). Black arrows show determined of dynamics (trajectories of the model for projection), gray arrows show movement of flows: if the trajectory of the flow reaches certain sphere (dashes), it could – with a certain degree of certainty – move to a random point of the route or to another flow.

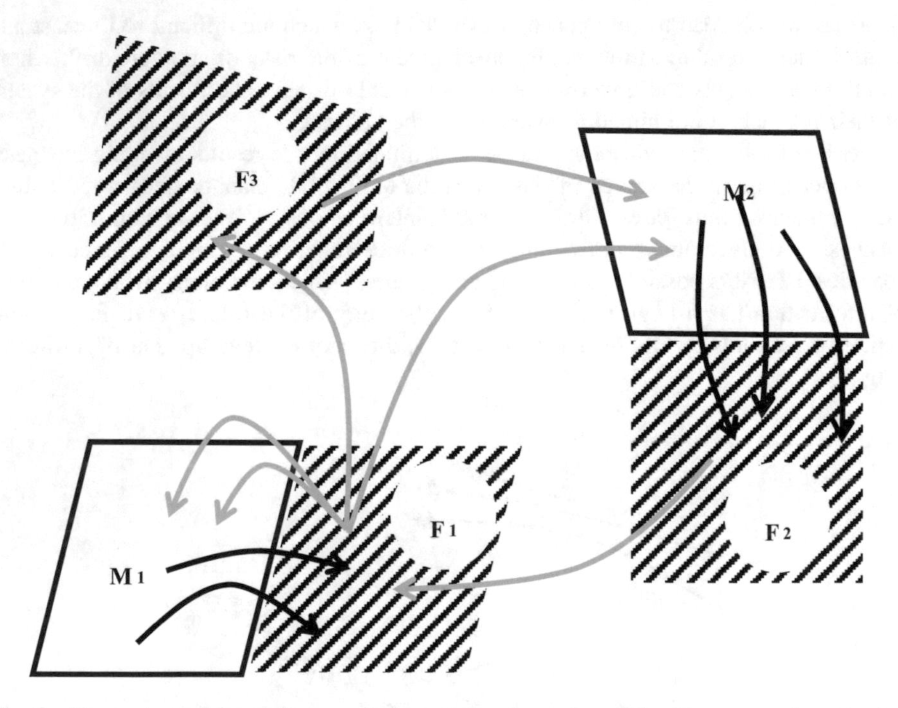

Fig. 3. The system of foresight research in complex dynamics of development and application of the tools of GR management for socio-economic development of territories. Source: compiled by the authors.

The system of foresight studies of the flow of information or actions becomes a tool of simplification of reality, when refusal from analysis of certain factors and strategies for transition to other possible alternatives takes place.

One the one hand, the perspectives of foresight study of the limits of development and application of the tools of GR management for socio-economic development of territories are obvious; on the other hand, any risk or opportunity leads to emergence of turbulence, violating the state of security.

5 Conclusions

It should be concluded that specific character of the modern world consists in a tendency of replacing the forward development by non-linear development, which creates the conditions of instability and unpredictability, as a result of which specifics of socio-economic development of territories changes and the territory receives a new impulse of growth or decline.

Foresight study is connected to the notion “socio-economic growth”. In the context of the authors’ evidences, foresight study of the limits of development and application of the tools of GR management should be used for socio-economic development of territories that is manifested in increase of favorable conditions for population: developed infrastructure, growth of territorial capital, rational usage of land (natural resources); balance of the ecological system, etc. The main tools are the mechanisms of leveling the risks in the system of interrelations of public authorities and business and the mechanisms of identification of opportunities for future socio-economic development of territories.

Acknowledgments. The reported study was funded by the Russian Foundation for Basic Research, grant No. 19-010-00103 “Cognitive modeling of innovative mechanism of partnership of business entities and public authorities (GR-management) in the conditions of digitalization”.

References

Ansoff, I.: Strategic Management. Ekonomika, Moscow (1989)

Aspects of security. Rules of inclusion into standards. [E-source]: GOST R 51898-2002: Decree of the State committee on metrology and standardization of the Russian Federation dated 5 June 2002, No.228-st. Access from the system Consultant Plus (2002)

Dibie, R.A., Dibie, J., Quadri, O.M.: Business and government relations in Kenya. Business and Government Relations in Africa. Taylor and Francis, pp. 340–370 (2017)

Gabsa, W.: Business and government relations in Cameroon. Business and Government Relations in Africa. Taylor and Francis, pp. 291–313 (2017)

Gadzekpo, L.: Government and business relations in Ghana. Business and Government Relations in Africa. Taylor and Francis, pp. 132–146 (2017)

Kayl, I.I., Lamzin, R.M., Epinina, V.S.: GR-management: realities and prospects of development. Fundam. Res. **9–3**, 597–600 (2016)

Makarova, N.N., Shokhnekh, A.V.: Turbulent approach to the system of provision of economic security of economic subjects. In: Makarova, N.N., Shokhnekh, A.V. (ed.) Audit and Financial Analysis 2012, no. 6, pp. 397–400 (2012)

Miropova, D.Y., Pavlova, E.A.: Influence of marketing and foresight research on competitiveness of innovative university developments. In: The Modern Problems of Science and Education 2013, no. 2 (2013). http://www.science-education.ru/108-8604. Accessed 25 Dec 2016

Moguev, B.D.: Development of foresight in the process of transition to the market relations with the help of benchmarketing. Russ. Entrepreneurship, no. 10(232), pp. 133–141 (2013)

Nur, Y.A.: Business and government relations in Somalia. Business and Government Relations in Africa. Taylor and Francis, pp. 147–161 (2017)

Pereslegin, S.: New maps of the future, or Anti-Rand. M.: AST: Moscow, SPb.: Terra Fantastica (2009)

Sokolov, A.V.: Foresight: a view into the future. Foresight 2007, no. 1 (2007)

Sidunova, G.I., Shokhnekh, A.V., et al.: Influence of the risks of strategic drift on foresight development of economic systems: monograph. Sidunova, G.I., Shokhnekh, A.V., Smykovskaya, T.K., Gomayunova, T.M. AETERNA, Ufa, 174 p. (2017)

Emirov, N.D.: Foresight technologies and perspectives of their application in the social sphere. Bulletin of South Ural State University 2012, no. 22, p. 130 (2012)

Clayton, A.: Technology Roadmapping for Developing Countries. Vienna: UNIDO5 Unido Technology Foresight Manual. United Nations Industrial Development Organization, Vienna, 2005, vol. I, p. 8 (2005)

Rogachev, A.F., Melikhova, E.V., Shokhnekh, A.V.: Information technology of cognitive modeling of industrial and investment self-development of the medium-sized and single-industry towns. Espacios **38**(27), 4 (2017)

Rogachev, A.F., Mazaeva, T.I., Shokhnekh, A.V.: Manufacturing and consumption of agricultural products as a tool of food security management in Russia. Revista Galega de Economia **25**(2), 87–94 (2016)

Shokhnekh, A.V., Skiter, N.N., Rogachev, A.F., Pleschenko, T.V., Melikhova, E.V.: The concept-strategy of ecosystem management through tax mechanisms of financial security. J. Adv. Res. Law Econ. **7**(7), 1854–1857 (2016)

Sidunova, G.I., Shokhnekh, A.V., Sidunov, A.A., Glinskaya, O.S., Sizeneva, L.A.: Approaches to modeling management and control processes in foresight management system taking into account expresentness conditions. Espacios **38**(24), 30 (2017)

Effective Technologies and Methods of GR Management that Stimulate Increase of the Level of Socio-Economic Development of Territories and Quality of Life

Iakow I. Kayl[1(✉)], Anna V. Shokhnekh[1], Veronica S. Epinina[2], Roman M. Lamzin[1], and Marina V. Samsonova[1]

[1] Volgograd State Socio-Pedagogical University, Volgograd, Russia
kailjakow@mail.ru, shokhnekh@yandex.ru, rom.lamzin@yandex.ru, marinasamsonova@yandex.ru
[2] Volgograd State University, Volgograd, Russia
econmanag@volsu.ru

Abstract. The article presents the main conceptual and methodological directions of development of effective technologies and methods of GR management, which stimulate the increase of the level of socio-economic development of territories and quality of life. Analysis of effectiveness of the methods of direct influence, influence through expert communities, support for political activities, media relations, and unethical GR technologies is performed. Participation in decision making by public authorities through expert and public councils allows for selection and application of the lobbying technologies. The article presents a four-stage technology of GR management; provides a functional model of GR management of socio-economic development of territories and population's well-being; offers a system of modeling of foresight development of GR management.

Keywords: Effectiveness · Technologies of GR management · Methods of GR management · Tools of GR management · Business and public authorities · Increase of the level of socio-economic development of territories · Increase of the level of population's quality of life

1 Introduction

New mechanisms and technologies of managing the interactions of public authorities and business form in the dynamic changes of economic relations. GR management, as an innovative mechanism that ensures the increase of effectiveness of cooperation of public authorities at all levels (local, regional, and territorial), is aimed at coordination of interests and efforts of entrepreneurship and the state at the macro- and micro-levels. Research and application of the modern methods and technologies of GR is very important in achieving success during implementation of business projects.

State has always been considered a stakeholder, with which mutually profitable cooperation is to be built, in view of public interests, for provision of successful development of business. In the 2nd half of the 20th century, the theory of stakeholders was

E. G. Popkova and B. S. Sergi (Eds.): ISC 2019, LNNS 91, pp. 191–207, 2020.
https://doi.org/10.1007/978-3-030-32015-7_23

developed (Akoff 1985; Freeman 1984). In the theory of stakeholders, business is defined as a center of the system of interconnected agents, which cannot be separate from public interconnections; and each participating agent contributes into the results of business's activities and expects profits from his contribution (Gadzekpo 2017; Igbokwe-Ibeto 2017; Rogachev et al. 2017). In these conditions, formation of the necessary interconnections with stakeholders and, especially, with the state, becomes the primary task of business – its solution determines successful development in the conditions of dynamic changeability of the market's situation (Kayl et al. 2016; Thomson et al. 2007). Positions of requirements of stakeholders lead business to understanding its role and responsibility in the system of development of social and ecological environment (Petrov 2004). In the modern management, the following groups of stakeholders are distinguished: public authorities at all levels (municipal, regional, and federal); society on the whole; shareholders and investors; credit organizations and banks; intermediaries (partners and suppliers; customers and potential clients); management of business; employees and unions; public organizations, etc. (Hussien et al. 2017; Rogachev et al. 2017; Shokhnekh et al. 2016; Sidunova et al. 2017).

Expectations of stakeholders are obvious, but they are directly connected to success, safe activities, and general well-being of business. However, certain interests of stakeholders have special significance for society. Thus, the state expects from successful and stable activities of business the increase of the level of socio-economic development of territories and quality of life. Tis expectation is obvious, as business does the following:

(1) creates jobs, reduces the level of unemployment, and increases the population's incomes;
(2) lifts off the load from the state regarding the payment of social payments, due to decent wages for population;
(3) pays taxes from incomes – as a taxpayer and agent;
(4) creates and improves infrastructure of the state, which is necessary for development of territory on which business conducts economic activities;
(5) raises quality of life of its labor resources;
(6) creates new markets;
(7) develops trans-regional and international relations;
(8) receives subsidies from the budget for socio-economic projects of business on a certain territory.

Effective technologies and methods of GR management, which stimulate the increase of the level of socio-economic development of territories and quality of life, could be studied on the basis of the national and global experience.

2 Materials and Method

The scientific material includes scientific publications on the issues of forecasting of the possibility of formation of partnership interactions of public authorities of all levels with business for socio-economic development of territories on the platform of foresight research. The authors use the methods of deduction, induction, comparative analysis, logical conclusions, and graphic conclusions of the results of the authors' concept, which allowed ensuring the necessary depth of elaboration of the scientific problem.

3 Discussion

In view of the interest of the state in development of entrepreneurship in the EU countries, GR management is used by large, medium, and small business. In Europe, GR is aimed not only at intra-national partnership but also at governing bodies of the EU.

The main institutional platform of interaction of the state and GR managers is multiple consultation committees and commissions, which exist under public authorities of the states and governing bodies of the EU. Large entrepreneurial associations, which synthesize interests and requests of business, are a mechanism of interaction of the state and entrepreneurial community. Active work is peculiar for the European Round Table of Industrialists; the Federal Union of German Industry; the Confederation of German Employers' Associations; the union of German Industry and Trade; the American Chamber of Commerce in Belgium, etc.

At present, entrepreneurial associations, which have the key position in relations between the state and business, set the balance between competitiveness of business in international, national, and territorial markets and significance of supporting social agreement and peace.

An example of successful application of the methods of GR management in Europe is the program of development of Airbus corporation. In 1967, French, British, and German manufacturers of aircraft formed the Airbus group. 1970 say conclusion of an agreement on creation of Airbus Industries, in which French Aerospatiale (government property of France) and German Airbus GmbH participated. Then, the following companies joined the project: Spanish CASA; British Hawker Siddeley; Dutch Fokker; Belgian Belairbus. For supporting the development of Airbus, European countries issued credits and guarantees. Also, governments of the European states insisted on using only European aircraft. Implementation of such GR project with the government support and wise formation of GR dialog made Airbus the world leader of civil aircraft engineering and the symbol of European integration (Smorgunova et al. 2012). In Russia, development of government relations has its specifics. The interrelations of public authorities and business were dominated by the tendency of strict control from the state, which hindered formation of conditions for effective dialog for formation of the mechanisms and tools of GR. However, after the market reforms of the early 1990's (privatization of state property and emergence of private business), there appeared preconditions for formation of a constructive dialog between the state, business, and non-profit organizations. During this period, large Russian corporations created departments for interaction with public authorities. GR departments appeared in oil, tobacco, and brewing companies – as their activities depended on state regulation; and in large monopolies - Joint Stock Company "Russian Railways"; Gazprom, etc. International corporations that work in Russia also create GR departments.

Since 2000's, professional GR management has become an inseparable part in the system of management of corporate business (Kokueva et al. 2009).

The main institute that forms the interaction between the state and business is associations and unions of entrepreneurs; the state actively participates in their work as an interested party (stakeholder) for building a dialog and interaction with business for accelerating the rates of socio-economic development of territories, growth of quality of life, and increase of population's well-being.

Associations of entrepreneurs are an "institute of development" and form positive general economic effects, which are manifested in improvement of the market: (1) protection of ownership rights, development of market infrastructure, stimulation of "transparency" of the market and the state; (2) possibility to change the behavior of business and develop the investment stimuli (Popandopulo 2011). Unions of entrepreneurs include sectorial directions and the whole business community. Business community is organized through the Russian union of industrialists and entrepreneurs and the Chamber of Commerce and Industry of the Russian Federation. The Russian union of industrialists and entrepreneurs formed as an association of large Soviet companies, which transformed into the organization that presented the interests of large capital in federal authorities. The Russian union of industrialists and entrepreneurs consists of sectorial commissions and committees. Also, the Chamber of Commerce and Industry lobbies the interests of entrepreneurs and organizes the dialog as a non-profit organization, unifying Russian large, medium, and small business. The Chamber of Commerce and Industry solves the following tasks: stimulates the development of the Russian economy and its integration into the world economic processes; forms the modern industrial and trade infrastructure; organizes favorable conditions for entrepreneurial activities; regulates the relations between entrepreneurs and their social partners; stimulates the development of various types of foreign economic relations between Russian companies and foreign partners. It should be noted that the Chamber of Commerce and Industry has been cooperating – with various organizational forms – since 19th century. According to Article 3 of the Law of the Russian Federation No. 5340-1 dated July 7, 1993 "Regarding the Chambers of Commerce and Industry in the Russian Federation", one of the main tasks of the Chamber is organization of interaction between the subjects of entrepreneurial activities and their interaction with government bodies and social partners. As a vertically integrated structure, the Chamber of Commerce and Industry effectively performs its functions at the federal, regional, and local levels (Regarding the Chambers of Commerce and Industry in the Russian Federation, the Law of the Russian Federation dated July 7, 1993, No. 5340-I).

Medium and small business is especially interested in membership in the Chamber of Commerce and Industry and Russian union of industrialists and entrepreneurs, as these platforms are the most effective means of organization of interaction with public authorities. These associations provide support for entrepreneurs, represent and protect their interests regarding the issues of economic activities; stimulate the diversification of entrepreneurial activities in view of the specifics and economic interests of Russia's regions; stimulate the development of continuous increase of business competencies of entrepreneurs; stimulate the training and advanced training of personnel for business; participate in development, implementation, and correction of government and international programs; provide information services; stimulate the creation of infrastructure; stimulate support for expansion of export; stimulate operations in the external market; take measures for preventing unfair competition and non-business partnership; stimulate regulation of arguments between subjects of business; provide the needs for services that are necessary for entrepreneurial activities. At present, sectorial associations and unions develop cooperation within the sphere and develop the sphere within the interaction with public authorities. Large sectorial associations in Russia are presented by the Association of Russian Banks; the Russian union of machine manufacturers; the Agro-industrial union of Russia; the Russian union of manufacturers; the Union of Russian car manufacturers; Association of Russian builders; Association of railway transporters “Union of transport companies”; the Union of Russian manufacturers of beer and soft drinks; the Union of timber industrialist and exporters of Russia; the Russian union of builders; the Union of railroad builders; the Russian union of tourism industry; the All-Russian union of insurers; the Association of companies of retail trade, etc. (Official web-site of the Russian union of industrialists and entrepreneurs [E-source] - Accessed [17.03.2019]: http://рспп.рф).

Also, there’s a system “Open government”, which develops for the purpose of developing the dialog between the state, business, civil society, and expert community. The purpose of open government is to increase transparency of activities and decisions of the public authorities at all levels and to build a mechanism of feedback between society, business, and the state. The system “Open government” includes the Ministry of the Russian Federation for cooperation with open government, the Expert council with the Government of the Russian Federation; projects “Open region”, “Open ministry”, and “Open data”, which are to increase the transparency of the activities of certain elements of the system of public authorities (Open government [E-source] - Accessed [17.03.2019] http://большоеправительство.рф/opengov/tasks/).

Development of the mechanisms of GR management in Russia was conducted during organization of “open government”. Formation of “Open government” began in February 2012 according to the Decree of the President of the Russian Federation and the work group for organization of the system “Open government”. The Ministry for cooperation with open government was created in 2012. The possibility for business’s participation in development of government decisions at the highest level of authorities is enabled by the institute of open government.

The Agency for strategic initiatives for promotion of new projects was created – as an institute of GR management – in August 2011 by the Decree of the Government of the Russian Federation. The government bodies are to provide support for the activities of the agency for strategic initiatives in promotion of new projects for conducting expertise of the projects of normative and legal acts that regulate the relations which participants are the subjects of the entrepreneurial and investment activities.

One of the mechanisms of the system of interaction of the government bodies and business is a new institute of entrepreneurs’ rights commissioner. The commissioner is authorized to approve the projects of legal acts regarding the rights and legal interests of the subjects of entrepreneurial activities and provide motivated offers on cancelling or termination of effect of legal acts of their provisions that complicate the entrepreneurial, including investment, activities. The federal entrepreneurs’ rights commissioner has a range of rights that are adopted in the federal law, which allow him to protect the entrepreneurs’ rights with great effect.

It is possible to state that GR management was created according to the requirements of the economic practice in the USA and the countries of Western Europe for large and medium business and non-profit organizations. The Russian infrastructure of Government Relations appeared in 1990's and is actively used in the practice of management of large and medium business.

4 Results

The main issue in development of GR management is development of the technologies and methods for practical implementation of formation of partnership between public authorities and business. Systematization of the methods and tools of GR and selection of the most effective practices becomes an important task of theoretical consideration of GR. The following GR technologies are distinguished:

- monitoring and forecasting of the results of activities of public authorities (local, regional, and federal);
- influence on the activities of government bodies;

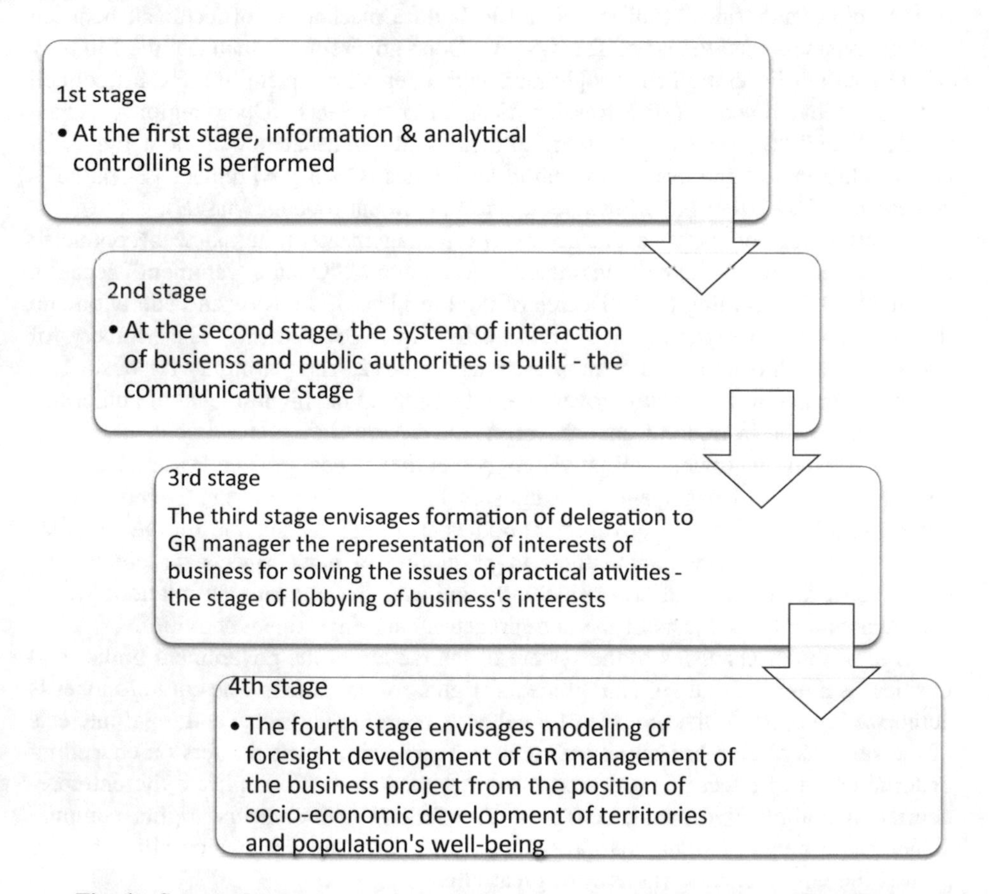

Fig. 1. Stages of technologies of GR management. Source: compiled by the authors.

- preparation of expert information and its provision for the employees of public authorities;
- creation of information events and organization of information campaigns;
- participation in social project (charity and volunteer activities for formation of a positive image in the state) (Mayorov 2008).
- The research showed that the technology of GR management could be implemented in four stages (Fig. 1).

1. At the first stage, information & analytical controlling is performed.

Considering the fact that GR manager has to be aware of possible changes in the external environment of business, it is necessary to perform diagnostics of the existing legal regulation and projects of the law. The main function of the state is regulation, and the task of a GR manager is to track and forecast certain changes of the "rules of the game" and state's interests in socio-economic development of territories. Also, a GR manager has to have knowledge of the government programs of socio-economic development of territories, for ensuring the business's participation in their implementation and receipt of government orders.

Specific features of the information & analytical work of GR managers of business depend on the specifics of the sphere and size of the business. The larger the business, the more important is this direction of work. It is necessary to study, analyze, and systematize various directions, including the trends of the global economy; problems of development of international relations; macro-economic indicators of the national economy; intra-political national problems; setting and ratio of powers of various political parties and groups; changes in the social dynamics, etc. Such organization of the information & analytical work will allow determining the main tendencies of development of society and forecasting possible actions for formation of business projects, with partnership participation of public authorities.

Collection of information and its analysis and synthesis are predetermined by the fact that attitude of public authorities towards large business depends on the changes of the economic and socio-political national dynamics of the country. Thus, natural monopolies in Russia are subject to state regulation of tariffs for services. As for regulation, public authorities use the cost of services and assets that are necessary for development of the sphere as the basis for regulation. However, natural monopolies are interested in increasing prices for provided services for the purpose of increase of profit. Realizing the importance of protection of population from the position of social demands from high tariffs of natural monopolies, the state has to deal with the challenges of social dissatisfaction. The price for the services of natural monopolies also influences the cost of all goods that are manufactured and moved in the national economy – and, therefore, the competitiveness of domestic products in the world markets.

Analysis of the information space, which is performed at the regional level, consists of analysis of the main directions of changes of socio-political dynamics of the territory. It is important to take into account the information on economic situation in the region and ratio of powers of various social and political groups.

2. The second stage envisages formation of the system of interaction between business and public authorities – the communicative stage.

It is possible to distinguish several typical methods of building a system of interaction of business and public authorities:

- interaction with public authorities through the platforms of the sectorial character;
- participation in the activities of consultation bodies of public authorities (in government establishments);
- participation in sessions, conferences, congresses, and exhibition activities, which are initiated by public authorities and business;
- participation in public-private partnership;

At present, sectorial associations exist almost in all spheres and sectors of economy. The purpose of the associations is protection of business, regardless of the types of activities (the Russian manufacturers union, the Russian association of managers, etc.). These organizations also stimulate the development of dialog between business, society, and public authorities. Sometimes, even public authorities initiate the creation of sectorial associations.

In the process of business's participation in sectorial associations, the organization could set before the public authorities the problem issues and start discussing and solving tasks. It should be noted that public authorities involve public organizations and interested parties in the process of development of the country's decisions. Before passing the legal projects they undergo public discussion (projects of budgets; infrastructural plans, projects of federal and regional laws, etc.). The government bodies also have consultation departments, which could influence the increase of awareness of public authorities, development of recommendations, and preliminary "elaboration" of government decisions.

3. The third stage envisages delegating the representation of interests of business to GR manager for solving the issues of practical activities – the stage of lobbying of interests of business.

Delegating representation of the interests of business in government bodies and direct interaction with them to a GR manager is a stage of implementation of practical results or a process of lobbying of interests of business. The problems of business that appear in the course of interaction with public authorities and that require solution could be classified in the following way:

- receipt of documents that allow conducting licensed business based on the Federal law dated 04.05.2011, No. 99-FZ "Regarding the licensing of certain types of activities";
- participation in self-regulatory organizations, which report to the state for quality of conducting sectorial types of activities based on the Federal law dated December 1, 2007, No. 315-FZ, "Concerning self-regulatory organizations";
- participations in contests for state order and grants for development of new (strategically important for the region, territory, or cluster) business;
- establishment of subsidized tax regimes and reduction of tax rates;
- adoption of a normative act that allows for amnesty for debt or violations.

The methods of direct interaction of business and public authorities could be used in the following directions:

- information and legal support for issues that have to be solved in government structures;
- negotiations with government structures;
- support – through sectorial associations – for regulation or correction of certain legislative acts;
- influence and negotiations with representatives of the government structures through mass media and the public.

An important method of lobbying of interests of business is expert and information & legal support, within which the purpose of a GR manager is provision of full and accessible information to representatives of government structures for decision making. Therefore, GR manager has to have a lot of tools, methods, and technologies for conducting his professional activities. Application of the methods and technologies of GR management at all stages of interaction with public authorities will allow GR manager to represent business's interests for solving the issues of practical activities and implementation of projects.

As of now, the following approaches to building the methods of GR management as the tools of growth of the level of socio-economic development of territories and quality of life could be distinguished, including (Volkova et al. 2006; Sokolov 2007; Higgins 1994).

- qualitative & quantitative approach (method of expert evaluations);
- qualitative approach to building the technologies and methods of GR management (morphological analysis, interview, literature overviews, "tree of goals", scenarios, role games, etc.);
- quantitative approach (analysis and forecast of indicators, methods of extrapolation, modeling, etc.);
- analytical and synthetic approach (Delphi method and brainstorm; road mapping; critical technologies, patent analysis; modeling, etc.).

The studies allow for essential characteristics of these technologies and methods of GR management as the tools of leveling the risks in the system of interaction between public authorities and business.

Another qualitative & quantitative method of GR management is the method of expert evaluations, which, based on qualitative features, provides qualitative evaluation for a possible choice of an effective alternative for the purpose of increasing the level of socio-economic development of territories and quality of life. The method of expert evaluations is used for formation of almost all government projects. The technology of this method consists in inviting competent experts from various sectors and spheres of activities for formation of the groups of 10–25 people, which have to consider – in the course of several months – the offered business projects that are aimed at socio-economic development of territory and present possible variants of the future on the set topic. The experts have to use the new analytical and information resources and developments (Sokolov 2007). The key advantage of the method of expert panels, aimed at increasing the level of socio-economic development of territories and quality of life, is as follows: communication and cooperation of specialists of various spheres of activities between each other and with public authorities and intermediaries. Communication with experts during the whole work process is requires for the formation of sufficient, substantiated, and reliable evidential basis. Experts outline the process of stimulation of realizing the potential of innovations or unexpected risks in business projects and joint public-private projects. The drawbacks include expensive, labor-intensive, and resource-consuming approach and possible mistakes due to subjectivism of experts (human factor), as the large number of experts in a group could influence the polarity of the obtained results in a negative way and stop the development of the necessary project for socio-economic development of territories.

As a qualitative method of GR management, it is possible to use the method of scenario approach for possible formation of directions and increase of the level of socio-economic development of territories. The method of scenario approach of GR management, as one of the most popular and effective technologies of forecasting the alternatives of the future, envisages preparation of detailed routes of implementation of the business project with government's involvement. Possible future alternative to the system of interaction between public authorities and business for implementation of an important project is implemented in case of meeting the corresponding conditions (Volkova et al. 2006; Sokolov 2007; Higgins 1994). The scenario approach for GR management envisages the selection of alternative texts, which are built around carefully selected points. The probability of an event is forecasted and designed simultaneously. Opportunities and advantages of the scenario approach are manifested in the following actions: maneuvering in the conditions of unexpected crises; building and developing a common understanding of the real problems in the system of interaction of public authorities and business for increasing the level of socio-economic development of territories; determining possible stimuli for development of the economic system; building the indicators and limits for experts; developing skills of group work with experts. The drawbacks of the method of scenario approach of GR management could be the following problems: (1) formation of a competent expert work group and high level of expert's subjectivity; (2) insufficient arguments of the results; (3) large labor-intensity; (4) large financial expenditures.

An analytical & synthetic method of GR management is the Delphi method – a tool aimed at attracting public authorities into the business project and, as a result, increasing the level of the territory's socio-economic development. The Delphi method of GR management is a technology that is used for forecasting and expertise (Higgins 1994). This method envisages structuring of the process of group communication of representatives of public authorities, business, and public organizations for working on a complex problem. This method uses independent surveys of the expert panels, which allow determining probability, values, and consequences of the factors and evaluating the tendencies and trends of the studied problem. The first stage envisages a survey – members of the expert panel receive all answers that are provided anonymously by other members, regarding the possible government's interest in the business project. This allows the experts to specify and correct their positions regarding the possible state subsidizing or investments. The key advantage of this method is the possibility of obtaining detailed, transparent, and objective results. Application of the Delphi method is aimed at leveling the influences of authoritative and active members on other experts, with absence of the necessity for joint expert meeting.

The analytical and synthetic methods of GR management also include brainstorm (Higgins 1994; Shokhnekh et al. 2016; Sidunova et al. 2017), which envisages effective solution of the problems, based on stimulation of the participants' creative activity.

The most adequate business projects, aimed at increasing the level of socio-economic development of territories and quality of life, are selected. This method allows for quick determination of the main potential possibilities and existing risks, connected to solving of the set task, formation of various options of its solution, and creation of alternative long-term strategies. This method does not require large financial expenditures, which allows forming non-standard variants of solutions, and has an approbated technique of forecasting. The drawbacks include a medium level of reliability in case of application of the method of brainstorm as the only approach to solving the problem of implementation of a business project.

The method of extrapolation of the trends of GR management is a quantitative approach (Ivanitskaya 2014) and is most popular in forecasting. This method consists in extrapolation (dissemination) of past achievements or failures for future events. The advantage of this method consists in its quick implementation and significant economy, with the necessary data of past indicators. The drawbacks of the method of extrapolation of trends from the position of GR management include high probability of a mistake in quantitative calculations and presentation of the results of extrapolation; complexity of analysis and evaluation of the existing factors of influence on the changes; insufficiency of information during analysis and evaluation of the limits of extrapolation, which determines the level of risk of a mistake in implementation of a business project.

One of the most popular methods that are used on the basis of the analytical & synthetic approach in GR management is the method of technological road map. The method of technological road map of GR management – as a tool aimed at increasing the level of socio-economic development of territories and quality of life – is used in the process of building the long-term foresight strategies of interaction of business and the state. Technological road maps are presented in the form of visual plan-scenario of development of technologies in view of opportunities (+) and risksв (−). Plan-scenario of technological road map reflects the expected scripts and point of critical decisions (Clayton 2005). The main opportunities and advantages of technological road maps for GR management could be presented in the form of possibility of assessment of future threats, existing potential, analysis, and selection of priorities; unification of the most important stimulating and negative factors into a consecutive strategic plan. As a result, the obtained map stimulates the determination of "narrow" spots, specification of "priorities of stimuli" in the sphere of investments into the innovative field of an economic subject, R&D, and HR policy. A drawback of the method is the necessity for the large number of resources – financial and time – for specialized training of experts that participate in road mapping.

Technological maps – as a method of GR management – allow recording the events that take place in business, for considering the three conceptual components (Clayton 2005):

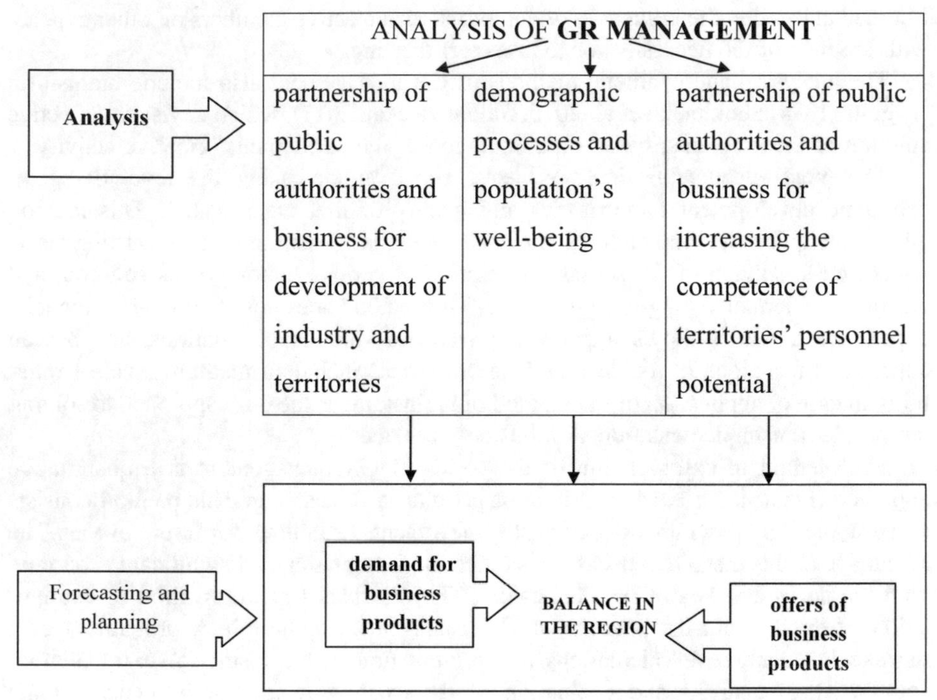

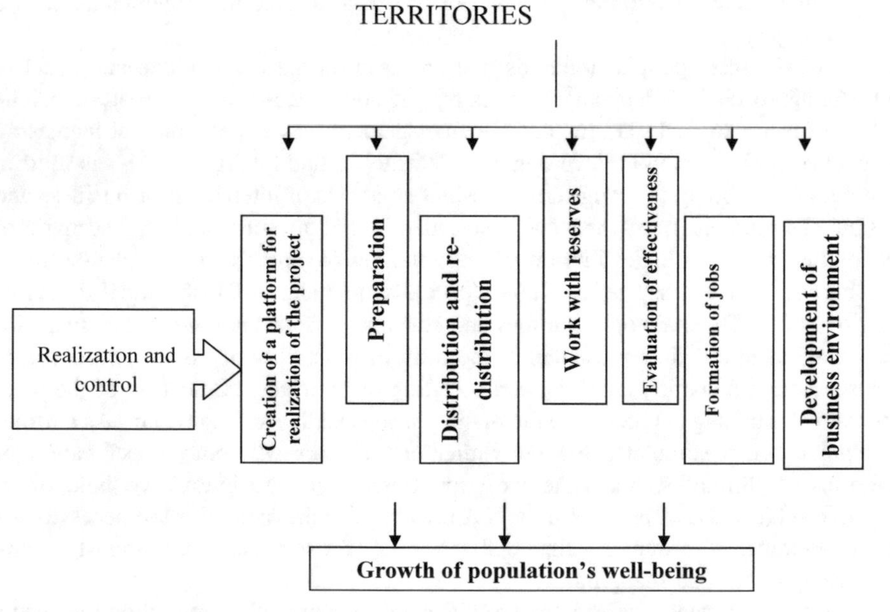

Fig. 2. The functional model of GR management of socio-economic development of territories and population's well-being.

(1) unambiguity of understanding the state and position of the business project that is lobbied by GR management;
(2) high utility and value of information, where usage of the method during solution of any tasks of GR management increases the quality and speed of receiving the result;
(3) accessibility of information for GR management, where users do not require special knowledge – logic and common sense are enough.

The quantitative approach could also be used on the basis of the method of modeling the process of lobbying of the business project in GR management. This method is the basis for the cyclic process, in which the schemes of functional models of formation of business projects are used (Fig. 2).

4. The fourth stage envisages modeling of foresight development of GR management of the business project from the position of socio-economic development of territories and population's well-being.

A modern method in the technologies GR management is foresight development. In the regional foresight development, functional determination of sectorial modeling is especially important.

Modeling of regional foresight development of GR management takes place in the process of transformation of strategic and current management of the region. Thus, it is necessary to establish a mechanism of "feedback" in the system of management, which would ensure formation of analytical & information resource that is built into an unbreakable chain of modern management of the future (Ermakova 2009).

Formation of the systemic approach to modeling of regional foresight development of GR management is predetermined by the necessity for order and integrity of life activities, namely: (1) studying the structural model of the system of relations of business and authorities in modern Russia; (2) studying the mechanisms of decision making in government bodies; (3) determining the specific features of GR management in Russia; (4) studying the interactions with sectorial associations and public organizations; (5) substantiating the role of public and expert councils with public authorities; (6) overview of the global trends of business communications; (7) analysis of the influence of changes in communications on the character of state and municipal management in Russia; (8) diagnostics of the declared policy of openness and transparency for GR communications with different government bodies; (9) building a full cycle of the process of interaction of business and public authorities: from monitoring to assessment of results; (10) diagnostics of the history of relations of business and public authorities/public organizations; (11) possibility for communication audit and communication SWOT analysis; (12) building the technologies of development of successful GR campaigns; (13) applying the means of mass information and PR strategies for solving business tasks.

Characteristics of the subjects of modeling of foresight development of GR management determines the circle of parties that have rights and obligations of the vertical and horizontal levels of management in business and government bodies.

The object of modeling of foresight development of GR management is the system of relations that is formed in various spheres and sectors during interaction of public authorities and business, which have a program of development, and business processes of strategic management and sub-processes for responsibility centers.

The objects of modeling of foresight development of GR management are as follows: foresight development of the territory; program of strategic usage of capital (material, non-material, and financial resources), land (all natural resources), and entrepreneurial and information resources.

The incoming analytical & information resource of modeling of foresight development of GR management is a detailed target resource of data for the users.

The tools of application of subject to subject and object of modeling of foresight development of GR management are a segment of methodology that determine the means, methods, and procedures.

Totality of the indicators of modeling of foresight development of GR management envisages establishment of the technique of collection of data and types and methods of the calculated indicators that characterize the state of the object, conditions, and sphere of the regional strategic management of the future.

Totality of parametric indicators of the scenario indicators of modeling of foresight development of GR management envisages building the precise criteria of the state of the object, item, conditions, and sphere of regional strategic management of the future for selecting the alternatives in the moment of action.

The outgoing analytical & information resource of the results of modeling of foresight development of GR management is organized for formation of directions of regional strategic management of the future.

The authors' studies show that the system of modeling of foresight development of GR management should have a process-oriented approach, which envisages establishment of interconnections in view of identification and systematization of the processes in the region's business. Implementation of the functions of the system of modeling of foresight development of GR management is based on the analytical & information resource, which determines the conditions of substantiated regional managerial decisions of the future socio-economic development of territories. Modeling of foresight development of GR management is performed in case of presence of the incoming analytical & information resource for transformation into the proper and sufficient information base: normative & legal information of the research sphere (allows initiative collection, storing, and accumulation of data); statistical, analytical and marketing.

Approaches to modeling of foresight development of GR management are determined by the process-oriented analytical & information resource of regional management (cluster approach). Unlike the linear & functional regional information resources, the cluster approach can form information according to the points of regional foresight development, determining weak aspects – namely, in the regional processes and subprocesses of the spheres of economy, which allows determining positive and negative effects of foresight development for the places of their emergence.

Effectiveness of application of the analytical & information resource of modeling of foresight development of GR management in the process-oriented approaches could be achieved during combination of the process and regional structural & functional direction. An important condition of formation of the analytical & information modeling of foresight development of GR management is information on regional revenues and expenditure for the centers of economic sectors. Orienting at the specifics of the sphere and sectors of the regional economy, it is expedient to determine the centers of formation of strategic development of business structures (including small and medium business), regional responsibility of administration and business, socio-economic contribution of business into regional product, share of business in the regional indicators, regional strategic parameters of the research sphere, and regional indicators in comparisons.

Modeling of foresight development of GR management envisages methodological means and methods that are regulated by the normative and legal order as to the levels of regional management and regional functional connections, as well as scientifically substantiated approach to evaluation of the system of regional budgeting of revenues and expenditures.

An important condition of effectiveness of modeling of foresight development of GR management is formation of working documents that reflect the detailed regulation and technology of strategic programs of development of the region and object of the research. Decision making of public authorities is influenced by the current GR methods and technologies that are to be accompanied by qualitative document paperwork in GR. It is necessary to understand the structure of writing of letters and requests, know typical errors during writing of letters to public authorities – for avoiding the risks of misunderstanding and negative influence on business and business projects from the consequences of the adopted normative acts at the federal, regional, and local levels. Therefore, it is necessary to perform a full cycle of the process of interaction with government bodies: from monitoring of GR model to evaluation of results.

5 Conclusion

An important issue is understanding the modern trends of development of the optimal structure of the system of interaction "business – public authorities": the notion of GR, its goals and main directions of activities of a specialist in the sphere of GRPR and GR, lobbyism: ratio of notions; determining the optimal functions in GR and business; specific features of GR management in Russia in the conditions of the digital economy. **Effective technologies and methods of GR management that stimulate the increase of the level of socio-economic development of territories and quality of life** include determining the criteria of work of GR departments in commercial organizations in view of the specifics of GR activities and its main goals; the system and structure of government bodies and local authorities; internal structure of the government bodies; determining the information flows in government structures; GR within a multi-stage process of decision making; full cycle of the process of interaction with public authorities; criterion of evaluation of effectiveness of GR-activities; legislative process; process of initiation and adoption of normative decisions and their types. It is necessary to mobilize public opinion for the future positive treatment of the decision of business and business projects. It is necessary to develop a strategic plan of media for smart interaction in GR activities, including the algorithm of development of the key messages in the communication with public authorities. It is possible to determine and apply the mechanisms of creation of an information event in GR during usage of mass media and PR strategies for solving business tasks. Monitoring of the existing opportunities for GR activities in the Internet space is also necessary and include determination of the classification of social networks from the position of the individual purpose as a resource for GR; determination and characteristics of possible risks of GR activities in social media.

Acknowledgments. The reported study was funded by the Russian Foundation for Basic Research, grant No. 19-010-00103 “Cognitive modeling of innovative mechanism of partnership of business entities and public authorities (GR-management) in the conditions of digitalization”.

References

Akoff, R.L.: A Concept of Corporate Planning. Progress (1985)
Volkova, V.N., Denisov, A.A.: The theory of systems. Vysshaya Shkola (2006)
Ermakova, N.A.: Regarding the accounting mechanism of interconnection of the systems of current and strategic managerial control (2009)
Ivanitskaya, I.P., Ivanitsky, A.Y., Gorbunov, A.V., et al.: Mathematical model of calculating entrepreneur’s income with changing taxation rates. Bulletin of Chuvash University, no. 4 (2014)
Kokueva, Z.M., Kulagin, V.E.: Development and the role of CRM technologies in the production sector. Management in Russia and abroad, no. 5 (2009)
Mayorov, R.A.: Technologies of corporate GR. Business key, no.6 (2008)
Regarding the Chambers of trade and commerce in the Russian Federation. The Law of the Russian Federation dated, no. 5340-I, 7 July 1993
Official web-site of the Russian union of industrialists and entrepreneurs. http://рспп.рф. Accessed 17 March 2019
Open government. http://большоеправительство.рф/opengov/tasks/. Accessed 18 March 2019
The Law “Regarding licensing of certain types of activities”, No. 99-FZ, Consultant Plus, Accessed 4 May 2011
The Law “Concerning self-regulatory organizations”, No. 315-FZ, Consultant Plus, 1 December 2007
Petrov, M.A.: The theory of interested parties: paths of practical application. Bulletin of SPbSU. Series 8, no. 2(16), p. 54 (2004)
Popandopulo, A.I.: Institutional forms of interaction of the state and business in the political system of Russia. State management, no. 29, December 2011 (2011). http://ejournal.spa.msu.ru/uploads/vestnik/2011/vipusk__29._dekabr_2011_g./pravovie_i_polititcheskie_aspekti_upravlenija/popandopulo.pdf
Smorgunova, L.V., Timofeeva, L.N.: GR relations with the state: theory, practice, and mechanism of interaction of business and civil society with the state. Study guide. In: Smorgunova, L.V., Timofeeva, L.N. (eds.) Russian political encyclopedia, p. 51 (2012)
Sokolov, A.V.: Foresight: a view into the future. Foresight, no. 1 (2007)
Clayton, A.: Technology roadmapping for developing countries. Vienna: UNIDO5 Unido Technology Foresight Manual. United Nations Industrial Development Organization, Vienna, 2005, vol. I, p. 8 (2005)
Freeman, R.E.: Strategic Management: A Stakeholder Approach. Pitman, Boston (1984)
Gadzekpo, L.: Government and business relations in Ghana. Business and Government Relations in Africa: Taylor and Francis, pp. 132–146 (2017)

Hussien, M., Dibie, R.A.: Business and government relations in Sudan. Business and Government Relations in Africa: Taylor and Francis, pp. 162–195 (2017)

Higgins, J.M.: 101 Creative Problem Solving Techniques: The Handbook of New Ideas for Business. New Management Publishing Company, Winter Park (1994)

Igbokwe-Ibeto, C.J.: Business and government relations in Tanzania. Business and Government Relations in Africa: Taylor and Francis, pp. 314–339 (2017)

Kayl, I.I., Lamzin, R.M., Epinina, V.S.: GR-management: realities and prospects of development. Fundam. Res. **9–3**, 597–600 (2016)

Rogachev, A.F., Melikhova, E.V., Shokhnekh, A.V.: Information technology of cognitive modeling of industrial and investment self-development of the medium-sized and single-industry towns. Espacios **38**(27), 4 (2017)

Rogachev, A.F., Mazaeva, T.I., Shokhnekh, A.V.: Manufacturing and consumption of agricultural products as a tool of food security management in Russia. Revista Galega de Economia. **25**(2), 87–94 (2016)

Shokhnekh, A.V., Skiter, N.N., Rogachev, A.F., Pleschenko, T.V., Melikhova, E.V.: The concept-strategy of ecosystem management through tax mechanisms of financial security. J. Adv. Res. Law Econ. **7**(7), 1854–1857 (2016)

Sidunova, G.I., Shokhnekh, A.V., Sidunov, A.A., Glinskaya, O.S., Sizeneva, L.A.: Approaches to modeling management and control processes in foresight management system taking into account expresentness conditions. Espacios. **38**(24), 30 (2017)

Thomson, S., John, S., Mitchell, G.J.: Public Affairs in Practice: A Practical Guide to Lobbying (PR in Practice). Kogan Page, London (2007)

Special Economic Zone of the Tourist and Recreation Type as a Tool of Regional Policy

Natalia S. Beskorovaynaya(✉), Elena V. Khokhlova, Ilya V. Ermakov, Zukhra S. Dotdueva, and Vitaly V. Lang

North Caucasus Federal University, Stavropol, Russia
agb20047@rambler.ru, Elena_kosh@mail.ru,
Ilya.ermakov@mail.ru, aibazova@yandex.ru,
Vitlang@yandex.ru

Abstract. Unsuccessful attempts to create special economic zones in several subjects of the Russian Federation, which have the corresponding natural conditions and resources, for the purpose of development of tourism and regional economy, actualize the problem of their economic substantiation and formation of the corresponding normative and methodological provision.

The purpose of the article is to determine the main factors that determine successfulness of implementing the projects for creation of special economic zones of the tourist and recreation type in Russian Federation and to improve organizational foundations of managing the development of territories that have tourist potential.

Methodology. In the recent years, the number of scientific works devoted to the problem of increase of effectiveness of functioning of special economic zones of various types in Russia has increased. However, most studies reflect peculiarities of development of successful zones – as a rule – technical and implementation, as well as theoretical aspects of their influence – on the territorial economy. At that, not enough attention is paid to the issues of economic substantiation of special zones of the tourist and recreation type as a tool of regional policy.

Analysis of the state of sanatorium-resort sphere in Stavropol Krai has been performed on the basis of statistical information, with the use of the methods of analysis, synthesis, comparison, systemic methods, etc.

Results. The resorts of the Caucasian Spas, most of which are located in Stavropol Krai, possess unique natural and geographic resources and the balneological basis. However, the quality of recreation services is low, and the municipal resort infrastructure is very old. The services' consumers are the Russian population (more than 90% of the general number of the customers). Thus, there's necessity for state support for development of the territorial resort complex.

Miscalculations and mistakes at the stage of substantiation of the project of a special economic zone in Stavropol Krai, which constituted in planning the construction of the resort infrastructure on new territories and orientation at new tourist products, which required large budget investments, became the reasons for its incompleteness. Besides, it is necessary to note the existing uncertainty of the order of actions for creation of a special economic zone, ignoring the

E. G. Popkova and B. S. Sergi (Eds.): ISC 2019, LNNS 91, pp. 208–218, 2020.
https://doi.org/10.1007/978-3-030-32015-7_24

dynamics of demand for recreation services, and the consequences of influence on the environment. However, such projects allow for receipt of large federal investments into modernization and reconstruction of old resort infrastructure of municipal entities.

Conclusions. It is concluded that special economic zone of the tourist and recreation type could be an effective tool of regional policy that implements a complex approach to management, oriented at activation of economic environment. At that, all efforts should be aimed at development of existing resort infrastructure and tourist business, and after that – at formation of new recreational territory.

Substantiation of the necessity for creation of special economic area should be based on criteria of sustainability of the project, its significance for development of the region, and economic effectiveness.

Keywords: Special economic zone · Regional policy · Effectiveness of functioning of special economic zone · Tourism · Sanatorium and resort activities · Tourist cluster

JEL Code: O 18 · R 12

1 Introduction

Special economic zones are treated as a means of development of depressive or underdeveloped territories and certain types of entrepreneurial activities. They are to attract foreign capital (as a rule, in developing countries) and national private capital (mostly in developed countries). A mandatory factor is large investments of state assets into creation of infrastructure and other forms of support for entrepreneurial activities. Large investments are required for organization of tourist and recreation special economic zones (SEZ).

Special economic zones stimulate the revival of entrepreneurial activity, economic and social development of the territory, and its integration into the global economy. However, their effectiveness, measures by comparing budget revenues and expenditures, is not always vivid, and in some cases the attempts to create SEZ end with a failure due to political, economic, and organizational reasons (in Sri Lanka, Guatemala, Liberia, Stockholm, etc.) (Prihodko and Volovik 2007). Also, negative influence of territories with a special status on the national and regional economy is observed, which leads to changes of prices and mechanisms of stimulation, growth of budget expenditures, using subsidies for money laundering, and aggravation of the environment (Beskorovaynaya and Rubchevskaya 2016).

Thus, actuality of studies devoted to analysis of the causes for failure to realize the SEZ projects in Russia and the issues of economic substantiation of expedience of their implementation and evaluation of effectiveness as a tool of regional policy grows.

Role of Special Economic Zone in Development of Region's Economy

Certain aspects of substantiation and evaluation of effectiveness of special economic zones, as a tool of regional policy in Russia, are an object of many scientific studies. In particular, the mechanism of state support for development of territories that possess tourist potential, is viewed in the works of A.I. Bolshakov, E.V. Selevanova, S.V. Prokhodko, N.P. Volovik, etc.

A lot of authors think that SEZ should be formed primarily in underdeveloped and depressive regions, thus creating "growth points" that stimulate development of adjacent territories (Peredkova 2013) and have to retranslate the achievements to other connected economic systems (Pavlov 2010). A specific model of a zone should be selected in view of the potential possibilities of the territory and existing competitive advantages.

In the foreign countries, SEZ began to be created in 1980's for the purpose of complex development of other underdeveloped territories, stimulating activation of entrepreneurial activities and reduction of unemployment. Development of tourist business should stimulate growth of the region's economy and bring additional budget revenues by means of increase of the tourist flow and involvement of small and medium business.

Topicality of creation of special economic zones of the tourist and recreation type is caused by high effectiveness of this sphere and its capability to influence the flow of investments, dynamics of development of other sphere on the local territory, and increase of population's employment.

P.V. Pavlov notes the important of pre-investment period of implementation of the SEZ project, which is to study the conceptual possibility of its creation and selection of the location, type, and size of the area. The risks of this stage are caused by uncertainty of the purposes of the project's participants, presence of preliminary title certificates, errors in financial and economic calculations, incompleteness, and lack of prevision of information on financial state and business reputation of the project's participants (Pavlov 2009).

The conceptual model of development of SEZ in Russia is based on the principles of creation, issues of formation of the system of state support and requirements to residents of the zone, and organization of economic activities. At that, the tools of state support should be chose in view of solution of political and social problems of the regions and implementation of comparative advantages, not as compensation of lacking factors of development (Pavlov 2010).

E.V. Erokhina considers the problems that emerge during creation of SEZ of the tourist and recreation type in Russia caused by passivity of small business, underdevelopment of the system of formation of demand, inaccessibility of services for most population, and low level of trust to public authorities from the society due to high level of corruption and crime (Erokhina 2015).

Russian law (Federal law "Concerning special economic zones in the Russian Federation" dated July 22, 2005 No. 116-FZ) allows for unification of several special economic zones within one cluster under the managing company- which is caused by the necessity for increase of their effectiveness.

Certain scholars also view SEZ TRT as a form of cluster (Bolshakov 2014). We think that there are differences between these two tools of regional policy, caused by the structure of participants and the level of state regulation. Thus, tourist cluster is formed of the companies of various sectors of economy, joint activities of which on a certain territory is aimed at issue of tourist product (Khokhlova and Pashayev 2017). At that, according to the Russian law, there are residents on the territory of SEZ TRT which conduct only the tourist and recreation activities, envisaged by the agreement with bodies of management of the SEZ and other individual entrepreneurs and commercial organizations. At that, the activities of residents are regulated and controlled by authorized bodies of executive power of various levels and by the managing company. Residents have the system of state support, which includes tax, administrative, and financial subsidies. Thus, organizations and individual companies of adjacent spheres are not a structural element of the system of zone's management.

We think that cluster approach to managing a territory is more preferable. It could be supplemented by creation of a SEZ of the federal or regional type, which allows receiving large federal investments for implementation of large-scale investment projects. However, such type of state regulation does not cover all problems of socio-economic development of municipal entities. The global experience shows the necessity for supporting economic development on the spot according to the endogenous economic theory (Beaumier 2002), based on active role of economic environment and its participation of management of territory.

Russian and foreign scholars studied the theoretical and methodolgocial foundations of creation of SEZ. However, low efficiency of this tool of regional policy in Russia, including the areas of the tourist and recreation type, determines the necessity for studying the conditions of their implementation and development of the corresponding normative and methodological provision of the necessity for their creation.

Normative and Legal Basis of Functioning of SEZ TRT in Russia

Federal law No. 116-FZ "Concerning special economic zones in Russian Federation" (2005) supposes the possibility of functioning of a SEZ for the purpose of development of tourism. A tourist and recreation zone is created on one or several areas of the territory that belongs to one or several subjects of the RF. Land plots that are given to SEZ should be in the state, municipal property or be a property of citizens or legal entities. A tourist and recreation zone may include lands and objects of protected territories, forest fund, and agricultural lands.

SEZ are created on the basis of application of regional authorities and municipal entities with substantiation of expedient of formation and its effectiveness for solving federal, regional, and local tasks.

SEZ are created for 49 years. For the purpose of effective management and control, the law supposes creation of a managing company and supervisory board.

The system of SEZ is a complex economic system which includes the territorial and organizational structure, administrative machine, tax and other preferences, and residents. The role of residents could be acquired by individual entrepreneurs and commercial organizations (excluding unitary), which are registered on the territory of the zone and conclude agreements on conduct of tourist and recreation activities with the bodies of management of the SEZ. For that, they have to provide a business plan.

At that, tourist and recreation activities are activities on construction, reconstruction, and exploitation of the objects of tourist industry, objects for sanatorium-resort treatment, medical rehabilitation and recreation of citizens, as well as tourist activities and the activities on development of sources of mineral water and other natural medical resources, including the activities on the sanatorium-resort treatment and prevention of diseases, etc. A resident cannot have branches beyond the SEZ.

Experience of Creation of SEZ in Stavropol Krai

The meeting of the Audit Chamber of the RF states that over ten years of implementation of the policy of creation of special economic zones, they have not become an effective tool for supporting the national economy. Moreover, eight special economic zones of the tourist and recreation type were liquidated in Russia in autumn 2016, one of them – in Stavropol Krai. No agreements on joint activities were concluded, and no residents were registered on these territories.

A SEZ in Stavropol Krai, which was created in 2007, supposed construction of 14 investment platforms, creation of more than 40,000 jobs, 17,000 of hotel spots, and attraction of additional 2.5 million tourists. Its purpose was to create a balneological resort of the international level with orientation at the European standards.

After the decision on creation of SEZ in Stavropol Krai, the planned financing was reduced to the value sufficient for only one investment project (Grand Spa Uza), with the volume of budget assets of RUB 8.7 billion, of which RUB 5.1 billion from the federal budget. Construction of the infrastructure should have been performed by means of the state, and tourist objects should have been constructed with the help of private investors.

Tax subsidies were planned for attracting residents:

– reduction of regional corporate tax to 13.5% (by 4.5%);
– exemption from corporate property tax for 10 years;
– exemption from land tax for 5 years;
– reduction of land rent fee to 0.01% for the period of the investment project;
– subsidizing the interest rate for credits in the volume of 2/3 of the refinancing rate of the RF Central Bank;
– free connection to the objects of engineering infrastructure.

Failure to realize the initial project of the SEZ could be explained by two reasons – reduction of the volume of budget financing and lack of interest of private investors. Obviously, at the stage of decision making on creation of SEZ it is necessary to use the scientifically substantiate conclusion on the volume of expected demand for the corresponding service.

In order to learn the reasons for unsuccessful realization of the SEZ TRT in "Concerning the criteria of creation of special economic zone" in Stavropol Krai, let us see how its characteristics conform to the criteria of creation of SEZ. These criteria were set in 2012 by the Decree of the Government of the RF No. 398:

– natural economic & geographical competitive advantages of the territory;
– correspondence to the priorities of the complex territorial development;
– expected growth of the volume of revenues of the budget system;
– presence of investment projects and potential investors with the volumes of investments that exceed the budget assets necessary for creation of the infrastructure;
– substantiation of project indicators of development of infrastructure and the indicators of effectiveness of functioning of the SEZ;
– availability of the organization that can perform the functions of a managing company;
– successful experience of implementing large investment projects in the region, etc.

The resorts of the Caucasian Spas have a competitive advantage that is expressed in unique natural and geographical resources: mineral waters and therapeutic muds, as well as proximity to ski resorts of the North Caucasus tourist cluster. The unique balneological basis of the Caucasian Spas consists of more than a hundred of sources of mineral waters of thirteen types and therapeutic mud of the Lake Tambukan.

The protected ecological region of the Caucasian Spas includes the territory of three subjects of the Russian Federation: Stavropol Krai, the Karachay-Cherkess Republic, and the Kabardino-Balkar Republic, which cover one artesian basin. 58% of the Caucasian Spas' territory is located in Stavropol Krai.

The balneological resort was created in 1803. In the Soviet times, 1 million people visited it, and as of now, more than 130 sanatorium-resort organizations can host 30,000 people.

At present, the following drawbacks of the resorts are observed: low quality of provided sanatorium services, problems with water supply and sewers, restoration of old funds of sanatorium organizations, and utilization of waste.

Development of tourism is one of priorities of the Strategy of socio-economic development of Stavropol Krai until 2020 and until 2025, set in 2009 by the Decree of the Government of Stavropol Krai No. 221-rp, which aimed at creation of a "leading Russian tourist and recreation and medical complex on the basis of the Caucasian Spas". The Strategy of development of tourism until 2030, which was adopted in 2015, is aimed at "provision of infrastructural and environmental development of the region in view of top-priority development of the industry of tourism". It sets the task of determining "the territories of Stavropol Krai that are favorable for tourism and formation of new tourist clusters and regional or local medical-recreational places and resorts".

The factor that reduces the interest of investors is high level of risks related to the security threats. At that, the ecological and sanitation-and-epidemiological state of the territory does not conform to the requirements set in Russia.

The most important criterion is availability of investment projects and investors. In order to answer the question of lack of interest to doing business on the territory of SEZ, it is necessary to view the state of the sanatorium-resort sphere in the region.

Analysis of the Modern Stage and Perspectives of Development of the Sanatorium-Resort Activities in Stavropol Krai

Let us view demand and offer for recreation services in Stavropol Krai. In 2015, the structure of the tourist services has the share of sanatorium-resort services at the level of 86.4%, tourist and recreation services and hotel services – 6.8% each (Territorial body of Federal state statistics service, 2016).

The consumers of these services are the population of the Russian Federation (93.9%). In the total number of consumers, the share of people from the CIS constituted 5.8%, outside of the CIS – 0.3% (Territorial body of Federal state statistics service, 2016).

Dynamics of demand for recreation services depends on the volume of revenues of the country's population. Thus, growth of real disposable income of the population in the RF until 2014 was accompanied by increase of the flow of vacationers and profitability of activities of the sanatorium-resort organizations (Fig. 1). Starting from 2015, all indicators reduced.

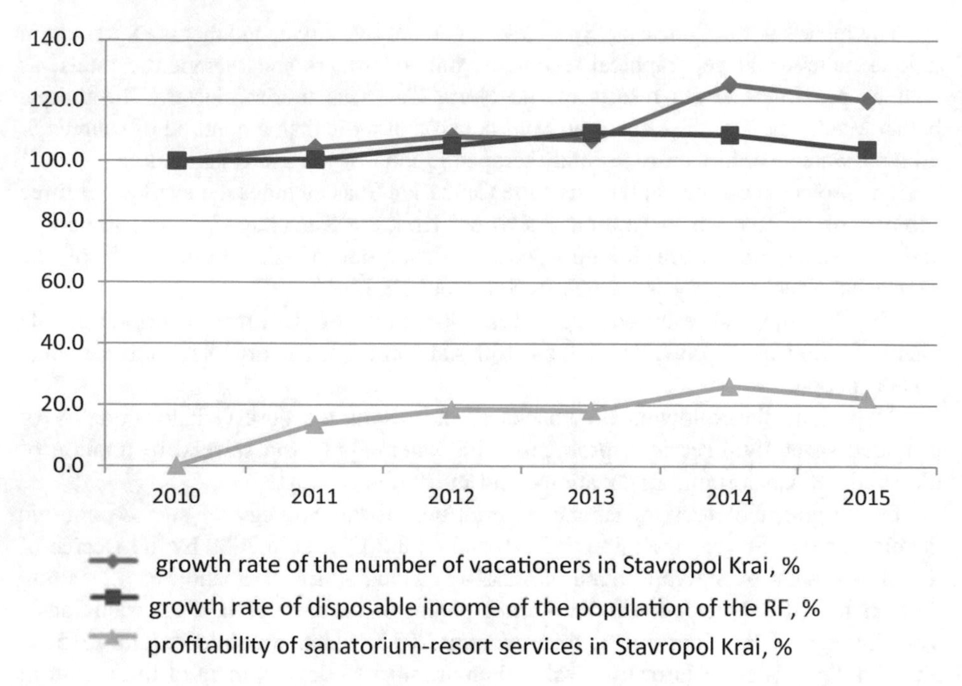

Fig. 1. Dynamics of profitability of the sanatorium-resort services in Stavropol Krai as compared to the growth rates of the number of their consumers and real disposable income of the population in the Russian Federation, %.

Compiled by the authors on the basis of (Territorial body of Federal state statistics service, 2016).

In 2015, the sanatorium-resort services were provided by 113 organizations of the corresponding profile (without micro-companies), of which 41.6% were state, 23.0% belonged to public associations (organizations) and trade unions, 18.6% – private, 10.6% – mixed property, 6.2% – foreign property. Despite the acceleration of the main funds of these organizations, the level of their wear is high – 39.7%. Financial result from exploitation was high in 2014, in the period of the highest inflow of the vacationers (Table 1).

Thus, the tourist and recreation sphere in Stavropol Krai requires increase of the tourist flow and restoration of the resort infrastructure. This requires financial support from the federal budget for those with low incomes that require sanatorium treatment and in the form of large subsidies for municipal budgets within the corresponding federal program.

Table 1. Indicators of activities of the sanatorium-resort organizations of Stavropol Krai

Indicator	2011	2012	2013	2014	2015	2015 in % to 2011
Sanatorium-resort organizations, quantity	116	114	115	118	113	97.4
places	30,012	30,729	30,011	31,033	30,578	101.9
consumers, thousand people	510.4	528.3	523.3	612.6	587.2	115.0
main funds as of year-end, RUB million	13,491.4	13,978.6	15,034.4	17,204.4	20,289.2	150.4
Level of wear, %	39.3	39.1	40.9	38.3	39.7	–
Restoration coefficient, %	19.1	5.1	8.2	13.2	6.5	–
Income from exploitation, RUB million	16,622.8	18,197.5	19,077.6	21,211.1	23,782,9	143.1
Expenditures for exploitation, RUB million	14,641	15,360	16,143.1	16,805.4	19,445.7	132.8
Financial result, RUB million	1,981.8	2,837.5	2,934.5	4,405.7	4,337.2	218.9

Reasons for unsuccessful implementation of SEZ TRT in Stavropol Krai

In order to substantiate the necessity for creation of SEZ TRT, it is necessary to take into account its formation:

- large expenditures for creation and reconstruction of infrastructure, conducted by means of the federal budget;
- dependence of the economic effect on risks of tourist business, primarily – on demand for services, which is determined by the influence of such factors as the volume of purchasing capacity of population, security and state of the transport infrastructure, and the system of state social help;
- large terms of projects' return;
- large share of small and medium entrepreneurship in the tourist services market;
- for recreation services – availability of a large number of highly-qualified personnel.

Thus, one of the main problems consists in attraction of investors, which, apart from the volumes of state support, are oriented at the offered demand for the tourist and recreation services.

The factors that caused the failure of SEZ TRT on the territory of Stavropol Krai could be divided into economic and organizational. The former include the following:

- low effective demand of the population of the Russian Federation, which does not allow – under the existing conditions – expecting a large growth of the vacationers' flow;
- low quality and high price for the provided sanatorium-resort services with large wear of the main funds of organizations and the municipal resort infrastructure, low level of professional preparation of working personnel;
- lack of large investors who are ready to invest assets, due to low investment attractiveness of territories and large payback time of the projects (over 12 years).

The factors of the organizational character are predetermined by miscalculations in planning and organization of SEZ:

- uncertainty of the order of stage-by-stage conduct of actions for creation of SEZ (lack of criteria and investment projects, perspective plans of development, indicators of effectiveness, and system of monitoring);
- orientation at construction of new tourist complexes with new products that require large capital investments with unclear perspectives of services' demand, with underfinanced reconstruction of the sanatorium-resort base;
- ineffectiveness of decisions on the choice of the state body that is responsible for implementation of SEZ;
- the necessity for complex development of the territory of the Caucasian Spas was not taken into account.

SEZ TRT is peculiar for non-applicability of the principle of innovativeness (i.e., creation of new productions), the expedience of which is obvious as to design of production and technical zones. This is caused by the necessity for large investments into creation of a new recreation territory and the efforts on formation of favorable image of a new product. There is also a plan for construction of new tourist recreation objects in Stavropol Krai, despite the presence of a sanatorium-resort complex in Stavropol Krai, which is a serious drawback of the SEZ project. The researchers note that a complex approach to territorial development should develop the mechanism of support for existing business and concentrate on creation of new companies and attraction of regional and foreign companies (Prihodko and Volovik 2007).

The drawbacks of organizations of SEZ in Stavropol Krai include lack of assessment of the character and intensity of influence on the natural eco-system, system of ecological monitoring and ecological certification, developed in view of international experience.

Organization of SEZ supposes participation of large investors, including foreign, which will be interested in receipt of commercial effect, not in development of territory. At that, local participants of the tourist complex of municipal entities, represented by subjects of small business, also require state support. Thus, it is necessary to determine the requirements set to the residents of the SEZ.

Economic Substantiation of Creation of SEZ TRT

We think that during decision-making on implementation of a SEZ project it is necessary to assess its sustainability, necessity for development of the region or sphere, and economic effectiveness.

Sustainability of the SEZ TRT project depends on demand for services and tourist potential of the territory. Changeability of demand and lack of proper control over exploitation of the natural and recreation resources in Stavropol Krai reduce the probability of the expected effect.

The necessity for creation of SEZ TRT is substantiated by supposed positive influence on the state of adjacent spheres and territories. However, in practice, expectations for creation of growth points do not come true, partially due to lack of necessary conditions and low level of entrepreneurial activity.

An important issue is selection of indicators of effectiveness of the SEZ functioning, at which the plans of their development are oriented and monitoring of results. At present, evaluation of effectiveness of SEZ functioning includes the indicators that characterize the results of activities of residents and management bodies of the SEZ and profitability of investment of budget assets, which are determined as compared to the planned values. We think that in view of the necessity for improvement of health and population's living standards due to growth of accessibility of tourist and sanatorium-resort services, it is necessary to envisage the indicators of the number of vacationers in the sanatorium-resort organizations of the zone.

Also, it is necessary to view the influence of risks caused by specialization of the territory, related to reduction of demand for tourist services or their seasonal character, which may lead to growth of unemployment or social tension. These and other problems of development of the territory of the Caucasian Spas could be solved in a complex, in view of interests of all subjects of economic activities and local population. In this aspect, effectiveness of the SEZ for development of the region seems to be limited. It is necessary to supplement it with other tools of state management of territories: cluster policy and public-private partnership.

State stimulation of development of the problem regions, on the basis of territorial justice, leads to reduction of economic effectiveness. However, implementation of such policy is necessary, as it stimulates fuller use of economic potential of underdeveloped territories, leveling of disproportions in the regional development, and stabilization of social relations.

2 Conclusions

The tools of regional policy – such as special economic zones, clusters, and public-private partnership – should be applied within the complex approach to management, oriented at activation of the region's economic environment. The strategy of development of territory should be aimed at development of existing resort infrastructure and tourist business, and at the next stage – at formation of new recreation territory, with new products.

Creation of SEZ TRT stimulates development of the territory and solving the problem of improvement of health and population's living standards. For a lot of municipal entities, it allows receiving large federal investments into reconstruction and formation of modern resort infrastructure.

The decision on creation of SEZ TRT should be based on criteria of sustainability of the project and its significance for development of the region and economic effectiveness.

Effectiveness of functioning of the territory with special economic status should be measures not only with investment activity of residents and payback of budget investments, as it is envisaged by the Russian normative tools. Special attention should be paid to dynamics of vacationers' flows, including those who require the sanatorium-resort treatment. There's also a necessity for the state subsidizing for demand for healthcare services for low-lived groups of population (retired, handicapped, and children).

References

Asian economic integration report 2015: How can special economic zones catalyze economic development? Asian Development Bank Publications, Mandaluyong City. https://www.adb.org/sites/default/files/publication/177205/asian-economic-integration-report-2015.pdf. Accessed 12 April 2017

Beskorovaynaya, N., Rubchevskaya, T.A.: Strategy for sustainable development of the region in conditions of globalization. In: Regularities and Tendencies of Development of Science in Modern Society Proceeding of the International Scientific-Practical Conference in Ufa, Russia, 2016, LLC Aeterna, pp. 42–44 (2016)

Bohmer, A., Farid, N.: Designing Economic Zones for Effective Investment Promotion (2010). http://www.oecd.org/mena/investment/44866506.pdf. Accessed 12 April 2017

Bolshakov, A.I.: Formation and development of cross-border tourism clusters, Moscow (2014)

Erokhina, E.V.: Influence of areas with special economic status on the spatial development of the regions. Nat. Interests Priorities Secur. **36**(321), 12–29 (2015)

Farole, T., Akinci, G.: Special Economic Zones Progress, Emerging Challenges, and Future Directions. The International Bank for Reconstruction and Development/The World Bank (2011). https://openknowledge.worldbank.org/bitstream/handle/10986/2341/638440PUB0Exto00Box0361527B0PUBLIC0.pdf. Accessed 12 April 2017

Federal State Statistics Service (2016). Russia in figures 2016, The concise statistical handbook, Moscow

Khokhlova, E.V., Pashayev, M.J.: Special economic zone in the Chechen republic. In: The Scientific Mechanisms for the Resolution of Problems of Innovative Development, Proceedings of the International Scientific-Practical Conference in Ufa, Russia, 2017, LLC Aeterna, pp. 211–217 (2017)

Kropinova, E.G., Mitrofanova, A.V.: Regional-geographical approach to the concept of tourism and recreation cluster, Vestnik Immanuel Kant Baltic Federal University, vol 1, pp. 70–75 (2009)

Kropova, A.A.: Evolution of special economic zones in Russia. Young Sci. **4**(108), 429–431 (2016)

McCallum, J.K.: Export processing zones: Comparative Data from China, Honduras. Nicaragua and South Africa (2011). http://www.ilo.org/wcmsp5/groups/public/—ed_dialogue/—dialogue/documents/publication/wcms_158364.pdf. Accessed 12 April 2017

Pavlov, P.V.: System of indicators for assessing the functioning of special economic zones. Finan. Credit **29**(365), 2–11 (2009)

Pavlov, P.V.: Institute of Special Economic Zones in the Russian Federation: Financial, Legal and Organizational-Economic Aspects of Functioning. Magistr, Moscow (2010)

Peredkova, I.V.: Organizational and economic mechanism of formation of special economic areas. Socio-Econ. Phenom. Process. **9**(55), 93–97 (2013)

Prihodko, S., Volovik, N.: Special Economic Zones. Consortium for Economic Policy Research and Advice, IET, Moscow (2007)

Rodrigues, A.B.: Turismo rural: práticas e perspectivas. Contexto, Sao Paulo (2003)

Selivanova, E.V.: International and Russian experience of creation of special economic zones for tourism and recreation: problems and prospects. Sci. Bull. SevKavGTI **1**(24), 46–50 (2016)

Territorial body of Federal state statistics service of the Stavropol territory (2016). The resort-tourist complex of the Stavropol territory for 2011–2015, Statistical compilation

The Problems of Modernization of the Social and Economic Development in the Light of the Scientific and Technological Revolution

Irina V. Novikova[1(✉)], Nataliia N. Muraveva[1], Lilianna Y. Grazhdankina[2], Nina Y. Zhelnakova[1], and Larisa A. Tronina[1]

[1] North-Caucasus Federal University, Stavropol, Russia
Iren-n@rambler.ru
[2] Stavropol State Medical University of the Ministry of Healthcare of the Russian Federation, Stavropol, Russia

Abstract. This research is aimed at identifying the indicators and criteria of modernization of the economy in the light of the scientific and technological revolution. The indicators of social and economic development of Russia and the most developed countries of the world: United States, Japan, United Kingdom, France and Italy were compared in the process of the study. The research methodology itself is based on a system approach that combines the methods of induction, deduction, comparative analysis, identification of the general, the special and the particular, etc. Various ratings, statistical data of the Federal State Statistics Service of the Russian Federation and its local offices, the Ministry of Economic Development of the Russian Federation, as well as the content from the Internet were used as an information base; author's own estimates. The basic results of research are as follows: the relevance of the development of criteria for modernization of the economy in the light of the scientific and technological revolution and modernization models has been revealed. It has been found that depending on whether modernization occurs on the basis of the scientific and technological revolution or on the basis of evolutionary changes, modernization can be either evolutionary or revolutionary. it has been proved that modernization is based on innovations, the ultimate goal of which consists in improving resource efficiency. Criteria for modernization of any economic system, which are relevant both for the scientific and technological revolution and for evolutionary modernizations, have been defined.

Keywords: Criteria · Indicators · Resources · Modernization · Scientific and technological revolution · Efficiency

JEL Code: O 110

E. G. Popkova and B. S. Sergi (Eds.): ISC 2019, LNNS 91, pp. 219–228, 2020.
https://doi.org/10.1007/978-3-030-32015-7_25

1 Introduction

Modernization is the process of the quality update of the object of modernization. Thematic justification of this problem lies in the fact that any economic system must constantly upgrade, that is, be updated, move to a different quality, innovate the basis of its existence, structure, mechanisms, methods etc. for increasing its competitive power. Only innovative countries can successfully develop in the contemporary world economic system, be successful in economic competition for market outlets and resources, provide their residents with a high and ever-increasing standard of living, and improve the quality of human capital. In turn, it is the latter that is currently one of the main factors of competitiveness. That is, a direct correlation can be observed here – the level of economic development determines the opportunities for the improvement of human potential, which, in turn, impacts the economic development. All of this is indicative of the fact that the problems of modernization appear to be extremely topical. Furthermore, at every stage of the time spiral, at every level of social and economic development, these problems will always differentiate, since the mechanisms for solving these problems will be subject to modernization as well. Moreover, permanent scientific and technological revolution which occurs differently in different countries, results in dependence of the factors of modernization, its goals and criteria on the stage of the scientific and technological revolution of a particular country, as well as its dominant technological mode.

The studies of the problems of modernization generally bump up against the identification of modernization model, and, accordingly, mechanisms and methods that are the most productive and reasonable in this economical system and at the current stage of the scientific and technological revolution. This is due to the fact that the need for modernization as such is not disputed; however, the structural components of the modernization process as such, its criteria and modernization models are still debatable. So, what are the most necessary aspects in this scientific problem, what are the tasks of research? In our opinion, the tasks are related to the fact what specimens in global development practices the directions of modernization should be supported by, and, accordingly, scientific and technological revolutions, which criteria in the development should be pursued by a particular country. And the second problem is how to achieve these criteria, what methods and mechanisms are used, what model of development is used to achieve them. Hence, the scientific challenge can be conveniently classified into two parts: the definition of the modernization criteria and the selection of modernization model at the current stage of the scientific and technological revolution.

The so-called model of the "catching-up development" gained the most currency in global practice. Its main goal consists in achieving the economic level of the most developed countries of the world by backward and less developed countries. This model has two varieties – neokeynesian and neoliberal. Both of them were formed by different countries of the world in the course of modernization policy. Notably, different countries achieved their modernization goals more or less successfully in the application of these models. Both models have their own advantages and disadvantages, as well as their own mechanisms of modernization. For example, neokeynesian model is essentially based on the active involvement of the state, every possible expansion of the state sector of the economy, various protectionist measures, foreign exchange regulation etc. This modernization model, without doubt, can significantly increase the pace of economic growth, the degree of industrialization of the economy, to catalyze the scientific and technological revolution, but, on the other hand, reduces competitiveness of national commodity producers, requires significant government expenditures, curtails the freedom of market agents, and, naturally, there is a fairly large number of the so-called "failures of the state", when only the development of the free market is able to efficiently solve certain modernization tasks.

The role of the state is also high in the neoliberal model, but the mechanisms are different – an indirect impact of the state on certain vectors of modernization of the economy for the purpose of stimulation of the scientific and technological revolution, achievement of high economic growth, progressive structural changes, an increased degree of freedom of economic agents, but also with active foreign exchange regulation. Public sector undertakings are increasingly being privatized, promoting competition, reducing the tax burden for national commodity producers, increasing the freedom of entrepreneurs, maintaining exports and a low exchange rate of the national currency, etc.

Both modernization models led to a certain success at different stages of the scientific and technological revolution throughout the world, but they also revealed various problems. In addition, new problems and new time challenges for development occur at every stage of historical development. This is why the identification of the essence of the concept of modernization as such, modernization model depending on the historical stage of development, actualization of goals and criteria of modernization calls for further scientific research and substantiation.

2 Methodology

The analysis of studies of Russian and foreign researchers in this area, aimed at actualization of methodological and methodical issues of modernization changes in economic complexes of various social and economic systems, served as the basis for the identification of the problems of modernization. The authors used the results of the study of statistical, informational methods, methods of content analysis for the verification of factors and directions of modernization of the economy in academic literature of Russia and rest of the world. The research methodology itself is based on a system approach that combines the methods of induction, deduction, comparative analysis, identification of the general, the special and the particular, etc. Various ratings, statistical data of the Federal State Statistics Service of the Russian Federation and its local offices, the Ministry of Economic Development of the Russian Federation, as well as the content from the Internet were used as an information base; author's own estimates.

3 Results

Modernization as a process was studied by the research workers of various scientific areas – social scientists, political scientists, cultural specialists, economists etc. For example, it was as far back as at the times of Karl Marx who believed that all countries undergo certain stages of their development, and the transition from one stage to another in this process is only made in the process of modernization.

Max Weber was one of the first researchers who became interested in the process of modernization. He considered this term from the perspective of political science as a process of transformation of one political system of the state to another political system.

One of the first theoretical studies of the modernization process was carried out by Black (1975). In his book "The Dynamics of Modernization" S. Black stated that modernization must be studied as an interdisciplinary area, which can be the object of study of a wide range of sciences: history, politology etc.

A wide range of researchers of the modernization process studied it from the perspective of changes in the social structure of society, increasing the degree of democratization, transition from the less developed technological modes to the more developed technological modes etc. In this list, we can highlight Eisenstadt (1966), Ward (1963), Bendix (1964), Vago (1989).

A wide range of foreign researchers considered modernization only as a growth factor, which can impact, among other things, the process of democratization. Thus, for example, Martin Lipset considered modernization as the process of democratization of the society (Lipset 1959). Samuel Huntington made a considerable contribution in modernization theory, considering it as an opportunity for economic growth, which, in turn, helps to establish the democratic political regime. Therefore, Huntington treated modernization as the factor of democratization which reduces the level of poverty rate of population and leads to the economic growth (Huntington 1991). In 1997, Adam Przeworski and Fernando Limongi based on a study of 224 countries with different political regimes proved that the higher is the level of economic development which can be reached by the modernization process, the higher are democratic values of the society (Przeworski and Limongi 1997). Scientists Boix and Stokes in 2003 (Boix and Stokes 2003) and Arat (1988) arrived at approximately the same conclusions in their empirical research. In other words, all these research workers considered modernization as a growth factor and democratization of the society, but did not study any models of modernization, its criteria, methods and mechanisms.

The papers of American political scientists Sidney Verba and Lucian Pye deal with the study of various modernization theories from the perspective of political science, in which its components are considered. They studied the modernization process from the perspective of politology and distinguished the structural components of this process, in a manner of speaking, the characteristics of the process as such.

Many papers in social sciences deal with the process of modernization as the means of social change of the society. They include papers by Smelser (1971), Gellner (1993), Sztompka (2002) et al.

The interest in the area of study of modernization is attracted by the research in the PhD program in the political center of the University of New York (Political Science of the City University of New York), the key data of which are published in articles by Ana (2008) and Arat (1988), as well as the papers by James (1989) and Gordon (1985). These authors paid special attention in their papers to the relation between modernization and the processes of democratization of the society, globalization of the world economic system and progressive improvements.

However, all these papers of the Western research workers are mostly focused not on the identification of criteria and models of modernization, but on the demonstration of economic and political modernization of the society, on the correlation between the level of economic development, which can be sharply increased by means of economic modernization and the level of democratization of the society.

The papers of economic scientists in the area of modernization, its goals, factors, criteria are of greatest interest for us. A wide variety of approaches can be observed here as well.

Much of the research in the area of economic modernization dates back to the mid-20th century. During this period, many papers were written dealing with the study of the experience of modernization changes in the countries of Southern and Eastern Europe.

There is a fairly large amount of papers dealing with the modernization theory, its various aspects and trends, by research workers of our country. Some papers of the Russian research workers deal with the study of modernization processes occurring throughout the world, identification of its factors, force application vectors, and results. It is quite natural that these processes are usually studied from the perspective of opportunities for using the best practices of modernization for our country. For example, Vinogradov, A.V, in contrast with the Chinese experience, considers the modernization process, on the one hand, as the type of historical development, and on the other hand, as a part of the world globalization process; hence, external factors associated with globalization directly impact the internal modernization.

A wide range of research workers analyze the modernization in Russia, study its special aspects, factors etc. For example, a fairly large amount of studies deals with the modernization process itself as an opportunity to increase competitive power of the national economy by means of its transition to the innovation-based growth. In other words, modernization in this case is treated as the opportunity of transition to the innovation-driven growth model. However, this brings up the question: should the modernization model be innovative for the whole world or for this country only? After all, all countries of the world are at different development stages; a particular technological mode is dominant in them – the higher is the degree of development, the higher is the mode. If for one country modernization, which leads to the transition to the fifth technological mode, will be innovative, for another country it will be a stagnation in development.

A fairly large amount of work is designed to identify the concept of modernization as such. The papers of Gelman, V. (modernizations as measures of social and economic policies), Yasin, E. (modernization as the means of achieving competitive power in the world economic system), Kostenko, M. and Nikolsky, A. modernization as a technology of "catching-up development"), Krasilshchikov, V. (modernization as a set of changes allowing to achieve the economic level of the most developed countries of the world) deal with this issue. The difference of modernization from economic development consists in the fact that development means gradual evolutionary changes, and the basis of modernization consists of radical transformations, qualitative transition from one technological mode to another.

Kovryzhko, V.V. obtained interesting results in the process of the study of the modernization theory. According to her opinion, modernization of the economy is a least-evil solution; it is the consequence of the availability of certain problems of social and economic development of the society; furthermore, modernizations always cover large time intervals and start as "catching-up" modernizations. She writes that "modernization can only be industrial, since the mobilization mechanisms are inefficient within the framework of postindustrial society; mobilization can only bring the society to the "launching site" for the formation of the information economy, but cannot create it". Naturally, we disagree with this statement, since it turns out that currently those countries that have already made a transition to postindustrial society, will not need modernization anymore. We shall give further consideration to this issue.

Several Russian research workers believe that economic modernization is a type of historical development which enables the transition from traditional to the contemporary society. Modernization has its national peculiarities due to the economic lag from the most developed countries of the world, accompanied by the intensification of production, the introduction of innovations, the improvement of production management, the growth of labor differentiation. In our opinion, this definition is too generalized, especially as regards the modern and traditional types of society. These concepts will not be identical at every stage of development for different countries.

When considering the theory of modernization in our country, Fedotova, V.G. uses the process approach; in particular, she subdivides modernization itself into several vectors, consisting in the intensification of production, introduction of new technologies, creation of specialized clusters, increase in production of consumer goods, and change of the production system as such by means of informatization.

In general, the analysis of the level of knowledge of the problems of modernization has shown that modernization theory is a fairly new school of thought, thus further development of this theory is required. In particular, the question about the system of criteria and indicators of modernization of the economy, technique, methods and institutions of modernization is of interest.

First, we must come up with the decision what we mean by "modernization". We shall start from the beginnings, from the root of the word itself. The root of the word – modern (lat.). The dictionary interprets this term as "an update, an improvement that meets modern requirements and tastes, as a process of transformation of social, economic and political life". According to other definition, modernization acts as the update of the object and bringing it to correspondence with new requirements and standards. Definitions given by other authors also indicate innovative transformations. Hence, we can state that modernization is the process of application or introduction of innovations. Any innovation is based on a permanent scientific and technological revolution which can be observed in almost all countries of the world. It is quite natural that Economic modernization is the process of introduction of innovations in production, unlike other innovations and modernizations (political life of the society, social, culturological etc.). This process is aimed at improvements. What does the society want improvements for? The ultimate goal of existence of any society is the satisfaction of increasingly growing needs of people under certain resource constraints, that is, the society must satisfy the growing needs, increasing the efficiency of utilization of limited resources. This means that modernization is necessary for the society for improving resource efficiency. Of course, when considering the concept of modernization, many authors claimed that it is aimed at achieving the level of developed countries of the world in terms of economic growth. But this goal is too abstract; moreover, as for the most developed countries of the world – why should they carry out modernization and what references they should be guided by? In our opinion, economic modernization of any country at any historical stage of its development is aimed at increasing the resource efficiency, which, in turn, will lead to the increased competitive power of the national economic system. However, we aim not for any modernization, but for that kind of modernization which will lead to the increase in the quality of living; hence, it should pursue not only the growing needs of the population for the amenities, but also the preservation of the environment, i.e. sustainable development. Therefore, we can conclude that economic modernization is the process of introduction of innovations in production with a view to improving resource efficiency, increasing the competitive power of the economy and its sustainable development.

Another question is what innovations serve as the basis for modernization, that is, what is a technological base of modernization. If innovations are evolutionary, gradually improving, such that do not affect the basic technological mode, this is an evolutionary modernization. If innovations represent a complete change of the technological mode, occur in the form of the scientific and technological revolution, this is a revolutionary modernization. Depending on what type of modernization occurs – evolutionary or revolutionary – modernization criteria should be selected. Thus, specific indicators indicating an increase in the resource efficiency or a change in the technological mode should act as the criteria for modernization for all countries of the world. Of course, the best global practices in the employment of resources should be used as the frame of reference.

The basic types of resources are labor, land, and capital. How can we assess the efficiency of their use compared to the developed countries of the world? The problem consists in the peculiarities of statistics of economic indicators in Russia and other countries of the world. Modernization criteria should be comparable, simple and easily understandable in the course of study.

Every country has certain land resources which can be represented by the total area of the territory and the area of farmlands. According to figures for the 2017, the total area of the territory of Russia is 17125.2 thousand square km, the area of farmlands is 1712.5 million hectares. In our opinion, the land utilization efficiency can be assessed as follows: gross return on investment in land, land requirement of GDP, return on investment in agriculture and agricultural requirement. Gross return on investment in land is determined as the quotient from the division of the total GDP by the total area of the territory; in turn, the land requirement of GDP shows: what amount of land is used for the generation of a unit of GDP, i.e. the quotient from the division of the total land area by the volume of GDP.

Return on investment in agriculture is determined as the quotient from the division of agricultural output by the total area of farmlands. Agricultural requirement shows: what volume of farmlands is needed for the production of a unit of agricultural products.

As for the next resource – labor – we can determine labor efficiency by in terms of GDP and personnel requirement of GDP. Productivity is the quotient from the division of the total GDP and the average annual number of employees, whereas personnel requirement of GDP is a reciprocal.

We can assess the capital utilization efficiency by the utilization of investments to obtain GDP. Investment requirement of GDP shows the ratio between the volume of investment and the volume of GDP, and the return on investment – the ratio between the volume of GDP and the volume of investment. Investment requirement of GDP shows efficiency of investments, in the country, from the perspective of their impact on the volume of GDP. And the indicator of the return on investment designates the volume of GDP that is required to receive the unit of investment.

The results of estimated modernization criteria put forward by us are presented in Table 1.

Table 1. Indicators for determining the criteria of modernization of the economy for Russia in 2017

Indicators	2017
Gross return on investment in land, thousand rubles per 1 km^2	5.37
Land requirement of GDP, square km per 1 billion rubles	0.19
Return on investment in agriculture, thousand rubles per ha	24.8
Agricultural requirement, ha per thousand rubles	0.04
Labor efficiency by in terms of GDP, thousand rubles per person	1.28
Personnel requirement of GDP, persons per thousand rubles	0.78
GDP return on investment, rubles per 1 ruble of GDP	0.17
Investment requirement of GDP, rubles	5.76

In order for these criteria to allow assessment of modernization processes, there must be reference values of these criteria. In our opinion, the criteria of global development practices should be taken as reference criteria. Of course, all monetary criteria should be recalculated in dollar terms for international comparisons. However, there are certain difficulties in terms of methodology, which prevent from making international comparisons by all modernization criteria put forward by us. For example, it is impossible to compare the indicators of yield on capital investments and capital/output ratio. This is due to the fact that regulatory framework of Russia according to the funds valuation procedure differs from international standards, in particular, for example, IFRS standards in Germany and GAAP in the United States.

The most developed countries of the world include United States, Japan, United Kingdom, France, Italy, Canada, etc. The results of modernization criteria put forward by us are presented in Table 2.

Table 2. Criteria of modernization of the economy compared to the developed countries of the world

Indicators	Russia	United States	Japan	United Kingdom	France	Italy
Gross return on investment in land, thousand dollars per 1 km^2	74.24	2025.86	11944.37	11784.17	4637.8	6308.5
Land requirement of GDP, square km per 1 million dollars	13.47	0.49	0.08	0.08	0.22	0.16
Return on investment in agriculture, thousand dollars per square km	39.11	49.64	1293.58	87.39	158.13	351.69
Agricultural requirement, square km per 1 million dollars	25.57	20.15	0.77	11.44	6.32	2.84
Labor efficiency by in terms of GDP, thousand dollars per person	16.79	124.49	69.14	90.44	85.65	75.20
Personnel requirement of GDP, persons per thousand dollars	0.06	8.03	14.46	11.06	11.68	13.30
GDP return on investment, dollars per 1 dollar of GDP	4.7	4.9	4.6	5.5	4.9	6.0
Investment requirement of GDP, dollars	0.21	0.20	0.22	0.18	0.20	0.16

The data that we have presented in Table 2, show that the efficiency of utilization of the basic types of resources in Russia is much lower compared to the developed countries of the world. The only exception is provided by investment the utilization efficiency of which is only worse in Japan as compared to Russia.

However, general conclusions clearly cannot be made on the basis of return on investment and investment requirement of GDP. These criteria clearly indicate the need of modernization of the country's economy. Specific mechanisms and models of modernization call for a separate research.

4 Conclusions

Hence, we have found that the process of modernization is the process of application or introduction of innovations. Any innovation is based on a permanent scientific and technological revolution which can be observed in almost all countries of the world. Economic modernization is the process of introduction of innovations in production with a view to improving resource efficiency, increasing the competitive power of the economy and its sustainable development. Depending on what innovations serve as the basis for modernization, it will be evolutionary or revolutionary, which leads to the change of technological mode. If innovations represent a complete change of the technological mode, occur in the form of the scientific and technological revolution, this is a revolutionary modernization. If innovations are evolutionary, gradually improving, such that do not affect the basic technological mode, this is an evolutionary modernization. Specific indicators indicating an increase in the resource efficiency or a change in the technological mode should act as the criteria for modernization for all countries of the world. The best global practices in the employment of the basic types of resources - land, labor and capital - should be used as the frame of reference. In order to enable comparison with the best global practices, modernization criteria should be comparable, simple and easily understandable in the course of study. We have put forward the criteria for modernization of the economy which are relevant for any stage of the scientific and technological revolution and any type of modernization – either evolutionary or revolutionary. They include such indicators of resource efficiency as: gross return on investment in land, thousand dollars per 1 km^2; land requirement of GDP, square km per 1 million dollars; return on investment in agriculture, thousand dollars per square km; agricultural requirement, square km per 1 million dollars; labor efficiency by in terms of GDP, thousand dollars per person; personnel requirement of GDP, persons per thousand dollars; GDP return on investment, dollars per 1 dollar of GDP; investment requirement of GDP, dollars.

Modernization criteria put forward by us present an example, a goal which should be pursued in the utilization of resources. Specific models of achievement of these goals depend on every historical stage of development, scientific and technological revolution, on the technological mode which is currently dominant in a particular country, on the condition of its infrastructure etc., and calls for further research.

References

Black, C.: The Dynamics of Modernization. A Study in Comparative History, NY (1975)

Eisenstadt, S.: Modernization: Protest and Change, Englewood Cliffs (1966)

Ward, R.: Modern Political Systems, NJ (1963)

Bendix, R.: Nation-Building and Citizenship, NY (1964)

Vago, S.: Social Change, NJ (1989)

Lipsct, S.: Some social requisites of democracy economic development and political legitimacy. Am. Polit. Sci. Rev. **53**(1), 69–105 (1959)

Huntington, S.: The Third Wave: Democratization in the Late Twentieth Century. University of Oklahoma Press, Norman (1991)

Przeworski, A., Limongi, F.: Modernization: theories and facts. World Polit. **49**(2), 155–183 (1997)

Boix, C., Stokes, S.: Endogenous democratization. World Polit. **55**(4), 517–549 (2003)

Arat, F.: Democracy and economic development: modernization theory revisited. Comparative Politics, Ph. D. Programs in Political Science. City University of New York, vol. 21, no. 1, pp. 21–36 (1988)

Smelser, N.J.: Sociological Theory: A Contemporary View, NY (1971)

Gellner, E.: The Coming of Nationalism, Storia d'Europa, vol. 1 (1993)

Sztompka, P.: Socjologia. Analiza społeczeństwa, Krakow (2002)

Ana, M.: Preemptive Modernization and the Politics of Sectoral Defense, Adjustment to Globalization in the Portuguese Pharmacy Sector. Comp. Polit. **40**(3), 253–272 (2008)

Zehra, F.: Arat democracy and economic development: modernization theory revisited. Comp. Polit. **21**(1), 21–36 (1988)

Digitalization of Kyrgyz Society: Challenges and Prospects

Chinara R. Kulueva[1(✉)], Mirlanbek B. Ubaidullayev[1], Kurmanbek I. Ismanaliev[1], Victor P. Kuznetsov[2], and Elena V. Romanovskaya[2]

[1] Osh State University, Osh, Kyrgyz Republic
ch.kulueva@mail.ru
[2] Minin Nizhny Novgorod State Pedagogical University, Nizhny Novgorod, Russia
kuzneczov-vp@mail.ru, alenarom@list.ru

Abstract. This scientific article is devoted to the issues of digitization of the Kyrgyz Republic, where problems and prospects of building a society in the new stage of development of the country in conditions of globalization. The main purpose of our work is aimed at theoretical coverage of opinions of scientists on the new technological way, its advantages, and disadvantages, as well as on the necessity of transition to digitalization. Digital society is researched and disclosed by scientific schools as a specific direction of artificial development of societies, thus they made certain clarifications in terminology and substantiated a new conceptual apparatus, which is the basis of expansion of further philosophical and economic researches of modern world processes. The role and significance of the stage of digital transformation, which will be implemented in our country in the presence of scales and more qualitative level of capacities of science, technology, and technosphere. Consequences of introduction of digitalization on socio-economic processes, where under the influence of high-tech and scientifically grounded productive forces there is not only gradual industrialization of the economy but also a radical change in the nature of technical and economic relations and adaptation of artificial conditions of society life. This study considers the government program "Digital Kyrgyzstan", which provides creation of favorable bases for the effective digitalization of the country within the framework of the Concept of the digital program transformation. It set ambitious goals and objectives aimed at complementing and expanding previously adopted programs and thereby defining the structure, management system and foundations of the digitalization process. In conclusion, suggestions and recommendations are given, relevant conclusions are drawn on the issues under consideration.

Keywords: Digitalization · Technological way · Technologies · Skills · Technosphere

E. G. Popkova and B. S. Sergi (Eds.): ISC 2019, LNNS 91, pp. 229–236, 2020.
https://doi.org/10.1007/978-3-030-32015-7_26

1 Introduction

The modern world in conditions of globalization is accompanied by the rapid development of mass communication and in our minds firmly entrenched concepts of "digitalization", "digital society", "informatization", "information technologies" etc., which at first glance bear some complexity in perception, although they prove to us their "simplicity" in their practical application, cardinally changing the social existence. Process of visualization through created and widespread personal computers, globalization of built digital and computerized systems, developed the latest technologies of social networks, virtual reality characterizes the formation of a new stage of development of society and economy in the XXI century. In turn, the new course taken on development of a digital society in the global community with universal integration and intercultural communication, in our view, should take into account implications of data development processes. Since cardinal changes in the processes of socialization, identification, communication allows us to speak about the significant changes in society and its cultural life affecting directly the self-development and introduction of innovative advanced technologies.

2 Methodology

The method of thematic research of processes in natural conditions was used as the methodological base of the research.

3 Result

According to the results of prospective studies, experts came up with the following forecasts, where the lion's share of the most significant events falls on the development of the Internet of Things (IoT): more than 50% of Internet traffic will be provided by mobile applications, including implementation of state-level programs "smart home" and "smart city"; 10% of the population will satisfy their clothing needs via the Internet, 10% of reading points will also be connected to the global network. Cloud technologies will acquire a wide scale, where almost 90% of the people on the planet will have a real possibility of unlimited and free storage of information data. About 90% of the population will have smartphones and permanent access to the Internet. With the development of 3D printing, there will be new potential opportunities for improving surgical medicine and medical engineering, as well as automotive, home building, consumer goods, development of robotization of many production processes, artificial intelligence and much more (Schwab 2017).

To date, this sphere is beginning to be viewed over all aspects of society as dominant and defining. Rapid acceleration of digitalization processes and the overthrow of human consciousness by various information greatly aggravates the process of digestion of changes taking place, and human brain in depending on acquired competencies, acquired skills, and human needs should "sort" them as needed. Hourly, unless necessary and unnecessary information is received every minute, therefore a person comes to the rescue of digitalization, with the help of which an individual has the opportunity to select necessary information using special software. Previously, to select necessary information used the method of linear programming (Lagrange, Euler), but times change and such serious things are done now with the help of digital technologies, what else time shows the exit of human consciousness from the usual framework of existing concepts, where long-term forecasts of social space were built taking into account analytical tools.

It should be noted that "technological development has accelerated significantly in recent decades. If earlier the change of technological ways occurred over the life of several generations, today during the lifetime of one generation several technological ways are replaced, based on which [5] lies the tightening of competition in the cost position and economic globalization.

There is no doubt and exaggeration that today digitalization and information technology play an important role in the life of each of us. Today on the person who doesn't have the mobile phone or the computer society considers to "eccentric" category of people. This proves once again the penetration of digital and information technologies into our daily lives that sometimes we have to abandon the outside world, replacing it with technical devices.

Modern economic science focuses its efforts on the study of modified economic relations, where the object is information nature and digitalization of society. Professionals and the public are influenced by continuous flows that generate new effects and are not yet explained by science because of their susceptibility to rapid change, which cannot be absolutized it as institutional theory, as this phenomenon is a venture, quite often variable and is in constant renewal. In other words, with the increase in the number of so-called "positive" behaviors that do not meet the established requirements of their reliability, relevance and completeness, the number of patterns of behavior that aim to deliberately distort information and use deliberately distorted products as a tool or mechanism (hackers, economic crimes in cyberspace, false information, anonymous letters, misinformation, etc.). As a result, economic entities: firms, companies, enterprises of various forms of ownership, even large corporations are forced to suffer large losses due to planned misinformation caused by unfair competition.

Based on the above, it can be noted that economic science cannot remain distant from the above problems, which determines a new round of research of science.

Kyrgyzstan in the Context of Adaptation to the Digital Society: Problems. In the context of globalization of the world community, Kyrgyzstan is intensifying its position every year in building a "digital community" as the basis for the development of the national economy in the twenty-first century. For this purpose, the National Development Strategy of the Kyrgyz Republic for 2018–2040 was approved by the Decree of the President of the Kyrgyz Republic on November 1, 2018, where within the framework of this strategy was also considered and approved the National Program of Digital Transformation "Digital Kyrgyzstan" - 2019–2023". In our view, the targeted creation of favorable foundations for an effective digitalization of Kyrgyzstan within the framework of the Digital Transformation Program Concept "Digital Kyrgyzstan" has set ambitious goals, aimed at complementing and expanding previously adopted programs and thus defining the structure, management system and the basis of the digitalization process to be generated in the following aspects:

- Development of non-digital and digital foundations of transformation;
- Construction of infrastructure of the international level;
- Universal broadband Internet access throughout the country;
- Ensuring the most efficient, reliable and inexpensive connection of the country with global networks and data networks;
- Focus on the development of infrastructure for data collection, processing and analysis;
- Use of cloud technology, regional data centers to achieve the level of driver international partnership, with counter offer reliable and inexpensive computational storage capacity [9].

In fact, it is evident that the modern digital infrastructure, adaptable and widely used by consumers, allows to form innovative platforms and platforms of interaction of three parties: the first - government authorities, the second - private sector, and the third - citizens and population (Fig. 1).

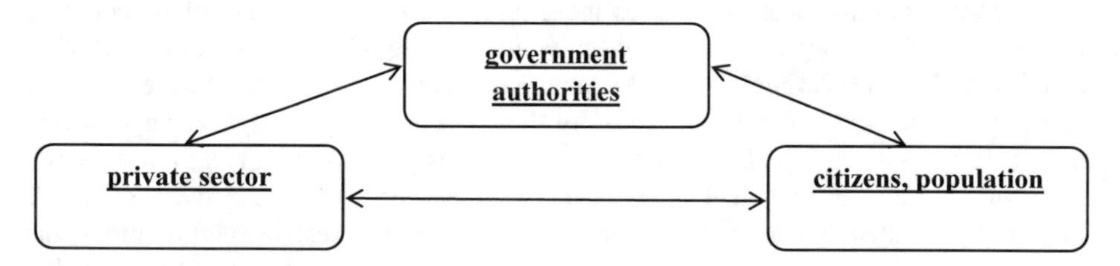

Fig. 1. Interaction of modern digital infrastructure of Kyrgyzstan

Today, according to our vision, digital development of the Kyrgyz Republic and the adaptation of associated digital technologies will contribute to improving public administration and improving the quality of services provided for citizens and community. Therefore, drastic measures are required to develop and utilize the information and digital potential.

Since 2019, the country has launched a company under the motto "Developing regions", where the main focus will be on supporting local business and introducing digital infrastructure at all levels and structures.

To achieve this goal, the goal is to provide the whole country with broadband Internet, where particular importance is attached to the issues of financing by economic entities of the cost of using and development of IT technologies, as well as digitalization processes (Fig. 2).

In Kyrgyzstan's banking sector, digital technologies offer a number of advantages to commercial banks, consisting of a much rapid transfer of funds to accounts, cost of carrying out all types of banking operations, both for the bank and for the client, thereby reducing the risk of not paying payments. The 24-h operation of the virtual bank creates a certain convenience, firstly, without visiting a financial institution, secondly, having credit cards to make small purchases, and thirdly, save time from order to delivery of goods.

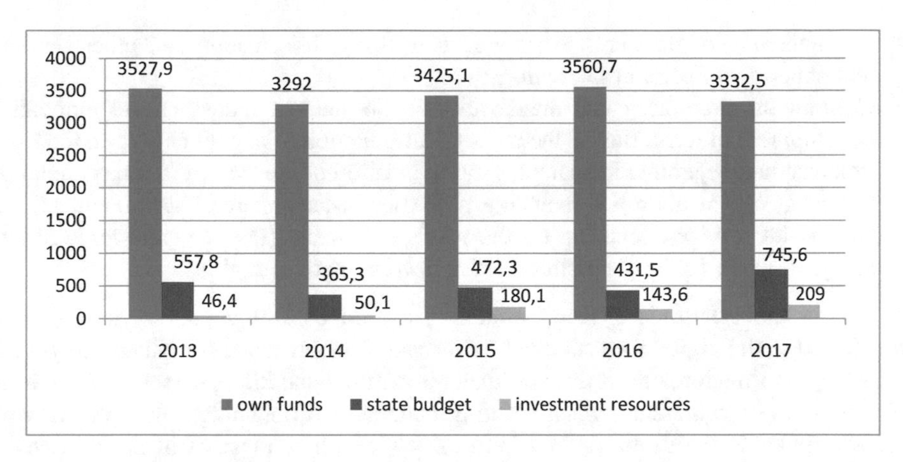

Fig. 2. Financing of expenses of RFS enterprises on use and development of IT-technologies, digitalization of processes (million soms *) **As of 15.02.2019, 1 dollar = 69.85 soms*

Consequently, with the development of digital and information technologies, the distance between producer and consumer of banking services is significantly reduced. Significantly increased interbank competition, thereby increasing the quality of banking services. This once again proves the rapid development of remote banking services for clients, such as Internet-banking services, mobile banking services, which can be used on the distance to make payments, pay for services and have access to other popular services of banks.

By the end of 2023, digital technologies will be introduced into the structures of education, health, tourism, agriculture, telecommunications, banking, light industry, and construction. Digitalization should reach a level where each farmer will have a real opportunity to get access to information about a state of markets and forecast prices for products in the target markets sales through Internet services [10].

Developed concept reflects 23 recommendations on digitalization sent to the government of the Kyrgyz Republic, the most significant of them are [9].

- Creation and implementation of unified digital platforms "Sanarip Kyrgyzstan" with a unified system of identification, electronic messages, digital payments, system of electronic interdepartmental interaction "Tunduk", portal and mobile application of electronic public services;
- Creation and implementation of the digital platform of the municipal management "Sanarip aimac" and regional centers (hubs) of competence;
- Wide introduction of information technologies in the field of education and health services with transfer of payments and payments to non-cash form in electronic format;
- Development of a system of classification of professions in the field of information technologies with the introduction of forecasting processes of promising innovative and high-tech professions, with development and introduction of training programs that meet international standards.

– Strengthening of measures to provide consulting, information and other services and support to SMEs in the regions;
– Continuous monitoring and diagnostics of the state of industrial and innovative development of territories on the basis of introduction of annual ratings on the main financial and economic indicators; - modernization of the system of transparency in planning, organization and financing processes in the sphere of social, educational and health services, focusing on the needs of citizens, demographic forecast and analysis, while creating possibilities for development of regions.

As for the electronic system "Tunduk", a total of 189 public services will be transferred to the digital format, which will make it much easier for citizens to get the necessary information, this or that certificate of criminal liability, payments for the birth of a child, upon retirement, when granting benefits and obtaining a passport, driving license, civil acts, registration of transport means, which as a result will reduce cases of direct contact with government officials in order to eliminate corruption elements.

In this case, we mean the creation of blocks of a digital state, such as "e-government", "e-parliament", "e-justice", "Digital economy", on which work competent specialists.

In many universities began to practice scientific and practical conferences of the High Technologies Park of Kyrgyzstan devoted to innovations in public administration and services, problems of acceleration process of digital transformation and digitization of society, discussion of the latest digital technologies, development of IT products, as well as new quality public services (Table 1).

Table 1. Proposed catalogue of public services for citizens of the Kyrgyz Republic

Digitalized Catalog of Public Services KR	
BLOCK: passport	Passport readiness
	Passport check
	All services
BLOCK: family and children	Check of marital status
	Marriage registration
	All services
BLOCK: address registration	Check of registration address
	Registration at the place of residence
	All services
BLOCK: transport	Inspection of fines
	Check on the date of passing the test
	All services
BLOCK: real estate	History of real estate rights
	Information on the technical characteristics of real estate
	All services
BLOCK: for legal entities	Extract from the Unified State Register
	All services

(continued)

Table 1. (*continued*)

Digitalized Catalog of Public Services KR	
BLOCK: education	Authentication of the diploma
	Certificate authentication
	All services
BLOCK: registration of foreign citizens	Extension of temporary registration
	All services

Compiled by the authors

4 Conclusion

In conclusion, when summarizing results of the main problems and conditions of development of digital technologies in the Kyrgyz Republic, it should be noted that scientifically based structural policy should make the program of digitization as the main nucleus of changing the structure of society, in particular the economic one, which consequently leads to formation of new factors and motivations for economic growth, requiring an appropriate level of preparation of analytical decisions and introduction of methods of analytical planning, requiring coordinated development of optimal solutions, their adoption and implementation in various development programs, which should take into account precisely calculated resources and their efficiency. In this case, the policy of project approach within the framework of specific programs and concepts, acting as the main method of planning. Makes it possible to assess the payback of implemented projects, which takes into account not only the invested budget, but also the degree of influence of new social functions, creation of new jobs, improvement of personnel skills, qualitative change in living standards, which in fact is provoked by information and digital technologies, requiring the production of settlement operations full economic efficiency of developed and implemented government programs for digitalization of society.

Therefore, the course taken on the digitization of society in Kyrgyzstan in parallel should be implemented:

- Active introduction of e-democracy technology;
- Involving citizens and business structures in the process of modeling digital management;
- Ensuring transparency in adoption of optimization decisions;
- Increase in digital education of the population;
- Facilitating access and quality of public services;
- Establishing a system of interdepartmental interaction “horizontally” and “vertically”;
- Raising the level of competence at all levels of government structures.

References

1. Dergacheva, E.A.: The Concept of Social and Natural Globalization: Interdisciplinary Analysis, 256 p. (2016)
2. Dergacheva, E.A., Baksansky, O.E.: Information Society and Challenges of the NBICS Convergence. Collection of articles I All-Russian scientific-practical conference on the theme "Challenges of the digital economy: conditions, key institutions, infrastructure", Bryansk, 21–22 March 2018, pp. 14–22 (2018)
3. Kaftan, V.V.: Philosophy of modern information society. Financial University under the Government of the Russian Federation. Bulletin of the Moscow Regional State University. Series: Philosophical Sciences, no. 2, p. 40 (2017)
4. Bell, D.: The Coming Post-Industrial Society. Experience of Social Forecasting, pp. 155–162 (1999)
5. Sadovaya, E.C.: The man in the digital society: dynamics of social and labor relations. South Russ. J. Soc. Sci. **19**(3) (2018)
6. Concept of the Kyrgyz Republic digital transformation "Digital Kyrgyzstan" - 2019–2023. http://ict.gov.kg/index.php?r=site%2Fsanarip&cid=27
7. Demidenko, E.S., Dergacheva, E.A.: Socio-philosophical analysis of formation and development of the concept of technogenic society. Modern Problems of Science and Education, no. 2 (2015). http://www.science-education.ru/131-23481
8. Roco, M., Bainbridge, W. (eds.): Converging Technologies for Improving Human Performance: Nanotechnology, Biotechnology, Information Technology and Cognitive Science, Arlington (2004). http://www.transhumanism-russia.ru/content/view/498/116/ednref1
9. https://ru.sputnik.kg/politics/20181214/1042430211/kyrgyzstan-sovbez-zasedanie-cifrovoe-razvitie.html
10. https://economist.kg/2019/01/11/zheenbekova-dal-23-rekomendacii-pravitelstvu-v-god-razvitiya-regionov-i-cifrovizacii-spisok/
11. https://academicimpact.un.org/ru/content/

Technological Revolution in the 21st Century: Factors, Conditions and Perspectives

Influence of Technological Revolution in the Sphere of Digital Technologies on the Modern Entrepreneurship

Elena B. Belik[1(✉)], Elena S. Petrenko[1], Georgy A. Pisarev[2], and Anna A. Karpova[3]

[1] Plekhanov Russian University of Economics, Moscow, Russia
belikilim@yandex.ru, petrenko_yelena@bk.ru
[2] Russian Presidential Academy of National Economy and Public Administration, Moscow, Russia
pisarev.georgi@gmail.com
[3] Volgograd State Agrarian University, Volgograd, Russia
anna-karpova-72@mail.ru

Abstract. Purpose: The purpose of the article is to determine the influence of technological revolution on modern entrepreneurship, which implements various strategies of using digital technologies and to determine the most preferable strategies.

Design/methodology/approach: Based on statistical information on usage of digital technologies in entrepreneurship, which was provided in open access by the National Research University "Higher School of Economics", the structure of entrepreneurship within the offered authors' classification according to the criterion of implemented strategies of using digital technologies in Finland, France, the Czech Republic, Sweden, the UK, and Russia in 2018 is offered. The authors also determine the consequences (with the help of analysis of variation) of selecting the strategy of using digital technologies for development of entrepreneurship, connected to efficiency, export activities, and provision of competitiveness.

Findings: The authors determine that activity and directions of usage of digital technologies in entrepreneurship in the conditions of technological revolution depend on a range of external (demand, infrastructure, state regulation, and competition) and internal (inclination for innovations) factors. In modern economic systems in entrepreneurship, four main strategies of usage of digital technologies are implemented: strategy of ignoring (refusal from using digital technologies), strategy of formal participation in technological revolution (usage of the Internet for certain business operations), strategy of expanded participation in technological revolution (usage of the Internet for purchases and sales), and strategy of leading technological revolution (usage of the Internet and the leading digital technologies).

Originality/value: It is proved that the most preferable strategies are strategies of leading technological revolution and expanded participation in technological revolution, as they allow increasing export and increasing competitiveness and efficiency of entrepreneurship. Contrary to this, separation from technological revolution (implementing the strategy of ignoring or strategy of formal usage of the Internet) leads to reduction of competitiveness and efficiency and to reduction

E. G. Popkova and B. S. Sergi (Eds.): ISC 2019, LNNS 91, pp. 239–246, 2020.
https://doi.org/10.1007/978-3-030-32015-7_27

of export. It is concluded that technological revolution in the sphere of digital technologies creates new opportunities for development of entrepreneurship, which have to be used.

Keywords: Technological revolution · Digital technologies · Entrepreneurship · Digital modernization

JEL Code: L26 · O31 · O32 · O33 · O38

1 Introduction

The modern entrepreneurship functions and develops in the conditions of technological revolution in the sphere of digital technologies. Revolutionary changes in the entrepreneurial environment are manifested, firstly, in expanded opportunities for communications and optimization of information management on the basis of formation of electronic data bases, which eliminate the necessity for storing physical carriers of information (e.g., with the help of cloud technologies) and their computer processing (e.g., with the help of Big Data).

Secondly, a manifestation of technological revolution in the sphere of digital technologies is new perspectives of authomatization of entrepreneurial processes: production (based on robototronics and "smart" conveyor technologies), distributive (based on online trade), accounting (based on electronic document turnover), financial (based on online payments), logistical (based on ERP-technologies), and marketing and sales (based on RFID- and CRM-technologies).

However, the influence of technological revolution in the sphere of digital technologies on modern entrepreneurship is ambiguous. Apart from the above opportunities for development and increase of effectiveness of entrepreneurial activities, digital technologies create additional entrepreneurial risks – e.g., risks of digital security, risks of return of investments into purchase and implementation of digital technologies into entrepreneurial activities, etc.

This causes a scientific and practical problem of determining the causal connections of development of modern entrepreneurship in the conditions of technological revolution depending on activity and directions (i.e., strategy) of usage of digital technologies. The purpose of the paper is to determine the influence of technological revolution on modern entrepreneurship, which implements various strategies of using digital technologies, and to determine the most preferable strategies.

2 Materials and Method

Specifics of development of modern entrepreneurship in the conditions of technological revolution are studied in the works Bezrukova et al. (2017), Pritvorova et al. (2018), Sergi et al. (2019), Sibirskaya et al. (2019), and Vanchukhina et al. (2018). The perspectives of using digital technologies in entrepreneurial activities are discussed in the works Morozova et al. (2018), Popkova (2019), and Bogoviz et al. (2019).

The literature overview showed that though separate issues of development of modern entrepreneurship in the conditions of technological revolution are studied in the scientific works of certain scholars and experts, the strategies of using digital technologies are poorly studied. Thus, it is expedient to study the influence of technological revolution on modern entrepreneurship through the prism of various strategies of using digital technologies.

As a result of considering the existing modern global practice of using digital technologies in entrepreneurship, we determined the four following strategies of this process:

- strategy of ignoring technological revolution, which envisages refusal from using digital technologies;
- strategy of formal participation in technological revolution, which envisages using the Internet for separate business operations;
- strategy of expanded participation in technological revolution, which envisages using Internet for purchases and sales;
- strategy of leading the technological revolution, which envisages using the Internet and the most leading (RFID, ERP, CRM) digital technologies.

Based on statistical information on usage of digital technologies in entrepreneurship, provided in open access by the National Research University "Higher School of Economics", the structure of entrepreneurship within the offered authors' classification according to the criterion of implemented strategies of using digital technologies in Finland, France, the Czech Republic, Sweden, the UK, and Russia in 2018 is offered. Also, the consequences of selecting the strategy of usage of digital technologies for development of entrepreneurship, connected to efficiency, export activities, and provision of competitiveness, are determined. The initial statistical and analytical indicators are shown in Table 1.

Table 1. Indicators of the influence of technological revolution in the sphere of digital technologies on modern entrepreneurship.

Indicator	Finland	France	Czech Republic	Sweden	UK	Russia
Initial statistical indicators						
Internet access in organizations, %	100	100	98	98	95	86
Usage of the Internet for purchases in organizations, %	39	29	57	28	27	18
Usage of The Internet for sales in organizations, %	21	17	24	29	20	12
Usage of cloud services in organizations, %	66	17	22	48	35	23
Usage of RFID-technologies in organizations, %	23	11	8	12	8	6
Usage of ERP-systems in organizations, %	39	38	28	31	19	19
Usage of CRM-systems in organizations, %	39	28	19	32	32	13

(*continued*)

Table 1. (*continued*)

Indicator	Finland	France	Czech Republic	Sweden	UK	Russia
Export of goods, USD million	68073.5	535188.4	182143	153109.51	441106.35	353548
GDP, USD million	252301.84	2582501	215914	535607.39	2637866.34	1578417
Efficiency of labor, USD per hour per 1 employee	55.45	59.82	35.3	61.74	52.5	24.1
Index of global competitiveness, points 1-7	5.49	5.18	4.77	5.52	5.51	4.64
Analytical indicators						
Entrepreneurship that does not use digital technologies	0	0	2	2	5	14
Entrepreneurship that uses the Internet for separate business operations	56	72	66	58	67	68
Entrepreneurship that uses the Internet for purchases and sales	21	17	24	28	20	12
Entrepreneurship that uses the Internet and the most leading (RFID, ERP, CRM) digital technologies	23	11	8	12	8	6
Export of goods, % of GDP	26.98	20.72	84.36	28.59	16.72	22.40

Source: compiled by the authors based on National Research University "Higher School of Economics" (2019), OECD (2019), World Bank (2019), World Economic Forum (2019).

3 Results

The determined structure of entrepreneurship according to the criterion of implemented strategies of using digital technologies in selected countries in 2018 is shown in Fig. 1.

As is seen from Fig. 1, the strategy of ignoring the technological revolution is not implemented in Finland and France. In the Czech Republic and Sweden, the share of companies that do not use digital technologies constitutes 2%, in the UK – 5%, and in Russia – 14%. The strategy of formal participation in technological revolution is the most popular one in the considered countries. The share of companies that use the Internet for separate business operations in Finland constitutes 56%, in France – 72%, in the Czech Republic – 66%, in Sweden – 58%, in the UK – 67%, and in Russia – 68%.

The strategy of expanded participation in technological revolution is the second popular strategy in the studied counties. The share of companies that use the Internet for purchases and sales in Finland constitutes 21%, in France – 17%, in the Czech Republic – 24%, in Sweden – 28%, in the UK – 20%, and in Russia – 12%. Strategy of leading the technological revolution is most popular in Finland, where the share of companies that use the Internet and the leading (RFID, ERP, CRM) digital technologies constitutes 23%, in Sweden – 12%, and in France – 11%. It has a very low level of implementation in the Czech Republic (8%), the UK (8%), and Russia (6%).

The consequences of influence of technological revolution on the companies that implement various strategies of using digital technologies are shown in Fig. 2.

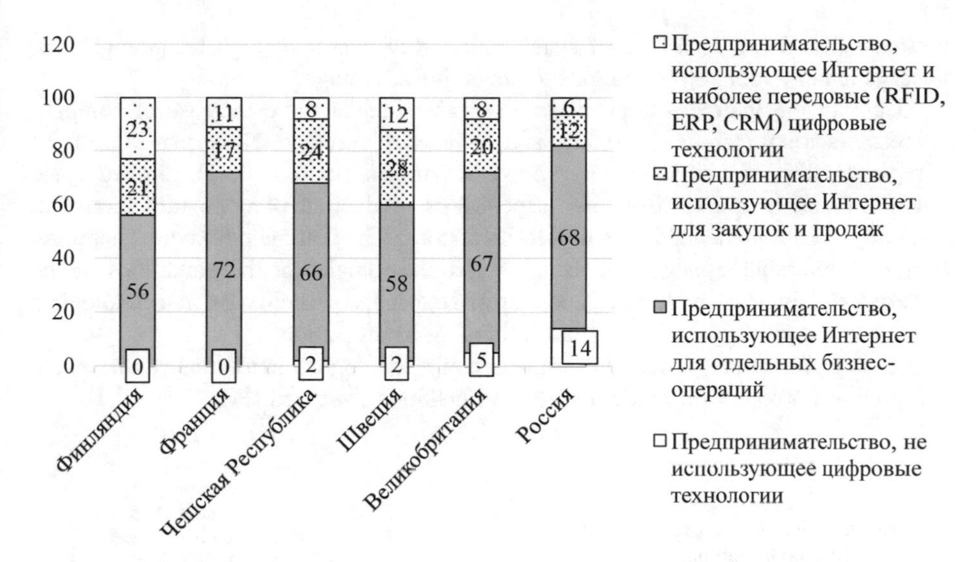

Fig. 1. Structure of entrepreneurship according to the criterion of implemented strategies of using digital technologies in selected countries in 2018. Source: calculated and compiled by the authors.

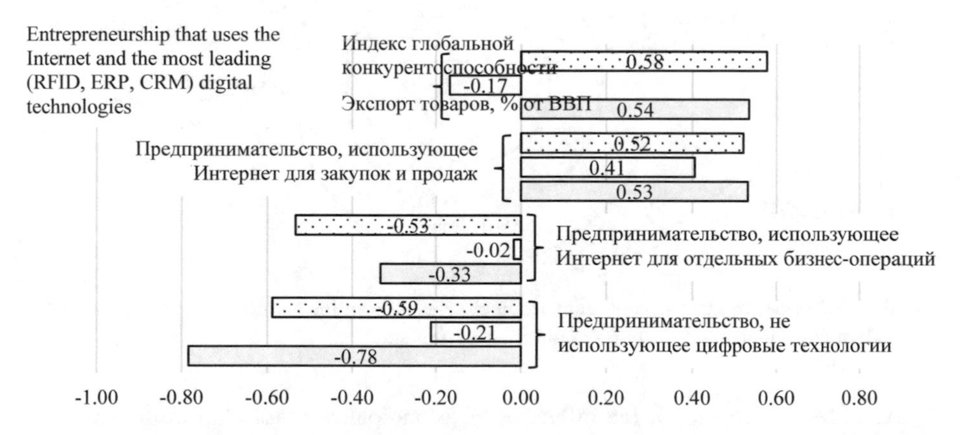

Fig. 2. Autocorrelation of competitiveness, export, and efficiency of entrepreneurship depending on the implemented strategy of using digital technologies in 2018. Source: calculated and compiled by the authors.

As is seen from Fig. 2, in the conditions of technological revolution, the most active and diverse the usage of digital technologies the more successful the entrepreneurship. Usage of the Internet and the most leading (RFID, ERP, CRM) digital technologies stimulates the increase of competitiveness (autocorrelation – 0.58) and efficiency (autocorrelation – 0.54) of entrepreneurship. At the same time, its export activity reduces moderately (autocorrelation – 0.17), which is probably caused by growth of internal demand and measures of state limitation of export of high technologies for preservation of national technological competitive advantages.

Usage of the Internet for purchase and sales stimulates the increase of competitiveness (autocorrelation – 0.52) and efficiency (autocorrelation – 0.53) of entrepreneurship, as well as increase of export (autocorrelation – 0.41). Limited usage of the Internet or refusal from its usage lead to reduction of competitiveness and efficiency of entrepreneurship, as well as reduction of the volume of export. Therefore, the most preferable strategies of using digital technologies in the conditions of the modern technological revolution are strategy of expanded participation in technological revolution and strategy of leading the technological revolution.

Based on the above, a model of influence of technological revolution in the sphere of digital technologies on modern entrepreneurship is compiled (Fig. 3).

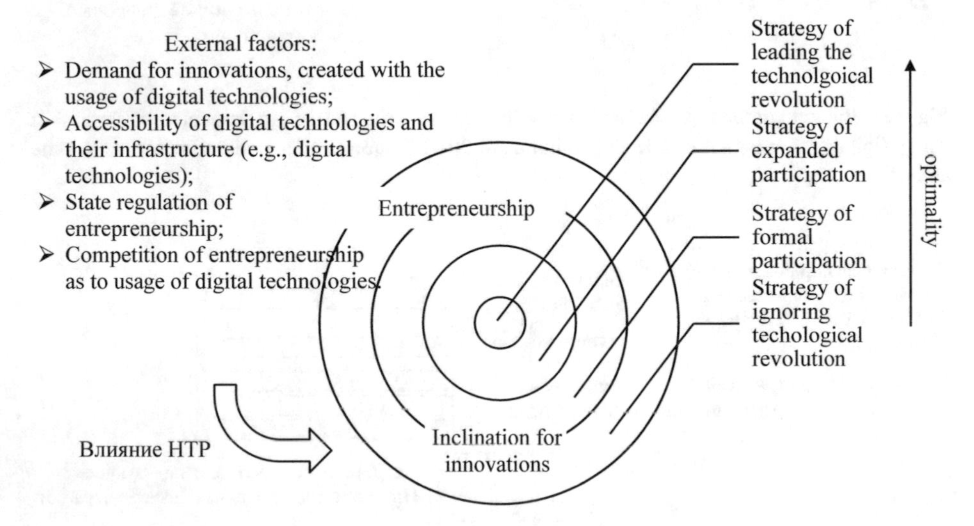

Fig. 3. The model of influence of technological revolution in the sphere of digital technologies on modern entrepreneurship. Source: compiled by the authors.

As is seen from Fig. 3, the influence of technological revolution in the sphere of digital technologies on modern entrepreneurship and selection of the strategy of using these technologies depend on the internal factor – inclination of each separate company for innovations (specifics of corporate culture, innovative activity of owners, management, and employees, sufficiency of financial and other resources, inclination for risk, etc.) – and external factors. They include demand for innovations that are created with the usage of digital technologies, in company's target sectorial market, accessibility of digital technologies and their infrastructure (e.g., digital personnel), state regulation of entrepreneurship (e.g., stimuli for using digital technologies), and competition of entrepreneurship as to usage of digital technologies.

4 Conclusion

Thus, activity and directions of using digital technologies in entrepreneurship in the conditions of technological revolution depend on a range of external (demand, infrastructure, state regulation, and competition) and internal (inclination for innovations) factors. In modern economic systems in entrepreneurship, four main strategies of usage of digital technologies are implemented: strategy of ignoring (refusal from using digital technologies), strategy of formal participation in technological revolution (usage of the Internet for certain business operations), strategy of expanded participation in technological revolution (usage of the Internet for purchases and sales), and strategy of leading technological revolution (usage of the Internet and the leading digital technologies).

The most preferable strategies are strategies of leading technological revolution and expanded participation in technological revolution, as they allow increasing export and increasing competitiveness and efficiency of entrepreneurship. Contrary to this, separation from technological revolution (implementing the strategy of ignoring or strategy of formal usage of the Internet) leads to reduction of competitiveness and efficiency and reduction of export.

Therefore, technological revolution in the sphere of digital technologies creates new opportunities for development of entrepreneurship, which should be used. Contrary to a popular belief that digital entrepreneurship will co-exist with pre-digital entrepreneurship in the near future, it is logical to expect pre-digital companies to be either ousted from the market or to expect their digital modernization. Future studies should dwell on this issue.

References

Bogoviz, A.V., Lobova, S.V., Ragulina, J.V.: Perspectives of growth of labor efficiency in the conditions of the digital economy. Lecture Notes in Networks and Systems, vol. 57, pp. 1208–1215 (2019)

Morozova, I.A., Popkova, E.G., Litvinova, T.N.: Sustainable development of global entrepreneurship: infrastructure and perspectives. Int. Entrep. Manag. J. **2**(1), 1–9 (2018)

OECD: GDP per hour worked (2019). https://data.oecd.org/lprdty/gdp-per-hour-worked.htm. Accessed 4 May 2019

Popkova, E.G.: Preconditions of formation and development of Industry 4.0 in the conditions of knowledge economy. Studies in Systems, Decision and Control, vol. 169, pp. 65–72 (2019)

Pritvorova, T., Tasbulatova, B., Petrenko, E.: Possibilities of blitz-psychograms as a tool for human resource management in the supporting system of hardiness of company. Entrep. Sustain. Issues **6**(2), 840–853 (2018). https://doi.org/10.9770/jesi.2018.6.2(25)

Sergi, B.S., Popkova, E.G., Bogoviz, A.V., Ragulina J.V.: Entrepreneurship and economic growth: the experience of developed and developing countries. In: Entrepreneurship and Development in the 21st Century, pp. 3–32. Emerald Publishing Limited (2019)

Bezrukova, T.L., Popova, E.V., Korda, N.I., Kuznetsova, T.E., Bezrukov, B.A.: Institutional traps of innovative and investment activities as an obstacle on the path to the well-balanced development of regions. Contributions to Economics (9783319606958), pp. 235–240 (2017)

Sibirskaya, E., Popkova, E., Oveshnikova, L., Tarasova, I.: Remote education vs traditional education based on effectiveness at the micro level and its connection to the level of development of macro-economic systems. Int. J. Educ. Manag. **33**(3), 533–543 (2019). https://doi.org/10.1108/IJEM-08-2018-0248

Vanchukhina, L.I., Leybert, T.B., Khalikova, E.A., Khalmetov, A.R.: New approaches to formation of innovational human capital as an element of institutional environment. Espacios **39**, 22–32 (2018)

World Bank: Merchandise exports (current US$) (2019). https://data.worldbank.org/indicator/TX.VAL.MRCH.CD.WT?view=chart&year_high_desc=true. Accessed 4 May 2019

World Economic Forum: The global competitiveness report 2017–2018 (2019). https://www.weforum.org/reports/the-global-competitiveness-report-2017-2018. Accessed 4 May 2019

National Research University "Higher School of Economics". Digital Economy (2019). https://www.hse.ru/primarydata/ice2019kr. Accessed 4 May 2019

The Transformation of Motor Vehicle Taxation in a Climate of Digitization of the Economy of the Russian Federation

Irina V. Gashenko(✉), Yulia S. Zima, and Elena N. Makarenko

Rostov State University of Economics, Rostov-on-Don, Russia
gaforos@rambler.ru, zima.julia.sergeevna@gmail.com

Abstract. Purpose: This paper is aimed at studying the trend of transformation of motor vehicle taxation in a climate of digitization of the economy of the Russian Federation and identifying the prospects for the development of this trend for the improvement of modern Russian practice of motor vehicle taxation.

Design/methodology/approach: The authors carry out the structural analysis of the fiscal systems of the regions of the Southern Federal District of the Russian Federation in 2018. It is used to determine the place of transport tax in the revenue structure, as well as the place of expenses for the maintenance and development of road transport infrastructure in the spending pattern of the consolidated budgets of the regions of the selected district.

Findings: The authors substantiate that digitization of the economy of the Russian Federation has intensified the trend of transformation of motor vehicle taxation, which serves to ensure the intended expenditure of tax levies, differentiate taxpayers, complicate the motor vehicle taxation structure, and increase the accuracy of tax computation and payment.

Originality/value: A mechanism of combined transport taxation based on the use of artificial intelligence has been proposed to enable the most complete use of the potential for the improvement of the practice of transport taxation that is opened in a climate of digitization of the economy of the Russian Federation. A mechanism has been developed to enable automatic determination of the amount of environmental and infrastructure transport taxes and charges, depending on a wide range of indicators, including specifications of motor vehicles, directions and activity of their use. This will enable the rise in transparency and efficiency of motor vehicle taxation in the Russian Federation.

Keywords: Taxation · Motor vehicles · Digitization · Economy · Russian Federation

JEL Code: E62 · H21 · H22 · H23 · L62 · L91 · O18 · O31 · O32 · O33 · O38 · R41

1 Introduction

Motor vehicle taxation plays a special part in the fiscal system of the Russian Federation, since it is characterized by two important features. The first feature is associated with the fact that transport tax levies go into the regional budgets and thus play a

E. G. Popkova and B. S. Sergi (Eds.): ISC 2019, LNNS 91, pp. 247–253, 2020.
https://doi.org/10.1007/978-3-030-32015-7_28

critical part in the economic decentralization and development of the regional economics in the Russian Federation. The second feature consists in exclusive standards of transparency of the motor vehicle taxation systems on the part of individual and corporate taxpayers (vehicle owners).

In this regard, there is an increased complexity of the management of the system of motor vehicle taxation in the Russian Federation. The transport tax is perceived by most taxpayers as a payment for the use of road infrastructure and thus there are severe requirements for its quality. Transport tax payment complicates the introduction of mechanism of public and private partnership for the development of road infrastructure based on tollways, since the taxpayers are interested in free access to this infrastructure or cancellation of a transport tax (are not willing to pay for the usage of infrastructure).

At the same time, the cost of maintenance and development of road infrastructure is so high that transport tax levies are unable to cover it, thus many infrastructure projects are financed from the federal budget. Thus, taxation of motor vehicles in the Russian Federation is contradictory and therefore inefficient, meeting neither the interests of the society (infrastructure development, equitable distribution of tax burden), or the interests of the state (increase in tax levies).

Therefore, a crucial task of economics is to find the ways of improvement of contemporary practice of motor vehicle taxation in the Russian Federation. In this research, it is hypothesized that the process of transformation of motor vehicle taxation was launched in a climate of digitization of the economy of the Russian Federation, making it possible to improve its contemporary practice. This paper is aimed at studying the trend of transformation of motor vehicle taxation in a climate of digitization of the economy of the Russian Federation and identifying the prospects for the development of this trend for the improvement of modern Russian practice of motor vehicle taxation.

2 Materials and Methods

Peculiar features of motor vehicle taxation have been studied in papers of Fung and Proost (2017), Gashenko and Zima (2019), Gashenko et al. (2019), Gashenko et al. (2018), Ližbetinová et al. (2017). Economy digitization processes, particularly in terms of their impact on taxation, have been studied at fundamental and empirical scientific levels in papers of Bezrukova et al. (2017), Bogoviz et al. (2019a), Bogoviz et al. (2019b), Pritvorova et al. (2018), Popkova et al. (2019), Sergi et al. (2019a), Sergi et al. (2019b), Sibirskaya et al. (2019), Vanchukhina et al. (2018).

The conducted literature review has revealed the lack of coordination between existing studies and publications on the selected topic and understudied research and practice problem of transformation of motor vehicle taxation in a climate of digitization of the economy, which determines the need for its further study. For this purpose, the authors of this paper carry out the structural analysis of the fiscal systems of the regions of the Southern Federal District of the Russian Federation in 2018. It is used to determine the place of transport tax in the revenue structure, as well as the place of expenses for the maintenance and development of road transport infrastructure in the spending pattern of the consolidated budgets of the regions of the selected district.

In 2018, there is a significant disbalance of nonoperating revenues, which includes transport tax levies and other expenses, the structure of which includes expenses for the maintenance and development of road transport infrastructure in the regions of the Southern Federal District of the Russian Federation. The ambiguousness of scope (the lack of accurate statistical information) and disproportionality of revenues and expenses is indicative of nontransparency of motor vehicle taxation in the Russian Federation and corroborates the need of its improvement.

3 Results

Following the results of the study of traditional practice of motor vehicle taxation in the Russian Federation from the early XXI century we have identified the trend of its transformation that was launched in connection with digitization of the economy of the Russian Federation. We have presented the comparative analysis of the traditional practice of taxation of motor vehicles in Russia and the new practice emerging in a climate of digitization of the economy in Table 1.

Table 1. Comparative analysis of the traditional practice of taxation of motor vehicles in Russia and the new practice emerging in a climate of digitization of the economy

Comparative characteristics	Practice of motor vehicle taxation	
	Traditional (before digitization of the economy started)	New (in a climate of digitization of the economy)
Transport tax collection principle	Compensation for damage caused by a motor vehicle to the environment of the region	Financial assurance of road infrastructure (usage charge)
The use of tax levies in regional economics	Off-target (depending on the fiscal policy pf the region)	Intended (for the maintenance and development of road infrastructure)
Structure of taxpayers	Uniform (unification of all types of transport)	Non-uniform (distinction between types of transport)
Structure of motor vehicles	Mono-structure (as simple as possible, includes the tax only)	Polystructure (complex, includes the tax as such and additional transport charges, for example, payment in the system "Platon")
Basis for tax payment (tax basis)	Engine power of a motor vehicle	Use of road infrastructure
Technological support	Traditional (pre-digital) tax computation and payment technologies	Digital technologies of tax computation and payment

Source: compiled by the authors.

Table 1 shows that the traditional practice of motor vehicle taxation has been used in the Russian Federation since the beginning of the XXI. According to it, transport tax is intended to compensate for damage caused by a motor vehicle to the environment of the region. Tax levies were not combined into an individual special-purpose fund, but went into the common regional budget and were spent depending on the fiscal policy of the region. Taxpayer structure is uniform, i.e. all types of transport were unified no distinction was made between them for purposes of taxation).

It should be noted here that tax exemptions were granted to certain categories of taxpayers. Nevertheless, unlike road infrastructure, there was no taxation of water infrastructure (for example, port infrastructure). Motor vehicle taxation was characterized by mono-structure, which was as simple as possible and included the tax only. The engine power of a motor vehicle served as the basis for transport tax payment. Taxes were computed and paid manually and did not involve the use of automation facilities.

A new practice of motor vehicle taxation emerges in a climate of digitization of the economy of the Russian Federation. The transport tax collection principle consists in financial assurance of road infrastructure (usage charge). As a result, the use of tax levies in regional economics becomes intended and implies the maintenance and development of road transport infrastructure of the region. Taxpayer structure becomes more complicated and non-uniform, distinction is made between the types of transport not only according to their power, but also, for example, according to their weight (laden weight).

A polystructure of motor vehicle taxation is being established, which includes not only transport taxation as such, but also additional transport charges, for example, payment in the system "Platon" (payment collected from cargo vehicles with a gross laden weight of more than 12 tons, according the right to receive tax rebate on transport tax). The use of road infrastructure becomes the basis for tax payment (tax basis). Technological support of new practice is provided by digital technologies of tax computation and payment (for example, technologies of electronic payment of charges – bank transfers, payment with fuel cards, payment via payment kiosks).

The revealed trend of transformation of motor vehicle taxation in a climate of digitization of the economy of the Russian Federation serves to increase the transparency of this practice, but provides no opportunity to overcome its inconsistency. Motor vehicles simultaneously cause damage to the environment and use road transport infrastructure, while the consideration of only one of these aspects causes distorted interpretation of the essence of motor vehicle taxation by the taxpayers and negates the effectiveness of this practice.

In this respect, from the standpoint of social justice in the determination of the tax basis of transport tax, the actual, not potential, damage to the environment and activity of the use of road infrastructure should be taken into account in the determination of the tax basis of transport tax. Thus, for example, currently in the Russian Federation the transport tax on a motor vehicle that is out of use is identical to the transport tax on a similar motor vehicle that is used on a regular basis.

Therefore, we believe that in the Russian Federation there is a need for the combined practice of motor vehicle taxation. The prospects for its emergence open in a climate of digitization of the Russian economy. In our opinion, the technology of

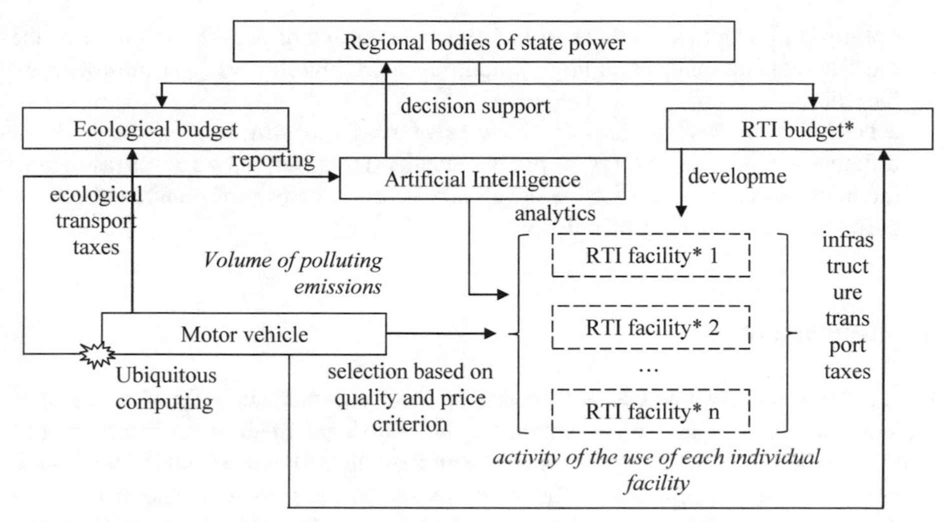

*RTI – road transport infrastructure.

Fig. 1. Mechanism of combined transport taxation based on the use of artificial intelligence Source: compiled by the authors.

Artificial Intelligence is one of the most in-demand technologies. We have developed a mechanism of combined transport taxation based on the use of artificial intelligence that is presented in Fig. 1.

Figure 1 shows that motor vehicle is equipped with ubiquitous computing – a technology which enables automatic determination of the activity of its use (fuel consumption, haul) and send data to Artificial Intelligence, which computes the amount of ecological transport taxes (according to the tax rates set by the state which may vary in different regions) based on specifications (for example, specific volume of polluting emissions characteristic of a particular motor vehicle).

Motor vehicles are also subject to the payment of infrastructure transport taxes, the rates of which are different for various road infrastructure facilities (depending on the quality) and various motor vehicles (depending on the weight). Artificial Intelligence analyzes the activity of the use of various road transport infrastructure facilities in regional economics and determines the most in-demand facilities, in favor of which are budgetary funds are reallocated. As a consequence of practical application of the proposed mechanism, the following advantages are achieved:

- drawing distinction between the ecological budget and the budget of road infrastructure in the region, increasing the transparency of motor vehicle taxation and guaranteeing intended budget expenditure;
- increasing the degree of social equity thanks to accurate computation of the amount of transport taxes and charges depending on specifications, directions (infrastructure facilities) and activity of the use of motor vehicles;

- optimizing budget expenditure in the region by means of precise definition of the most in-demand (and, accordingly, requiring more investment) road infrastructure facilities;
- establishing market relations in the area of road infrastructure on the basis of different tax assessment (as a pricing element) for various road infrastructure facilities opens prospects for the development of mechanism of public and private partnership and creation of tollways.

4 Conclusion

Hence, the hypothesis that was put forward has been substantiated – the digitization of the economy of the Russian Federation has intensified the trend of transformation of motor vehicle taxation, which serves to ensure the intended expenditure of tax levies, differentiate taxpayers, complicate the motor vehicle taxation structure, and increase the accuracy of tax computation and payment. This transformation is most vividly manifested in the launch of the system "Platon" which uses the technology of electronic payment of charges depending on the number of motor vehicles.

A mechanism of combined transport taxation based on the use of artificial intelligence has been proposed to enable the most complete use of the potential for the improvement of the practice of transport taxation that is opened in a climate of digitization of the economy of the Russian Federation. A mechanism has been developed to enable automatic determination of the amount of environmental and infrastructure transport taxes and charges, depending on a wide range of indicators, including specifications of motor vehicles, directions and activity of their use. This will enable the rise in transparency and efficiency of motor vehicle taxation in the Russian Federation.

References

Bezrukova, T.L., Popova, E.V., Korda, N.I., Kuznetsova, T.E., Bezrukov, B.A.: Institutional traps of innovative and investment activities as an obstacle on the path to the well-balanced development of regions. Contributions to Economics (9783319606958), pp. 235–240 (2017)

Bogoviz, A.V., Alekseev, A.N., Ragulina, J.V.: Budget limitations in the process of formation of the digital economy. Lecture Notes in Networks and Systems, vol. 57, pp. 578–585 (2019a)

Bogoviz, A.V., Lobova, S.V., Ragulina, J.V.: Shift of the global investment flows in the conditions of formation of digital economy. Lecture Notes in Networks and Systems, vol. 57, pp. 1216–1223 (2019b)

Fung, C.M., Proost, S.: Can we decentralize transport taxes and infrastructure supply? Econ. Transp. **9**, 1–19 (2017)

Gashenko, I.V., Zima, Y.S.: Risk-oriented approach to tax control for stabilization of financial systems of developing countries by the example of modern Russia. Lecture Notes in Networks and Systems, vol. 57, pp. 932–938 (2019)

Gashenko, I.V., Zima, Y.S., Davidyan, A.V.: Tax system of a state: Federal, regional and local taxes and fees. Studies in Systems, Decision and Control, vol. 182, pp. 13–21 (2019)

Gashenko, I.V., Zima, Y.S., Stroiteleva, V.A., Shiryaeva, N.M.: The mechanism of optimization of the tax administration system with the help of the new information and communication technologies. Advances in Intelligent Systems and Computing, vol. 622, pp. 291–297 (2018)

Ližbetinová, L., Fábera, P., Jambal, T., Caha, Z.: Road tax as an economic tool of the support for the development of multimodal transport in selected EU states. In: MATEC Web of Conferences, vol. 134, p. 00031 (2017)

Popkova, E.G., Zhuravleva, I.A., Abramov, S.A., Fetisova, O.V., Popova, E.V.: Digitization of taxes as a top-priority direction of optimizing the taxation system in modern Russia. Studies in Systems, Decision and Control, vol. 182, pp. 169–175 (2019)

Pritvorova, T., Tasbulatova, B., Petrenko, E.: Possibilities of blitz-psychograms as a tool for human resource management in the supporting system of hardiness of company. Entrep. Sustain. Issues **6**(2), 840–853 (2018). https://doi.org/10.9770/jesi.2018.6.2(25)

Sergi, B.S., Popkova E.G., Vovchenko, N., Ponovareva, M.: Central Asia and China: financial development through cooperation with Russia. In: International Symposia in Economic Theory and Econometrics, vol. 26. Emerald Publishing Limited (2019a)

Sergi, B.S., Popkova, E.G., Bogoviz, A.V., Ragulina J.V.: Entrepreneurship and economic growth: the experience of developed and developing countries. In: Entrepreneurship and Development in the 21st Century, pp. 3–32. Emerald Publishing Limited (2019b)

Sibirskaya, E., Popkova, E., Oveshnikova, L., Tarasova, I.: Remote education vs traditional education based on effectiveness at the micro level and its connection to the level of development of macro-economic systems. Int. J. Educ. Manag. **33**(3), 533–543 (2019). https://doi.org/10.1108/IJEM-08-2018-0248

Vanchukhina, L.I., Leybert, T.B., Khalikova, E.A., Khalmetov, A.R.: New approaches to formation of innovational human capital as an element of institutional environment. Espacios **39**, 22–32 (2018)

Federal State Statistics Service of the Russian Federation: Regions of Russia. Socio-economic indicators - 2018 (2019). http://www.gks.ru/bgd/regl/b18_14p/Main.htm. Accessed 7 May 2019

Teleology of Scientific and Technical Progress

Igor V. Astafyev[1(✉)] and Dmitry P. Sokolov[2]

[1] Civil Society Development Foundation, Kostroma, Russia
iastafjev@mail.ru
[2] Financial University under the Government of the Russian Federation, Moscow, Russia
dpsokolov@fa.ru

Abstract. Scientific and technical progress as a complex systemic phenomenon is one of the main factors of socio-economic development. As any element of a system of relations, STP has a complex objective function, which, on the one hand, defines its tasks, on the other, it predetermines results and consequences.

The teleology of scientific and technical progress has two clearly expressed directions - cognitive process and the growth of the quantity and quality of consumption. These areas are closely interrelated.

The article focuses on the features of scientific and technical progress in terms of its goals and objectives at the present stage of civilizational development.

Keywords: Civilization · Society · Scientific and technological progress · Macroeconomic system · Work objectives · Market · Commodity goods · Science · Marginal utility

JEL Code: O400

1 Introduction

Any considerable human activity has a goal. Even knowingly idle activities (rest, distraction) are aimed at something. Another question is that goals from different points of view can be both constructive and deconstructive. Only the behavior of mentally challenged people makes no sense, and even then conditionally - because they are confident that their actions are logical.

Scientific and technical progress (STP) is directly related to cognition, science which is a complex phenomenon. One of the main, if not the most important, factors of STP is fundamental science. And in this regard, a strange and somewhat amusing analogy with mental illness is becoming quite serious and reasonable (it was not for nothing that many famous scientists were first considered crazy).

2 Methodology

Designer and thinker R.O. di Bartini suggested that "it would be much easier to build for the beginning not a model of the development of science and technology, but a model of a person capable of developing science and technology" (Buzinovsky 2003).

E. G. Popkova and B. S. Sergi (Eds.): ISC 2019, LNNS 91, pp. 254–263, 2020.
https://doi.org/10.1007/978-3-030-32015-7_29

This is logical, since neither the process nor the phenomenon can have goals, but people themselves, not even groups of people, can be carriers of goal-setting, and.

The system of goals, which includes economic aims, has a level structure in which scientific and technical progress is one of the benefits of the highest level. “Along with commodity and public goods, the highest benefits should also be highlighted. … These are goods-factors, goods-conditions, goods-circumstances, … not distributed, and not consumed in the traditional economic sense at all. These are knowledge, justice, honesty, progress.” (Astafyev 2018).

What are the goals of scientific and technical progress? For obvious reasons, the answer will not be easy. First of all, it is necessary to distinguish what kind of progress and in what sphere is considered.

- Is it the progress of fundamental scientific knowledge?
- Is it the progress of machinery and technology? What exactly, productive or serving the ultimate needs? Is it the progress of military equipment?
- Is it the progress of individual final consumption?

Each of the listed items corresponds to a specific goal, which are not always the same, but do not contradict each other. However, it is possible to divide the goals of STP into two major directions: cognition and consumption. The public goods (science, education) mainly have knowledge goals, and commodity goods serve the consumption purposes.

At the same time, distinction is made between applied science and fundamental science. Products of applied science may to a large extent have the properties of goods than the fundamental scientific knowledge, the properties of which completely coincide with the properties of public goods (non-exclusivity and lack of rivalry in consumption, irreducibility to commodity units). “Repeated and still ongoing attempts to make commodity goods out of knowledge are constantly undergoing regular failures. … But it is impossible to trade in knowledge and ideas in the same way as introducing into the economic circulation areas of cosmic bodies far from the Earth or trading letters of the alphabet or sounds of speech, having patented them.” (Astafyev 2018).

Obviously, it is unlikely that anyone will succeed, even if it is possible to document this, to sell for personal use to someone even at a reasonable price, even declaring a grand discount, at least part of Ohm’s Law, Pythagorean Theorem or Kirchhoff’s Laws. “It’s hard to imagine that, say, the Schrödinger equation from quantum mechanics, the Potryagin-Kuratovsky theorem from graph theory, the Arrow-Debre and Mackenzie models from mathematical economics can be consumed by individual people and be a part of their utility function”. (Greenberg 2008)

When trying to prioritize between the above goals, we get the following situation. Applied research is impossible without fundamental knowledge, it is obvious. In other words, without public goods commodity does not appear (if there is no state, at least in the role of a “night watchman”, then trade becomes impossible, robbery takes its place).

Consequently, scientific and technical progress has two teleological components: cognition and consumption.

3 Results

3.1 Economic Properties of Scientific and Technical Progress

How relevant is this question? In the macro- and mega-economic systems of the free market, it is, paradoxically, less relevant. Without funding, practically no research of either fundamental or applied nature can take place. (Certain exceptions may be made by mathematics and philosophy.) The difference between the principles of their resource provision is that basic research, unlike applied research, cannot have neither a clearly defined result, nor a guaranteed success fixed by an economic agreement.

This implies a difference in the principles of financing: either predominantly from the state (whether it is grants or permanent funding of scientific organizations); or mainly from commercial entities.

But in this case we are more interested in another question: how to link the economic value of knowledge to its social significance. Labor assessment of knowledge does not make sense, as well as the market, because the marginal utility in this case is hard to realize. It is possible, with a certain degree of reliability, to determine the cost of resources for obtaining this or that fundamental scientific knowledge, although this will be a deliberately conditional estimate without taking into account unsuccessful attempts and the costs of obtaining related knowledge. But this assessment is of little importance, because the cost of resources in different countries and different macroeconomic systems is incomparable.

Let us define the elemental composition of scientific and technical progress.

The philosophical interpretation of STP is known as the process of the materialization of scientific knowledge or as the improvement of technologies and means of production. Such interpretations are (a) not exhaustive; (b) controversial, since the STP covers all aspects of cognition and its manifestations in practical activities. The components of scientific and technological progress can be represented as follows (Fig. 1).

The given interpretation of the elements of the STP, it seems, quite successfully and fully describes the elements of the scientific and technical process as a socio-economic phenomenon.

It can be seen that the elemental structure of the STP is, at same time, a synthesis of goals of different levels and different subjectivity. There is a problem of determining the goal-setting of scientific and technical progress - in what form, to what extent and for whom exactly it is necessary.

It would be a mistake to assert that the STP is constantly in demand, useful to everyone and always. A paradox of development is known when, for example, the opening of a new mass source of energy, an alternative to the combustion of hydrocarbons, is not beneficial for the beneficiaries of this segment of the economy, since it will change the structure of the world market beyond recognition. The same applies to the discovery of new types of drugs, methods of construction, communications, etc.

Goals define constructiveness or lack of it. There is an interesting approach to the STP in terms of teleology. Does the problem of overpopulation of the Earth (the ratio of the maximum number of people to the total capabilities of the planetary resource support) which Malthus was one of the first to set, exist? Yes, it does. It is objective.

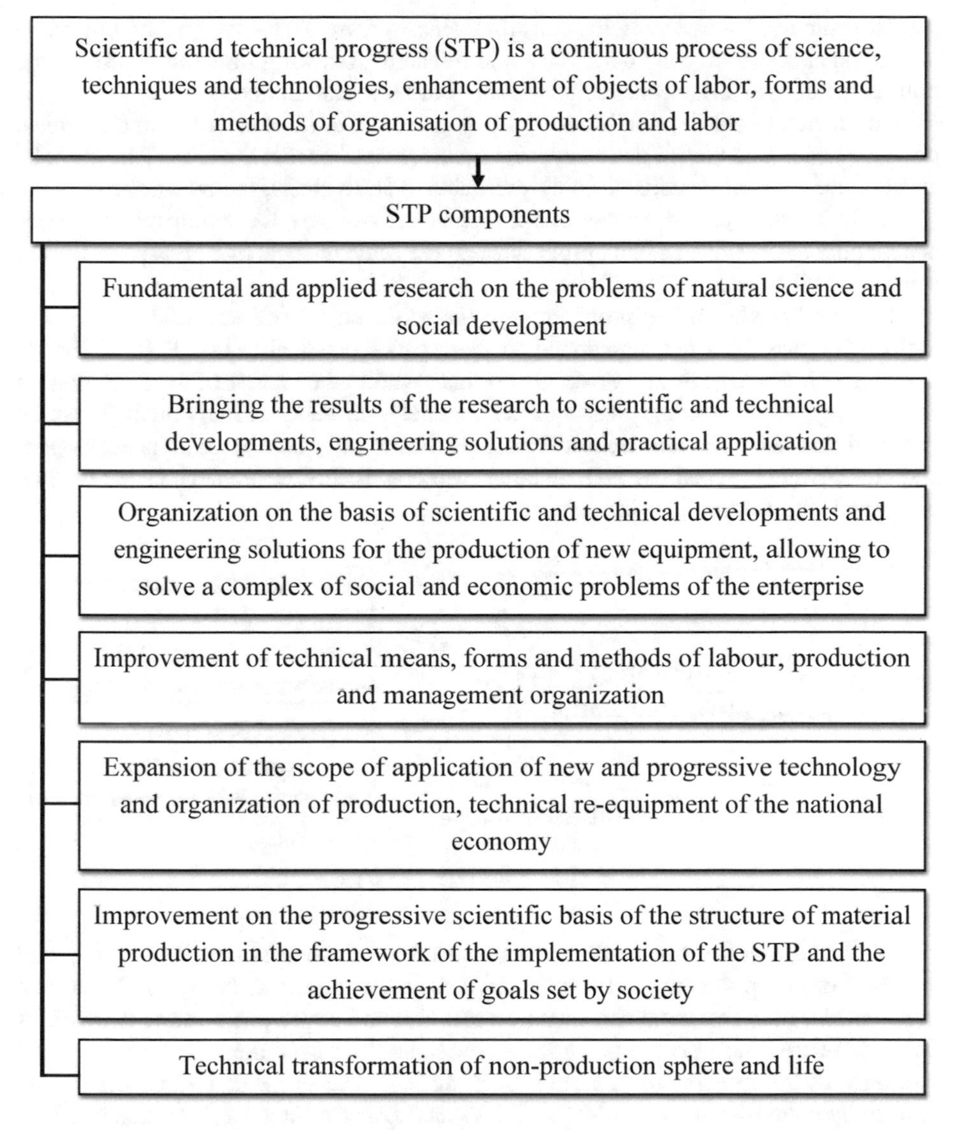

Fig. 1. The components of scientific and technical progress. Source: (Nechaev 2009, p. 59).

Is there a task of providing food in adverse natural conditions? Certainly there is one.

But both of these real and undeniable problems can be perceived in a completely opposite way from different perspectives. What does overpopulation mean? Is it a speech on birth control (sterilization)? Is it about the extermination of the population using wars? What does genetic modification of plants and animals mean, the threat of existence, or vice versa, the rescue option?

The criterion for assessing the constructiveness of the STP is the goals followed by one or other research. Moreover, the same research, following different goals, can be both constructive and deconstructive. Goals can be social and commercial.

Commercial goals a priori do not correspond to the tasks of scientific and technical progress, since knowledge generally, as a public good, benefits the society as a whole, and the commercial benefits are only profitable to individuals, the beneficiaries.

It can be said otherwise: scientific and technical progress for commercial purposes is generally impossible. Making profit through the achievements of STP is possible, but it is not at all the same (Fig. 2).

The person who makes profit through the achievements of scientific and technological progress does not correspond to the type of personality that R.O. di Bartini talked about. Because in this case, it is not man who is the creator of the STP, but, on the contrary, the STP is the creator of the material well-being of individuals. It can be assumed that scientific and technical progress is not only an economic phenomenon, and not so much social, as civilizational. Moreover, the teleological factor of STP determines the future of civilization.

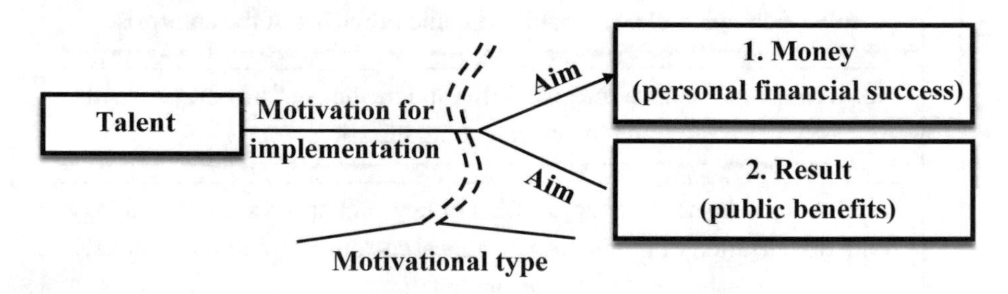

Fig. 2. Teleology of talent

It is interesting to compare the share of spending on science in relation to GDP and the share of non-productive services in GDP itself. According to the World Bank data in the world, only two countries spend on research and development more than 4% of GDP, these are South Korea and Israel. At the same time, even the leaders of the world economy direct to scientific research only up to 3.6% of GDP ("Expenditures on science by countries of the world" 2017): Japan 3.6%; Sweden 3.2%; Germany 2.9%; United States 2.7%; China 2%. Based on the Table 1 and World Bank data, world GDP, the total expenditure on science and the cost of services of primarily non-productive character can be compared.

In 2016, the total expenditure on science in the world amounted to about 1,650 billion dollars. The world GDP is about $75,278 billion (Dynkin and Baranovsky 2018, p. 16.) According to the World Bank, the total volume of the world services market was $31,843.2 billion. Thus, in comparison with the world GDP, the total expenditure on science is 2.19%, and the services market - 42.3%.

Table 1. The main indicators of science in the twenty countries-leaders in terms of domestic expenditure on research and development based on the purchasing power parity of national currencies: 2016

Country	Internal expenses on research and development						Developers in the equivalent of full occupation			
	In terms of purchasing power parity of national currencies		Per one developer		As a percentage of gross domestic product		Position of the country	Thousand people	Per 10000 employed	
	Position of the country	Billion of US dollars	Position of the country	Thousands of US dollars	Position of the country	%			Position of the country	Number of people
Russian	**10**	**39,9**	**47**	**93,0**	**34**	**1,10**	**4**	**428,9**	**34**	**60**
The U.S.A	1	511,1	2	359,9	11	2,74	2	1380,0	18	91
China	2	451,2	8	266,6	15	2,12	1	1692,2	48	22
Japan	3	168,6	9	253,4	6	3,14	3	665,6	14	100
Germany	4	118,5	6	295,6	8	2,94	5	400,8	16	92
The Republic of Korea	5	79,4	17	219,6	2	4,24	6	361,3	5	138
France	6	62,2	16	220,6	13	2,25	9	277,6	12	101
India	7	50,1	24	177,1	44	0,62	8	283,0	57	6
The UK	8	47,2	31	162,1	21	1,69	7	291,4	17	92
Brazil	9	41,1	13	229,1	28	1,28	10	183,9	51	20
Taiwan	11	35,8	11	242,5	5	3,16	12	147,7	6	131

Source: (Higher School of Economics 2018)

Moreover, if we compare the costs of science in the world with the advertising market, the situation looks even less optimistic. In 2016, according to expert estimates (tadviser.ru›index.php…Реклама_(мировой_рынок); sostav.ru›publication/reklam-nyj-rynok-budet-rasti; pwc.ru and others) the global advertising market ranged from 579 billion US dollars to 965 billion. according to other estimates. There is no exact method for assessing the advertising market, since this sphere is dispersed between practically all types and subjects of economic activity, and not just the mass media. Moreover, part of the advertisement is in the shadow sphere. But nevertheless, it is clear that only the advertising market (excluding the entertainment market) is already comparable to the global expenditure on science, and taking into account similar related services, it even surpasses them.

This suggests that the goals of consumption substantially and even fatally dominate the cognition goals.

3.2 The Interconnection Between Economics, Teleology and STP

It is obvious that only the knowledge goals in the scientific and technical progress correspond to the task of the possible preservation of civilization in case of possible planetary cataclysms. The goals of increasing individual end consumption correspond to the complete dependence of humanity on environmental conditions and in fact contribute to the final consolidation of the status of the planet's endemic in modern civilization. In fact, a civilization, taken as a whole, in this case will not differ in anything from a large primitive tribe with a relatively high level of consumption.

"The changes that we have been experiencing over the past three decades are nothing compared to what awaits us in the next 30 years." With the current trend of energy consumption, "tens of thousands of nuclear power plants will not be enough for a consumer society". Only in the USA, Canada and Western Europe, where less than 12% of the world's population lives, accounts for over 60% of the global consumption of goods and services. In 1972, 0.8 of the planet was enough for people, but nowadays 1.35 of it is not enough. (Meadows 2007)

It is known that the organization of the macroeconomic system on the principles of the free market and private capital, as K. Marx rightly pointed out, as a result of the impact of scientific and technical progress corresponds to the constant growth of the so-called "organic composition of capital" (Marx 1890 and (1960), pp. 638, 657), that is, its permanent part, the "past labor" embodied in machines. At present, this trend has not just intensified, but moved to a new stage - the growth of the information and digital structure of capital, its "knowledge capacity" according to Bodrunov (Bodrunov 2018, p. 231) This leads to a relative reduction in GDP in the so-called "real sector" of production, but an increase in the intellectual component in goods (Fig. 3).

But this either leads to imbalances in the cost estimates, or to the drift of commodity goods towards the public ones.

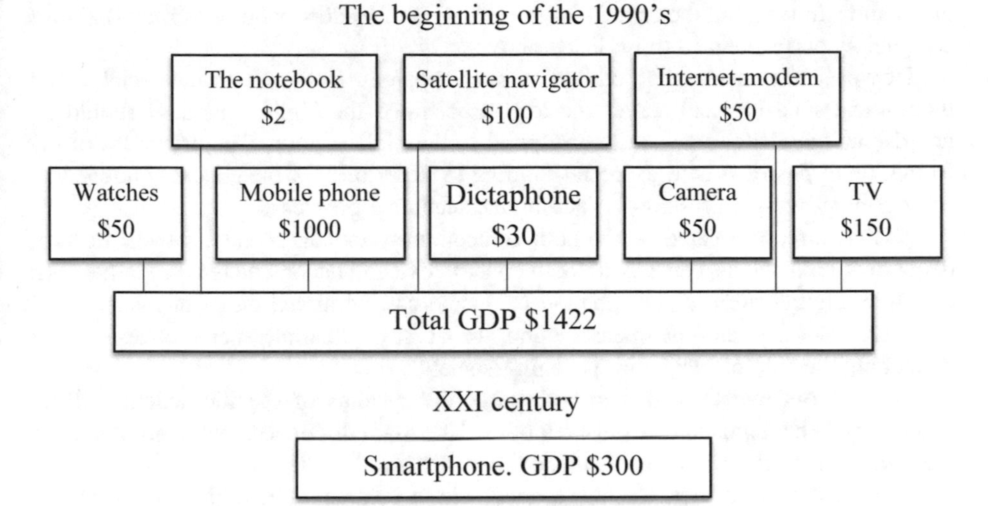

Fig. 3. Example of GDP reduction due to the progress of information technologies. Source: (Bodrunov 2018, p. 97)

4 Conclusions

The close interconnection between the scientific and technical progress and the system of authority organization and governance at macro- and mega levels. The goals of growth of individual final consumption correspond to the principles of the formation of government bodies acting primarily in the interests of certain groups of people who manage the processes of wealth distribution in the economic system (it can be said of the ruling classes or situs groups).

The goals of knowledge in the framework of scientific and techncфд progress correspond to the principles of the formation of power, one way or another focused on meritory administration in the interests of the whole society.

Meritors can be understood as different strata and/or representatives of different strata under the meritors, but common properties, characteristic of the representatives of this myriad, can be outlined:

- High cultural, educational and intellectual level;
- Fundamental voluntary refusal of commercial activity in any form, not only for oneself, but also for close relatives (otherwise it automatically transforms into lobbyism);
- The presence of public recognition in one form or another.

So far, such principles of organization and management at the macrolevel of recent history are unknown. The system of the so-called "Soviet power" (this name is not quite correct, rather, it can be described as a multi-stage democratic formation of the legislative power by the population) and, strange as it may seem, control in theological states was closest to them. In the latter case, it is understood, of course, not religious goals and attitudes, but the criteria for the selection of leaders, who (the criteria) almost completely correspond to those listed above.

However, the mentioned "oddity" is only apparent. In fact, both the socialist and theological state has and serves the idea proclaimed the highest good. It should be noted that knowledge may well correspond to the highest good. The principles of the formation of power structures, corresponding to the priority of the goals of knowledge, fully comply with the tasks of scientific and technical progress.

STP is directly related to the communication system in society, which, in turn, depending on its goals, can acquire both constructive and deconstructive properties. For example, digitalization of information exchange can be aimed either at total control over civil society and its manipulation, or at strengthening social cohesion and enhancing the role of justice in managing society.

The authors underline that in both cases the systems of specific actions will be practically indistinguishable from each other. The main differences will consist only in the fact with what ultimate goals this or that structure is organized.

In particular, chipping, electronic document circulation, electronic voting, remote judicial proceedings and interaction with government bodies, etc. can serve both the tasks of total control and the suppression of the collective will, and the absolutely opposite - the construction of a perfect civil society.

The next aspect of scientific and technical progress is its connection with the economic principles of organizing activities. The marginal utility, which is the basis of the market turnover of goods, is in a certain contradiction with the interests of society, because it forces to create, apply and continuously improve high-cost technologies that minimize the possible consequences of opportunistic behavior.

Exact data on the costs of resources to perform these tasks is obviously not available, but indirectly, their volume can be judged by the fact that "the costs of certifying the correctness, for example, of banking operations, account for about half of all costs of the banking system." (Bodrunov 2018, p. 280)

Technologies of trust, however, is not fully compatible with commodity and is completely incompatible with the market turnover of goods inseparable from the goal of maximizing individual final consumption. In fact, recognition and identification technologies based on fingerprints, face, iris, etc., preventing unauthorized access, surveillance, video surveillance and video recording, etc., require a lot of time and resources while (according to one of the Chinese stratagems "Taking the brushwood out of the fire") the need for all of this is eliminated with just one, while unrealistic, condition - honesty and trust.

At the same time, it should be noted that honesty and trust are becoming possible only when the absolute majority of individuals working together have common goals with which they agree.

Thus, the teleology of scientific and technical progress is the main factor determining its final result or consequences.

References

Astafyev, I.V.: The System of Motivation as a Key Element in the Evolution of Macroeconomic Systems, Philosophy of Economy, Almanac of the Center for Social Sciences and Faculty of Economics of the Lomonosov Moscow State University, No. 3, p. 128 (2018)
Bodrunov, S.D.: Noonomics, Moscow - St. Petersburg - London, Cultural Revolution (2018)
Buzinovsky, S., Buzinovskaya, O.: The Mystery of Woland, Barnaul, p. 10 (2003)
Dynkin, A.A., Baranovsky, V.G.: Russia and the World: 2018 Economy and Foreign Policy. Annual forecast. Project head – A.A. Dynkin, V.G. Baranovsky - M.: IMEMO RAN, 2017 (2018)
Expenditures on Science by Countries of the World, 27 November 2017. https://theworldonly.org/rashody-na-nauku-po-stranam/. Accessed 05 Apr 2019
Greenberg, R.S.: Basis of a Mixed Economy. Economic Sociodynamics, Moscow, Institute of Economics, Russian Academy of Sciences, p. 160 (2008)
Higher School of Economics: Ranking of the Leading Countries of the World in Science Costs, Science, Technology, Innovations, Institute for Statistical Studies and Economics of Knowledge of the Higher School of Economics, 24 July 2018. issek.hse.ru›mirror/pubs/share/221869863. Accessed 07 Apr 2019
Marx, K.: Capital. A Critique of Political Economy, vol. I (1890)
Marx, K., Engels, F.: Essays, 2nd edn., vol. 23, Moscow (1960)
Meadows, D.H.: Tens of Thousands of Nuclear Power Plants are not Enough for a Consumer Society. Ecol. Law. **2**(26), 42 (2007)
Nechaev, V.I., Paramonov, A.F., Khalyavka, I.E.: Economics of Enterprises of Agroindustrial Complex, Part III, Krasnodar (2009)

Potential Horizons of Technological Development of Russia

Lyudmila V. Oveshnikova[1(✉)], Elena V. Sibirskaya[1], Evgeniya P. Tenetova[1], Natalya P. Kuznetsova[2], and Mariya O. Grigoryeva[1]

[1] Plekhanov Russian University of Economics, Moscow, Russia
{Oveshnikova.LV, Sibirskaya.EV, Tenetova.EP, Grigoryeva.MO}@rea.ru
[2] Ufa State Aviation Technical University, Ufa, Russia
Natalerk1977@yandex.ru

Abstract. At present, a significant feature of a country's economy is intensification of innovative processes and their transformation into the vector of economic growth. That's why the issues of assessment of a country's technological development with the usage of comparative analysis of the indicators that are formed with application of methodological foundations of studying the innovative development of various countries become very important.

The purpose of the paper is to study the indicators of technological development of Russia and foreign countries, to perform structural & dynamic analysis, build a complex of economic & mathematical models, analyze the innovative processes, and apply complex integral indices that reflect the level of development of technologies and innovations.

The scientific novelty of the research consists in substantiation of theoretical provisions and methodological recommendations for studying the economic processes in the sphere of implementation of technologies and innovations with application of descriptive, component, and factor methods. The scientific novelty is proved by applicability of the research methods to the Russia and foreign economies. The performed research allows solving the tasks of determining the potential horizons of Russia's technological development.

The structural & dynamic analysis, coefficients of innovative activity, and complex integral indices that reflect international level of development of technologies and innovations in various countries of the world are used for evaluating the level of technological parameters in Russia. Special attention is paid to interconnection of the frequency of emergence of innovations and appearance of new companies based on application of multiple regression models. The performed models has a multi-aspect character, including analysis of the Russian experience of intensification of innovative processes, and inter-country comparisons, according to the indicators of the level of information and communication technologies, research activity, demand for technologies, and expenditures for technological innovations in different countries.

Keywords: Technologies · Innovations · Statistics · Analysis · Modeling · Entrepreneurship · Hi-tech business

E. G. Popkova and B. S. Sergi (Eds.): ISC 2019, LNNS 91, pp. 264–274, 2020.
https://doi.org/10.1007/978-3-030-32015-7_30

JEL Code: C 10 • O 33

1 Introduction

In the modern world, the contribution of science, innovations, and breakthrough technologies is one of the most significant factors of social and economic development of a country. Economic development is based on innovative activity with functioning of economic systems and usage of the innovative policy in market relations (Naumenko 2012). Innovations are a strategic aspect of growth, influence the structure of public production, reform the country's economy, and normalize social policy in the country.

Speaking of a country's innovative development, it is necessary to study the hi-tech sphere (Zavyalov 2017). Self-provision of the Russian economy with hi-tech products is at a rather low level. In developed countries – e.g., Japan, the USA, the UK, Germany, and France – the level of self-provision with products of the hi-tech complex is around 80%.

International experience of implementation of technologies and innovations could be analyzed according to the level of the Global Innovation Index (2018), which contains the data of comparative analysis of innovative systems of 127 countries. As of 2017, the rating's leaders were Switzerland, Sweden, the Netherlands, the UK, and the USA. Russia went down two positions, ranked 45th.

2 Research Methodology

The methodology of the research is based on complex study of the indicators of technological development of Russia and foreign countries and is based on structural & dynamic analysis, building a complex of economic & mathematical models, analysis of the innovative processes, and application of complex integral indices that reflect the level of development of technologies and innovations.

In this research, descriptive, component, and factor methods are used for studying the economic processes in the sphere of implementation of technologies and innovations. The descriptive methods include analysis of dynamics and structure, graphical method, and method of analytical grouping. The component methods include the index method; and the factor methods include creation of multiple regression models. There's also a group of non-parametric methods, which are based on calculation of various coefficients. Various methods should be combined for achieving the best effect.

The information and empirical basis of the research includes legislative and normative documents of the federal level, the data of the Federal State Statistics Service, information web-sites of the National Programs, and the data of Eurostat. The results are also based on the data that were obtained in the process of independent authors' research and conclusions.

3 Results

3.1 Structural Analysis of Technological Production in Russia

In the modern progressive economy, one of the main elements that from the structure of the national innovative system is hi-tech company (Kotov 2011), (Kuznetsova 2018). This group of companies is oriented at mastering, creation, and implementation of the totality of innovations (Zuev 2018). The performed research showed that in 2010–2017 the share of products of the hi-tech and science-driven spheres in the total volume of GDP grew by 13%, constituting 21.7% in 2017. On average, the indicator was growing by 1.8% each year (Fig. 1).

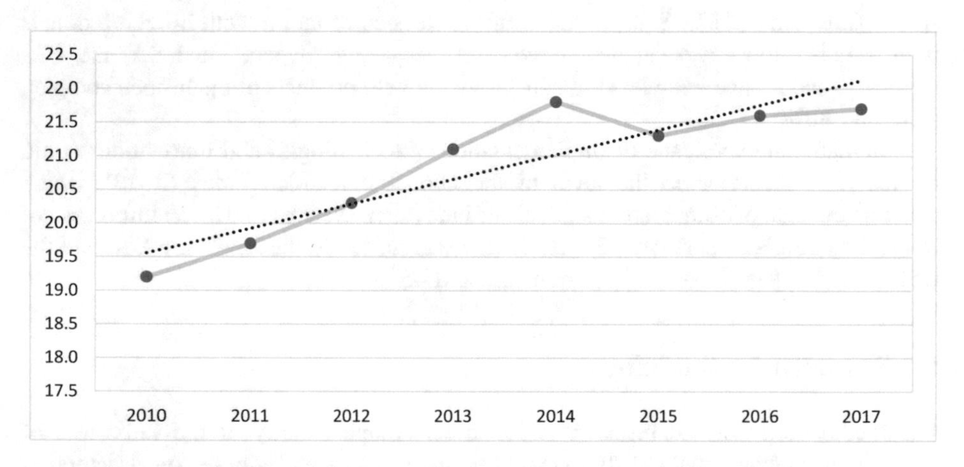

Fig. 1. Dynamics of the share of the hi-tech products in GDP, %.

Figure shows that this indicator has a positive tendency, which is a positive factor for development of the country's economy.

Studying the level of innovative activity in Russia showed that in 2017, as compared to 2010, the share of organizations that implemented innovations dropped to 8.5%. In 2017, the highest share in the total number of innovative activity of organizations belonged to the sphere of scientific R&D (29.8%), and the lowest – to combined agriculture (1.3%) (Fig. 2).

Comparing the developed leading technologies in 2017 to the used technologies for the main types of economic activities, it is possible to see that the highest share in the total volume of the used leading technologies belonged to the sphere of processing production (157,881) and scientific R&D (23,628), and the lowest – to technical maintenance and service (709). It is possible to conclude that the highest number of developed and used leading technologies belongs to the same spheres.

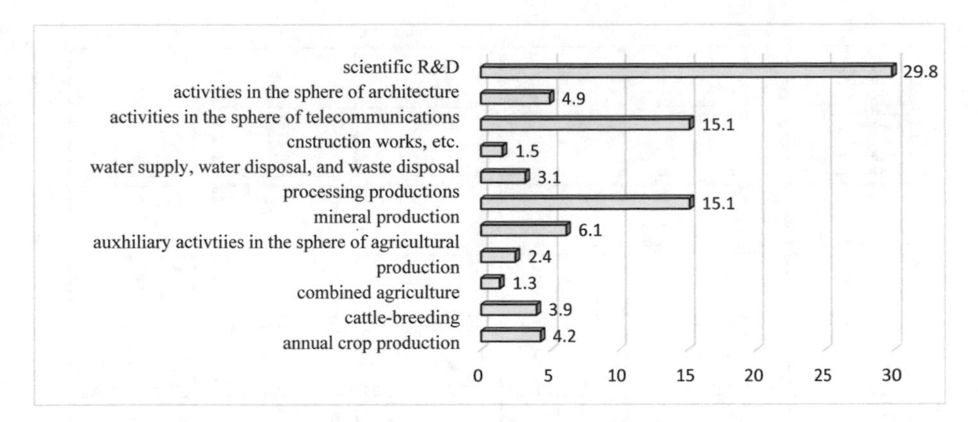

Fig. 2. The structure of innovative activity of organizations for the main types of economic activities in 2017, %.

3.2 Factor Analysis of Emergence of Innovations and Creation of New Companies in Russia Based on Multiple Regression Models

Very often, emergence of innovations is connected to appearance of new companies. Thus, let us forecast the values of birth rate of organizations for 2018 and 2019. Let us build a model, using multiple regression.

For building a multiple regression model of the indicator of birth rate of companies we selected the following characteristics:

Y – birth rate of organizations, per 1,000 organizations;

X_1 – number of population with income below the subsistence level, million people (the more people with low income, the more often they think of opening their own business for improving their material well-being);

X_2 – credits, deposits, and other assets that are given to organizations, individuals, and credit organizations, RUB trillion (people use credits for starting entrepreneurial activities);

X_3 – entrepreneurial intentions, % (people who are going to start business participate in creation of new organizations);

X_4 – general level of innovative activity, % (stimulates creation of a company);

X_5 – number of people that received traumas at production, thousand people (after such cases people often think about opening their own, safe, business).

Graphs and calculations of the model's parameters are obtained in environment R (CRAN). The graph of paired scatter diagrams is presented in Fig. 3.

Table 1 shows a matrix of paired coefficients of correlation of the model's initial parameters.

The regression model is built with the help of the method of backward selection of factors. As a result, the obtained regression equations, their standard errors and results of evaluation of their significance are presented in Table 2.

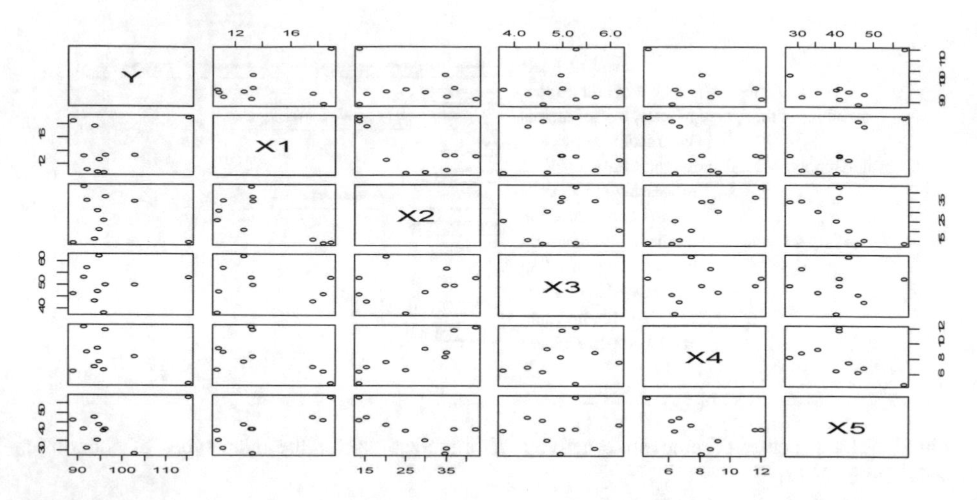

Fig. 3. Scatter diagrams of dependence of the initial parameters of the model and birth rate of companies.

Table 1. Matrix of paired private coefficients of correlation.

	Y	X1	X2	X3	X4	X5
Y	1.000	0.163	−0.032	0.142	−0.207	0.053
X1	0.163	1.000	−0.719	−0.169	−0.496	0.777
X2	−0.032	−0.719	1.000	0.319	0.887	−0.794
X3	0.142	−0.170	0.319	1.000	0.303	−0.230
X4	−0.207	−0.496	0.887	0.303	1.000	−0.490
X5	0.053	0.777	−0.794	−0.230	−0.490	1.000

Table 2. Evaluation of the parameters of the multiple regression model for birth rate of organizations for significant factor features.

	Regression coefficient	Standard error	t-criterion	p-Value	Assessment of p-value	Evaluation of significance
Intercept	61.1015	9.0341	6.763	$1.21 * 10^{-7}$	$p < 0.001$	Significant
X1	0.6662	0.2892	2.303	0.028	$P < 0.05$	Significant
X2	1.3696	0.2572	5.324	$7.73 * 10^{-6}$	$P < 0.001$	Significant
X4	−4.7917	0.8761	−5.469	$5.07 * 10^{-6}$	$P < 0.001$	Significant
X5	0.6906	0.1993	3.465	0.00153	$P < 0.01$	Significant

Standard error of regression residues: 2.981
Multiple R-square: 0.5065, Adjusted R-square: 0.4448
F-criterion: 8.211, p-value: 0.0001127

The obtained equation of the model of birth rate of companies will have the following form:

$$Y = 61.1 + 0.67X1 + 1.37X2 - 4.79X4 + 0.69X5,$$

The estimate value of F-criterion equals 8.211, p-value = 0.0001 < 0.001; the equation is very significant according to F-criterion at the trust level of 95%. Based on Table 2, all regression coefficients in the model are significant at the trust level of 95%, which allows using the obtained model for forecasting the birth rate of a company. Standardized regression coefficients (b-coefficients) of the obtained model are shown in Table 3.

Table 3. Standardized (b-coefficients) of the multiple regression model of organizations' birth rate.

(intercept)	β1	β2	β3	B4
0.0000000	0.475	3.228	−2.280	1.130
Influence of factors	Population	Credits and deposits	Innovative activity	Number of people with traumas

Thus, it is determined which factor influence the birth rate of organizations the most, which allows for more precise forecasting of the dynamics of emergence of new organizations, the level of innovative activity, and frequency of emergence of innovations in Russia.

3.3 Comparative Analysis of Russia and Foreign Countries as to the Level of Technological Development

Study of the international innovative position of countries with the index method, performed by different expert communities, led to the following results. The index of development of information and communication technologies shows countries' achievements from the point of view of development of countries as to the level of information technologies. It is an objective indicator of the country's well-being both in the technological and socio-economic aspects. In this rating, Russia is ranked 45th, far behind the leading countries.

The level of research activity is one of the key indicators characterizing the state of the country's technological development. It is calculated as a total number of research articles that are published in peer-reviewed scientific journals from the system of scientific citation indexing: Science Citation Index (SCI) and Social Sciences Citation Index (SSCI). The rating of countries as to the level of research activity in the international statistics is one of the key indicators that characterizes the state of a country's technological development. Russia's position here is more perspective, but the underrun from the USA is very large. Besides, the country of author's origin is not taken into account, and a lot of leading Russian and European authors have their papers published in the European and American journals in English, for attracting wider audience, increasing the interest, and obtaining grants. In this aspect, the USA is the most attractive platform for scholars.

Human development index (HDI) includes evaluation of the living standards and healthcare and education standards – i.e., the main significant aspects of public life (Anichin 2012). Russia is among the countries with low HDI, ranked 43rd in the world (Rating 2017) (Table 4).

Table 4. Rating of countries as to development of the ICT, RA, and HDI, 2017.

Index of development of information and communication technologies (ICT)			Level of research activity (RA)			Index of human development (HDI)		
Position	Country	Value	Position	Country	Value	Position	Country	Value
1	Iceland	8.93	1	USA	212,384.9	1	Norway	0.953
2	South Korea	8.88	2	China	89,894.4	2	Switzerland	0.944
3	Switzerland	8.86	3	Japan	47,105.7	3	Australia	0.939
4	Denmark	8.71	4	Germany	46,258.8	4	Ireland	0.938
5	UK	8.65	5	UK	46,035.4	5	Germany	0.936
6	Hong Kong	8.61	6	France	31,685.5	6	Iceland	0.935
…	…	…	…	…	…		…	
45	Russia	6.91	15	Russia	14,150.9	43	Russia	0.816

Comparing the factual demand for technologies and innovations from the countries, let us note that it is at a low level in Russia – as to evaluations of business and as to the level of investments into non-material assets (Fig. 4).

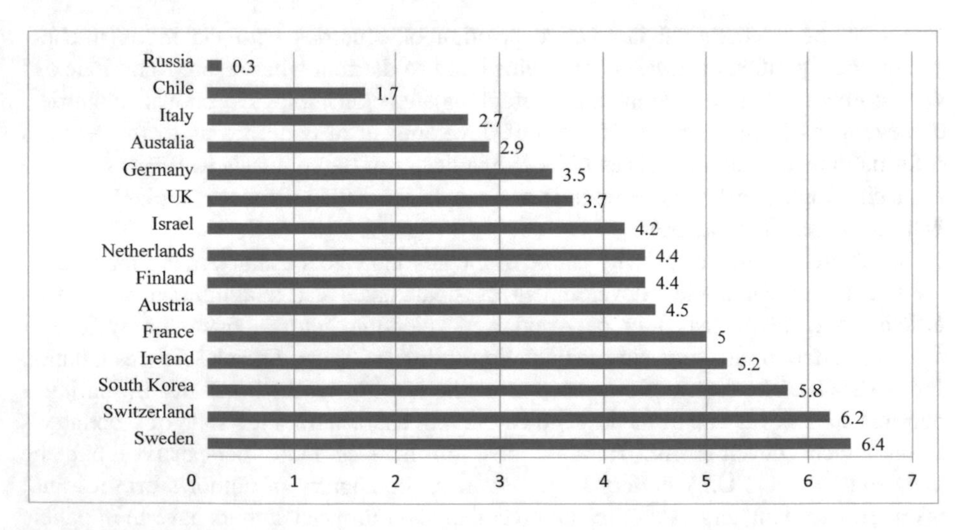

Fig. 4. Demand for technologies in countries of the world (as to investments into non-material assets, % of GDP).

For a more detailed comparison of certain aspects of the level of technological production of Russia, we selected European countries with similar state programs - Germany and France. These countries are members of the EU, which conducts a large-scale policy on transforming the "innovative union" by stimulating the innovative development of its members, establishing key landmarks, and determining top-priority directions.

Studying the dynamics of expenditures for technological innovations in Russia, Germany, and France showed that in 2010–2017 expenditures in Russia grew by EUR 13,688.5 million (3.57 times), constituting EUR 19,009.5 million (or RUB 1,404,985.3 million). In Germany, expenditures in the same period grew by EUR 26,139.1 million (by 37%), France – by EUR 7,735.6 million (by 18%) (Table 5).

Table 5. Expenditures for technological innovations in Russia and countries of Europe, EUR million

	Russia	Germany	France
2010	5,321	70,014.208	43,468.832
2011	9,742	75,569.073	45,111.514
2012	12,009	79,110.378	46,519.037
2013	14,769	79,729.508	47,362.045
2014	16,089	84,246.766	47,918.737
2015	15,936	88,781.819	49,839.13
2016	17,054	92,419.184	50,099.33
2017	19,009.5	96,153.347	51,204.413

In Russia, in the studied period of time, expenditures were growing by EUR 1,955.5 million (1.2 times), or RUB 143,454.5 million each year (Fig. 5).

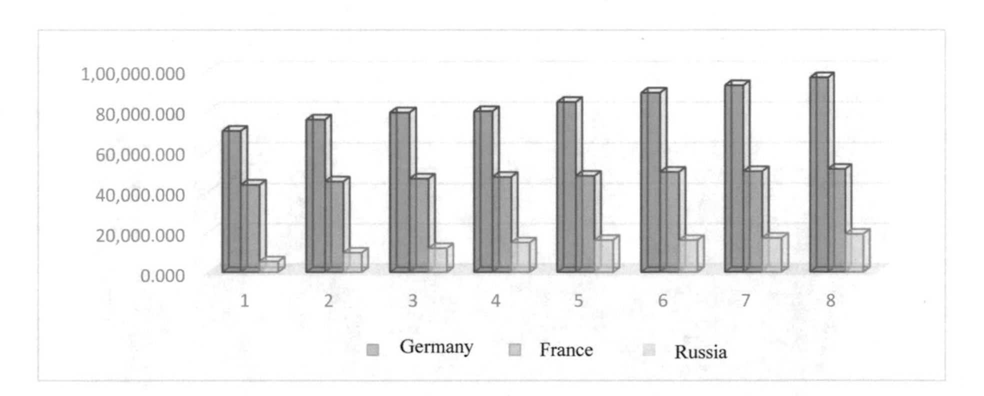

Fig. 5. Dynamics of expenditures for technological innovations, EUR million.

We see that a vivid leader on expenditures for development of the innovative activities is Germany. It is largely due to the leading position of Germany in most spheres. Germany is followed by France, with its stable economy, and Russia. It should be noted that each year the expenditures for development of innovative activities in Russia are growing, which is a positive factor of development of separate spheres of production and of the Russian economy on the whole. Let us study the dynamics of expenditures for technological innovations in the total volume of supplied goods, performed works, and provided services in the Russian Federation (Fig. 6).

Figure 6 shows that in 2010–2013 the share of the used technological innovations in the volume of production was growing, and in 2014–2017 it decreased by 0.3%.

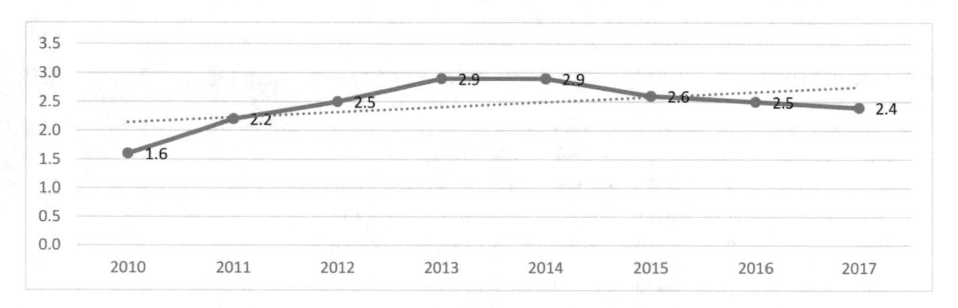

Fig. 6. Share of expenditures for technological innovations in the total volume of supplied goods, performed works, and provided services in the Russian Federation, %.

In 2010–2017, the share of organizations that implement innovative activities in the Russian Federation decreased by 1%, while in Germany it grew by 13% and in France by 13.8%. The share of the leading organizations in European countries is much higher than the share of organizations in Russia – by 3 times in Germany and by 2 times in France (Fig. 7).

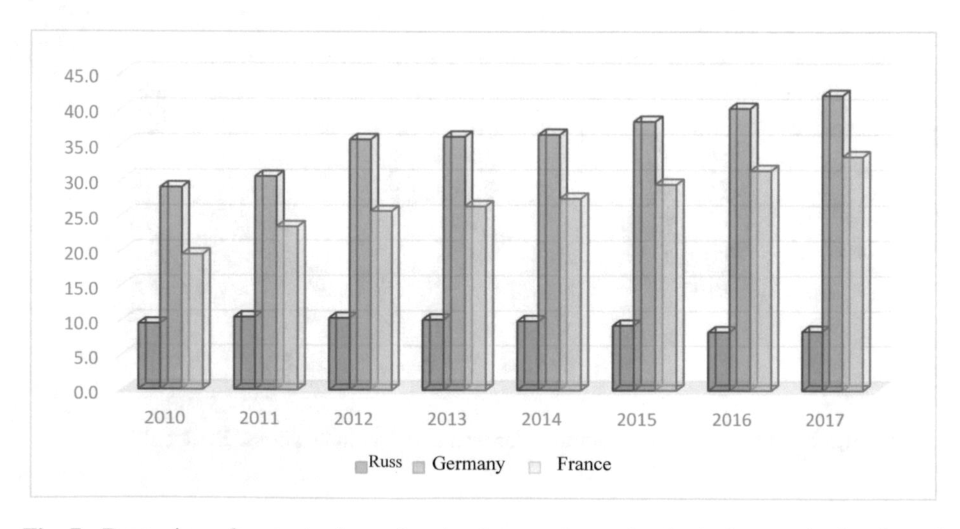

Fig. 7. Dynamics of organizations that implement the technological, organizational, and marketing innovations in the total number of studied organizations.

4 Conclusion

It is possible to conclude that Russia's position in the sphere of technological production and innovations strengthened and acquired an upward tendency. In recent years, expenditures for innovations and investments into development of technological products and services grew. Russia increased export of hi-tech products, increasing its positions in the global market. The multiple regression model helped to determine the factors that influence birth rate of organizations the most, which allows for more precise forecasting of the dynamics of emergence of new organizations and the level of innovative activity and frequency of appearance of innovations in Russia.

However, a comparative analysis of Russia and countries and Europe shows that Russia is behind the leading countries according to most indicators, which shows the necessity for accelerating such directions as development of human capital, increase of effectiveness of expenditures for R&D, and usage of the possibilities of international cooperation in the scientific, technological, and innovative spheres.

References

Anichin, V.L.: VRP: advantages and disadvantages of the indicator, No. 6. pp. 33–47 (2012)

European Statistical Commission (2018). https://ec.europa.eu/eurostat. Accessed 25 Nov 2018

Federal State Statistics Service (2018). http://www.gks.ru/. Accessed 20 Nov 2018

Global Innovative WBC Index (2018). http://www.globalinnovationindex.org. Accessed 20 Nov 2018

Innovative platform of France (2018). https://www.innovationpolicyplatform.org/. Accessed 21 Nov 2018

Kotov, D.V.: Evaluation of the innovative development of the national economy. Current issues of economics and management: materials of the Intern. scientific conf. Russia, 2011, Moscow, pp. 29–33 (2011). https://moluch.ru/conf/econ/archive/9/238/. Accessed 04 Dec 2018

Kuznetsova, N.P., Nasyrova, A.D.: Progressive approaches to introducing innovations in modern management. Forum Young Sci. **2**(18), 344–347 (2018)

Naumenko, I.N.: Problems of innovative development of the economy (2012). http://www.be5.biz/ekonomika1/r2012/2300.htm

Portal of the RF State Program (2018). https://programs.gov.ru/Portal/. Accessed 20 Nov 2018

Porter, M., Stern, S.: The Global Competitiveness Report 2001–2002. New York: Oxford University Press (2001). http://www.isc.hbs.edu/econ-innovative.htm. Accessed 15 Nov 2018

Rating of the countries of the world in terms of the development of information and communication technologies (2017). https://gtmarket.ru/ratings/ict-development-index/ict-development-index-info

Senotrusova, S., Svinukhov, V., Vavilova, E., Makarova, I., Demenko, O.: Economic consequences of Russian embargo on meat and meat products. WSEAS Trans. Bus. Econ. **15**, 93–98 (2018)

Spirin, I., Zavyalov, D., Zavyalova, N.: Globalization and development of sustainable public transport systems. Globalization and its socio-economic consequences. In: 16th International Scientific Conference Proceedings, pp. 2076–2084 (2016)

Zuev, V.M., Manakhov, S.V., Gagiev, N.N., Demenko, O.G.: Problems and prospects of convergence of higher education systems in countries of the eurasian economic union. Espacios **39**(29), 21 (2018)

Zavyalov, D.V., Saginova, O.V., Zavyalova, N.B.: The concept of managing the agro-industrial cluster development. J. Environ. Manage. Tour. **8**(7), 1427–1441 (2017)

Technology of Secure Remote Voting via the Internet

Marina L. Gruzdeva(✉), Zhanna V. Smirnova, Elena A. Chelnokova, Olga T. Cherney, and Zhanna V. Chaikina

Nizhny Novgorod State Pedagogical University named after K. Minin, Nizhny Novgorod, Russia
grul234@yandex.ru, z.v.smirnova@mininuniver.ru, chelnelena@gmail.com, o.t.chernej@pochta.vgipu.ru, jannachaykina@mail.ru

Abstract. *Relevance*

In this article described the technology of remote voting using biometric authentication methods, protected from unauthorized interception of personal data of voters and forgery results, which allows quickly and with minimal cost of material resources to conduct public opinion polls among different groups of the population of the Republic of Kazakhstan, and also conduct elections and referendums at levels from enterprises and regions to country as a whole.

Methods

When conducting research methods of system analysis, construction of computer networks, modern theory of information security, neural networks, biometrics, and cryptography were used. Application of these research methods will allow solving the main scientific and practical tasks of the project, to obtain new theoretically right results. The explanation in favor of the chosen research methods is generally accepted world practice of using them to solve scientific and technical problems and authors' experience in using these methods.

Results

Implementation of remote voting systems, based on introduction of the original concept of security core, allows for comprehensive remote mass control of all procedures at all stages by any interested persons to guarantee the protection of results. Introduction of results will allow a widespread use of remote voting systems, which significantly save material costs for voting procedures, and speed up procedures. Also, they are convenient to participants and generate confidence in results of their implementation.

Discussion

Authors of this article describe the technology of remote voting using biometric authentication methods, protected from unauthorized interception of personal data of voters and forgery results of expression of will, which allows to quickly and with minimal cost of material resources to conduct public opinion polls among different groups of the population of the Republic of Kazakhstan, and also conduct elections and referendums at any levels from enterprises and regions to the country as a whole.

Conclusion

Authors of this article believe that creation of remote voting technology using biometric authentication methods, protected from unauthorized interception of

E. G. Popkova and B. S. Sergi (Eds.): ISC 2019, LNNS 91, pp. 275–281, 2020.
https://doi.org/10.1007/978-3-030-32015-7_31

personal data of voters and falsification of the results of the will, allows to conduct polls of public opinion among different groups of the population of the Republic of Kazakhstan quickly and with minimal cost of material resources, and also conduct elections and referendums at any level from enterprises and regions to the country as a whole. Such systems significantly save material costs of voting procedures, speed up implementation of these procedures, are convenient to the participants and generate confidence in final results.

Keywords: Information security · Risk analysis · Informatization of education · Computer networks

JEL Code: K390

1 Introduction

Problem of creating a perfect system of remote voting through the internet has become more and more urgent over the past two decades, although in different countries among different segments of the population have different ideas about criteria for the perfection of such systems. Among reasons for the lack of significant progress in remote voting technologies in their present form, it is stated that risks and negative consequences of the electoral fraud can be significantly greater than, for example, in electronic commerce. Therefore, additional efforts are required in the development of technology of the data.

2 Theoretical Basis of Research

Problem of creating a perfect system of remote voting through the internet has become more and more urgent over the past two decades, although in different countries among different segments of the population have different ideas about criteria for the perfection of such systems. In works (Jefferson, Sokov, Lessons from the EVOTE 2014 Internation Conferens, Schneier 2004, Vishnyakov 2014, Dodis and Reyzin 2014) considered criteria of systems perfection and compared criteria for voting systems with electronic commerce systems, where Internet technologies found a wide application and have shown their effectiveness. Among reasons for the lack of significant progress in remote voting technologies in their present form, it is stated that risks and negative consequences of the electoral fraud can be significantly greater than, for example, in electronic commerce. (Kuznetsov 2018). Consequently, additional efforts are needed in development of this type of technology.

In European countries, Internet voting is becoming increasingly common: each year an increasing number of countries are using the electronic voting system. At present, the world has accumulated enough experience of voting with remote technologies. For example, in Norway, Switzerland, Estonia, Internet technologies are used in various forms, while in Austria, Brazil and a number of other countries voting is conducted using electronic systems (Table 1).

Table 1. Percentage of voting using remote technologies worldwide

Number of countries using Internet voting	2010	2015	2018
	3 (1,55%)	22 (11,4%)	29 (15%)

During remote voting, experts identified the following problems:

– Difficulty of ensuring the anonymity of voters;
– Inability to provide universal access to remote voting channels;
– Difficulty in tracking the number of votes by voters;
– Possibility of falsification of data or unauthorized interference of the voting system;
– Complexity of control and surveillance by the public, etc.

Computer voting by definition is fraught with programming errors, human errors, technical failures and malicious interference by third parties. Because of the complexity of the technologies used, which are very few available, it is essential that electronic voting systems include a mechanism to verify and track votes. At the same time, an indelible record must be created for each vote, accuracy of which each voter could verify before casting his vote, and which cannot be modified after verification. This must be achieved without violating the secrecy and authenticity of the voting procedure.

Experience of implementation of modern voting technologies in foreign countries allows to formulate requirements for remote voting systems in the work (Schneier 2004), where it is said:

1. Only those who have the right to vote can vote.
2. Everyone can vote no more than once.
3. No one can find out how a particular participant in a voting procedure voted.
4. No one can vote for anyone else.
5. No one can secretly change someone's vote.
6. Each voter can verify that his vote was taken into account when summing up results.

In addition, some voting schemes may require the following requirement:

7. Everyone knows who voted and who didn't.

Most experts conclude that none of the existing systems for voting over the Internet fully meets requirements set out above.

One of the main reasons that prevent the compliance of existing systems with these requirements, as noted in the works (Sokov, Lessons from the EVOTE 2014 Internation Conferens, Schneier 2004, Vishnyakov 2014), is a difficulty of gaining confidence of voters about impartiality when counting votes and preserving secrecy of will expression. In other words, voting system must be designed in such a way as to ensure that there is no possibility of distorting of results or violating secrecy in any way by anyone, including system administrators, who have the highest rights of access to information resources of the system. Only full transparency and control of all procedures of the voting system without exception of any interested entity, regardless of its status, can be a condition for ensuring voters' trust. Presence of at least one non-transparent procedure can cause distrust and discredit voting system (Kuznetsov 2017).

Therefore, functional profile of information resources protection of the remote voting system should include in addition to services of protection against violation of confidentiality and integrity, as well as services related to ensuring full public control. Full public control means such control application of which should leave no doubt about the absence of any deviation from the normal configuration and normal functioning of voting system.

In almost all countries, including Kazakhstan, the main obstacle to remote voting systems in practice of public relations is lack of trust. The fact that results of vote will be objective, without any distortion, reflect the true will of people. Often voting subjects are not sure that organizers of voting procedure (for example, in elections, contests, referendums, polls, etc.) are able to ensure secrecy of the will, neutralize possible attempts to distort voting results, effectively counteract possible illegal actions of malefactors.

In the case of remote voting, a voter is outside of controlled zones of security domains, which significantly increases potential of malefactors to violate the freedom of will expression. In addition, confidential data exchange between voters and server is carried out through open communication channels, which also increases possibility of attacks. Existing systems do not solve the problem of distrust of voters. Current project removes the problem of distrust of remote voting systems over the Internet by mass users and removes the main obstacle to their wide range of use in practice.

In the voting method developed by authors, synthesis of the model of remote voting via the Internet is carried out and formal evidence of crypto stability of protection mechanisms is presented. In case of implementation of the presented development organizers of various types of elections, competitions, polls, referendums, examinations and other events, where possible use of remote voting procedures, will receive a modern, efficient and user-friendly tool for holding such events, ensuring reliable and convincing for their participants guarantees for protection personal data.

Implementation of remote voting systems, based on the introduction of the original concept of security core, allowing for comprehensive remote mass control of all procedures at all stages of voting by any interested party, while ensuring the necessary levels of security of information, will enhance the possibilities of using these systems in practice and will have a positive impact on the development of remote voting technologies.

The main expectation of potential consumers of research results is guaranteed protection of protected data (in particular, personal data of voters, as well as voting results), which are generated, processed, transmitted through open communication channels and stored in remote voting systems.

Implementation of results will allow widespread use of remote voting systems. Such systems significantly save material costs of voting procedures, speed up implementation of these procedures, are convenient to the participants and generate confidence in final results.

3 Research Methodology

When conducting research methods of system analysis, construction of computer networks, modern theory of information security, neural networks, biometrics, and cryptography were used. Application of these research methods allowed to solve the main scientific and practical tasks of the project, to obtain new, theoretically right results. The explanation in favor of the chosen research methods is the generally accepted world practice of using them to solve scientific and technical problems and authors' experience in using these methods.

4 Analysis of Results of the Study

The main expectation of potential consumers of research results is guaranteed protection of protected data (in particular, personal data of voters, as well as voting results), which are generated, processed, transmitted through open communication channels and stored in remote voting systems.

Means of biometric identity authentication of voters are based on both statistical methods (fingerprints, iris, curse geometry, etc.) and dynamic methods (assessment of emotional state and mimics of a person) through constant hidden monitoring with the help of a webcam (Fig. 1).

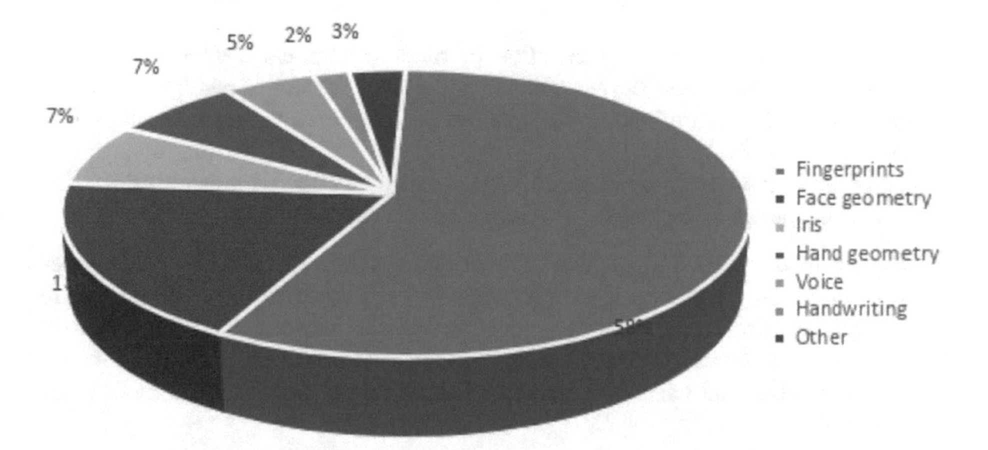

Fig. 1. Percentage of use of different biometric authentication methods

In carrying out researches in the field of development of modern methods of biometric authentication perspective is development of mathematical apparatus, information, and software in this area (Cavoukian and Stoianov 2007).

In the course of this study, following procedures were developed:

- Procedure for setting the minimum and maximum threshold of biometric characteristics:

- Procedure for determining the stability of biometric characteristics of a user depending on stability of his emotional state and mimics;
- Procedure of establishing minimum and maximum limits, in which an algorithm of increasing security of information will work;
- Procedure for calculating "pass corridor" and average value for each biometric characteristic of a user;
- Procedure of an algorithm to increase the security of information based on assessment of emotional state and mimics;
- Procedure for creating a biometric user key.

The method of voting results protection developed by authors was tested on the basis of Zhetysu State University named after Zhansugurov. Comments and recommendations of students and teachers of the university participating in approbation served as the basis for improvement and debugging of the system.

Participants of the approbation noted the advantages of proposed technology of safe remote voting:

- use new protection system, based on the implementation of original concept of the security core, involving remote mass control by any interested persons, number of which is not limited, behind the processes and files on the server of the voting system, which does not allow for substitution of software or hidden (including malicious) interfering of server process. In this case, all remote voting software is open and accessible for any advance checks by anyone wishing to perform such tests;
- use of new mechanisms of generation of random numbers realized by standard means of Internet access devices that do not require the use of additional software and hardware, which promotes the massive use of remote voting technology for a wide range of tasks;
- use of new cryptographic means of information protection with proven durability, as well as means of biometric authentication of voter's identity.

Implemented in this study, new remote voting systems, based on original concept of the security core, allows for a comprehensive remote mass control of all procedures at all stages of voting on the part of any interested parties, ensuring the necessary levels of durability of information protection, will increase possibilities of the use of these systems in practice and will have a positive impact on development of remote voting technologies.

5 Conclusion

Ensuring secrecy of voting is currently one of the main problems in voting through the Internet. Implementation of results of this study will allow widespread use of remote voting systems. Such systems significantly save material costs of voting procedures, speed up implementation of these procedures, are convenient to the participants and generate confidence in final results.

References

Jefferson, D.: If I Can Shop and Bank Online, Why Can't I Vote Online? https://www.verifiedvoting.org/resources/internet-voting/vote-online

Sokov, B.B.: Development and analysis of methods of multifactor authentication using biometric characteristics of a person. Scientific community of students of the XXI century. Technical science: Collection of articles on materials. LX International student scientist - practical conference No. 12 (59). https://sibac.info/archive/technic/12(59).pdf

Lessons from the EVOTE 2014 InternationConferens http://e-lected.blogspot.com/search?updated-min=2014-01-01T00:00:00-08:00&updated-max=2015-01-01T00:00:00-08:00&max-results=50

Schneier, B.: What's Wrong With Electronic Voting Machines? https://www.schneier.com/essays/archives/2004/11/whats_wrong_with_ele.html

Vishnyakov, V.M.: Open system secret voting. In: Vishnyakov, V.M., Prigara, M.P., Voronin, O.V. (eds.) Management of warehousing systems. Collection of science articles, Vip. 20, pp. 110–115 (2014). http://urss.knuba.edu.ua/files/zbirnyk-20/22.pdf

Dodis, Y., Reyzin, L., Smith, A.: Fuzzy Extractors: How to Generate Strong Keys from Biometrics and Other Noisy, Data 13 April 2004

Verbitskiy, E., Tuyls, P., Denteneer, D., Linnartz, J.-P.: Reliable Biometric Authentication with Privacy Protection. In: Proceedings of 24th Benelux Symposium on Information Theory (2003)

Cavoukian, A., Stoianov, A.: Biometric Encryption: A Positive-Sum Technology that Achieves Strong Authentication, Security AND Privacy, March (2007). http://www.ipc.on.ca

Daugman, J.: Probing the uniqueness and randomness of iris codes: results from 200 billion iris pair comparisons. In: Proceedings of the IEEE, vol. 94, no. 11, pp. 1928–1935, November 2006

Bragina, E.K., Sokolov, S.S: Modern methods of biometric authentication: review, analysis and definition of development prospects. Bulletin of ASTU. 2016 No. 1 (61). https://cyberleninka.ru/article/v/sovremennye-metody-biometricheskoy-autentifikatsii-obzor-analiz-i-opredelenie-perspektiv-razvitiya

Kuznetsov, V.P., Garina, E.P., Andryashina, N.S., Romanovskaya, E.V.: Models of Modern Information Economy Conceptual Contradictions and Practical Examples, p. 361. Emerald Publishing Limited, Bingley (2018)

Kuznetsov, V., Kornilov, D., Kolmykova, T., Garina, E., Garin, A.: A creative model of modern company management on the basis of semantic technologies. Communications in Computer and Information Science (2017)

Transformation of the Industrial Potential in the Digital Economy

Elena V. Dudina[1(✉)], Ekaterina I. Mosina[2], Elena E. Semenova[2], Maria A. Stepanova[2], and Nataljya A. Baturina[2]

[1] Orel State University, 95 Komsomolskaya Str., Orel, Russia
orel-osu@mail.ru

[2] Orel State University of Economics and Trade, Orel, Russia
super-ya-57@mail.ru, osuet@mail.ru, orel-osu@mail.ru,
1278orel@mail.ru

Abstract. A formalized approach to the study of the industrial production demonstrates the need to transform key criteria. One of these criteria is the industrial potential. The industrial potential declares a set of methods and means of production. The digital environment changes the significance of this concept. This framework transforms the industrial potential in the digital economy. The purpose of the scientific article is to determine the role of industrial potential in the digital economy. The main objectives of the study are: to consideration of the structuring of industrial potential; to determination of the transformation processes of the industrial potential on the basis of manifesting factors; to grouping the levels of the industrial potential development within the digital economy; to proposal for the transformation of industrial potential in the digital economy. Methodological tools of research include: the method of scientific definitions, the method of external reflection of essential circumstances, the method of decomposition components, the method of factorization of the whole, the method of aggregate differentiation, the method of scientific presentation, the method of deduction and induction, the method of analysis and synthesis, the method of scientific substantiation and analysis, the method of formalization, the generalization method and the abstraction method. Dedicated tools form the theoretical part of the study. The theoretical part of the research is aimed at determining the internal and external components of the industrial potential. The practical significance of the study determines the importance of the use of digital tools in the context of the transformation of industrial potential.

Keywords: Industrial potential · Digital economy · Transformation · Changes · Factors · Life cycle

1 Introduction

The production of change in society leads to transformational processes. Adjustment of priorities occurs. The potential of tools and tools reflects the aspect of change. Potential models intersectoral changes occurring. Potential forms a certain number of conditions. This series of conditions is associated with emerging external transformation processes.

E. G. Popkova and B. S. Sergi (Eds.): ISC 2019, LNNS 91, pp. 282–290, 2020.
https://doi.org/10.1007/978-3-030-32015-7_32

One of the characteristic priorities is the industrial potential. The industrial potential reflects a set of industry characteristics, regulated in the context of emerging production areas. Let us highlight some priority aspects of the industrial potential.

The mining nature of the industrial potential states the significance of the resource. The industrial potential regulates the group of minerals. Minerals are the basis for the formation of this potential. The processing nature of the industrial potential determines the instrumental basis of the resource component. The processing nature of the industrial potential allows us to distinguish the technological base of the industrial production. These characteristics determine the nature and significance of the industrial potential.

Another aspect of the study of the industrial potential is the sectoral focus. The sectoral focus of the industrial potential produces the importance of separate enterprises producing single products based on similar technologies. Mining and processing characteristics for the selected enterprises are the same. The sectoral focus of industrial potential classifies industry trends. Destinations include the metallurgy, the fuel industry, the electric power, the metalworking, the light and food industries and other typical the industrial production.

The industrial potential transformed in accordance with environmental conditions. The emergence of the conceptual framework of the digital economy leads to the need to take into account these conditions in the formation of the industrial potential. Signs of the use of breakthrough technologies within the industrial potential are put forward in the first place. Secondary are the models of the organizational-market nature of the industrial production. Within the framework of the thesis put forward, the significance of the topic being studied is determined. The significance aspect is directed at examining the industrial potential within the digital economy.

The purpose of writing a scientific article is to determine the role of the industrial potential in a digital economy. The goal indicates a multitasking theme. Multitasking solution implies the following:

- to consider the structuring of the industrial potential;
- to determine the transformation processes of the industrial potential on the basis of the manifesting factors;
- to group the levels of development of the industrial potential within the digital economy;
- to propose the transformation of the industrial potential in the digital economy.

Multitasking research reveals theoretical and methodological characteristics. Firstly, the study assesses the internal component of the industrial potential. Secondly, the influence of the factor component on the regulation of the industrial potential is highlighted. Thirdly, the levels of the industrial potential are used in the digital economy.

2 Methods

Methodical tools are defined in the context of a scientific nature. The unified system of methods reveals the essential elements of the subject under study. The main methods are:

– the method of scientific definitions is the essence and internal components of the industrial potential;
– the method of external reflection of essential circumstances is the idea of the external manifestations of the industrial potential;
– the method of decomposition components is a set of elements of the industrial potential as a single subject of scientific research;
– the method of factoring the whole is the factorial components of the priority elements of the object under study within the framework of the selected criteria;
– the method of aggregate distinctions is the implication of circumstances on the basis of level components;
– the method of scientific presentation is the formation of proposals in the framework of the studied topic.

Formed methods are narrowly focused. The study of this subject requires the application of general scientific methods. General scientific methods include: the method of deduction and induction, the method of analysis and synthesis, the method of scientific substantiation and analysis, the method of formalization, the method of generalization and abstraction. Let's apply the methods within the framework of scientific research on the transformation of the industrial potential in the digital economy. Initially, we consider the structuring of the industrial potential.

3 Main Part

The study of the industrial potential is based on the identification of definitions and scientific studies. The main definitions are:

– the industrial potential is a set of production characteristics reflecting the production properties of goods produced by industrial sectors (Oakey 1991);
– the industrial potential is a differentiated set of goods and enterprises within a single direction of the industrial sector (Ershova and Karakulina 2018);
– the industrial potential is the material base of social reproduction of products of the industrial sector characterized by a combination of productive capital (Smetanov 2016).

Highlighted directions cause some characteristic features. Firstly, the industrial potential acts as a production characteristic. The production function of the industrial potential reflects the synopsis of the means of producing differentiated products, industrial orientation. Secondly, the industrial potential reflects a differentiated set of goods and products. Industry is inherently demarcated into a large number of segments. Each of the industrial segments forms products for production in the industry. Thirdly, the structuring of the industrial potential reflects a single view of this direction through various conditions of the external and internal state. Consider this thesis in more detail (Fig. 1).

The industrial potential is structured within the internal component and external manifestation. The internal component of the industrial potential reflects the structure of this definition, revealing its internal characteristics.

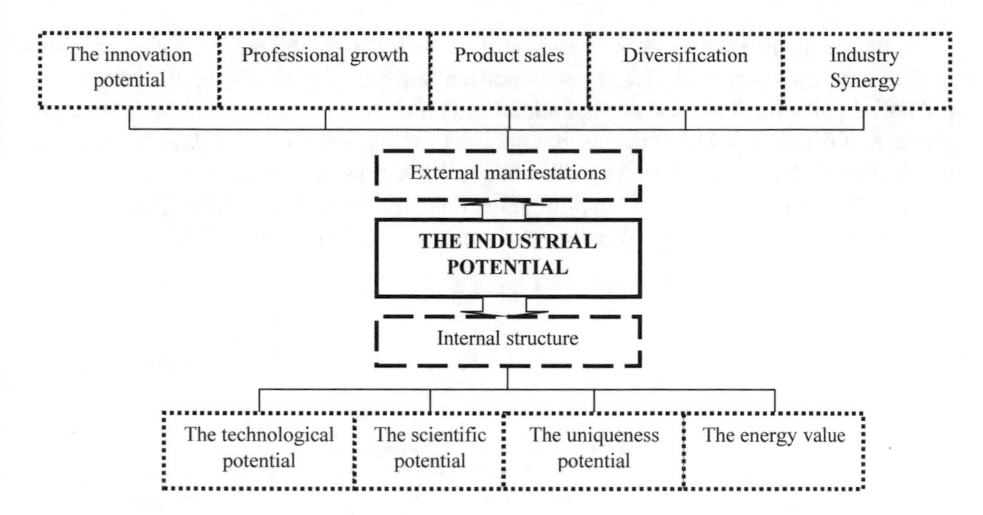

Fig. 1. Structuring of the industrial potential

Based on the research, the following internal structure of the industrial potential was identified:

- the technological potential is a set of tools, methods, tools and technologies in production activities with the aim of forming the final product that satisfies consumer preferences;
- the scientific potential is a set of educational and resource capabilities, accumulated in the process of organizing the solution of economic problems;
- the potential of uniqueness is a set of exceptional characteristics differentiating the result depending on the singularity of application of the methods, means, technologies, tools of this definition;

The energy value is the positive emergence of properties and characteristics of non-intrinsic, studied definitions (Gafiyatullina et al. 2015).

External manifestations reflect the manifestations of the industrial potential. These methods isolate the importance and necessity of using the industrial potential. The external manifestations of the industrial potential include:

- the innovation potential is a set of resource components for the organization of innovation activities in the framework of the creation of the final product (Vagner 2009);
- professional growth is a characteristic, skills and competencies necessary to increase the criterion level of actors;
- product realization is a set of operations that help reduce the flow of creation and full satisfaction of the consumer's needs with a product;
- diversification is a complete variety of properties, types and characteristics of a single set of products, goods, services (Saltykov 2012);
- sectoral synergy is an increase in the effect of two or more adjacent sectors (industries).

External manifestations and internal structure determine the essential characteristic of the industrial potential. These areas concentrate the importance of the use of the industrial potential. The industrial potential is a criterion and property of the production process. It is important to consider the process of transformation of industrial potential under the influence of factors. As part of this task, we consider the transformation of the industrial potential. The transformation of the industrial potential includes the life cycle, the scientific and technological revolution, government incentives/restrictions (Fig. 2).

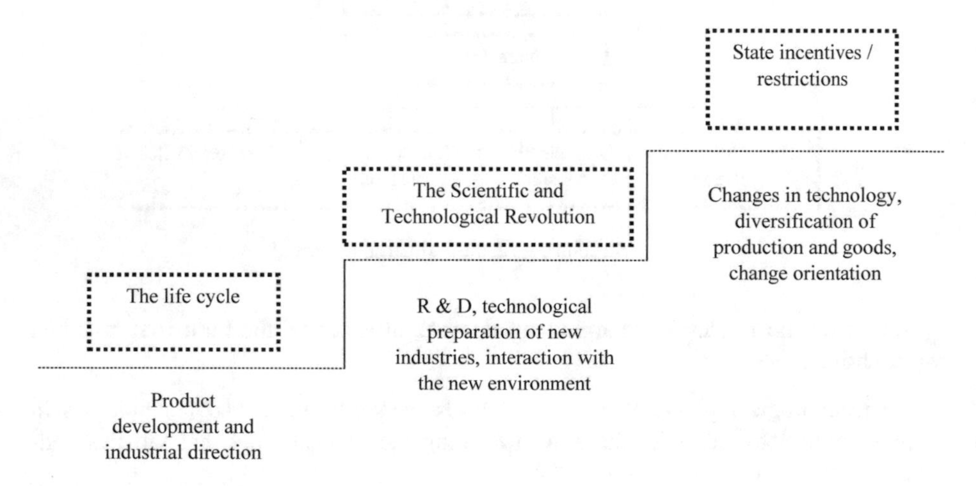

Fig. 2. Transformation of industrial potential based on emerging factors

The identified factors express organizational transformation (the life cycle), state transformation (state stimulation/restriction), country transformation (the Scientific and Technological Revolution). The life cycle is a range of natural phenomena by changing ontogenetic generations (Salimov and Mingaleev 2012). The life cycle of the industrial potential has an impact on the development of goods and industrial areas. This characteristic is manifested in the process of changing ontogenestical generations. A little differently factorized industrial potential in the context of the scientific and technological revolution. The scientific and technological revolution has an impact on the production of R&D. The development of R & D contributes to the emergence of a new environment-competitors. This circumstance provokes the need to match the industrial potential of the characteristics of research and development. Government incentives/ restrictions imply a transformation process. Initially, the transformation of the industrial potential implies a change in technology with their possible diversification. The main characteristic of this transformation is the orientation towards changes, focused on the internal structure and external manifestation of the industrial potential.

The essence and factorial features of the industrial potential allow us to form a current understanding of this concept. Digitalization conditions do not take into account the industrial potential as a means of reflection. To exclude this aspect from the study, we will form the levels of development of the industrial potential within the digital economy.

4 Result

The levels of development of industrial potential in a digital economy reflected in Fig. 3.

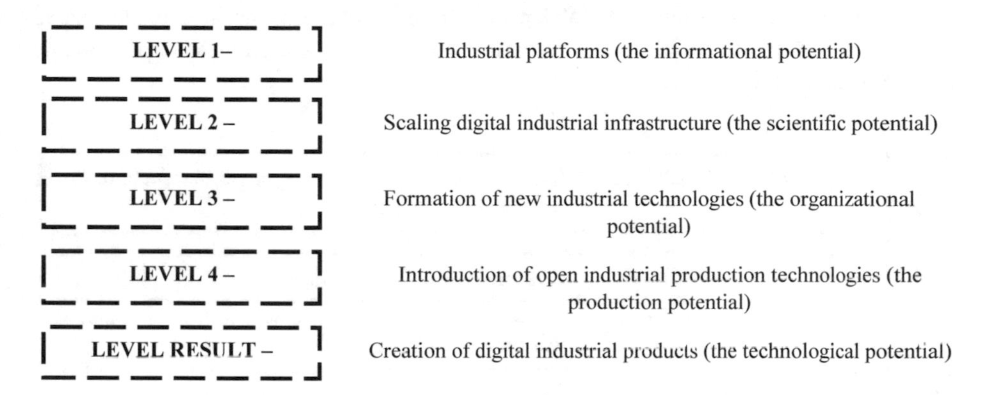

Fig. 3. Levels of the industrial potential development within the digital economy

The first level of digitalization is associated with biased information potential and industrial potential. The priority of information flows and technologies corrects the industrial potential towards the formation of the industrial platform. The industrial platform is a technological platform that implements the result of industrial activity. The industrial potential reflects the marketing component of industrial activity. The industrial platform is formed on the basis of the electronic industrial market. The industrial platform correlates flows between agents providing industrial interconnections. Level 1 allows you to reflect the industrial potential through a technological platform.

The second level of the industrial potential digitalization based on the interaction with scientific potential. The context of the scientific component connected with the elaboration of the issue of scaling the digital industrial infrastructure. The scale of the digital industrial infrastructure is driven by the need for strategic and software. The program approach provides a digital composition of the industrial potential. The industrial potential is considered as a software product. This product aimed at scaling and visualization of industrial infrastructure (Stroeva et al. 2017).

Level 3 digitalization regulates the process of transforming organizational potential. The organizational potential is involved in the formation of new industrial technologies. The introduction of new industrial technologies doesn't state their innovativeness. Innovative aspect refers to software products. The industrial technology is a versatile product that has additional, related characteristics that allow ascertain its unique character. The industrial potential defined as a means, product, formed in the regulations of the new industrial technology.

The adjustment of level 3 on the development of the industrial potential produces a relationship with the production potential. The production potential is transparent. The factor of transparency reflects the process of introducing technologies of open industrial production. This fact testifies to the constant technologization of the industrial potential. The industrial potential is determined through the introduction of technology into industrial production based on the concentration of external changes.

The final level of the industrial potential development is the creation of digital industrial products. The key role is played by the technological potential. The technological potential determines the totality of innovative connections. The technological potential is unique. The advanced properties approximate the need to transform the industrial potential (Fig. 4).

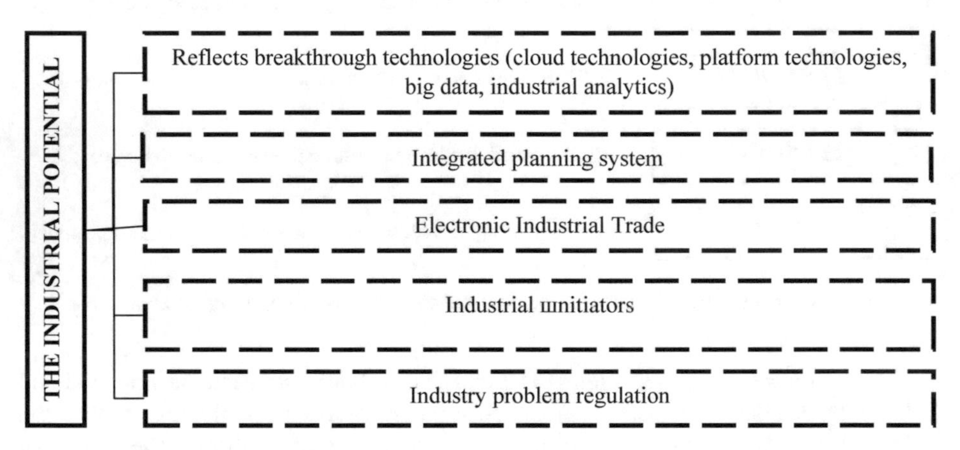

Fig. 4. Transformation of the industrial potential to the conditions of the digital economy

The transformation of the industrial potential is an addition to the digital economy. The transformation of the industrial potential allows to form a clear regulation of the parameters of this phenomenon. Based on the author's generalizations, five aspects of the transformation of the industrial potential are proposed. Consider them in more detail.

The industrial potential reflects the importance of breakthrough technologies (the cloud technologies, platform technologies, big data, industrial analytics). Summarizing the final level of industrial potential development, the importance of creating digital industrial products been established. The creation of digital industrial products due to the availability of breakthrough technologies. Breakthrough technology is an innovative range of technological tools that create products, goods, services, production activities. Industrial potential takes into account the presence of breakthrough technologies in production activities.

The industrial potential reflects an integrated planning system. Integrated planning reflects a set of industrial production tools. The system of integrated planning of the industrial production determines the links of industrial supplies, the levels and horizon of planning, methods for reducing riskiness.

Full electronic industrial trade is comparable to the level of the development of the industrial production. This transformation component defines the basis of digitalization virtualization. The actors of this transformation are industrial initiators. Industrial initiators are persons involved in the organization, production, and implementation of industrial activity. The industrial potential forms a collection of data on industrial initiators.

On the second plan in the process of transforming the industrial potential goes the current planning. The digital economy is prioritizing problem-based industry regulation. The aspect of eliminating problem situations is the basis of strategic planning. Problem regulation of industry allows for the accommodation of negative processes.

Selected aspects of the transformation of the industrial potential change the essence of this component. Currently, industrial potential forms a system of aggregate interconnections and production elements. Transformation of the industrial potential requires modification of this component. Breakthrough technology is given priority transformation.

5 Conclusion

The conducted scientific research led to the following conclusions.

1. The structuring of the industrial potential revealed two key patterns. Firstly, the internal structure of the industrial potential forms an idea of the essence of this component. Secondly, the external manifestations of the industrial potential determine the response to changes. These features allow a comprehensive assessment of the definition of «the industrial potential» .
2. Selected factors allowed to imitate the idea of transformation processes of industrial potential. This transformation occurs of three links. The links are organizational element, country element, international element. Organizational element determines the characteristics of the development of goods in the context of industrial activity. The country link is focused on the implementation of measures to support/limit industrial activity. The international link reflects the impact of scientific and technological progress on changes in the industrial potential.
3. Grouped levels of the industrial potential development include an industrial platform, digital infrastructure, new industrial technologies and manufacturing, digital industrial products. The industrial potential correlates with the information potential, the scientific potential, the organizational potential, the production potential and the technological potential.
4. The main proposals for the transformation of the industrial potential associated with the development of this component. Emphasis is placed on the reflection of the importance of breakthrough technologies, an integrated planning system, electronic industrial trade, industrial initiators, and problem-based industry regulation. Dedicated postulates contribute to bringing the industrial potential to the conditions of the digital economy.

References

Ershova, I.G., Karakulina, K.N.: Methodical approach to the management of the industrial potential of the enterprise. Sci. Bus. Ways Dev. **12**(90), 90–93 (2018)

Gafiyatullina, A.Z., Pavlova, A.V., Vesloguzova, M.V., Takhaviev, R.K., Kashirina, I.B., Ashmarina, S.I.: Structure and development of the intellectual potential of the industrial enterprise personnel. Rev. Eur. Stud. **7**(1), 117–122 (2015)

Oakey, R.: High technology small firms: their potential for rapid industrial growth. Int. Small Bus. J. **9**(4), 30–42 (1991)

Salimov, R.I., Mingaleev, G.F.: The regional policy of industrial IPM services for the development of knowledge potential in Russia. Implementing International Services: A Tailorable Method for Market Assessment, Modularization, and Process Transfer, pp. 425–435 (2012)

Saltykov, N.A.: Modeling the process of managing the innovative potential of an industrial enterprise. Bulletin of Udmurt University. Ser. Econ. Law **2**, 64–69 (2012)

Smetanov, A.Y.: The development of the industrial potential of the metropolis in the market. Russian Eng. **1–2**(48–49), 76–79 (2016)

Stroeva, O., Lyapina, I., Mironenko, N., Lukyanchikova T., Polyanin, A.: Formation of Organizational Directions for Implementing Road Mapping into Activities of Industrial Enterprises. Overcoming Uncertainty of Institutional Environment as a Tool of Global Crisis Management. Ser. Contributions to Economics, pp. 143–150 (2017)

Vagner, O.V.: Management of innovative and technological potential of an industrial enterprise. Scientific and technical statements of the St. Petersburg State Polytechnic University. Economics **4**(81), 260–270 (2009)

Scientific Discussions of FES Russia on Digitalization of Economy of the 21st Century

Konstantin V. Khartanovich(✉), Tatyana V. Starikova, Alexander V. Milenky, and Natalya E. Tikhonyuk

Vladimir Branch of RANEPA, Vladimir, Russia
kohart@yandex.ru

Abstract. The world economic science in the 21st century faced a digital challenge. Unfortunately, it was not ready to accept it.

Scientific circles are actively searching for answers to numerous questions of development of modern and future digital economy. A lot of opinions presented. But, as we know, truth is born in discussions.

Purpose of the work: analyze important aspects of scientific discussions of scientists and practitioners of FES Russia on problems of digitalization of the economy, which are still being conducted at the most primary level. Scientists are trying to reach consensus on at least the basic concepts of the digital economy, starting with the identification of it.

The methodology for studying a content of discussions consists of: development of conceptual apparatus, method of interviewing and opinions of experts, statistical (quantitative) methods, methods of forecasting, estimation methods of state policy in the field of digital economy.

The results of the study: revealed common points of ground on the conceptual apparatus, generalized trends of digitalization, revealed the remaining differences. Found that the government is a powerful driver of digitalization.

Recommendations: give the status of a separate National project to a set of measures for the digitalization of the economy, Russian own technologies. Defense industry is a multiplier of digitalization of the national economy and also the attraction of mass investment in specific projects, as well as training of highly professional specialists.

Keywords: Economy · Digital economy · Digitalization · E-government · New industrialization · Government technology policy

JEL Code: O14 · O33 · O38

1 Introduction

The cardinal economic innovations in the 21st century have rapidly entered Russian and world political and socio-economic reality. In the book by Klaus Schwab, the world-famous economist, founder of the Davos Economic Forum, "The Fourth

E. G. Popkova and B. S. Sergi (Eds.): ISC 2019, LNNS 91, pp. 291–297, 2020.
https://doi.org/10.1007/978-3-030-32015-7_33

Industrial Revolution", published in Russian in 2017, talks about the upcoming changes in human life.

K. Schwab is convinced that the fourth revolution will be the most difficult in the history of the existence of mankind. He believes that the fourth industrial revolution will begin in 2025 (Schwab 2017).

Relevance of investigated process is caused by scientific and practical necessity of analysis of the formed base of the digital economy.

Largest public association of economists in the country is the Free Economic Society of Russia, the first institute of civil society, created by decree of Empress Catherine II in 1765 (FES 2018).

Association makes a huge contribution to the development of various aspects of digitalization of the economy. It brings together representatives of many economic directions and views.

It has developed a common vision of the main drivers of development:

1. New industrialization (analogue of the Western concept of the fourth technological revolution). 2. Digital economy. 3. The policy of import substitution, relying on its own technological power.

The President of the Russian Federation used the term "own" several times in his message to Federal Assembly of the Russian Federation, showing the importance of maintaining independence of the domestic information market and technological security of our economy. He noted that it is necessary to create own digital platforms compatible with the global information space. It is important to ensure our own development and localization of key technological solutions. (The main content of the message of President of the Russian Federation 2018).

The most detailed debate is about the digital economy. These details cover definition, goals and objectives, mechanisms and tools, social and entrepreneurial benefits, and, of course, effectiveness and consequences. Of course, it is impossible to focus on all aspects of the discussion under this article. Let's highlight some important in our opinion.

2 Methodology

2.1 Scientific Discussions on the Term "Digital" Economy

In 2017, the first mass discussion on "Technological components of the new economy" took place in FES of Russia. Vice-President of the FES, Corresponding Member of the Russian Academy of Sciences, Doctor of Economics, and Professor D.E. Sorokin recalled that many definitions to the basic category - "ECONOMY" - over the past decades tried to use.

Dmitry Evgenyevich refreshed memories that in 2001 the course on "innovative economy" which would have developed in 2007–2010 was proclaimed. "Then we formulated the priority of modernization of the economy. Now we have formulated the priority of digital economy, although the transition to which everyone understands differently, but it is possible that it is the shift in productive forces that may change the system of all socio-economic relations." (Sorokin 2017)

There are many discussion platforms in FES. One of them very respected by authors was on an annual forum "Abalkinsky Readings", in memory of the great economist Leonid I. Abalkin. on March 6, 2018, with the theme "Digitalization and National security."

At the readings were discussed a lot of relative "synonyms" - "digital", "web economy", "digital economy", Internet economy, etc.

During discussion, the most pressing problems of digitalization of the economy were discussed. A lot of questions have no answers yet.

The authors of this article believe that the digital economy is an innovative part of information technology that allows for the provision of a complex, interrelated operation and development of industries, infrastructure, government, social institutions, and citizens.

2.2 Interviewing Method and Expert Opinions

Sergey Bodrunov, President of FES Russia, put forward the most topical issues for scientific discussions:

"What kind of economy are we digitizing? The raw economy? Or what kind of economy are we going to put to digitalization? Will our digitalization contribute to accelerating the development of product vector? Should we start digitalization now, when there is no new economic model and there are no clear benchmarks in the economy, what do we want to get in general? When we change priorities, we risk losing what we have, and not buying new ones." (Bodrunov 2017b)

Nobel Prize laureate Zhores Ivanovich Alferov expressed even sharper: "It is absurd to talk about some digital economy without having an element base." (Alferov 2017). Academician Alferov is right, most IT technologies are imported.

Member of the Presidium of FES Russia, First Deputy Chairman of the Committee of the Federation Council of the Federal Assembly of the Russian Federation on Economic Policy, Doctor of Economics, Professor S. IN. Kalashnikov suggested that the digital economy is an inevitable future but there is no clear vision of this future and that "we are in our misunderstanding of what will happen tomorrow, misunderstanding that we are on a cardinal turning-point. The vast majority of countries, even the most actively financing various modern information technology production, economically, that is, in economic theory did not understand these processes. And we are here on equal terms with them, and we are not more foolish than others…" (Kalashnikov 2017).

2.3 Statistical (Quantitative) Methods in Discussions and Their Assessment

In the context of economic instruments, we can talk about the volume of the digital economy in GDP. The statistics are as follows: Russia −3.9%; USA 10.9%; China 10%; EU 8.2% (Greenberg 2017). Strong positions in the world have: Great Britain, Denmark, Finland, Singapore, South Korea and other countries.

Statistics are always relative indicators, sometimes contradict some figures to another. Executive Director of the Board of Huawei Wu Xu complains about the lack of a unified statistical methodology. He mentions such figures that in 2016 digital economy was 15.5% of world GDP (11.5 trillion dollars). In 2025, according to him, digitalization will give 24.3 of world GDP (Wu Xu 2017).

Scientists of the Higher School of Economics published the latest statistical materials on the state of digital economy of Russia and world data. Statistics are provided for all three subjects of market relations—public sector, business sector, and home economy (Digital Economy 2019).

These statistics show that the digital economy is already in existence and is actively developing. That's a real fact.

2.4 Methods of Forecasting Dynamics of Digitalization of Economy

Methods of economic forecasting, especially strategic ones, have always been and will be the most difficult. A wide variety of options for prospects of the digital economy are presented. At least 3% of world GDP by 2024. Maximum—30%.

MasterCard and Tufts University presented the Digital Evolution Index rating for 60 countries in 2017. Russia was included in the group of promising countries, which "despite relatively low level of digitalization, are at the peak of digital development and demonstrate steady growth rates. It is expected that within 10 years Russia will be able to increase the volume of the digital economy by one and a half or two times." (Bodrunov 2017c)

By 2025, Huawei and Oxford Economics forecasts 23.3% of the world economy will be digital economy with a total volume of 23 trillion dollars. The equivalent additional income per average worker will be about $500. Digital economy will be a major factor in global economic growth (Report 5 September 2017).

2.5 Method of Assessing the Effectiveness of Government Policy in the Sphere of Digital Economy

It should be noted that government is actively introducing digital technologies at all territorial levels—federal (e-government with a portal of public services), regional and municipal (have their own pages on the Internet).

The digital economy is an economic activity in which the key factor of production is data in digital form, processing large volumes and using results of analysis of which compared to traditional forms of management allow to significantly increase the efficiency of various types of production, technologies, equipment, storage, sale, delivery of goods and services (including number and state) (Kuznetsov 2018).

3 Results of the Study

3.1 Common Points of Contact on Conceptual Apparatus

Most scientists consider "digital" economy as a way to find a way to answer HOW to produce. It is a purely innovative technological mechanism of production (Garina 2017).

However, many researchers still take the term “digital economy” in quotation marks. It seems to be necessary to take quotation marks not two words, but one —“digital” economy, which shows not a change in the basic essence of the concept of economy, but focuses on innovative technology of production economic product.

In other words, the “digital” economy is part of economic relations, which is mediated by the Internet, cellular communication, information, and communication technologies. It is a virtual environment that radically changes the economic way of life of society.

At the same time, the main goal of the development of the “digital” economy, as noted by all researchers, is to increase the competitiveness of the industry in conditions when: (1) country cannot compete in labor costs with developing countries and even with some developed countries; (2) software development and software platforms will become a powerful basis for digitalization (Kuznetsov et al. 2017).

3.2 Synthesis of Digitalization Trends

Among many different trends of digital economy, in the opinion of FES, it is necessary to consider in the first place:

1. Ensuring cyber security. Natalia Kasperskaya, in her speech at the St. Petersburg Digital Forum in 2017, noted: “It is not necessary to digitalize for the sake of digitalization or because the world simply moves forward. Risks of digitalization are clear: any modern technology has remote control. If you completely put the government administration on a digital system, it can be turned off at some point.” (Kasperskaya 2017).
2. Digitalization of public services. The digital economy is a mediator in relations between population and public administration.
3. Establishment of common platforms for cross-border cooperation.

3.3 Remaining Disagreements

Opinions of scientists are very different. Starting from complete non-perception of term itself.

Thus, Head of the Department of Economics and Management of Enterprises and Production Complexes of St. Petersburg State University of Economics, Doctor of Economics, Honored Scientist of the Russian Federation A. E. Karlik generally claims that he does not like the term “digital economy”. He states: “In my opinion, there is an economy—and that is it. And then, when adjectives begin to be attached to it: digital, import-substituting, innovative—we are not talking about the economy. It is clear that here is an innovative economy, and digital is an infrastructure of the economy first of all. It’s not the economy itself, in any case. This is the infrastructure and innovative provision of the economy.” (Karlik 2017).

Opinions of representatives of technical sciences who are members of FES were summarized by Roman Meshcheryakov, Doctor of Technical Sciences, Professor of the Russian Academy of Sciences, who singled out two approaches to concept of “digital” economy:

1. IT based economy.
2. The use of digital technologies in the production of an economic product (Meshcheryakov 2017).

"Digital" economy is a tool, technology, a means of achieving the main economic goal—ensuring the growth of welfare of citizens. There can be no other sense in its establishment.

3.4 Government Is a Powerful Driver of Digitalization of Economy and Management of Society

All over the world there is a significant lag of government regulation from the rapid pace of technological development of the economy.

To solve these problems in Russia was proposed the concept "Government as a platform", presented at the Gaidar Forum 2019 in the RANEPA. It provides for the end-to-end interdepartmental digitalization of processes and creation of an integrated organizational and technical infrastructure both for provision of public services and activities system of public administration. It should bc based on data collection and analysis, as well as on establishment of standardized interfaces with end-users and other information systems. An important element of the concept is the ability of independent vendors to create applications for end-users running on a platform.

Adoption of the national program "Digital Economy" at the end of 2018 was the first step in development of this direction in Russia. Program includes six federal projects: "Regulatory regulation of the digital environment", "Information infrastructure", "Personnel for the digital economy", "Information security", "Digital technologies" and "Digital government".

4 Conclusion and Recommendations

Of course, discussions about a pace, methods, positive and negative consequences of digitalization will continue for more than a decade and possibly centuries.

The first results of a discussion and recommendations were summarized by the President of FES of Russia S.D. Bodrunov in an interview with "Rossiyskaya Gazeta" in April 2019:

1. To give the status of a national project to the National Technology Initiative (NTI), which is based on the idea of digital economy.
2. Development of own technologies.
3. The defense industry should become a multiplier of digitalization of all sectors of the economy.
4. Attracting investments in digital projects (Business, banks, population, budget, foreign investors).
5. Digitalization of public administration, professional development of government workers.
6. Reducing dependence on foreign technologies.
7. The basis of success is "knowledge and developed into high-tech products.

As the main conclusion of S.D. Bodrunov stated: Sustainable GDP growth leading to improvement of the quality of life and big role of the country in the world is possible only with fast technological development" (Bodrunov 2019c).

References

Alferov, Zh.I.: About the digital economy. From the conversation with G.G. Malinetsky. https://regnum.ru/2388789.html

Bodrunov, S.D.: Questions for discussion. Conversations about the economy. Scientific-Popular Edition. FES Russia, pp. 241–242 (2017a)

Bodrunov, S.D.: Digital Evolution Rating. Conversations about the economy. Scientific-Popular Edition. FES Russia, p. 227 (2017b)

Bodrunov, S.D.: Interview of the Rossiyskaya Gazeta. Rossiyskaya Gazeta, no. 72(7830) from 02 April 2019 (2017c)

Free economy. Editorial topic of the issue. "Digitalization of the whole country", Free economy, no. 3, p. 17 (2017)

FES of Russia. Information-presentation material, p. 2 (2018)

Greenberg, R.S.: The volume of the digital economy in GDP. Conversations about the economy. Scientific-Popular Edition. FES Russia, p. 246 (2017)

Garina, E.P., Kuznetsova, S.N., Garin, A.P., Romanovskaya, E.V., Andryashina, N.S., Suchodoeva, L.F.: Increasing productivity of complex product of mechanic engineering using modern quality management methods. Acad. Strateg. Manage. J. **16**(4), 8 p. (2017)

Kalashnikov, S.V.: To realize that the world is different. Conversations about the economy. Scientific-Popular Edition. FES Russia, pp. 243–244 (2017)

Karlik, A.E.: Digitalization as a push. Conversations about the economy. Scientific-Popular Edition. FES Russia, pp. 222–223 (2017)

Kasperskaya, N.: Trends of digital economy (2017). https://ruscoins.info/fag/cifrovaya-tcjnomika-v-rossii

Meshcheryakov, R.: What is the digital economy? (2017). https://ruscoins.info/fag/cifrovaya-tcjnomika-v-rossii

Program "Digital Economy of the Russian Federation". http://static.government.ru/media/files/9gFM4FHj4PsB79I5v7yLVuPgu4bvR7M0.pdf

Kuznetsov, V.P., Garina, E.P., Andryashina, N.S., Romanovskaya, E.V.: Models of Modern Information Economy Conceptual Contradictions and Practical Examples. Emerald Publishing Limited, Bingley, 361 p. (2018)

Sorokin, D.E.: Every vegetable has its own time. Conversations about the economy. Scientific-Popular Edition. FES Russia, pp. 251–252 (2017)

Sui, U.: + Intelligence: motor of digitalization of the industry. Associated effect of digitalization. Report of HUAWEI and OXFORD ECONOMICS, p. 2 (2017)

Kuznetsov, V., Kornilov, D., Kolmykova, T., Garina, E., Garin, A.: A Creative Model of Modern Company Management on the Basis of Semantic Technologies. Communications in Computer and Information Science. Springer, Cham (2017)

Risk Management System in the Engineering Design Practice

Ekaterina P. Garina[1(✉)], Natalia S. Andryashina[1], Bolotbek M. Seitov[2], Alexander P. Garin[1], and Gulzat T. Tashmurzaeva[3]

[1] Minin Nizhny Novgorod State Pedagogical University, Nizhny Novgorod, Russia
e.p.garina@mail.ru, natali_andr@bk.ru, rp_nn@mail.ru
[2] Osh Technological University, Osh, Kyrgyz Republic
ch.kulueva@mail.ru
[3] Osh State University, Osh, Kyrgyz Republic
ch.kulueva@mail.ru

Abstract. Relevance of the research topic is due to the existence of sustainable risks in the economic activity of industrial enterprises, including risks of financial losses arising from a manufacturer at implementation of the overall strategy, which necessitates: implementation of the management support function in corporate and administrative practice; formation of measures to prevent damage; minimization (exclusion) of loss risks. The aim of the study is to develop recommendations for improving the risk management system as a tool for effective management of development at the engineering enterprise. Where the risk management system (hereinafter referred to as RMS) is considered as a set of processes carried out by management within the overall strategy of the enterprise. Methodology of project studies was used as a methodological basis and heuristic basis of the research process. Subject of research—risk management system of a machine-building enterprise. In the course of the study it is determined that: in the practice of domestic enterprises, the RMS is most often three-tier and minimizes risks of damage are formed in terms of prevention: failure of counterparties to perform their obligations; improper performance of contractual obligations by counterparties. It is proved that at present in operating activities of domestic enterprises most often represented: sustainable risks of disruptions of supply resources, and as a consequence—growth in the level of expenses; risks of financial losses. Based on what solutions are proposed for the future.

Keywords: Project activity · Risk management · Prevention actions

1 Introduction

Relevance of the research topic is due to the existence of sustainable risks in the economic activity of industry enterprises, including risks of financial losses arising from a manufacturer at implementation of the overall strategy of an enterprise, which necessitates: implementation of the management support function in corporate and administrative practice; formation of measures to prevent damage; minimization

E. G. Popkova and B. S. Sergi (Eds.): ISC 2019, LNNS 91, pp. 298–305, 2020.
https://doi.org/10.1007/978-3-030-32015-7_34

(elimination) of risks of losses. The aim of the study is to develop recommendations for improving the risk management system as a tool for effective management of development at the engineering enterprise. Where the risk management system (hereinafter referred to as RMS) is a set of processes carried out by management within the overall strategy of an enterprise, designed to provide:

- timely detection and reduction of events that negatively affect achievement of indicators of business plans of producers;
- efficient and economical use of resources;
- sufficient protection of resources of a manufacturer (including assets, personnel, information systems and databases, business reputation);
- completeness and reliability of financial, accounting, statistical, managerial and other reporting;
- compliance by managers and employees of an enterprise with the legislation of the Russian Federation, industry and internal regulatory documents;
- establishment of personal responsibility for minimizing risks and improving skills of employees, for understanding consequences of decisions (Gruzdeva et al. 2018; Markova and Narkoziev 2018);

The RMS, as one of the elements ensuring the achievement of manufacturer's objectives, is provided by:

- compliance by each executor of corporate standards in the area of its responsibility and related areas (Mizikovsky et al. 2018);
- assessment of the main potential threats of loss of values or failure to achieve the goals (risks) in terms of possible losses and probability of occurrence;
- development and implementation of control procedures and measures for all significant risks within areas of responsibility, minimizing these risks to an acceptable level;
- implementation of continuous assessment and improvement of control procedures and measures for risk management.

2 Methodology

Methodological basis and heuristic basis of the research process were a project research methodology, case study method of natural processes, and method of the pilot case study.

3 Results

Purpose of implementation of the RMS is to create common methods of risk management inherent in all types of activities of a manufacturer. In the practice of domestic enterprises, the RMS is most often three-tier (Fig. 1).

The first is the level of the Group (holding, corporations); this level assesses the risks of the Divisions and Enterprises exceeding limits set for them, as well as risks associated with the Group's activities in general. The second level is a division level; at this level, the risks of enterprises exceeding the maximum level established by their management body, as well as the division's own risks are considered. The third level is the enterprise level; each enterprise manages risks inherent in its activities within the maximum level established by its management.

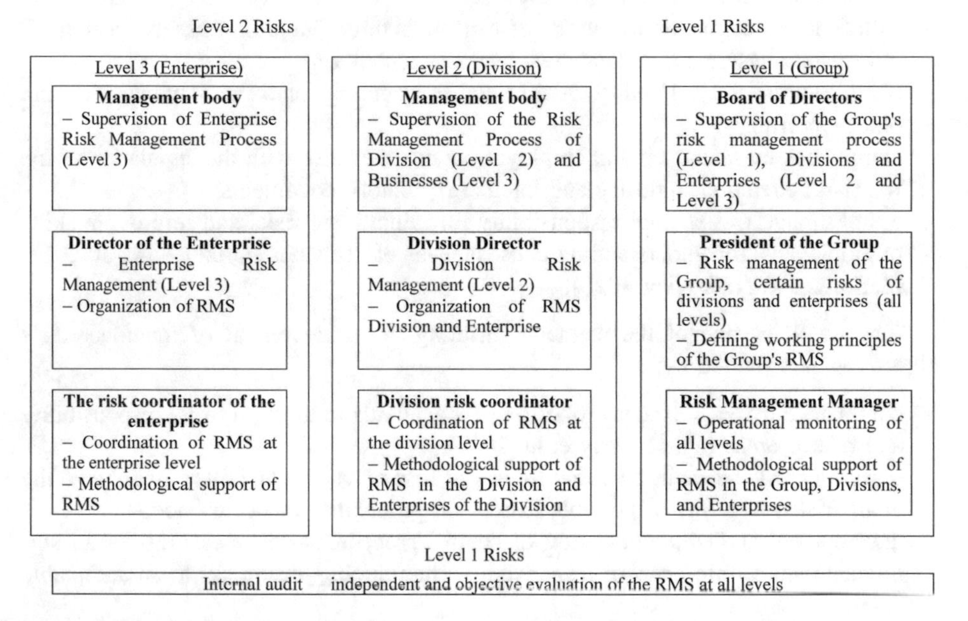

Fig. 1. Scheme of organization of risk management system in GAZ Group (Garina et al. 2016)

Considering formation of the risk management system of industrial enterprises, let's dwell on the design practice of a separate enterprise of GAZ Group—Gorky Automobile Plant (GAZ PJSC), which is a manufacturer of light commercial vehicles producing medium tonnage cars for small and medium businesses. The key indicators of the company's activity are presented in Tables 1, 2 and 3:

Table 1. Technical and economic indicators of production (Druzhilovskaya 2018)

Indicators	Years		Absolute deviation	Growth rate, %	Increment rate, %
	2016	2017			
1. Revenue from sales, goods, products, works, services (mln. rub.)	20 803.3	21,619.3	816.0	103.9	3.9
2. Cost of production (mln. rub.)	18 092.7	19,549.3	1 456.6	108.1	8.1
3. Gross profit, million rubles	2 710.6	2,070.0	−640.6	76.4	−23.6

(*continued*)

Table 1. (*continued*)

Indicators	Years		Absolute deviation	Growth rate, %	Increment rate, %
	2016	2017			
4. Average number of employees (list of employees + external part-time workers + working under contracts) (people)	7 813	7 643	−170.0	97.8	−2.2
6. Average annual cost of fixed assets (million rubles)	5256	5099	−157	97	−3
7. Average annual working capital balance (mln. Rub.)	703	860	157	122	22

Table 2. Indicators of the financial result of the enterprise (Mizikovsky et al. 2018)

Indicators	Years		Absolute deviation	Growth rate, %	Increment rate, %
	2016	2017			
1. Revenue from sales	20 803 266	21 619 254	815,988	103.9	3.9
2. Cost of sales	18 092 675	19 549 320	1 456 645	108.1	8.1
3. Gross profit	2 710 591	2 069 934	−640 657	76.4	−23.6
4. Management costs, Commercial expenses	1 971 596	2 010 874	39 278	102.0	2.0
5. Profit (loss) on sales	738,995	59 060	−679 935	8.0	−92.0

Table 3. Indicators of profitability of the company (Mizikovsky 2011; 345)

Indicators	Years		Absolute deviation	Growth rate, %	Increment rate, %
	2016	2017			
Profitability of sales	4.0	0.27	−3,73	−6,75	–
Return on assets of the enterprise, %	0.10	0.28	0.18	280.0	180.0
Profitability of core activities	0.44	1.01	0.57	250	150
Return on equity, %	0.37	1.06	0.69	286.5	186.5

The main technical and economic indicators of the company for 2016–2017 show mixed dynamics. On the one hand, there are certain positive decisions on the development of operating activities (revenue growth from 20,803.3 to 21,619.3 million rubles, output from 2,369.6 to 2,629, 2 rubles per person). On the other hand, there is either low efficiency of activity (average value of capital return 4 rubles., cost per ruble—98.5 rubles., profitability of core activity is less than 1%), or decrease in the indicator in a study period (decrease in gross profit from 2,710.6 to 2,070.0 million rubles, a decrease in profitability of sales from 4 to 0.27%). This indicates the need for significant changes in a management system at the operational level, including through the improvement of the RMS.

Implementation of the management support function in corporate and administrative practices, as well as risk management in order to bring them to a level acceptable to shareholders, consider on the example of a separate investment project of the enterprise (Tables 4 and 5) in the context of: goal-setting, rationalization of performance indicators, assessment of risk probability and expected actions to prevent them.

Table 4. Initial characteristics of the project (Garina et al. 2017)

Indicators	Basic parameters of the project
1. Project objectives	Organization of production of heavy drivetrains for a perspective line of machines of JSC "AZ "URAL" and LLC "LiAZ" within the "GAZ Group" Substitution of unit import for the perspective line of machines of JSC "AZ "URAL" and LLC "LiAZ"
2. Customer	JSC "AZ "URAL" and LLC "LiAZ"
3. Initiator and executor	PJSC "GAS"
4. Terms of implementation: – Investment phase – Operation phase	August 2018–April 2021 September 2020—December 2029
5. Project budget: – R & D costs, thousand rubles including VAT – Investments, thousand rubles	135,730 585,510, where the source of financing is own funds

Table 5. Economic performance indicators

Indicators	Value
NPV (net present value, thousand rubles)	583,370
IRR (internal rate of return, %)	47,6%
PP (payback period, years)	5.1
DPP (discounted payback period, years),	6.2
PI	2.2

In the strategic plan, an analysis of project risks shows a significant probability of increasing the project budget in the forecast period, associated with the increase in investment costs in equipment, and decrease in planned output (Table 6).

Table 6. Main risks (expert assessment) (Andrashina 2014)

Description	Probability	Implications	Measures to neutralize
Increase in investment costs	Average	Increase a payback period of the project	Adjustment of the direction of investment costs. Increasing the price of components
Reduction of production volumes of the customer division	Average	The decline in profitability of production	Expansion of the product line

4 Conclusion

At the operational level, it is defined:

1. Existence of sustainable risks of disruptions of supply resources, and as a consequence—growth in the level of expenses (Mizikovsky 2011). In this connection, at the initiative of the Risk Monitoring Department of PJSC GAZ (hereinafter referred to as "DMR"), work is carried out in comparison with prices of deliveries and their reduction in comparison with previous delivery conditions. For example, a contract is concluded on the volume of 2 350 m^3, where the price by the efforts of performers is reduced by 1 167 rubles for each cubic meter. As a result, the economic effect under a separate contract amounted to 2,742,45 thousand rubles. (excluding VAT). Thus, risks of unjustified costs under most contracts are excluded. What is the potential for quality project implementation? According to open sources, the total amount of excluding risks of unreasonable costs in the reporting 2018 year amounted to 25,890,4 thousand rubles.
2. Risks of financial losses business - units in the conditions of crisis phenomena. Experience of the DMR shows that risks from the external environment are primarily related to the conclusion and execution of contracts with counterparties, suppliers, or contractors. These risks can be divided into two types (Yashin et al. 2018):

   - First: risks of non-fulfillment by counterparties of their obligations. These risks are most often expressed in "non-delivery" of products, non-provision of services and non-performance of works, when under the terms of the contract the products, works, and services are paid. As a result, receivables are generated. In order to eliminate a threat of receivables, when concluding delivery contracts, as a rule, 100% advance payment by counterparties is provided.
   - Second: risks of improper performance of contractual obligations by counterparties. Risk of overstatement of the volume of work performed, use of cheap materials instead of expensive or technology-provided works, use of not contractor labor, but customer one (business—units). Risk of delivery of goods with low-quality characteristics under the guise and price of goods with high-quality characteristics.

Taking into account the existing risk gradation, in order to reduce economic risks when interacting with suppliers, customers and partners in the forecast period of the DMR business unit it is necessary to:

1. Pay special attention to the study of counterparties and contracts, when there are at least one of the following features:
   - non-compliance with the regulations;
   - accelerated procedure of registration of contracts;
   - conclusion of contracts without holding a tender commission;
   - entering a tender of the same firms (on the same issue), as a result of which a decision of tender commission is won by the same firm;
   - lack of alternative suppliers/contractors, etc. (Schätz 2016; 89);
   - "fresh" registration of a company;
   - mismatch of the actual location with the presented, the address is not specified.
   - refusal of companies to provide necessary information (questionnaire, copy of the balance sheet, copy of the certificate of registration, licenses, etc.).
2. Take measures to minimize the risks of damage:
   - examination of tender materials by DMR specialists for completeness of submission of documents;
   - on the counterparty—check potential winner for the contractual relationship with an enterprise, if any—the status of accounts payable and receivables, the presence of negative information in the with respect to the activity of the counterparty; collect data on founders (main shareholders), analysis of financial stability;
   - with the involvement of a financial unit of business to analyze prices for similar goods, works, services from third-party counterparties, i.e., in the market.
3. Also, in accordance with the stress-scenario for the current year, quarterly to determine measures to minimize (exclude) risks of financial losses of business units, among which may be:
   - strengthening the collection of receivables;
   - increase in accounts payable to the maximum;
   - organization of negotiations with suppliers on deferment of payment;
   - ensuring the termination of advance of suppliers;
   - sale suspension of finished products or increase of sales prices in accordance with the growth of the dollar. Currently at least 65% (Schätz 2016);
   - fixation of foreign currency purchases at the level of "ruble/dollar";
   - shipment of products only on advance payment;
   - refusal of loans at new rates.

Exclusion of risks of financial losses of a business unit in this case is possible through:

- constant monitoring by DMR over the implementation of the regulatory framework in the selection of counterparties and performance of contractual obligations on their part (Romanovskaya 2011);

- constant monitoring of conclusion and execution of contracts in close cooperation with lawyers, financiers and those responsible for the implementation of relevant activities in the areas (construction, supply, etc.) by managers of business units;
- carrying out inspections by DMR specialists of the performance of works, according to the concluded contracts: control of completeness of the volume and quality of work performed, compliance of the actual expenses spent by a contractor and material expenses under the act of performed works (rendered services);
- plan activities carried out by DMR to control all areas of financial and economic activity of business units, taking into account all economic risks.
- intensify the work of DMR on the study of information on schemes for working with crisis phenomena.

References

Andrashina, N.S.: Analysis of best practices of development of domestic machine-building enterprises. Bull. Saratov State Socio-Econ. Univ. **1**(50), 24–27 (2014)

Garina, E.P., Garin, A.P., Efremova, A.D.: Research and generalization of design practices of product development in the theory of sustainable development of production. Humanit. Socio-Econ. Sci. **1**(86), 111–114 (2016)

Garina, E.P., Kuznetsova, S.N., Garin, A.P., Romanovskaya, E.V., Andryashina, N.S., Suchodoeva, L.F.: Increasing productivity of complex product of mechanic engineering using modern quality management methods. Acad. Strateg. Manage. J. **16**(4), 8 p. (2017)

Gruzdeva, M.L., Smirnova, Zh.V., Tukenova, N.I.: Application of Internet services in technology training. Vestnik of the Minin University, vol. 6, no. 1(22), 8 (2018)

Druzhilovskaya, T.Yu.: Problems of practical application of fair value for estimation of accounting objects. International accounting, vol. 21, no. 9(447), 1086–1099 (2018)

Kuznetsov, V.P., Andrashina, N.S.: Innovative activity of industrial enterprises: problems and prospects. Izvestiya Penza State Pedagogical University named after V.G. Belinsky, no. 28, pp. 408–410 (2012)

Markova, S.M., Narkoziev, A.K.: Production training as a component of professional training of future workers. Vestnik of the Minin University, vol. 6, no. 1 (2018). https://doi.org/10.26795/2307-1281-2018-6-1-4. Accessed 12 Dec 2018

Mizikovsky, I.E.: Harmonization of indicators of internal control. Audit statements, no. 12, pp. 62–66 (2011)

Mizikovsky, I.E., Druzhilovskaya, T.Yu., Druzhilovskaya, E.S.: Accounting and Reporting of Non-credit Financial Organizations: Textbook. Nizhny Novgorod, 511 p. (2018)

Romanovskaya, E.V.: Formation of organizational and economic mechanism of restructuring of engineering enterprises: dissertation for the degree of candidate of economic sciences. Nizhny Novgorod State University. N.I. Lobachevsky. Nizhny Novgorod, 162 p. (2011)

Schätz, C.: A Methodology for Production Development: doctoral thesis. Norwegian University of Science and Technology, 126 p. (2016)

Yashin, S.N., Trifonov, Y.V., Koshelev, E.V., Garina, E.P., Kuznetsov, V.P.: Evaluation of the effect from organizational innovations of a company with the use of differential cash flow. Advances in Intelligent Systems and Computing, vol. 622, pp. 208–216 (2018). https://doi.org/10.1007/978-3-319-75383-6_27

Problems and Objectives of the Regional Socio-Demographic Potential Development

Yuri N. Lapygin(✉), Denis Yu. Lapygin, and Pavel Yu. Makarov

Russian Presidential Academy of National Economy
and Public Administration of Vladimir Branch, Vladimir, Russia
lapygin.y@gmail.com, lapygin.den@gmail.com,
makarovpu@ya.ru

Abstract. The elaboration of development prospects as a model for achieving goals presumes an orientation towards solving the most significant problems and attaining the desired state in the future. According to this, the purpose of this study is to design the model of regional goal development of systematization. Article presents the results of the Vladimir region's sociodemographic potential analysis, which is considered as a factor in elaboration of a socio-economic development strategy for the region. The presented analysis is carried out on the basis of statistical data of Russian Federation Federal State Statistics Service. It is shown that the demographic situation continues to deteriorate and that this problem is differently treated on regional and municipal levels, so there is a demand on regional goals systematization. Conclusions about the key problems of the Vladimir region in this area have been made and an algorithm has been constructed for systematizing goals at all stages of regional development.

Keywords: Socio-demographic potential · Region · Trends · Problems · Goals · The federal subjects of Russia · Strategy · Factors

JEL Code: R58 · J11 · J18

1 Introduction

According to the Federal Law "On Strategic Planning in the Russian Federation" from June 28, 2014. N 172-Fl among tasks of strategic planning is a definition of internal and external conditions, trends, restrictions, imbalances, opportunities of socio-economic development.

Implementation of this task of strategic planning at the regional level involves a comprehensive analysis of the potential of the constituent entity of the Russian Federation, among which the socio-demographic potential because the population of the region is both a key resource and stakeholder (Balakhina and Lapygin 2017) of socio-economic development. In this regard, one of the directions of the initial analysis in the development of a strategy of the region is an analysis of socio-demographic potential.

Based on the above, purpose of this work is to assess a socio-demographic potential of the Vladimir region and to identify circumstances that require attention in the development of the strategy development of the region.

E. G. Popkova and B. S. Sergi (Eds.): ISC 2019, LNNS 91, pp. 306–315, 2020.
https://doi.org/10.1007/978-3-030-32015-7_35

Next, we will consider theoretical basis of the research, results of analysis and conclusions.

2 Theoretical Basis of Research

First of all, we will define the concept that characterizes the object of research—the socio-demographic potential of the region.

In literature on demographic and social issues of regional development, this concept is collective and is interpreted as a "balanced system of quantitative and qualitative characteristics of population development" (Zvereva 2011; Lagerev et al. 2012). Due to composite nature of this concept, a number of authors operate with separate components. For example, work of Polkova (2014) presents the concept of "demographic (reproductive) potential", which means a set of observed and measured parameters reflecting a process of reproduction of the population.

Speaking about methods of assessing the socio-demographic potential, we can state the existence of quite a large number of author's models, among which it is difficult to distinguish the most acceptable. In this regard, it is proposed to give preference to analysis of data on those components of the socio-economic potential for which most authors agree and do not use any more narrowly specialized model.

Thus, in mentioned works and a number of other studies in particular (Kostromina 2017; Oborin et al. 2017) models of socio-demographic potential of the region can be concluded and characteristics should be considered:

Population dynamics;

- Factors contributing to this trend (fertility, mortality, migration);
- Economic aspect of population dynamics (demographic load, etc.);
- Social and economic situation of the population (income level, their distribution, etc.).

Based on the above directions, we will consider the assessment of the socio-demographic potential of the Vladimir region.

3 Results of the Study

Based on official statistics, it is possible to state a decrease in the population of Vladimir region during the last decade (Fig. 1A). At the same time, there is no positive dynamics of the rate of loss of the population (Fig. 1B).

A decrease in the population is taking place against the background of increasing demographic pressure: proportion of the population above and below working age is increasing (see Fig. 2).

Speaking about reasons for this situation, we can state that in reduction of the population begins to play an increasingly significant role the outflow of population from the region (Fig. 3A) to other subjects of the Russian Federation (Fig. 3B).

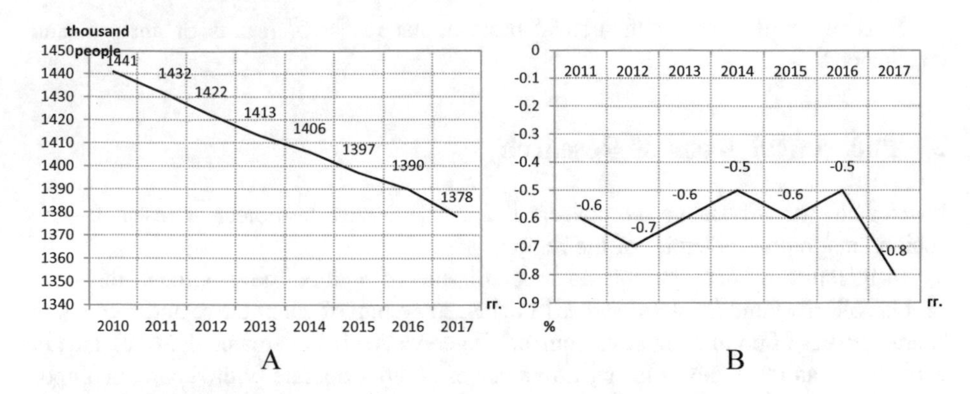

Fig. 1. Dynamic population of the Vladimir region in 2010–2017: absolute indicators (A) and growth rates (B)

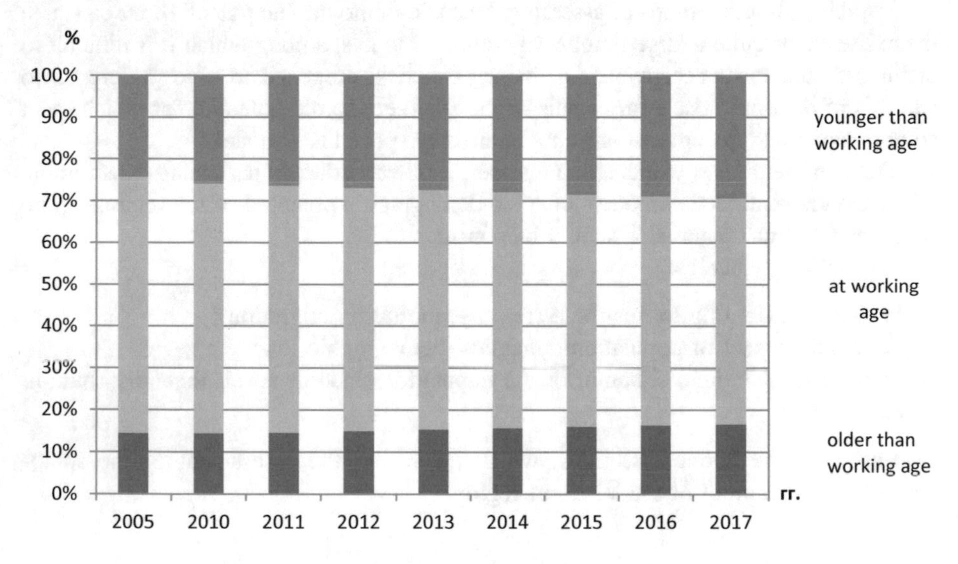

Fig. 2. Changes in the age composition of the population of Vladimir region in 2005–2017.

Such changes appear to be a serious problem against the steady excess of mortality over fertility (see Fig. 4) observed in all Central Federal District entities.

With regard to the social aspect of life in the region, there is a predominantly negative trend in the income of the population (see Fig. 5A), which is typical for most of the Central Federal District and country as a whole. At the same time, income distribution (see Fig. 5B) is at an acceptable level relative to other regions: Vladimir region is on the second place among regions of the Central Federal District in terms of uniformity of income distribution, having values of fund coefficient of 10.2 and Gini coefficient 0,352.

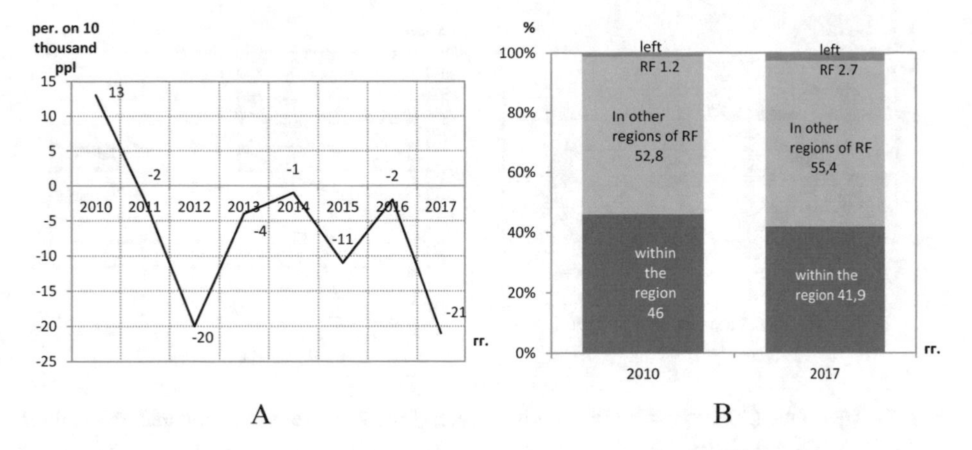

Fig. 3. Migration changes in the population of Vladimir region in 2010–2017: values of growth (decrease) (A) and direction of outbound migration (B)

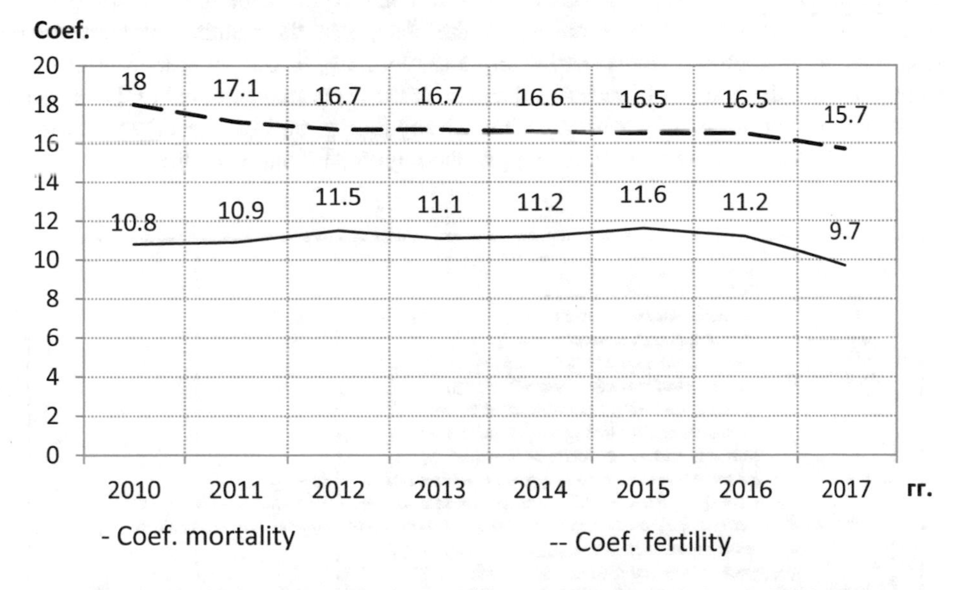

Fig. 4. Fertility and mortality rates of Vladimir region population in 2010–2017

In view of the limited volume of the publication, we do not provide a detailed analysis of important social indicators such as unemployment and proportion of the population with incomes below minimum. According to Rosstat data, Vladimir region has similar values to other regions of these indicators and is characterized by a tendency to their gradual improvement, which gives a reason not to consider as significant distinguishing features of the potential of the region.

Therefore, it can be noted that the trend of population decline, which has been developed in recent decades, continues.

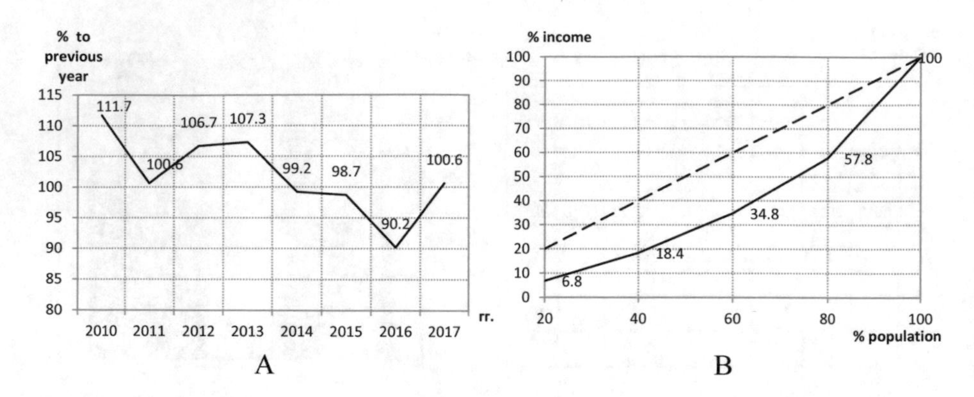

Fig. 5. Dynamics (A) and uniformity of distribution (Lorenz curve) (B) of income of Vladimir region population in 2010–2017

At the same time, the main socio-demographic problem requiring attention in the short term is a problem of population outflow from the region. Although, in the view of some authors, this factor does not affect the change in the number of permanent residents in the region as much as fertility and mortality, in our view the outflow of population, on the one hand, leads to a decrease in the economic potential of the region. It leaves the region, mainly with able-bodied population, and on the other - reflects presence of problems with attractiveness of the region as a place for life and work.

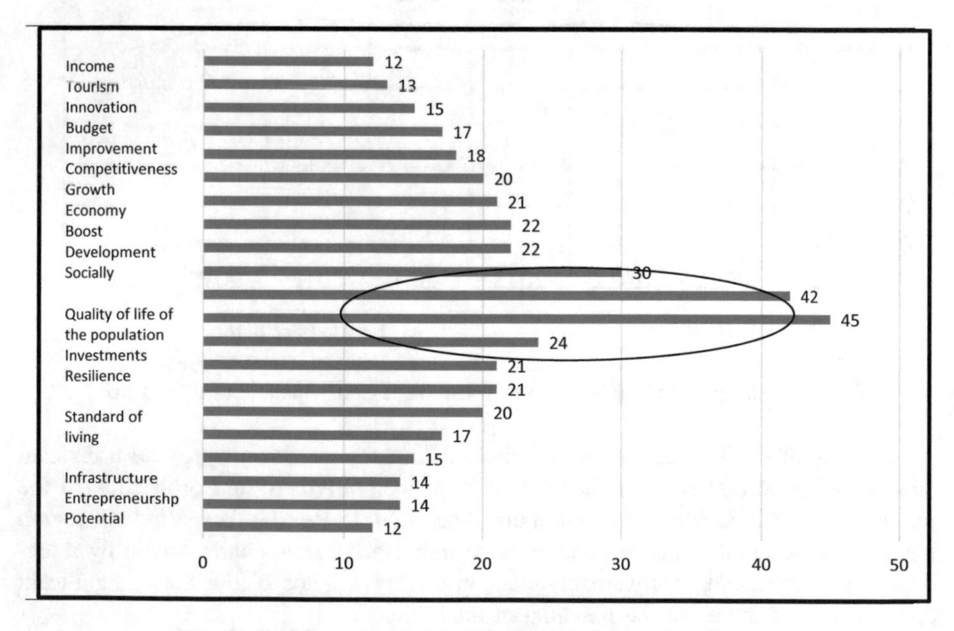

Fig. 6. Problems of municipal formations of Vladimir region

In turn, depopulation due to the excess of mortality over fertility appears to be a problem, firstly, requiring a long - term solution across the country as a whole, and secondly for other regions, and therefore does not determine the competitive advantages or disadvantages of the Vladimir region (Kuznetsov et al. 2018).

We will find out how problems of socio-demographic plan are reflected in the strategy of Vladimir region and strategies of municipalities of the region.

Results of the content analysis carried out in relation to the problems presented in the development strategies of municipal formations of Vladimir region, allow to construct the distribution reflected in Fig. 6.

Distribution built on results of the content analysis of goals reflected in the strategies of municipal entities of the Vladimir region is presented in Fig. 7.

The marked distributions allow to distinguish the main problem and strategic goal of development of municipalities by building slogans, including the most significant words and collocation of the considered formulations in the strategies of municipalities.

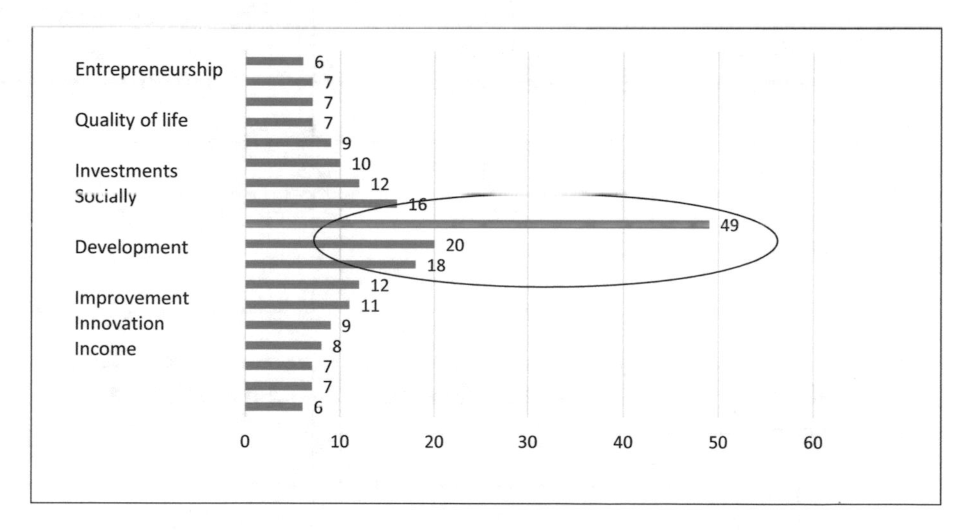

Fig. 7. Development goals of municipalities of the Vladimir region

Consequently, the main problem of municipalities may sound like this: "The authorities do not ensure the attraction of investments in infrastructure and social sphere of the municipality, which leads to the outflow of qualified personnel." The strategic objective would be as follows: "The growth of the quality of life on the basis of development of the social and economic system of the municipality by investing innovations in infrastructure and development of entrepreneurship, which ensures competitiveness and sustainability of budget revenues" (Kuznetsov et al. 2017). Now, if you turn to formulation of the "global goal" development of the Vladimir region, recorded in the Strategy of socio-economic development of the Vladimir region until 2030 ("improving the quality of life population to a level not below the average in the Central Federal District on the basis of the use of geopolitical advantages of the region, the realization of its industrial, scientific and recreational potential"), it becomes clear that simple generalization of goals of development of municipalities of the region is not enough to build a strategy. Goals differ in their content and orientation.

Most municipal development strategies in the region are formulated prior issuance of the Federal Law on Strategic Planning and can be adjusted to provisions of the law, however, appropriate tools are needed to solve such problem (Lapygin 2014; Lapygin 2008; Tulinova 2018), allowing and correctly perform the analysis, as well as to formulate goals and development strategy.

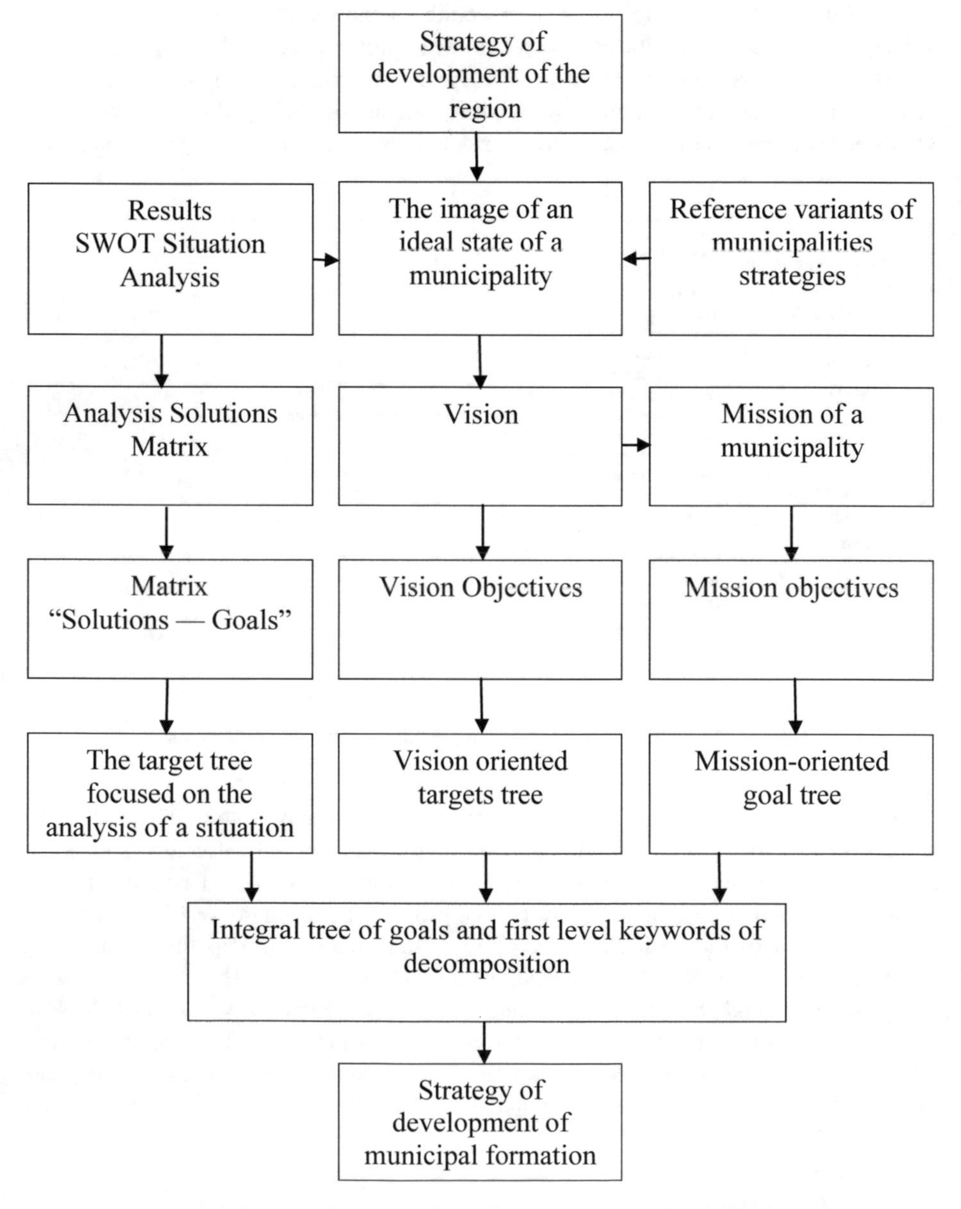

Fig. 8. Algorithm of systematization of development goals

As a procedure of transition from results of an analysis to construction of goals and strategy itself, it is possible to propose the algorithm of systematization of goals of development of municipal formation, presented in Fig. 8.

The algorithm is based on the idea of building an integral tree of municipal formation development goals (Tulinova and Lapygin 2018), which highlights the first level of decomposition goals. Keywords which constitute the slogan of the municipal development strategy.

In this case, the integral target tree is an ordered hierarchy of goals derived from three target blocks reflecting:

- goals formulated by results of construction of morphological matrices (Resolutions-Goals matrix (Tulinova 2015) and matrix of decisions (Lapygin and Lapygin 2012));
- goals formulated by results of a construction of morphological matrices (Balakhina et al. 2017) and focused on content of the Vision, which built on results of the formulation of the ideal state municipality, which is possible under favorable conditions;
- goals formulated by results of the construction of morphological matrices (Balakhina et al. 2017) and focused on the content of Mission.

The image of the ideal state of a municipality is formed not only on the basis of development strategy of the region and exemplary strategies of other municipalities but also taking into account results of factors of external and internal environment.

Described algorithm makes systematic in the process of development strategies of municipal formations by reflecting goals of all stages of strategy implementation:

- the first stage—implementation of objectives focused on results of the analysis of factors reflecting current situation (available opportunities and threats, significant problems and available potential of the municipal education);
- the second stage—achievement of goals that ensure implementation of Mission of the municipality;
- third stage—as close as possible to goals of the ideal state of municipality (ensuring implementation of Vision).

In addition to the fact that an algorithm allows to determine the set of goals at each stage of strategy implementation, it also provides a formalization of procedure for building a hierarchy of strategically significant development goals municipality and strategy itself.

The algorithm is also applicable for the development of strategically significant goals of the region, with the only adjustment that the place of the "Strategy of development of the region" will be taken by the "Strategy of development of Russian Federation" (set of goals designated in the Decree of the President of the Russian Federation (Balakhina et al. 2017)), and a place of municipalities will take the region.

In this case, it is necessary to adjust the development strategy of the region (or develop again), and strategies of municipalities of the region to form again or adjust, focusing on the goals of development of the region.

4 Conclusion

Summarizing results of this work, it is possible to identify the following conclusions obtained in the course of implementation:

- Socio-demographic potential of the region is an important factor in the development of a strategy because it reflects the state of the population as its key resource and stakeholders;
- Vladimir region, analyzed in this article, demonstrates the deterioration of socio-demographic potential. This problem needs to be reflected in plans of socio-economic development;
- This problem is reflected at the level of strategies of municipalities of the Vladimir region, which is linked to the problem of insufficient investment in infrastructure and social sphere;
- in turn, this is not reflected in the objectives of the current strategy, which emphasizes the geopolitical situation, industrial, scientific and recreational potential.

Results indicate the need for following recommendations on organization of the regional development process:

- There is a need to systematize and harmonize the analysis and regional development objectives;
- For this purpose, as a tool of transition from the results of analysis to construction of goals, an algorithm of systematization of development goals, which is based on the idea of building an integral tree of goals is proposed.

References

Balakhina, A.A., Lapygin, Yu.N.: Strategic goals of municipal development. Manage. Issues (2), 74–81 (2017)

Balakhina, O.V., Lapygin, Yu.N., Tulinova, D.V.: Issues of development strategies for the development of the towns. Uchenye zapiski: Sci. J. (2), 67–71 (2017)

Kostromina, E.V.: Structural and dynamic analysis of demographic processes of the Republic of Mari El. Stat. Econ. **14**(2), 70–78 (2017)

Kuznetsov, V.P., Garina, E.P., Andryashina, N.S., Romanovskaya, E.V.: Models of Modern Information Economy Conceptual Contradictions and Practical Examples. Emerald Publishing Limited, Bingley, 361 p. (2018)

Kuznetsov, V., Kornilov, D., Kolmykova, T., Garina, E., Garin, A.: A Creative Model of Modern Company Management on the Basis of Semantic Technologies. Communications in Computer and Information Science. Springer, Cham (2017)

Lagerev, A.V., Sazonov, A.S., Filippov, R.A.: Model for assessing the 2009-demographic potential and its impact on the structure of higher and postgraduate education in the region. Inf. Syst. Technol. **3**(71), 72–77 (2012)

Lapygin, D.Yu.: Tools for the Formation of a Strategy for the Development of Cities. Publishing House of the Vladimir branch of the RANEPA, Vladimir (2014)

Lapygin, D.Yu., Lapygin, Yu.N.: Logic of strategic matrices. In: Dynamics of Complex Systems, no. 4, pp. 12–25 (2012)

Lapygin, Yu.N.: System problem solving, Eksmo, Moscow (2008)

Oborin, M.S., Sheresheva, M.Yu., Ivanov, N.A.: Justification of strategic guidelines for the socio-economic development of small cities in Russia. Perm Univ. Bull. Ser. Econ. **12**(3), 437–452 (2017)

Polkova, T.V.: Demographic potential as a component of quality of life. Econ. Reg. (3), 118–130 (2014)

Tulinova, D.V.: Strategic Planning Tools in Cities. Publishing House of the Vladimir Branch of the RANEPA, Vladimir (2018)

Tulinova, D.V.: Methods of building strategic maps. Voronezh Univ. Bull. Ser. Econ. Manage. (3), 159–165 (2015)

Tulinova, D.V., Lapygin, Yu.N.: Integral tree of strategic goals. Bull. VlSU Ser. Econ. Sci. (1), 35–41 (2018)

Zvereva, N.V.: Human potential for sustainable innovative development of Russia. In: Innovative Development of the Russian Economy: Institutional Environment Proceedings of the IV International Scientific Conference in Moscow, Russia, pp. 771–780 (2011)

Selection of Preferred Information Support of the Management Enterprise in the Sphere of Services Using the Method of Analysis of Hierarchies

Sergey N. Yashin[1(✉)], Sergey A. Borisov[1], Arthur E. Ambartsoumyan[2], Natalia A. Shcherbak[3], and Yana V. Batsyna[4]

[1] Nizhny Novgorod State University named after N.I. Lobachevsky, Nizhny Novgorod, Russia
jashinsn@yandex.ru, ser211188@yandex.ru
[2] Advertising agency "MOST", Nizhny Novgorod, Russia
adm@ra-most.ru
[3] Moscow State University of Technology and Management named after K.G. Razumovsky (First Cossack University), Moscow, Russia
[4] Institute of Food Technologies and Design — the branch of Nizhny Novgorod State University of Engineering and Economics, Nizhny Novgorod, Russia
alx_jn@sinn.ru

Abstract. Subject. The subject of this study is the problem of choosing the most preferred information support in order to improve the management of interaction with customers for enterprises of the service sector. In conditions of digitalization and growth in the market of various tools solutions, heads of enterprises and heads of IT - services need formalized methods for choosing information systems, which can increase the competitiveness of an enterprise by improving customer interaction.

Goals. The goal is formation of a complex methodology of selection of information systems for improvement of management of relationships with clients and a further selection of the most preferred solutions based on a combination of methods of strategic management in the field of information technology (MacFarlan matrix) and qualitative mathematical method - analysis of hierarchies.

Methodology. Proposed methodology consists in the complex application of the strategic matrix of "information intensity" and the method of analysis of hierarchies to determine the area of the most preferred information systems and then select the most appropriate solution for an enterprise based on its performance requirements.

Results. As the results of this research the authors propose a technique that allows to choose the most preferred options of information support for enterprise management in the field of improvement customer relationship management. A concrete example shows the possibility of applying this technique and demonstrates how to use it to determine first the class of the most preferred information system, and then select the most effective solution in the field of

E. G. Popkova and B. S. Sergi (Eds.): ISC 2019, LNNS 91, pp. 316–324, 2020.
https://doi.org/10.1007/978-3-030-32015-7_36

informatization for a particular enterprise, taking into account stated interests, preferences, and limitations.

Conclusions. The results presented by the authors can be useful for directors of services enterprises, heads of IT departments, as well as other top managers of medium and large enterprises. According to the authors, presented methodology can be used by enterprises of any scale and organizational and legal forms, and can also be adapted for automation of enterprises, engaged in other economic activities.

Keywords: Information system · Hierarchy Analysis Method · Digitalization · CRM—system

JEL Code: M31 · O33

1 Introduction

At the moment, due to increasing role of information and digital solutions, automated information systems are beginning to play an increasingly important role in managerial decision-making (Skinner 2018). They eliminate many routine transactions by entrusting storage, processing, and analysis of data to machine systems (McAfee and Brynjolfsson 2017). At the same time, the decision - making function remains for the person making a decision. The goal is formation of a complex methodology of selection of information systems for improvement of management of relationships with clients and a further selection of the most preferred solutions based on a combination of methods of strategic management in the field of information technology (MacFarlan matrix) and qualitative mathematical method - analysis of hierarchies.

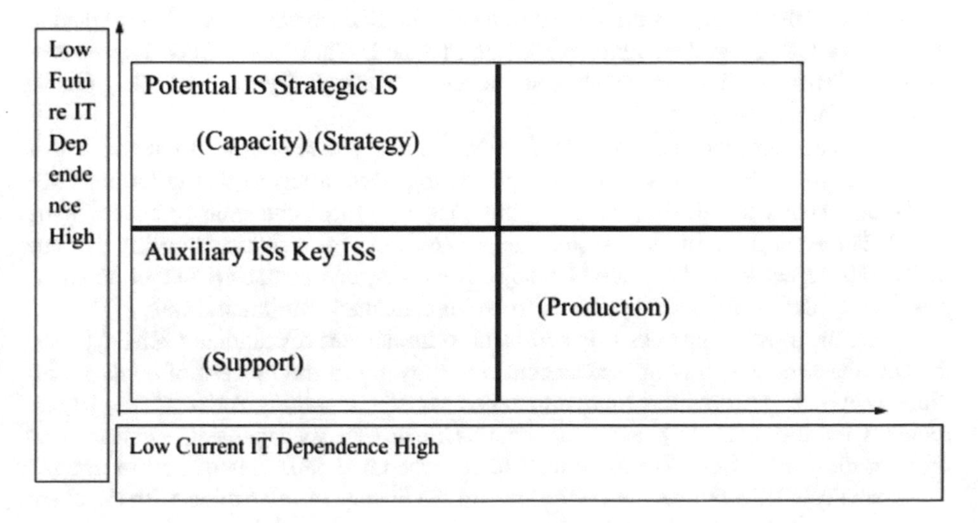

Fig. 1. Information Intensity Matrix

One of the most important areas of application of modern information systems (IS) is the sphere of services. On the one hand, IS in this area is generally not considered to be conditions for obtaining a key advantage at the current time, but it can be seen as a promising tool for enhancing the competitiveness of a firm, so they are called in accordance with the McFarlan matrix "potential IS" (Fig. 1). On the domestic and foreign IS market there is a huge number of software products that allow to automate business processes in the field of services, including trade. Since trade is mainly related to work with clients, for information support of this activity IS, called CRM systems, are mainly used.

With the use of CRM systems, it is possible to keep automated records of transactions and counterparties (Pohludka and Štverková 2019). They can also remind about scheduled meetings, important calls, and letters. Consequently, these systems allow for the automation of routine processes and, ultimately, lead to increased profits.

Methodology. Proposed methodology consists in the complex application of the strategic matrix of "information intensity" and the method of analysis of hierarchies to determine the area of the most preferred information systems and then select the most appropriate solution for an enterprise based on its performance requirements.

Results. Based on research conducted by Capterra, it was determined that 65% of companies implement CRM systems during the first five years of their existence, with 20% of companies implementing them in during 1–2 years of operation, 23% of companies implement such systems in 3–5 years (Teslenko 2019). It should also be noted that 10% of surveyed companies are ready to implement CRM systems within 6–10 years of work on the market, and another 10% of companies are ready to implement these systems after 10 years of work.

Presented on the market CRM - systems differ in their functionality and cost - it can be both small budget systems and huge IS for automation of activities on large enterprises of the service sector. An example of domestic IS, which can be attributed to the class of CRM—systems, are: 1C. CRM Standard, ClicK and others. Representatives of CRM of Western manufacturers are: Microsoft Dynamics CRM, Oracle Siebel CRM, and SAP CRM.

When choosing the most preferred CRM system, you should also remember that they differ in the level of information processing. According to this criterion, three types of systems are distinguished: operational, analytical and collaborative (Payne 2006). The description of these systems is presented in Fig. 2. When choosing the type of IS, it is necessary to be guided by objectives, task, and characteristics of an enterprise where the information system is to be implemented (Molineux 2002).

Thus, for most companies it is sufficient to implement operational CRM systems, the main task of which is to increase customer loyalty in the process of contact with him. They can perform the following tasks: register incoming traffic (calls, letters, requests for the site, etc.), automate internal document management, set tasks and monitor the work of employees, remind about scheduled calls, letters, and meetings.

Analytical CRM allows not only to record the history of interaction with the client but also to trace patterns, for example, how customers are distributed on the "sales funnel". They can: segment a customer base, determine the value of a client, analyze and forecast sales volume, show the distribution of transactions on the sales funnel.

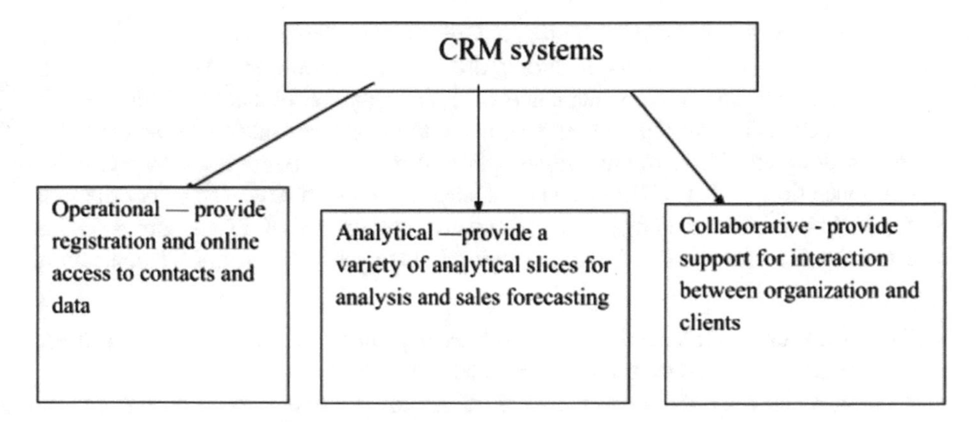

Fig. 2. Types of CRM—systems by level of information processing

Another type of CRM systems are collaborative systems, which allow communication with customers for feedback (Shmidt and Cochen 2014). With their help, the top management of a company can make decisions on the correction of assortment, decisions in the field of price policy, etc.

Currently, the combined CRM systems, which are a combination of the types of systems shown in Fig. 2, come to the fore. As a rule, operating CRM systems are used, to which analytical functions are added (sales reports, managers performance, etc.). Also, modules that provide the ability to communicate with customers are added to composition of collaborative systems.

Thus, we see that the existing systems differ in a number of parameters, among which it is worth highlighting: available functionality, cost, level of data processing, etc. (Nguyen et al. 2007). It is sometimes difficult for a modern entrepreneur in the wide range of IS offers available on the market to choose a software solution (Kesthong et al. 2007). In this case, different analytical tools for selecting the most preferred IS should be available to assist in selection. One such tool, as presented above, is the Information Intensity Matrix - it allows to select the class that is most preferable for an enterprise IS. However, the application of this tool is not enough to select one or two or three IS's that can be recommended for implementation in the enterprise. Accordingly, it can be considered as a step in selecting preferred, and most efficient IS.

The next step in selection of a system should be as narrow as possible, and ideally the choice of the only system to be recommended for implementation. Tools of this stage, as a rule, are based on qualitative and quantitative parameters that characterize the efficiency of application of the system to solve problems of an enterprise, and automation of its most significant business processes. Expert evaluation methods are actively used in the process of selecting the most preferred IS. Among these methods, an important role is played by the efficient IS selection method using a pairwise comparison method, also known as the Hierarchy Analysis Method (MAI).

Consider an application of this method to select the most preferred IS to improve the competitiveness of the retail business.

As the key problems facing a trading enterprise, consider the following: low cost of processing data of enterprise IS, including due to imperfection of POS-terminals, high cost of improvements and maintenance of operability, redundancy of functionality installed on enterprise system, inconvenience of the interface and functional content of the report designer. Therefore, a simple option of replacing the current version with a newer generation system of the same company is not considered. To select the most preferred information system, management and employees of IT department of an enterprise we consider the following criteria, divided into two groups - technical and cost.

- **Technical criteria:** speed of data processing, interface convenience, technical requirements for the system, openness and scalability;
- **Cost criteria:** cost of a license, cost of training of personnel, value for money, cost of consulting and implementation, cost of maintenance of information system.

To select a system, we propose to use the method of analysis of hierarchies (Saaty 1980). The method makes it possible to find a solution that is best consistent with an understanding of a problem and requirements for its solution.

The algorithm of the method:

1. Identification of the problem and definition of the purpose;
2. Highlighting the main criteria and alternatives;
3. Hierarchy construction: from goal through criteria to alternatives;
4. Construction of a matrix of pairwise comparisons of criteria by goal and alternatives by criteria;
5. Application of the method of analysis of obtained matrices;
6. Determining weights of alternatives according to the methodology of hierarchy

The tree of alternatives (fragment) in our case may look as shown in Fig. 3 (Borisov et al. 2017).

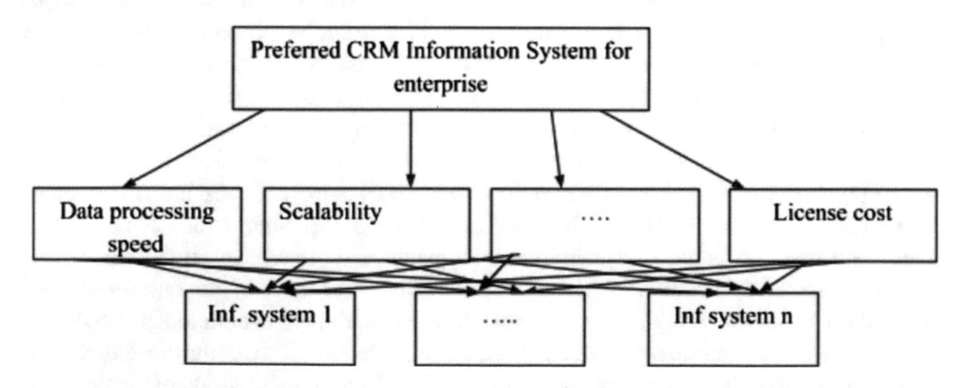

Fig. 3. Tree of alternatives for choosing the most preferred CRM information system—class

Also, for the selection of an information system based on the Saati principle, it is necessary to use the scale of relative importance, which is presented in Table 1.

Table 1. Relative importance scale (T.Saati)

Importance	Definition
1	Equal importance
2	Intermediate solution between adjacent
3	Moderate superiority
4	Intermediate solution between adjacent
5	Significant superiority
6	Intermediate solution between adjacent
7	Considerable superiority
8	Intermediate solution between adjacent
9	Very strong superiority

To use a backward comparison between features, reverse scores are applied. If we present this in the form of a formula, we get the coefficient a_{ij} when comparing objects directly, and the coefficient of $1/a_{ij}$ when comparing objects.

The following information systems were identified as objects for comparison (alternatives): 1C, Trade X, Oracle, Microsoft Dynamics CRM, Norbit. Experts, who were managers, employees of IT - services, as well as consultants from higher educational institutions of Nizhny Novgorod, the following initial parameters were defined for comparison of systems (Table 2).

Table 2. Initial expert data for choosing your preferred CRM system

System name/Name parameter	1 C	Trade X	Oracle	Microsoft Dynamics CRM	Norbit
Processing speed	0.3	0.6	0.7	0.8	0.2
User-friendly interface	0.6	0.8	0.7	0.9	0.4
Technical requirements	0.4	0.5	0.6	0.7	0.2
Openness	0.6	0.5	0.4	0.8	0.1
Scalability	0.4	0.6	0.5	0.8	0.2
License price	0.5	0.4	0.2	0.9	0.9
Staff training price	0.2	0.8	0.1	0.7	0.9
Price/quality ratio	0.4	0.6	0.5	0.8	0.3
Consulting price	0.4	0.8	0.6	0.4	0.6
Price of maintenance	0.6	0.3	0.4	0.7	0.8

In the future, all parameters of a system for analysis were divided into two groups: parameters that should be minimized and that should be maximized.

Parameters that require minimization and make up the vector P include: technical requirements for installation and maintenance of the system; price of purchase of the system; cost of training to work with the system; cost of consulting services; cost of maintenance.

The parameters that require maximization and make up the Q vector should include: speed of data processing; usability; openness; scalability; value for money.

In order to form integral criteria P and Q it is necessary to initially perform the analysis using the method of paired comparisons. Consider an example of using MAI in relation to the parameter "Data processing speed" (Table 3).

Table 3. Applying MAI to Data processing speed

	1C	TradeX	Oracle	Microsoft	Norbit	SGO	VP
1C	1	1/4	1/4	1/6	2	0.5	0.07
TradeX	4	1	1/2	1/2	4	1.3	0.20
Oracle	4	2	1	1/2	6	1.9	0.28
Microsoft	6	2	2	1	6	2.7	0.40
Norbit	1/2	1/4	1/6	1/6	1	0.3	0.05
Amount						6.7	1.0

On the basis of Table 2 and guided by the scale presented in Table 1, experts shall make appropriate direct and inverse estimates for this parameter for each alternative to be compared (in this case—information system).

Then, for each system parameter, the geometric mean estimate (CDS) shall be calculated using the following formula (1):

$$\text{average geometric score (EGS)} = \sqrt[n]{x1^{*}x2^{*}[?][?][?]^{*}xn} \qquad (1)$$

$$\sqrt{\Pi}$$

$$\sqrt{\Pi}$$

where X1, X2, Xn—relative expert systems obtained by the method of paired comparisons; n—the number of compared systems, CDF—the geometric average estimate.

The priority vector is defined by the following formula (2):

$$\text{VP} = \text{S/RSGO} \qquad (2)$$

The priority vector for the remaining parameters is calculated in the same way. On the basis of the parameters aspiring to the maximum, the priority vector P is formed by the method presented in Table 4.

Table 4. Forming the priority vector P for parameters aspiring to the maximum

	Speed	Convenience	Openness	Scalability	Price/quality	P (Sum)
1C	0.07	0.11	0.24	0.10	0.10	0.63
TradeX	0.20	0.27	0.16	0.24	0.24	1.11
Oracle	0.28	0.15	0.12	0.16	0.14	0.84
Microsoft	0.40	0.42	0.45	0.45	0.45	2.17
Norbit	0.05	0.05	0.03	0.05	0.06	0.25

Table 5 presents the Q vector, for the parameters that aspire to a minimum.

Table 5. Formation of Q priority vector for minimizing parameters

	Technical requirements	Price	Training	Consulting	Maintenance/	Q (amount)
1C	0.11	0.12	0.05	0.10	0.18	0.55
TradeX	0.19	0.05	0.30	0.40	0.06	1.00
Oracle	0.25	0.05	0.03	0.20	0.09	0.62
Microsoft	0.40	0.39	0.22	0.10	0.28	1.39
Norbit	0.06	0.39	0.40	0.20	0.40	1.44

Having specified vectors P and Q, we form Table 6, which is the final vector of priorities.

Table 6. Final priority vector

	P	Q
1C	0.63	0.55
TradeX	1.11	1.00
Oracle	0.84	0.62
Microsoft	2.17	1.39
Norbit	0.25	1.44

Based on the results presented in Table 6, using method B. Pareto (Yashin et al. 2017), we can conclude that Norbit has not entered the field of effective solutions, and Microsoft has but is losing to solutions offered by Trade X and Oracle. Thus, the most effective options for switching from the existing version of Axapta system are two options: Trade X and Oracle. Taking into account the high cost of implementing and maintaining the Oracle system, small number of specialists in this system in Russia and the fact that Oracle functionality redundancy is the most preferred option is TradeX.

2 Conclusion

In conclusion, it should be noted that the considered example of an application of the method of paired comparisons to a solution of the problem of choosing preferred information system demonstrates the effectiveness of an application of this approach in practice and can be recommended to enterprise analysts when deciding on the transition from one information support to another.

References

Skinner, Ch.: Digital Human: The Fourth Revolution of Humanity Includes Everyone. Marshall Cavendish International (Asia) Pte Ltd., Singapore (2018)

McAfee, A., Brynjolfsson, E.: Machine, Platform, Crowd: Harnessing Our Digital Future. W.W. Norton and Company, New York (2017)

Pohludka, M., Štverková, H.: The Best Practice of CRM Implementation for Small- and Medium-Sized Enterprises. Adm. Sci. **9**(1), 22 (2019)

Teslenko, E.: Types of CRM systems (2019). https://salesap.ru/vidy-crm-sistem/. Accessed 03 Mar 2019

Payne, A.: Handbook of CRM: Achieving Excellence in Customer Management. Elsevier Butterworth-Heinemann, Oxford (2006)

Molineux, P.: Exploiting CRM: Connecting with Customers. Hodder & Stoughton, London (2002)

Shmidt, E., Cochen, J.: The New Digital Age: Transforming Nations, Businesses, and Our Lives Paperback, Reprint edition. Vintage, New York, NY (2014)

Nguyen, Th., Sherif, J., Newby, M.: Strategies for successful CRM implementation. Inf. Manage. Comput. Secur. (2007)

Kesthong, W., Wei, J., Liu, L., Koong, K.: Standardization of internet-retail e-business solution. Inf. Syst. Educ. J. (2007)

Saaty, T.L.: The Analytic Hierarchy Process. McGraw-Hill, New York (1980)

Borisov, S., Usov, N., Novikova, V.: The analytic hierarchy process application selecting the most preferred solutions to improve information management enterprise. Working Paper, Economy and Entrepreneurship, Moscow (2017)

Yashin, S., Boronin, O., Sukhanov, D.: Formation of tools for multi-criteria evaluation of the effectiveness of innovation projects in the field of industrial safety of enterprises. Health Saf. (2017)

Motives Behind GCC Sovereign Wealth Funds' Investment in Foreign Securities and Their Role in Regional Capital Markets' Integration

Ibrahim A. Onour(✉)

Department of Business Administration,
University of Khartoum, Khartoum, Sudan
onour@uofk.edu, ibonour@hotmail.com

Abstract. To capture the long-run association between major GCC stock markets and foreign exchange reserves of Saudi Arabia, UAE, and Qatar at post-international financial crisis of 2008, we employed multivariate cointegration analysis (As there is no foreign debts payable by these countries, change in foreign exchange reserves of these countries can be a good proxy for change in their sovereign wealth funds (SWFs)). Our findings suggest an evidence of long run association between the foreign exchange reserves and the stock markets' prices. To assess risk attitude of GCC Sovereign Wealth Funds (SWFs) we further investigated cointegration between one of the biggest GCC SWF (SAMA fund) investment in foreign securities, and Dubai and Qatar stock markets prices before and after the global financial crisis. Our results indicate, while there is no significant evidence of common co-movements between change in the stock markets prices and SAMA fund before the financial crisis, there is a strong evidence of co-movements at the post-financial crisis era. This result implies an evidence of risk aversion attitude of SAMA fund, as opposed to the finding of other studies that associate primary motive of GCC SWFs with political influence on host countries.

Keywords: GCC · Sovereign wealth funds · Capital markets

JEL Code: G1 · G11 · G21 · G32

1 Introduction

The Gulf Cooperation Council (GCC) for the Arab States launched in 1981 with the objective of attaining, integration, coordination and cooperation among member GCC countries in various economic issues. With limited progress realized in the first twenty-five years of its beginning, GCC economic goals gained unprecedented drive since Muscat summit of GCC leaders in 2001. In Muscat summit of leaders an economic accord was signed with the objective of speeding up the process of economic integration among GCC countries. The new agreement called for equal treatment of all GCC nationals in all investment activities in GCC countries, including equity ownership and new business establishment, and free mobility of capital in member states. The agreement also urges for integration of financial markets, and for synchronization

E. G. Popkova and B. S. Sergi (Eds.): ISC 2019, LNNS 91, pp. 325–336, 2020.
https://doi.org/10.1007/978-3-030-32015-7_37

of all investment laws and regulations. One important vehicle to achieve these goals was GCC sovereign wealth funds, that if managed properly can boost economic unity and capital markets integration. Since the global financial crisis of 2008, the economic strategy, mood and financial management perception has been changed in the region (Barnett and Sergi 2018a, b; 2019). GCC authorities have not only been searching for safer heavens for their massive investment wealth, but also have been working towards becoming the financial center of the region by competing with each other, and adopted sophisticated financial investment strategies including the creation of various funds to sustain economic development. But more recently, with significant increase in the size of these funds, GCC countries investment abroad faced mounting skepticism from governments in Western countries over their concern that foreign investors may take over assets with strategic value, be they manufacturing plants, banks, industrial enterprises, or high-tech companies (Adekola and Sergi 2016). One noticeable case in this respect is blocking of Dubai Ports World's acquisition of P & O, that operated six major ports in the Eastern United States. Similar worries have been raised with regard to Dubai's offer to buy the Auckland international Airport and the Swedish stock exchange OMX. Such a trend led many GCC SWFs to diversify geographically and channel some of their surplus capital to GCC member states and Middle East countries. It is apparent that these skepticisms have been augmented by the emerging research that raise a number of doubts about the real motives behind sovereign wealth funds in general. Bernstein et al. (2013) explain the presence of government officials or politicians inside the board of SWFs cannot rule out the search for strategic political objectives in the hosting countries. Similarly, Knill et al. (2012) indicate some SWFs are more biased towards investing in countries in which they have weaker political links, revealing the political motives behind some of SWFs. However, Dyck and Morse (2011) conclude that despite investment decisions of SWFs are distorted by political considerations, a major goal of SWFs is the development of the homeland capital markets. Chhaochharia and Laeven (2009) explain, SWFs display strong tendencies towards countries they share a common culture with and industry similarities as the case of oil producing countries' SWFs investment in oil company stocks. Murtinu and Scalera (2016) indicate a large or a majority acquisition cases of SWFs in foreign industries reveal willingness to engage in effective corporate governance activities, and it can be part of a general industrial development strategy for SWFs as it facilitates the development of joint ventures and help strengthening expertise or industrial complementarities with the home country industries.

As a result of the mounting skepticisms about foreign sovereign wealth funds in general and to avoid political risk abroad (e.g., Masood and Sergi 2008; Hsing and Sergi 2010), GCC countries felt to direct a portion of their wealth to acquisitions and investment in homeland capital markets, by investing into assets of GCC banks, financial institutions and other sectors such as real estate and even retail markets. As indicated in the financial report of Kuwait Financial Centre (2008: 1) there are 36 GCC SWFs owning 131 GCC listed companies with a value of $300 billion, which represent about 27% of market capitalization in GCC capital markets. These figures indicate SWFs have played increasing role in GCC capital markets development in the past decade. While such returning capital expected to have implications of deepening GCC capital markets, the judicious role of GCC SWFs in boosting the linkage between GCC capital markets has not been studied so far.

As the literature on SWFs focus mainly on SWFs motives abroad, to date, there is not much attention paid to the role of SWFs on homeland capital markets development. The current paper addresses this issue by looking at the role of GCC SWFs in GCC capital markets integration. The paper contributes to the existing literature by addressing the following issues. Since GCC capital markets' integration is a major goal in GCC union agreement, signed four decades ago, this paper investigates the role of SWFs in achieving such an objective, and how the international financial crisis of 2008 have influenced such a role to reveal their risk behavior to conjecture their motives in investing in foreign securities.

The paper also reveals short term effect of SWFs on GCC capital markets.

The remaining parts of the paper are structured as follows. Section 2 highlights the research methodology. Section three include the empirical findings and statistical analysis. The final section concludes the study.

2 Data and Methodology

Data employed in this study includes monthly observations on foreign exchange reserves and stock markets indices for the main GCC countries, Saudi Arabia, UAE, and Qatar during the sample period from January 2012 to December 2018 (84 observations) and Saudi Arabian Monetary Authority (SAMA) fund during the period April 2005 – Sept 2018 (163 observations). To assess long term association between SWFs and stock markets we employed a number of diagnostic and validation tests that represent prerequisite for Johansen and Jusilus (1991) multivariate cointegration test. To assess volatility spillover across SWFs and stock markets we employed dynamic conditional correlation (DCC) GARCH model. The general specification of DCC GARCH model, developed by Engle (2002), can be shown as:

$$y_t = \delta X_t + e_t \tag{1}$$

$$e_t = H_t^{1/2} v_t \tag{2}$$

$$H_t = D_t^{1/2} R_t D_t^{1/2} \tag{3}$$

$$R_t = diag(Q_t)^{-1/2} Q_t diag(Q_t)^{-1/2} \tag{4}$$

$$Q_t = (1 - \mu_1 - \mu_2)R + \mu_1 e_{t-1} e'_{t-1} + \mu_2 Q_{t-1} \tag{5}$$

Where

y_t is (mx1) vector of dependent variables
δ is (mxk) matrix of parameters
X_t is (kx1) vector of independent variables
$H_t^{1/2}$ is the Cholesky factor of the conditional covariance matrix H_t
v_t is (mx1) vector of independent and identically distributed innovations

D_t is a diagonal matrix of conditional variance, in which each diagonal term $\sigma^2_{i,t}$ evolves according to GARCH specification:

$$\sigma^2_{j,t} = \alpha_j + \sum\nolimits_{i=1}^{p} \beta_i e^2_{j,t-i} + \sum\nolimits_{i=1}^{q} \pi_i \sigma^2_{j,t-i}$$

R_t is a matrix of conditional correlation, e_t is a vector of standardized residuals, and μ_1 *and* μ_2 are parameters that govern the dynamics of conditional correlation, that satisfies $0 \leq \mu_1 + \mu_2 \leq 1$.

$h_t = vech(H_t)$; the vech ( ) function stacks the lower diagonal elements of a symmetric matrix into a column vector, A and B each is a matrix of parameters.

3 Results and Discussion

3.1 Basic Statistical Analysis

Summary statistics for stock returns and change in GCC FX reserves are presented in Table 1, and show there is a considerable difference in stock returns among the three countries, as Saudi stock market exhibit monthly decline of 0.02% and Dubai and Qatar show positive monthly returns of 0.17 and 0.27% respectively. The skewness and kurtosis coefficients indicate the distributions of returns for all six variables characterized by peakness and fat tails relative to a normal distribution. The Jarque-Bera (JB) test for joint normal kurtosis and skewness reject the normality hypothesis for all variables[1]. The descriptive statistics of change in foreign exchange reserves indicate UAE foreign exchange reserves exhibit highest mean annual growth rate, as it grows on average by annual rate of 0.88%, whereas Saudi foreign exchange reserves show a declining annual rate at −0.23%. Skewness coefficient also indicate similar conclusion of negative skewness for Saudi foreign exchange reserves implying the likelihood of negative returns. It seems there is a positive association between stock markets performance and foreign exchange reserve growth rates as reflected by the mean growth and skewness coefficients.

To test the order of integration for each market, the augmented Dickey-Fuller (ADF), and Phillips and Perron unit root tests has been employed. Since methodologies of these tests are well documented in the literature, we report in Table 1 tests results for each market. The lag length parameter for ADF test is determined using the Akaike information criteria (AIC). Results of unit root tests indicate that the levels of all variables are non-stationary, while their first differences exhibit stationarity behavior. This result suggest that the order of integration of these variables is unity, I(1), which implies the first differenced series will be used in the upcoming cointegration and causality analysis. To assess the causality direction between foreign exchange reserves and stock markets prices we reported Granger causality test results in Table 2 indicating that Dubai stock market and Qatar stock markets are caused or influenced by the foreign exchange reserves of the three countries, but not the opposite causal direction,

[1] Green, W.; 1993, (page 310) explains that the Normality test based only on kurtosis and skewness is essentially non-constructive, as a finding of Normality does not necessarily suggest what to do next, and failing to reject it does not confirm normality; it is only a test of symmetry and mesokurtosis.

Table 1. Basic statistics

	Dubai stock M	Saudi stock M	Qatar stock M	UAE FX reserves	Saudi FX reserves	Qatar FX reserves
Mean (%)	0.17	−0.02	0.27	0.88	−0.23	0.65
St. deviation (%)	6.93	5.94	6.24	6.03	5.58	6.41
Skewness	0.46	−1.1	0.61	0.31	−4.33	0.20
Kurtosis	2.43	4.58	14.5	2.79	24.32	4.94
JB	28.78	124	610	70.9	1297	91.7
p-value	0.00	0.00	0.00	0.00	0.00	0.00
DW	1.98	1.92	2.00	2.18	2.00	1.97
ADF test						
- Level	1.25	0.96	1.02	4.0	3.2	5.31*
-1st diff	10.21*	16.59*	10.98*	18.16*	9.85*	12.0*
PP test						
- Level	1.31	1.30	0.86	3.85	3.15	5.32*
-1St diff	24.1*	20.98*	26.12*	72.5*	20.4*	20.3*

*significant at 5% significance level (critical value = 4.68).

Table 2. Granger causality tests

H0: GCC FX reserves	F test
Do not Granger causes	
- Saudi stock market	1.34
- Dubai stock market	2.25*
- Qatar stock market	2.21*
H0: GCC stock markets	F test
Do not Granger causes	
- UAE FX reserves	1.59
- Saudi FX reserves	2.03
- Qatar FX reserves	0.55
H0: SAMA do not Granger cause	
- Saudi stock market	2.08*
- Dubai stock market	0.52
- Qatar stock market	0.62
H0: Saudi stock market do not Granger	
Cause SAMA	1.66
H0: Dubai stock market do not Granger	
Cause SAMA	0.28
H0: Qatar stock market do not Granger	
Cause SAMA	0.29

*Reject the null-hypothesis at 5% significance level.

where as GCC foreign exchange reserves has no significant causal effect on Saudi stock market. To investigate this result further we employed SAMA fund, instead of GCC foreign exchange reserves, which indicate evidence of significant causal effect on Saudi stock market.

3.2 Long-Term Analysis

Multivariate cointegration analysis investigates long term association between the foreign exchange reserves of the three countries (UAE, Saudi Arabia, and Qatar) and their stock markets to capture the common forces driving the long-run co-movements between the two sets of the data. Co-integration of a number of variables implies that some linear combination of two or more variables yield stationary series even though each of the series is non-stationary and some long-run equilibrium relation ties the individual series together. In the following, co-integrating relationships are investigated using the multivariate approach of Johansen and Juselius (JJ) (1991). Since parameter stability and absence of structural break in the data is a prerequisite for Johansen cointegration validity, Tables 3 and 4 report Hansen tests results for structural break hypothesis in the data. Results of Hansen test reject the null hypothesis of structural break for all six variables (jointly and separately) for the sample period (2012 January to 2018 September). Figure 1 portrays a common trend in the three stock markets. More formally, Johansen test in Table 5 report the results of cointegration test and suggest that there is at most one cointegrating vector, or analogously there is one independent common stochastic trend combining the set of stock indices with the foreign exchange reserves. To assess the impact of the international financial crisis of 2008, on Sovereign Wealth Funds we investigate cointegration between SAMA investment in foreign securities and Dubai and Qatar stock markets before and after August 2008[2]. Results in Tables 6 and 7 indicate while there is no significant evidence of long term association between Dubai and Qatar stock markets and SAMA fund before the international crisis date of August 2008, but

Table 3. Hansen test for joint parameter instability

	Test stat	Critical value	Decision
GCC FX reserves Granger cause			
- Saudi stock market	1.86	3.26	Stable
- Dubai stock market	2.60	3.26	Stable
- Qatar stock market	2.50	3.26	Stable

Note: A test statistic that exceeds the critical values indication of rejection of the null-hypothesis of parameter stability. The joint stability test includes the variance (see Green 2000, chapter 7).

[2] SAMA is the only fund that has lengthy time series available before the international financial crisis of 2008.

Table 4. Hansen test for Structural break*

Stock markets			Foreign exchange reserves		
Coefficient	Test stat	Result	Coefficient	Test stat	Result
Saudi market			Saudi FX reserves		
β_0	0.056	Stable	α_0	0.110	Stable
β_1	0.065	Stable	α_1	0.144	Stable
β_2	0.10	Stable	α_2	0.162	Stable
Dubai market			UAE FX reserves		
β_0	0.040	Stable	α_0	0.077	Stable
β_1	0.062	Stable	α_1	0.068	Stable
β_2	0.360	Unstable	α_2	0.29	Stable
Qatar market			Qatar FX reserves		
β_0	0.117	Stable	α_0	0.032	Stable
β_1	0.113	Stable	α_1	0.085	Stable
β_2	0.256	Stable	α_2	0.109	Stable

Note: A test statistic values that exceeds the 10% critical value of 0.353 support an evidence of rejection of the null-hypothesis of parameter instability.
*To check structural break we employed AR(2) model: $\Delta y_t = \pi_0 + \pi_1 \Delta y_{t-1} + \pi_1 \Delta y_{t-2} + v_t$ and test stability of parameters.

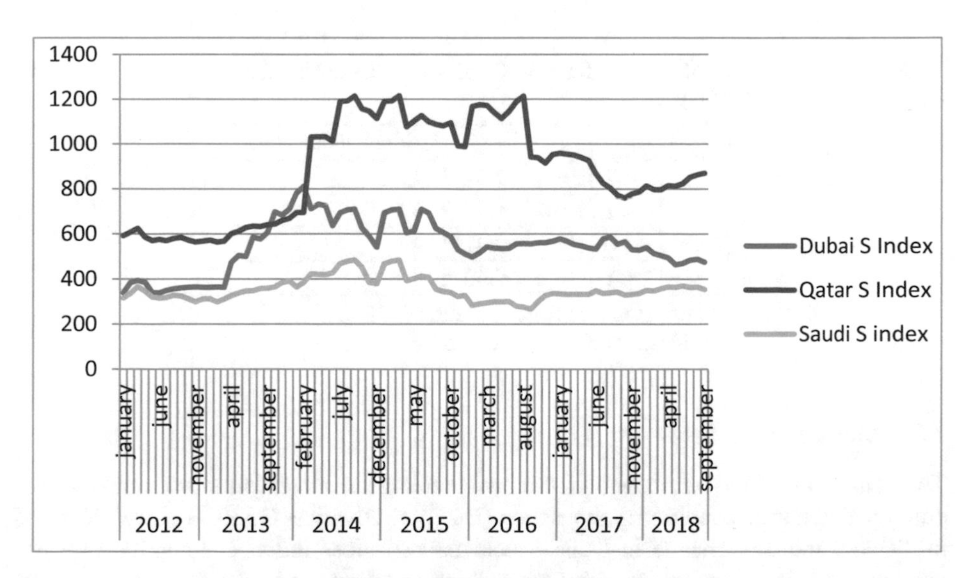

Fig. 1. GCC stock markets

there is strong evidence of cointegration in the period after the financial crisis. This result, as contrary to the findings in existing literature which associate the primary motives of GCC SWFs with politics and non-profit objectives, support an evidence of risk aversion behavior of SAMA fund.

Table 5. Cointegration of GCC FX reserves & stock markets

H_0	H_1	L_{max}	Critical values (5%)
$r = 0$	r = 1	60.13*	33.3
$r \leq 1$	r = 2	26.92	27.3
$r \leq 2$	r = 3	21.91	21.3
$r \leq 3$	r = 4	11.85	14.6
$r \leq 4$	r = 5	4.74	8.1
$r \leq 5$	r = 6	1.19	6.2

All critical values from Hamilton (1994).
*significant at 5% significance level.

Table 6. Cointegration of SAMA fund & Dubai stock market (before & after international financial crisis –Aug 2008)

Before	H_0	H_1	L_{max}	Critical values (5%)
	$r = 0$	r = 1	10.53	14.6
	$r \leq 1$	r = 2	5.46	8.1
After	$r \leq 0$	r = 1	20.0*	14.6
	$r \leq 1$	r = 2	0.43	8.1

All critical values from Hamilton (1994).
*significant at 5% significance level.

Table 7. Cointegration of SAMA fund & Qatar stock market (before & after international financial crisis –Aug 2008)

Before	H_0	H_1	L_{max}	Critical Values (5%)
	$r = 0$	r = 1	7.34	14.6
	$r \leq 1$	r = 2	6.57	8.1
After	$r \leq 0$	r = 1	69.47*	14.6
	$r \leq 1$	r = 2	0.40	8.1

All critical values from Hamilton (1994).
*significant at 5% significance level.

3.3 Short Term Effect

To assess the short-term effect of foreign exchange reserves on stock markets we employed dynamic conditional correlation (DCC) multivariate GARCH model. Results in Table 8 indicate change in Dubai stock market prices influenced significantly by change in foreign exchange reserves of the three countries, and volatility in this market is associated with the previous month own volatility. Change in Qatar stock market prices are influenced by changes in UAE and Saudi foreign exchange reserves, and volatility in this market is influenced by its own last month volatility. However, Saudi stock market is only influenced by change in Saudi foreign exchange reserves and its past month volatility. These results indicate even in the short term there is a significant

Table 8. The mean equation

Variables	Coefficient	Std. err	p-value
Dubai Stock Index			
UAE FX Reserves	0.0021*	0.0005	0.000
Saudi FX Reserves	0.0002*	0.0001	0.004
Qatar Fx Reserves	0.0027*	0.0010	0.007
Constant	77.31	65.13	0.235
ARCH (1)	0.558*	0.109	0.000
GARCH (1)	0.47*	0.067	0.000
Constant	410*	197.86	0.038
Qatar Stock Index			
UAE FX Reserves	0.0037*	0.0012	0.003
Saudi FX Reserves	0.0002	0.0001	0.134
Qatar FR Reserves	0.0058	0.0031	0.061
Constant	178.90*	86.92	0.040
ARCH (1)	0.87*	0.197	0.000
GARCH (1)	0.25*	0.117	0.031
Constant	722	485	0.137
Saudi Stock Index			
UAE FX Reserves	0.0003	0.0002	0.159
Saudi FX Reserves	0.00009*	0.00003	0.010
Qatar FR Reserves	−0.0001	0.0005	0.843
Constant	272*	30.6	0.000
ARCH (1)	0.66*	0.12	0.000
GARCH (1)	0.36*	0.05	0.000
Constant	95.86*	47.49	0.044
Log likelihood = −1398; Wald Chi2 (9) = 156			0.000

*significant at 5% sig. level; **significant at 10% sig. level

evidence of foreign exchange reserves influence on the stock markets of Dubai and Qatar. However, in the short-term Saudi stock market influenced by change in Saudi reserves only, but not the reserves of the other two countries.

4 Concluding Remarks

To investigate co-movements between change in foreign exchange reserves of Saudi, UAE, and Qatar and change in stock markets prices of Saudi, Dubai, and Qatar, we employed multivariate cointegration analysis to capture the common forces driving the long-run co-movements using monthly data during the post-international financial crisis era, January, 2012 to September, 2018.

Results of cointegration test suggest there is at most one cointegrating vector, or analogously there is one independent common stochastic trend combining the set of stock indices with the foreign exchange reserves. To investigate the impact of the international financial crisis of 2008, on GCC SWF we searched association between SAMA fund's investment in foreign securities and Dubai and Qatar stock markets before and after August 2008. Results indicate while there is no significant evidence of long-term association between GCC stock markets and SAMA fund, but there is a strong evidence of association in the period after the financial crisis. This result implies evidence of risk aversion attitude of SAMA fund, as contrary to findings of other studies that link the primary purpose of GCC SWFs with political motives and non-profit objectives.

Analysis of short-term effect of foreign exchange reserves on stock prices indicate change in Dubai stock market prices influenced by change in foreign exchange reserves of the three countries, and volatility in this market is associated with one month lagged own market volatility. Change in Qatar stock market prices are influenced by changes in UAE and Saudi foreign exchange reserves, and volatility in this market is influenced by its own past month volatility. However, Saudi stock market is only influenced by change in Saudi foreign exchange reserves and its past month volatility.

Data Appendix

Saudi FX reserves (Millions of SAR)
https://tradingeconomics.com/saudi-arabia/foreign-exchange-reserves

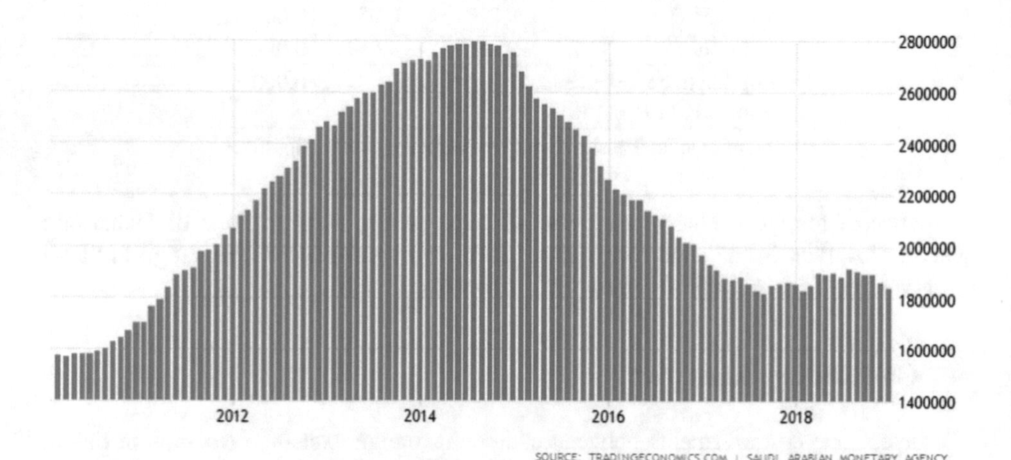

Qatar FX reserves (Millions of QAR)

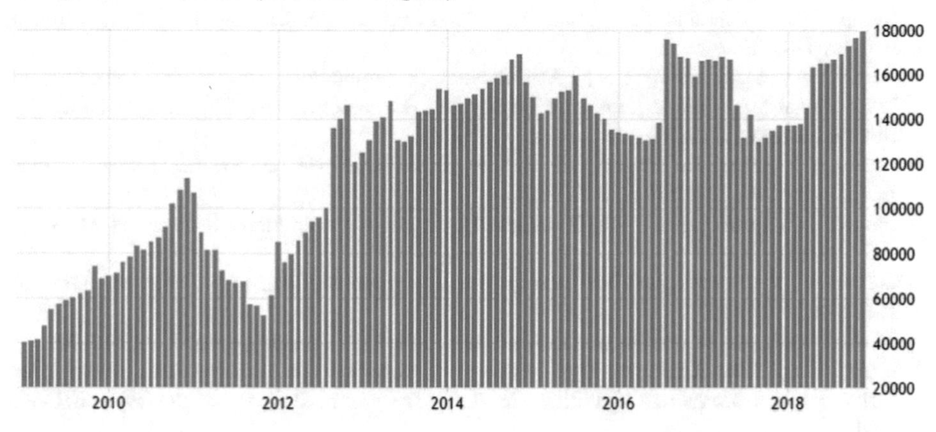

UAE FX reserves (Millions of AED):
https://tradingeconomics.com/united-arab-emirates/foreign-exchange-reserves

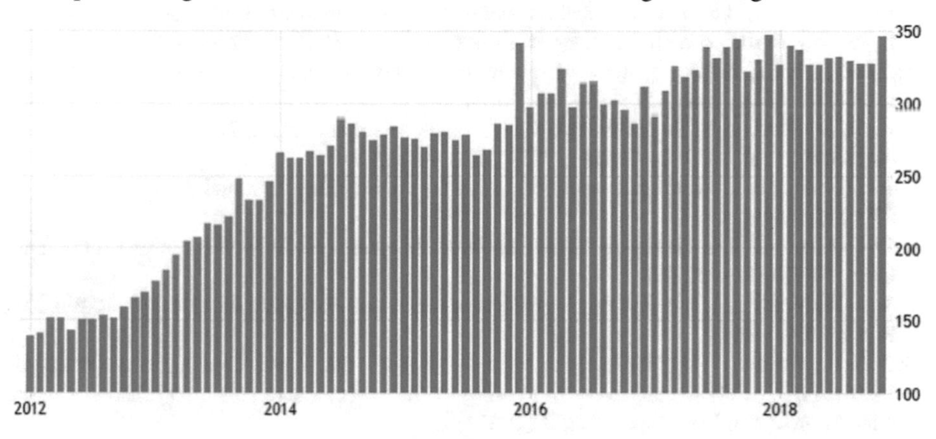

References

Adekola, A., Sergi, B.S.: Global Business Management: A Cross-Cultural Perspective. Routledge, New York (2016)

Aizenman, J., Glick, R.: Sovereign wealth funds: stylized facts about their determinants and governance. Int. Financ. **12**, 351–386 (2009)

Barnett, W.A., Sergi, B.S.: Introduction. In: Barnett, W.A., Sergi, B.S. (eds.) Banking and Finance Issues in Emerging Markets, pp. 1–8. Emerald Publishing Limited, Bingley (2018a)

Barnett, W.A., Sergi, B.S. (eds.): Banking and Finance Issues in Emerging Markets. International Symposia in Economic Theory and Econometrics, vol. 25. Emerald Publishing Limited, Bingley (2018b)

Barnett, W.A., Sergi, B.S. (eds.): Asia-Pacific Contemporary Finance and Development. International Symposia in Economic Theory and Econometrics, vol. 26. Emerald Publishing Limited, Bingley (2019)

Bernstein, S., Lerner, J., Schoar, A.: The investment strategies of sovereign wealth funds. J. Econ. Perspect. **27**, 219–238 (2013)

Carpantier, J.F., Vermeulen, W.N.: Emergence of sovereign wealth funds. J. Commod. Mark. **11**, 1–21 (2018)

Chhaochharia, V., Laeven, L.: Corporate governance norms and practices. J. Financ. Intermed. **18**(3), 405–431 (2009)

Dyck, A., Morse, A.: Sovereign wealth fund portfolios. MFI Working paper 2011 (2011)

Hsing, Y., Sergi, B.S.: Responses of real output to changes in euro exchange rates, stock prices, and other macroeconomic conditions in Italy. Transform. Bus. Econ. **9**(1(19)), 101–108 (2010)

Amar, J., Carpantier, J.-F., Lecourt, C.: GCC Sovereign Wealth Funds: Why do they Take Control? Working paper AMSE 2018-35 (2018)

Knill, A., Lee, B.-S., Mauck, N.: Bilateral political relations and sovereign wealth fund investment. J. Corpor. Finance **18**(1), 108–123 (2012)

Masood, O., Sergi, B.S.: How political risks and events have influenced Pakistan's stock markets from 1947 to the present. Int. J. Econ. Policy Emerg. Econ. **1**(4), 427–444 (2008)

Megginson, W.L., Fotak, V.: Rise of the fiduciary state: a survey of sovereign wealth fund research. J. Econ. Surv. **29**, 733–778 (2015)

Murtinu, S., Scalera, V.G.: Sovereign wealth funds' internationalization strategies: the use of investment vehicles. J. Int. Manag. **22**(3), 249–264 (2016)

Onour, I., Sergi, B.S.: GCC stock markets: how risky are they? Int. J. Monet. Econ. Finance **3**(4), 330–337 (2010)

Qcrimi, Q., Sergi, B.S.: Development and social development in the global context. Int. J. Bus. Glob. **14**(4), 383–407 (2015)

Sachs, J.D., Warner, A.M.: Natural resource abundance and economic growth. NBER Working Paper, 5398 (1995)

Sachs, J.D., Warner, A.M.: The curse of natural resources. Eur. Econ. Rev. **45**, 827–838 (2001)

Sala-i-Martin, X., Subramanian, A.: Addressing the natural resource curse: an illustration from Nigeria. IMF Working Paper 03/139 (2003)

Seznec, J.-F.: The Gulf sovereign wealth funds: myths and reality. Middle East Policy **15**(2), 97–110 (2008)

Smith, B.: Oil wealth and regime survival in the developing world, 1960–1999. Am. J. Polit. Sci. **48**(2), 232–246 (2004)

Truman, E.M.: Sovereign Wealth Funds: the Need for Greater Transparency and Accountability. Peterson Institute for International Economics (2007). http://www.iie.com/publications/pb/pb07-6.pdf. Accessed 29 Sept 2015

Ziolo, M., Sergi, B.S. (eds.): Financing Sustainable Development: Key Challenges and Prospects. Palgrave Macmillan, London (2019)

Artificial Intelligence: Reality in the 21st Century

Technological Revolution in the 21st Century: Digital Society vs. Artificial Intelligence

Elena G. Popkova[1(✉)] and Kantoro Gulzat[2]

[1] Plekhanov Russian University of Economics, Moscow, Russia
210471@mail.ru
[2] Kyrgyz National University named after J. Baiasagyn, Bishkek, Kyrgyzstan
gulzat.tashkulova@mail.ru

Abstract. Purpose: The purpose of the research is to determine the scenarios of socio-economic development of Russia under the influence of technological revolution in the 21st century, until 2030.

Design/methodology/approach: The authors determine the causal connections of the technological revolution in the 21st century with the help of the materials of a statistical collection of the National Research University "Higher School of Economics" "Digital economy – 2019", which reflect the factors that restrain the usage of the Internet (as the most revolutionary technology of modern times) by the population.

Findings: It is substantiated that the technological revolution in the 21st century is a very complex and variable process, which could take place according to one of the three scenarios, depending on formation of digital society and adoption of artificial intelligence. Scenario of deintellectualization is most probable in modern Russia, and its signs are seen as of now. It envisages low level of development of the digital society and low interest of consumers in artificial intelligence due to insufficient marketing support. That's why investment flows will be redirected in favor of conventionally revolutionary (not providing completely new opportunities) intellectual technologies. Scenario of segmental intellectualization is also rather probably in Russia, as it is traditionally peculiar for the processes of innovative development of the Russian economy (e.g., it could be realized during dissemination of mobile communications). This scenario envisages elitism of artificial intelligence until 2030, due to its high cost. Scenario of intellectual breakthrough is most probable, but least probable in Russia. It is connected to mass dissemination of artificial intelligence – in entrepreneurship, consumption, and state management. The key condition of its implementation is active marketing support for artificial intelligence.

Originality/value: The described scenarios reduce uncertainty and open wide opportunities and perspectives for state management of technological revolution in modern Russia.

Keywords: Technological revolution · Digital society · Artificial intelligence · Russia

JEL Code: D74 · M31 · O31 · O32 · O33 · O38

E. G. Popkova and B. S. Sergi (Eds.): ISC 2019, LNNS 91, pp. 339–345, 2020.
https://doi.org/10.1007/978-3-030-32015-7_38

1 Introduction

The 21st is marked by a new technological revolution, which is peculiar for the global scale, having covered most of the participants of the global economy, and is much faster that previous revolutions. Specifics of the new technological revolution determine, firstly, accelerated processes of society's transformation, which complicate social adaptation to the current changes. Digital competencies are completely new for representatives of most of the modern professions and, moreover, for consumers. Their mastering is hindered by psychological unpreparedness of interested parties and large duration of the process of preparation of the educational and methodological programs.

Secondly, the new technological revolution causes high social risks. The opportunities of revolutionary technologies are so high that they allow for total automatization. At present, usage of AI during solving a lot of organizational and managerial tasks is preferable. In the course of development of AI it might compete with human intellect and even form sustainable (unachievable for human) competitive advantages that are connected to effective processing of Big Data, multitasking, and absolute rationality (objectivity).

For managing AI, a manager has to possess the necessary digital competencies, which is much simpler that managing humans as the sources of intellectual resources – which requires correct selection of personnel, team-building, development and implementation of complex systems of motivation, and stimulation of labor. Multiple advantages of AI cause justified worries of humans and ambiguous reaction of the society to the technological revolution, as there are supporter and opponents.

Thus, in the technological revolution in the 21st century an important role belongs to social factors, which have to be taken into account during managing this revolution in the interests of its acceleration and development in the target direction and for leveling its social negative consequences and maximization of advantages for society. From the scientific and practical point of view, studying the causal connections of mutual influence of the digital society and artificial intelligence is especially topical. The purpose of this research is to determine the scenarios of socio-economic development of Russia under the influence of technological revolution in the 21st century, until 2030.

2 Materials and Method

The technological revolution in the 21st century is the object of research and is studied in a lot of modern scientific publications. However, the existing studies on the topic of the technological revolution in the 21st century have two drawbacks. 1st – narrow character of the compiled forecasts of the revolution. Most modern works consider only one selected scenario development of technological revolution in the 21st century. For example, the works da Silva et al. (2019), Masood and Egger (2019), Popkova (2019), Popkova (2017), Popkova and Sergi (2019), and Sergi et al. (2019) connect the consequences of the Fourth industrial revolution to transition to Industry 4.0, while only its algorithm and duration in time are changeable.

2nd – the technological revolution in the 21st century is studied fragmentarily. In most of the existing publications certain aspects of this revolution are studied – either formation of the digital society or formation of AI. For example, the works Ansong and Boateng (2019), Bogoviz et al. (2019a), Bogoviz et al. (2019b), Hodžić (2019), Hrustek et al. (2019), and Mueller and Grindal (2019) consider the opportunities and perspectives of digital modernization of certain economic processes.

These drawbacks do not allow forming a systemic vision of the technological revolution in the 21st century and determining its causal connections and scenarios of development. This causes uncertainty and hinders the management of this revolution, increasing its economic and social risks. That's why this papers aims at overcoming the determined drawbacks and establishing the causal connections of the technological revolution in the 21st century. For this, the authors use the materials of a statistical collection of the National Research University "Higher School of Economics" "Digital economy – 2019", which reflects that factors that restrain the usage of the Internet (as the most revolutionary technology of modern times) by the population (Fig. 1).

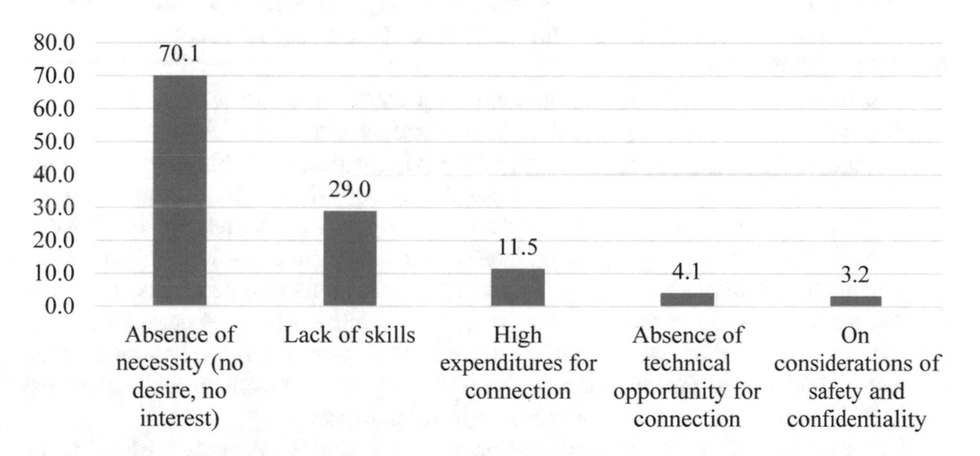

Fig. 1. Factors that restrain usage of the Internet in Russia. Source: compiled by the author based on National Research University "Higher School of Economics" (2019).

Figure 1 shows that the most significant factor (70.1% of Russian population) is absence of interest in revolutionary technologies. That's why there's a necessity for marketing support for technological revolution, which is aimed at attraction of attention and interest, as well as explaining the advantages of revolutionary technologies for potential consumers. The second factor (29%) is lack of skills. Thus, there's a need for stimulation of mastering of digital competencies. Other significant factors that restrain usage of revolutionary technologies in Russia are their high cost (11.5%), inaccessibility (4.1%), insufficient reliability, and low protection of data (3.2%). Therefore, there's a need for expanded financial provision of technological revolution in Russia, development of its infrastructure, and provision of safety of revolutionary technologies.

3 Results

The determined factors of using revolutionary technologies in Russia will be probably topical also for AI. State management of these factors will be determining the terms of socio-economic development of Russia under the influence of technological revolution in the 21st century until 2030 and its results. This is the basis for compiling three scenarios of socio-economic development of Russia under the influence of technological revolution in the 21st century until 2030, which qualitative treatment is shown in Table 1.

Table 1 shows that the first distinguished scenario envisages deintellectualization – i.e., refusal from the revolutionary breakthrough in the sphere of AI and limitation of the current plans on its usage. The key reason for implementation of this scenario is insufficient marketing support for AI, which hinders its commercialization. There's no stimulation of mastering of digital competencies, due to which ("market gap") digital competencies are a rarity. Financing of the revolution is provided from the state (more than 90% from the federal) budget and budgets of large corporations. Development of the revolution's infrastructure is delayed. Provision of safety of revolutionary technologies is insufficient.

The result of the revolution is mass social protests in the form of strikes of the employees of companies that perform automatization on the basis of AI and in the form of low demand for goods and services that envisage usage of AI. Instead of this, revolutionary technologies in which the possibilities of AI are used at the minimum level – "smart" devices and intellectual decision support – are widely disseminated.

The second scenario envisages fragmented intellectualization – i.e., elitism of AI. The key reason of implementation of this scenario will be fragmented (only in narrow circles) marketing support for AI. Stimulating the mastering of digital competencies is insufficient, due to which only a half of the society possess the digital competencies. Financial provision of the revolution is based on separate investment and innovative projects, including by the terms of public-private partnership.

Development of the revolution's infrastructure is timely. Provision of safety of revolutionary technologies is sufficient, but elite – e.g., high effectiveness and high cost of anti-virus software. The result of the revolution is neutral reaction of the society or restrained approval. Fragmentary automatization of consumer and entrepreneurial processes by the terms of elite access to AI is performed.

The third scenario envisages intellectual breakthrough – i.e., mass dissemination of AI. The key reason of implementation of this scenario is full-scale and systemic marketing support for AI, which stimulates its commercialization. Stimulation of mastering of digital competencies is sufficient – they are mastered by most of the society.

Financial provision of the revolution envisages attraction of investments from a lot of private companies that implement their own innovative projects on realization of AI. Development of the revolution's infrastructure is rapid. Provision of safety of revolutionary technologies is sufficient and widely accessible. The result of the revolution is wide social support, expressed in high demand for AI. Ubiquitous automatization on

Table 1. Qualitative characteristics of scenarios of socio-economic development of Russia under the influence of technological revolution in the 21st century until 2030.

Characteristics of the scenario		Scenario of socio-economic development		
		Deintellectualization	Fragmented intellectualization	Intellectual breakthrough
State management of the factors of the revolution	Marketing support for revolution	Insufficient	Fragmented	Full-scale, systemic
	Stimulation of mastering of digital competencies	Absent: digital competencies is a rarity	Insufficient: more than a half of the society possess digital competencies	Sufficient: mass mastering of digital competencies
	Financial provision of the revolution	From state budget and budgets of large corporations	Certain projects, including public-private partnership	Attraction of investments from private companies
	Development of the revolution's infrastructure	Delayed	Timely	Rapid
	Provision of safety of revolutionary technologies	Insufficient	Sufficient, but elite	Sufficient and accessible
Results of the revolution	Social reaction to the revolution	Mass social protests	Neutral reaction or restrained approval	Mass social support
	Directions of usage of AI	"smart" devices, intellectual decision support	Fragmented automatization of consumer and entrepreneurial processes	Ubiquitous automatization on the basis of AI
	Scale of dissemination of revolutionary technologies	Mass dissemination of conventionally revolutionary technologies	Elite access to artificial intelligence	Mass dissemination of AIe

Source: compiled by the author.

the basis of AI is performed. Quantitative characteristics of the considered scenarios is shown in Fig. 2.

Figure 2 shows that within the scenario of deintellectualization and with the current budget of technological revolution in Russia (the budget of the program "Digital economy of the Russian Federation"), which constitutes RUB 1.63 trillion (1% of GDP in 2018) the share of digital business (that uses revolutionary technologies) will remain at the current level (86.1%), as well as the share of digital society (73%). Within the

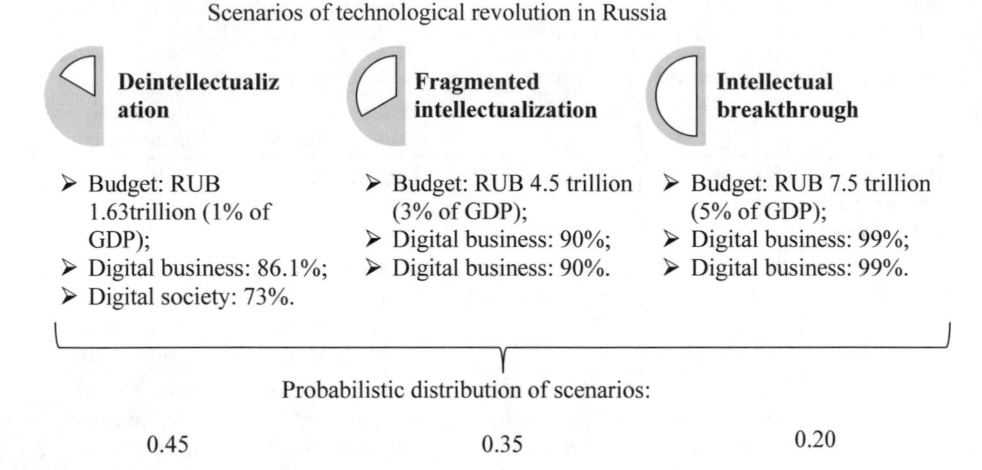

Fig. 2. Quantitative characteristics of scenarios of socio-economic development of Russia under the influence of technological revolution in the 21st century until 2030. Source: compiled by the author based on National Research University "Higher School of Economics" (2018).

scenario of fragmented intellectualization, increase of the budget by 2.5 times (up to RUB 4.5 trillion - 3% of GDP) by means of private investments will lead to increase of the share of digital business and digital society to 90%.

Implementation of the scenario of intellectual breakthrough will require higher increase of the budget of technological revolution in Russia – up to RUB 7.5 trillion (5% of GDP) also by means of private investments – which will allow achieving the 99% share of digital business and digital society. The most probable (0.45%) scenario is the scenario deintellectualization, as this is a forecast "with all other conditions being equal". The scenario of fragmented intellectualization is also rather probable (0.35). The least probable scenario is the scenario of intellectual breakthrough (0.20) – though this scenario might enable significant growth of global digital competitiveness and achievement of the level of developed countries as to the criterion of socio-economic development.

4 Conclusion

Thus, the technological revolution in the 21st century is a complex and variable process, which could take place according to one of the three scenarios, depending on formation of the digital society and adoption of AI. The scenario of deintellectualization is most probable in modern Russia, and its signs could be observed as of now. It envisages low level of development of the digital society and low interest of the consumers in AI due to insufficient marketing support. That's why investment flows will be redirected in favor of conventionally revolutionary (not providing completely new opportunities) intellectual technologies.

The scenario of fragmented intellectualization is also rather probable in Russia, as it is traditionally peculiar for the processes of innovative development of the Russian economy (e.g., it was implemented during dissemination of mobile communications). This scenario envisages elitism of AI until 2030 due to its high cost. The scenario of intellectual breakthrough is the most popular one – but it is least probable in Russia. It is connected to mass dissemination of AI – in entrepreneurship, consumption, and state management. The key condition of its implementation is active marketing support for AI. The described scenarios reduce uncertainty and open expanded opportunities and perspectives for state management of technological revolution in modern Russia.

Acknowledgments. The research has been performed with financial support from the Russian Foundation for Fundamental Research within scientific project No. 18-010-00103 A.

References

Ansong, E., Boateng, R.: Surviving in the digital era – business models of digital enterprises in a developing economy. Digit. Policy Regul. Gov. **21**(2), 164–178 (2019)

Bogoviz, A.V., Alekseev, A.N., Ragulina, J.V.: Budget limitations in the process of formation of the digital economy. Lect. Notes Netw. Syst. **57**, 578–585 (2019a)

Bogoviz, A.V., Lobova, S.V., Ragulina, J.V.: Distortions in the theory of costs in the conditions of digital economy. Lect. Notes Netw. Syst. **57**, 1231–1237 (2019b)

da Silva, V.L., Kovaleski, J.L., Pagani, R.N.: Technology transfer in the supply chain oriented to industry 4.0: a literature review. Technol. Anal. Strateg. Manag. **31**(5), 546–562 (2019)

Hodžić, S.: Tax administrative challenges of the digital economy: the Croatian experience. eJournal Tax Res. **16**(3), 762–779 (2019)

Hrustek, N.Ž., Mekovec, R., Pihir, I.: Developing and validating measurement instrument for various aspects of digital economy: E-commerce, E-banking, E-work and E-employment. Int. J. E-Serv. Mob. Appl. **11**(1), 50–56 (2019)

Masood, T., Egger, J.: Augmented reality in support of Industry 4.0—Implementation challenges and success factors. Robot. Comput.-Integr. Manuf. **58**, 181–195 (2019)

Mueller, M., Grindal, K.: Data flows and the digital economy: information as a mobile factor of production. Digit. Policy Regul. Gov. **21**(1), 71–87 (2019)

Popkova, E.G.: Economic and Legal Foundations of Modern Russian Society: A New Institutional Theory. Advances in Research on Russian Business and Management. Information Age Publishing, Charlotte (2017)

Popkova, E.G.: Preconditions of formation and development of industry 4.0 in the conditions of knowledge economy. Stud. Syst. Decis. Control **169**, 65–72 (2019)

Popkova, E.G., Sergi, B.S.: Will industry 4.0 and other innovations impact Russia's development? In: Exploring the Future of Russia's Economy and Markets, pp. 34–42. Emerald Publishing (2019)

Sergi, B.S., Popkova, E.G., Bogoviz, A.V., Ragulina, J.V.: Entrepreneurship and economic growth: the experience of developed and developing countries. In: Entrepreneurship and Development in the 21st Century, pp. 3–32. Emerald Publishing Limited (2019)

National Research University "Higher School of Economics": Digital economy – 2019: a short statistical collection (2019). https://www.hse.ru/data/2018/12/26/1143130930/ice2019kr.pdf. Accessed 09 May 2019

Modernization of Modern Entrepreneurship on the Basis of Artificial Intelligence

Mansur F. Safargaliev[1(✉)], Yuliana A. Kitsay[2], Elena N. Egorova[3], and Lilia V. Ermolina[4]

[1] A.N. Tupolev Kazan National Research Technical University, Kazan, Russia
MFSafargaliev@kai.ru

[2] I.Kant Baltic Federal University, Kaliningrad, Russia
Juliana_kn666@mail.ru

[3] Russian State Social University, Moscow, Russia
elegni@yandex.ru

[4] Samara State Technical University, Samara, Russia
Ermolina@mail.ru

Abstract. Purpose: The purpose of the article is to determine the perspectives of modernization of various components of entrepreneurial activities on the basis of AI and to develop a structural and logical scheme of systemic modernization of a company that would take into account the specific features of each component.

Design/methodology/approach: The authors determine the readiness of entrepreneurship to modernization on the basis of AI by the example of modern Russia, which is among top 10 countries as to most of the indicators of the digital economy. For this, analysis of activity of Russian companies' using digital technologies in various entrepreneurial processes - production, marketing, and sales – is performed.

Findings: It is concluded that perspectives of automatization of different components of entrepreneurial activities during modernization of entrepreneurship on the basis of AI are different. Modern Russia is peculiar for a contradiction - in management, where the perspectives of modernization on the basis of AI are the lowest, activity of automatization is the highest. At the same time, in production, where perspectives of modernization on the basis of AI are the highest, activity of automatization is the lowest. The determined contradiction shows that institutionalization of the observed practice of digital modernization of entrepreneurship in Russia might lead to advantages (e.g., growth of efficiency, increase of service, and rationalization) and risks (e.g., reduction of the level of competitiveness and corporate social responsibility).

Originality/value: For supporting high effectiveness and sustainability entrepreneurship in the process of modernization of on the basis of AI, it is recommended to use the developed structural and logical scheme of this process.

Keywords: Modernization · Entrepreneurship · AI · Modern Russia · Automatization · Digital modernization

JEL Code: L26 · O31 · O32 · O33 · O34

E. G. Popkova and B. S. Sergi (Eds.): ISC 2019, LNNS 91, pp. 346–352, 2020.
https://doi.org/10.1007/978-3-030-32015-7_39

1 Introduction

Entrepreneurship should become the main driving force of digital modernization of the modern economy, as in the process of entrepreneurial activities new technologies are implemented and innovative goods and services are created. AI is one of the most progressive technologies of modern times. Most of the other developed technologies envisage using AI in cyber-physical systems.

The existing approaches to studying the perspectives of implementing AI into the economic practice envisage sectorial overviews or consideration of certain companies. Generalized consequences of modernization of modern entrepreneurship on the basis of AI are considered. The key advantage is systemic integration of entrepreneurial processes on the basis of the unified and effective management with the help of AI.

The most popular and probable macro-economic drawbacks of mass usage of AI in entrepreneurship is high dependence on technical devices, which require technical maintenances, repairs, and system administration – without which the risks of failures in the work of equipment and crisis of entrepreneurship are high. Also, there's high risk of energy collapse – depletion of energy resources due to quick increase of the volume of their consumption. Uncertainty as to usage of AI entrepreneurship and limitations of its usage remain.

We offer a hypothesis that there's a necessity for a new approach to studying the perspectives of implementing AI into the economic practice, which envisages studying this process through the prism of the components of entrepreneurial activities, perspectives of modernization of which are different. This predetermined the purpose of the research – determining the perspectives of modernization of various components of entrepreneurial activities on the basis of AI and developing a structural and logical scheme of systemic modernization of a company, which would take into account specific features of each component.

2 Materials and Method

In the existing scientific literature, the issue of modernization of modern entrepreneurship on the basis of AI is widely discussed. Bogoviz et al. (2019), Li and Wang (2017), Sergi et al. (2019), Popkova (2019), Swan (2018), Ye (2017) note large potential of AI in creation of competitive advantages for modern entrepreneurship and expedience of its mass dissemination for increasing the created value (benefit) for the consumers.

Mingaleev (2012), Pritvorova et al. (2018), Salimov and Mingaleev (2012), Bezrukova et al. (2017), Sibirskaya et al. (2019), Valitov et al. (2015), Vanchukhina et al. (2018) note that implementation of AI is connected to large expenditures for entrepreneurship (e.g., expenditures for new equipment and technologies, personnel training, marketing, etc.), which are not always returned and are justified only for companies that have sufficient financial resources and function in competitive markets, at which consumers show high effective demand for innovative and technological products.

The literature overview showed that though the perspectives of modernization of modern entrepreneurship on the basis of AI are studied in the existing works, the specific features of implementing AI into entrepreneurial processes are not elaborated sufficiently – which complicates practical implementation of modernization measures in the modern entrepreneurship.

Let us determine readiness of entrepreneurship to modernization on the basis of AI by the example of modern Russia, which is in top 10 countries of the works as to most indicators of the digital economy. For this, analysis of activity of Russian companies' using digital technologies in various entrepreneurial processes - production, marketing, and sales – is performed (Fig. 1).

Table 1 shows that software in management are used by 52.7%. In the process of marketing and sales, various means of automatization (in particular, online financial accounting, SCM-, ERP-, CRM- and RFID-technologies, cloud services, and the Internet) are used by 19% of Russian companies. Though automatization of industrial production on the basis of robototronics is a new phenomenon for Russia, potential readiness for creation of cyber-physical systems are shown by 7.1% of Russian companies, which have access to broadband Internet with the speed over 100 Mb/s (mandatory condition of remote control in production).

Therefore, the highest opportunities of modernization on the basis of AI in Russia in 2018 are peculiar for management. Then come marketing and sales, and production has the lowest opportunities.

Table 1. Activity of Russian companies' using digital technologies in various entrepreneurial processes in 2018, %.

Share of companies that use software in management	52.7
Share of companies that perform financial payments in electronic form	53.7
Share of companies that use SCM technologies	7.1
Share of companies that use ERP technologies	13
Share of companies that use CRM technologies	19.2
Share of companies that use RFID technologies	6.2
Share of companies that use cloud services	22.6
Share of companies that use the Internet for sales	12.1
Share of companies that use the Internet for purchases	18.1
Share of companies that have broadband Internet access (more than 1000 Mb/s)	7.1

Source: compiled by the authors based on National Research University "Higher School of Economics" (2019).

3 Results

Evaluation of perspectives of modernization of various components of entrepreneurial activities on the basis of AI is performed in Table 1.

Table 2 shows that during modernization of production on the basis of AI the objects of automatization could be planning of production (tactical), organization of production, sorting, branding, and packaging, as well as quality control of manufactured products. The accompanying technologies are industrial Internet and other technologies of wireless communications, robototronics, and sensors.

Advantages of automatization are growth of efficiency, scale effect, reliability (precision and continuity), and safety. Limitations of automatization are caused by the fact that automatization should have systemic character, and communications with consumers have to be avoided. Risks of automatization consist in the fact that in case of

Table 2. Evaluation of perspectives of modernization of various components of entrepreneurial activities on the basis of AI.

Characteristics	Object of modernization (component of entrepreneurial activities)		
	Production	Marketing and sales	Management
Objects of automatization	– planning of production (tactical); – organization of production; – sorting, branding, packaging; – quality control of manufactured products	– receipt of orders; – calculations and document turnover; – modeling of products; – promotion; – logistics and shipment	– strategic planning; – implementation of innovations; – accounting; – solving arguments
Accompanying technologies	– Industrial Internet and other technologies of wireless communications; – robototronics and sensors	– distributed register (including blockchain); – 3D-print; – RFID, CRM, ERP, SCM	– quantum computers; – Big Data; cloud services
Advantages of automatization	– growth of efficiency; – scale effect; – reliability (precision and continuity) and security	– increase of service; – expansion of sales markets; – growth of scale of marketing	– absolute rationality; – quick making of optimal decisions
Limitations of automatization	– automatization should have systemic character; – communications with consumers should be avoided	– All sub-processes have to be standardized, which reduces flexibility of entrepreneurship	– impossibility or incompleteness of accounting of social factors during decision making
Risks of automatization	– in case of defects, the scale of defects is very high; – full stop of production in case of failures of machine communications	– errors during communication; – reduction of competitiveness and demand	– "narrow" thinking; – reduction of the level of corporate social responsibility

Source: compiled by the authors.

its defects, the scale of the defects is very large, and production is fully stopped in case of failures of machine communications.

During modernization of marketing and sales on the basis of AI, the objects of automatization could be receipt of orders, calculations, document turnover, modeling of products, promotion, and logistics and shipment. The accompanying technologies could be distributed register (including blockchain), 3D-print, RFID, CRM, ERP, and SCM. The advantages of automatization include expansion of service, expansion of sales markets, and growth of the scale of marketing. Limitations of automatization are caused by the fact that all sub-processes have to be strictly standardized, which reduces flexibility of entrepreneurship. The risks of automatization are connected to errors during communication and reduction of competitiveness and demand.

During modernization of management on the basis of AI, the objects of automatization could be strategic planning, implementation of innovations, accounting, and solving arguments. The accompanying technologies could be quantum computers, Big Data, and cloud services. The advantages of automatization include absolute rationality and quick making of optimal decisions. Limitations of automatization are connected to impossibility or incompleteness of accounting of social factors during decision making. Risks of automatization consist in "narrow" thinking of AI, which does not allow it to consider social factors during management and in reduction of the level of corporate social responsibility.

Thus, the largest perspectives of modernization of a modern company on the basis of AI are peculiar for production, then come marketing and sales, and management. Based on this, a structural and logical scheme of modernization of a modern company on the basis of AI is compiled (Fig. 1).

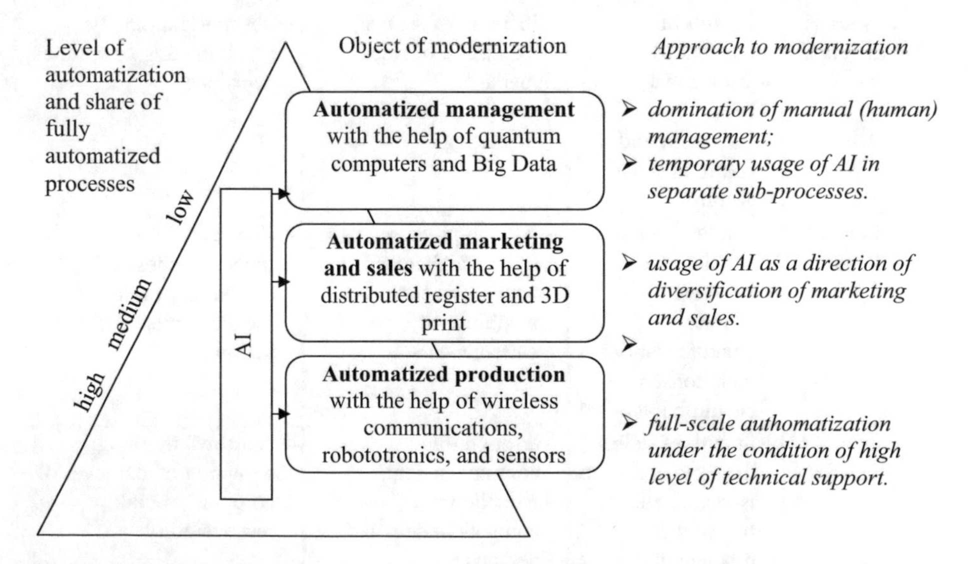

Fig. 1. Structural and logical scheme of modernization of a modern company on the basis of AI. Source: compiled by the authors.

Figure 1 shows that the company that is modernized on the basis of AI is based on automatized production. In this entrepreneurial process, it is recommended to conduct full-scale automatization under the condition of high level of technical support (level of automatization is high). The central element of a company is automatized marketing and sales, in which AI is to be used as a direction of diversification (as an addition to the existing practice, the level of automatization is medium).

Management with low level of automatization is at the top. For supporting high effectiveness and sustainability of company, manual labor and temporary usage of AI in separate sub-processes (if necessary) are required. This will allow obtaining advantages from modernization of entrepreneurship on the basis of AI and overcoming its limitations and risks.

4 Conclusion

Thus, as a result of the research, the offered hypothesis is confirmed. The perspectives of automatization of different components of entrepreneurial activities during modernization of entrepreneurship on the basis of AI are different. Modern Russia is peculiar for a contradiction - in management, where the perspectives of modernization on the basis of AI are the lowest, activity of automatization is the highest. At the same time, in production, where perspectives of modernization on the basis of AI are the highest, activity of automatization is the lowest.

The determined contradiction shows that institutionalization of the observed practice of digital modernization of entrepreneurship in Russia might lead to advantages (e.g., growth of efficiency, increase of service, and rationalization) and risks (e.g., reduction of the level of competitiveness and corporate social responsibility). For supporting high effectiveness and sustainability of entrepreneurship in the process of its modernization on the basis of AI, it is recommended to use the developed structural and logical scheme of this process.

References

Bogoviz, A.V., Lobova, S.V., Ragulina, J.V.: Perspectives of growth of labor efficiency in the conditions of the digital economy. Lect. Notes Netw. Syst. **57**, 1208–1215 (2019)

Li, Z., Wang, J.: An optimization application of artificial intelligence technology in enterprise financial management. Boletin Tecnico/Tech. Bull. **55**(11), 83–89 (2017)

Mingaleev, G.F.: The economic aspects of the development of car manufacturing and after-sales services in the republic of Tatarstan. In: Implementing International Services: A Tailorable Method for Market Assessment, Modularization, and Process Transfer 2012, pp. 437–449. Gabler Verlag, Germany (2012)

Popkova, E.G.: Preconditions of formation and development of industry 4.0 in the conditions of knowledge economy. Stud. Syst. Decis. Control **169**, 65–72 (2019)

Pritvorova, T., Tasbulatova, B., Petrenko, E.: Possibilities of blitz-psychograms as a tool for human resource management in the supporting system of hardiness of company. Entrep. Sustain. Issues **6**(2), 840–853 (2018). https://doi.org/10.9770/jesi.2018.6.2(25)

Salimov, R.I., Mingaleev, G.F.: The regional policy of industrial IPM services for the development of knowledge potential in Russia. In: Implementing International Services: A Tailorable Method for Market Assessment, Modularization, and Process Transfer 2012, pp. 425–435. Gabler Verlag, Germany (2012)

Sergi, B.S., Popkova, E.G., Bogoviz, A.V., Ragulina, J.V.: Entrepreneurship and economic growth: the experience of developed and developing countries. In: Entrepreneurship and Development in the 21st Century, pp. 3–32. Emerald Publishing Limited (2019)

Bezrukova, T.L., Popova, E.V., Korda, N.I., Kuznetsova, T.E., Bezrukov, B.A.: Institutional traps of innovative and investment activities as an obstacle on the path to the well-balanced development of regions. Contrib. Econ. (9783319606958), 235–240 (2017)

Sibirskaya, E., Popkova, E., Oveshnikova, L., Tarasova, I.: Remote education vs traditional education based on effectiveness at the micro level and its connection to the level of development of macro-economic systems. Int. J. Educ. Manage. **33**(3), 533–543 (2019). https://doi.org/10.1108/IJEM-08-2018-0248

Swan, M.: Blockchain for business: next-generation enterprise artificial intelligence systems. Adv. Comput. **111**, 121–162 (2018)

Valitov, S.M., Mingaleev, G.F., Khadeev, N.R., Antonova, N.V.: Methods to assess economic loss while implementing energy saving programs in oil-production enterprises. Mediterr. J. Soc. Sci. **6**(3), 766–769 (2015)

Vanchukhina, L.I., Leybert, T.B., Khalikova, E.A., Khalmetov, A.R.: New approaches to formation of innovational human capital as an element of institutional environment. Espacios **39**, 22–32 (2018)

Ye, S.: Research on the enterprise accounting statement evaluation and financial management optimization based on computer artificial intelligence method. Boletin Tecnico/Tech. Bull. **55** (20), 208–215 (2017)

National Research University "Higher School of Economics" (2019). Digital modernization: 2019. https://www.hse.ru/primarydata/ice2019kr. Accessed 24 May 2019

Intellectual Production and Consumption: A New Reality of the 21st Century

Aleksei A. Shulus[1(✉)], Elena S. Akopova[2], Natalia V. Przhedetskaya[2], and Ksenia V. Borzenko[2]

[1] Plekhanov Russian University of Economics, Moscow, Russia
shulus@bk.ru
[2] Rostov State University of Economics, Rostov-on-Don, Russia
kafedra_kil@mail.ru, k_cherry@mail.ru, nvpr@bk.ru

Abstract. Purpose: The purpose of the research is to develop a conceptual model of intellectual production and consumption.

Design/methodology/approach: For scientific substantiation of the necessity for a concept of intellectual production and consumption, the authors perform analysis of the existing experience of functioning of the Internet economy in Russia. Based on the accessible statistical information, the authors compile a generalized macro-economic added value chain of intellectual production and consumption in Russia in 2018.

Findings: It is determined that intellectual production and consumption are a new reality of the 21st century in the conditions of the Internet economy. Despite the existence of the technical opportunity for intellectual production and consumption with most of the population and entrepreneurial structures of Russia, less than 30% of economic subjects perform it. It confirms the necessity for a scientific concept of organization of this process, which will allow involving all potentially interested parties, most of which are not interested in it. At present, Russia is peculiar for implementation of separate (scattered) practices of intellectual production and consumption, which does not allow obtaining advantages from them in full – as, together with these practices, traditional (with low level of automatization) production and consumption are conducted.

Originality/value: It is substantiated that perspectives of development of the Internet economy are connected to ousting the traditional practices and mass dissemination of practices of intellectual production and consumption. This is stimulated by the developed conceptual model of intellectual production and consumption. It is to become a scientific and methodological platform for organization of production and distribution processes in the modern economy, thus reducing uncertainty and risk and increasing the attractiveness of intellectual production and consumption.

Keywords: Intellectual production · Intellectual consumption · AI · Intellectual decision support · Internet economy · Russia

JEL Code: D11 · D12 · E23 · L86 · P46 · O31 · O32 · O33

E. G. Popkova and B. S. Sergi (Eds.): ISC 2019, LNNS 91, pp. 353–359, 2020.
https://doi.org/10.1007/978-3-030-32015-7_40

1 Introduction

AI (as an analog of human intellect, which possesses a creative component and is capable of self-education and full autonomy) is in the process of development, but intellectual technologies of decision support are widely used – so intellectual production is a new reality of the 21st century. As decision making and conclusion of deals in the market have to take place in real time based on the current information, intellectual production and consumption are conducted on the Internet – i.e., they belong to the sphere of the Internet economy (structural component of the national economy that envisages conduct of economic operations with the usage of the Internet).

In developed countries and most developing countries, including Russia, the digital infrastructure is already formed, which allows developing the Internet economy. However, in practice, its development is very slow. For example, the share of Internet purchases in Russia grew in 2018 by 1.26 times, as compared to 2016; the share of Internet purchases of companies in this period grew by 1.08 times; the share of Internet sales of companies reduced by 4%. According to a sociological survey by the specialists of the National Research University "Higher School of Economics" (2019), the most important restraining factor on the path of increasing the activity of using the Internet by Russia's population is absence of the necessity (desire, interest), which is peculiar for 70.1% of consumers who have Internet access but do not use it.

The primary reason of low interest of population in intellectual consumption and of entrepreneurial structures in intellectual production is unclear organization of this process and potentially obtained advantages by its participants. That's why an important task of the modern economic theory is to form the conceptual foundations of intellectual production and consumption that would reflect the perspective directions of applying the technologies of intellectual decision support by economic subjects in the Internet economy, general logic of this process, and advantages of all its participants. The purpose of the research is to develop a conceptual model of intellectual production and consumption.

2 Materials and Method

The scientific and methodological provision of formation and management of the Internet economy is presented in the works Bezrukova et al. (2017), Bogoviz et al. (2019), Calzada and Tselekounis (2018), Chang and Dai (2018), Pritvorova et al. (2018), Sergi et al. (2019), Sibirskaya et al. (2019), Sukhodolov et al. (2018a), Sukhodolov et al. (2018b), Sukhodolov et al. (2018c), and Vanchukhina et al. (2018).

Certain issues of implementation of AI and intellectual technologies of decision support into the modern economic practice are discussed in the works Belkadi et al. (2019), Chen et al. (2018), Jemmali et al. (2018), López (2017), Muñoz and Capón-García (2019), Vidyasagar et al. (2018), and Zhao et al. (2018).

The comprehensive concept as a systemic scientific idea of the organization of intellectual production and consumption is not yet formed and requires further elaboration. For scientific substantiation of the necessity for this concept, let us perform analysis of the existing experience of functioning of the Internet economy in Russia. A generalized macro-economic added value chain of intellectual production and consumption in Russia in 2018 is shown in Fig. 1.

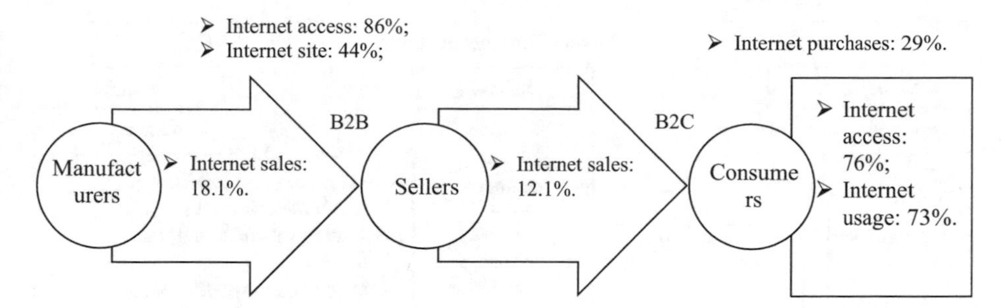

Fig. 1. Generalized macro-economic added value chain of intellectual production and consumption in Russia in 2018. Source: compiled by the authors based on the materials of the National Research University "Higher School of Economics" (2019).

According to Fig. 1, added value chain of intellectual production and consumption consists of three links. 1st link: intellectual production (manufacturers). 2nd link: intellectual mediation and sales (sellers). 3rd link: intellectual consumption (consumers). Only 86% of companies in Russia have Internet access, of which only 44% (less than half) have their own Internet site (i.e., the possibility of independent product sales via the Internet).

Internet purchases (i.e., Internet sales in the form B2B – business to business) are performed by 18.1% of Russian companies. Internet sales (in the form B2C – business to consumers) are performed by 12.1% of companies. 76% of Russians have Internet access, but only 73% of the population use the Internet, and only 29% perform Internet purchases. Therefore, despite the existence of the technical possibility for intellectual production and consumption with most of the population and entrepreneurial structures of Russia, less than 30% of economic subjects perform it. This confirms the necessity for a scientific concept of organization of this process, which will allow involving all interested parties which do not have interest in it as of now.

3 Results

As a result of studying the capabilities of intellectual technologies of decision support, we compiled a conceptual model of intellectual production and consumption (Fig. 2).

The model in Fig. 2 shows that consumer sets demand for the product of the studied market of the Internet economy. He visits the web-site either of this market (if any) or the Internet store, which contains the products of the market, and sets the criteria of product selection:

- technical features of products, which could satisfy the need – e.g., form, color, size, weight;
- consumer preferences – e.g., manufacturing country or delivery time;
- preferred terms of payment and price range.

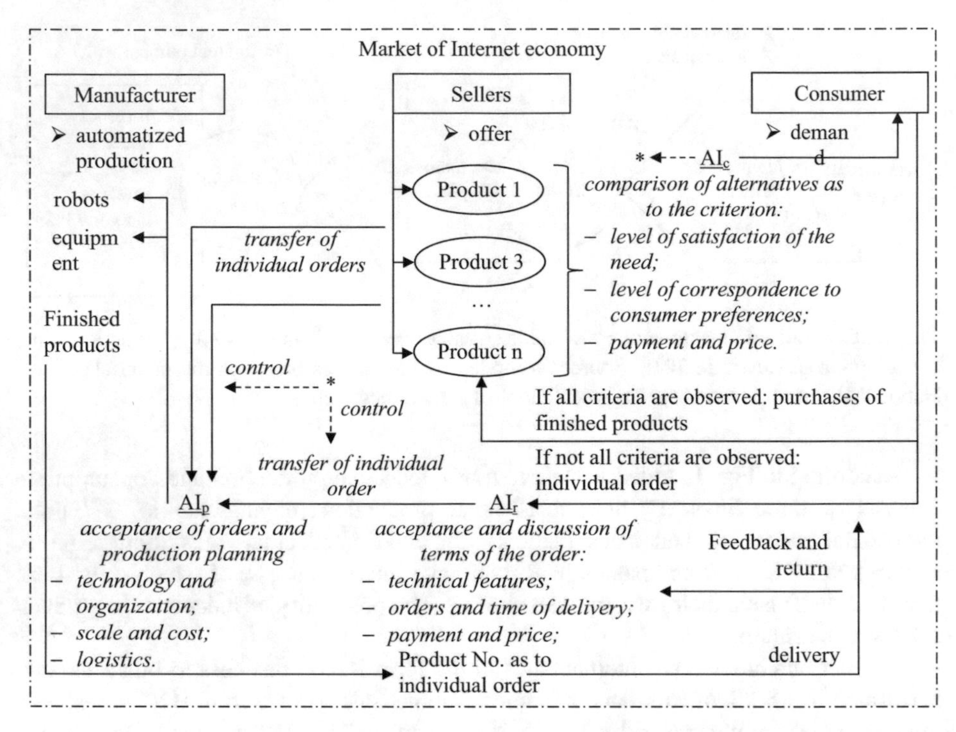

Fig. 2. The conceptual model of intellectual production and consumption. Source: compiled by the authors.

According to the set criteria, AI (AI_c – intellectual technology of decision support regarding the purchase) compares the presented alternative market offers and provides a report to the consumer. If all criteria are met, the consumer purchases the finished product (e.g., product n). If not all criteria are met, the consumer may place an individual order. Retail seller uses AI (AI_r – intellectual technologies of autonomous collection of orders) accepts the order and discusses its terms: technical features of the products, order and time of delivery, payment methods and price.

At the same time, retail sellers accept individual orders from consumers, which could coincide (with low level of originality). All orders are transferred to the manufacturer. Its AI (AI_p – namely, intellectual technology of production management) accepts all orders and plans production:

- determines the technology and organization of production in view of the possibilities of its standardization;
- selects the optimal scale and cost of production in view of the possibilities of gaining the "scale effect";
- determines the optimal variant of logistics.

Intellectual technology of production management performs practical implementation of the developed plan, issuing orders to robots and equipment within automatized production. Finished products N are sold to intermediaries, which then deal with the retail sales. The consumer controls the manufacturer and seller's activities in the course of the whole period of his individual order and is notified as a result of its production and supply.

Advantages of intellectual production for sellers are connected to the following:

- reduction of costs: product sales and collection of orders and automatized and are conducted in the Internet, which allows reducing expenditures for labor payment and fixed capital;
- increase of the service level: consumer could obtain all necessary information and apply for the seller's consultation (messaging or "call back") in a convenient place and time;
- expansion of sales market: collection of orders with the help of AI envisages usage of template forms for communication with consumer, which allows guaranteeing high quality of transfer and precision of the passed information.

The advantages of intellectual production for manufacturers are as follows:

- optimization of production: AI makes rational decisions on planning and organization of production, reducing the costs and increasing the efficiency;
- automatization of production: collection of orders by AI forms an information data base for further automatized manufacture of products, while human collection of orders requires their additional transfer into the electronic form.

Consumers obtain advantages from intellectual consumption, which consist in the following:

- optimal satisfaction of needs: AI allows saving time on decision making regarding purchases and guarantees the most rational decisions;
- controllability of the process of execution of orders: consumer could receive all necessary information on accessibility of the selected product in the seller's storage and the course of execution of the individual order.

4 Conclusion

Thus, intellectual production and consumption are a new reality of the 21st century in the conditions of the Internet economy. At present, Russia is peculiar for implementation of separate (scattered) practices of intellectual production and consumption, which does not allow obtaining advantages from them in full – as, together with these practices, traditional (with low level of automatization) production and consumption are conducted.

The perspectives of development of the Internet economy are connected to ousting the traditional practices and mass distribution of the practices of intellectual production and consumption. This is stimulated by the developed conceptual model of intellectual production and consumption. It is to become a scientific and methodological platform for organization of production and distribution processes in the modern economy, thus reducing uncertainty and risk and increasing the attractiveness of intellectual production and consumption.

References

Belkadi, F., Dhuieb, M.A., Aguado, J.V., Laroche, F., Bernard, A., Chinesta, F.: Intelligent assistant system as a context-aware decision-making support for the workers of the future. Comput. Ind. Eng. **2**(1), 139–145 (2019)

Bezrukova, T.L., Popova, E.V., Korda, N.I., Kuznetsova, T.E., Bezrukov, B.A.: Institutional traps of innovative and investment activities as an obstacle on the path to the well-balanced development of regions. Contrib. Econ. (9783319606958), 235–240 (2017)

Bogoviz, A.V., Lobova, S.V., Ragulina, J.V.: Perspectives of growth of labor efficiency in the conditions of the digital economy. Lect. Notes Netw. Syst. **57**, 1208–1215 (2019)

Calzada, J., Tselekounis, M.: Net Neutrality in a hyperlinked Internet economy. Int. J. Ind. Organ. **59**, 190–221 (2018)

Chang, X., Dai, D.: Analysis on marketing combination and mode transformation under internet economy. J. Adv. Oxid. Technol. **21**(2) (2018). 201804588

Chen, X., Huang, R., Shen, L., Chen, H., Xiong, D., Xiao, X., Liu, M., Xu, R.: Intellectual production supervision perform based on RFID smart electricity meter. In: IOP Conference Series: Earth and Environmental Science, vol. 128, no. 1 (2018). 012050

Jemmali, M., Alharbi, M., Melhim, L.K.B.: Intelligent decision-making algorithm for supplier evaluation based on multi-criteria preferences. In: 1st International Conference on Computer Applications and Information Security, ICCAIS 2018 (2018). 8441992

López, W.L.: Can scientrometrics contribute to the assessment of intellectual production? Universitas Psychologica **16**(4), 1–2 (2017)

Muñoz, E., Capón-García, E.: Intelligent mathematical modelling agent for supporting decision-making at industry 4.0. Adv. Intell. Syst. Comput. **865**, 152–162 (2019)

Pritvorova, T., Tasbulatova, B., Petrenko, E.: Possibilities of blitz-psychograms as a tool for human resource management in the supporting system of hardiness of company. Entrep. Sustain. Issues **6**(2), 840–853 (2018). https://doi.org/10.9770/jesi.2018.6.2(25)

Sergi, B.S., Popkova, E.G., Bogoviz, A.V., Ragulina, J.V.: Entrepreneurship and economic growth: the experience of developed and developing countries. In: Entrepreneurship and Development in the 21st Century, pp. 3–32. Emerald Publishing Limited (2019)

Sibirskaya, E., Popkova, E., Oveshnikova, L., Tarasova, I.: Remote education vs traditional education based on effectiveness at the micro level and its connection to the level of development of macro-economic systems. Int. J. Educ. Manage. **33**(3), 533–543 (2019). https://doi.org/10.1108/IJEM-08-2018-0248

Sukhodolov, A.P., Popkova, E.G., Kuzlaeva, I.M.: Internet economy: existence from the point of view of micro-economic aspect. Stud. Comput. Intell. **714**, 11–21 (2018a)

Sukhodolov, A.P., Popkova, E.G., Kuzlaeva, I.M.: Modern foundations of internet economy. Stud. Comput. Intell. **714**, 43–52 (2018b)

Sukhodolov, A.P., Popkova, E.G., Kuzlaeva, I.M.: Perspectives of internet economy creation. Stud. Comput. Intell. **714**, 23–41 (2018c)

Vanchukhina, L.I., Leybert, T.B., Khalikova, E.A., Khalmetov, A.R.: New approaches to formation of innovational human capital as an element of institutional environment. Espacios **39**, 22–32 (2018)

Vidyasagar, S., Ghosh, M., Deka, D., Mondal, S.: Intellectual inverter for efficient unit consumption. J. Adv. Res. Dyn. Control Syst. **10**(5), 427–431 (2018)

Zhao, W., Wu, J., Shi, P., Wang, H.: Intelligent sensing and decision making in smart technologies. Int. J. Distrib. Sens. Netw. **14**(11), 38–46 (2018)

National Research University "Higher School of Economics" (2019). Digital economy: 2019. https://www.hse.ru/primarydata/ice2019kr. Accessed 24 May 2019

Imagination, Invention and Internet: From Aristotle to Artificial Intelligence and the 'Post-human' Development and Ethics

Qerim Qerimi[1,2(✉)]

[1] University of Antwerp, Antwerp, Belgium
Qerim.Qerimi@uantwerpen.be
[2] University of Prishtina, Pristina, Kosovo

1 Introduction

The notion that perennial questions of human welfare and future life defined or dictated by technological advances are as ancient as Aristotle's thought might be intellectually curious, yet factually accurate and discursively valid. It is true that one of the fundamental tenets of Aristotle's peripatetic philosophy is that the goal of life is to maximize happiness by living virtuously, fulfilling your own potential as a human, and engaging with others—family, friends and fellow citizens—in mutually beneficial activities. In his conception, purposively imagining a better, happier life is feasible since humans have inborn abilities that allow them to promote individual and collective flourishing. These include the inclinations to ask questions about the world, to deliberate about action, and to activate conscious recollection (Hall 2018). There is but one element in the life maximization equation that plays a critical role, namely imagination.

Thanks to imagination, we almost see in the once utopian Aristotle the anticipator of modern developments in artificial intelligence. He imagines the possibility that everyone will one day be able to realize his potential and make full use of all his faculties (the distinctive 'Aristotelian principle' according to the political philosopher John Rawls). Aristotle envisages a futuristic world in which technological advances would render human labor unnecessary. He recalls the mythical craftsmen Daedalus and Hephaestus, who constructed robots that worked to order (Aristotle 1988, book I, part IV):

> *for if every tool could perform its own work when ordered, or by seeing what to do in advance, like the statues of Daedalus in the story, or the automatic tripods of Hephaestus which the poet says 'enter self-moved the company divine,' if shuttles could weave like this, and plectrums strum harps of their own accord, master-craftsmen would have no need of assistants and masters no need of slaves.*

This passage could be taken as a testament to the constant quest present in the history of ideas, imaginary or actual, seeking to enlarge and express human potential to its fullest. Indeed, the task of this chapter is not so much about any historical scrutiny of inventive thinking as much as about the contemporary results of such a process, thus investigating the present achievements made in the field of artificial intelligence and its revolutionary effects and also ensuing ethical challenges posed for humanity.

E. G. Popkova and B. S. Sergi (Eds.): ISC 2019, LNNS 91, pp. 360–371, 2020.
https://doi.org/10.1007/978-3-030-32015-7_41

The next first section will look at how imagination is mobilized, including the pertinent conditioning factors. Unlike hard sciences in some relative sense, the scope or scale of imaginative and innovative thinking in social sciences is rather need and problem-based. Examples will be presented from such areas as human rights and environmental protection, both critical composite components of the process of development. Variations of the way other spheres of socio-economic life are interacting with technological advances will also be reviewed; however, the list is hardly finite. The subsequent second section will look more specifically at the impact of technological advances on development process, as well as related challenges, expressed chiefly in terms of globally uneven access to such technologies. Next, the third section, will engage in a more concerted fashion with present realities and prospective developments of artificial intelligence and superintelligence. The presentation of a conceptual framework will be offered here, same as a discussion of the ethical implications inherent in this 'post-human' conundrum. One specific operational sample could be obtained from the development and deployment of technological inventions in the area of security or military more specifically, which is subject of discussion in section four of the chapter. Section five then deals with the imperative of inventing and designing adequate and effective regulatory regimes that both enable use and prevent abuse of artificial intelligence. Finally, there is a concluding section that sums up the key messages and findings.

2 Imagination in Operation: Rights, Freedoms and Environment

Triggering and employing imagination is context-sensitive and might be also deferential to particular disciplines. For instance, in the arena of human rights, its development and evolvement can be credited to imagination as much as to problems and needs thereof. The same could be true of environmental concerns and increased awareness for action and protection. Imagination has been rather activated and mobilized by the needs ensuing from, or identified on the basis of, real-world problems. Think of Raphael Lemkin, who coined the term genocide and initiated the Genocide Convention during World War II. Or Jody Williams and the International Campaign to Ban Landmines, culminating in the Convention on the Prohibition of the Use, Stockpiling, Production and Transfer of Anti-Personnel Mines and on their Destruction (known also as the Mine Ban Treaty or Ottawa Treaty). Al Gore's Oscar-winning documentary *An Inconceivable Truth* moved climate change onto the list of public policy debates in the US and across continents. This was also based on a problem.

In any event, the origin of the invention in the broader arena of human development is to a great extent linked with the individual. The adoption of major international human rights instruments was often a result of processes that were initiated by one or a small group of individuals who clarified their own objective, then used the resources of modern information and communication to mobilize other non-governmental groups, which in turn influenced key national governments to pursue and ultimately adopt a particular program. The same process increased the pressure on more reluctant members of the international community. The initial international human rights instruments, same as those concerning the environment and environment and development, were indeed issued as Declarations rather strictly legally-binding treaties. And that was precisely because the States concerned were initially unwilling to lock themselves into a network of binding obligations (Reisman et al. 2004). Conceptually, the challenge has been to compromise on the notion of State sovereignty and in favor of individual empowerment.

However, in process, dictated by advances in new technologies, the role and functions of the individual are not only transformed, but in certain segments now and more in the future, they will be replaced by smart machines or at the very least unmanned digitally-operated functions, as in the Artistotle's quoted passage.

The underlying approach adopted so far in regard to classic human rights framework and digital technologies has been the simple transfer of human rights and fundamental freedoms applied offline to the online context, in particular in connection with the right to freedom of information and expression. For instance, a recent resolution adopted by the UN Human Rights Council (2016), entitled *The promotion, protection and enjoyment of human rights on the Internet*, affirms that 'the same rights that people have offline must also be protected online, in particular freedom of expression, which is applicable regardless of frontiers and through any media of one's choice.' Back in 2011, the Special Rapporteur on the promotion and protection of the right to freedom of opinion and expression reasoned that Article 19 of the Universal Declaration of Human Rights and the International Covenant on Civil and Political Rights—the right to freedom of opinion and expression—was drafted with the foresight to include and to accommodate future technological developments through which individuals can exercise their right to freedom of expression. Therefore, the framework of international human rights law remains relevant and is applicable to the internet (Human Rights Council 2011).

In the context of the relationship between the rights to privacy and to freedom of expression, throughout history, debating controversial issues in the public sphere—in offline platforms—has always implied the possibility to do so anonymously, a practice acknowledged and approved by the judiciary (U.S. Supreme Court 1999, 1995 and 1960). Therefore, the UN Special Rapporteur has also called upon States to ensure that individuals are able to express themselves anonymously on the internet and to refrain from adopting real-name registration systems (Human Rights Council 2011, para. 84). Although there may be legal exceptions that warrant limiting the right to privacy, such as for the purposes of administering criminal justice or preventing crime, these measures ought to comply with the international human rights framework.

Privacy or the protection of personal data on the internet could also be challenged by another competing right, that of copyright protection. The tension between the two or between privacy and other competing rights has led to judicial proceedings and decisions both before national courts and regional judicial bodies, such as the Court of Justice of the European Union (CJEU) and the European Court of Human Rights. For instance, the CJEU that internet users' IP addresses are 'protected personal data,' and that the installation of a 'filtering system [to enable copyright holders to institute legal proceedings] would involve systematic analysis of all content and the collection and identification of users' IP addresses' (CJEU 2011, para. 51). Requiring installation of the contested filtering system could undermine freedom of information 'since that system might not distinguish adequately between unlawful content and lawful content,' and could therefore 'lead to the blocking of lawful communications' (CJEU 2011, para. 51). In addition, such a system would not respect the requirement that a fair balance be struck between copyright protection and the freedom to conduct business since it would require the internet service provider 'to install a complicated, costly, permanent computer system at its own expense' (CJEU 2011, paras. 48–49). Ultimately, a fair balance ought to be struck between the right to intellectual property, on the one hand, and the freedom to conduct business, the right to protection of personal data and the freedom to receive or impart information, on the other (Qerimi 2017).

In the long list of areas being affected by the use of digital technology is the electoral process or indeed the overall democratic process. In a case concerning the collection of digital signature in an electoral process, the Venice Commission and the OSCE Office for Democratic Institutions and Human Rights (ODIHR) recommended the clarification of exact requirement for digital signatures, 'with a view to ensuring equal treatment between the collection of classical and digital signatures, and easy access to the process in both cases' (Venice Commission 2007, para. 40). In a different case, a mechanism for electronically collecting the fingerprints of voters at polling stations was provided, so as to check for cases of potential multiple voting. In this case, the Venice Commission and OSCE/ODIHR stated that: 'should any new technologies be introduced in the electoral process, a number of issues should be thoroughly considered, including a risk assessment of the costs, benefits and challenges of introducing such technologies, harmonization of new provisions with existing data protection laws and standards, but also ensuring trust in the process, necessary check-ups and pilot procedures, proper procedures for procurement, public testing and certification of the equipment, contingency planning if the technology fails, sufficient efforts for training electoral staff, and effective awareness-raising among voters and political parties' (Venice Commission 2018, para. 33). In a related context, the Committee of Ministers of the Council of Europe has highlighted in its Recommendation (2004)11 on legal, operational and technical standards for e-voting that 'e-voting shall be as reliable and secure as democratic elections and referenda which do not involve the use of electronic means.'

Beyond these segments of democratic functioning, this chapter also looks somewhat briefly at the way technological advances impact socio-economic development.

3 Specific Impact on Development and Mounting Challenges

There are countless other dimensions that are and will be affected, often in paramount ways, by the employment of artificial intelligence and, generally, digital technologies.

The effects of AI will not always be—least not automatically—beneficial for all societies or sectors of societies. The overall effects of the AI and digital technologies, however, in the longer run, will on the whole be more beneficial (Sachs 2019) and, in some aspects, such as sustainable finance and banking development (Chen and Sergi 2018), labor productivity growth (Sachs 2019; Sergi et al. 2019; Wamboye et al. 2015) and overall sustainable development (Popkova and Sergi 2019; Qerimi 2012a), are expected to have major impacts.

In any event, artificial intelligence will exert impact on all key sectors of the economy, including agriculture, mining, manufacturing, business services (including transportation, warehousing, and finance), trade, healthcare, education, finance, tourism, transportation, and governance (Sachs 2019). As Sachs (2019, p. 162) put it, 'In short, digital technologies will enable a massive increase in productivity for a wide range of goods and services. Overall production will become more capital intensive, including three kinds of capital: human, infrastructure, and business. The demand for skilled workers will rise while the demand for unskilled workers will decline'. Further and in order to reap the benefits of the digital revolution and manage rapid urbanization —another major effect of digital technologies—'the key challenge for developing countries will be to finance a rise in capital investments while simultaneously incurring a loss in labor-intensive manufacturing export earnings' (Sachs 2019, p. 163).

The underlying challenge in the near to medium-term would be to collapse the divide that persists between countries with varying degrees of economic development. In order to extend the benefits of digital revolution also to low-income countries, digital-based development strategies are an acute imperative. Development institutions such as the IMF, the World Bank and other multilateral development banks, UN organs and development agencies, national governments and key private-sector leaders and institutions should prioritize 'concerted efforts to transfer digital skills and technologies as rapidly as possible to the low-income nations to enable them to build the digital-based industries, skills and jobs of the future' (Sachs 2019, p. 166).

The list of issues involving new technologies or areas of life, private and public, is growing to being almost virtually *ad infinitum*. This makes the tackling of every possible aspect that is being influenced or modified by technological advances impossible. However, through a select list of issues that pose significant ethical challenges and risks, the fundamental thrust of future research projects or operations is to explore the adequacy and extent to which the existing international legal and institutional regime, largely conceived in the absence of contemporaneous technological innovations, can be applied to an increasingly dominant *online* world. In process, limitations and associated implications of the classical human rights framework should be identified and assessed, with a view to identify potential solutions to real or *imaginary* problems.

However, understanding the underlying structure and logic of artificial intelligence, as well as anticipating its probable or actual future development, remains the key to envisioning the proper parameters for its governance. The dialogue, national and transnational, ought to include actors at the intergovernmental, nongovernmental, and industry levels.

4 Manifestations of Artificial Intelligence/Artificial General Intelligence and Ethical Implications

It should come as no surprise that there is no consensus on a definition of Artificial Intelligence (AI), as distinct from a related, yet different category, of superintelligence, better known as Artificial General Intelligence (AGI).

The notion 'intelligence' is in itself contentious and complex, and not only because of the various types of known intelligence, be it emotional intelligence, musical intelligence, or analytical skills, spatial visualization abilities, and so on (Lin and Allhoff 2019). Thus, 'how we define, and eventually measure, intelligence in natural beings such as humans is notoriously difficult' (Roff 2019, pp. 129–130). A simple, brief conception of AI would be that of a computational system preconceived to *automate* decisions intelligently or with the appearance of intelligence, however the term is defined (Lin and Allhoff 2019).

AI is often equated and/or used simultaneously with 'machine learning' and 'deep learning'. In the case of machine learning, algorithms are designed to identify relationships, develop particular predictive models, and ultimately make decisions. Image classification would be a typical example. Deep learning, on the other hand, uses data to find patterns used to make predictions about new data, independently of humans. Data are thus directly fed into the deep learning algorithm, in turn predicting the occurrence of objects. Deep learning systems are used in such instances as speech recognition and natural language processing. Finally, there is the 'reinforcement learning', whose task is to improve the performance of an algorithm relating to multiple engagements with a problem, adapting or adjusting action based on constant feedback from previous actions (Livingston and Risse 2019).

The term Artificial General Intelligence or AGI, as introduced earlier, is often used interchangeably with the terms 'general AI' and, simply, 'superintelligence'. Livingston and Risse (2019, p. 142) conceptualize AGI as 'an algorithm or set of algorithms that are, minimally, as capable as a typical human across multiple problem domains.' While there are no known examples of AGI currently in existence, rapid advances are being made to expand the adaptability of algorithms to multiple problem domains through the initiatives of the like of Google's DeepMind and Google Brain.

With current technological advances in mind, one cannot rule out—at least by way of imagination—the plausible creation of AGI. In this connection, however, it might be prudent to distinguish between an entirely autonomous or *pure AGI* and a human-extended AGI or *hominin AGI.* An example of the latter would be Elon Musk's Neuralink enterprise, which seeks to enhance human intelligence by surgically implanting high brain-bandwidth brain-computer interfaces. This model would exert significant moral implications, in that an AGI achieved by way of enhancing human beings differs drastically from an autonomous AGI conceived and built exclusively by computer engineers (Eady 2018). Indeed, neural implants are already in use by people with Parkinson's disease to steady tremors, same as epilepsy patients have implanted electronic monitors to detect signs of impending seizures and release electric pulses to prevent them. These cases only reinforce the understanding about the benefits and opportunities presented by smart machines in improving 'our ability to prevent, to diagnose and to treat serious diseases' (Brownsword and Harel 2019, p. 111). Neuralink's ultimate goal, however, extends further, aiming a 'symbiosis with artificial intelligence.' Neuralink is obviously not alone in this quest, a trend that demands both action and imagination over 'a blended world of machine and human' (Livingston and Risse 2019, p. 147), and ensuing implications, ethical and practical, of the moral standing and status of superintelligent machines and more complex variations of enhanced human intelligence, which might be referred to as *humanized AGI*.

As this question merits deeper scrutiny and its examination would go beyond the narrower scope of this chapter, it would suffice to say for now that it is within the reach for future intelligent and especially superintelligent machines to be accorded with some sort of moral status. This statement is made for situations extending beyond the intriguing context of a possibly materialized neural link—which would appear to be on course of its materialization—and which as well presents challenges of unprecedented scale to seek to understand where does the individual begin and the machine end, the otherwise existence of what Livingston and Risse (2019, p. 151) call a 'human-machine mélange.' One could in the present context also make a distinction between the possible accordance of moral status of sorts to AI and the consideration of AI as a 'moral agent', the latter possibility already ruled out by some scholars (see Roff 2019). As Roff (2019, p. 128) puts it, 'when it comes to normative considerations, AI is not and will never be a "moral agent," and thus cannot determine what is morally right.' Others remark that 'If we create AI that is smarter than us we have to be open to the possibility that we might actually lose control to them' (Tegmark 2018). The alternative, short of losing control to, or even being exterminated by, artificial superintelligence—as presented by Harari (2015)—would be the possibility of gradual creation of superhumans defined by an increased resistance to disease, greater intelligence, and heightened physical endurance. As Harari (2015, p. 4) notes, 'Homo sapiens is not going to be exterminated by a robot revolt. Rather, Homo sapiens is likely to upgrade itself step by step, merging with robots and computers in the process … In pursuit of health, happiness and power, humans will gradually change first one of their features and then another, and another, until they will no longer be human.'

Whatever the outcome and gradual shape of it, there is one certain decisive element, and that is all would depend on human decision and human-dictated shape of events. One specific aspect of artificial intelligence in operation could be observed in the field of weaponry (next section), which while still developing, offers us a glimpse of the nature of its operation. Another particularly opportune moment for the utilization of artificial intelligence or superintelligence could be space and the exploration of celestial bodies, in search of finding or creating conditions for life in other planets, hence transforming human beings into transplanetary beings.

5 Artificial Intelligence and Human Security

In such areas as defense and security, new technologies have already been used. Examples include unmanned aerial vehicles (UAV)—alternatively referred to as 'drones', but also remotely piloted vehicles (RPV)—meaning aircrafts without a human operator on board. Their existence as such is not inherently illegal, as their operation does not make it impossible to observe the requisite rules of the relevant bodies of law, which in this case is international humanitarian law and international human rights law. Violations of both bodies of law have nevertheless occurred as such violations have indeed occurred also in contexts defined by the absence of use of such technologies. The implication is however the demand for a scrutiny of adequacy and applicability of existing rules and principles of international law and the need for their potential modification.

A more complex scenario is presented by the existence and use of a different type of technology, which is the autonomous weapon system, which is a weapon system that can select (search for, detect, identify or track) and attack (use force against, neutralize, damage or destroy) targets without human intervention. After initial launch or activation by a human operator, it is thus the weapon system itself—using its sensors, computer programming (software) and weaponry—that takes on the targeting functions that would otherwise be controlled by humans.

The relevant body of law would in this case be international humanitarian law (IHL). The IHL rules on the conduct of hostilities—notably the rules of distinction, proportionality and precautions in attack—are addressed to those who plan, decide upon and carry out an attack in armed conflict. The lawful use of autonomous weapon systems will therefore require that combatants retain a level of human control over their functioning in carrying out an attack.

The classic body of law would still remain valid as far as the principles governing the conduct of hostilities are concerned, as well as those of command and control responsibility. The most important conclusion is the ultimate imposition by IHL to ensure compliance with its rules and principles. This implies in practice a necessity for supplementing the current framework in ways that determine the type and degree of human control required in the use and operation of autonomous weapon systems to carry out attacks, so as to ensure compliance with international humanitarian law. As also suggested in the existing literature, 'binding norms will eventually be necessary to manage AI's international security implications', however in the meantime 'a good start can be made through a principles and standards-driven approach to governance' (Gill 2019, p. 169).

The above considerations demand anticipatory action in the form of debate and legal regulation.

6 Regulating Artificial Intelligence

The need for regulation artificial intelligence would be warranted both for purposes of its better use and also prevention of abuse (Qerimi 2012b). In the best scenario, for instance, smart machines, smart policing and smart cities might be effective tools to fight or end crime. *Au contraire*, in the worst, we can imagine various dystopian features where the threats presented by AI are realized (Joh 2015). As Hawking (2018, p. 188) famously remarked 'the advent of super-intelligent AI would be either the best or the worst thing ever to happen to humanity.'

The hitherto unforeseen magnitude of smart phones and other imaging applications and platforms, including a significant number of high-resolution satellite imagining, offer fertile grounds for an abundance of surveillance. There is a high degree of probability that, once something happens, it stands a good chance of being recorded in one or another way either by a camera on earth or a satellite in orbit. For instance, in 2017, the International Criminal Court, the permanent criminal court established by the Treaty of Rome in 1998, issued an indictment for the arrest of a Libyan warlord based on satellite images and videos taken of the executions ordered or conducted by him. Those satellite imagery and videos were posted on social media by his followers (Cluskey 2017). Other related spheres of life positively affected by the use of AI include forensic anthropology. One such example is the DNA sequencing, whose application holds a much greater degree of scientific accuracy—in addition to being more efficient and of lower cost—in forensic investigations (D'Agaro 2018).

On the other pole of the AI utilization, namely its abuse or potential abuse, one could point to the AI's use by governments to monitor citizens and intrude into their privacy. In China, for instance, the social credit score system uses observational surveillance, facial recognition software, and social media surveillance to monitor citizen behavior and assign scores based on how well people comply with the country's rules (Kobie 2019; Human Rights Watch 2017). Even scarier perhaps is the phenomenon of 'deepfakes', an AI application that effectively alters visual content; for instance, digitally inserting the face and voice of one person into a video of another person, creating in turn 'highly realistic but entirely synthetic video content' (Livingston and Risse 2019, p. 144). In this specific scenario, AI is used 'to delude, obfuscate, and humiliate targeted individuals and organization' (Livingston and Risse 2019, p. 144).

Preventing abuse and making the best of the benefits the AI has to offer, an effective legal framework has no serious alternative candidate. International cooperation is a fundamental prerequisite. It might be unrealistic, in the short run at least, to see legally-binding agreements in place on the norms to regulate or govern specific aspects of AI. As an initial step, easier to agree on, states and the wider international community could resort to guiding principles, standards and codes short of binding law. However, steps could be taken with a view of codifying parts of relevant rules and principles, especially those of a less controversial nature. One such example is the work recently initiated by the Council of Europe on a binding international treaty on AI, with the involvement of Governments, technology experts, and civil society.

Other alternative avenues to be pursued, in the form of defining guiding principles, are presented by, for example, the European Commission High-Level Expert Group on Artificial Intelligence (2018). This EU Expert Group has identified five guiding principles in the function of trustworthy and ethical development of the AI, namely: beneficence, nonmaleficence, autonomy of humans, justice, and explicability (European Commission High-Level Expert Group 2018). In the more specific context of the military use of AI, in 2018, the Group of Governmental Experts of the Convention on Certain Conventional Weapons on emerging technologies in the area of lethal autonomous weapon systems (LAWS), identified ten guiding principles on emerging technologies in the sphere of LAWS. These principles are accompanied by common understandings on definitions and the nature of human intervention required at various stages of development and deployment, so as to ensure compliance with international law and, more specifically, international humanitarian law. The Group of Governmental Experts currently comprises 125 states, including all those thought to be pursuing national security applications of AI (Group of Governmental Experts 2018). Finally, the United Nations, through its principal and subsidiary organs and agencies, should pursue a more proactive policy in the field of international legal regulation of AI.

7 Conclusion

This chapter has sought to examine and expose the revolutionizing effects of imagination and innovative thought from early history to the most contemporary realities and prospective developments in the arenas of artificial intelligence and superintelligence. From basic rights and freedoms to environmental and developmental concerns, and from health to security and to potentially making the human race a transplanetary race, the development and deployment of artificial intelligence is observed both in terms of its benefits and several of its challenges, ethical and operational. The chapter reveals various expressions in artificial intelligence and superintelligence, calling for an inescapable imperative of international dialogue and regulatory initiatives about how to make best use of the benefits smart machines have to offer, as well as to define their limits.

References

Aristotle: The Politics. Cambridge University Press, Cambridge and New York (1988)

Brownsword, R., Harel, A.: Law, liberty and technology: criminal justice in the context of smart machines. Int. J. Law Context **15**, 107–125 (2019)

Chen, K., Sergi, B.S.: How can fintech remake Russia's development? In: Sergi, B.S. (ed.) Exploring the Future of Russia's Economy and Markets: Towards Sustainable Economic Development, pp. 1–11. Emerald Publishing, Bingley (2018)

Cluskey, P.: Social media evidence a game-changer in war crimes trial. Irish Times, October 3, 2017 (2018). https://www.irishtimes.com/news/world/europe/social-media-evidence-a-game-changer-in-war-crimes-trial-1.3243098. 1 Aug 2019

Court of Justice of the European Union (CJEU) 2011: Case C-70/10, Scarlet Extended SA v. Société belge des auteurs, compositeurs et éditeurs SCRL (SABAM). http://curia.europa.eu/jurisp/cgi-bin/form.pl?lang=EN&Submit=Submit&numaff=C-70/10. 30 July 2019

D'Agaro, E.: Artificial intelligence used in genome analysis studies. EuroBiotech J. **2**(2), 78–88 (2018)

Eady, T.: Does recent progress with neural networks foretell artificial general intelligence? (2018). https://medium.com/protopiablog/does-recent-progress-with-neural-networks-foretell-artificial-general-intelligence-9545c17a5d8b. 31 July 2019

European Commission High-Level Expert Group on Artificial Intelligence 2018: Draft: Ethics Guidelines for Trustworthy AI. www.ec.europa.eu/digital-single-market/en/news/draft-ethics-guidelines-trustworthy-ai. 1 Aug 2019

Gill, A.S.: Artificial intelligence and international security: the long view. Ethics Int. Aff. **33**(2), 169–179 (2019)

Group of Governmental Experts of the High Contracting Parties to the Convention on Prohibitions and Restrictions on the Use of Certain Conventional Weapons Which May be Deemed to Be Excessively Injurious or to Have Indiscriminate Effects 2018: Report of the 2018 Session of the Group of Governmental Experts on Emerging Technologies in the Area of Lethal Autonomous Weapons Systems. https://www.unog.ch/80256EDD006B8954/(httpAssets)/20092911F6495FA7C125830E003F9A5B/$file/CCW_GGE.1_2018_3_final.pdf. 1 Aug 2019

Hall, E.: Aristotle's Way: How Ancient Wisdom Can Change Your Life. Penguin Press, New York (2018)

Harari, Y.N.: Homo Deus: A Brief History of Tomorrow. Harper, New York (2015)

Hawking, S.: Brief Answers to the Big Questions. John Murray, London (2018)

Human Rights Council: Resolution 32/13. A/HRC/RES/32/13, 18 July 2016

Human Rights Council: Report of the Special Rapporteur on the promotion and protection of the right to freedom of opinion and expression. U.N. HRC, 17th Sess., U.N. Doc. A/HRC/17/27, 16 May 2011

Human Rights Watch 2017: China: Police DNA Database Threatens Privacy, 15 May 2017. www.hrw.org/news/2017/05/15/china-police-dna-database-threatens-privacy. 1 Aug 2019

Joh, E.E.: The new surveillance discretion: automated suspicion, big data, and policing. Research Paper No. 473. UC Davis Legal Studies Research Paper Series (2015)

Kobie, N.: The complicated truth about China's social credit system. Wired, 21 January 2019. www.wired.co.uk/article/china-social-credit-system-explained. 1 Aug 2019

Lin, P., Allhoff, F.: Arctic 2.0: how artificial intelligence can help develop a frontier. Ethics Int. Aff. **33**(2), 193–205 (2019)

Livingston, S., Risse, M.: The future impact of artificial intelligence on humans and human rights. Ethics Int. Aff. **33**(2), 141–158 (2019)

Popkova, E.G., Sergi, B.S.: Will industry 4.0 and other innovations impact Russia's development? In: Sergi, B.S. (ed.) Exploring the Future of Russia's Economy and Markets: Towards Sustainable Economic Development, pp. 51–68. Emerald Publishing, Bingley (2019)

Qerimi, Q.: Bridge over troubled water: an emerging right to access to the internet. Int. Rev. Law **2017**(1), 1–22 (2017)

Qerimi, Q.: Development in International Law: A Policy-Oriented Inquiry. Martinus Nijhoff Publishers, Leiden and Boston (2012a)

Qerimi, Q.: Technology and Development: Universalizing Access to and Protection of Information and Communication Technology. Int. J. Soc. Ecol. Sustain. Dev. **3**(1), 1–21 (2012b)

Reisman, W.M., Arsanjani, M.H., Wiessner, S., Westerman, G.S.: International Law in Contemporary Perspective. Foundation Press, St. Paul (2004)

Roff, H.M.: Artificial Intelligence: Power to the People. Ethics Int. Aff. **33**(2), 127–140 (2019)

Sergi, B.S., Popkova, E.G., Bogoviz, A.V., Litvinova, T.N.: Understanding Industry 4.0: AI, the Internet of Things, and the Future of Work. Emerald Publishing Limited, Bingley (2019)

Sachs, J.D.: Some Brief Reflections on Digital Technologies and Economic Development. Ethics Int. Aff. **33**(2), 159–167 (2019)

Tegmark, M.: Speaking in 'Do You Trust This Computer?', YouTube video, 1:18:03, posted by Dr. Caleb Cheung (2018). www.youtube.com/watch?v=DVprGRt39yg. 1 Aug 2019

U.S. Supreme Court: Buckley v. Am. Constitutional Law Found. 525 U.S. 182 (1999)
U.S. Supreme Court: McIntyre v. Ohio Elections Comm'n. 514 U.S. 334 (1995)
U.S. Supreme Court: Talley v. California. 362 U.S. 60 (1960)
Venice Commission: Joint Opinion on the Albania's Draft to the legislative Initiative of the Citizens. Adopted at its 116th Plenary Session, 19–20 October 2018
Venice Commission: Joint Opinion on the Draft Law on State Register of Voters of Ukraine. Adopted at its 71st Plenary Session, 1–2 June 2007
Wamboye, E., Tochkov, K., Sergi, B.S.: Technology adoption and growth in sub-Saharan African Countries. Comp. Econ. Stud. **57**(1), 136–167 (2015)

Application of Smart-Contracts When Using the Exclusive Rights to Results of Intellectual Activity

Agnessa O. Inshakova[1(✉)], Tatyana V. Deryugina[2], and Evgeny Y. Malikov[3]

[1] Volgograd State University, Volgograd, Russia
gimchp@volsu.ru, ainshakova@list.ru
[2] Moscow University of the Ministry of Internal Affairs of the Russian Federation of V.Ya. Kikot, Moscow, Russia
sofija96@mail.ru
[3] Volgograd Humanitarian Institute, Volgograd, Russia
imeuufo@gmail.com

Abstract. In the article is considered problems of registration of contractual legal relationship on use of the result of intellectual activity using the smart-contract. The Russian, foreign and international standards of the right are investigated. The conclusion about the lack of a uniform and consistent system of legal regulation specified legal relationship is drawn. It is specified the reasons for such a situation. Also, the bad points connected with the possibility of application of the smart-contract when using results of intellectual activity come to light positive. Possible mechanisms of legal regulation are investigated. A conclusion about the need for the use of a possibility of positive regulation of the developing legal relationship is proved. The detailed analysis of the existing Russian legislation which elements it is possible to use by analogy at regulation of the relations with the use of the smart-contract for results of intellectual activity is carried out. The attention to the last projects approved by the State Duma of the Russian Federation is paid. Detailed analysis of innovations is carried out; it is specified positive and negative the moments, offers on improvement of rules of law are made. The analysis of international legislation supports drawn conclusions. The research of the doctrine demonstrates an ambiguous approach to the definition of the legal nature of the smart-contract. The analysis made it possible to systematize the points of view of Russian and foreign scientists, to highlight the technological and legal approaches to determining the nature of a smart-contract, within the latter to determine the broad and narrow approaches. It carried out the analysis allowed to formulate specifics of the smart-contract, to point out its key features and also to draw a conclusion on the need of fixing of the third (along with written and oral) transaction forms – program. Proceeding from the key feature of the smart-contract consisting in a separation of the will be directed to the performance of the contract from the actual execution the position connected with the expression of will in out of is proved and also characteristic is given to features of the offer and acceptance for the conclusion of the smart-contract.

E. G. Popkova and B. S. Sergi (Eds.): ISC 2019, LNNS 91, pp. 372–384, 2020.
https://doi.org/10.1007/978-3-030-32015-7_42

Materials - The legal basis for carrying out the research made the Civil Code of the Russian Federation defines the general requirements to the conclusion, change and termination of the contract and also to form and content; The Federal law No. 34-FL adopted by the State Duma of the Russian Federation on March 12, 2019 "About modification of parts the first, second and article 1124 parts of the third of the Civil code of the Russian Federation" establishing a new concept the "digital" right and features of transactions in electronic form; The project of a Federal Law No. 419059-7 "About digital financial assets" (the edition accepted by the State Duma of Federal Assembly of the Russian Federation in the I reading on May 22, 2018), which for the first time in the Russian legislation formulates a definition of the smart contract; Federal Law draft No. 419090-7 "About alternative ways attraction of investments (crowdfunding)", pointing out features of smart contracts at crowdfunding.

The analysis of foreign sources allowing to study the positive and negative experience of legal regulation of these or those elements of legal reality is of great importance for carrying out research. It is, in particular, Arizona Amendment Act (2017), Uniform Electronic Transaction Act (2018), Contracts in civil law of foreign countries (2018), Decree of the President of the Republic of Belarus No. 8 (2017) and others.

The basis for a definition of the legal nature of smart contracts and the possibility of its application in the international commercial turnover the Principles of the global commercial treaties UNIDRUA (2010), the Principles of the European contract law (the added and revised version of 1998), et al. were acting.

The essential doctrinal basis of the conducted research concerning the use of technical means when using an author's right were the works of Kuzevanova (2018), Grin (2017), Svechnikova (2009). In the analysis of the legal nature of smart-contracts the works of Nikiforova and Smirnova (2017) are investigated; Dyadkin et al. (2018), Novoselova (2017), Bagaev (2018), Saveliev (2016), Amuial et al. (2016), that allowed to allocate various approaches to understanding smart-contracts and to draw conclusions about their legal nature.

The research of specific signs of the smart-contract was conducted through a prism of the theory of will and a expression of will of Pokrovsky (1998), Ioffe et al. (2003), Oygensicht (1983), Nemov (2003), Kechekian (1958), Novitsky (1954), Krasavchikov (2005) and others.

Methods - General scientific methods of knowledge made the methodical basis of the research: the historical, allowed to carry out a comparative analysis of the legislation and doctrine in a historical retrospective. The analogies, as a result of which application have concluded the possibility of implementation of international law to the Russian reality. The modelling, which allowed to create a design of the relations between the author and the person, using the result of its intellectual work, using the smart contract. The analysis and synthesis which were a basis for the scientific conclusions drawn in research. When carrying out research also particular and proper methods are used (comparative and legal, legalistic, etc.).

1 Introduction

Development of digital technologies and the Internet allows using results of intellectual activity widely. With regret, it should be noted that along with lawful use, new technologies create unlimited opportunities for illegal use of results of intellectual work. Because the author's right belongs to the absolute rights and the circle of subjects which can violate this right is unlimited and is not defined that creates additional difficulties for protection of author's rights. Also, if until recently, there were no technical means, and together with them and legal ways of prevention of such violations, then the development of blockchain technology, the emergence of smart-contracts (smart-contract) allows not only to create the structured database. However, also to fix a legal mechanism of the automated conclusion and performance of the license contract with the person using the result of intellectual activities for means of the smart-contract.

Now legal regulation of the relations with the use of smart-contracts in Russia and international law is in a formation stage that is caused, including, some unresolved theoretical problems. Demand fundamental permission questions concerning the identification of the subjects entering a legal relationship, their legal personality; the moment of placement of the offer and obtaining the acceptance and communication with it is the moment of signing of the contract; maintenance of the contractual relations; performance of obligations and so forth.

The particular specifics of the arising legal relationship are also caused that the smart-contract, represents a set of certain systematised information (database) and the computer software that does it by the independent object of intellectual activity (Articles 1260, 1261 Civil Code of the Russian Federation). Owing to what, undoubtedly, questions of protection of the rights of the author who created the database applied by other subjects of civil circulation including, and for the generation of profit are particularly acute.

Use of smart-contracts in the intellectual right, on the one hand, is hugely perspective, and with another – causes the numerous difficulties connected with security author's rights.

It is necessary to carry the simplified procedure of the introduction in a contractual legal relationship to a number of good points; lack of need of development of contractual conditions; lack of subjective control of execution of the obligation (in fact, control is exercised by the computer program); minimizing of a possibility of breaches of contract and so forth.

Along with pluses of the introduction of such systems, in scientific literature also possible minuses are noted. In particular, the specifics of use of the smart-contract are that exchange of information is carried out passing the person of law, in fact, interaction happens between right objects (electronic devices), in this regard, created by centuries of the theory of the right (in particular, the theory of transactions) cannot adequately be applied and demands cardinal revision. As a result, there is a problem of definition of the subject of responsibility behind infliction of harm by the action of the automated systems.

In literature also the problem of the impossibility of definition of the author of a part of the program code created by the self-training computer program (the technical person) (Gurko 2017) is noted. It should be noted that in this specific case, the classical right proposes to us the solution, proceeding from a position, as robotic systems, and computer programs are objects, but not persons of law. Owing to what, the person who created these systems has rights of the author (by analogy with an ownership right where fruits belong to the owner of a thing).

Causes concern and the blockchain technology used in smart contracts, communications with a possibility of entering into the program code of illegal changes (Nikiforova and Smirnova 2017). The absence legislative regulation does not allow to use smart contracts as proofs (Dyadkin et al. 2018).

2 Part I. Emergence of a Legal Relationship on the Use of Results of a Particular Activity Using the Smart-Contract

According to scientific (Kuzevanov 2018; Grin 2017), the most popular ways of copyright violation on the Internet is illegal copying, reproduction, use by different ways of results of an intellectual activity without instructions of the author, placement of the reference to the website and so forth. At the same time, it should be noted that the legislator attempted to solve problems of illegal access to results of intellectual activity on the Internet.

In particular, in Article 1299 of the Civil Code of the Russian Federation (further the Civil Code of the Russian Federation) points to a possibility of use of technical means of protection of author's rights. The legislator allocates technical means on control of access to results of intellectual activity; the technical means preventing unauthorised use; the technical means limiting unauthorised use. Thus, the given norm solves the problem connected with copyright violation but does not solve a problem of the use of objects of intellectual activity in civil circulation.

The scientists investigating problems of use of technical means of protection of author's rights as the purpose of creation and use of such means see creation of obstacles for illegal use of objects of an author's right (I.V. Svechnikova of 2009), or prevention and suppression of illegal actions concerning intellectual property (Kuzevanov 2018).

In all given examples, authors recognise a possibility of copyright violation and need for prevention of offences in the field that, indeed, genuinely. However, in this case, it is necessary to use, according to us, possibilities of positive regulation. It is necessary to provide specific incentives in the right permitting legally without excess encumbrances to increase lawful use of objects of intellectual activity in civil circulation. Introduction of the smart-contract allowing in the automated mode without the commission of any difficult implied actions to conclude the bargain will allow to solve the specified problem and also will order collecting a royalty, or single payments for the use of objects of intellectual activity.

Professional solutions for the creation of a possibility of use of smart-contracts exist both in Russia and in other countries. The most widespread means is the establishment of the digital watermark – the unique code which marks an object of intellectual activity and information on his author, the owner, and so forth. The unique program recognises these codes at the appeal to these objects, and watermarks remain when rewriting data. Such technologies are used in various programs for the establishment of illegal loans in author's texts (anti-plagiarism). Computer programs can notify users on possible unauthorised use of objects of the author's rights.

Identification of the watermark marking the object of intellectual activity testifies to the appeal to this or that object of the intellectual rights and has to attract the conclusion of the smart-contract for the use of the work of the author. In this case, placement of an object of an author's right on the Internet, in the absence of restrictions (from the author) set for full use, is the offer directed to the signing of the contract.

3 Part II. Russian and Foreign Legislation About Smart-Contracts

Unlike technical means, legal mechanisms of regulation of the use of smart-contracts in Russia does not exist. The Federal Law No. 34-FL adopted by the State Duma of the Russian Federation on March 12, 2019 "About modification of parts the first, second and article 1124 parts of the third of the Civil code of the Russian Federation" enter a new object of the civil rights "the digital right". However at the description of use in the civil circulation of a new law of estate the term the smart contract is not used, unlike Federal Law draft No. 424632-7 "About modification of parts the first, second and fourth the Civil code of the Russian Federation" where definition and the smart-contract was initially entered. At the same time, the smart-contract was not considered as an independent type of the transaction and was defined as a way of automatic execution of the obligation that, in our opinion, is the single approach lending the sphere of use of the smart-contract. However, already in edition of the Federal Law No. 424632-7 project, accepted in the first reading by the State Duma of the Russian Federation on May 22, 2018, the term "smart-contract" was already absent, and the possibility of performance of obligations without separately expressed will with the use of information technologies is reflected in Article 309 of the Civil Code of the Russian Federation.

In turn the Federal Law N 419059-7 project "About digital financial assets" (the edition accepted by the State Duma of Federal Assembly of the Russian Federation in the I reading on May 22, 2018), formulates definition of the smart-contract (Article 2) as contracts made in an electronic form by which execution happens in an automatic order. Thus, the smart-contract is considered as the independent type of the contract having specifics of execution. Attracts attention that the State Duma considers both projects in one day, and nevertheless, a real understanding of the smart contract in them is not offered.

Federal Law draft No. 419090-7 "About alternative ways attraction of investments (crowdfunding)" does not define the smart-contract, referring to use of the definitions which are available in the legislation (Paragraph 2 of Article 2 of the Federal Law).

However, at the same time in Paragraph 3 of Article 9 of the Federal Law indicates possibility, changes and the terminations of the legal relationship of participants of the investment platform based on the smart-contract. Them what directly follows that it is about the independent contractor as the legal fact, but not about a way of execution of the obligation.

Thus, it is necessary to summarise that the attempt of the legislator to legally issue a concept of the smart-contract was not crowned with success now.

In states Arizona (Arizona Amendment Act (2017)), Tennessee (2018), Ohio (Uniform Electronic Transaction Act (2018)) of the USA, since 2017, is published in the legislation a mention of smart contracts. However, if in the legislation of Arizona definition of the smart-contract as the decentralised register on means of which transfer of assets is possible, then in Tennessee and Ohio it is only pointed out impossibility to nullify the relations is formulated if at their registration were used blockchain technology or the smart-contract. As for the Law on the introduction of amendments to laws of the State of Arizona (2017), the mention of a possibility of transfer of assets using the smart contract suggests an idea of independence of the smart-contract in the system of contracts. And not just about the legalisation of an electronic form of the contract what it was initially being talked during the creation of the bill.

Perhaps, the most advanced legislation concerning the smart-contract is the legislation of Belarus now (Decree of the President of the Republic of Belarus No. 8 of December 21, 2017). In particular, in it the possibility of use of the smart-contract and as the bases of the emergence of legal relationship and a way of performance of obligations is noted (item 5.3. The decree) that demonstrates the broadest approach to an understanding of the smart-contract. Moreover, considering the smart contract as the program code in the register of blocks, the legislation of Belarus allows commission with its use not only transactions but also other legally significant actions (paragraph 9 of Appendix 1 to the Decree of the President of Republic of Belarus No. 8).

4 Part III. Civilistic Approaches to the Legal Nature of the Smart Contract

In scientific literature, the legal nature of smart-contracts is treated ambiguously. Analysing the civil doctrine, it is also possible to allocate technological and legal approaches to the definition of the nature of the smart-contract.

In particular, T.S. Nikiforova and K. M Smirnova believe that the smart-contract is a way of automation of execution of terms of the contract. The contract as the legal document has to exist irrespective of use further of the smart-contract (Nikiforova and Smirnova 2017). As the element of the computer code realised with the use of blockchain technology considers smart contract G. Greenspan (Greenspan 2016).

Considering the legal approach, it is traditionally possible to allocate narrow and broad approaches to a solution.

D.S. Dyadkin, Yu.M. Usoltcev, N.A. Usoltceva, representatives of the narrow direction, believe that the smart-contract can be considered or as a contract form or as a way of ensuring the performance of obligations (Dyadkin et al. 2018). Only as the way of signing of the contract treats the smart-contract D.I. Skvortsov. L. Novoselova believes that the smart-contract represents "special type of registration and implementation of civil transactions, which has several specific characteristics" (Novoselova 2017). According to I.A. Drozdov (Bagaev 2018), the smart-contract is a unique way of execution of the obligation.

Representatives of broad approach M.Yu. Yurasov and D.A. Pozdnyakov believe that the smart-contract has a dual legal nature: it can be considered as the additional agreement and as the independent contract (Yurasov, Pozdnyakov). From a position of the developed theory of a contract law A.I. Savelyev (Saveliev 2016) proves the point of view about the independence of the smart contract in the system of contracts. In particular, the author points to such signs of the smart contract as a possibility of transfer of values on means of the smart-contract; action existence by the will and in the interest; a duty of performance of the contract according to its conditions.

As the independent transaction with a suspensive condition of execution V.N. Boyarkin considers the smart-contract, at the same time specifying, the performance of the contract does not depend on the will of the parties of the agreement.

In foreign literature, there is also no unity of opinions on nature, the arising legal relationship. At an initial stage, the smart-contract was considered as a way of execution of the obligation (Szabo 1994).

The later broader approach was created. Amuial S., Dewey J. N., Seul J. considers the smart-contract as the legal document created and executed using blockchain technology (Amuial et al. 2016).

C. Sargeant, C. Lim, T. Saw specify that comparison with the traditional contract allows to mark out only one similarity, both the smart-contract and the usual contract, are the main of the relationship of subjects of the transaction. The distinction is that the smart-contract chooses the behaviour options which are in advance programmed and no subjective factors; additional lead conditions can influence this choice (Sargeant, Lim, Saw). However and at such approach, understanding of the smart-contract is beyond an easy way of execution of the obligation. It is considered not only as of the legal fact, which is the legal relationship emergence basis but also as the legal form establishing its contents.

Foreign researchers at the same time note that ultimately it is impossible to automate the relations (Autor et al. 2003), at the same time conditions which assume programming and those who cannot be described using the program code are allocated (Clack, Bakshi, Braine).

Considering a smart-contract design in terms of the signs of the civil contract put in the Civil Code of the Russian Federation, it should be noted that it completely meets the established requirements. The smart-contract represents the coordinated will (agreement) of two and more parties; its purpose – establishment, change and the termination of the civil rights and duties (taking into account the specified signs the smart-contract acts as the legal fact mediating emergence of the civil relations). The conclusions of the smart-contract result in the legal relationship, which is characterised by the existence of obligations (contractual obligations). Thus, the smart-contract can be characterised and as the transaction which defines the rights and duties and as the contractual legal relationship which resulted from the transaction comprising certain obligations.

The smart-contract can mediate as property legal relationship (the property contract), providing transfer of material benefits, and organisational (the organisational contract). In the light of told, the smart-contract can be classified and as the preliminary contract and as the main. Application of the smart-contract is possible after public contracts, option agreements. As a rule, the smart contract is the contract of accession; however, there are no bases to believe that the parties cannot agree using the smart-contract on individual conditions. At the same time, it is necessary to remember that each smart-contract represents an object of an author's right (from the right side) and complicated technological model (from technical aspect) that draws natural a conclusion about the larger quantity of economic and time expenditure to its creation. Thus, the perspective direction is consideration of the smart-contract as contracts of the accession which is not excluding at the same time an opportunity to provide individual conditions which have to be reflected in the computer program at a stage of its creation or changes.

By nature, smart-contracts can be both typical, and mixed, unilateral and bilateral, paid and gratuitous.

Proceeding from the given characteristics of the smart-contract, it is possible to conclude that it is about the contract in that sense which was put in it by the legislator. The similar conclusion can be drawn, having analysed both the legislation of foreign countries and the international acts.

In particular, the Civil code of Germany (German civil code) contains the signs of the contract (§ 145 157 of GGU) allowing to draw a conclusion that it is necessary to understand the coordinated will of the parties corresponding to each other, at the same time will (the acceptance and the offer) as the contract are independent transactions.

In the USA, in the majority of states, the contract is considered as the agreement, and only in some – as communication (Contracts in civil law of foreign countries 2018). It should be noted that in the legislation of the USA there is a smart contract prototype called the contract of access, mediating access to information bases (Law on the electronic form of transactions 1999) using electronic means.

Signs of the coordinated will are also noted in the Principles of the international commercial treaties UNIDRUA (UNIDROIT principles of international commercial contracts 2010), in Article 2:101 of the Principles of the European contract law (the added and revised version of 1998), et al. acts.

Thus, considering the smart-contract as the coordinated will of its parties directed to emergence, change and the termination of the legal relationship, we can speak about a classical kind of contracts.

The main difference of the smart-contract from other contracts consists in order and a way of signing of the contract which cannot be described as an exchange of electronic documents transferred on communication channels (a simple written form). The smart-contract is a unique computer program with the use of databases that does it by an independent object of an author's right. Describing the smart-contract, we cannot use the existing contract form designs (oral and written), it is a unique, new form which is valuable not only as of the legal fact generating legal relationship and as the act regulating this legal relationship but also as an independent object of the right. Therefore in this case introduction of amendments to the existing Russian legislation, by addition of the third form of the transaction to – a program form is necessary (that is the conclusion of the transaction by the addressing the particular computer program).

5 Part IV. The Specificity of the Smart-Contract, Mediating Intellectual Property Relations, from the Standpoint of the Theory of will and a Expression of will

The specifics of the smart-contract, consist, in features of will. In legal literature, there is a point of view that smart-contracts have one essential feature distinguishing it from other contracts at an execution stage strong-willed activity of the parties of legal relationship is absent (Bagayev 2018). This point of view is interesting by the clear theoretical message connecting will, will and the smart contract in uniform whole.

The set of scientific works (Pokrovsky 1998; Ioffe et al. 2003; Oygensicht 1983) is devoted to the research of will and a expression of will. Without going into details, so it is not a research objective, we will specify that the concept of will formulated in philosophy, psychology, sociology cannot unconditionally be applied in jurisprudence (Flume 1975). A consequence of understanding of will an internal act of awareness of requirement, decision-making and efforts for its achievement (Nemov 2003) that is characteristic of psychology and philosophy, is the emergence of a legal concept of will. Otherwise, it is impossible to prove the theory of invalidity of transactions when they will and a expression of will not coincide.

Division of categories "will" and "will" where the will is considered as a mental process (a subjective element) is a consequence of such approach, and its expression in out of is will (an objective element) (Kechekian 1958; Novitsky 1954). In due time, O.A. Krasavchikov considered a ratio of will and a expression of will as a ratio of content and a form (Krasavchikov 2005). At such approach, the will contents are reflected in a form will, therefore, there is a legal consequence. However at such approach, the concept of will is identical to a concept of the purpose (causa) as concluding the bargain, we seek for the achievement of a particular result which is reflected in statements of the purpose and is the transaction basis.

From other positions, A.A. Panov considering will as deciding on a transaction which is expressed in out of it will (Panov 2011) approaches an understanding of a ratio of will and a expression of will. At the same time, the decision is included in acceptance, both awareness of requirements, and the statement is more whole also the development of ways of achievement of the goal. Legal consequences arise only in the case of objectification of the will.

All theories of will and a expression of will, it is possible to divide into those which proceed from a will priority before will (Pokrovsky 1960), wills before will (Novitsky 1954), the equivalence of concepts of will and a expression of will (Krasavchikov 2005; Ioffe et al. 2003; Andreev 2017).

Accepting a position that will have to reflect the will of the subject in the valid transaction, concerning the smart-contract signed for the use of an intellectual property item, it is possible to note the following.

Any agreement represents a set of the wills directed to emergence, execution and the termination of contractual obligations. At the same time, it is necessary to agree with O.A. Krasavchikov that the contract cannot be considered as the sum of unilateral wills, it is always the coordinated will of subjects of the contract (Krasavchikov 2005). Thus, in the smart-contract, as well as in any contract it is possible to allocate at least two wills of the parties, the first – is directed to the signing of the contract, the second – to its execution. However, in the smart-contract (at least the smart contract on the use of results of intellectual activity) the introduction moments in a contractual legal relationship and its execution coincide. Any addressed an intellectual property item at the same time enters a contractual legal relationship and expresses will on its execution. At the same time, one fact of the appeal to an intellectual property item should not be considered as the will directed to the introduction in the contractual legal relationship. It is necessary to provide a commission of the individual actions indicating the voluntary will of the addressed party to the introduction in the contractual legal relationship.

For the owner of the author's right allowing using an object of an author's right on the Internet, the moment of placement of the offer to sign the contract (offer) coincides with the will moment on its execution according to the conditions which are contained in the smart-contract. For the contractor (the person wishing to use results of intellectual activity) this moment also coincides. However, for the signing of the contract and its execution he needs to make specific actions (for example, to enter the data, or the unique code, directed by communication, or to confirm the introduction in a contractual legal relationship on means of pressing of a combination of keys et al.). That expression of the acceptance by inaction in a situation with the smart-contract is inadmissible. Otherwise, there is an opportunity to challenge the fact that they will be directed to the signing of the contract.

6 Conclusion

The conducted research allows us to draw the following conclusions:

1. It is necessary to provide the incentives in the right permitting legally without encumbrances to increase lawful use of objects of intellectual activity in civil circulation. Introduction of the smart-contract allowing in the automated mode without the commission of difficult implied actions to conclude the bargain will allow to solve the specified problem and also will order collecting a royalty, single payments for the use of objects of intellectual activity.
2. Placement of an object of an author's right on the Internet, in the absence of the restrictions (from the author) set for full use, is the offer directed to the signing of the contract.
3. The smart-contract as the coordinated will of its parties directed to emergence change and the termination of a legal relationship is an independent kind of contracts. The main difference of the smart-contract from other contracts consists in order and a way of signing of the contract which cannot be described as an exchange of electronic documents transferred on communication channels (a simple written form).

4. The smart-contract is a unique computer program with the use of databases that does it by an independent object of an author's right. Describing a form of the smart-contract it is impossible to use the existing contract form designs (oral and written), it is a unique, new form which is valuable not only as of the legal fact generating legal relationship and as the act regulating this legal relationship but also as an independent object of the right. Introduction of amendments to the existing Russian legislation, by addition of the third form of the transaction to – a program form is necessary (the conclusion of the transaction by the addressing the particular computer program).
5. For the owner of the author's right allowing using an object of an author's right on the Internet, the moment of placement of the offer to sign the contract (offer) coincides with the will moment on its execution according to the conditions which are contained in the smart-contract. For the contractor (the person wishing to use results of intellectual activity) this moment also coincides. However, for the signing of the contract and its execution he needs to make specific actions (for example, to enter the data, or the unique code, directed by communication, or to confirm the introduction in a contractual legal relationship on means of pressing of a combination of keys et al.). That an expression of the acceptance by inaction in a situation with the smart-contract is inadmissible.
6. The specifics of the performance of the smart-contract are in what the will directed to the performance of the smart-contract has to coincide with the introduction moment in the contractual legal relationship. It does not coincide with the moment of the actual execution of the smart contract which is carried out on means of functioning of an object of the right (the computer program) without the participation of the subject.

Acknowledgments. The reported study was funded by RFBR according to the research project No. 18-29-16132.

References

Amuial, S., Dewey, J.N., Seul, J.: The Blockchain. A Guide for Legal and Business Professionals, p. 123. Thomson Reuters, Danvers (2016)

Andreev, Yu.N.: Treaty in the civil law of Russia: a comparative legal study: Monograph, p. 213. NORM, INFRA-M, Moscow (2017)

Arizona Amendment Act. https://forklog.com/gubernator-arizony-podpisal-istoricheskij-zakon-o-smart-kontraktah-i-tehnologii-blokchejn/. Accessed 27 Mar 2019

Autor, D.H., Levy, F., Murnane, R.J.: The skill content of recent technological change: an empirical exploration. Q. J. Econ. **118**(4), 1279–1333 (2003)

Bagaev, V.: It is worth thinking about what can be simplified in the existing legislation if people solve their economic tasks in alternative ways [Interview with I.A. Drozdov]. Zakon **2**, 6–17 (2018)

Boyarkin, V.N.: Smart Contract Decompilation: Scheme, Form and Control. https://zakon.ru/blog/2017/11/4/dekompilyaciya__smart-kontraktov_eschyo_odna_statya#ssil2. Accessed 05 Mar 2019

Clack, C.D., Bakshi, V.A., Braine, L.: Smart contract templates: foundations, design landscape and research directions. https://arxiv.org/pdf/1608.00771.pdf. Accessed 05 Mar 2019
Decree of the President of the Republic of Belarus No. 8 of December 21 (2017). http://president.gov.by/ru/official_documents_ru/view/dekret-8-ot-21-dekabrja-2017-g-17716/. Accessed 05 Mar 2019
Dyadkin, D., Usoltsev, Yu.M., Usoltseva, N.A.: Smart contracts in Russia: prospects for legislative regulation. Universum Econ. Law Electron. Sci. J. **5**(50) (2018). http://7universum.com/ru/economy/archive/item/5806. Accessed 03 Mar 2019
Flume, W.: Allgemeiner Teil des Burgerlichen Rechts. Zweiter Band: Das Rechtsgesch ft. Berlin, 52 (1975)
German Civil Code: Introductory Law to the Civil Code, 4th edn., pp. VIII–XIX, 1–715. Infotropik Media, Pererab.- M (2015)
Grin, Ye.S.: Registers of complex intellectual property rights (using the example of audiovisual works). Actual Prob. Russ. Law **8**, 99–105 (2017)
Gurko, A.: Artificial Intelligence and Copyright: a look into the Future. Information system. Copyright Relat. Rights **12**, 7–18 (2017)
Ioffe, O.S.: Relationships under Soviet civil law. Selected Works on Civil Law, Moscow, p. 575 (2003)
Kechekian, S.F.: Relationship in a socialist society, Moscow, p. 42 (1958)
Krasavchikov, O.A.: Categories of the science of civil law. Selected works: in 2 tons, Moscow, vol. 2, pp. 145–147, 274 (2005)
Kuzevanov, A.I.: Technical means of protection of copyright and related rights. Information system. Copyright Relat. Rights **1**, 21–28 (2018)
Nemov, R.S.: Psychology: V 3 t. Moscow, Volume 1: General principles of psychology, pp. 425–427 (2003)
Nikiforova, T.S., Smirnova, K.M.: Will lawyer robots be left without work? Zakon **11**, 110–123 (2017)
Novitsky, I.B.: Transactions. Limitation of actions, Moscow, pp. 22–23 (1954)
Novoselova, L.: “Tokenization” of civil law objects. Econ. Law **12**, 29–44 (2017)
Oygenzicht, V.A.: Will and will (essays of theory, philosophy and psychology of law), Dushanbe, p. 24 (1983)
Panov, A.A.: On the issue of categories of will, will and vice of will in the theory of a legal transaction. Civ. Law Bull. **1**, 55–56 (2011)
Pokrovsky, I.A.: Basic problems of civil law, p. 245. Statute, Moscow (1998)
Rabinovich, N.V.: Invalidity of transactions and its consequences. Leningrad **6**, 245 (1960)
Saveliev, A.I.: Contract Law 2.0: “smart” contracts as the beginning of the end of the classic contract law. Civ. Law Bull. **3**, 32–60 (2016)
Sargeant, C., Lim, C., Saw, T.: Smart contracts: bridging the gap between expectation and reality. https://www.law.ox.ac.uk/business-law-blog/blog/2016/07/smart-contracts-bridging-gap-between-expectation-and-reality. Accessed 28 Feb 2019
Svechnikova, I.V.: Copyright: Textbook, p. 164. Dashkov and K, Moscow (2009)
Skvortsov, D.I.: Smart contracts in the system of contract law of the Russian Federation. https://icoreview.ru/smart-kontrakty-v-sisteme-dogovornogo-prava-rossijskoj-federacii/. Accessed 03 Mar 2019
Szabo, N.: Smart contracts in essays on smart contracts, commercial controls and security (1994). http://szabo.best.vwh.net/smart.contracts.html. Accessed 28 Feb 2019
The law on the electronic form of transactions in 1999 (adopted by all states) in the collection: The Treaty as a general legal value: Monograph, 381 p. IZISP, Statute, Moscow (2018)
Lando, O., Beale, H. (eds.): The Principles of European Contract Law. Nijhoff, Dordrecht (1995)

Gaydayenko Sher, N.I., Grachev, D.O., Leshenkov, F.A., et al.: Treaties in the civil law of foreign countries: Monograph. Resp. ed. S.V. Solovyov, p. 344. IZISP, Norma, INFRA-M, Moscow (2018)
UNIDROIT Principles of International Commercial Contracts (2010). International Commercial Transactions, 4th edn. (2011)
Uniform Electronic Transaction Act. https://letknow.news/news/shtat-ogayo-uzakonil-hranenie-dannyh-na-blokcheyne-8696.html. Accessed 27 Mar 2019
Yurasov, M.Yu., Pozdnyakov, D.A.: Smart contract and prospects for its legal regulation in the era of blockchain technology. https://zakon.ru/blog/2017/10/9/smart-kontrakt_i_perspektivy_ego_pravovogo_regulirovaniya_v_epohu_tehnologii_blokchejn. Accessed 03 Mar 2019

The Role and Value of Intellectual Resourccs in thc 21st Century and the Scientific and Educational Platform for their Training and Development

The Flagship University's Model in Terms of Digitalization: The Case of Industrial University of Tyumen as a Center of Strategic Decisions in the Field of Smart-City, IoT/IIoT and Big Data

Viktoria A. Lezer(✉), Liubov N. Shabatura, and Igor A. Karnaukhov

Industrial University of Tyumen, Tyumen, Russia
vika_64@list.ru, lnshabatura@mail.ru,
ikharnauhov@gmail.com

Abstract. The authors seek to consider the concept and model of a digital university in this article. The Industrial University of Tyumen (IUT) activity is the case of current research resulting from the specifics of software and hardware solutions that allow implementing the strategic project of developing smart technologies. The main idea of the project is to make a center for initiatives to develop and implement solutions for the effective management of urban resources and the environment in the Tyumen region.

The objectives of the strategic project include: developing and implementing software and hardware solutions, organizing consulting, and examining regional projects for developing smart technologies. In addition, it helps to the IUT to form reputation as a leader in the areas of competence of Smart-City, IoT/IIoT and Big Data in the Tyumen region. Therefore, the IUT can be involved in the solutions the difficult tasks of effective city management by local communities and professionals. Moreover, the strategic project opens the ways to developing educational programs and project-based learning forming the specific skills in implementation of smart technologies.

Keywords: Digital university · Transformation · Stakeholders · Supporting platforms · Digital marketing · Smart-technologies

JEL Code: I230 · I250

1 Introduction

In the conditions of the digital revolution, the modern Russian education system needs radical modernization and competitiveness enhancement to ensure a technological and economic breakthrough in the international world stage. Institutional, financial, managerial, technological fields that are sufficiently capable for reflection the interests of the educational system stakeholders are subject to improvement. Any university, developing a strategy for its development at the present time, takes into account the paradigm content of digital transformation. It means the implementation of information

E. G. Popkova and B. S. Sergi (Eds.): ISC 2019, LNNS 91, pp. 387–396, 2020.
https://doi.org/10.1007/978-3-030-32015-7_43

technologies and information technology solutions, optimization of management processes, improvement of the results, development of the organizational level and corporate culture. The world scientific and educational map is rapidly changing as a part of the growing process of economic globalization and dynamic technicalization. Therefore, the first reason for digitalization is to increase the competitiveness of universities. Competition begins with the struggle for the most cultural, the most intelligent applicant-student, not within a cluster or even a country, but in the international stage as a whole. In this regard, there are fundamental changes including the ability to ensure flexible and speedy alterations, effective management, interaction, innovation and socio-cultural transformations of universities. It can preserve their competitive advantages, improve the scientific and educational environment, create a system of a new level, a new scale, and new meaning. The meaning lies not so much in the technological parameters, levels and scales, but in the sociocultural effectiveness of these parameters and technologies. The tasks are to prepare highly cultured skilled specialists, responsible, promising employees and socially competent citizens; to make optimal conditions for academic staff for continuous development and introduction of innovations in the scientific and educational environment; and to create the most effective management system for administrative staff. The extra task is to meet the employers' and stakeholders' needs and requirements as much as possible. The synergistic effect of this meaning should form a hypothetically more developed humane society with the harmonization of the relationship between social classes, where an individual is the aim, not the mean to achieve other goals.

The purpose of this article is to consider the concept and model of a digital university. The IUT activity is the case of current research resulting from the specifics of software and hardware solutions that allow implementing the strategic project of developing smart technologies. It forms reputation of the IUT as a leader in the areas of competence of Smart-City, IoT/IIoT and Big Data in the Tyumen region. Therefore, the IUT can be involved in the solutions the difficult tasks of effective city management by local communities and professionals. The strategic project opens the ways to developing educational programs and project-based learning forming the specific skills in implementation of smart technologies.

The article is using the following scientific papers:

Scholl and Scholl (2014). Smart governance: A roadmap for research and practice; Albino et al. (2015). Smart cities: Definitions, dimensions, performance, and initiatives; Federici et al. (2015). Gentlemen, all aboard! ICT and party politics: Reflections from a mass-eParticipation experience; Gil-Garcia et al. (2015). What makes a city smart? Identifying core components and proposing an integrative and comprehensive conceptualization; R.G. Hollands, Will The Real Smart City Please Stand Up? Intelligent, Progressive, or Entrepreneurial?; Caragliu et al. (2011). Smart Cities in Europe; Kuznechov and Engovatova (2016). Universities 4.0: growing-points of knowledge economy in Russia; Kondakov (2017). Education in the era of the 4th industrial revolution; Barabanova et al. (2018). Digital economy and university 4.0, and etc.

2 Methodology

Theoretically, the conceptual model of a digital university consists of five levels or supporting platforms. We shall consider the components of each level.

The dominant first level is represented by internal and external university stakeholders: students, academic staff, university partners, graduates and applicants.

The second level reflects a single information space created in the university with basic and additional information services that are able to effectively provide digital interaction within the university by using flexible tools.

The third level is represented by services for students and academic staff providing educational, design, and research activities. This level has any time access to scientific literature and other necessary resources for carrying out scientific work, and also creating a higher level of comfort for students and university teachers. It has a positive effect on the image of school. The same level comprises digitization of scientometrics, including accumulation, monitoring, and analytics of scientometrical data. This is one of the most important and promising university fields, as it provides and builds two strategic goals for the university. The first is to develop and identify promising basic research areas that are most relevant to the school. The second is to determine the levels, indicators of publication activity and citation of universities.

The fourth level reflects:

- optimization of internal processes in the use of new information technology methods and approaches leading to the simplification of the interaction between stakeholders and to increasing the net effect;
- tracking technological innovations, ongoing training and consulting on options for their implementation and use;
- improving incentive policy.

From the perspective of implementation, this level is the most resource-intensive, but at the same time it allows the university to get the highest added value. It consists of the following services: research project management, procurement management, digital marketing, and interaction with applicants and students.

As the innovation for Russian universities, digital marketing aims at solving the following tasks:

- organization of effective interaction with educational support staff, academic staff, all-level-managers, students, applicants, graduates using all modern tools and technologies, and the entire spectrum of digital communication channels;
- monitoring of changes in the perception of the university brand in the target markets based on the results of researching social networks; forming the positive image of the university;
- encouraging the creation of new digital communities and innovations at all stages of the educational cycle, as well as communication of the educational programs content and features of student activities for applicants;
- development of personalized marketing materials for target audiences based on analysis of data from different sources.

Interaction with applicants and students includes the following tasks:

- using digital technologies to interact with applicants and inform them about the stage of processing applications for admission;
- using analytics to determine the most promising applicants and increase their enrollment rate;
- use of various communication channels, both digital and conventional, to provide applicants with the most complete information about the university. This task is most relevant for foreign applicants who cannot visit the university and want to form an idea about it using the necessary materials and information from the Internet;
- analytics to identify the quality of training level and, the most and least successful students;
- student's office automation.

The fifth level consists of digital technologies, which will be probably used in the academic setting in the period from 2019 to 2020. Such technologies, for example, include drones. According to an estimate from a recent PwC study, the global market for potential application of drone-based solutions in 2015 amounted to $127 billion. Undoubtedly, we see it quite logical that universities, especially technical ones, will want to participate in the development of this market. In this context, as a first step, universities will actively introduce drones technologies into the internal educational and research areas, purchasing of equipment, setting up laboratories, encouraging students and researchers to test and work with the new technology.

The transformation to a digital university is impossible without supporting activities focused on introducing changes at the university. Such events may include:

- the development of optional or core modules in training programs aimed at improving digital literacy among students;
- supporting the academic staff who set trends in the development of digital skills and are engaged in the development of innovative teaching methods;
- encouraging the advanced use of training platforms by the academic staff in order to ensure higher student learning outcomes, and improve the efficiency of the university work;
- rendering assistance to academics who are less advanced and do not have the skills to use digital technologies.

In our opinion, in order to move to the modern level, the university must adequately complete all levels of the digital university model described above, and constantly maintain feedback with key stakeholders – students, academic staff, industry, academic partners, graduates, applicants and their parents.

3 Results

At the premises of the Industrial University of Tyumen and in accordance with the development program of the flagship university, the strategic project «Creating a Regional Innovation Cluster in the SMART-City, IoT/IIoT and Big Data Competence Field» has been implemented since 2017.

The main idea of the project is to make a center for initiatives to develop and implement solutions for the effective management of urban resources and the environment in the Tyumen region.

The objectives of the strategic project include: developing and implementing software and hardware solutions, organizing consulting, and examining regional projects for developing smart technologies. In addition, it helps to the IUT to form reputation as a leader in the areas of competence of Smart-City, IoT/IIoT and Big Data in the Tyumen region. Therefore, the IUT can be involved in the solutions the difficult tasks of effective city management by local communities and professionals. Moreover, the strategic project opens the ways to developing educational programs and project-based learning forming the specific skills in implementation of smart technologies.

The project «Creating a Regional Innovation Cluster in the SMART-City, IoT/IIoT and Big Data Competence Field» is included in the Strategic Development Program of the flagship university until 2020. The subject of the project covers more than 50 areas. Among them are «Smart City», «Smart Road», «Smart Home», «Smart Production», ecological, sociological and many other projects. The center of this research data in IUT is the department of Road transport, construction and road machines. Nowadays 16 structural university units are involved in the development of Smart City, including Departments of building structures, business informatics and mathematics, heat gas supply and ventilation, operation of road transport, cybernetic systems, design and urban planning, power economy, water supply and drainage, applied geophysics, construction management, housing and utilities, architecture and design, roads and aerodromes, marketing and municipal management, Multidisciplinary College, etc.

In 2018, the university took part in the development of a transport demand management concept in the territory of Tyumen, commissioned by MKU Tyumengortrans. The strategic project is being implemented with financial support from the Government of the Tyumen Region, as well as with the active participation of the industrial partners of the university PJSC Rostelekom, MKU Tyumengortrans, and IBM.

Oleg Danilov, a head of the strategic project Smart City IUT, noted that «the leading engineering university in the region is ready to become the foresight center of the Smart City project, engage in analytics, expertise, coordination of the activities of the Smart City in the external information field».

Governor of the Tyumen region Alexander Moor at one of the meetings stated «we do not just know that our flagship university is working on the Smart City Tyumen concept, we are actively participating in its formation, being a principal of a document aimed at creating a fundamentally new comfortable living environment in the city and fundamentally new conditions for resource management».

So, one of the most important stakeholders of the strategic project are the city administration and the government of the Tyumen region, which opens up certain prospects.

In the period from 2017–2019 the project achieved the following results:

- in the field of science and innovations, 20 prototypes of «smart» software, hardware and technological solutions in five relevant areas in the region (traffic management, energy saving, industrial monitoring, IT infrastructure, automation of construction processes) were transferred to pilot industrial operation at city-owned premises);
- in the field of education, a new model of It-education has been created: practical training for bachelors is carried out on software-hardware solutions, project-based learning is carried out in accordance with the standards of the international initiative CDIO;
- in the area of promoting regional development, the project contributed to the formation of a communication platform at the university (roundtable discussions with the Tyumen region authorities, with the city administration, and with the business community; the Hackathon and Hackathon kids marathons, the Internet of Things in the Tyumen Region).

All prototypes developed within the project are responsible for improving the quality of citizens' life and reducing the costs of municipalities, the region and enterprises. For instance, at the moment, IUT specialists, commissioned by Rostelekom, have developed and created an electronic screen that displays the time of arrival of buses and waiting times. This screen is already available to citizens on the Melnikayte Street. Soon these information screens will be installed at other bus stops in the city.

Smart traffic lights will be in Tyumen. They will help on the loaded sections of roads with traffic jams. There are cameras at intersections where an adaptive system is installed. It collects statistics during the month: about the number of cars passing through it at different times. Thus, the traffic lights distinguish the usual situation from the unusual ones.

IT-park and intelligent stopping complex will be constructed on the Lunacharskogo street close to the buildings of the IUT. The working title of the project is «Yamskaya Sloboda». In winter, the complex will be warm due to the heating system, and in summer it will be cool due to the installed air conditioners. At the bus stop on-line information about the arrival of buses will be displayed. Facing the complex is planned from metal frames with wooden inserts. Panoramic windows will be convenient for a good view of those who are waiting for their bus. The complex will be equipped with stands with useful information for passengers, television screens with schedule of urban transport.

At the moment, one electric bus and several buses on natural gas have already started moving around the city. In the future, their number will increase. All of them will be equipped with sensors to track the route.

In modern houses, built-in heat and water supply sensors will appear, information from which will be sent to a single base substation. The information will be automatically sent to dispatching services of the water and energy services. It will significantly save the time of citizens.

Defining the role of external project stakeholders, it can be said that at this moment agreements have been reached with a large number of companies. We divide them into three groups. The first one is urban engineering services, water supply, water drainage, heat, gas, electricity, housing and public utilities. The city should be interested in the development of digital technology firstly. Smart City provides him an enormous economic effect. The second group is the corporations that supply the complete and finished product. We call them industrial partners. And the third group comprises universities, training organizations that can become partners in the implementation of this work.

4 Discussion

As noted by Scholl and Scholl: «smart interaction with stakeholders is a broader field of interest in smart governance research that has emanated from traditional electronic government research» (Scholl and Scholl 2014).

To create a smart city, you need to create a smart management system. «smart governance means various stakeholders are engaged in decision-making and public services» (Albino et al. 2015). In addition, smart management means «new technologies – that is, social media, the internet, open data, citizen sensors, and serious games – are used to strengthen the collaboration between citizens and urban governments» (Federici et al. 2015).

Gil-Garcia, Pardo and Nam supplemented the image of a smart city with the following components: «(1) technology and data, (2) government, (3) society, and (4) physical environment» (Gil-Garcia et al. 2015). Technology is a special matter, spreading to other elements, connecting and expanding each of them.

European researcher Hollands identified a number of essential characteristics of the «smart city» concept:

1. «The utilization of networked infrastructure to improve economic and political efficiency and enable social, cultural, and urban development;
2. An underlying emphasis on business-led urban development;
3. A stress on the crucial role of high-tech and creative industries in long-run urban growth;
4. Profound attention to the role of social and relational capital in urban development;
5. social and environmental sustainability as a major strategic component of smart cities» (Hollands 2008).

The city becomes «smart» when significant investments in human capital and traditional (transport) and modern (ICT) communication infrastructure fuel sustainable economic growth and a high quality of life, with a wise management of natural resources, through participatory governance» (Caragliu et al. 2011).

The author and the head Oleg Danilov, professor, doctor of technical sciences, comments that «the concept of Smart City is different for everyone. There are world leaders in the development of smart cities. Among them, the most successful is IBM, which has implemented dozens of similar projects in North America, Europe, and Korea».

A «smart city» is a huge system that includes many subsystems: energy, ecology, roads, passenger transport, safety, education, medicine and much more. And all of them need to be managed. Management in the modern world is put on digital, or smart technologies. They provide simultaneous work of all city services. It shows a degree of cities in smart fields.

A specific example: somewhere in the system there was a leak of water, it was fixed by special sensors that can give a signal to the computing cluster. The computer decides to close the valve. The valve is closed, then it recommends the operator how to eliminate the accident. The operator either sends a repair crew to the site, or corrects this decision, or accepts an additional one. This is the model of «smart city».

In one way or another, smart workers are now engaged in power engineering, public utilities and safety. But they do it initiatively and separately. Information is processed on their servers, they make some decisions. The goal is to combine all these efforts. At the same time, an industrial university plays the role of a *connector* between all project participants.

Nowadays, a co-working center is being created at the university. It is a platform for communication, where active, enthusiastic people can meet each other, where a start-up movement may be born, and new associations can be established. It should be noted that the Industrial University of Tyumen represents the Smart city project as the engineering project.

«Smart City» are planned to be shifted into curricula, work programs, which can be studied by students. It is planned to move to project-based learning, the most advanced technology in higher education

In the future, all best practices for the «Smart City» are planned to be shifted into curricula, work programs, which can be studied by students. It is planned to move to project-based learning, the most advanced technology in higher education. Everything will work on improving the training quality of our graduates.

5 Conclusions

1. The implementation of the strategic project of the Industrial University of Tyumen «Creating a Regional Innovation Cluster in the SMART-City, IoT/IIoT and Big Data Competence Field» is based on universal and European values, and aims to create a comfortable urban environment for harmonious human existence and the development of its creative and professional abilities.
2. Tyumen, having studied the positive experience of other cities, took the first steps towards the «smart city». On solving this problem, experts of the Industrial University of Tyumen are currently working. Thanks to the efforts of the government, business, citizens, the educational and engineering community, much has

already been done to make Tyumen as a «smart city». Tyumen is most successfully developing in the field of transport, communications, and infrastructure.

3. The Smart City project at the premises of the Industrial University of Tyumen is more perceived as engineering. In the future, all best practices for the «Smart City» are planned to be shifted into curricula, work programs, which can be studied by students. It is planned to move to project-based learning, the most advanced technology in higher education.
4. Distance education and e-learning marked the beginning of a new global trend – «smart education». This is not so much about technology as about the philosophy of education. Smart learning is flexible learning in a vibrant and ever-changing educational environment. Maximum availability of knowledge ensures that all information is freely available. At the same time, the educational process becomes more interactive and it has a variety of approaches.
5. «Smart education» is a transition from passive content to active, online, digital education. E-learning provides two-way communication between teachers and students, allows you to share knowledge, and it does not matter how far the interlocutors are from each other. Thus, e-learning is embedded in the structure of the digital society and is even its central, fundamental element. «Smart education» is an association of students, teachers and knowledge from around the world.

References

Scholl, H.J., Scholl, M.C.: Smart governance: a roadmap for research and practice. In: iConference 2014 Proceedings (2014)

Albino, V., Berardi, U., Dangelico, R.M.: Smart cities: definitions, dimensions, performance, and initiatives. J. Urban Technol. **22**(1), 3–21 (2015)

Federici, T., Braccini, A.M., Sæbø, Ø.: Gentlemen, all aboard! ICT and party politics: Reflections from a mass-eParticipation experience. Govern. Inf. Q. **32**(3), 287–298 (2015)

Gil-Garcia, J.R., Pardo, T.A., Nam, T.: What makes a city smart? Identifying core components and proposing an integrative and comprehensive conceptualization. Inf. Polity **20**(1), 61–87 (2015)

Hollands, R.G.: Will the real smart city please stand up? Intelligent, progressive, or entrepreneurial? City **12**, 3, 303–320 (2008)

Caragliu, A., Del Bo, C., Nijkamp, P.: Smart cities in Europe. J. Urban Technol. **18**(2), 65–82 (2011)

Kuznechov, E.B., Engovatova, A.A.: Universities 4.0: growing-points of knowledge economy in Russia. Innovation **5**, 3–9 (2016)

Kondakov, A.: Education in the era of the 4th industrial revolution, Vesti obrazovanija (2017)

Barabanova, M.I., Trofimov, V.V., Trofimova, E.V.: Digital economy and university 4.0. J. Legal Econ. Res. **1**, 178–184 (2018)

Zhurkin, M.Y., Karnaukhov, I.A.: IOP Conf. Ser. Mater. Sci. Eng. **463**, 042067 (2018)

Robinson, C.M.: Improvement in city life. Atlantic Monthly **83**, 771 (1899)

Belogolovskij, V.: Green house, Moscow, Tatlin, p. 198 (2009)

Beevers, R.: The Garden City Utopia, p. 70. St. Martin's Press, New York (1988)

Lezer, V., Muratova, I., Korpusova, N.: Issues of transport security and human factor (2019). https://www.e3s-conferences.org/articles/e3sconf/abs/2019/17/e3sconf_tpacee2019_08068/e3sconf_tpacee2019_08068.html

Shabatura, L.N., Martseva, L.M., Tarasova, O.V., Yatsevich, O.E.: Strategy of the social state in early 21-st century Russia, Moscow, Sociological Studies, pp. 171–173 (2017)

Investments in Human Capital as a Key Factor of Sustainable Economic Development

Angelika P. Buevich(✉), Svetlana A. Varvus, and Galina A. Terskaya

Financial University under the Government of the Russian Federation, Moscow, Russia
buanpet@mail.ru

Abstract. The modern approach to human capital consists in a comprehensive vision hereof as a crucial socio-economic resource and as a factor of public goods' creation. Human capital acts simultaneously as a factor and purpose of development for a person, family, and society. Jointly with financial, natural and physical resources, it lay the basis of the national wealth. The article discusses the main approaches and criteria for measuring human capital, its importance in different countries, as well as ways of efficient investments in human capital.

Keywords: Investments · Human capital · Sustainable economic development · Poverty · Education · Human Development Index · Knowledge-based economy

JEL Code: D31 · E22 · E24

1 Introduction

The concept of sustainable development recognized by the UN as a consensus official paradigm of the development of the 21st economy determined obligatory participation of all the countries in the resolution of various social and environmental problems, the transformation of current business models and strategies for society, nature, and man.

In this context, an understanding of the topical issue related to the movement of Russian economy towards a sustainable economic growth should be displayed primarily in attracting civil society and business to resolution of global problems and establishment of a favorable inclusive environment with regard to scientific, creative, technological, intellectual, and resource capacity under critical role of human capital.

2 Methodology

Benchmarking method, time series, analogies.

The Human Development Index is used to assess human capital. Based on Updated Statistic Data on Human Development Indicators Indices and Indicators within the United Nations Development Program, the authors discuss the main problems restraining the world progress.

E. G. Popkova and B. S. Sergi (Eds.): ISC 2019, LNNS 91, pp. 397–406, 2020.
https://doi.org/10.1007/978-3-030-32015-7_44

3 Results and Discussion

The goals of sustainable development declared by the United Nations are aimed at improving the welfare and protection of our planet. The UN stresses that measures on poverty elimination should be taken jointly with efforts on enhancement of economic growth and resolution of a wide range of education, healthcare, social safety, employment, and environmental protection issues. More than half of seventy goals declared are connected with human development, such as elimination of poverty and starvation, good health and well-being, high-quality education, a decent job, and economic growth, industrialization, innovation and infrastructure development, sustainable cities and settlements, responsible consumption and production, partnership for sustainable development.

3.1 Sustainable Development of the Economy and Human Capital

According to the definition suggested by the OECD and approved by the United Nations (Guide for Human Capital Measurement 2016), human capital is a combination of knowledge, skills, abilities and other qualities embodied in people and contributing to their personal, social and economic well-being. The state of human capital affects the use of all other development resources.

The combination of measures to ensure the innovation development of the economy will make it possible to achieve sustainability of the socio-economic development of both international and domestic system. At the same time, it is the accumulation and reproduction of human capital that acts as a determinant of innovative development of the economy.

The relationship between human capital (HC) and investments is determined as follows:

$$\mathrm{HC} = \mathrm{f}\,(\mathrm{j}, \mathrm{q}, \mathrm{I}, \mathrm{X}) = \mathrm{cI},$$

where j is the labor quality index in a broad definition;
q is a quality index of accumulated human capital;
I is an investment in human capital;
X are other variables, which human capital depends on, including the accumulated human capital (Korchagin 2004).

A continuous circulation takes place: human capital assists the growth of economic efficiency, and production efficiency, in turn, opens opportunities for investment in the development of human capital. American scientist K. Brunner stated the conversion of non-economic capital into the economy: a resourceful, evaluative, and maximizing man (REMM) (Brunner 1993).

All measurements show that rich countries have bigger human capital against poor ones. Therewith, the gap widens when the concept of human capital becomes broader and goes beyond education.

Under the support of the UN, in 1990 a technique for calculation of the Human Development Index (HDI) was designed. It has been applied for a comparative assessment of countries in terms of human capital development and monitoring of changes hereof during time periods. In 2013, the index became the "human development index" with the following main components of the measurement technique:

- measurement of the human development index based on the assessment of life expectancy that depends on the state of the healthcare system and social security of citizens;
- assessment of the level of education and literacy of the adult population, as well as the coverage hereof by various levels of education, which make it possible to reveal the peculiarities of the country's educational system;
- evaluation of the gross national product per capita.

The country rating made up in accordance with the Human Development Index 2018 is included in the UN Development Program Report. In 2018, Norway had the highest level of human development (out of 189 countries). Switzerland and Australia are also the top three countries in terms of this indicator (Human Development Indices and Indicators 2018). Ireland, Germany, Iceland, Hong Kong, Sweden, Singapore, and the Netherlands were in the top ten. The Central African Republic, South Sudan, Chad, and Burundi round out the rating. Russia is 49th in the rating of 2008 (see Fig. 1). In accordance with the report, since the 1990s the rate of human development at the global level has been 22%. This fact determined the growth of life expectancy, wider education opportunities, an increase in the population income.

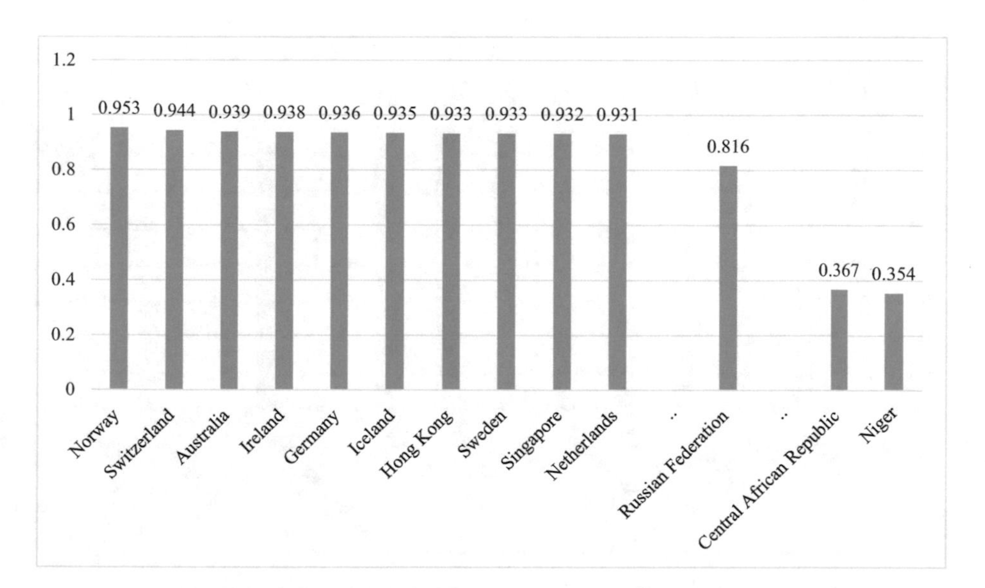

Fig. 1. HDI data in 2018 (fragmentarily) (Source: developed by the authors based on Human Development Indices and Indicators 2018).

HDI data research discovers some problems:

- unequal human development. New reports of Oxfam International showed that eight people have the same wealth as 3.6-billion poorest population; 82% of overall global wealth in 2018 passed to 1% of rich;
- gender inequality (the world average HDI for women (0.705) is 5.9% lower than that for men);
- the level of deprivation of people is still high (in countries with low human development level, 47.5% of the adult population is illiterate and only 17.1% have access to the Internet) (Human Development Indices and Indicators. Updated Statistical Data 2018).

Another approach to the measurement of human capital was proposed by the World Bank. It is based on a discount method for assessment of human capital through the value, which is calculated as the difference between the value of overall national wealth and expression of natural and physical capital in monetary terms (Parushina et al. 2017).

According to the World Bank data (Human Capital Index and Components 2018), Singapore became the first in terms of human capital out of 157 countries in 2018. Japan and Korea also were in the top three. Top ten was represented by Hong Kong, Finland, Ireland, the Netherlands, Canada, Australia, Sweden. Russia was the 34th. South Sudan and Chad rounded out the rating (see Fig. 2). From the point of human development, true progress can be achieved through ensuring high-quality education, healthcare and other areas of human activity.

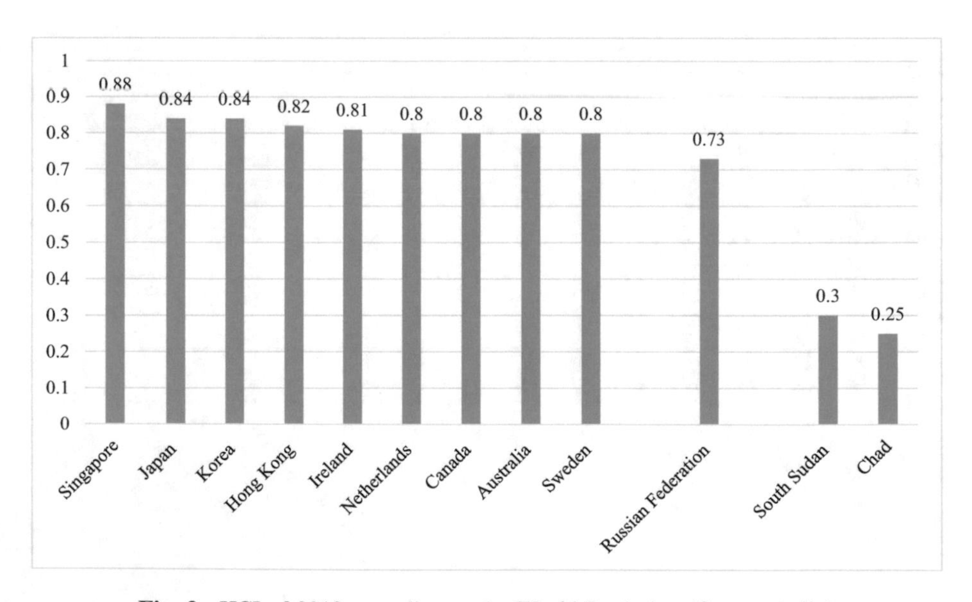

Fig. 2. HCI of 2018 according to the World Bank data (fragmentarily)

The problem is that the investments in education and healthcare are not instantly cost-effective and risky, and thus restrain the authorities from investing. The obvious reasons for deviations from the expected profit are possible death, loss of working ability, and various qualities of the person, such as energy, luck, etc. According to the World Bank data, nowadays a quarter of the world's young people don't manage to entirely fulfill themselves due to chronic malnutrition and growth-retarding diseases, which affects the child's cognitive development, school progress and, consequently, and future income.

If today's children in adulthood are not able to meet the demands of the labor market, the country will not be able to provide people with job and production growth on its own and will not be fully equipped for economic competition. The project of the World Bank for HCI calculation emphasizes the importance of solving challenging issues by the government, such as accounting of inadequate or cost-ineffective expenses, management and service provision, population and demography, infrastructure development, etc. (World Bank. The Human Capital Project 2018).

As we can see from the data presented in Fig. 2, HCI Russia is 0.73. This means that children born in Russia today will be effective in adulthood only by 73% due to their insufficient health and education. According to World Bank statistical data, the average graduate of the Russian school will not complete studies for over two years. Survival in Russia of 2018 is 78%, i.e. only such number of today 15-year-old compatriots will live to 60 years (World Bank. The Human Capital Project 2018). This means that not all children will live to the moment when they get an education and start making a contribution to social development, not everyone will be able to get a decent education. Besides, there are issues in education and healthcare systems.

3.2 The Main Points and Peculiarities of Investing Human Capital in the Modern Russian Economy

According to the World Bank data (The Changing Wealth of Nations 2018), human capital is the biggest component of global wealth, its share account for two-thirds of overall global wealth, while natural capital – only 9% hereof (see Table 1).

According to Table 1, in the countries-leaders of the HCI World Bank rating 2018 the share of human capital is 55–75% of the country's overall wealth. In Russia, human capital is slightly more than 48% thereof, which is closer to the countries-outsiders of the World Bank rating.

In general, human capital accounts for 70% of the wealth of high-income countries, and only 40% accrue to low-income countries. This trend will persist. In this context, we agree with the findings of the World Bank report that a skilled employee is key to the future (The Changing Wealth of Nations 2018).

According to the calculations of the World Bank, the introduction of various kinds of reforms by developing countries in 2018–2027 that contribute to growing number of graduates from secondary school on average (5%), increasing coverage of population by secondary, vocational and higher education (7%), and higher estimated life expectancy (by 2,5 years) will allow increasing GDP by nearly 0.2 p.p. and mitigating population's income inequality.

Table 1. The proportion of human capital in the overall wealth of the country

Country	Overall wealth	Production capital	Natural capital	Human capital	The share of human capital in the overall wealth of the country, percent
Singapore	775.196	186.017	56	466.119	60.13
Japan	571.927	179.277	3.741	365.157	63.85
Korea	424.052	126.650	4.013	291.748	68.80
Ireland	627.256	189.309	15.912	473.656	75.51
Netherlands	792.396	234.415	9.528	516.543	65.19
Canada	1016.593	229.999	52.438	730.832	71.89
Australia	1046.785	311.442	180.792	585.737	55.96
Sweden	886.129	285.792	27.890	576.521	65.06
…					
Russian Federation	188.715	48.807	46.921	90.812	48.12
…					
Chad	20.077	1.619	9.973	9.099	45.32

We believe that these recommendations should be applicable to our country, where low funding of key components of human capital, such as education and human health, is one of the main obstacles towards economic development.

Thus, the budget expenditures on education in Russia over 2013–2018 accounted for 3.6–3.9% of GDP (see Fig. 3). Russia was 112nd out of 189 countries by total current public expenditures on education.

As for the total expenditures on health, in 2016, Russia was 122nd out of 189 countries (see Fig. 4). Investments in fixed assets aimed at developing health care amounted to 1.2 of total investments in fixed assets in the economy of the country in 2016. We suppose that insufficient funding of education and healthcare determined Russia underdevelopment by number and quality of accumulated human capital per capita. The development of these areas should be taken into account within priority national projects for the period up to 2024 (Decree of the President of the Russian Federation 2018).

Talking about investments in human capital, it is necessary to pay due regard to the effectiveness hereof. Education and health care need structural reforms. In our opinion, it's worth discussing the rejection of the model of compulsory medical insurance in Russia, because it's invalid, detrimental to people's health, worsening the position of medical staff, and return to public funding of the USSR times. Many mistakes were also made in education; the expected results expressed in higher-quality of knowledge and affordability of education were not achieved. It is necessary to reconsider all previous experience and choice of our own model for the development of education and healthcare systems. At the same time, it is necessary to toughen the social

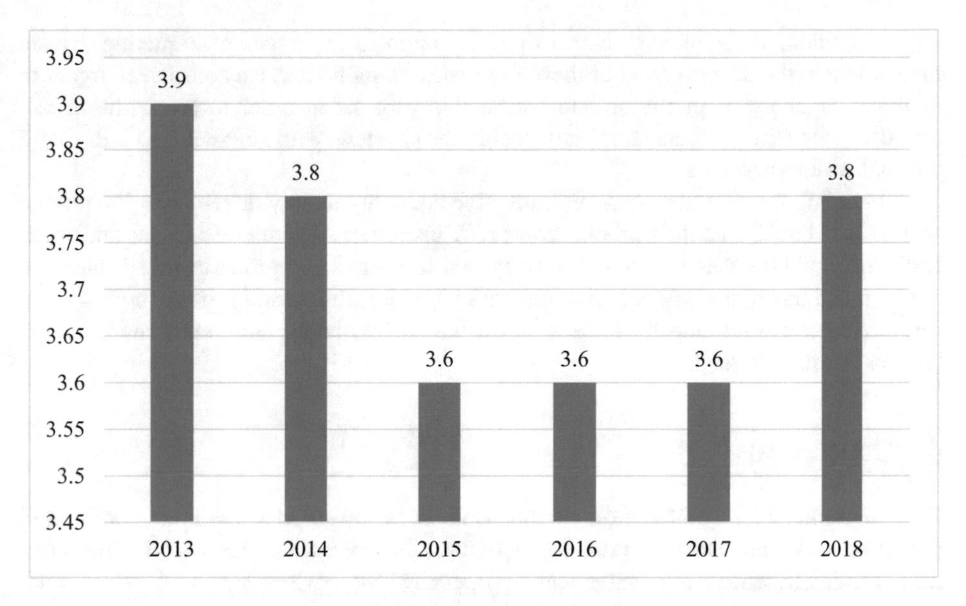

Fig. 3. Budget expenditures on education, percent of GDP (Source: developed by the authors based on Russia and the World 2018).

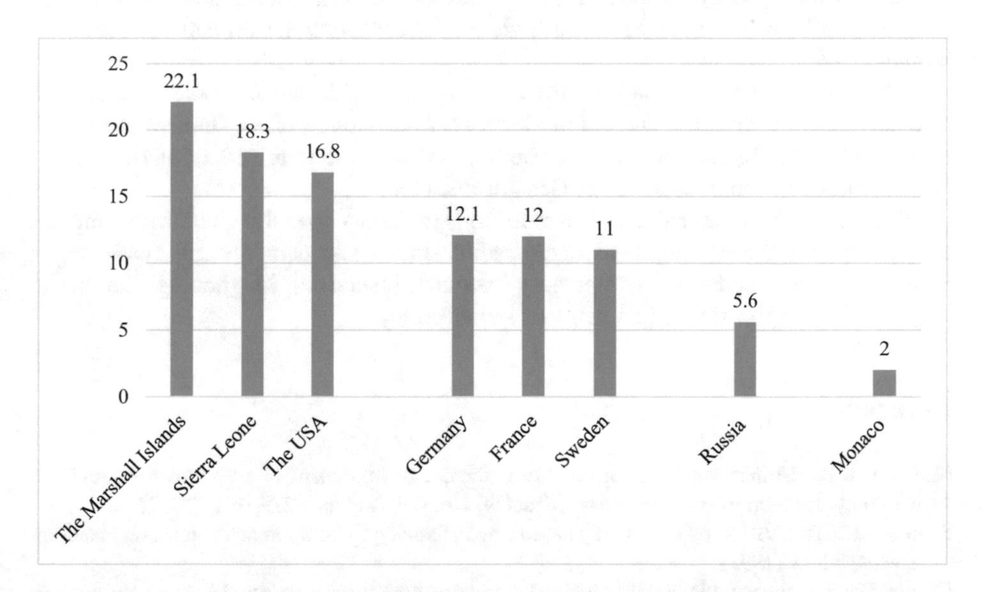

Fig. 4. Total health expenditures in 2016, percent of GDP (fragmentarily) (Source: developed by the authors based on The World Data Atlas 2019)

responsibility of the state as an actor of social policy in education and healthcare, also by means of effective supervision over the fulfillment of social, financial and other obligations established in accordance with law by all participants of investment activities, as well as provision of free education and healthcare.

In addition, it seems expedient to pay attention to the issues of attracting private investment in the development of these industries. In particular, we consider it urgent to increase the share of public-private partnership projects in order to cover the infrastructure deficit in the education and health care system with reference to successful international experience.

Regarding the fact that modern innovative economy mainly develops in the private sector, we should establish private non-profit institutions (both educational and medical), the goal hereof is not to gain money, but to protect and promote social interests, develop and introduce new forms of education and healthcare and provide high-quality services in education and healthcare that make society better and create more opportunities for the citizens.

4 Conclusions

Human capital is a combination of innate abilities, acquired knowledge, skills, and expertise in the process of production activities. In present conditions, it is characterized by intelligence, innovativeness, propensity for partnership, and educational mobility.

The human capital is developing over life through continuous investment at personal, company, and government levels. From the point of human development, true progress can be achieved by ensuring high-quality education, health and other areas of human activity.

Investments in human capital are the most profitable against other forms. The studies confirm a sizeable and a long-term socio-economic effect. Innovative development of both Russia and other modern countries is determined primarily by the development of human capacity as a major resource.

Structural education and health care reforms in Russia should include: choosing our own model for the development of education and health care systems, including a revision of the model of compulsory medical insurance; toughening the social responsibility of the state as an actor of social policy.

References

Buevich, A.P.: Human capital as a condition for the development of an innovative regional economy. Economics and Business, Moscow, No. 4-1 (81), pp. 279–291 (2017)

Brunner, K.: The Vision of the Man. The concept of society: two approaches to understanding. Thesis, No. 3 (1993)

Human Capital: theory, historical experience, and development prospects. Monograph/ number authors; Project Leader Shapkin I.N. M.: Ruscience (2017)

Chuvakhina, L.G., Terskaya, G.A., Buevich, S.Y.: Innovations as a factor of public improved performance in the world economic system. Revista Espacios, vol. 39 (# 04), p. 22 (2018)

Indices and Indicators of Human Development. Updated Statistical Data 2018. http://hdr.undp.org/sites/default/files/2018_human_development_statistical_update_ru.pdf. Accessed 14 Apr 2019

Federal State Statistics Service. http://www.gks.ru/wps/wcm/connect/rosstat_main/rosstat/ru/statistics/databases/. Accessed 17 Apr 2019

Human Capital Index and Components (2018). https://www.worldbank.org/en/data/interactive/2018/10/18/human-capital-index-and-components-2018. Accessed 14 Apr 2019

Korchagin, Y.: The Effectiveness of National Human Capital: A Measurement Technique. CIRE, Voronezh (2004)

Krasyuk, I.A., Kobeleva, A.A., Mikhailushkin, P.V., Terskaya, G.A., Chuvakhina, L.G.: Focus on economic interests as a basis of investment policy development. Revista Espacios, vol. 39 (# 31), p. 19 (2018)

Mailyan, F.N.: Human capital: measurement problems. Bull. NSU Ser. Socio-Economic Sci. **10**(3) (2010). http://www.nsu.ru/rs/mw/link/Media:/22733/06.pdf. Accessed 14 Apr 2019

Maryganova, E.A., Dmitrievskaya, N.A.: Human capital as a factor of sustainable development. Economics, Statistics and Computer Science. Bulletin UMO, No. 6, pp. 73–78 (2013)

World Data Atlas (2019). https://knoema.ru/atlas/topics. Accessed 17 Apr 2019

Parushina, N.V., Lytneva, N.A., Semidelikhin, E.A.: Methods of measuring and evaluating human capital (scientific review). Economics, No. 2, pp. 89–99 (2017)

Pogodina, T.V., Terskaya, G.A., Chuvakhina, L.G.: Development of Innovative Economic Area: Russian and International Experience. Perot Publishing House, Moscow (2017)

Russia and the World: Stat. Book. Rosstat. M., 375 p. (2018)

Forecast of the long-term socio-economic development of the Russian Federation for the period up to 2030 (developed by the Ministry of Economic Development of Russia. http://www.consultant.ru/document/cons_doc_LAW_144190/. Accessed 11 Apr 2019

Draft Federal Law "On Public Regulation of Greenhouse Gas Emissions and on Amendments to Certain Legislative Acts of the Russian Federation" (developed by the Ministry of Economic Development of Russia). http://www.consultant.ru/cons/cgi/online.cgi?req=doc;base=PRJ;n=178575#046177848216803863. Accessed 11 Apr 2019

Guide for Human Capital Measurement 2016. The United Nations Economic Commission for Europe, VIII, 138 p. UN, New York and Geneva (2016)

Schulz, T.: Investing in people: the economy of population quality, M. (1999)

Schumpeter, J.: Theory of Economic Development. Capitalism, Socialism, and Democracy. Foreword by V.S. Avtonomova, 864 p. EKSMO, M. (2007)

Terskaya, G.A.: The role of human capital in the development of regional innovation system. The phenomenon of the market economy: from the beginnings to the present day. Business, innovation, information technology, modeling. In: Sidorova, V.A., Yadgarova, Y.S. (eds.) Proceedings of the 7th International Research and Practice Conference, pp. 374–383. Electronic Publishing Technologies LLC, Maikop

Terskaya, G.A.: Development of the regional innovation capacity. Self-management **2**(1), 155–158

The Changing Wealth of Nations (2018). https://www.worldbank.org/en/news/feature/2018/01/30/the-changing-wealth-of-nations-2018. Accessed 13 Apr 2019

Presidential Decree No. 204 as of May 7, 2018 (as amended on 07.19.2018) "On the national goals and strategic objectives of the development of the Russian Federation for the period up to 2024" (2018). http://www.consultant.ru/document/cons_doc_LAW_297432/. Accessed 11 Apr 2019

Varvus, S.A.: Poverty measurement: an analysis of current approaches. Econ. Taxes Right **9**(6), 26–34 (2016)

World Bank: The Human Capital Project. World Bank, Washington D.C. (2018). https://openknowledge.worldbank.org/handle/10986/30498. Accessed 13 Apr 2019. License: CC BY 3.0 IGO

Zajtseva, N.: The research findings – responsible consumption: the space of new business opportunities and the experience of Russian companies (2018). http://ecounion.ru/wp-content/uploads/2018/06/Zajtseva-N.pdf. Accessed 11 Apr 2019
Healthcare in Russia 2017: Stat. Book. Rosstat. M., Z-46, 170 p. (2017)

Functional Model of the Head of the Third-Generation University Department

Irina V. Neprokina(✉)

Togliatti State University, Togliatti, Russian Federation
ivneprokina@rambler.ru

Abstract. This article is dedicated to the development of a functional model of the head of a third-generation university department and introduction thereof into the activity of a modern university. The relevance of this model is determined by the fact that it allows universities efficiently respond, integrate and, the most important, manage the processes of accelerated technological development that drastically change all global economic and social landscapes today.

The purpose of this study is to develop a functional model of the head of university structural unit; in the University of a third generation it's not a chair, but a department. Departments are established in accordance with the principles of project and process approaches. Departments are subdivisions that unite larger academic communities than ones existing within the chairs and provide lecturers with higher academic freedom.

We showed that universities seeking to be ranked "third-generation" need a new model of a leader who is able to be a director general for a dynamic and business institution, which meets the challenges of the 21st century, and an experienced research scholar who can develop the intellectual authority of the university.

We revealed the factors affecting the qualification of the head of the university department for management activities in terms of a third-generation university: the strategic goals of the institute; quality of work performed; business qualities; leadership skills.

We developed the process of building a leader's qualification: it is a complex of organization and management bodies that provide an assessment of the management process, the results thereof and identification of factors affecting their quality.

As a result of the study, we discovered the factors affecting the activity of department heads and proved their low efficiency using the method of expert assessments and diagnostic tools, such as Ishikawa and Pareto charts, and also developed requirements for the leader of new generation.

The content of the study is of practical value, the conclusions and recommendations can be used to improve the current system of training the heads for management activities.

Keywords: Changes in university management · Third-generation university · Requirements to the head of the university · Factors affecting the qualification of the head · Functional model of the head of the university department

E. G. Popkova and B. S. Sergi (Eds.): ISC 2019, LNNS 91, pp. 407–414, 2020.
https://doi.org/10.1007/978-3-030-32015-7_45

1 Introduction

The higher education sector is a highly competitive market where educational institutions directly compete: education has become a commodity today.

In this regard, the role of the head in an educational institution is growing. Nowadays, the head of an educational institution should act both as a scholar and an administrator and a strategic thinker and financial expert, as well as have political knowledge and diplomatic skills [4].

The leader's importance is obvious for the intellectual, strategic, and financial state of the institution in general, but the issues of selection of the most suitable candidate, criteria and procedures used are rarely discussed in the literature [8].

Having analyzed modern management trends and requirements for a manager, we proposed a functional model of a third-generation university manager.

The manager of a third-generation university should be a trans-professional. According to the opinion of the British social historian G. Perkin, the term "trans-professionalism" means "global professional revolution" [10].

From a historical point, there are three revolutions.

The first revolution allowed choosing professional activities and created a means of social mobility.

The second one is the industrial revolution that gave rise to the massification of professions, established professional standards and vocational training system, unified training.

Many authors call the third revolution "global". It happened in an era of uncertainty and generated virtual networks [6, 9]. The environment becomes multilingual and IT – saturated, which leads to the development of a set of basic, obligatory high-level skills. A specialist should be able to solve any complex problems, organize teams, forge complicated communication systems, i.e. to be a trans-professional. Obligatory requirements are lifelong learning, high social activity within the company and in various professional environments, creativity, innovative knowledge, moral development. The trans-professional does not see the usual vertical control systems; he solves professional problems from the point of efficiency. In this regard, this specialist needs natural and human science knowledge, basic legal and economic training, team building, command of several languages [8].

We agree with N.I. Kaligin who says that "a critical management function is the encouragement of the teaching staff of an educational institution, i.e. the possibility to influence the people, inspire them to effective activities to achieve the goals. Researchers noted the probable enhancement of motivation under a clear and precise goal setting. It is also believed that difficult goals compared to the easy ones have greater motivating power. As a rule, a desired psychological mood is created by the very person but can be tuned from the outside through the special means of influence used by a manager" [5].

We have earlier noted that "well-established management teams of educational institutions can become a real driving force of the renewal, but this does not automatically guarantee the progress of the transformation. The performance of the educational institution is finally determined by the human factor. In a team without coordinated actions aimed at a common result and constructive settling of disputes, usually, have low performance and attractiveness of the educational institution. It becomes obvious that the desired quality of educational results and the effectiveness of organizational changes is unattainable without a management system uniting the isolated groups of employees into a team of like-minded people" [7].

Training successful global leaders is a competitive advantage for companies. In addition to basic leadership skills, global leaders face special challenges that require additional skills. A global leader is defined as a person who develops a company on the foreign market, sets a strategy at the global level and manages diverse and isolated groups. Some of the problems include managing a diverse group of employees and business processes; flexible problem solving; adaptation to new values and cultures, various types of business and personal stress factors [9].

In accordance with the Concept of the University 3.0, the third-generation university deals with three main challenges: training, research, commercialization of scientific research. Moreover, the development of business skills, processes and culture are of primary importance.

For example, innovative and business activity of students in the United States has become a key factor of the universities' competitiveness. The landmarks of this education model are the American universities, such as the Massachusetts Institute of Technology, Stanford and Harvard Universities, as well as European role models, such as the University of Cambridge and the Catholic University of Leuven. Currently, the vast majority of international scholars work in interdisciplinary teams focusing on specific areas of research.

The objective of the research is to determine the right path of development based on the establishment of organization and management conditions for training the department head-manager of a third-generation university.

In the era of monodisciplinary studies, the perfect form of their organization were departments. It becomes a problem for interdisciplinary teams because they have to look for new organizational forms. These changes also concern the authorities of a modern university who should create new jobs for management of know-how profit-making, as well as be ready to be effective in the face of growing scale and complexity of the challenges.

2 Methodology

Within the university's transition to a third-generation model, various universities transform into university centers for innovation and technological development. This process also touched Togliatti State flagship university. The objectives of the University 3.0 are education, research, commercialization, i.e. the university should become an institutional representation of the student. The university is moving to a new model of teaching and management: from the departmental model to the management of educational programs, from isolated educational programs to a modular development of educational programs, invariability and unification of the disciplines and their modules. Departments are established in accordance with the principles of the project and process approaches. Departments are subdivisions that unite larger academic communities than ones existing within the chairs and provide lecturers with higher academic freedom [2, 3].

The relevance of this model is determined by the fact that it allows universities efficiently respond, integrate and, the most important, manage the processes of accelerated technological development that drastically change all global economic and social landscapes today. And although there are no well-established criteria and parameters for this model (it is quite flexible, because University 3.0 already get transformed into 4.0 stage), one fact has become generally recognized: universities that actively grow start-ups within business model develop much faster than ones without start-ups.

Togliatti State flagship University experimentally established three departments: the Bachelor's Department, the Master's Department (Business Programs) for the Management of Innovative Projects and Further Certification (Project Management Expert), Business Department, which trains students for project management based on international standards with a further study in master's program [2, 3].

Research and training in the departments are transdisciplinary or interdisciplinary and their goal is unity and creativity. All this will create conditions for the most talented students and teachers. Departments this way can perform the function of mass education, create particular conditions for training academic staff as a researcher or conduct research to enhance the development of knowledge and achieve one of the main goals of the University 3.0 – business and commercialization of ideas.

Let us present the main list of functional responsibilities of heads of the bachelor's department, the master's department and the business department, which served the basis for the development of a management model of a third-generation university department.

For targets of educational, scientific and management activities to be achieved, the head of the bachelor's department should:

- use the facilities of communication office equipment for receiving and transmitting information;
- keep information confidentiality;
- apply modern facilities of collecting, processing and transmitting information;
- evaluate the results within the framework of objectives;
- work with computer and office equipment;
- hold business correspondence.

The criteria for the performance of functional responsibilities are:

- implementation of the plan to improve the performance of the financial and economic activities of the department, including the rate of internal return, the execution of the department's budget;
- the compliance of the department's executive discipline, execution of orders and decrees of the rector and vice-rectors in due time, instructions of the direct manager, as well as the execution of other local regulations;
- monitoring of students' satisfaction with the quality of training not lower the established level;
- completion of the plan for the integration of curricula, the content of modules with professional standards and foreign partner universities; plan on the number of

published monographs; plan to meet license and accreditation performances, to involve applicants and students in competitions and contests at various levels, as well as in research activity.

The criteria for the performance of functional responsibilities by the head of the master's department to achieve the targets of educational, scientific and management activities are as follows:

- implementation of the master's department media plan in due time;
- completion of the plan on publication of the articles in academic periodicals indexed by foreign and Russian institutions (Web of Science, Scopus, Russian citation index), as well as in Russian peer-reviewed journals; maintaining the positive changes in the number of publications indexed in the information-analytical system of scientific citation Scopus and Web of Science;
- well-timed organization of the training of postgraduate students, doctoral students and candidates for thesis defense and promotion of growing number of graduate students in the department;
- enhancement of the media rating of the Master's Department staff (business programs);
- ensuring the achievement of research activity performances by the Master's Department (business programs);
- an increasing number of modules of disciplines for invariant use in master programs and securing an appropriate competitive advantage hereof. And finally, the head of the Business Department should:
- select university projects developed in the framework of the project activities, upgrade them, package, and launch;
- provide students with the opportunity to make money while studying;
- create a business environment at the university to attract talented young people and develop their skills in the field of start-up technologies through the launch of real projects.

3 Findings

Guided by the factors required for successful management in a third-generation university, we determined the level of qualification of current heads for management activities in new condition. For that we used the statistical methods of quality assurance:

- checklist;
- bar chart;
- scatter diagram;
- Pareto chart;
- stratification (interleaving);
- Ishikawa chart (cause-and-effect diagram);
- control chart.

The listed tools can be considered both as individual methods and a system of methods that provide comprehensive control of quality performances. They are an integral part of the control system of Total Quality Management [1].

Talking about seven statistical methods of quality assurance, we should emphasize that their main purpose is to control the occurring process and provide the participant hereof with information to adjust and improve it. The knowledge and practical application of seven quality assurance tools underlie one of the indispensable requirements of TQM –continuous self-monitoring [1].

However, in our study, we employed only two assessment methods: Ishikawa and Pareto chart, which brought 80% of the result.

The Pareto chart was applied to determine the critical factors affecting the occurrence of inconsistencies in management activities. This makes it possible to prioritize the actions required for problem-solving. In addition, the Pareto chart and the Pareto rule allow dividing major factors from minor ones.

Ishikawa cause-and-effect chart was constructed to assess the quality of managers' activity. We analyzed four groups of factors that have an impact on the quality of the manager's activity: the strategic goals of the department, the quality of the work performed, business and leadership skills.

As a result of the research, we revealed that the crucial factors in the activity of the bachelor's department head are knowledge of the regulatory documents, labor productivity, and basic leadership skills. For the head of the master's department – knowledge of the regulatory documents, the achievement of financial and economic performances, meeting the terms of financial and economic activities.

For the head of the business department, these are strong interpersonal skills, command of a foreign language, publicity.

In addition to the above-described methods, we applied the method of expert assessments.

According to the results of the expert assessment, we revealed the factors that are insufficiently developed by the heads of departments (average level). They require the measures to be improved.

Thus, the heads of the bachelor departments (economic and management programs), the master department (business programs) and business ones need to raise knowledge of regulatory documents, the head of the master department (business programs) should increase financial and economic performance, and the head of the business department should enhance command of English.

For weak points of a university department managers to be eliminated and the performance of their activity to be increased, it's necessary to take the following measures:

- to improve the knowledge of regulatory documents, deliver employees to a practice-oriented workshop of the National Accreditation Agency;
- to draw up a media plan with the purpose of department recognition to fully reach the performances of financial and economic activities, which will allow finding new consumers of the department's scientific research;

– to improve the level of a foreign language and arrange language courses for staff. Knowledge of a foreign language at a professional level will allow you to participate in international Start-up sites.

4 Conclusion

As a result of the study, we found out that third-generation universities include commercial and innovative activities on the implementation of the know-how created by university scholars. This trend defines the university as some independent unit that is able to pursue policy in competitive education, research, and development, which are used in the manufacturing industry, even in the conditions of insufficient public funding [4]. Today, the requirements for universities become higher: as providers of educational services, they should provide the maximum possible number of people with access to education; as research centers, they should face the global challenges of our time; as regional engines of innovation, they should apply research results, create jobs, and also tightly cooperate with the economy, culture, and society. Within the framework of this strategy, it is required to redetermine the position of a leader who should be a trans-professional: both an academic leader and a director general, as well as a specialist in other fields.

Following these requirements when appointing the head of a third-generation university will allow the latter to be competitive and take a leading position among universities that have proved themselves as a center for efficient know-how.

Universities seeking to be ranked "third-generation" need a new model of a leader who is able to be a director general for a dynamic and business institution, which meets the challenges of the 21st century, and an experienced research scholar who can develop the intellectual authority of the university.

References

1. GOST R ISO 9001-2008: Quality Management System. Requirements
2. Development Program and Transformation Program at the University Center for Innovation and Technological Development of the Samara Region. https://www.tltsu.ru/about_the_university/transformation/Programma_razvitiya_i_transformacii.pdf. Accessed 13 Jan 2018
3. Regulations on the Department of the Institute of Finance, Economics and Management of Togliatti State University (Resolution of the Academic Council of TSU No. 109 as of February 22, 2018. Togliatti: TSU, 2018.7 p.)
4. Wissema, J.G.: University of the Third Generation: University Management in Transitional Period. M.: Sberbank, 422 p. (2016)
5. Kaligin, N.A.: Principles of organizational management. Tolyatti: Printing house of Front Office on Information Systems of "AVTOVAZ" AO, 269 p. (2001)
6. Levitskii, Yu.V., Varaksa, A.M.: Self-development of education and science as a result of economic globalization. In: Levitskii, Yu.V., Varaksa, A.M. (eds.) Topical Issues of the Human and Natural Sciences, no. 11-2, pp. 105–108 (2014)
7. Neprokina, I.V., Shestova, E.I.: Change Control in Preschool Educational Institution. In: Neprokina, I.V., Shestova, E.I. (eds.) Symbol of Science, vol. 1, no. 1, pp. 212–215 (2017)

8. Farrington, D., Ismaili, D.: Finding the right person to lead a third-generation university: a new approach in the Republic of Macedonia. Procedia Soc. Behav. Sci. **15** 2083–2087 (2011). https://www.sciencedirect.com/science/article/pii/S1877042811006045. Accessed 17 Apr 2019
9. Third Generation University Plan Launched. Financial Tribune (2016). https://financialtribune.com/articles/people/43749/third-generation-university-plan-launched. Accessed 17 Apr 2019
10. Perkin, G.: The third revolution: Professional society in the international perspective. L. (1996)

Priority of Intellectual Resources for Development of Digital Entrepreneurship

Irina V. Mukhomorova[1(✉)], Aydarbek Giyazov[2], Gulzat K. Tashkulova[3], and Nurgul K. Atabekova[4]

[1] Russian State Social University, Moscow, Russia
mukhomorova@mail.ru
[2] Kyrgyz National University named after J. Balasagin, Bishkek, Kyrgyzstan
aziret-81@mail.ru
[3] Kyrgyz National University, Bishkek, Kyrgyzstan
gulzat.tashkuliva@mail.ru
[4] Kyrgyz State Law Academy, Bishkek, Kyrgyzstan
nur-aika@mail.ru

Abstract. Purpose: The purpose of the article is to develop an algorithm of development of a digital company on the basis of intellectual resources, which emphasizes their priority and reflects the logic of their usage.

Design/methodology/approach: The empirical part of the research is connected to studying the experience of modern Russia in using intellectual resources in the activities of digital entrepreneurship. For this, the structural analysis, analysis of variation, and correlation analysis are used. The sectorial structure of digital entrepreneurship in Russia (determined according to the criterion of using the Internet for purchases and sales) in 2018 is determined.

Findings: As a result of logical analysis of the modern Russian practice of using intellectual resources in entrepreneurship, the authors determine an "institutional trap" – innovations and automatization are performed in Russian entrepreneurship for obtaining state support. In a lot of cases, intellectual resources are not in high demand in entrepreneurship, and thus they are used formally for bringing the indicators of economic activities of a company in accordance with the state's requirements. That's why instead of acceleration, state stimulation of using intellectual resources leads to restraint of development of digital entrepreneurship in Russia. Overcoming this "institutional trap" envisages adoption of a new strategy of state stimulation of development of digital entrepreneurship, which has to be oriented not at setting of plans and financial support (direct stimulation) but at creation of market stimuli for usage of intellectual resources in the interests of increase of competitiveness of entrepreneurship (indirect stimulation).

Originality/value: An algorithm of development of a digital company on the basis of intellectual resources is developed – it reflects the perspectives of improving the management of its internal processes and allows using its possibilities, solving its current problems, and using the potential of its innovative development of and automatization.

Keywords: Intellectual resources · Development of digital entrepreneurship · Automatization · AI · Russia

E. G. Popkova and B. S. Sergi (Eds.): ISC 2019, LNNS 91, pp. 415–421, 2020.
https://doi.org/10.1007/978-3-030-32015-7_46

JEL Code: L26 · L53 · O15 · O31 · O32 · O33 · O38

1 Introduction

Digital entrepreneurship is a growth vector of the digital economy, in which the new (Fourth) technological mode ensures increase of effectiveness of economic activities, increase of society's quality of life, strengthening of positions in the world markets, and expansion of capabilities for the export activities. Specific features of digital entrepreneurship determine priority of intellectual resources for its development.

The peculiarity of connection between the digital company and external (market) environment is connected to usage of digital information and communication technologies (of which the Internet is the most popular one) for communications and deals with intermediaries and consumers. In particular, digital financial and tax accounting is conducted, reports are provided for the regulating bodies of public authorities, digital financial operations are performed, and digital document turnover with intermediaries is conducted.

The peculiarity of the internal organization of a digital company consists in the high level of automatization of its economic activities. For using digital information and communication technologies and the means of automatization it is necessary to have intellectual resources: digital managerial personnel that are capable of using these technologies, innovators that adapt new technologies to the company's economic practice and create innovations for formation and strengthening of technological competitive advantages, and AI that ensures flexibility and control over the automatization means.

For acceleration of transition to the digital economy, most governments, including the Russian government, perform stimulation of development of digital entrepreneurship. This support is aimed at overcoming the deficit – i.e., increase of accessibility of intellectual resources. Financial provision of implementation of the national program "Digital economy of the Russian Federation" in 2018 constituted RUB 7.726 billion (TAdviser 2019), and expenditures for R&D in Russia constituted 1.096% of GDP, or RUB 1,827.62 billion (World Bank 2019). Aggregate state expenditures for stimulation of development of digital entrepreneurship in 2018 constitute RUB 1,832.35 billion (1.1% of GDP).

According to the data of the National Research University "Higher School of Economics" (2019), the share of companies in Russia that use the Internet for purchases increased in the studied period of time from 16.7% to 18.1%, and the share of companies that use the Internet for sales reduced from 12.6% to 12.1%. The given statistical data show low effectiveness of spending of budget assets and a tendency for development of digital entrepreneurship in Russia with slow rate, which does not allow forecasting quick transition to the digital economy (until 2024, according to the National Program).

Thus, the problem of the search for alternative means of stimulating the development of digital entrepreneurship becomes important. The hypothesis that is offered here is that a barrier on the path of development of digital entrepreneurship in Russia is not deficit but non-target usage of intellectual resources. This article is to verify the offered hypothesis and aims at developing an algorithm of development of a digital company on the basis of intellectual resources, which emphasizes their priority and reflects the logic of their usage.

2 Materials and Method

Significance of intellectual resources for development of entrepreneurship is noted in the works Dereń et al. (2019), Morris and Snell (2011), Popkova et al. (2017a, b), Popkova and Parakhina (2019), Sibirskaya et al. (2019), and Vanchukhina et al. (2018). The conceptual and applied issues of functioning and development of digital entrepreneurship as a subject of the digital economy are discussed in the works Ansong and Boateng (2019), Bezrukova et al. (2017), Bogoviz et al. (2019), Pritvorova et al. (2018), Sergi et al. (2019), and Shah et al. (2019).

However, despite the high level of elaboration of separate components of the problem of this research, specifics of using intellectual resources in the activities of a digital company in the interests of its development is insufficiently studied in the existing sources of economic literature.

The empirical part of the research dwells on experience of modern Russia in usage of intellectual resources in the activities of digital entrepreneurship. The authors use structural analysis, analysis of variation, and correlation analysis.

The highest share of digital entrepreneurship in Russia is observed in the sphere of energy supply (the share of companies that use the Internet for sales constitutes 29.7%, for purchases – 10.1%), in the sphere of telecommunications (the share of companies that use the Internet for sales constitutes 29.1%, for purchases – 26.0%), and the tourist sphere – i.e., in the activities of hotels and catering companies (the share of companies that use the Internet for sales constitutes 27.6%, for purchases – 20.1%).

The share of digital entrepreneurship in Russia in the total structure of entrepreneurship constitutes at least 30%. Variation of the share of companies that use the Internet for purchases constitutes 31.26%, for sales – 55.17%. This show heterogeneity of digital entrepreneurship in Russia.

The highest activity of using intellectual resources is observed in the sphere of telecommunications (the share of companies that implement innovations constitutes 12,2%, the share of companies that use program means in management – 72.1%), the processing industry (the share of companies that implement innovations constitutes 13.3%, the share of companies that use program means in management – 66.4%), and the sphere of professional, scientific, and technical activities (the share of companies that implement innovations constitutes 30.7%, the share of companies that use program means in management – 52.0%).

The highest profitability of sold products is observed in the sphere of minerals extraction (25.9%), telecommunications (18.4%), and agriculture (17.3%). In the sphere of the processing industry, profitability of sold products is medium – 11.5%, in the sphere of professional, scientific, and technical activities – 12.3%. Correlation of profitability of sold products in the Russian entrepreneurship in 2018 with implementation of innovations constitutes 0.2365, with automatization – 0.0988.

Variation of the share of companies that use program means for solving organizational, managerial, and economic tasks constitutes 20.38%. Variation of the share of companies that implement innovations constitutes 94.43%. Variation of profitability of sold products constitutes 52.98%. Therefore, intellectual resources are used in different ways in different spheres of the modern Russian economy and do not have statistically significant influence on profitability of sold products – i.e., they do not ensure the development of digital entrepreneurship.

3 Results

As a result of logical analysis of the modern Russian practice of using the intellectual resources in entrepreneurship we determined an "institutional trap" – innovations and automatization are implemented in the Russian entrepreneurship in the interests of obtaining state support. In a lot of cases, intellectual resources are not popular in entrepreneurship, and thus they are used formally for bringing the indicators of a company's economic activities in accordance with the state's requirements.

That's why instead of acceleration, state stimulation of using intellectual resources leads to restraint of development of digital entreprencurship in Russia. Overcoming this "institutional trap" envisages adoption of a new strategy of state stimulation of development of digital entrepreneurship, which has to be oriented not at setting of plans and financial support (direct stimulation), but at creation of market stimuli for usage of intellectual resources in the interests of increase of competitiveness of entrepreneurship (indirect stimulation).

As the object of this research is not state stimulation but digital entrepreneurship here we focus on the perspectives of improvement of the modern Russian practice of using the intellectual resources in entrepreneurship. For this an algorithm of development of a digital company on the basis of intellectual resources is developed (Fig. 1).

According to Fig. 1, the first stage of the developed algorithm is automatic management (with high level of automatization), monitoring, and control over the production and distribution processes of a company by AI (accessible in the current technological mode). At the second stage, AI sends report to the subjects of company management – top manager (regarding new opportunities) and middle managers (regarding current problems), and to innovators.

At the third stage, innovators conduct R&D and at the fourth stage inform the subjects of management on accessible innovations. At the fifth stage, middle managers perform "brain storm" and select tactical problems, and the top manager selects strategic decisions. At the sixth stage, they send application for analytical support from AI via computer programming. At the seventh stage, AI performs comparative analysis of alternative decisions; at the eighth stage, it provides intellectual support for decision making.

At the ninth stage, subjects of management make decisions; at the tenth stage, they pass orders to AI on execution of the decisions and specify the priorities of automatized management and the following analytics. As a result, highly-effective strategic and tactical management of a digital company is achieved – it allows using its capabilities, solving the current problems, and using the potential of its innovative development and automatization.

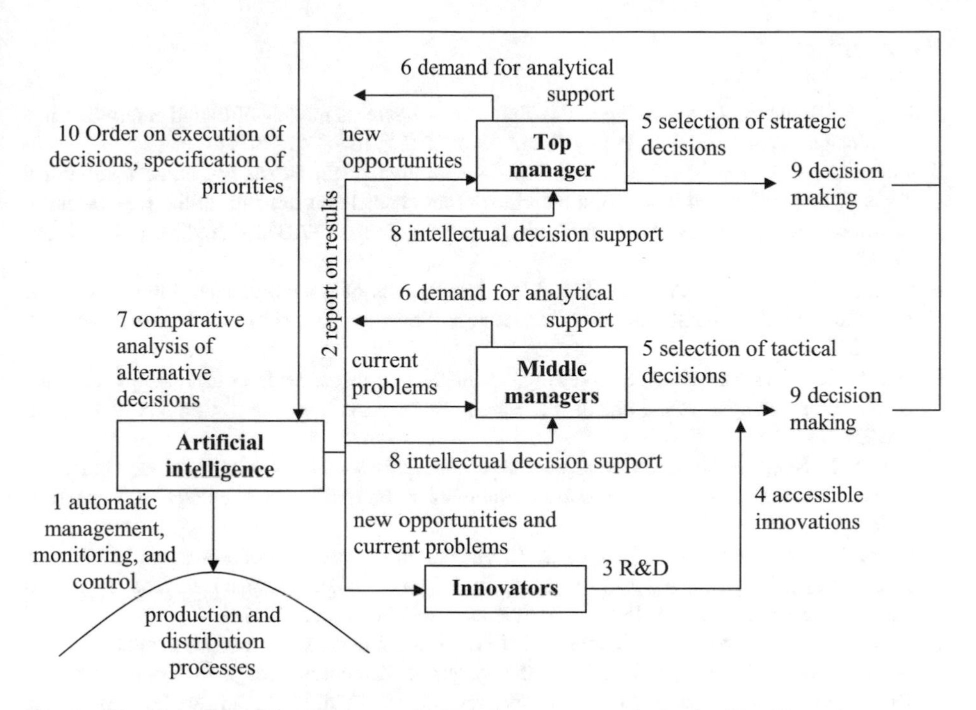

Fig. 1. The algorithm of development of a digital company on the basis of intellectual resources. Source: compiled by the authors.

4 Conclusion

The performed research showed that when using digital technologies entrepreneurship of modern Russia prefer external communications, as it allows corresponding to the state norms and apply for state financial (including tax) support. This confirms the offered hypothesis on non-target usage of intellectual resources in digital entrepreneurship of Russia. For solving this problem, an algorithm of development of a digital company on the basis of intellectual resources is developed – it reflects the perspectives of improving management of its internal processes.

Market stimuli for practical application of this algorithm in modern Russia are very weak. That's why the state's efforts on stimulating the development of digital entrepreneurship on the basis of usage of intellectual resources should be aimed at supporting such competitive environment in sectorial markets that would ensure stimuli for innovative development and automatization of the Russian companies but prevent dumping and other means of monopolization. The strategy of anti-monopoly regulation of sectorial markets in the interests of using the potential of usage of intellectual resources for development of digital entrepreneurship should be developed in further studies.

References

Ansong, E., Boateng, R.: Surviving in the digital era – business models of digital enterprises in a developing economy. Digit. Policy Regul. Gov. **21**(2), 164–178 (2019)

Bezrukova, T.L., Popova, E.V., Korda, N.I., Kuznetsova, T.E., Bezrukov, B.A.: Institutional traps of innovative and investment activities as an obstacle on the path to the well-balanced development of regions. In: Contributions to Economics, (9783319606958), pp. 235–240 (2017)

Bogoviz, A.V., Lobova, S.V., Ragulina, J.V.: Perspectives of growth of labor efficiency in the conditions of the digital economy. In: Lecture Notes in Networks and Systems, vol. 57, pp. 1208–1215 (2019)

Dereń, A., Seretna-Sałamaj, D., Skonieczny, J.: Intellectual resources in polish enterprises in the SME sector - analysis of the obtained research results. Adv. Intell. Syst. Comput. **854**, 49–59 (2019)

Morris, S.S., Snell, S.A.: Intellectual capital configurations and organizational capability: an empirical examination of human resource subunits in the multinational enterprise. J. Int. Bus. Stud. **42**(6), 805–827 (2011)

Popkova, E.G., Bogoviz, A.V., Lobova, S.V.: Vacuum in the structure of human capital: A view from the position of the theory of vacuum. In: Human Capital: Perspectives, Challenges and Future Directions, pp. 163–181. Nova Science Publishers, Inc. (2017a)

Popkova, E.G., Morozova, I.A., Litvinova, T.N.: New challenges for human capital from the positions of its infrastructural role in the system of entrepreneurship. In: Human Capital: Perspectives, Challenges and Future Directions, pp. 257–275. Nova Science Publishers, Inc. (2017b)

Popkova, E.G., Parakhina, V.N.: Managing the global financial system on the basis of artificial intelligence: possibilities and limitations. In: Lecture Notes in Networks and Systems, vol. 57, pp. 939–946 (2019)

Pritvorova, T., Tasbulatova, B., Petrenko, E.: Possibilities of blitz-psychograms as a tool for human resource management in the supporting system of hardiness of company. Entrepr. Sustain. Issues **6**(2), 840–853 (2018). https://doi.org/10.9770/jesi.2018.6.2(25)

Sergi, B.S., Popkova, E.G., Bogoviz, A.V., Ragulina J.V.: Entrepreneurship and economic growth: the experience of developed and developing countries. In: Entrepreneurship and Development in the 21st Century, pp. 3–32. Emerald Publishing Limited (2019)

Shah, S., Shah, B., Amin, A., Al-Obeidat, F., Chow, F., Moreira, F.J.L., Anwar, S.: Compromised user credentials detection in a digital enterprise using behavioral analytics. Future Gen. Comput. Syst. **93**, 407–417 (2019)

Sibirskaya, E., Popkova, E., Oveshnikova, L., Tarasova, I.: Remote education vs traditional education based on effectiveness at the micro level and its connection to the level of development of macro-economic systems. Int. J. Educ. Manag. **33**(3), 533–543 (2019). https://doi.org/10.1108/IJEM-08-2018-0248

TAdviser.: Financial provision of implementation of the national program "Digital economy of the Russian Federation" (2019). http://www.tadviser.ru/index.php/Статья:Финансирование_программы_Цифровая_экономика. Accessed 25 May 2019

Vanchukhina, L.I., Leybert, T.B., Khalikova, E.A., Khalmetov, A.R.: New approaches to formation of innovational human capital as an element of institutional environment. Espacios **39**, 22–32 (2018)

World Bank (2019). https://data.worldbank.org/indicator/GB.XPD.RSDV.GD.ZS. Accessed 25 May 2019

National Research University "Higher School of Economics" (2019). Digital economy: 2019. https://www.hse.ru/primarydata/ice2019kr. Accessed 24 May 2019
Federal State Statistics Service of the Russian Federation (2019). Russia in numbers: statistical collection. http://www.gks.ru/bgd/regl/b18_11/Main.htm. Accessed 25 May 2019

The Role of Managerial Competence of an Executive in Improving the Quality of Pre-school Educational Organization

Natalia V. Belinova(✉), Irina B. Bicheva, Larisa V. Krasilnikova, Tatyana G. Khanova, and Anna V. Hizhnaya

Minin Nizhny Novgorod State Pedagogical University, Nizhny Novgorod, Russia
belinova@mail.ru, irinabicheva@bk.ru, larvladkr@mail.ru, tanyaha10@mail.ru, xannann@yandex.ru

Abstract. Modern trends in the development of preschool education are conditioned by requirements of the Federal State Educational Standard and are aimed at the processes of its modernization and quality improvement. Authors emphasize that the quality of preschool education is studied from the point of view of actual conformity of conditions, organization, methods of evaluating the properties and results of education, an educational process to necessary requirements of the standard. The dynamism of quality characteristics of preschool education implies construction and development of quality management system in conditions of preschool educational organization in order to achieve optimal balance of internal and external conditions of its activity, harmonization of relations with teachers and social environment, effective implementation of norms of professional ethics, orientation for the future development and improvement. Development and implementation of such a system place high demands on managerial competence of an executive, which is considered by authors as a complex of value-motivational, analytical and design, organizational and executive, communication and activity, reflexive and assessment competencies. Components and indicators of achievement of the proposed competencies are allocated. Article presents the results of managerial competence formation of executives of preschool educational organizations of Nizhny Novgorod, which revealed a significant need for managers to increase their level of constituent competencies. Prospect of further research assumes the design of the quality management system of preschool educational organization on the basis of the real level of formation of managerial competencies and individual trajectories of their improvement.

Keywords: Preschool education · Quality of preschool education · Executive of preschool educational organization · Quality management system · Managerial competence · Managerial management competence

JEL Code: I2—M5

E. G. Popkova and B. S. Sergi (Eds.): ISC 2019, LNNS 91, pp. 422–429, 2020.
https://doi.org/10.1007/978-3-030-32015-7_47

1 Introduction

Modern conditions for development of preschool education, reflecting the correspondence of real results of activities of a preschool educational organization to requirements of the Federal State Educational Standard as a normative document. Initiate development of an effective system of quality of preschool education in conditions of its modernization (Boguslavskaya 2017; Gorina 2014; Edakova 2013; Safonova 2015).

This article solves the following task: determine the content of managerial competencies of the executive of preschool educational organization, allowing to carry out professional activity and take into account the fundamental principles of quality management of preschool educational organization.

2 Methodology

The research is carried out on the basis of methods of analysis and generalization, which allowed offering a set of managerial competencies of preschool educational organization executive, ensuring construction of a quality management system. Diagnostics of the level of managerial competence of managers was carried out in preschool educational organizations of Nizhny Novgorod. About 120 leaders participated in the study.

3 Results

In scientific and pedagogical research, concept of "quality of education" is studied from the point of view of the actual conformity of conditions, organization, methods of evaluating the properties and results of education, educational process to necessary requirements of the standard.

Study of publications on problem allows distinguishing the most significant directions of scientific research, revealing various meanings of the quality of preschool education: methodological (I.B. Edakova, M.V. Krulecht, S. M. Markova, M.M. Potashnik, N.V. Fedina, etc.), management (K.Yu. Belaya, L.V. Gorina, Z.A. Klimentyeva, O.L. Knyazeva, L.G. Loginova, E.G. Yudina, etc.), criteria-evaluative (T.N. Boguslavskaya, N.A. Veraksa, N. A. Vinogradova, N.V. Miklyaeva, T.A. Nikitina, O.A. Safonova, O.A. Skorolupova, L.I. Fishman, etc.).

A special role is acquired by development of quality management system in order to achieve the optimal balance of internal and external conditions of preschool educational organization activity, harmonization of relations with teachers and socio-professional environment. (Kaznacheeva et al. 2016) (Kuznetsov, et al. 2018) Development and implementation of such a system reinforces the importance of targeted, integrated and coordinated interaction between management and managed subsystems, which puts high demands on managerial competence of an executive of preschool educational organization.

Managerial competence is determined by experience, motives, individual abilities of managers (Kuznetsov et al. 2017) continuously self-education and self-improvement in the field of management of preschool educational organization, manifests itself in the creative implementation of management function (Klimenteva 2013). In accordance with this provision, the most important and relevant in development of quality management system of preschool educational organization are personal and professional characteristics of an executive system, normative, constructive, activity attitude to management (Kaznacheeva et al. 2017). In this regard, managerial competence of the head of a preschool educational organization can be considered as a complex of the following competencies and forming them:

- Value-motivational (purposefulness, motivation, awareness, conviction, strategic vision, etc.) (Gutsu et al. 2016);
- Analytical (planning and correction of strategic and tactical ways of development of preschool educational organization, management decisions based on analysis, etc.);
- Organizational and executive (responsibility, initiative, strategic vision, effectiveness in achieving results, adherence to standards, etc.);
- Communication (influence, cooperation, interaction, respectful and friendly attitude, etc.);
- Reflexive-evaluative (objectivity, adaptability, openness to changes, orientation to achieve a qualitative result, etc.).

Evaluation of the formation of these competencies involves a definition of indicators of their achievement.

Value-Motivational Competences:

- Clearly represents the image of desired future preschool educational organization, priorities (mission), system of values, traditions, culture;
- Sets specific, achievable goals and objectives of innovative development necessary to increase the level of provision of educational services, including additional and variable;
- Applies various ways of motivation to achieve the planned result, taking into account the professional interests of teachers, needs of society and strategy of development of preschool educational organization;
- Owns methods of persuasion, argumentation of position, conducting polemics; etc.

Analytical competences:

- Owns diagnostic methods of research, including monitoring of parental activity in the development of a complex of educational services and program of development of preschool educational organization, selection of content and organizational forms of intersubject interaction (Smirnova and Krasikova 2018; Gutsu and Kochetova 2017);
- Timely notices and analyzes external and internal changes in resources (information, administrative, social, motivational, methodical, material-technical, etc.) (Kuznetsov et al. 2018);
- Interprets analytical results, establishes their compliance with regulatory requirements for design of management tasks and decisions;

- Determines managerial impact, objectively performs setting and distribution of tasks, delegation of authority and responsibility; etc.

Organizational and executive competencies:

- Acts in accordance with established regulatory requirements and rules, norms of professional ethics (Belinova et al. 2017); and pedagogical culture (Lyashenko and Mineeva 2018);
- Carries out the exchange of information between teachers, parents and social environment on key indicators and actual values of activity of preschool educational organization, with Internet opportunities;
- Creates an information database to design possible improvements based on professional requests and needs in a specific and prospective situation (Gruzdeva et al. 2018);
- Retains and demonstrates professional productivity in conditions of uncertainty, constant load, resource constraints and changes;
- Expands scientific, methodological and research capabilities of the preschool educational organization through creation of an innovative educational cluster with institutions of higher education; Science, Culture, Business Community, Board of Trustees (Bogorodskaya et al. 2018).

Communication and activity competence:

- Uses methods of program and target management, technologies of cooperation for professional support, expansion of participation of teachers, parental community in the activities of preschool educational organization, and building an effective team of like-minded people;
- Controls manifestation of their emotions in conflict situations;
- Owns ways of resolving professional conflicts; etc.
- Reflexive and evaluation competences:
- Carries out self-knowledge as a continuous study and understanding of its individual characteristics, properties, ways of activity in various situations, including moral choice (Bicheva and Filatova 2018);
- Evaluates real educational situation in the socio-professional space to compare results of their work and competitive educational organizations, identify their own competitive advantages, best and effective practices, and adaptation of good practices;
- Objectively assesses the available level of development of professional competence of each teacher, educational process; etc.

We conducted a study to determine the level of managerial competence of executives of pre-school educational organizations of Nizhny Novgorod. 120 executives participated in the study. They were asked to assess the degree of manifestation of proposed managerial competencies from 1 to 10 points. Based on results of the study, levels of their formation were determined: high, medium and low (Fig. 1).

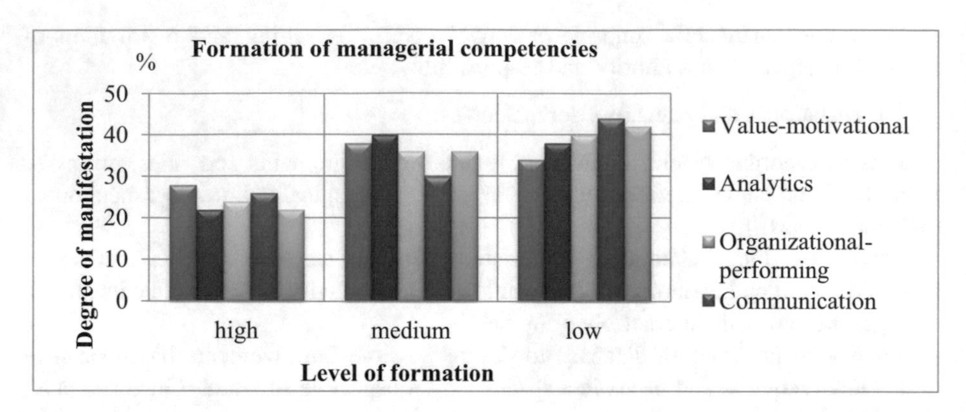

Fig. 1. Level of formation of managerial competencies of executives of preschool educational organizations

High level (8–10 points) is characterized by an innovative orientation in manifestation of managerial competencies. Managerial activity is characterized by efficiency and pronounced individual style. Average level (5–7 points) is associated with manifestation of the majority of managerial competencies, management activity is carried out quite steadily. Low level (2–4 points) indicates a partial manifestation of managerial competencies, managerial activity is built as standard, formally, there is a lack of flexibility and ability in performing professional functions.

Summarizing obtained data, it follows, the level of formation of managerial competence of executives of preschool educational organizations is characterized, first of all, as low (39,6%) and average (36%) for all components. 24.4% of managers have a high level (Fig. 2).

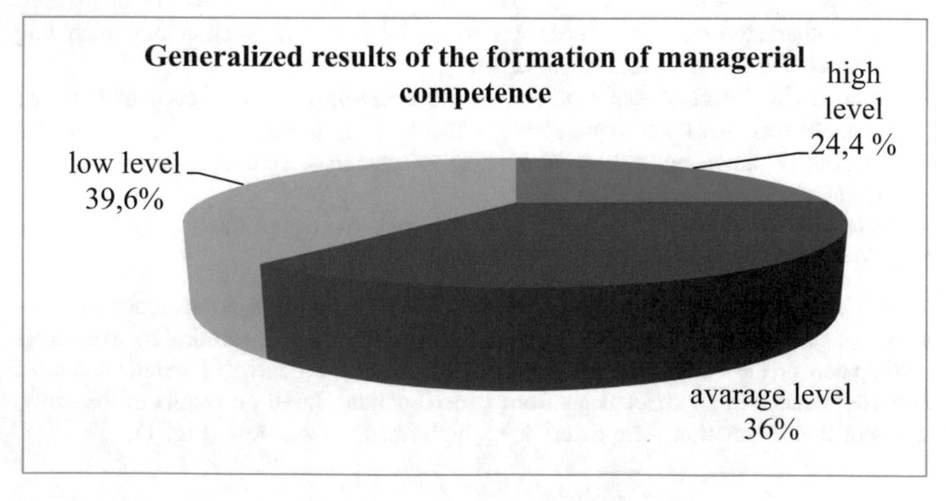

Fig. 2. Level of formation of managerial competence among executives of preschool educational organizations

Analyzing results highlighted the difficulties, which indicated the majority of managers.

Understanding specifics of professional management associated with formation of individual managerial style and impact on the collective in the solution of promising problems of innovative development, managers find it difficult to objectively analyze a professional situation, set professional goals, they show a rather low level of independence. Despite general understanding of the role of information of preschool education, managers do not have sufficient knowledge of information tools to develop projects/programs in the field of implementation of professional activities with methods of design. In situations of management practice the problems of professional interaction are highlighted: low sensitivity in perception of the professional environment, unawareness of own managerial position in the team, uncertainty in situations of communication, etc. Management activity is considered as strict implementation of normative and instructional documents. According to managers, this presupposes the priority of authoritarian methods of leadership without taking into account individual characteristics of teachers, leads to categorical judgments, frequent intolerance, especially in conflict situations, which prevents effective interaction with professional environment. Quite rarely they show such qualities as logic and consistency in the process of explanation and statement of a problem, do not always know how to reduce tension in the team, and objectively assess professional activities.

Thus, results of the study revealed a significant need for executives of preschool educational organizations increase their level of development of managerial competence.

4 Conclusions and Recommendations

Achieving the necessary quality of pre-school educational organization is a significant condition for increasing its competitiveness and social status in ensuring the effectiveness of provided educational services. An important role in this process is assigned to the quality management system as a factor of modernization of the management system of preschool educational organization based on the principles of adequacy, objectivity, reliability, reliability of diagnostic and evaluation procedures.

High level of development of managerial competence allows carrying out a qualitative assessment of efficiency of its activity to achieve necessary the degree of guaranteed educational services that meet the government standards, expectations, and demands of consumers. This implies taking into account the diverse perspectives of all participants: children (from the perspective of interest in learning), parents (from the perspective of learning efficiency), educators (from the perspective of evaluation of work, individual achievements and successful results of professional activity, state of mental and physical health of pupils), executive (from the point of view of efficiency of professional growth, individual progress of pupils, high assessment from parents and educational authorities).

Prospect of further research assumes design of the quality management system of preschool educational organization on the basis of the real level of formation of managerial competencies and individual trajectories of their improvement.

References

Bicheva, I.B., Filatova, O.M.: Prospects of professional training of teachers: axiological approach. Bulletin of Minin University, vol. 6, no. 2(23), p. 3 (2018). https://doi.org/10.26795/2307-1281-2018-6-2-3

Boguslavskaya T.N.: Assessment of the quality of preschool education in the conditions of its standardization. Problems of Modern Education, no. 4, pp. 142–149 (2017)

Gorina, L.V.: Professional growth of teachers as a factor of improving the quality of preschool education. Izvestiya of Saratov University. New series. Acmeology of Education. Psychology of development, vol. 3, no. 4, pp. 372–375 (2014)

Gutsu, E.G., Kochetova, E.V.: Management of development of educational organization on the basis of self-assessment results. Nizhny Novgorod education, no. 2, pp. 30–35 (2017)

Edakova, I.B.: On the question of assessment of quality of preschool education in conditions of introduction of FSES. Primary school plus before and after, no. 12, pp. 17–22 (2013)

Kaznacheeva, S.N., Bicheva, I.B., Yudakova, O.V.: Improving the efficiency of personnel management: trends and features of management. Modern science-intensive technologies. Regional application, no. 2, pp. 44–50 (2016)

Klimenteva, Z.A.: On development of managerial competences of future teachers and executive of preschool educational organizations. Education and self-development, no. 4(38), pp. 134–139 (2013)

Safonova, O.A.: Technology and effects of independent assessment of quality of preschool education. Management of preschool educational institution, no. 4, p. 16 (2015)

Smirnova, Zh.V., Krasikova, O.G.: Modern means and technologies of evaluation of learning results. Bulletin of Mininsky University, vol. 6, no. 3, p. 9 (2018). https://doi.org/10.26795/2307-1281-2018-6-3-9

Belinova, N.V., Bicheva, I.B., Kolesova, O.V., Khanova, T.G., Khizhnaya, A.V.: Features of professional ethics formation of the future. Espacios, vol. 38, no. 25, p. 9 (2017)

Bogorodskaya, O.V., Golubeva, O.V., Gruzdeva, M.L., Tolsteneva, A.A., Smirnova, Z.V.: Experience of approbation and introduction of the model of management of students' Independent work in the university. In: Advances in Intelligent Systems and Computing, vol. 622, pp. 387–397 (2018). https://doi.org/10.1007/978-3-319-75383-6_50

Gutsu, E.G., Demeneva, N.N., Kochetova, E.V., Mayasova, T.V., Belinova, N.V.: Studying motivational-axiological component of professional competence of a college teacher. International Journal of Environmental and Science Education, vol. 11, no. 18, pp. 12650–12657 (2016)

Gruzdeva, M.L., Prokhorova, O.N., Chanchina, A.V., Chelnokova, E.A., Khanzhina, E.V.: Post-graduate information support for graduates of educational universities. In: Advances in Intelligent Systems and Computing, vol. 622, pp. 143–151 (2018). https://doi.org/10.1007/978-3-319-75383-6_19

Kaznacheeva, S.N., Chelnokova, E.A., Bicheva, I.B., Smirnova, Z.V., Lazutina, A.L. Worldwide management problems, Man in India, 97(15), pp. 191–199 (2017)

Kuznetsov, V.P., Garina, E.P., Andryashina, N.S., Romanovskaya, E.V.: Models of modern information economy conceptual contradictions and practical examples, 361 p. Emerald publishing limited (2018)

Kuznetsov, V., Kornilov, D., Kolmykova, T., Garina, E., Garin, A.: A creative model of modern company management on the basis of semantic technologies. In: Communications in Computer and Information Science (2017)

Lyashenko, M.S., Mineeva, O.A.: Methodological approaches to teaching culture understanding in teaching research. Perspektivy Nauki i Obrazovania, 5(35), pp. 10–17 (2018). https://doi.org/10.32744/pse.2018.5.1

Markova, S.M., Sedykh, E.P., Tsyplakova, S.A., Polunin, V.Y.: Perspective trends of development of professional pedagogics as a science. In: Advances in Intelligent Systems and Computing, vol. 622, pp. 129–135 (2018). https://doi.org/10.1007/978-3-319-75383-6_17

Implementation of the Division Model of Pedagogical Labor in the Teacher Training System of a New Type

Tatyana K. Belyaeva(✉), Evgeniy E. Egorov, Tatyana K. Potapova, Tatyana L. Shabanova, and Mikhail Y. Shlyakhov

Minin Nizhny Novgorod State Pedagogical University, Nizhny Novgorod, Russia
btk66@yandex.ru, eeegorov@mail.ru, konctantinka@gmail.com, shabanovatl@mail.ru, mik-shlyakhov@yandex.ru

Abstract. The article analyzes the intermediate results of the implementation of the strategic initiative "The teacher of the future" developed by the project team of teachers within the framework of the Development Strategy of Minin University for 2014–2023. The aim of the study is to obtain a new quality of teacher training in the context of dynamic socio-cultural transformations in the country and world due to the differentiation of pedagogical work and emergence of new pedagogical positions. The novelty of the research is the author's model for the division of pedagogical work in school and system for training teachers in a pedagogical university built on the basis of this model. The methodological basis for the design of a new training system was the idea of meta-competence formation. Experience of teaching students with the use of the latest educational technologies on the basis of educational institutions is analyzed. Difficulties and problems that have arisen in the course of implementation of the project which need to be studied and solved are formulated.

Keywords: Division of pedagogical work · New pedagogical professions · Subject expert · Moderator · Tutor · Corrector · Meta-competence

JEL Code: I21 · I23

1 Introduction

Dynamic changes in society, connected with the processes of globalization, informatization, cause the systemic renewal of education. Modernization processes also define new requirements for the competence of a teacher. Today, the effectiveness of a teacher depends not only on knowledge in the subject area but also on the presence of new professional abilities to moderate the educational process, build dialogue communication, mediation of conflict and stress situations (Loughran 2006).

The transformation of education requires not only a qualitative change of teachers who have fallen into new conditions but also creates a demand for new pedagogical professions. Therefore, the experience of introducing new pedagogical positions in schools is used in the world educational practice. Thematic editor of online lessons,

E. G. Popkova and B. S. Sergi (Eds.): ISC 2019, LNNS 91, pp. 430–438, 2020.
https://doi.org/10.1007/978-3-030-32015-7_48

content director of the educational platform, developer of educational interfaces—this is not a complete list of vacancies posted, for example, in the jobs section on the American edtech-edition EdSurge (Hanushek and Woessmann 2008).

Minin University (NGPU named after K. Minin) also strives to implement its strategy of training new specialists in demand on the labor market and adequate requirements of modern conditions. To this end, an innovative educational project "Teacher of the Future" is being implemented, aimed at "modeling the personality of a future teacher, involving building a system of promising over-subject competences in preparing bachelors and masters by changing the structure and content of basic educational programs and developing individual educational trajectories" of students (Fedorov 2017).

2 Methodology

The methodological basis of the project is ideas of cultural, historical and activity approaches of Russian psychologists and teachers L.S. Vygotsky, A.N. Leontiev. Theory of developing and personal - oriented training, (A.G. Asmolov, V.V. Davydov, D.B. Elkonin,), humanistic pedagogy and psychology (G.Olport, K.Rogers, P.Lengrand, etc.), LLL-model of teaching, the concept of pedagogical abilities of V.A. Krutetsky, G.N. Malkovskaya, V.A. Slastenina, et al. *The theoretical basis* of experimental activity is the conceptual provisions of the draft model of the National System of Teacher Growth, and Professional Standard of the Teacher.

3 Results

The main *objective of the project is*: obtaining a new quality of teacher training, in the context of dynamic socio-cultural transformations in the country and in the world in connection with the differentiation of pedagogical work and emergence of new pedagogical positions. Teacher training takes place on the basis of the model of division of pedagogical labor developed by a group of teachers of the Minin University (Egorov 2016). Figure 1.

This picture requires some explanation. "External system-forming factors" are understood as:

1. Social order and world trends in education development.
2. Federal state standard of higher education.
3. The professional standard of the teacher and the requirements of the employer.
4. Key changes in pedagogical education. Open pedagogical education. Formation of the future image of a teacher.

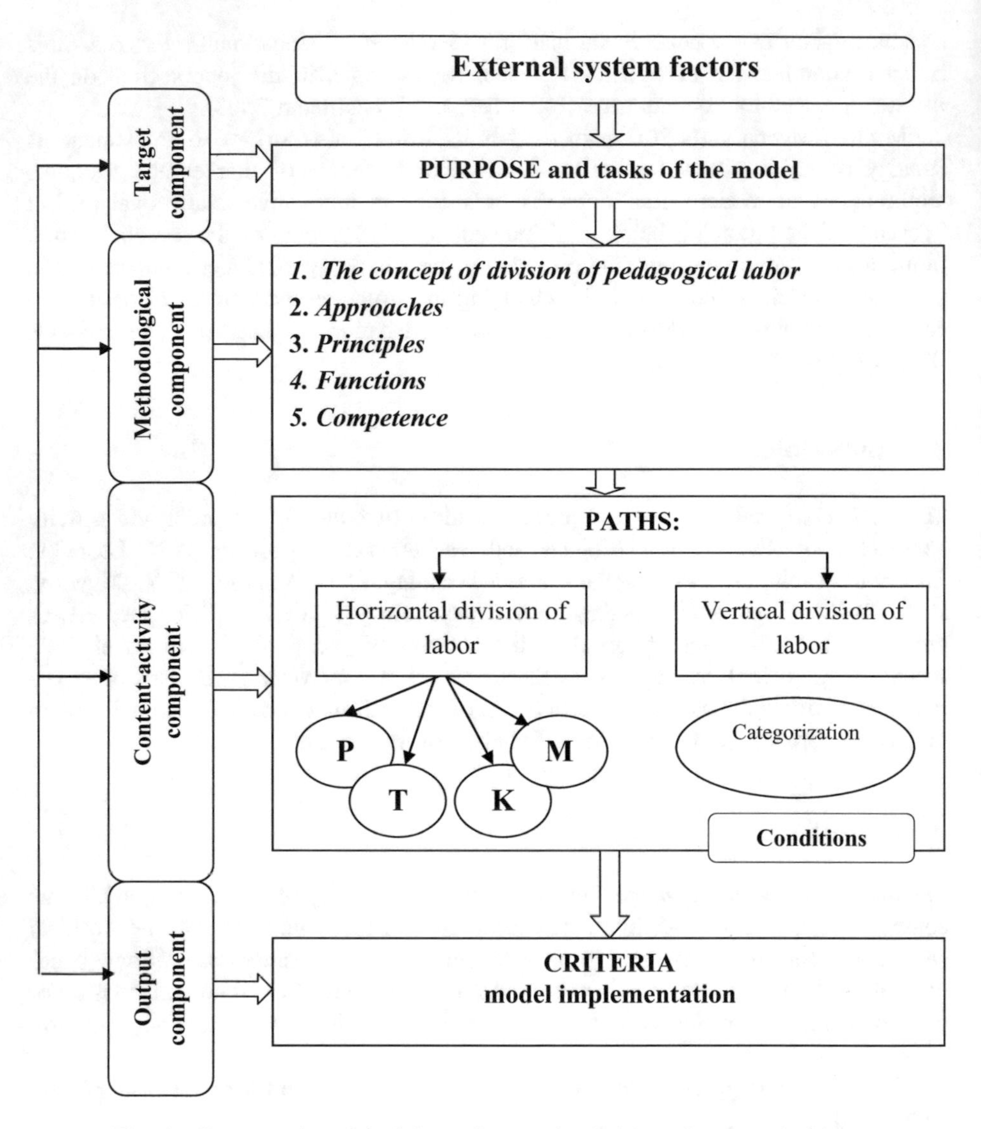

Fig. 1. Conceptual model of the project on the division of pedagogical labor

Model purpose: substantiation of theoretical and practical necessity and the possibility of creating a system to increase the effectiveness of educational activities based on the division pedagogical work.

Criteria for effective implementation of the model are:

1. Formation of a scientific-methodological style of thinking, growth of professionalism and creative potential of the teacher.
2. Change of structure and content of organization of educational process.
3. Formation of comfortable educational environment.

4. Valuable attitude to pedagogical activity.
5. Improving the effectiveness of pedagogical activity.

The key idea of creating the model is the formation of four new additional directions of activity: *subject teacher*, endowed with different functions of the expert, able to navigate in the subject area, *teacher tutor*, performing the formation of individual educational trajectories of students by studying their motives, interests, inclinations; *teacher-moderator*, forming group educational trajectories of students, projecting and managing educational processes that organize distance learning and professional development of teachers (Kuznetsov 2018), and *an educator* who carries out activities to identify and improve teachers' qualifications (Kuznetsov 2018) elimination of deviations in the educational process, its harmonization through diagnostics and comprehensive analysis of educational results.

At first glance, there is nothing new in these professions. But we draw attention to the fact that in the Russian education system nobody has managed to bring them into a single harmonious working system (Egorov 2016; Ilaltdinova et al. 2015). The system of division of pedagogical labor proposed in the project "The teacher of the future" offers a new approach to understanding the content of teacher training.

The subject teacher remains the central figure in pedagogical professions. In the planned system, there is such a provision that all students must go through this profession. This passage can be stepped through, the position of assistant, internship, passing the professional exam and finally obtaining the status of a Teacher. In the future, the professional and career trajectory of a particular teacher can be different, taking into account all professions. Ensuring the multiplicity of professional trajectories in the future at the stage of teacher training is possible through the formation of meta-competence.

We proceeded from the principle that each new pedagogical profession is built around one leading meta-competence. Thus, there were four main meta-competences, and two integrative competences, which are fundamentally important in any accentuation of the future teacher's training.

Scientific and methodical meta-competence, understood as readiness to carry out educational activities on the basis of modern scientific research in the subject area, is inherent mainly to the teacher - to the subject expert. *Managerial* meta-competence, defined as the ability to effectively manage educational processes, projects, programs should be implemented as much as possible in the activities of the moderator. In the work of the tutor, the key importance is *diagnostic* meta-competence, formulated as the ability to diagnose elements of the educational system (participants, processes, results) for ensuring and improving the quality of the pedagogical process. *Psychological and pedagogical* meta-competence defined as the ability to carry out psychological and pedagogical support, correction and harmonization of the educational process on the part of all interested participants, is implemented in the activities of the corrector. Among the two important integrative meta-competences, we consider information technology and communication (Kuznetsov et al. 2017). The change in the system of meta-competences and the transition to a new model of preparing the teacher of the future led to change in the basic educational program and relevant curricula (Kuzminov et al. 2011; Potapova 2016). The modular curriculum planning approach developed by us makes it possible to preserve the subject component of each profile and direction of training. At the same time, student has the opportunity to get additional specialization in the field of fast-changing pedagogical work.

A single module "Introduction to pedagogical activity" was introduced for all students, forming the basic psychological and pedagogical competencies for any teacher. The development of the introductory module allowed students to understand the basics of pedagogical work and to form a desire to master special competencies. This allowed them to become a narrower specialist in the field of division pedagogical work.

After mastering module students were given the opportunity to study profile modules containing disciplines of a narrower orientation of one of the future professions: subject expert, moderator, tutor, and corrector. Students were trained according to the modules of specialization in the system of division of pedagogical labor. Students mastered the basics of knowledge on the formation of individual and group trajectories, on pedagogical diagnostics and monitoring, management and marketing of educational organizations, modern IT—technologies in education. Another module "Personal potential of the teacher" included disciplines that were aimed at the formation of integrative competencies, previously communicative, reflexive, characterizing it as a highly organized, communicative, creative leader of society. Further, students studied the discipline of the division of pedagogical labor of choice, for example: "Purpose setting and choice in the professional environment", "Subject expert activity of the teacher", "Moderator activity in the educational process", "Activity of the tutor in education", "Pedagogical" corrective and developing activities".

Students developed research skills and analytical thinking that form the basis of scientific and methodological meta-competence (Taylor 1985), (Shabanova 2015). Psychological and pedagogical meta-competence of students was formed in the process of designing and implementing corrective work with students. During class time, they mastered the skills of developing correctional and developmental programs aimed at improving intellectual processes, volitional qualities, and emotional self-regulation among students (Dubrovina and Lubovsky 2017; Shabanova and Novik 2015).

Management meta-competence was formed in the process of development and implementation of research and creative projects by students.

Formation of meta-competence is impossible without the latest educational technologies. The main ones in the training process were the following: case studies, visualization, trainings, discussions, simulation and business games, video-fixing of educational processes, preparation of presentations, sparring partnership, research and creative design. The use of such technologies contributed to the stimulation of educational motivation of students, the formation of creative and cognitive activity through the organization of quasi-professional activities in learning conditions.

During the implementation of the project, we carried out psychological and pedagogical diagnostics in order to identify deficits and resources of students to assess the effectiveness of activities and plan for further strategy. We used a psycho-diagnostic techniques package: Spilberger's self-assessment scale of anxiety, developed by us questionnaire "A look at the teacher of the future" and the methodology of personal SWOT analysis of educational and professional activities.

We studied the degree of emotional well-being of students, level and content of professional motivation, indicators of self-assessment of personal qualities and resources (Fig. 2).

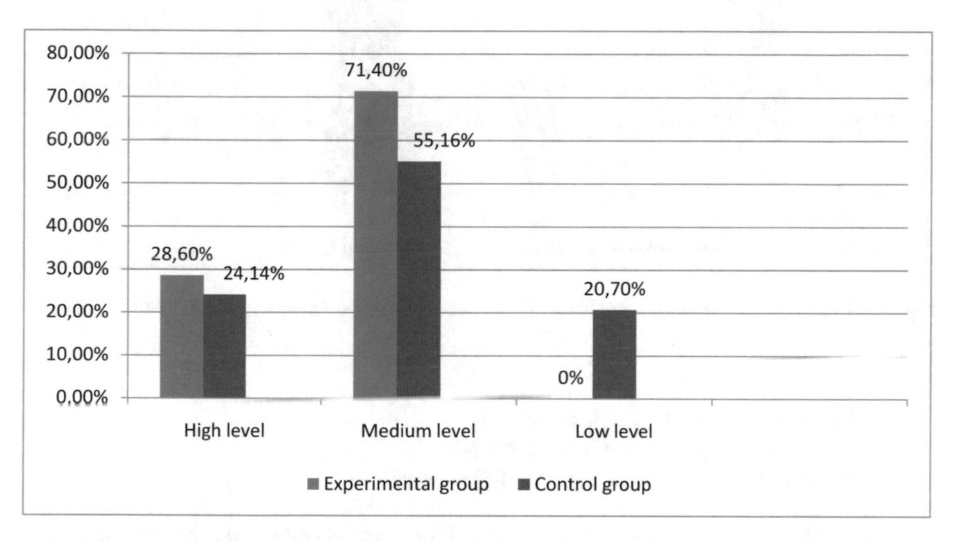

Fig. 2. Results of comparative analysis of personal anxiety indicators in experimental and control groups

Students of the experimental group are less afraid than the students of the control group of new unknown events, unpredictable situations of communication with students. In the future, they do not predict their professional lack of demand, do not exaggerate the probability of possible failures, conflicts, and unprofessionalism. Comparative analysis of the results of measurement of the general level of anxiety showed that moderate anxiety, having a constructive, motivating orientation, is characteristic of most (71.4%) students of the experimental group and only half of the students (55,16%)—control group not participating in the experiment.

Participation in the project contributed to increasing professional motivation among students of the experimental group (Fig. 3).

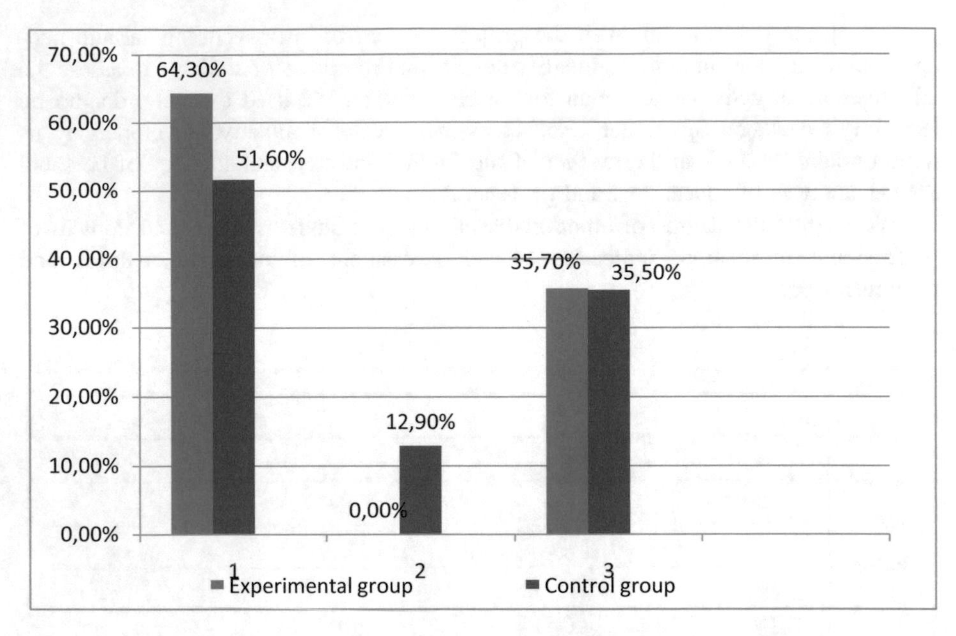

Fig. 3. Results of comparative analysis of indicators of professional motivation of experimental and control groups

1—"Yes, I'm going to work in school"
2—"No, I'm not going to work in school"
3—"I do not know, I have not yet determined"

Participation in the project contributed to increasing professional motivation among students of the experimental group. The percentage of students who in the future are going to work as teachers in the experimental group is higher than in the control group 64.3%. All the students studied showed interest in the innovative project implemented by us and obtaining additional pedagogical professions.

The results of the comparative analysis of self-assessment of professionally significant qualities showed that participation in the project contributed to the development of such qualities among students of the experimental group as responsibility, kindness, responsiveness, and sociability (Fig. 4).

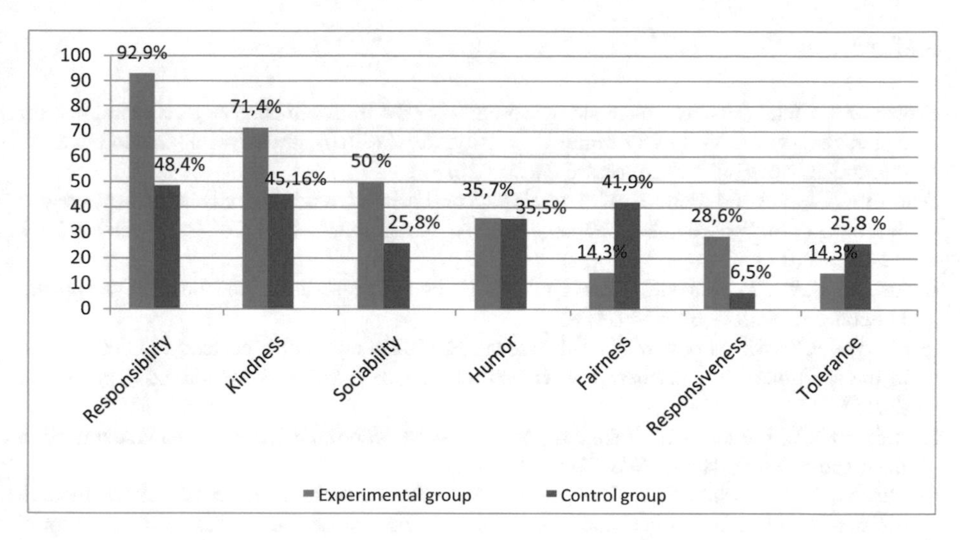

Fig. 4. Results of comparative analysis of self-assessment of professionally significant qualities of students of experimental and control groups

The above results of experimental work in the framework of the project "The teacher of the future", prove the correctness of the guidelines formulated by the authors about the necessity and effectiveness of changing the training teachers based on the model of division of pedagogical labor.

4 Conclusion

In the course of realization of the developed model of division of pedagogical labor certain results were achieved. Internal professional motivation of students is strengthened, professional confidence is improved, psychological and communicative culture is developed. Students have acquired additional special competencies that will allow them to successfully realize themselves in the future. However, there were also problems that require further study and resolution.

First, there is a need to fill the teacher's professional standard with specific competencies. Secondly, it is necessary to address the issue of a clearer differentiation of the functions of teachers already existing in the school and activities of teachers of new types of professions. Thirdly, for school to accept teachers of new professions, it is necessary to motivational readiness for these changes of all participants of the educational process. Fourth, there are a number of problems related to changing the content and technology of training teachers of new professions. Fifthly, it is necessary to consider the possibility of combining both the best traditional educational technologies and innovative in the training of teachers of a new type.

References

Egorov, E.E.: Intendiffia as a methodological basis for the transformation of teacher training for general education. Vestnik of Minin University, no. 2 (2016). https://vestnik.mininuniver.ru/jour/article/view/188/189. Accessed 24 Dec 2018

Fedorov, A.A., et al.: Portrait of a teacher. The basis of modeling educational programs: monograph. In: Fedorov, A.A., Paputkova, G.A., Novgorod, N. (eds.) Minin University, 2017 - 202 p. (2017)

Hanushek, E.A., Woessmann, L.L.: The role of cognitive skills in economic development. J. Econ. Lit. **46**(3), 607–668 (2008)

Ilaltdinova, E.Y., Shlyakhov, M.Y., Shlyakhova, M.M.: Pedagogical expediency. Differentiation of the functions of the teacher. New as well forgotten old. Public education, no 1, pp. 80–84 (2015)

Kuzminov, Y.I., Frumin, I.G., Zakharov, A.B.: Russian school: an alternative to modernization from above. Educ. Issues **2011**(3), 5–53 (2011)

Kuznetsov, V.P., Garina, E.P., Andryashina, N.S., Romanovskaya, E.V.: Models of modern information economy conceptual contradictions and practical examples, 361 p. Emerald publishing limited (2018)

Kuznetsov, V., Kornilov, D., Kolmykova, T., Garina, E., Garin, A.: A creative model of modern company management on the basis of semantic technologies. In: Communications in Computer and Information Science (2017)

Lenskoy, E., Pinskaya, M.: Russian teachers in the mirror of the international comparative study of the pedagogical corps (TALIS 2013) [Text]. National researches University "Higher School of Economics", Institute of Education. M.: Publishing House of the Higher School of Economics, 36 p. Modern education analyst, vol. 1 (2015). ISBN 978-5-7598-13262

Loughran, J.J.: Developing a Pedagogy of Teacher Education: Understanding Teaching and Learning About Teaching. Routledge, London (2006)

Potapova, T.K.: The challenges of the time, the reform of education and training of future teachers in the model of the division of pedagogical labour. Vestnik of Minin University, no. 3 (2016). https://vestnik.mininuniver.ru/jour/article/view/241/242. Accessed 24 Dec 2018

Shabanova, T.L.: Formation of psychological and pedagogical meta-competence among students in the process of studying the author's courses of the innovative project of the university "Teacher of the Future". Concept. no. 06, June 2015. http://e-koncept.ru/2015/15201.htm. ISSN 2304-120X. Accessed 24 Dec 2018

Schleicher, A.: Building a high-quality teaching profession: Lessons from around the world. OECD Publishing (2011). http://dx.doi.org/10.1787/9789264113046-en. Accessed 24 Dec 2018

Sundli, L.: Mentoring – a new mantra for education? Teaching and Teacher Education, vol. 23, pp. 201–214 (2007)

Taylor, C.W.: Cultivating multiple creative talents in students. J. Educ. Gifted **8**, 187–198 (1985)

Van Der Merwe, L., Verwey, A.: Leadership meta-competencies for the future world of work. SA J. Hum. Res. Manag. **5**(2), 33–41 (2007)

Subjective Representation Study of University Teachers About the Significance of Changes in Higher Education

Elena G. Gutsu(✉), Nadezhda N. Demeneva, Elena V. Kochetova, Oksana V. Kolesova, and Tatyana V. Mayasova

Minin Nizhny Novgorod State Pedagogical University, Nizhny Novgorod, Russian Federation
elenagytcy@mail.ru, nndemeneva@yandex.ru, evkoch@mail.ru, a-m-kolesov@yandex.ru, vip.mayasova@mail.ru

Abstract. The effectiveness of innovative processes in higher education largely depends on how these innovations will be accepted by those who depend on their direct implementation. The aim of this article is to study the subjective attitude of teachers of higher education to some changes and innovative processes taking place in the system of higher education. The following changes are highlighted for analysis: introduction of competence approach, reduction of the share of classroom time, technologization of training, introduction of an effective contract for teachers, increase of the share of distance learning, and tightening of discipline. The study was conducted using the author's methods. Of particular interest are the results of a pilot study of subjective attitude of university teachers to changes in modern higher school. The article assesses the impact of these changes in three directions: on content, effectiveness, and specifics of work. The empirical data obtained in the course of the study may be of interest to direct managers in the system of higher education (heads of departments or heads of educational programs), as well as for the administration of universities for development and implementation of personnel management programs in the educational organization.

Keywords: University lecturer · Modernization of higher education · Competence approach · Innovations · Changes in the system of higher education · Subjective attitude

JEL Code: I230

1 Introduction

The system of domestic higher education is at the stage of reformation, which is taking place against the background of significant changes in Russian society. The beginning of the 2000s is marked by the adoption of an extensive package of official documents related to the modernization of the education system. However, their actual implementation in practice has not yet produced desired effects. The government has not yet

E. G. Popkova and B. S. Sergi (Eds.): ISC 2019, LNNS 91, pp. 439–445, 2020.
https://doi.org/10.1007/978-3-030-32015-7_49

developed a clear and well-thought-out strategy for the modernization of higher education, although it has realized the significance of this problem (Protasova 2017).

Modern educational system is the most complex form of social practice, its role is unique and exceptional. Education, as noted by V.I. Slobodchikov, is the most significant social institution that translates and embodies the basic values and goals of Russian society (Slobodchikov 2005). Consequently, requirements for modern education are higher and more complex than before.

Implementation of these tasks, according to many authors, is possible on the basis of introduction of competence-based approach, orienting education on systematic development of learners of knowledge and skills of practical activities that ensure the successful functioning of a person in the main spheres of life, both in interests of himself, society and the government (Stelzer-Rothe 2005; Kuznetsov 2018).

Many researchers note that the teacher is a key figure in education reforms, on which the pace and nature of modernization depend. (Vasilenko and Wielz 2016). As noted by V.V. Serikov, creation of any serious and deep theory requires the simultaneous reflection of the model of a teacher who is able to implement this theory (Serikov 2010).

With this in mind, the problem of the personality of the teacher as a subject of professional pedagogical activity, capable of self-development, has now taken one of the central positions in educational psychology.

The level of development of the personality of an educator, more than in any other field, determines the success of his professional activity (Gorelova 2004). If the pedagogical process is primarily the interaction of individuals, then the main means of influence becomes the teacher himself as a person, and not only as a specialist who owns a set of necessary knowledge and skills (Smirnov and Fadeev 2006).

There is, perhaps, not a single researcher who would not speak about the importance of the teacher's personality in the educational process. But this problem was especially acute in connection with the implementation of the competence-based approach. "For the traditional model of teaching, the main thing is what they teach and how they teach. For competence—who teaches!" (Serikov 2005). Culture, which penetrates into the consciousness of the pupil, always passes through the prism of the personality of the teacher (Serikov 2005). The personal sphere is responsible for the value-semantic component of the content of education. Therefore, if the teacher's personality is not taken into account, the most important resource for improving the quality of education is missing.

Problems of the development of the personality of a higher education teacher, his socio-psychological characteristics and value-semantic orientations, self-realization and self-actualization are discussed in the works of E.A. Zaluchenova, L.E. Pautova, L.A. Povornishenoy, O.S. Rudenko, L.V. Khazova, (Ilaltdinova et al. 2018).

Problems of professional deformation and professional burnout of a teacher are studied in researches of (Zhagalina 2006), A. Kozlova, E.G. Ozhogova, N.V. Prokoptseva, et al.

The study of professionally important qualities, as well as the conditions of their development, is undertaken in the works of S.M. Batashova, G.U. Matushansky, M.N. Strikhanov and others.

The issues of motivation for the professional activities of a university teacher (Gutsu 2018; Sorokoumova et al. 2014; Gutsu et al. 2016; Minaeva et al. 2018; Kochetova et al. 2017a, b) are becoming increasingly relevant in recent years, among which studies on the specifics of the teacher's activities in modern conditions and in the context of innovation (Armstrong Lloyd 2017; Malkina et al. 2018; Efendiyev 2008; Kornilov, D.).

According to (Serikov 2005), personal "contribution" of the teacher to the educational process consists:

- in the formulation of objectives of professional activity on the basis of subjectively understood pedagogical situation;
- in value-semantic interpretation of the content;
- to use their personal experience as a source of content of education;
- in their own style and forms of support and psychological support of cognitive activity of pupils;
- in the specifics of reflection and self-evaluation of their activities on the basis of their own ideas, which do not always coincide with the official standards.

All of the above demands to treat the teacher not as a passive performer of ready-made instructions, but as a subject of his pedagogical activity. The subject of professional activity is determined by the values, meanings, and motives that guide a particular teacher. Therefore, the proposed concept of "subjective pedagogical reality" is undoubtedly valuable. It is understood by the author as an ideal form of realization in the consciousness of the teacher of educational goals and means of their achievement; project of educational paradigm; as well as system of professional and personal reflections (Serikov 2010).

Y.V. Senko and M.N. Frolovskaya introduce the concept of "professional image of the teacher's world." According to the authors, it includes the values and meanings of education, style of pedagogical thinking, which actively participate in the analysis and personal experience of existing socio-cultural situation, allocating a pedagogical task from it, searching for ways of its realization and solution, and analyzing the obtained results. "The professional image of the world is a "transformed form"of competence." An important point for understanding the real processes in education is the realization of how the processes occurring in the system of higher education are subjectively perceived and experienced by teachers themselves.

2 Purpose of the Study

The aim of the work is to study the subjective representation of university teachers about the importance of some changes in higher education and their impact on professional pedagogical activity in three directions: on the content and specificity of teachers' work, on the quality of work and on professional improvement.

3 Methods of Research

In order to identify the subjective attitude of teachers to some changes in the system of higher education, we developed a "Questionnaire for studying the motivation of professional activity of the teacher University" (Gutsu 2017).

The following changes in higher education have been identified for compiling diagnostic tools: introduction of a competence-based approach, reduction of classroom instruction time, technologization of education, the introduction of an effective contract for teachers, an increase in the share of distance learning, and stricter discipline.

Teachers were asked to evaluate the changes proposed above in terms of their impact on:

(1) content and specificity of teachers work,
(2) quality of teacher's work (quality of students' training),
(3) for professional improvement of a teacher.

The evaluation ranged from "−5" to "+5". The "−5" rating corresponded to a significant negative impact of the estimated change, and the "+5" rating corresponded to a significant positive impact. The results were calculated in average values.

In our study, 80 university professors took part, ranging in age from 34 to 72.

4 Results of the Study

We conducted a pilot study of subjective representation of teachers of higher education about the impact of changes taking place in higher education on content and specifics of professional work, quality of their work and the professional improvement. Results of the study are reflected in Table 1.

Table 1. Subjective perception of teachers about the importance of changes in education (in average values, min = −5/significant negative changes, max = +5/significant positive changes)

Changes	Direction of influence		
	On the content and specifics of the teacher's work	On the quality of the teacher's work	Professional improvement
Implementation of competency-based approach	+1.5	+0.2	+1.5
Reduction of classroom learning time	−0.5	−1,5	+1.8
Technologization of learning	+2.7	+0.2	+1.8
Introduction of an effective contract for teachers	+1.1	−0.3	−0.9
Increase in the share of distance learning, including full-time study	+1.9	−1.1	+2.3
Tighter discipline	−3.8	−0.3	−0.6

5 Discussion of Results

The obtained data allowed to establish that a large part of the teachers who took part in the survey relate to changes in the system of higher education indifferent. In their view, the importance of change is extremely small. The impact of most changes on the content and specifics of the teacher's work, on the quality of the teacher's work (the quality of the student's training), as well as on professional improvement is assessed survey participants in the range from "−1" (minimum negative changes) to "+1" (minimum positive changes). This shows that most of the changes, according to the subjective opinion of the teachers who took part in the study, do not have a significant impact on the real situation.

It is quite optimistic that the very introduction of competence approach is evaluated by the majority of teachers in a positive range in all three directions of influence, although in mean values, this change is evaluated as having a slightly positive effect. This shows that at this stage the competence approach is generally accepted by teachers and does not cause resistance.

The greatest negative values were recorded for the "toughening discipline" and "reducing classroom time" changes. The greatest positive estimates are for the changes "Technologization of learning" and "Increase of the share of distance learning, including full-time education". It should be noted that the positive impact of these changes was noted only on the content and specificity of the teacher's work and on professional improvement. In other words, according to teachers, the introduction of new technologies in education has a positive impact on the professional development of the teacher, allowing him to acquire new competencies, but paradoxically, does not affect the result of the teacher's work (quality of the student's training).

The teachers' assessment of such a change as "Introduction of an effective contract for teachers" deserves special attention. The significance of this change was defined as "causing minor positive changes" in relation to the content and specificity. But at the same time as negative—for productivity and professional self-improvement.

The fact that the effectiveness of teachers' work, understood as the quality of students' training, according to the teachers who took part in the study, is in the least degree depends on changes in higher education.

6 Conclusion

Thus, the data obtained from the study provide the following preliminary conclusions.

The processes of modernization taking place in the modern system of higher education are refracted through the prism of subjective understanding and experience of teachers of changes taking place.

Administrative system of stimulation without analysis and taking into account the specifics of a particular professional group has extremely insignificant influence on the real professional activity of university teachers. Increasing technologization of the educational process, toughening of disciplines and even intensification of teachers' work taking into account progressive material remuneration (due to the introduction of effective contract) are poorly combined with the creative nature of scientific and pedagogical work.

A significant part of the changes introduced in higher education through the administrative resource, from the point of view of teachers, do not have a pronounced positive impact on the real educational process, especially in terms of the quality of education. Moreover, the impact of some changes is assessed by teachers in a negative range.

Undoubtedly, for a more objective understanding of the real situation, a more detailed and thorough analysis of the obtained data and a more detailed study of subjective perception by university teachers is necessary processes taking place in higher education. It seems that the solution to this problem is possible interdisciplinary basis combining approaches developed in psychology, sociology, and management. This may be the purpose of independent research.

References

Vasilenko O.Yu., Veltse, V.: University lecturer: motivation and stimulation of labor activity (Review of sociological research) [Electronic resource]. Access Mode: http://www.vevivi.ru/best/Prepodavatel-vuza-motivatsiya-i-stimulirovanie-trudovoi-deyatelnosti-Obzor-sotsiologicheskogo-issledovaniya-ref88191.html. Accessed 27 Oct 2016

Gorelova, G.G.: Crisis and pedagogical profession. Moscow Psychological and Social Institute, p. 320 (2004)

Gutsu, E.G., Nyagolov, M.D., Runov, T.A.: Research of motivation of labor activity of a university teacher. Bulletin of Minin University, vol. 6. no. 3 (2018). Access Mode: https://vestnik.mininuniver.ru/jour/article/view/874/686. Accessed 12 Nov 2018

Gutsu, E.G.: Diagnostics of professional competence of the teacher of higher education. In: Gutsu, E.G. (eds.) Methodical Manual for University Teachers. M., "Flinta" (2017)

Zhalagina, T.A.: Motivational aspect of prevention of professional deformation of the teacher's personality. Human factor: problems of psychology and ergonomics, no. 2, pp. 11–14 (2006)

Myalkina, E.V., Zhitkova, V.A.: System of complex assessment of administrative and managerial personnel in the university: practice and features. Bulletin of Minin University, vol. 6, no. 1 (2018). Access Mode: https://vestnik.mininuniver.ru/jour/article/view/751/642. Accessed 25 Oct 2018

Protasova, I.I.: Modernization of the system of management of motivation of teachers in the sphere of higher education. Abstract for the degree of Candidate of Economic Sciences. Krasnodar (2013)

Serikov, V.V.: Nature of pedagogical activity and features of professional education of the teacher. Pedagogy **5**, 29–37 (2010)

Slobodchikov, V.I.: Essays of psychology and education, 2nd edn. Revised and Supplemented, p. 272. Birobidzhan (2005)

Smirnov, D.G., Fadeev, A.M.: Teacher and scientist: models of university synthesis. Intelligentsia and Mir **4**, 112–128 (2006)

Sorokoumova, S.N., Gutsu, E.G.: Development of a motivational and valuable component of professional competence of the university teacher in the system of intra-university professional development. Privolzhsky Sci. J. **4**, 305–309 (2014)

Gutsu, E.G., Demeneva, N.N., Kochetova, E.V., Mayasova, T.V., Belinova, N.V.: Studying motivational-axiological component of professional competence of a college teacher. Int. J. Environ. Sci. Edu. **18**, 12650–12657 (2016). T. 11

Ilaltdinova, E.Y., Frolova, S.V., Lebedeva, I.V.: Top qualities of great teachers: national and universal. Adv. Intell. Syst. Comput. **677**, 44–52 (2018). https://doi.org/10.1007/978-3-319-67843-6_6

Kochetova, E.V., Gutsu, E.G., Demeneva. N.N., Mayasova. T.V., Fedoseeva. O.I.: Psychological mechanisms of future pedagogues' professional individualization formation during their studies in a higher educational institution. J. Fundam. Appl. Sci. **2**, 1484–1493 (2017a). T. 9

Kuznetsov, V.P., Garina, E.P., Andryashina, N.S., Romanovskaya, E.V.: Models of Modern Information Economy Conceptual Contradictions and Practical Examples, p. 361. Bingley, Emerald Publishing Limited (2018)

Kuznetsov, V., Kornilov, D., Kolmykova, T., Garina, E., Garin, A.: A creative model of modern company management on the basis of semantic technologies. In: Communications in Computer and Information Science (2017b)

Stelzer-Rothe, T.: Competencies in higher education. A tool for good teaching and learning at universities, p. 400, Rinteln (2005)

Development of Professional Creativity of Teachers in the System of Professional Safety Culture of Children in Transport

Galina S. Kamerilova(✉), Marina A. Kartavykh, Elena L. Ageeva, Irina A. Gordeeva, and Marina A. Veryaskina

Minin Nizhny Novgorod State Pedagogical University, Nizhny Novgorod, Russia
{kamerilova-galina, mkartavykh}@rambler.ru, lenaageeva2015@yandex.ru, iku09@mail.ru, veryaskina_ma@mail.ru

Abstract. Relevance of the research. Transformation of fundamental bases of education actualized the problem of teachers' professional creativity development, ensuring the application of innovative approaches in the system of qualification for the formation of a culture of safety in transport. The analysis revealed a contradiction between understanding the role of creativity in the system of postgraduate pedagogical education and insufficiency of this issue in theory and pedagogical practice. The purpose of the study: theoretical substantiation and practical realization of the development of professional creativity of teachers within the framework of the project of the professional development of education workers on the formation of children transport security culture.

Methods of research. The research was carried out on the basis of scientific-pedagogical methods of analysis, comparative generalization and systematization of theoretical and empirical sources, questionnaires determining the necessary objectivity and heuristic conclusions.

Results of the study. The work establishes the essence, indicators, conditions and levels of development of professional creativity of pedagogical staff in the system of advanced training in developing a safety culture for children based on innovative educational technologies. A developed cluster of innovative technologies implemented in the digital educational environment of the University of Minin (MOOC), which represents the novelty of the research, is proposed.

Discussion and conclusion. The cluster was based on culturological, competency-based, problem-situational, environmental, and contextual principles. In the development of a creative teacher the main role is played by following groups of pedagogical technologies: technologization and goal-setting, focusing on the meaning and legitimacy of the upcoming activities, understanding the importance of new relevant competencies; management of the process of assimilation of content with demonstration of examples and mastering methods of creative activity in the digital educational environment; organization of creative independent activity with the use of electronic educational resources; reflexive technologies of evaluation of the results of training, combining quantitative and qualitative methods; technologies of pedagogical communication and support, delicate management of educational activities, tutor assistance, supervision, and moderation in group discussions.

E. G. Popkova and B. S. Sergi (Eds.): ISC 2019, LNNS 91, pp. 446–451, 2020.
https://doi.org/10.1007/978-3-030-32015-7_50

Keywords: Professional creativity · Safety culture · Pedagogical technology

JEL Code: I230

1 Introduction

Sociocultural conditions of modern social development make special requirements for the formation of a creative personality of the teacher, ready and capable of innovative educational activities (Kuznetsov 2018). In this regard, modernization of postgraduate pedagogical education carried out in the system of professional development, as a priority direction is put forward the creativity of pedagogical staff. N.B. Bogoyavlenskaya defines creativity as a deep personal property, which is expressed in the original statement and solution of the problem-filled with personal meaning. Creativity is considered as the leading criterion of professional competence of pedagogical workers.

Federal Law No. 273-FZ of 29.12.2012 "On Education in the Russian Federation" provides for systematic training of teaching staff through the development of additional professional programs as exemplified by the programs of advanced training courses for teachers in the field of education developed at the University of Minin and tested in various regions of Russia. The research was carried out within the framework of the development and implementation of state contracts with the Ministry of Education and Science of the Russian Federation from 19.09.2017 No. 07.P61.11.0034 and 0034 and No. No. No. 07.P61.11.0008 from 07.05.2018, where in two years 2618 teachers of general education organizations, organizations of additional education and preschool educational organizations were trained, as well as teachers of regional institutes of professional development and institutes from 61 regions of the Russian Federation. Training was conducted in the system of remote training, as well as in-person on the basis of the reference sites of the Federal districts. The effectiveness of management was ensured by systematically conducted installation, thematic and final webinars, forums, and chat rooms.

The competence format of the developed and implemented copyright programs provided the development of professional creativity, activity of the original performance of the proposed assignments related to the development of the author's portfolio, creation of unparalleled electronic educational resources in the field of transport security, and initiatives in the discussion questions.

Each educational program was accompanied by detailed methodological recommendations, which revealed theoretical aspects of educational innovations and examples of best practices.

Targeted competence orientation of programs and project management of courses on the basis of digital educational platform MOOC provide modern training of teachers of regional institutes, teachers of general education organizations, preschool and additional education to innovative professional activity and necessary changes in the field of formation of a culture of safety in transport. Creativity developed on their basis, being the main feature of the creative personality (J. Gilford, D.B. Bogoyavlenskaya, Ya.A. Ponomarev, N.V. Kuzmina, M.F. Morozov, L.M. Friedman, Y.N. Kulyutkin, K.A.

Abulkhanov-Slavskaya) [4, 8, 9, 11], provides overcoming the usual stereotypes, productive transformations in pedagogical activity, search for fundamentally new approaches, original ideas that contribute to progress of educational systems (Kuznetsov and Kornilov 2017).

The performed analysis showed a variety of approaches to indicators of creativity and conditions of its development (D.B. Bogoyavlenskaya, T.A. Barysheva, J. Gilford, I.N. Dubrovina, M.M. Kashapov, A.I. Popov) [2, 3, 7]. Generalization of scientific views allowed to establish that professional creativity implies divergence of pedagogical thinking, fluidity, flexibility, originality, and intellectual initiative. Now teacher is open to innovation, is actively involved in research, and showing improvisation. Development of professional creativity is associated with the nature of communications, the free choice of creative activity, possibilities of the educational environment, and sequence of passing through the stages of formation of creativity in ontogenesis [1]. The issue of technological support for the development of professional creativity [10] while pedagogical technologies have enormous creative potential.

The purpose of this work is: study of possibilities of development of professional creativity of pedagogical personnel at courses of qualification improvement on the formation of the safety culture of children in transport with innovative pedagogical technologies.

2 Methodology

A pedagogical study of the problem of developing creativity among working teachers in the field of transport safety of children was carried out in the logic of developing additional educational programs and electronic educational and methodological complexes for advanced training courses held at the University of Minin, in 2017 and 2018. The initial review and comparison of best practices and applied methods, theoretical analysis were carried out scientific and pedagogical literature on the nature and methods of developing creativity in professional pedagogy, a generalized approach to creativity indices creative potential of pedagogical technologies, and assess their effectiveness in the electronic learning environment.

3 Results

The cardinal changes in education require a teacher of a new style of scientific and pedagogical thinking aimed at humanistic principles associated with creativity (V.A. Slastenin, L.S. Podymova) In this regard, the ongoing project of professional development of pedagogical workers was built in the logic of creative approach, presupposing a variety of innovations in the field of acquisition by listeners of relevant competencies for the formation of a culture of safety in transport. Ideas of development of professional creativity are laid in the procedure of project management; competence, goal setting; allocation of axiological, cognitive, praxiological, personal-professional aspects of the content related to the structural heterogeneity of justified competencies; the stage of the educational process organized in the digital educational environment;

intermediate and the final diagnosis of the results considered by us earlier [6]. Great importance was paid to motivation and search for meanings that make up the value-semantic platform of creative activity, intellectual initiative, and search for original solutions of pedagogical problems on safe behavior of children on the roads.

An indispensable condition for the development of professional creativity of listeners is the cluster of pedagogical technologies developed by us. The following approaches have been adopted in the sphere of formation of a culture of safety in transport for children: (1) cultural studies (E. Bondarevskaya), which provides educational activities in the coordinates of modern culture with the ideas of post-non-classical, humanization, dialogue, integration; (2) competence (V.A. Bolotov, E.F. Zeer, I.A.Zimnyaya, G.A. Paputkova, J. Raven, A.P. Tryapitsyna), the nature of which means the unity of values, knowledge, skills, thinking, initiatives, reflection, as well as a deep interest in solving a pressing problem: ensuring the safety of children in transport; (3) problem-situational (I.Ya. Lerner, M.I. Makhmutov, M.N. Skatkin, Z.Y. Yuldashev), oriented creative independence and leading to formation of subject-subject interaction, opening of new knowledge and ways of pedagogical activity by students; allowing the use of specific traffic situations that are actually found in children in life; (4) environmental (A.V. Mudrik, V.I. Slobodchikov, I.V. Robert), meaning comfort of business communication, active work of teachers in the digital educational environment, saturated with electronic educational resources and involves the development of various Information and communication technologies; (5) contextual (A.A. Verbitsky), which determines the choice of technologies in accordance with real pedagogical activity in the field of formation of children's culture of safe behavior in road traffic. The cluster structure, based on the ideas of interactivity, represents a set of interrelated technologies of education:

1. Technologies of motivation and goal-setting, focusing on the meaning and legitimacy of the upcoming activities, understanding the significance of new competencies. It is assumed to understand the relevance of the problem of transport safety of children, a departure from traditional but ineffective methods, installation on wide possibilities of visualization in the digital educational environment. 2. Technologies of management of the process of assimilation of content with the demonstration of examples and mastering methods of creative activity in the digital educational environment: information search (compilation of annotated lists and catalogs of Internet resources on transport security), problem training (lectures with problem presentation, heuristic workshops, case learning by party-search methods), "brainstorming", simulation of game activity (business game "Organization and conduct of the webinar "Formation of children's skills of safe participation in road traffic" for teachers); creative projects of prevention of children's road traffic injuries. Technologies of organization of creative independent activity in the digital educational environment, determining individuality in the formation of professional creativity through the performance of creative tasks: interactive participation in master classes (demonstration business portfolio of web design studio "Organization of teachers" activities in the sphere of formation of children's skills of safe participation in the road traffic"); author's presentations ("Bank of effective technologies in the field of teaching children the rules of safe behavior on the roads"); SWOT analysis as a method of strategic planning in the field of transport safety of children in a specific educational

organization. 4. Reflective technologies of evaluation of learning results, combining quantitative and qualitative methods: rating, test diagnostics, essay "My idea of professional competence of teachers in the sphere of formation of children's culture of safe behavior on the roads", representing its kind perspective model author's reflexive vision of learning results; analysis of creative products activity (the final project, author's e-educational resource, posted on the website "System of dynamic formation of cross-platform electronic educational resources", diary, questionnaires). Generalized analysis of results showed significant positive dynamics in the development of professional creativity of teachers from stimulus-productive to heuristic level of creativity and further to creativity, characterized by stable internal motivation, intellectual initiative, divergence thinking, attitude to the research nature of the activity. 5. Technology of pedagogic communication and support, reflecting the subject-subjectivity of the relationship: expansion of access to information resources (verbalized forms, hypertext, multimedia, "virtual reality"), an increase of contact interaction through dialogue and cooperation [5], synchronous and asynchronous interactivity, and training interpersonal communication. Pedagogical support involves delicate management of educational activities of listeners, tutor assistance in drawing up and implementation of personal educational route, supervision as support of professional improvement and development of creativity. moderation, and also ensuring the productivity of the group discussion.

4 Conclusions and Recommendations

The importance of professional creativity at advanced training courses on formation of safety culture in transport led to the need to search for means of its development. The educational resource with high creative potential is innovative pedagogical technologies developed and implemented in the digital educational environment of the University of Minin, the cluster which implements culturological, competence, problem-situational, environmental, contextual principles of modern continuous professional and pedagogical education.

Acknowledgments. The work was performed as part of the Government contract Ministry of Education of the Russian Federation from 09.19.2017, No. 0034 and 07.R61.11.0034 and No.07. R61.11.0008 from 07.05.2018.

References

1. Barysheva, T.A.: Psychological structure and development of creativity of adults: Doctor of Psychological Sciences, p. 38. T.A.Barysheva. SPb (2005)
2. Bogoyavlenskaya, D.B.: Psychology of creative abilities, p. 320. M. "Academy" (2002)
3. Gilford, J.: Three sides of intelligence. Psychology of thinking. In: Matyushkina, A.M. (ed.) Moscow (1965). (translation of the article)
4. Golubova, V.M.: Research of the nature of creative thinking and creativity of the personality. Fundam. Res. no. 2–5, 1067–1071 (2015). http://www.fundamental-research.ru/ru/article/view?id=36985. Accessed 17 Nov 2018

5. Kamerilova, G.S., Kartavykh, M.A., Ageeva, E.L., Gordeeva, I.A., Astashina, N.I., Ruban, E.M.: Communicative teaching models: The formation of the professional pedagogy competence among health and safety school teachers. Espacios **39**(29), 7 (2018)
6. Kartvykh, M.A.: Scientific and methodical support of additional professional training programs of teaching staff in the sphere of formation of children's skills of safe participation in road traffic [Electronic resource]. In: Kartavykh, M.A., Kamerilov, G.S. (eds.) Bulletin of Mininsky University, vol. 6, no. 1 (2018). http://vestnik.mininuniver.ru/jour/article/view/752/643. Accessed 07 Oct 2018
7. Popov, A.I.: Development of professional creativity in the system of higher education. Continuing pedagogical education: problems and searches of continuing pedagogical education: problems and search (2016). http://kpfu.ru/psychology/programma-razvitiya-ipo-kfu. No. 1 (1) 50.S
8. Markovich, E.V.: Pedagogical Creativity as a Leading Component of the Structure of Pedagogical Gifted. https://nsportal.ru/blog/nobshcheobrazovatelnaya-tematika/all/2012/11/08/pedagogicheskaya-kreativnost-kak-vedushchiy
9. Tarakanov, A.V.: Development of creativity of students with the help of active teaching methods in the university Tarakanov. Sib. Pedagogical J. **2**, 65–68 (2012)
10. Aizikovitsh, E., Amit, M.: Evaluating an infusion approach to the teaching of critical thinking skills through mathematics. In: Aizikovitsh, E., Amit, M. (eds.) Procedia Social and Behavioral Sciences. vol. 2, pp. 3818–3822 (2010)
11. Botcheva, L., Shih, J., Huffman, L.C.: Emphasizing cultural competence in evaluation: a process-oriented approach. In: Botcheva, L., Shih, J., Huffman, L.C. (eds.) American Journal of Evaluation, vol. 30, no. 2, pp. 176–188 (2009)
12. Boyatzis, R.E.: The Competent Manager: A Model for Effective Performance. Wiley, New York (1982). http://www.cbsd.ru/center/31evel.php?PHPSESSID=uij22ebi9nac24fd
13. Hutchby, I., Moran-Ellis, J., Wach, L. (eds.): Children and Social Competence. Arenas of Action. Falmer Press, London (1998)
14. Dronkers, J., Robert, P.: Has educational sector any impact on school effectiveness in Hungary? In: Dronkers, J., Robert, P. (eds.) European Societies, vol. 6, no. 2, pp. 205–236 (2004)
15. Kuznetsov, V.P., Garina, E.P., Andryashina, N.S., Romanovskaya, E.V.: Models of Modern Information Economy Conceptual Contradictions and Practical Examples, p. 361. Bingley, Emerald Publishing Limited (2018)
16. Kuznetsov, V., Kornilov, D., Kolmykova, T., Garina, E., Garin, A.: A creative model of modern company management on the basis of semantic technologies. In: Communications in Computer and Information Science (2017)

Forecasting the Development of Professional Education

Svetlana M. Markova(✉), Svetlana A. Tsyplakova, Catherine P. Sedykh, Anna V. Khizhnaya, and Olga N. Filatova

Nizhny Novgorod State Pedagogical University named after K. Minin, Nizhny Novgorod, Russia
cveta-ts@yandex.ru

Abstract. *Relevance.* The article deals with the problems associated with determining the prospects for the development of vocational education. Forecasting in the system of vocational education is determined by the factors of social development, which include the volume of the total national product, pace of production automation, number of skilled workers in the total amount employed in social production, changes in occupations over the course of a person's life, emergence of new areas in the economic sphere that require special educational training, and growth rate of knowledge.

Methods. The study uses general scientific methods of cognition; analysis, synthesis, historical and logical, and systematic approach, as well as special methods related to specific tasks of research: extrapolation and morphological analysis, modeling for construction of prognostic models of professional education development.

Results of the study. Forecasting of pedagogical objects is considered as a specially organized process of research aimed at obtaining perspective information on the future state of pedagogical phenomena and processes, with the aim of modernizing the content, methods, means and organizational forms of professional and pedagogical activity.In the result of generalization of the available approaches to the development of the basic provisions of prognostic activity, principles of prognosis relating to professional and pedagogical activity: connection of predictive object, consistency, polyvariability, continuity, verifiability, necessity, consistency, and subordination.

Information-dynamic model of a professional educational institution is constructed.

Discussion and conclusion. Thus, the use of principles of forecasting and facts of economic growth allowed to determine prognostic directions of development of professional education: integration of science, technology, education and production, interaction of personal and professional development of workers and specialists, technologization and humanization of education, interaction of pedagogical, information and production technologies; integration and universalization of the content of education; training of professionally competent and competitive specialists for the innovative economy.

Keywords: Professional education · Forecasting · Forecasting principles · Professional and pedagogical activity

JEL Code: I210

E. G. Popkova and B. S. Sergi (Eds.): ISC 2019, LNNS 91, pp. 452–459, 2020.
https://doi.org/10.1007/978-3-030-32015-7_51

1 Introduction

The concept of "forecast" is associated with the prediction of the future of a particular subject, phenomenon, or process. A forecast is a scientifically based judgment about the prospects, possible states of a particular phenomenon in the future, which carries probabilistic character. The forecast is considered as an expression of alternative ways of development of the process and the timing of its implementation.

Forecasting covers all spheres of scientific and social life. The object of forecasting is real phenomena and processes reflected in the public consciousness of the future state or the course of its development. Forecasting is considered as a process of studying probable future situations, as well as the patterns of organization and functioning of forecasting systems.

The most developed direction of forecasting is the field of scientific and technical forecasts. It has the most defined, clearly defined objects. A more complex process of forecasting in the social environment, aimed at creating theoretical models of the future.

The purpose of this article is a system representation of forecasting and professional education, which process of determining the prospects of development of professional and pedagogical objects. Also allocation of factors of social development, and influencing the definition of prospects of professional education development on the example of building an information-dynamic model of a professional educational institution.

2 Scientific Novelty of Research

On the basis of logical and methodological analysis the essence of pedagogical forecasting is defined, consisting in synthesis of socio-economic, scientific, technical, professional knowledge at the level of general scientific interdisciplinary integration. Revealed the system-forming role of professional activity as a condition of successful functioning of professional educational development. Information-dynamic model of a professional educational institution as a system basis of forecasting, providing information on dynamics of parameters and characteristics of a professional educational institution is constructed.

3 Theoretical Basis of Research

The study of the relationship between professional, pedagogical and prognostic activities is associated with the study of prospects for the development of economic, managerial, technical and technological, cultural, sociological professional.

For the development of professional education there is a great importance of foresight (N.A.Berdyaeva, A.P.Bogdanova, V.I.Vernadsky, N.D.Kondratieva, I. Prigozhinaidr.) and forecasting theory (D.M.Gvshiani, V.A.Lisichkin, Belyaeva, B.S. Gershunsky, M.V.Clarinidr.), which is considered as social and technical integration of education, science, and production.

For the study of the essential characteristics of forecasting in the system of professional education, the study of educational and pedagogical prognostication, providing the way to receive advanced knowledge of the development of educational systems (B.S. Gershunsky), to develop professional education (A.P. Belyaev, G.A.Ivanov, G.I.Lukinidr.), perspectives of development of educational process (I.P. Podlasy), identification of personal and professional development (A.M. Gendin).

4 Methodology of Research

The study uses general scientific methods of cognition; analysis, synthesis, historical and logical, and systematic approach, as well as special methods related to specific tasks of research: extrapolation and morphological analysis, modeling for constructing prognostic models of professional education development (Kuznetsov 2018).

Vocational education is considered as a complex system. For the development of the research is subject to the following issues: strategic initiatives for the development of vocational education, integration of science, education of production, interconnection of pedagogical and production processes, educational and production activities; interconnection of general, polytechnic and professional education.

5 Results of Research

Predicting the system of professional education is preceded by a perspective of the process of social development. The most important economic indicators of the development of the system of professional education are:

- amount of aggregate national product;
- the rate of automation of production;
- the number of qualified workers in the total volume of production involved;
- change of professions in the course of all human life;
- emergence of new areas in the economic sphere requiring special educational training (Kuznetsov and Kornilov 2017);
- growth rate of knowledge.

Thus, by 2020, skilled workers from vocational schools will increase to 2.5 million, compared with 2.0 million in 2005; growth rate of 29%; average annual growth of 1.2%. This corresponds to the needs of the national economy for new skilled workers and the growth of resources of young people in the forecast period. In this regard, there are tasks of active career guidance for young people to work professions.

Direct contracts of professional educational institutions with enterprises that will pay for training, retraining and professional development will be the priority.

It is necessary to actively develop professional centers in small and medium cities by creating branches to train workers for the production and social infrastructure of rural areas.

The structure of professions needs to be changed. It is necessary to move to the training of skilled workers in complex professions for the scientific and industrial type of production.

Prediction of professional education associated with the prediction of preparing skilled workers with the oscillations in the dynamics of population waves. They occur as a consequence of changes in birth and mortality rates. For example, the dynamic numbers of 15th and 18th years old had a minimum in 1962, and a maximum in 1979 for 18-year-old boys and girls. In the future, the growth of the number of young people will increase.

The rapid technical development is accompanied by an increase in the requirements of a button of professional competence. The growing pace of technological transformation will lead, that one basic professional education in human life will not suffice.

The study has shown that the objects of forecasting in the system of professional education can act as follows:

- social and economic phenomena;
- complex objects, including different interconnections between changes;
- objects of quantitative and qualitative characteristics.

It is possible to conclude that predicting objects act as a complex single system, for which the system approach is used.

The following methodological principles should be read in the following methodological principles:

- consistency (the need to consider the prediction object for the complex system);
- natural specificity (need to take into account the specificity of nature of the object, laws of internal development, and possible growth limits);
- description optimization of the object.

For the description of the social system, it is important to consider a single property, namely: elements may have different nature. Another property is the purposefulness of the system. The system continues to function for the purpose. The purpose of the stability of the system is characterized by the interaction of the internal self-development of the social laws of the system.

Exploring professional education as a complex system can highlight properties:

- The complex system has a hierarchical structure;
- complex system is an open system conducive to development;
- complex system is an autonomous system that has spatial-temporal characteristics;
- a complex system has a set of models, each of which reflects the defined part of the entity (simulability of the system).

The complex system is mandatory organizationally and manageable, has a large number of information links. “Organization, ordering system—the ability of the system to predetermine its future.”

Increasing ordering means the dependence of factors that determine the behavior of the system. The views of external random factors, it means the existence of a system of interconnection, establishing a correspondence between the properties of the environment and functions of the system" (V.V.Druzhinin, D.S.Konstorov).

This property is based on the creation of predictive models of complex systems, such as information-dynamic model of a professional educational institution. Prognosis of complex systems (of the system of professional education) is defined as an integrative property.

With the help of extrapolation and morphological analysis, an information-dynamic model of a professional educational institution that has passed all stages of development was built. Thus, it can be used as a systematic basis for forecasting.

Information-dynamic model of a professional educational institution shows the most important sign of statistical stability in the process of interaction with the external environment.

It can be used to draw conclusions about the dynamics of the parameters and characteristics of professional institution of the future and, on this basis, draw conclusions about the main directions of scientific research in the field of professional education.

Stages of development

1921	1926	1933	1941	1949	1954	1959	1966	1976	1986	2005
1925	1932	1940	1948	1953	1958	1965	1975	1986	2005	2020

period of instability period of own development (decline in educational institutions) (increase in the number of educational institutions)

The system approach allows you to identify the main trends of behavior of complex systems, determine general patterns, develop development pathway, which makes it possible to choose the development strategy of the object.

Forecasting of pedagogical objects is considered as a specially organized process of research aimed at obtaining perspective information about the future state of pedagogical phenomena and processes in order to modernize content, methods, means and organizational forms of professional and pedagogical activity (B.S.Gershunsky).

In the system of professional education prognostic is a field of professional and pedagogical knowledge, in which the basic ideas of determining the prospects of development of various objects of professional education are studied.

The study of the principles of development, pedagogic forecasts in the system of professional education allowed to determine the basic principles of development of forecasts:

- consistency that requires interrelationship and subordination of the elements of prediction object and the forecast background;
- coherence requiring appropriate coordination of regulatory and search forecasts of different nature;
- variability, requiring a strict sequence of forecast development, coming from the forecast background;
- continuity requiring changes in forecasts in the field of new forecast objects;

- verifiability requiring the determination of the credibility and validity of the predictions;
- similarity, which presupposes a continuous comparison of the object in a given area similar to objects and models, the purpose of defining analog use of analysis and forecasting of individual elements.

These principles are universal, used in the prognosis process in the system of professional education. Gershunsky B.S. has allocated the system of prognosis of pedagogical objects, which are the following:

- principle of holistic consideration of the object of pedagogical forecasting, revealing the informative approach of forecasting;
- principle of a complex approach of prognosis and research of all component of the educational process characterizing procedural aspects of forecasting;
- principle of collective development of the forecast and the adoption of the corresponding managerial decisions;
- principle-experimental verification of pedagogical forecasts;
- interconnection of the prognostic subject of the prognostic background.

In the process of development of the forecast, following conditions should be observed: patterns of development of objects; ensuring continuity of the process of forecasting, with each forecast due to the previous, constant correction of the information (V.A. Chabrovsky).

A.A. Kupriyanov distinguishes the following special principles of forecasting in the system of professional education: practical orientation, social determination, logical continuity, and hierarchical coordination.

These principles are of internationality character, which express universal statements reflecting the general approach to scientific research.

In the result of generalization of the available approaches to develop the principles of forecasting, it is possible to highlight the principles of forecasting in the system of professional education:

- interconnection of the predictive object (environment);
- consistency of regulatory and search approaches to forecasting;
- polyvariant predictions resulting from the maximum possible number of options for the future state of the object can be developed;
- continuity, ensuring continuity of the process of forecasting, future state of the object and its adjustment;
- verification, requiring the reliability of the forecast;
- the need, in agreement with which the statement of encouragement is sufficient reliability;
- systemic character, requiring to consider the object of forecasting (system of predicting formation) as a larger complex system;
- subordination, which the private objectives of forecasting are subordinated to the general purpose of the management of the system of professional education.

6 Conclusions

The study of factors of social development and prospects of development of society allowed revealing that innovative development of the economy is accompanied by introduction of achievements of scientific and technical process in production, informatization of production equipment, use of innovative control technologies, development of automation of production, and big role of the human factor.

Thus, the use of principles of forecasting and facts of economic growth allowed to determine prognostic directions of development of professional education: integration of science, technology, education and production, interaction of personal and professional development of workers and specialists, technologization and humanization of education, interaction of pedagogical, information and production technologies; integration and universalization of the content of education; training of professionally competent and competitive specialists for the innovative economy.

References

Bogorodskaya, O.V., Golubeva, O.V., Gruzdeva, M.L., Tolsteneva, A.A., Smirnova, Z.V.: Experience of approbation and introduction of the model of management of students' Independent work in the university. In: Advances in Intelligent Systems and Computing, vol. 622, pp. 387–397 (2018). https://doi.org/10.1007/978-3-319-75383-6_50

Bulaeva, M.N., Vaganova, O.I., Koldina, M.I., Lapshova, A.V., Khizhnyi, A.V.: Preparation of bachelors of professional training using moodle. In: Advances in Intelligent Systems and Computing, vol. 622, pp. 406-411 (2018). https://doi.org/10.1007/978-3-319-75383-6_52

Bystrova, N.V., Konyaeva, E.A., Tsarapkina, J.M., Morozova, I.M., Krivonogova, A.S.: Didactic foundations of designing the process of training in professional educational institutions. In: Advances in Intelligent Systems and Computing, vol. 622, pp. 136–142 (2018). https://doi.org/10.1007/978-3-319-75383-6_18

Gruzdeva, M.L., Prokhorova, O.N., Chanchina, A.V., Chelnokova, E.A., Khanzhina, E.V.: Postgraduate information support for graduates of pedagogical universities . In: Advances in Intelligent Systems and Computing, vol. 622, pp. 143–151 (2018). https://doi.org/10.1007/978-3-319-75383-6_19

Khizhnaya, A.V., Kutepov, M.M., Gladkova, M.N., Gladkov, A.V., Dvornikova, E.I.: Information technologies in the system of military engineer training of cadets. Int. J. Environ. Sci. Edu. **13**, 6238–6245 (2016)

Kuznetsov, V.P., Garina, E.P., Andryashina, N.S., Romanovskaya, E.V.: Models of modern Information Economy Conceptual Contradictions and Practical Examples, p. 361. Emerald Publishing Limited, Bingley (2018)

Kuznetsov, V., Kornilov, D., Kolmykova, T., Garina, E., Garin, A.: A creative model of modern company management on the basis of semantic technologies. In: Communications in Computer and Information Science (2017)

Markova, S., Depsames, L., Tsyplakova, S., Yakovleva, S., Shherbakova, E.: Principles of building of objective-spatial environment in an educational organization. EJME: Math. Educ. **11**(10), 3457–3462 (2016)

Markova, S.M., Sedykh, E.P., Tsyplakova, S.A., Polunin, V.Y.: Perspective trends of development of professional pedagogics as a science. In: Advances in Intelligent Systems and Computing, vol. 622, pp. 129–135 (2018)

Aleksandrova, N.M., Markova, S.M.: Problems of development of professional and pedagogical education. Bull. Minin Univ. **1**(9), 11 (2015)

Markova, S.M.: Forecasting as a strategic direction of development of engineering and pedagogical education. Sci. Sch. **6**, 18–21 (2010)

Markova, S.M., Narkoziev, A.K.: Polytheoretical bases of professional and pedagogical education. Bull. Minin Univ. **3**(16), 2 (2016)

Markova, S.M., Tsyplakova, S.A.: Professional bases of professional and pedagogical education. Psychol. Pedag. Res. **9**(1), 38–43 (2017)

Tsyplakova, S.A.: Modeling project activity in the system of professional education. Probl. Mod. Pedag. Educ. **58–3**, 274–277 (2018)

Tsyplakova, S.A.: Professional and pedagogical education and trends and of its development. Probl. Mod. Pedag. Educ. **56–8**, 274–280 (2017)

Project Activities of University Students by Means of Digital Technologies

Elvira K. Samerkhanova(✉), Lyudmila N. Bakhtiyarova, Aleksandr V. Ponachugin, Elena P. Krupoderova, and Klimentina R. Krupoderova

Nizhny Novgorod State Pedagogical University named after K. Minin, Nizhny Novgorod, Russia
samerkhanovaek@gmail.com, l_bach@rambler.ru, sasha3@bk.ru, krupoderova_ep@mininuniver.ru, kklimentina@gmail.com,

Abstract. **The goal** is to analyze possibilities of the project method using digital technologies in the process of teaching students.

Relevance of the research lies in the wide application of the project method in the process of teaching university students, on the one hand, on the other hand, the potential of using this method in combination with digital technology is not fully disclosed.

Methods—general scientific methods were used in the course of research: analysis, including, domestic and foreign literature, synthesis, analogy; special: pedagogical experiment, method of studying products of students' activity, etc.

The main results and novelty of the research—the analysis of digital technologies in the project activity of university students was carried out. It was revealed that digital technologies take on several functions: tool for project creation and presentation, information environment of project operation, information space of project presentation and storage.

Practical significance of research—demonstrated the use of various digital tools in the development of business plans, organization of network project activities using different Internet tools. The impact of the project method using digital technologies as a means of implementing the activity method on the student in the process, interaction with social and virtual environments. Examples of the impact of project activities on the formation of personal and professional qualities of future graduates are presented.

Keywords: Project method · Project activity of university students · Digital technologies · Network services · Professional competence · Personal qualities of the student

JEL Code: I230

E. G. Popkova and B. S. Sergi (Eds.): ISC 2019, LNNS 91, pp. 460–467, 2020.
https://doi.org/10.1007/978-3-030-32015-7_52

1 Introduction

At present, great importance is attached to project activities in the educational process in general and in the educational activities of university students in particular. The project method has a deep history and a sufficient number of examples of its practical implementation.

The principle of personal interest of students laid down in the method of projects by its founders J. Dewey and V.H. Kilpatrick, allows us to speak about the conditions of formation of professional competences within the educational process. Safonova and Podolsky note that the essence of the project method is to solve a specific practical problem in the field of future professional activity. Project method, according to authors, represents learning technology, in which students acquire new knowledge in the process of step-by-step solution of the problem (Safonova and Podolsky).

The use of digital technologies has greatly expanded the scope of application of the project activity and its functionality. Identification of the potential for the implementation of the project method by means of digital technologies will make it possible to push its boundaries of influence on the formation of professional competencies of future specialists.

2 Theoretical Basis of the Study

The interest in the student's personality manifested at the end of the 19th at the beginning of the 20th centuries, and revision of views on the educational system led to emergence of the project method. The term "project method" was widely used in pedagogical literature in 1911. Further understanding and theoretical substantiation of the project method was presented in V. Kilpatrick (theoretical basis), E. Killings (advantages of the project method over the traditional learning system), E.G. Kagarova (analysis of the experience of the method), G. Meandrova (stages of work on the project), etc. The first attempts to apply the project method of training in Russia were realized at the beginning of the 20th century by a group of employees led by S.T. Shatsky.

The modern approach of using the design method is reflected in the works of J.K. Jones, G.M. Kojaspirova, P.F. Kaptereva, M.V. Krupenina, E.S. Polata, N.Y. Pakhomova, V.N. Shulgina and others. For example, J.K. Johnson proposed design analysis, which is universal in nature, and brainstorming as a method of new ideas.

The use of modern information technologies in educational activities is revealed in many works of Russian and foreign scientists (L. Dawley, C.N. Gunawardena, M.S. McIsaac, O.A. Kozlov, I.V. Robert, E.K. Samerkhanov, A. Saavedra, V. Opfer, et al.).

Project activity of students with the use of information and communication technologies is presented in the works of E.S. Polat, I.V. Robert, E.K. Samerkhanova, et al.

Most authors note the interdisciplinarity, deep integration of knowledge, increasing the degree of independence and the desire for self-education of students in the implementation of the project method by means of digital technologies.

3 Research Methodology

Analysis of domestic and foreign literature on topic of research showed that, despite the widespread use of modern information technologies of the project method, the potential of method and digital technologies are not fully disclosed. Systematization of the possibilities of the project method implemented by means of digital technologies will allow to have a positive impact on the quality of the educational process (Kuznetsov et al. 2018).

In many works, it is noted that information and communication technologies act in relation to the project method of learning as a tool (Kuznetsov et al. 2017). For example, today for the organization of project activities of students many teachers use the possibilities of network services. Zabelina writes that capabilities of the electronic information and educational environment of a higher education institution allow significantly diversify forms and types of organization of project activities, also adapt them to the capabilities and needs of each student (Zabelina 2014).

Didactic capabilities of modern ICT tools based on Web 2.0 services and e-learning environments for organizing students' project activities are discussed in the works of many authors (Krupoderova 2013; Krupoderova 2014; Ponachugin and Lapygin 2018, Samerkhanova and Izhmarova 2018). It should be noted that digital technologies in this case also act as an information space for project presentation and storage, which requires integrative knowledge from students: on the one hand—the subject area, which is reflected in the project, including the phased creation, on the other hand—the information space as a means of presenting and storing.

The diversity of digital technologies in project activities of university students is quite wide. In practice-oriented projects, such as the development of design or economic-oriented investment projects, business projects, modern information technologies, on the one hand, act as a tool for creating and presenting a project, on the other hand, as the environment for its operation (Bakhtiyarova 2014).

Thus, it can be stated that digital technologies in the project activities of university students perform three main complex functions:

- tool for creating and presenting a project;
- information environment;
- information space for presentation and storage.

4 Analysis of the Results

The project method of digital technologies is used as a tool for creating and presenting a project and an information environment for the project at the Minin University implemented in many disciplines, including "Information Technologies in Management".

In accordance with the Federal State Educational Standards of Higher Education, the area of preparation 38.03.02 Management (undergraduate level) should have skills of quantitative and qualitative analysis of information when making decisions, building economic, financial, organizational and management models, and business planning skills.

In the course "Information Technologies in Management", students are in small groups (2–3 people), using computer technologies for creating business plans. For example, "Business plan for creating a furniture manufacturing company" or "Business plan for creating studio on tailoring "etc." This project is practice-oriented. The methodology for preparing a business plan by means of computer technology includes a set of tasks:

- creation of economic dynamic business model in spreadsheets;
- preparation of a text document "Business Plan" by means of a text editor;
- creation of a database of employees by means of the database management system;
- Preparation of presentation of a business created by means of presentation graphics (Bakhtiyarova 2014).

Spreadsheets, text and graphic editors, database management system act as a tool for creating a project. In addition, spreadsheets and database management systems are the operating environments of a project: a business plan created in spreadsheets, as an electronic business model of a business, can be modified by making changes to key positions in a business plan, presence of source and dependent cells in spreadsheets make the business plan model dynamic.

E.S. Polat writes: "The solution to the problem inherent in any project always requires involvement of integrated knowledge" (Polat and Bukharkina 2007). Developing a business plan for a company requires students to have integrated knowledge that includes the subject area (for example, furniture production or car repair), economics (business plan development), digital technologies (office software, the central part of which are spreadsheets as a means of creating an economic model business).

Project activity of students in microgroups involves not only the implementation of the entire range of activities, but such managerial functions as distribution of responsibilities for collecting and analyzing information, developing an economic business model, including the organizational structure of the company, text document, advertising product, database, and a business plan presentation. One of the problems inherent in the project is the creation of effective business, the payback of which should come at the end of 1–2 years of its operation. In the project, students analyze information gathered about the subject area, discuss the structure of the business plan and form of its presentation. Using the presence in spreadsheets of functions of source and dependent cells, students optimize the economic model of the business created in the electronic environment.

An example of a fragment of one of the main tables "Income and Expenses Plan" of the business plan for creating a hairdressing salon is shown in Fig. 1.

	A	B	C	D	E	
3						
4	План доходов и расходов					
5						
6	Наименование статей	Условное обозначение	Формула, комментарий	Сумма по статье в месяц, руб.	Сумма по статье в год, руб.	Гс
7	1.Чистая выручка от продаж(нетто)	Вч	Вч=Впр*100/(100+20)	292 500	3 510 000	
8	1.1. Выручка от реализации продукции (выручка от продаж), включая НДС	Впр	см. лист "Объем продаж"	351 000	4 212 000	
9	2. Переменные затраты (издержки прямые)	Ип		32 704	392 448	
10	2.1. Стоимость продукции		см.лист " Прямые издержки"	32 704	392 448	
11	3. Валовой доход	Дв	Дв=Вч-Ип	259 796	3 117 552	
12	4.Косвенные издержки (постоянные затраты)	Ик		237 008	2 844 091	
13	4.1. Фонд зароботной платы (фонд оплаты труда)	ФОТ	см. лист "Оплата труда"	142 016	1 704 192	

Fig. 1. Fragment of the Business Plan Table "Revenue and Expense Plan"

Electronic educational-methodical complex "Information technologies in management" is located in the electronic information and educational environment of the university. Within the course, manuals, methodological recommendations are presented, elements of course—tasks as separate algorithmic units of a project are placed. Fragment of the electronic educational and methodical complex "Information technologies in management", including the development of a business plan, is shown in Fig. 2.

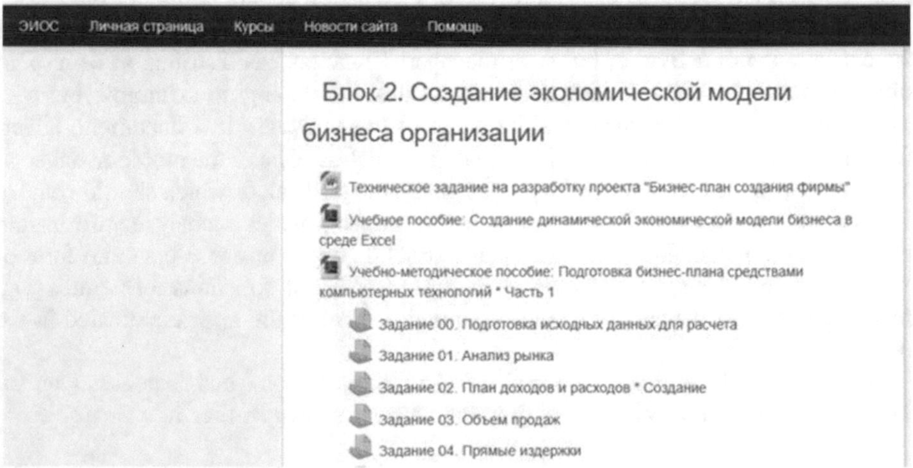

ЭИОС Личная страница Курсы Новости сайта Помощь

Блок 2. Создание экономической модели бизнеса организации

Техническое задание на разработку проекта "Бизнес-план создания фирмы"

Учебное пособие: Создание динамической экономической модели бизнеса в среде Excel

Учебно-методическое пособие: Подготовка бизнес-плана средствами компьютерных технологий * Часть 1

Задание 00. Подготовка исходных данных для расчета

Задание 01. Анализ рынка

Задание 02. План доходов и расходов * Создание

Задание 03. Объем продаж

Задание 04. Прямые издержки

Fig. 2. Fragment of "Information Technologies in management"

Project activity of students, according to the presented methodology, is carried out with students of full-time and part-time forms of education for several years. The results of the experiment were collected, summarized and ranked in accordance with the complex functions of digital technologies and the project training activities of students (Table 1).

Table 1. Results of a business plan activity of students

A comprehensive digital function	Software tool	Design activity of students	Percentage of students of the total number of in groups that received the maximum number of points (or close to it) for completing tasks
Project creation and presentation tool	Search engines	Collecting information for the execution	~82%
	MS Excel	Creating an economic business model (business plan)	~73%
	MS Access	Creating a database of employees	~75%
	MS Word	Create a Business Plan text document	~85%
	MS PowerPoint	Creating a business plan presentation	~88%
Information environment of operation	MS Excel	Business plan optimization	~62%
	MS Access	Replenishment and optimization of the database	~78%
Information space for presentation and storage of the project	Moodle	Work in the electronic educational environment of the university	100%

For the third year, the University of Minin has been working under universal baccalaureate conditions. As part of the training of bachelors and masters of pedagogical direction, the formation of a set of competencies ensuring project activities takes place in the context of the development of the information and educational environment of the university (Samerkhanova et al. 2016a, b).

All first-year students study the module "Information technologies", within which the project activity is provided. In preparing their research papers, students used various network services such as wiki, joint document editing services, online mental maps, time tapes, infographics, virtual boards, and online presentations. These services allow you to organize joint activities; create products that contain information presented in the form of text, graphics, video or combined. The opportunity for a student to receive tutor advice and perform tasks, discuss the performance of the assignment not only with the teacher but also with each other (Samerkhanova et al. 2016a, b).

The results of project activity of students within the module “Information technologies” are presented in Table 2.

Table 2. Results of project activity of students in the framework of the module “Information technology”

A comprehensive digital function	Software tool	Design activity of students	Percentage of students of the total number of in groups that received the maximum number of points (or close to it) for completing tasks
Project creation and presentation tool	Search engines, online questionnaires, co-editing tables	Collecting information for the execution	~90%
	Wiki, joint document editing services, online mental maps, time tapes, infographics, virtual boards, online presentations	Creation of joint products	90%
Information environment of operation	Wiki, blogs, virtual boards, co-editing documents	Participation in brainstorming, self-assessment and mutual evaluation, reflection	~85%
Information space for presentation and storage of the project	Moodle	Work in the electronic educational environment of the university	100%
	Wiki, Google sites, co-editing documents	Presentation of projects through relevant Internet services	~95%

5 Conclusions

As practice shows, application of the project method using digital technologies in the educational process of a university with focus on productive learning, as well creates conditions not only for the successful formation of a system of basic knowledge and professional competencies but also generates motivation for permanent acquisition of new knowledge, contributes to the disclosure of personal qualities of students, bordering on professional characteristics, such as readiness for social dialogue and partnership. The method develops creative abilities, initiative, and independence in decision making.

References

Bakhtiyarova, L.N.: Creation of a Dynamic Economic Model of Business in the MSExcel Environment: Textbook, 148 p. NGPU n.a. K. Minina, N. Novgorod (2014)

Zabelina, H.A.: Application of the project method for the formation of professionally significant qualities of students-informatics. Materials of the International conference-exhibition “Information technologies in education” (2014) [Electronic resource]. http://ito.su/main.php?pid=26&fid=3670. Accessed 10 Apr 2018

Krupoderova, E.P.: Preparation of students for project activity in the information educational environment of the XXI century (2013). http://www.mininuniver.ru/scientific/scientific_activities/vestnik/archive/. Accessed 12 Sept 2018

Krupoderova, K.R.: Patriotic education of youth through network project activity, no. 1. Bulletin of Minin University (2014) [Electronic resource]. http://www.mininuniver.ru/scientific/scientific_activities/vestnik/archive/. Accessed 11 Jan 2018

Polat, E.S., Bukharkina, M.Yu.: Modern Pedagogical and Information Technologies in the Education System: Textbook for Students of Higher Educational Institutions, 368 p. Publishing Center “Academy” (2007)

Ponachugin, A.V., Lapygin, Yu.N.: Organization of interactive interaction in e-learning [Electronic resource]. https://vestnik.mininuniver.ru/jour/article/view/696. Accessed 18 Dec 2018

Samerkhanova, E.K., Imzharova, Z.U.: Organizational and pedagogical conditions of formation of readiness of future teachers for project activity in the conditions of digitization of education, no. 2. Vestnik Minin University (2018). https://vestnik.mininuniver.ru/jour/article/view/805. Accessed 15 Nov 2018

Safonova, K.I., Podolsky, S.V.: Project activity of students in the university: the principle of selection of projects and the principles of formation of project groups [Electronic resource]. https://doi.org/10.24158/spp.2017.9.11. Date of appeal 28 Apr 2018

Dawley, L.: Social networking knowledge construction: emerging virtual world pedagogy. OntheHorizon **17**(2), 109–121 (2009)

Gunawardena, C.N., McIsaac, M.S.: Distance education. http://ocw.metu.edu.tr/file.php/118/

Kuznetsov, V.P., Garina, E.P., Andryashina, N.S., Romanovskaya, E.V.: Models of Modern Information Economy Conceptual Contradictions and Practical Examples, 361 p. Emerald Publishing Limited (2018)

Kuznetsov, V., Kornilov, D., Kolmykova, T., Garina, E., Garin, A.: A creative model of modern company management on the basis of semantic technologies. Communications in Computer and Information Science (2017)

Week10/Gunawardena-McIsaac-distance-ed.pdf. Accessed 17 Oct 2018

Saavedra, A., Opfer, V.: Learning 21st-century skills requires 21st-century teaching. Phi Delta Kappan **94**(2), 8–13 (2012)

Samerkhanova, E.K., Krupoderova, E.P., Krupoderova, K.R., Bahtiyarova, L.N., Ponachugin, A.V.: Students’ network project activities in the context of the information educational medium of higher education institution. Int. J. Environ. Sci. Educ. **11**(11), 4578–4586 (2016a)

Samerkhanova, E., Krupoderova, E., Krupoderova, K., Bahtiyarova, L., Ponachugin, A.: Networking of lecturers and students in the information learning environment of higher school by means of cloud computing. IEJME Math. Educ. **11**(10), 3551–3559 (2016b)

Information-Project Technology for the Formation of General Competencies of Students by Means of Electronic Information and Educational Environment

Alexandra A. Tolsteneva[1(✉)], Valeria K. Vinnik[2], Marina V. Lagunova[3], Anna A. Voronkova[2], and Natalia D. Zhilina[3]

[1] Minin Nizhny Novgorod State Pedagogical University, Nizhny Novgorod, Russia
tolstenev25@yandex.ru

[2] Nizhny Novgorod State University n.a. N.I Lobachevsky, Nizhny Novgorod, Russia
vinnik@yandex.ru, anavoronkova@mail.ru

[3] Nizhny Novgorod State University of Architecture and Construction, Nizhny Novgorod, Russia
mvlnn@mail.ru, zhilina@nngasu.ru

Abstract. The relevance of presented research is determined by the need for development and implementation of training technologies that provide high-quality professional training of students in the context of increasing the amount of time devoted to independent work. The goal of this work is to develop and test pedagogical technology aimed at the formation of general competences of students with the effective use of electronic information – environment (EIE). Methods of research used by authors: analysis of scientific literature, modeling, pedagogical experiment, methods of mathematical statistics, etc. The article considers experience and new possibilities of organization of independent work of students in conditions of informatization. Also, analysis of scientific and pedagogical researches devoted to the study of methods of organization of independent work is carried out. Theoretical and methodological substantiation of new author's information and project technology of independent work of students with the use of electronic educational-methodical complexes implemented, including methodological approaches and principles. Authors proposed the original system, which includes principles: cumulative, modular and integrative. All stages of interaction between teachers and students in the implementation of technology are described, namely: motivational, cognitive, integrative activity, evaluative and reflexive. Estimates of efficiency of implementation of the author's teaching technology are offered. The results of approbation and project technology are presented. The efficiency of the developed technology was assessed, and possible directions for the further development of EIE due to the expansion of methodological opportunities for the systemic use of interdisciplinary methodical complexes are presented. The result of the pedagogical experiment was a statistically significant increase in the components of general competences of students: motivational, cognitive, activity, evaluative and reflexive. Thus, the information and project technology

E. G. Popkova and B. S. Sergi (Eds.): ISC 2019, LNNS 91, pp. 468–476, 2020.
https://doi.org/10.1007/978-3-030-32015-7_53

proposed by authors of formation of general competences of students can be proposed for use in educational organizations.

Keywords: Electronic information and educational environment (EIE) · Independent work of students · Information and project technology · Electronic educational-methodical complex

JEL Code: I230

1 Introduction

At the present stage of development of the digital society should be the most important task of the university is training demanded and hence competitive specialists who are ready to continuous professional development, and self-education (Kuznetsov 2018). Integral characteristic of the graduate's readiness is a complex of formed general and professional competencies. General competencies allow independently learning, predicting, and improving professional competence throughout life. Innovative transformations of the modern educational system significantly changed the organization of educational work. The emphasis in the preparation of students is transferred to the process of independent knowledge (Kuznetsov and Kornilov 2017). One of the modern tools for the organization of educational process is the EIE of organization. Currently, EIE is a set of separate, independently studied electronic courses, which limits the possibilities of training of students and makes it relevant search for new effective approaches to the organization of independent work of students based on new opportunities provided by EIE (Lagunova and Yurchenko 2011).

2 Methodology

In recent years, significant changes have been made to the organization of study work of the university, namely the redistribution of the academic load: reduction of classroom hours and increase the proportion of students' independent work. In the context of this study, the analysis of certain directions of organization of self-educational activity of students is significant (Gibbons and Phillips 1982; Reeve and Jang 2006 etc.). Of particular importance for our research are the works of recent years, revealing various aspects of self-education activities of students (Klentak 2017; Strekalova 2017; Rashid and Asghar 2016; Kim et al. 2014).

Based on the research of students' independent work (Zimnyaya 2002; Klentak 2017; Strekalova 2017; Schunk and Mullen 2012, etc.), we define the concept of "independent work" as a kind of internally motivated trainee activity work aimed at the formation of general and professional competencies, which is interdisciplinary, integrative in nature, implemented at all stages of educational activities under the guidance of teachers through interactive interaction with students in terms of information and communication technologies (Tolsteneva and Winnick 2016).

Changes in the informatization of society as a whole and the system of vocational education, we note the appearance of a number of contradictions caused by: insufficient development of pedagogical technologies using modern means of informatization; constant increase in the volume of hours devoted to independent work of students; non-system study of certain disciplines of the curriculum.

A significant resource for resolving contradictions is, in our opinion, a rationally organized independent work, forming a positive motivation of the doctrine, as well as the most important personal qualities—independence, cognitive creative activity, and responsibility.

Here are the main pedagogical approaches that create conditions of self-development and self-realization of the student in the course of independent work in the conditions of EIE (Fig. 1).

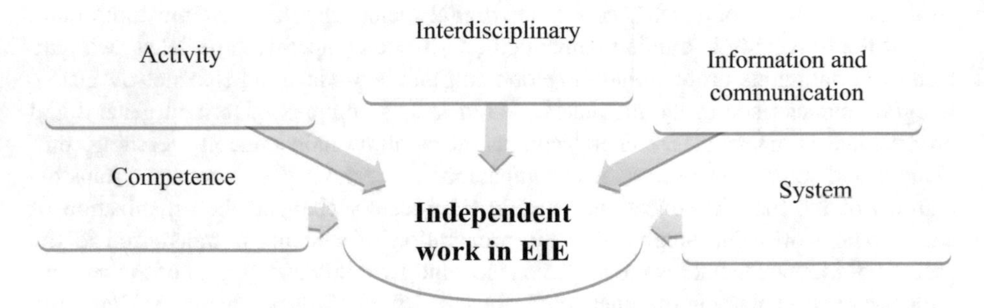

Fig. 1. The set of pedagogical approaches that make up the methodological basis of organization of independent work

Let us characterize the selected approaches.

Competence—provides the formation of general competences of students; aims to develop the ability of students to independently solve didactically adapted professional tasks, using EIE resources.

Activity—represents independent work as a way of organization of educational and cognitive activity, in which students are active participants of cognitive process; orients all methodological possibilities for creating conditions for purposeful, intensive, constantly increasing complexity of professional-oriented activities.

Interdisciplinary—involves the creation and use of EIE, in which the student can solve professionally significant tasks both within individual disciplines and at the level of discipline cycles; using the information resources of various courses;

System approach—represents EIE as a set of interrelated elements - electronic courses, which, in turn, have a single ordered structure. Relatively independent components of EIE are considered not in isolation, but in their relationship, which allows to obtain integrative system properties and new qualitative characteristics, which are missing from the individual components of the system.

Information and communication—involves expansion of the information field in the process of communication and exchange of information; It shifts the emphasis from the informational role of the teacher as an informant to the role of the organizer of the communicative process, helping to learn independently.

Each of these approaches is implemented through a system of principles, including along with general pedagogical author's principles, reflecting the specifics of the work carried out and possibilities of EIE (Fig. 2).

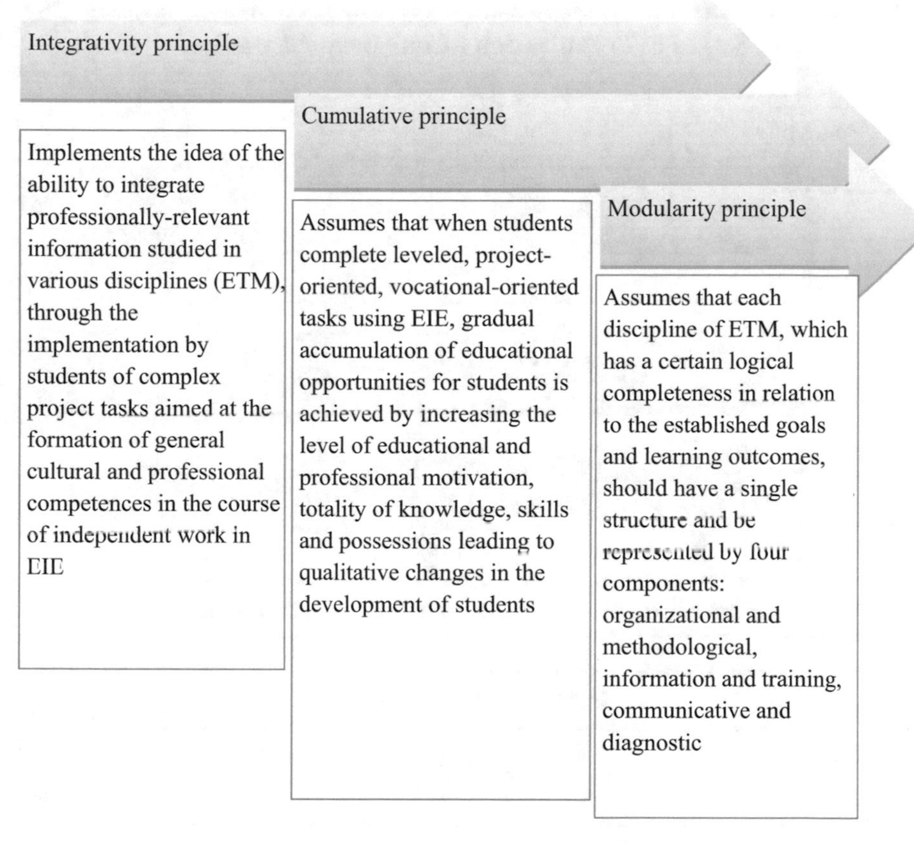

Fig. 2. The author's principles determining the specifics of independent work in conditions of EIE

In the course of independent activity of students within the framework of EIE was developed information and project technology of teaching. The essence of project technology is the management of students' independent work, in which general competencies are formed in the process of planning and performing progressively more complex professionally significant tasks - projects carried out using information and communication technologies. Project technology allows to form general competencies, develop professional and educational motivation of students, form a system of universal knowledge, skills, and possessions of learners, as well as to navigate in information space.

The scheme (Fig. 3) presents the stages of organization of independent work in EIE, determining the joint activity of a teacher and student through the project technology of training.

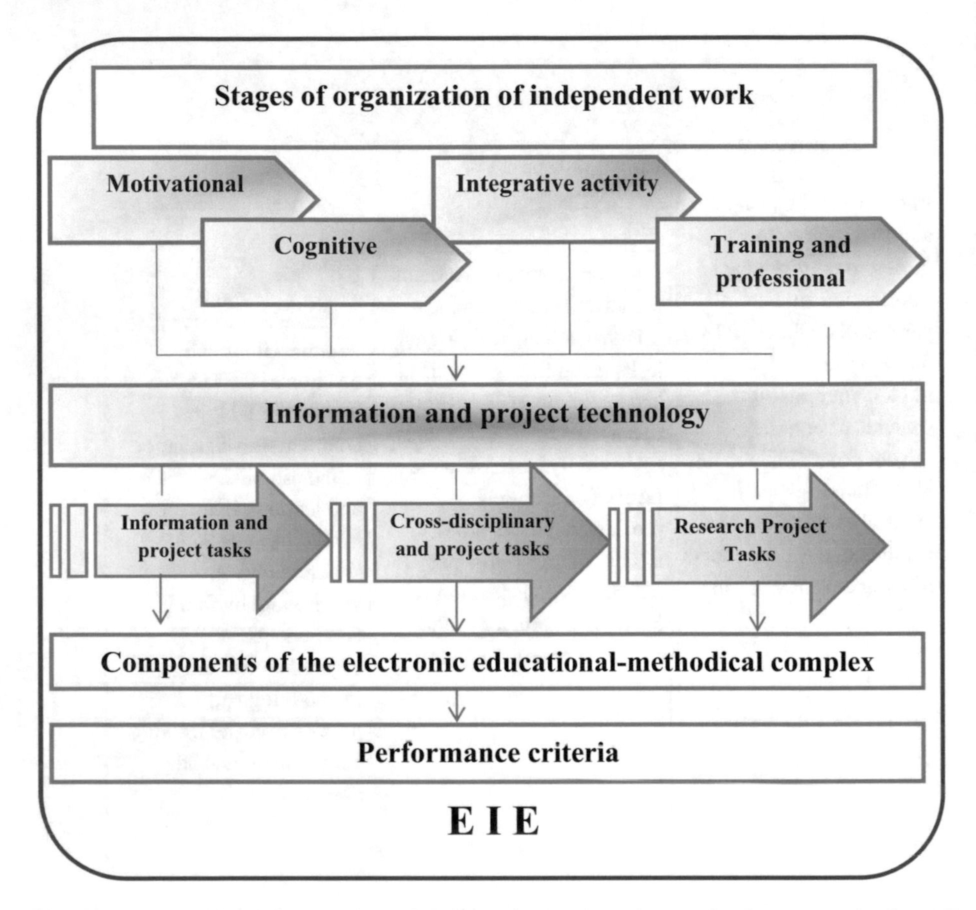

Fig. 3. Scheme of implementation of information and project technology organization of independent work in EIE

Motivational stage—motivation to study electronic courses as elements of preparation for future professional activity is formed. The implementation of tasks at this stage is aimed at achieving self-organization of students, efficiency, success and self-confidence.

The cognitive stage is aimed at the formation of the system of basic concepts within the framework of the development of certain disciplines by students of the project tasks of the first level (information project tasks), aimed at collecting information, assessing quality, depth, analyzing and synthesizing it.

Stage of integration presupposes presence of the developed system of concepts of disciplines, ability to search and systematize professionally significant information, and EIE skills. At this stage, the implementation of project tasks of the second level in groups - interdisciplinary projects; students organize both their own activities and the work of the team for resolution of standard situations in solving professional tasks.

Training and professional stage involves the implementation of research project tasks aimed at making decisions in a non-standard professional situation; forms decision - making skills, responsibility for their work and team members. Research project tasks represent independent educational and research work, which requires in-depth study of individual problems of professional activity. The result is the creation of new products, objects, models or the improvement of existing ones.

Criterial-diagnostic apparatus for assessing the effectiveness of the developed technology assumes the evaluation of following indicators:

- *motivational:* expression of interest in the independent study of disciplines as an element of preparation for future professional activities, implementation of design tasks using ICT, and perseverance in the organization of independent work;
- *cognitive:* determines the amount and quality of self-acquired theoretical knowledge in the course of development in EIE;
- *activity:* ability to timely and independent performance of professionally significant project tasks requiring an in-depth study of individual issues of professional activity on the basis of the system knowledge generated during the implementation of project activities in EIE;
- *evaluative reflexive.* ability to plan, organize and manage their work in the EIE, objectively assess their and other's results in projects.

For the implementation of the technology, a structure and content of the complex on disciplines of general education, general professional and interdisciplinary cycles were developed. By complex we mean a set of educational and methodical materials united by means of EIE, providing a full didactic cycle of studying disciplines, including the following components:

- *organizational and methodical*—contains information on the organization of the educational process on discipline;
- *informative*—reflects a content of the discipline;
- *communicational*—contains various elements of a course, providing contact communication between teacher and student;
- *diagnostic*—contains elements of the course "Tests", "Task", etc. The "Task" element allows students to download completed tasks and submit them to teacher in various forms.

3 Results

To test the effectiveness of the developed technology, a pedagogical experiment was conducted on the basis of UNN n.a N.I. Lobachevsky when preparing students for programs of secondary vocational education of 02.19.10 "Technology of catering products" and 09.02.04 "Information systems (by industry)" during 2015–2017 years. The pilot group was 87, the control group - 78.

To determine the motivation index, the following methods were used: observation, analysis of student activity results, questionnaire (complex of diagnostic methods of assessment of motivation of educational activity of V.A. Yakunin (Badmaeva 2004); Zamfir's methodology of "Study of motivation and professional activity").

To assess the cognitive component, the method of determining the quality coefficient of education by Grabar and Krasnyanskaya (1977), adapted to this study was used, as well as tests on topics and sections of academic disciplines.

Evaluation of activity component was carried out by the analysis of project level tasks, method of determining the level of independence of Tretyakov and Sennovsky (2001).

The method of A.V. Karpov "Diagnostics of Reflection" (Karpov 2003) was used to measure the evaluative and reflexive component.

Dynamics of levels of formation of the selected indicators are presented in Fig. 4.

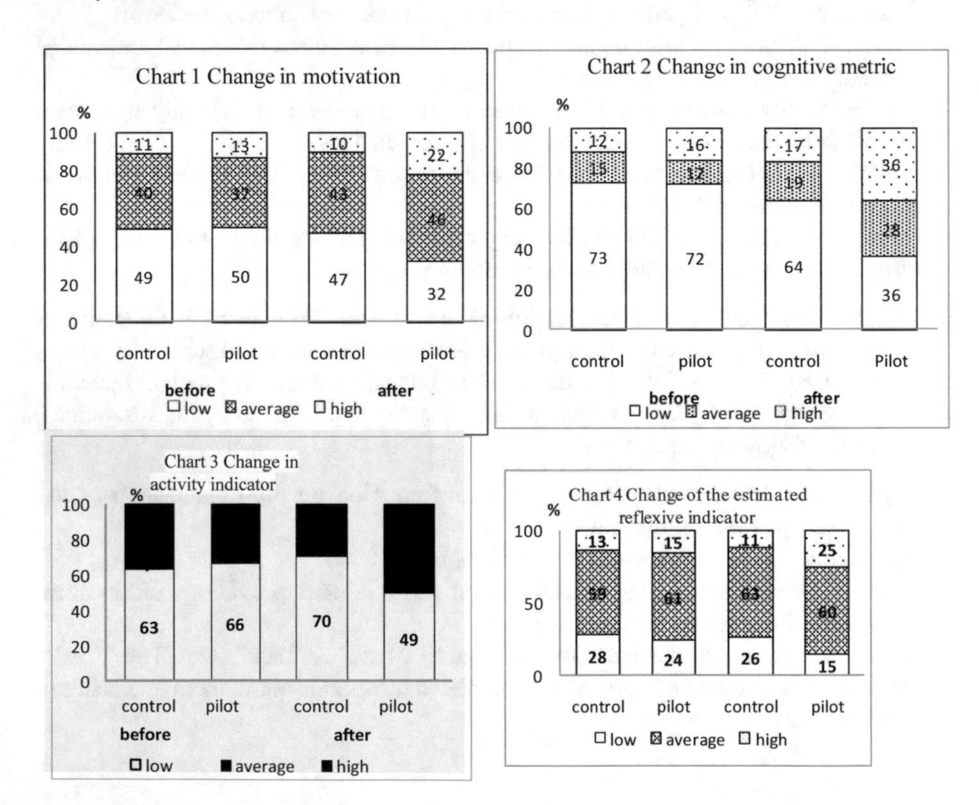

Fig. 4. Dynamics of levels of formation of general competences

Empirical values $\chi 2$ for the distribution of students control and experimental groups by the level of formation of indicators of formation of general competences at the end of the experiment with the level of values of 0.05 for the number of degrees of freedom 2 (v = 3 − 1) revealed the inequality of ^{2}pcs. ^{2}criteria for all indicators.

This leads to the conclusion that the obtained results of the experiment are significant and reliable with an accuracy of 95%.

Results indicate more effective formation of general competencies in the experimental group, where the learning process was built with the use of information and project technology.

4 Conclusion

The obtained results testify to the pedagogical effectiveness of proposed information and project technology of training. In addition, the use of the proposed technology allows expanding the methodological capabilities of EIE due to systematic application of interdisciplinary links of educational program, and providing integrative professional training with independent work.

References

Badmaeva, N.C.: Methodology for diagnostics of educational motivation of students (A.A. Rean and V.A. Yakunin Modification of N.T. Badmaeva). Badmaeva, N.T.: Influence of motivational factor on the development of mental abilities: Monograph, Ulan-Ude, pp. 151–154 (2004)

Grabar, M.I., Krasnyanskaya, K.A.: Application of mathematical statistics in pedagogical research. Non-parametric methods. Pedagogy, p. 136 (1977)

Gibbons, M., Phillips, G.: Revue canadienne de l'éducation. Can. J. Educ. **7**(4), 67–86 (1982)

Zimnaya, I.A.: Pedagogical psychologists. Series: Textbook of the XXI century. M.s Logos, 304 p. (2002)

Karpov, A.V.: Reflectivity as a mental property and a technique of its diagnosis. Psychol. J. **24** (5), 45–57 (2003)

Kim, R., Olfman, L., Ryan, T., Eryilmaz, E.: Leveraging a personalized system to improve self-directed learning in online educational environments. Comput. Educ. **70**, 150–160 (2014)

Kuznetsov, V.P., Garina, E.P., Andryashina, N.S., Romanovskaya, E.V.: Models of Modern Information Economy Conceptual Contradictions and Practical Examples, 361 p. Emerald Publishing Limited (2018)

Kuznetsov, V., Kornilov, D., Kolmykova, T., Garina, E., Garin, A.: A creative model of modern company management on the basis of semantic technologies. Communications in Computer and Information Science (2017)

Klentak, L.S.: Formation of the ability to self-organization of independent work of students of technical university: thesis of Candidate of Pedagogical Sciences: 13.00.08; Klentak Lyudmila Stefanovna, Place of protection, Samara, 188 p. (2017)

Lagunova, M.V., Yurchenko, T.V.: Management of cognitive activity of students in the information and educational environment of the university: monograph, 160 p. NNU, N. Novgorod (2011)

Strekalova, N.B.: Quality management of students' independent work in the open information and educational environment: Doctor thesis of Pedagogical Sciences: 13.00.08; Strekalova Natalia Borisovna, Place of protection: national research academy Koroleva, Samara, 588 p. (2017)

Schunk, D.H., Mullen, C.A.: Self-efficacy as an engaged learner. In: Christenson, S., Reschly, A., Wylie, C. (eds.) Handbook of Research on Student Engagement, Self-Education: The Process of Life-Long Learning, pp. 219–235. Springer US, Boston (2012)

Rashid, T., Asghar, H.M.: Technology use, self-directed learning, student engagement and academic performance: examining the interrelations. Comput. Hum. Behav. **63**, 604–612 (2016)

Tolsteneva, A.A., Vinnik, V.K.: Organization of independent work of students based on information and communication technologies: monograph, 102 p. Minin University, Nizhny Novgorod (2016)

Tretyakov, P.I., Sennovsky, I.B.: Technology of modular education in school: Practically-oriented monograph. 352 p. (2001)

Intellectual Resource as a Factor of Ensuring National and Cultural Security in the Conditions of the Training Course "Teacher of Russian as a Foreign Language"

Elena M. Dzyuba(✉), Victoria T. Zakharova, Natalia M. Ilchenko, Anna L. Latuhina, and Tatyana N. Sheveleva

Minin Nizhny Novgorod State Pedagogical University, Nizhny Novgorod, Russia
dsjubannov@list.ru, victoriazaharova95@gmail.com, ilchenko2005@mail.ru, alatuhina@yandex.ru, tatyana.n.sheveleva@gmail.com

Abstract. The aim of this study is to analyze methods of translating the national cultural code in the context of exporting education in Russian, which formed the basis for the concept of training a teacher of Russian as a foreign language (RFL). Methodological guidelines define the concepts: national cultural code, intellectual resource, national and cultural security. The national cultural code is understood by authors of the article as a didactic unit in conditions of creating an educational resource for training specialist in the field of teaching RFL. Current state of exports of Russian education is analyzed including a humanitarian sphere. Intellectual resource in the field of pedagogical education is understood as a set of professional competencies acquired by a graduate.

Export of education also involves understanding the problem of national and cultural security, condition of which is formation of a concept of education in Russian, taking into account the peculiarities of mentality, Russian linguocultural specifics.

Results of this study suggest the need to include an educational program of teacher training of the RFL disciplines containing national and cultural specificity, plan of lessons, elective courses, and writing of qualification work in the framework of linguocultural and sociocultural, regional specifics, correlated with the basic axiological vectors of Russian society life.

It is concluded that the methodologically well-implemented national concept of teaching Russian as a foreign language and preparing a teacher of Russian as a foreign language will ensure national and cultural security of an educational concept.

Keywords: National cultural code · Concept of RFL teaching · Intellectual resource

JEL Code: I230

E. G. Popkova and B. S. Sergi (Eds.): ISC 2019, LNNS 91, pp. 477–483, 2020.
https://doi.org/10.1007/978-3-030-32015-7_54

1 Introduction

Among the priority federal projects of the Russian Federation is the project "Development of export potential of the Russian education system" (Passport of the priority project "Development of export potential of the Russian educational system" 2017), implementation of which is designed to increase the competitiveness of Russian education in the international market of educational services. This component of the so-called "non-raw export" is characterized as a strategic resource of the Russian economy (Kuznetsov 2018). Its potential is focused on the long term and is associated with the assessment of quality and demand of specialists, including foreign specialists, who have received education in higher education institutions Russian Federation in Russian.

An important role in the strategy of educational export plays an intellectual resource, which is transposed by a graduate of a certain educational program of a Russian university and mastered in a potential professional environment of a graduate. As a result, it becomes part of an "important, intangible asset of an enterprise" (Gosteva and Akentieva 2015), (Kuznetsov 2017).

In the field of humanitarian education, including the subjects of program teacher of Russian as a foreign language and/or programs that train foreign students in Russian language, intellectual resource is the competence of a graduate in the field of pedagogical, research, managerial and aesthetic activity, as well as the product of their scientific activity (Vladimirova et al. 2018).

It should be noted that the economy of development of intangible, intellectual resource in conditions of production is focused on measurability of result, including conditions of an educational institution. From the point of view of a modern researcher the intellectual resource is: "resource that characterizes knowledge", "intellectual potential of the organization", "its competitiveness"; "value of the set of alienated and inalienable intellectual assets involved in the economic activity of a legal entity, including intellectual property (B.B. Leontiev)"; "intangible material values that increase the market value of a company" (Astafieva 2007).

Humanitarian sphere as a subject of development in educational activities is not clearly measurable. That is why there are no relevant indicators that allow taking into account value/axiological characteristics of an intellectual resource. But they are largely related to a specificity of the exporting country of education, its culture and tradition. These indicators are also reflected in specifics of the export of education in the Russian language.

At the same time, an economic evaluation of the intellectual resource needs to be clarified, from the point of view of the philosophy of educational export, its ideological component, which includes issues of national and cultural security and understanding of the role of national and cultural code in the process of education in Russian.

The aim of our research is to identify ways of translating national values/national cultural code into educational activities in Russian and for education in Russian as factor of ensuring national security.

Achievement of research goals requires a sense of a number of direct research tasks, among which:

– National and cultural code as a key methodological concept within the framework of export of Russian education and education in Russian;
– Educational intellectual resource of a foreign student and teacher of Russian as a foreign language;
– Potential of the scientific and educational center as a driver of research activities in the field of definition of value vectors.

2 Methodological Installations of the Study

In the research of recent years, a discussion of English-language training programs in domestic universities is relevant. In the article of G. A. Krasnova the thought of attractiveness and perspective of "English educational existence", its prospects and mobility, especially in relation to master's programs: "It is not a question of substitution or replacement of Russian language with English. We are talking about the other —Russian-English bilingualism in truly modern domestic education, education as a unity of teaching and learning. Which is immersed also in the widest social context" (Krasnova 2013) "The lack of good educational programs in English" leads, according to experts, to a loss of share participation of Russian education in the world market of educational services.

This kind of analytics is symptomatic at the present stage of reforming domestic education in higher education, as evidenced, for example, by the detailed commentary on content and prospects development of the English language teacher training program at the Nizhny Novgorod State Pedagogical University named after Kozma Minin (Shamov and Kim 2018).

In the article, developers of the program, first of all, emphasize academic benefits that the university receives in the field of academic mobility, professional scientific communication of teachers and "export of available educational services at different levels of employees training (bachelor, master and postgraduate)" (Shamov and Kim 2018).

It will ensure from the point of view of a number of researchers attractiveness of Russian education not only in the capital's universities of Moscow and St. Petersburg but also in province, still does not give up the position of fundamental education in fields of physics, computer technology, medicine, and linguistics. Problems of education of foreign citizens in English, increasing the number of hours for teaching English are also in the center of research attention in work of Pimonova and Fomina (2018).

In this case, the national Russian language becomes an important translator of the national cultural code, and its development becomes a means of eliminating intercultural conflict. According to the fair observation of S.G. Ter-Minasova, "behind the word is the concept, behind the concept—object or phenomenon of reality of the world, and this is the world of another country" (Ter-Minasova 2000). Export of education in Russian, translation and development of intellectual resources assume mastering of cultural (conceptual) picture of the world, which we understand as "reflection of the real picture through the prism of concepts formed on the basis of representations of man obtained by means of senses and passed through his consciousness, both collective and individual" (Ter-Minasova 2000).

The term "cultural picture of the world" is also related to the basic concept of "cultural code" or "national cultural code", methodological relevance of which for our research we consider to be the most significant. By culture code we mean a set of mentefacts connected with cultural senses phenomena belonging to the same type and/or sphere of existence, forming a certain picture fragment of the world) (Krasnih 2016).

3 Results

The decryption of the national cultural code of community is achieved only by approaching its linguistic culture, mastering the key meanings of a particular culture in the national language. Cultural centricity as a feature of modern humanitarian education is reflected in the educational paradigm of teaching methods of RFL: culture becomes the goal of learning, language is a learning tool for culture. Therefore, language and culture at classes are an inseparable unity, where priority is given to culture, the assimilation of which in the process speech of communication helps to reveal national cultural codes not only of the country but also to learn the peculiarities of their own mentality. That is why the presence of educational programs of humanitarian profile of preparation of disciplines based on national values, seems to be very important. "Country studies of Russia", "Main plot in Russian literature", "Archetypes of Russian culture", "Philological analysis of text" and others can be included as a variable component in a training program as foreign students receiving specialized education "Russian language", "Teaching Russian as a foreign language", and as a part of training programs for masters and future teachers of RFL these disciplines are designed to form a number of professional competencies of the future teacher: general humanitarian, linguistic, methodological, vocational and communicative (Molchanovsky 1999). The formation of these competencies gives an opportunity to develop students an increased susceptibility of national cultural values, ability to apply certain communicative strategies and act as mediator in intercultural contacts (Beyer 2010; Wiseman and Koester 1993). At the end of development of an educational program, we get an intellectual resource: availability of necessary knowledge, ability to use them in pedagogical activity, competitiveness in the market of educational services.

The product of scientific and methodical activity of graduate students - future teachers of RFL are the program of author's elective courses, developed within the framework of production (project) and partially tested during pedagogical practices: "Welcome": formulas of acquaintance and rules of behavior in Russia; "What is my name in you?" (Russian onomastics for foreign students), "Secrets of the Russian soul (concept sphere of the Russian nation)", "In our life, beautiful, and strange …", "Russian soul in B. Okudzhava's poetic text", "Peoples of Russia: peculiarities of everyday life, character, culture", "Laughter in a foreign language is a mystery"; "Literature of Russian emigration", etc.

A plan of elective courses involves the active inclusion of the regional cultural component: for example, elective course "M. Gorky in Nizhny Novgorod"; "Texts about Nizhny Novgorod as a didactic material within the course of language studies of foreign students".

Within the framework of research and scientific-methodical activity of students on teacher training programs of RFL it is supposed to write final qualification work, content and the problems of which also reflect the national and cultural specifics of Russian speech behavior, customs, rules, social conventions, rituals, country studies: "Features of perception of Russian mentality of foreign students in the analysis of Russian names of their own (on the examples of the XIX century): theoretical and methodological aspect", "Language image of the family in the Russian saying: materials for the special course for foreign students", "The text of the author's song as a source of the formation of linguistic and cultural competence of foreign students in RFL classes", "Conceptual analysis in the teaching Russian as a foreign language (on the example of the concept "happiness"), "Anecdote at RFL: sociocultural, linguocultural and communicative aspects".

A number of studies are traditionally devoted to axiological vectors of Russian literature: "Romantic story as a source of knowledge about Russian culture in classes with foreign students", "Formation of socio-cultural competencies in RFL classes (on the example of concept "nobility's nest" in works of I.C. Turgenev).

The concept of the educational program of teaching Russian as a foreign at different levels of its existence is supported by scientific and educational centers, which exist at the specialized departments (in our case NOC "Axiology of Slavic Culture"), accumulate experience of scientific activity, become bases of research practice for students. Here they get the first experience of acquaintance with Russian science, the possibility of publishing in collections of young researchers international scientific conferences, which are held in Nizhny Novgorod State Pedagogical University named after M. K. Minin (Axiology of Slavic Culture 2015, 2016, 2017).

4 Conclusions

The analysis allows us to confirm the need for

(1) Formation of general humanitarian, methodical and professional-communicative competences of future teachers of RFL (undergraduates), linguocultural competence of foreign students (students of RFL at the bachelor level), subject to the inclusion of a broad national cultural context in an educational program;
(2) Practice of planning elective courses that implement different aspects of teaching RFL, on the basis of precedent texts representing basic concepts of national culture.
(3) Practice for writing scientific work: from an article to final qualifying work, focused on the sociocultural (historical and modern) space of Russian reality.

Therefore, it can be said that components of the RFL teacher training program at the baccalaureate and magistracy levels allow:

- Methodically competently implement the national concept of teaching Russian language and Russian culture;
- Use it as a variant of "soft export" of education in Russian and a means of promoting Russian government and national interests abroad;
- Ensure the national and cultural security of the educational concept.

References

Astafieva, N.V.: Intellectual capital in the system of higher education. Bulletin of Saratov State Technical University, Saratov, pp. 147–155 (2007)

Dzyuba, E.M. (ed.): Axiology of Slavic Culture: international collection of scientific articles of young scientists, graduate students, students. Minin University, Nizhny Novgorod (2015, 2016, 2017)

Bedniy, A.B., Yerushkina, L.V.: Nizhny Novgorod through the eyes of foreign students: sociological analysis, p. 40, 42, 50. NNGU named after N.I Lobachevsky, Nizhny Novgorod (2011)

Gosteva, O.V., Akentieva, E.I.: Intellectual resources as the basis of intellectual capital of the organization. Int. Res. J. no. 6, pp. 34–35 (2015). Yekaterinburg

Krasnova, G.A.: Export of Russian education: myth or reality. University Management: practice and analysis, Ekaterinburg, UrFU, no. 6, pp. 70–73 (2013)

Krasnykh, V.V.: Dictionary and Grammar of Linguoculture: Fundamentals of Psycholinguological Culturology, p. 552. Gnosis, Moscow (2016)

Molchanovsky, V.V.: Composition and content of professional competence of a teacher of Russian as a foreign language: abstract of dissertation for Phd: 13.00.02 A.S. Pushkin. Moscow, 42 p. (1999)

Passport of the priority project. Development of export potential of the Russian education system (2017). http://government.ru/news/28013/ (in the version of the protocol dated May 30, 2017 No. 6)

Pimonov, S.A., Fomina, E.M.: Improving the attractiveness of Russian universities for foreign students. University Management: practice and analysis. Ekaterinburg, UrFU, No. 4, pp. 97–109 (2018). http://umj.ru/index.php/pub/inside/2029/

Ter-Minasova, S.G.: Language and Intercultural Communication, p. 624. Word, Moscow (2000)

Shamov, A.N., Kim, O.M.: Corporate training program "English for professional activity" in the modern educational space of goals, content, and results'. Bulletin of Minin University, T. 6, No. 3 (2018). https://elibrary.ru/download/elibrary_36350745_21143929.pdf

Beyer, S.: Intercultural Communication. A literature Review. Research Paper (postgraduate) (2010). https://www.grin.com/document/153743

Kuznetsov, V.P., Garina, E.P., Andryashina, N.S., Romanovskaya, E.V.: Models of modern information economy conceptual contradictions and practical examples, 361 p. Emerald publishing limited (2018)

Kuznetsov, V., Kornilov, D., Kolmykova, T., Garina, E., Garin, A.: A creative model of modern company management on the basis of semantic technologies. Communications in Computer and Information Science (2017)

Novikov, A., Novikova, I., Shkvarilo, K.: Psychological and linguistic specifics of the foreign language acquisition by multinational university students. Language and Super-diversity: Explorations and Interrogations: Abstracts. Jyvaskyla, Finland, pp. 179–180 (2013)

Vladimirova, L., Safin, R., Ivanova, D., Litvina, T.: Succeeding in Foreign Language Study: Teachers and Students Standpoints. In: The International Scientific and Practical Conference Current Issues of Linguistics and Didactics: The Interdisciplinary Approach in Humanities and Social Sciences (CILDIAH-2018). SHS Web Conference, vol. 50, p. 01196 (2018). https://www.shs-conferences.org/articles/shsconf/pdf/2018/11/shsconf_cildiah2018_01196.pdf

Wiseman, R.L., Koester, J.: Intercultural communication competence. International and Intercultural Communication Annual. SAGE Publications (1993)

The Role of Scientific and Educational Platform in Formation of the Innovative Economy of Kyrgyzstan: Foreign Experience, Realities, and Prospects

Chinara R. Kulueva[1(✉)], Inabarkan R. Myrzaibraimova[1], Gulbar B. Alimova[1], Victor P. Kuznetsov[2], and Elena V. Romanovskaya[2]

[1] Osh State University, Osh, Kyrgyz Republic
ch.kulueva@mail.ru
[2] Minin Nizhny Novgorod State Pedagogical University, Nizhny Novgorod, Russia
kuzneczov-vp@mail.ru, alenarom@list.ru

Abstract. The article is devoted to issues of disclosure of the role of scientific and educational platforms in realization of prospects for building innovative economy in the territory of the Kyrgyz Republic, where the realities and prospects for a new stage of development of the country in the context of globalization. In this work an attempt is made to highlight the opinions of domestic and foreign scientists on the innovative way of market development of relations, and on the need to build an innovative economy. Foreign experience of Finland in construction of innovative economy, their problems, features, key factors and stages of development are studied in detail. By the method of observation and theoretical research, the role and importance of the scientific and educational platform in the formation of a new innovative model of economic development were reliably determined. Article analyzes the role of government regulation in the issues of a revival of production and economic activity, in support of business activity through activation of innovation in the country and a change in the financing of R & D. This article examines the role of a regional university, its development model in the provision of innovative educational services based on the formation of an innovative production infrastructure, taking into account regional characteristics; expanding and exploring sources of funding and governance mechanisms for the development of the university; contribution to the development of business by creating joint ventures with an educational institution, companies in the implementation of research projects. In conclusion, suggestions and recommendations are given, relevant conclusions are drawn on the issues under consideration.

Keywords: Innovations · Digitalization · Modernization · Technological way · Human resources potential

E. G. Popkova and B. S. Sergi (Eds.): ISC 2019, LNNS 91, pp. 484–491, 2020.
https://doi.org/10.1007/978-3-030-32015-7_55

1 Introduction

The National Concept of the Kyrgyz Republic "Digital Kyrgyzstan" – 2019–2023. (ict.gov.kg), it is noted that the main mechanism of technological modernization of the economy and increase its competitiveness by development of an innovative economy. However, introduction of innovative processes in our country by economic entities is perceived inert, remains extremely low and does not meet requirements of sustainable growth. The main reason lies, in our view, firstly, in absence or in some cases the lack of key innovation resources (Negroponte N., 1995). Secondly, an insufficient number of carriers of innovative elements, that is human resources resulting from its quality labor resources, which are largely associated with intellectual resources. It is safe to note that the government has taken a course of modernization. Economic renewal should rely on creative and not afraid of radical changes young people and generation. Therefore, the main purpose of this article is a theoretical substantiation of the necessity of transition of the Kyrgyz model of development of market relations. Change innovative economy through improvement of education policy in regions of Kyrgyzstan, where the preparation and development of a scientific and educational platform is important (Castells 2004).

2 Methodology

Transition to an innovative economy, and it's technological way—digitalization is faced with the problem of a quality of human resources, which requires a policy of formation in all links of human resources potential, who has professional knowledge of new competencies of digital economy (Roco 2004). Refusal from Soviet methods of training and development of staff, formation of personnel reserve fund to date has not decided the problems that have been marked in the early 90s and continues to this day. Therefore, in our opinion the words "Personnel solves everything!" remain relevant.

3 Results

We would like to dwell on foreign experience in formation of an innovative system, in this case we were attracted by experience of Finland. Today, Finland is growing fast, and it is impossible to believe that once, it was an exporter of raw materials and remained a country with a predominantly rural population until the 1960s. Industrialization began in 1960. At the end of the 1970s, one of the largest state committees was created - the "Technology Committee", which developed a *nationwide vision of the country's* technological future. The goal is to expand the national base of research resources, where regional universities that were part of this structure were the main agents of technology policy, who were given the right to expand vocational training and implement certain structural changes.

In 1982, a government decree on technical policy was adopted, in 1983. A National Technology Agency (Tekes) was established to manage the program by Technology Committee, where the Government agreed to a continuous increase in funding for R & D. At same time development of science and technology policy was supported by society.

A key element in networking and clustering is triangular cooperation between firms, universities and research institutes (Fig. 1).

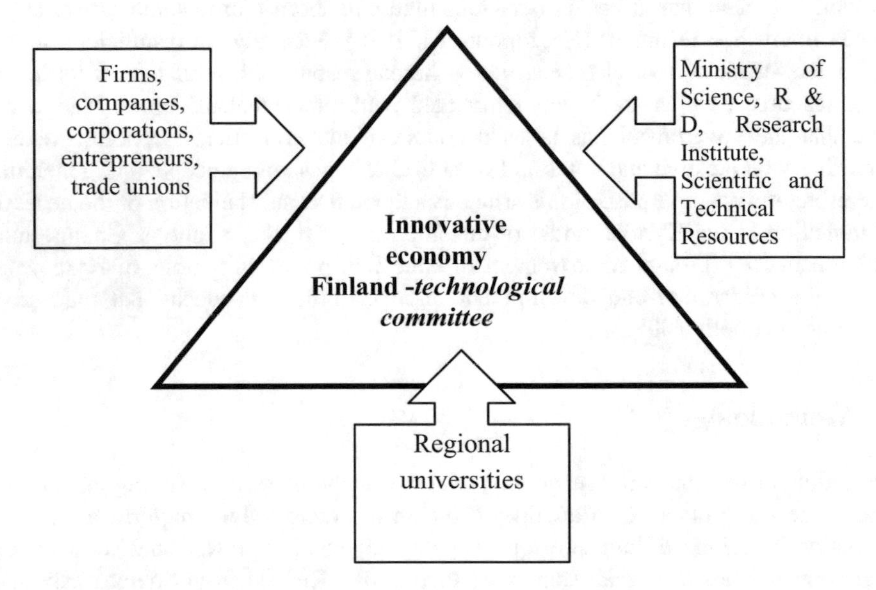

Fig. 1. The peculiarity of Finland's innovation system is tripartite cooperation between firms, universities and research institutes

In 1998, the Finnish University Act was adopted, which, together with amendments, aims to encourage universities to promote education, personnel, and research, as well as facilitating the dissemination and practical use of their results by commercial and business entities.

The National Fund for Research and Development (Sitra), established in 1967, made an important contribution to the process of creation and implementation of a creative model of innovative economy. Its purpose as a government venture fund, unlike Tekes, is to finance new companies (start-ups) (Man and Labor 2010), through which tangible structural changes in the - industry and Finnish exports due to their high-tech nature.

Therefore, summing up the Finnish experience, we can confidently note its role in creating world-class technologies, which are grouped into following clusters: information, communication, metallurgical, metalworking, forestry, social security, biotechnology, chemical, environment, energy, real estate, construction, food. Today's Finland is attractive for foreign capital due to favorable innovation climate—it speaks about the right path chosen by the country back in the 30–40 s of the last century.

The state of transition of Kyrgyzstan innovation policy: facts. Innovative development of Kyrgyzstan involves the use of competitive advantages of a domestic economy not only in traditional (energy, transport, agricultural sector, tourism) but also in relatively new knowledge-intensive sectors, thus making innovation a major source of economic growth.

Analysis of the level of innovative products to the total volume of manufactured industrial products showed an annual decrease from 15.3% in 2004 to 1.7% in 2010 and 1.2 2016. In Kyrgyzstan, the volume of innovative products decreased from 4822.8 million soms in 2004 to 1243.7 million soms in 2016 or decreased 4 times.

There has also been a reduction in the number of innovative-active enterprises from 50 to 32 over the same period. The proportion of innovative-active enterprises to the number of surveyed has a slight decrease from 8.82% to 5.5% with an increase in the number of surveyed enterprises from 567 to 709, where the level of completed innovation is also quite low. If in 2004 the share of completed innovations to the number of innovative-active economic entities amounted to 76.5%, this indicator in 2016 decreased to 48%.

Today Kyrgyzstan sets itself an ambitious goal—by 2035 the development of industry should be based on involvement of mainly new resources in the economic turnover, increasing production capacity through expansion, modernization, restoration of new technical and technological bases. The question is rhetorical but quite interesting and intriguing. Firstly, if Finland went to the innovation economy for almost half a century, we will not have time, secondly, if we use the experience of the Finns and other developed countries, we will develop the implemented Strategy primarily with a restoration of former glory of our country in a short time (5–7 years) thirdly, if we rely on educated young people, carriers of new types of specialties, fourthly, if universities switch from classical forms to production once, then an innovative economy is possible. This once again proves the need for close communication and interaction between the government and educational platform.

As for the government, no one cancels its role. it must take care of the revival of production and economic activity, create favorable conditions for firms and entrepreneurs, develop business activity in the development of industrial production and all other spheres of the national economy (Kulueva 2018). Due to the fact that our country is agrarian, where over 65% of the population live in rural areas, about 30% of GDP is produced, we think, to form a new structure in the Government of Kyrgyzstan—Ministry agro-industry and innovative technologies (Musakojoev, 2016).

To activate innovation in the country, we consider it necessary to change the process of financing applied R & D (Musakojoev 2016) by government according to the following scheme.

In order to enhance innovation activity, it is necessary to determine an algorithm of financing applied R & D by the government with determination of their sources. It should be borne in mind that the process of transition to science, technology and innovation policies should be carried out with the direct participation and support of individual firms and organizations engaged in research and development as well as their clustering to market entry.

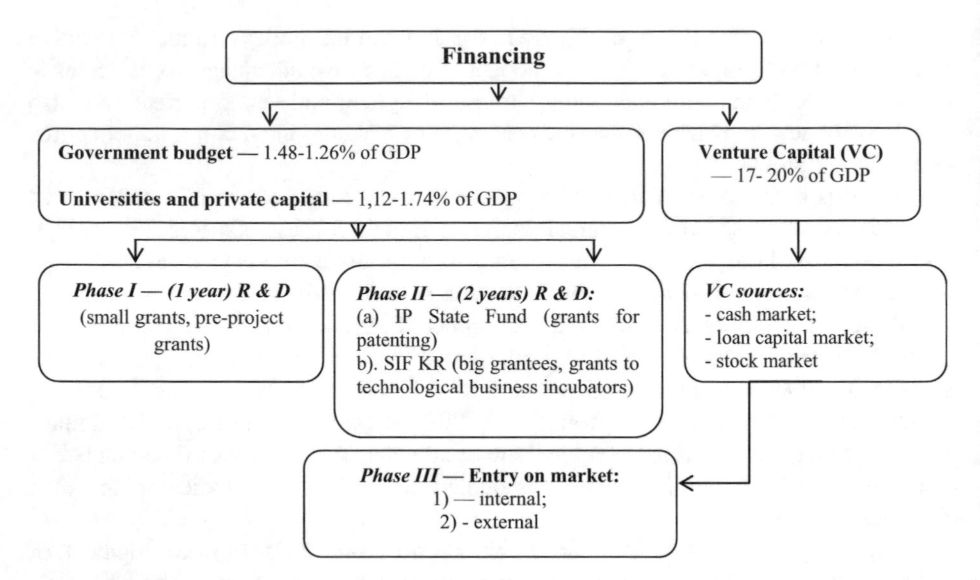

Fig. 2. The procedure for financing innovative and venture activities from public expenditures in the amount of 2.6–3.0% of GDP of the Kyrgyz Republic

Active promotion of development and support of small and medium-sized businesses, by training relevant personnel, employees and specialists by improving organizational skills, the introduction of teaching modules in schools, it is possible to change a situation that can directly affect a motivation to start their own business, increase their knowledge in business planning, marketing, finance, etc. It is safe to say that young professionals with knowledge and skills are the basis of an innovative economy. Although, at present, few domestic employers are satisfied with their level of training (Fig. 3).

In continuation, we would like to note that Osh State University as the third year in its educational-methodical, research, social-educational and career guidance activities is guided by adopted by the staff Development Concept (oshsu.kg), where a special line is the Strategy of active involvement of students in practical activities in learning process in the form of internships in a real sector, project, and teamwork, as well as their participation in R & D.

Today Osh State University develops according to the ***model "3-I (I)—Innovation-Investment-Integration"*** on the basis of the adopted Concept of development of Osh State University for 2019–2024, where the focus is on transition from the classical type of university to the Innovation and Research University with entry into the TOP-700 world universities, for which perspective benchmarks are defined. The main task of the university is transition to an innovative way of development at the expense of additional sources of financing by providing, a kind of symbiosis of education, research and industrial activity (Wattenhofer 2016), which requires formation and improvement of innovative infrastructure.

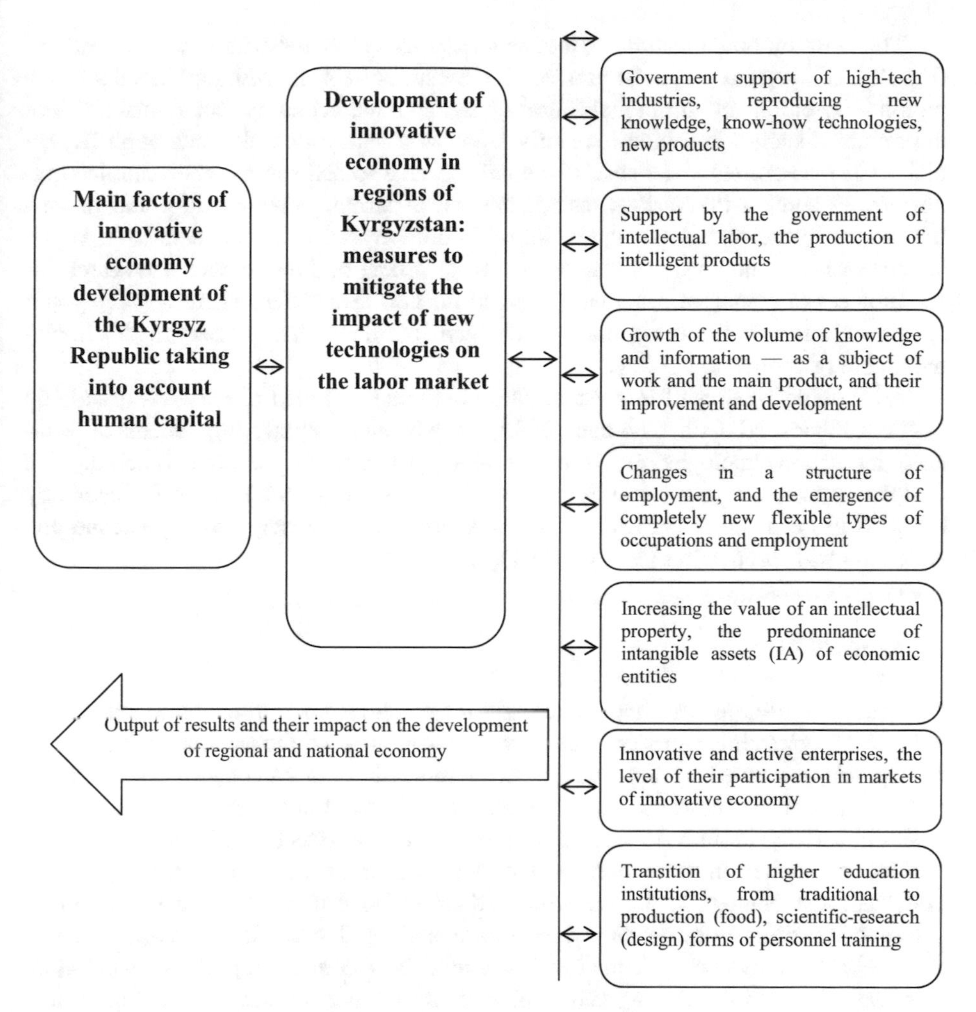

Fig. 3. Interaction of development of innovative economy of Kyrgyzstan with state of human capital and labor market

We consider it necessary to develop and implement the government target program that stimulates the involvement of students, graduates, enterprising young specialists in the activities of innovative companies that should be sufficient in Kyrgyzstan to form a new way of a market economy (Roos 1998). This, in turn, will bring a tangible positive effect for the domestic economy. In this way, companies will be able to release funds and direct them to R & D and create brand new products. Attraction of creative young people, promising employees will give a certain opportunity to increase the overall professional level of personnel, as well as to maintain some continuity of knowledge within labor collective.

The most important features of the knowledge-based digital economy are as follows (Fig. 2). The Kyrgyz Republic crossed the threshold of 2019 with ambitious goals—regional development and digitalization of the national economy (economist.kg). For the republic, such a transition is mainly associated with economic integration (Rykunich 2013), as currently, no state alone will be able to achieve macroeconomic equilibrium, to combat the various manifestations of current potential risks and threats. Innovation in the economic context should be addressed as a new economic value-added, such as a new way of organizing work, a new business model, the search for new sources of resources, where the demand for new specialties, where creative youth act as carriers. They are the main mechanism of innovation process in the implementation of innovation policy.

As forming spheres of innovative enterprises can be noted physics, electrical and IT-technologies (Hafedh Chourabi 2012), mechanical engineering, chemical technologies, nanotechnologies, etc. In recent years, Osh State University is following this innovative path, trying not only to train specialists for different areas of the economy but also to be a direct participant in this sphere, more closely linking science and education with production (Sakaya 1991).

4 Conclusion

In modern conditions, the importance of higher education institutions increases not only in an educational process but also in a structure of the national economy of Kyrgyzstan. As noted by the study in proposed article, in recent years the trend of development of the country is accompanied by an increase in scientific, innovative and business activity of universities, some increase in the effectiveness of their use of intellectual capital in formation of the domestic innovation economy, where an important role belongs to the scientific and educational platform, in our case. Innovative universities with developed scientific and production laboratories, where a generation of innovators is formed and new models of competent professionals can be provided, aimed at developing and implementing innovative ideas, technologies, and products for all spheres of the national economy.

References

Bontis: (1998); MERITUM Project Guidelines for Managing and Reporting on Intangibles (Intellectual Capital Report). (2002). http://www.pnbukh.com/files/pdf_filer/MERITUM_Guidelines.pdf. Sveiby, K.E.: The New Organizational wealth: managing and measuring knowledge-based assets. Berrett Koehler, San Francisco (1997)

Bradburn, Coakes 2004; Knowledge Management: Classic and Contemporary Works, 2002; Volkov, Garanina 2008; Sveiby, K.E.: The New Organizational wealth: managing and measuring knowledge-based assets. Berrett Koehler, San Francisco (1997)

Bezrukova, T.L., Gyyazov, A.T., Popkova, E.G., Kulueva, Ch.R., Rayymbaev, Ch.K.: Innovative problems of economic growth in modern market conditions (article). Int. J. Econ. Policy Emerg. Countries. Italy, Bruno, Medical University, no. 5 (2017)

Zhang, W.V.: Capital and Knowledge. Dynamics of Economic Structures with Non-Constant Returns. Springer, Heidelberg (1999)

Kulueva, Ch.R., Kupuev, P.K., Ubaidullaev, M.B.: Informatization as a Mechanism of Fighting Tax Evasion. "Optimization of the Taxation System: Preconditions, Tendencies and Perspectives", First Online: 29 September 2018, Part of the Studies in Systems, Decision and Control book series (SSDC, volume 182), pp 143–149 (2018)

Pigou, F.C.: Industrial Fluctuations. Macmillan, London (1927)

Saint-Onge, 1996; Sveiby, K.E.: The New Organizational wealth: managing and measuring knowledge-based assets. Berrett Koehler, San Francisco (1997)

Slepov, V.A., Gerzeliyeva, Zh.I.: Intellectual capital of the university and indicators of its evaluation. Creative Econ. **9**(8), 995–1008 (2015). https://doi.org/10.18334/ce.9.8.579

Robertson, D.N.: A Study of Industrial Fluctuations. Aldwich, London (1948)

Roos, J., Roos, G., Dragonetti, N.C., Edvinsson, L.: Intellectual Capital: Navigating the New Business Landscape. Macmillan Press, London (1998). Sveiby, K.E.: The New Organizational wealth: managing and measuring knowledge-based assets. Berrett Koehler, San Francisco (1997)

Schumpeter, J.A.: The Theory of Economic Development. Harvard University Press, Cambridge (1934). English translation: Schumpeter, J.: Theory of economic development (1982)

Schumpeter, J.A.: Business Cycles: A Theoretical, Historical and Statistical Analysis of the Capitalist Process, 2-vols. McGraw-Hill, New York (1939)

Success of the Innovative Economy in the 21st Century: Expectations and Reality

Scientific and Methodological Foundations of an Innovative Company Management

Olga V. Konina(✉)

Moscow State Pedagogical University, Moscow, Russia
koninaov@mail.ru

Abstract. Purpose: The purpose of the paper is to determine the scientific and methodological foundations of an innovative company management, determine the drawbacks in the existing Russian practice of management of innovative companies, and to develop recommendations for overcoming them in the interests of improving the existing practice.

Design/Methodology/Approach: For empirical purposes of the research, the authors analyze the statistics of the activities of modern Russian companies (by the example of industry, for which the most extensive statistical accounting is present) in 2018, in three directions of managing an innovative company: receipt of innovations, implementation of innovations into entrepreneurial processes, and taking innovative products to the market.

Findings: It is substantiates that the practice of managing innovative companies in modern Russia has a range of drawbacks, which cause disproportions in the process of commercialization of innovations and reduce innovative activity of entrepreneurship. For example, high level of the investment and risk load onto Russian companies in the process of implementation of innovative activities.

Originality/Value: For overcoming the determined drawbacks, the authors offer a perspective model of managing an innovative company in Russia, which has the following advantages: expanded investment provision of innovative activities due to usage of venture investments; reduced risk component of innovative activities due to moving the risks to suppliers of innovations; growth of investment activity of personnel due to higher motivation and stimulation; achievement of long-term (sustainable) competitive advantages due to production of essentially new products and marketing support for their sales.

Keywords: Management · Innovative company · Competitive advantages · Innovative development · Modern Russia

JEL Code: G32 · G34 · D81 · O31 · O32 · O33

1 Introduction

Innovative development is one of the key priorities of the modern economy, as it sets foundations for provision of its global competitiveness and supporting sustainability. In modern Russia, the issues of innovative development of economy are given a lot of attention from the state, which is confirmed by the adoption of the Strategy of

E. G. Popkova and B. S. Sergi (Eds.): ISC 2019, LNNS 91, pp. 495–502, 2020.
https://doi.org/10.1007/978-3-030-32015-7_56

innovative development of the Russian Federation until 2020 (Decree dated December 8, 2011, No. 2227-r. At present (2019), implementation of this strategy is near its finish, which opens opportunities for its scientific analysis and reconsideration.

Social plan, which is presented in this strategy and which is brought down to increasing the level of education of the Russian population, could be considered achieved – but the research and entrepreneurial plan has not been fully implemented. For example, factual value of the share of innovative goods, works, and services in the aggregate volume of supplied goods, performed works, and provided services of organizations of industrial production (8.4%), according to the Federal State Statistics Service (2019), is three times lower than the planned one (25%) (Government of the Russian Federation 2019). The macro-economic manifestation of moderate investment activity of entrepreneurship in the Russian economy is Russia's underrun from other developed countries as to the value of innovative index (37.90 points out of 100, 46th position in the world) (WIPO 2019).

The working hypothesis of the research is that the reason of the moderate level and rate of innovative development of the modern Russian economy and incomplete achievement of the research and entrepreneurial plan, which was set in the Strategy of innovative development of the Russian Federation until 2020, is imperfection of management of innovative companies. The purpose of the paper is to determine the scientific and methodological foundations of an innovative company management, determine the drawbacks in the existing Russian practice of management of innovative companies, and to develop recommendations for overcoming them in the interests of improving the existing practice.

2 Materials and Method

There are three conceptual approaches to treatment of the essence of an innovative company in the modern economic literature; they determine the specifics of its management. The representatives of the first approach - Chen et al. (2015), Yang et al. (2013), Kumar and Dwivedi (2019) – define innovative company as an economic subject that conducts R&D and shows formalized (positive) results of their conduct – publication of scientific works, registration of rights for the objects of intellectual property, etc.

Within the second approach, Popkova et al. (2018), Morozova et al. (2018), Butorin and Bogoviz (2019), Morozova et al. (2019), Tekin and Konina (2019), Ostrovskaya et al. (2017) treat innovative company as an economic subject that implements innovations into their entrepreneurial processes (e.g., marketing and production) for their optimization.

According to the third approach, Bezrukova et al. (2017), Pritvorova et al. (2018), Sergi et al. (2019), Sibirskaya et al. (2019), Vanchukhina et al. (2018) consider innovative company to be an economic subjects, which offers innovative products (goods, works, and services) in the market.

According to the authors of this paper, all existing approaches are narrow, as they cover separate processes of the stages of the process of commercialization of innovations. In practice, innovative company often combines all the above types of activities. That's why from the positions of management of an innovative company it is offered to distinguish three directions of management:

1. Receipt of innovations (in any accessible and preferable way) – management is to fully satisfy the company's need for innovations with smallest expenses;
2. Implementation of innovations into entrepreneurial processes – management is to ensure effective organization of these processes and to prevent opposition to innovations from the company's employees;
3. Taking innovative products to the market, including its marketing and sales – management is to ensure maximization of the company's profit.

The above directions are enumerated in the succession of the stages of the process of commercialization of innovations; however, from the positions of management their order could be different (e.g., conducting preliminary marketing studies for determining potential demand for planned innovative products, as well as further marketing of finished innovative products) and, in particular, they could be implemented simultaneously.

For empirical purposes, let us analyze the statistics of activities of modern Russian companies (by the example of industry as a sphere for which the fullest statistical accounting is available) in the three above directions in 2018 (Fig. 1).

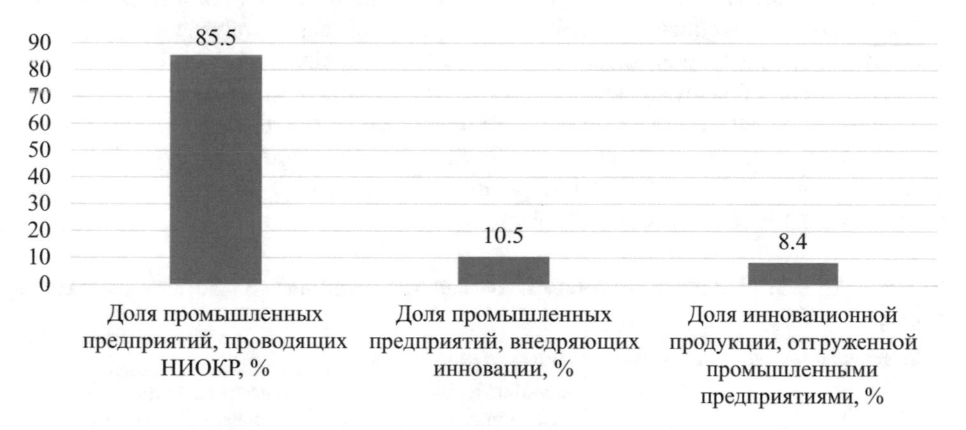

Fig. 1. Statistics of innovative entrepreneurship in Russia in 2018 (by the example of industry). Source: compiled by the author based on Federal State Statistics Service (2019).

Figure 1 shows that own R&D (on the basis of research and engineering departments) were conducted by 85.5% of Russian industrial companies in 2018. However, the share of industrial companies that implement innovations into their activities is 10.5%, and the share of innovative products that are supplied by industrial companies is 8.4%. The determined disproportions show the imperfection of management of innovative companies in modern Russia.

3 Results

Within each distinguished direction of management of an innovative company, the universal (common for all directions) objects of management are as follows:

- Investments: the task of management is financial provision of the company's innovative activities;
- personnel: the task of management is personnel's involvement into the company's innovative activities;
- risks: the task of management is to reduce the risks of the company's innovative activities.

There are also specific objects of management within each direction. Within receipt of innovations, the object of management is innovations – the decision what is the source of innovations for the company – own R&D or finished innovations from specialized organizations (e.g., research institutes) – is made, and the process of receipt of innovations is organized.

Within implementation of innovations in entrepreneurial processes, the object of management is innovative processes – they are organized, and the decision is made, how they could serve the purpose of increasing the company's competitiveness. Within movement of innovative products to the market, the object of management is the created innovative products – a decision on the level of their novelty is made, and the process of their production, marketing, and sales is organized.

As a result of studying the existing Russian practice of managing innovative companies its drawbacks were determined; they could be the reasons of disproportion of efficiency of the stages of commercialization of innovations. For overcoming the determined drawbacks, an alternative practice of management of an innovative company in Russia is recommended (Table 1).

Table 1. The scientific and methodological foundations of managing an innovative company in Russia.

Direction of management	Object of management	Method of management	
		According to the modern Russian practice	According to the recommended practice
Receipt of innovations	Innovations	Conducting own R&D	Outsource of R&D
	Investments	Foundation of own investments	Attraction of venture investments
	Personnel	Considering team results	Considering individual results
	Risks	Moving risks to employees	Moving risks to suppliers of innovations

(*continued*)

Table 1. (*continued*)

Direction of management	Object of management	Method of management	
		According to the modern Russian practice	According to the recommended practice
Implementing innovations into entrepreneurial processes	Innovative processes	Striving for growth of effectiveness (for increasing the competitiveness)	Striving for production of innovative products
	Investments	Foundation on own investments	Attraction of venture investments
	Personnel	Standardization and norming of investment activity	Motivation and stimulation for manifesting investment activity
	Risks	Independent risks	Moving risks to suppliers of innovations
Taking innovative products in the market	Created innovative products	Creation of products that are new for this company	Creating completely new products
	Investments	Foundation on own investments	Using the possibilities of integration
	Personnel	Foundation on own personnel	Outsource of personnel
	Risks	Moving risks to intermediaries	Independent risks

Source: compiled by the authors.

Table 1 shows that management of an innovative, according to the modern Russian practice, is conducted in the following way:

- In the process of receipt of innovations, Russian companies conduct their own R&D, which are peculiar for high level of risk (e.g., risk of receipt of negative results). The companies use their own investments and thus have a deficit of financial provision of R&D. Team results are taken into account during personnel management, which destroys motivation of personnel for innovative activities. Risks of R&D are moved to the employees, who bear responsibility for the results of innovative activities. For example, requirements in the sphere of publication and patent activities are set to them;
- In the process of implementation of innovations in the entrepreneurial processes, Russian companies strive for growth of effectiveness (increase of efficiency and reduction of cost), while production of innovative products is a not a goal in itself. The companies use their own investments. Personnel management envisages standardization and norming of innovative activities. The company bears the risks independently;
- In the process of taking innovative products to the market, Russian companies create the products that are new for them, though very often these products are not new for the market. The companies use their own investments and personnel (employees on marketing and sales), which complicates the organizational structure. At the same time, risks of sales are moved to intermediaries – they will suffer losses in case of low demand for products in the market.

The recommended practice is presented in the model of managing an innovative company in Russia in Fig. 2.

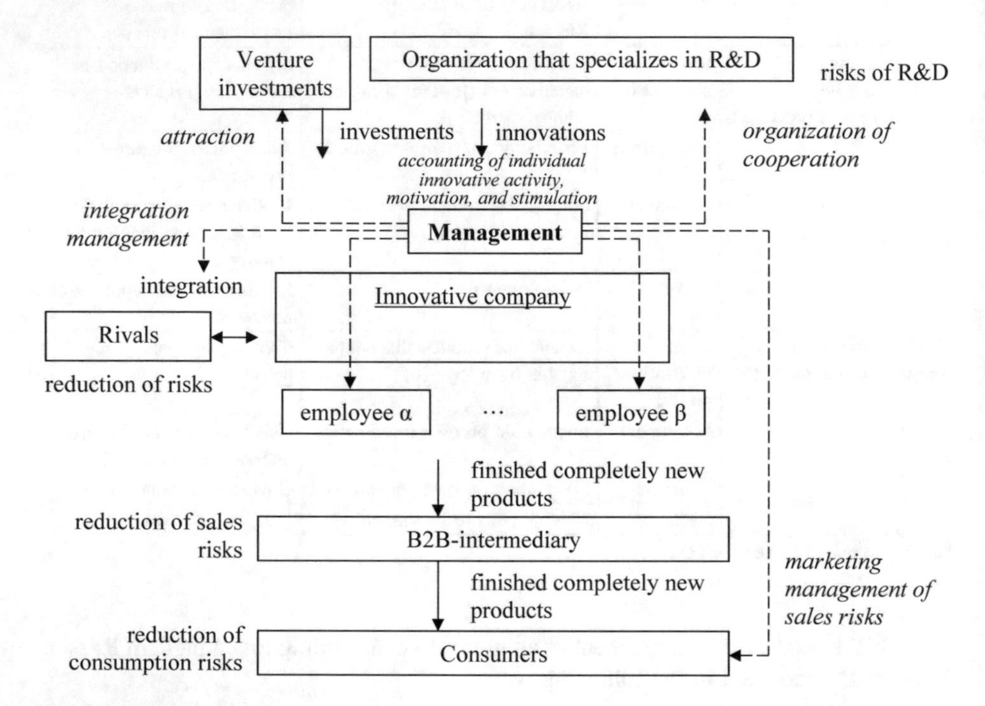

Fig. 2. A perspective model of managing an innovative company in Russia. Source: compiled by the authors.

As is seen from Fig. 2, according to the offered perspective model, management of innovative company is conducted in the following way:

- in the process of receipt of innovations, it is offered to use outsource, purchasing ready innovations from specialized organizations which will bear the risks of R&D. If the employees show investment activity (during creation of know-how), it is necessary to consider individual results for strengthening the motivation. Also, it is recommended to attract venture investments for full-scale financial provision of innovative activities;
- in the process of implementation of innovations в entrepreneurial processes, it is necessary to strive for production of innovative products, as sustainable competitive advantages could be formed only in this case. Attraction of venture investments is expedient. Personnel management should be conducted on the basis of motivation and stimulation (including by individual and flexible terms) of investment activity. The risks should be moved to suppliers of innovations, which provide active support for implementation of innovations at a company;

- In the process of taking innovative products to the market, it is recommended to create completely new products for increasing the global competitiveness. During management of investments, it is recommended to use the possibilities of integration with rivals – e.g., with the help of clustering. Personnel outsource – i.e., external marketing – is offered. Sales risks are to be borne by the company (passing the products for sales with return of unsold products); the company also has to manage them with the help of marketing.

4 Conclusion

Thus, as a result of the research, the working hypothesis is proved – the practice of management of innovative companies in modern Russia has certain drawbacks, which cause disproportions in the process of commercialization of innovations and reduce innovative activity of entrepreneurship. An example could be high level of investment and risk load onto Russian companies in the process of implementation of innovative activities. For overcoming the determined drawbacks, a perspective model of managing an innovative company in Russia is offered; it has the following advantages:

- expanded investment provision of innovative activities due to usage of venture investments;
- reduced risk component of innovative activities due to moving the risks to suppliers of innovations;
- growth of investment activity of personnel due to higher motivation and stimulation;
- achievement of long-term (sustainable) competitive advantages due to production of completely new products and marketing support the sales.

References

Bezrukova, T.L., Popova, E.V., Korda, N.I., Kuznetsova, T.E., Bezrukov, B.A.: Institutional traps of innovative and investment activities as an obstacle on the path to the well-balanced development of regions. Contributions to Economics (9783319606958), pp. 235–240 (2017)

Butorin, S.N., Bogoviz, A.V.: The innovational and production approach to management of economic subjects of the agrarian sector. In: Advances in Intelligent Systems and Computing, vol. 726, pp. 758–773 (2019)

Chen, J., Lin, S., Meng, Z.: Enterprise technological innovation and market innovation of innovative management. Metall. Min. Ind. **7**(9), 980–985 (2015)

Kumar, S., Dwivedi, A.K.: Innovative business model that creates nano-curcumin-based enterprise (with respect to sustainable enterprise management). Smart Innov. Syst. Technol. **135**, 143–150 (2019)

Morozova, I., Litvinova, T., Mordvintsev, I.A., Konina, O.V.: Using public-private partnership for stimulating innovational developments in the infrastructural sphere. J. Adv. Res. Law Econ. **9**(4), 1382–1386 (2019)

Morozova, I.A., Popkova, E.G., Litvinova, T.N.: Sustainable development of global entrepreneurship: infrastructure and perspectives. Int. Entrepr. Manag. J. 1–9 (2018)

Ostrovskaya, V.N., Tyurina, Y.G., Konina, O.V., Przhedetskaya, N.V., Pupynina, E.G., Natsubidze, A.S.: Perspectives of elimination of "institutional gaps" in foreign economic activities of subjects of SME within the global crisis management. Contributions to Economics (9783319606958213218), 213–218 (2017)

Popkova, E.G., Sozinova, A.A., Grechenkova, O.Y., Menshchikova, V.I.: Deficiencies in the legislative support of innovative activities in contemporary Russia and ways of addressing them. Russ. J. Criminol. **12**(4), 515–524 (2018)

Pritvorova, T., Tasbulatova, B., Petrenko, E.: Possibilities of blitz-psychograms as a tool for human resource management in the supporting system of hardiness of company. Entrepr. Sustain. Issues **6**(2), 840–853 (2018). https://doi.org/10.9770/jesi.2018.6.2(25)

Sergi, B.S., Popkova, E.G., Bogoviz, A.V., Ragulina, J.V.: Entrepreneurship and economic growth: the experience of developed and developing countries. In: Entrepreneurship and Development in the 21st Century. Emerald publishing limited, pp. 3–32 (2019)

Sibirskaya, E., Popkova, E., Oveshnikova, L., Tarasova, I.: Remote education vs traditional education based on effectiveness at the micro level and its connection to the level of development of macro-economic systems. Int. J. Educ. Manag. **33**(3), 533–543 (2019). https://doi.org/10.1108/IJEM-08-2018-0248

Tekin, A.V., Konina, O.V.: The role of information and communication technologies in the process of strategic management of entrepreneurial structures activities: the budget and financial aspect. In: Advances in Intelligent Systems and Computing, vol. 726, pp. 269–278 (2019)

Vanchukhina, L.I., Leybert, T.B., Khalikova, E.A., Khalmetov, A.R.: New approaches to formation of innovational human capital as an element of institutional environment. Espacios **39**, 22–32 (2018)

WIPO. Global innovation index 2018 (2019). https://www.wipo.int/publications/en/details.jsp?id=4330. Accessed 09 May 2019

Yang, Q., Hu, J., Yao, J., Li, J.: On the innovative countermeasures to the incentive management of the high-tech enterprise employees. Adv. Sci. Lett. **19**(4), 1176–1179 (2013)

Government of the Russian Federation. Strategy of innovative development of the Russian Federation until 2020, adopted by the Decree dated December 8, 2011, No. 2227-r (2019). https://www.garant.ru/products/ipo/prime/doc/70006124/. Accessed 09 May 2019

Federal State Statistics Service. Russia in numbers 2019 (2019). http://www.gks.ru/bgd/regl/b18_11/Main.htm. Accessed 09 May 2019

Transformation of Business Models in Terms of Digitalization

Olga B. Digilina[1(✉)] and Irina B. Teslenko[2]

[1] RUDN University, Moscow, Russia
o.b.digilina@mail.ru
[2] Vladimir State University named after Alexander and Nikolay Stoletovs, Vladimir, Russia
iteslenko@inbox.ru

Abstract. In the article, the authors set a goal to analyze the changes in business models that occur in the process of digitalization of economic processes in Russia. The authors emphasize that on digital platforms, economic agents have the opportunity to effectively interact not with several separate enterprises, but with a whole ecosystem of companies. This interaction has a significant synergistic effect. Modern enterprises in order to maintain their competitive position should move from traditional ways of doing business to modern business models through business restructuring, business process reengineering.

Keywords: Business model · Sharing economy · Car sharing · Crowdfunding · Digital insurance technologies

JEL Code: M 20

1 Introduction

The development of information technology opens up new incredible opportunities. This is not only fast communication at a distance via mobile communication, social networks, not only the ability to buy goods, sitting at home via the Internet, not only the ability to store a huge amount of information on compact media, etc., it is also the formation of new models of people's relationships.

Business models are the ways organizations use to create value and generate profit [17]. Digital economy at the expense of new unique technologies simplifies and accelerates traditional business processes, customizes the service using large amounts of data, ensures the emergence of virtual businesses, opens up opportunities for flexible individualized pricing, etc.

New business models cover all areas of social and economic life: from business to government structures. They transform the activities of companies and entire industries, encouraging more and more new market players to reconsider their attitude to business models.

The emergence of new business models associated with the changing needs and requirements of customers. With the advent of mobile devices and the development of

E. G. Popkova and B. S. Sergi (Eds.): ISC 2019, LNNS 91, pp. 503–509, 2020.
https://doi.org/10.1007/978-3-030-32015-7_57

mobile applications, so-called "platform" business models are becoming increasingly popular. Let us dwell on the characteristics of some of them.

2 Background and Methodology

The object of this article was the process of introducing digital technologies into the business structure of Russia. The subject of the study was changes in business models in the process of digitization of economic sectors and their new look. The methodological basis of this study was a set of techniques, methods and cognitive attitudes, which are currently known to scientific research and which have been adapted to the specificity of the object of study of this article.

The empirical basis for the preparation of this article was the socio-economic practices of the Russian Federation, large enterprises, their best practices in the field of transformation of business models in terms of digitalization. In addition, this study relies on academic statistical and analytical reviews, government documents that reflect the impact of the digital economy on business, and economic articles in the media.

3 Discussion and Results

One of the modern and highly sought-after business models is the business model of sharing, or, as it is often called, the sharing economy.

The economics of sharing (eng. Option - sharing economy) in its most general form is the exchange of goods and services between an unlimited number of people worldwide without intermediaries. This is a new socio-economic model of interaction, which is based on a conscious preference for collective ownership instead of private ownership. The choice of such a model is associated not only with a lack of money, but with the desire to expand its capabilities and take advantage of the opening advantages.

Sharing of goods is rooted in the distant past. The modern concept of ESP was proposed by economists R. Botsman and R. Rogers in the book "What's Mine Is Yours: The Rise of Collaborative Consumption" (2010) [8]. Their idea was that it was easier and more profitable for consumers to pay for temporary access to a product than to own it.

The formation of an economic sharing model led to two factors: technological and value. As for the first, the emergence of platform technologies allows you to quickly and easily connect economic agents with each other. The essence of the value factor is the rejection of established habits, in the new perception of many things, behaviors that previously seemed unshakable.

Not so long ago, the car was considered an indicator of human status, but now it is perceived by many as a means of transportation, which should be convenient to use. However, the increase in the number of cars on the roads, many hours of traffic jams, the lack of parking spaces, the expensive maintenance of their own car initiated the creation of special services for the new model of interaction. This, for example, Uber (application for ordering a taxi) or Carsharing (car sharing).

Already today, thousands of people in the world use the services of a shared-use economic model, such as BlaBlaCar - search for travel companions; eBay - online auction; Airbnb - rental housing service; and etc.

Thanks to the concept of sharing, a market has appeared even for those things that were never considered potentially profitable. A dozen square meters of driveway can be profitable as a parking space through Parking Panda. The room in the house can become a dog house through the service DogVacay. With the SnapGoods service, an idle drill in a garage also becomes a source of income [7].

In Russia, due to the fairly rapid spread of digitalization processes, ESP is also developing. Thus, in 2018, the country increased the volume of operations in the ESP markets by 30% compared with 2017. The largest share in the volume of ESPs was electronic sales between consumers (C2C - Consumer-to-consumer - 72%). Freelance services (19%), car sharing (5%) and rental of premises for a short period (2%) lag far behind them.

The Association of Tour Operators of Russia (ATOR) predicts that in 2019 tourist flow will grow by 20% compared with 2018. Since, as you know, renting accommodation through the Airbnb service allows you to save travelers money, it is quite obvious that an increasing number of tourists will use this particular service when traveling in Russia and abroad.

According to Forbes, the income received through ESP and coming directly to its participants will grow at a very rapid pace. ESP is not just a profitable model, but a real revolution in the economy.

Despite the advantages and growing popularity, the ESP raises a number of problems and has certain disadvantages.

The first and most important problem is the possible reduction of jobs in the implementation of ESP services.

The next problem is tax regulation. For the new economic model, new methods and methods of tax collection should be developed.

The third problem is connected with the introduction of additional measures to ensure the security of information platforms.

There is the problem of organizing fraudulent schemes.

At the same time, the main advantages of EPS are: convenience, trusting relationships of participants, maximum use of available resources, and obtaining permanent income from property that the owner does not use constantly.

The development of crowd-based technologies has led to the emergence of such a business model in a digital economy such as crowd-trusting. Crowd technologies involve the use of the abilities, knowledge, and creative skills of a large number of people to solve various issues and problems during their joint online work on a specialized Internet platform.

The scope of crowd technology is very extensive. With their help, a business can, for example: promote its brand; improve the design of the site, organize interaction with customers through online surveys and forums; attract resources, which in modern conditions is particularly important.

It's no secret that the performance of an organization depends on its employees. Each organization seeks to find the best employees.

Now the recruitment of personnel in organizations and enterprises engaged in personnel departments, as well as specialized organizations - employment services and recruitment agencies. Recruiting is a business process for the selection and selection of specialists for the client company [18]. And online recruiting is the involvement of qualified personnel or finding the desired vacancy by the applicant using the websites: zarplata.ru, superjob.ru, hh.ru, job.ru, rosrabota.ru, rabota.ru, bankir.ru, etc. (Simanova [14]).

In recent times, overseas and in Russia have started talking about the technology of crowd racking.

This is not just a way to hire workers online. This is something more. Crowd recruiting (from the English. "Crowd" - the crowd and recruiting - recruiting for certain positions, recruitment) is a way to find and hire an employee by assessing his qualities, real capabilities and abilities in carrying out the project proposed by the company [16].

A characteristic difference in crowdfunding from other ways of selecting candidates for work is the full involvement of project participants in solving the tasks, so that they have the opportunity in a short time to show themselves from completely different sides, maximizing their potential (Bobko [3]).

In the course of the work of the candidate on the project, the employer can learn a lot about the abilities of applicants, their skills or inability to act in different situations.

Crowd recruiting reduces the workload of employees of the HR service of the enterprise, eliminating the need to attract third-party recruitment agencies, reduces the cost of selecting each of the candidates [4].

At the end of the project, the employer will only have to send invitations to the final interview or offer to take the vacant place to those who are interested in it [5].

In Russia, crowd racking technology was applied in 2012 by Witology on the order of Rosatom State Corporation during the implementation of the unique project TempP 2012. The project was aimed at selecting 100 specialists for various areas of Rosatom's activities. With traditional recruiting, this would take about half a year. This project based on crowd recruiting took only a month.

Along with identifying the most creative and knowledgeable candidates, crowd rucking allowed us to form a specific talent pool of young professionals, to introduce many applicants to the company and the specifics of its work [4].

Crowd recruiting has its advantages and limitations. In addition to the advantages of both the speed of selection and cost savings, it should be noted that it is the online format that is very much in demand by modern young people, it is clear to them. Therefore, graduates of the challenge of different specialties - this is the main audience of the cruise crewing projects.

However, this technology is hardly suitable for selecting a driver, secretary, candidates for rare professions, highly qualified specialists.

Crowd recruiting involves the use of Internet platforms (for example, BrainForce or FuturUS), so its cost is high enough, which means that not all employers, especially small and medium business owners can use this technology.

Crowd recruiting is not so widespread in Russia, according to the Research Center of the portal Superjob.ru, because quite often employers want personal contact with the candidate, because for them are important different things, such as appearance, style of dress, manner of behavior, etc. (Gorunova 2015).

The further development of the crowdfunding technology is to improve the feedback, since the questions: how selected candidates cope with the work, how many of them “got accustomed” to the company, how many quit and how quickly - remain open. From this point of view, the effectiveness of the technology has not yet been determined [6].

Crowd recruiting can be used not only in the selection of candidates for vacancies, but also when it is necessary to solve some problem within the company. Employees of the company have the opportunity to express themselves and thereby begin to treat their work differently, and later, perhaps, to move up the career ladder.

As for the company’s customers and third-party respondents, they get the opportunity to become participants in the creation of a product or service that interests them [12].

In general, the further development of crowdfunding can make this technology effective and very promising in the selection of personnel in the 21st century.

The use of big data - Big Data - opens up great opportunities for the development of new business models.

Big Data is either huge amounts of structured or unstructured digital information that cannot be processed with traditional tools (in world practice), or data processing technologies (domestic approach).

Some researchers believe that the next information revolution will occur unnoticed precisely because of the large-scale implementation of Big Data technology. In a few years it will not be something fantastic anymore if in all the coffee shops of the world a person will always be offered the same coffee brewed as soon as he likes it.

All channels through which information is received are divided into digital ones - these are various sensors and software - these are various programs. Devices installed in homes, cars, hand bracelets, as well as programs (from Facebook to Google Maps) collect all information about the population.

Using the sensor, you can estimate the standard activity of a person and monitor abnormalities. A person himself can call for help using such devices.

Big Data provides personalized communication with customers. The technology is used in many areas, is no exception and insurance.

Insurers began tracking data from fitness bracelets and other smart devices that help monitor the health and lifestyle of customers. So, the American insurance company John Hancock Financial offers free fitness trackers from Fitbit or Apple, if the client reaches certain sports goals.

Some track how well their customers brush their teeth. In particular, the insurance company Beam Technologies, using a smart toothbrush, controls the duration of tooth brushing, and if it is at least two minutes, the client regularly receives free brush tips and a discount on insurance services [7].

As for health insurance, insurers make discounts (or surcharges) on some health insurance products, depending on what kind of lifestyle a person leads. Such products, for example, are in Scandinavia, they began to be used in Estonia [2].

Insurers began tracking data from fitness bracelets and other smart devices that help monitor the health and lifestyle of customers. So, the American insurance company John Hancock Financial offers free fitness trackers from Fitbit or Apple, if the client reaches certain sports goals.

Some track how well their customers brush their teeth. In particular, the insurance company Beam Technologies, using a smart toothbrush, controls the duration of tooth brushing, and if it is at least two minutes, the client regularly receives free brush tips and a discount on insurance services [11].

Russian insurers are not yet using the capabilities of Big Data as widely as foreign ones. However, the first steps in this direction have already been made. This is a "smart hull" using telematics devices. These are devices (sometimes just a program in a mobile phone) that are installed in a car and collect various data related to driving style: speed, number of maneuvers, braking, etc. However, while this information is used by insurers only for the preliminary calculation of the tariff for insurance hull insurance.

The domestic market can boast only 50 thousand personalized offers that have been implemented by insurance companies over the past few years using telematics. Statistics show that these sales make up no more than 1% of the total number of all CASCO policies.

There may be many reasons, but the main problem is that there is no specialized legal framework in Russia that could standardize such systems and the way they work with insurance companies [10].

The company in the field of telematics "Smart Driving Laboratory" has begun cooperation with key players in the insurance market: Alfastrakhovanie, Rosgosstrakh, Soglasie, Ingosstrakh.

The company launched the first telegram bot in Russia that allows you to remotely control a car, developed the first in Russia system of recognition and early warning of accidents based on artificial intelligence - Crash AI (the technology is able to distinguish a real accident from a false one and the severity of damage and other important characteristics of the accident) [15].

As for property insurance, built-in sensors help to reduce risks, which allow detecting unauthorized penetration into the premises, gas leakage, breakthrough in the water supply system or fire in time. Such devices are of interest both from the point of view of minimizing losses, and also serve as the basis for the formation of a more favorable insurance offer in combination with statistics and the accumulated history of insured events.

Over time, the use of the Internet of Things (IoT) will lead not only to the formation of innovative types of insurance, but also change the interaction model within the classical types of insurance.

4 Conclusions

So, the digitization of recent years has influenced almost all spheres of life and activities of society.

Gone are the days when the business was mainly focused on the price and quality of the product. Now more important are the speed and convenience of its receipt. At the present time on digital platforms it is possible to interact not with several enterprises, but with a whole ecosystem of companies. Improving technology, automating business processes becomes the basis of successful competition,

Modern enterprises, in order to develop and generate revenues, to maintain their competitive position, must move from traditional ways of doing business to modern business models by restructuring business, reengineering business processes to meet the needs of their customers and generating additional income.

References

1. Big data analysis will help insurers understand customer needs in healthcare and insurance. [Electronic resource]. https://forinsurer.com
2. Big data technologies in life insurance. [Electronic resource]. https://lifeinsurance.kz/ekspert/tehnologii-big-data-v-strahovanii-zhizni
3. Bobko, A.: Crowdrilling. Search for people through the search for ideas. [Electronic resource] (2018). https://witology.com/blog/company/264/
4. Crowdrekruting - a new technology of mass talent search. [Electronic resource]. http://iinsider.biz/
5. Crowdrebring - crowdsourcing model of hiring labor [Electronic resource]. http://forumbusiness.net/showthread.php?t=27323
6. Crowdrekruting as a method of finding candidates. [Electronic resource]. http://hr-portal.ru/blog/kraudrekruting-kak-metod-poiska-kandidatov
7. Economy sharing. [Electronic resource]. https://forbes.kz/finances/markets/ekonomika_sovmestnogo_polzovaniya
8. Everywhere sharing: what is a co-consumption economy. What is sharing economy: the history of the term. [Electronic resource]. https://rb.ru/story/share-it/
9. Goryunova, O.: In one fell swoop to kill two birds with one stone - this is the result that promises a new technology of recruitment. [Electronic resource] (2015). https://bankir.ru/publikacii/20150316/prizhivetsya-li-u-nas-kraudrekruting-10006171/
10. How to apply “big data” in insurance: ITMO University projects. [Electronic resource]. https://habr.com/ru/company/spbifmo/blog/329762/
11. How clever technology. IoT and Big data will help insurance companies and how will they affect tariffs in the future? [Electronic resource]. https://forinsurer.com
12. Idea management and crowdfunding. [Electronic resource]. http://www.sergiev-posad.ru/useful/?ID=18134
13. Why insurers need telematics, fitness trackers and smart toothbrushes. [Electronic resource]. https://allinsurance.kz/articles/analytical/5813-kak-umnye-tekhnologii-iot-i-big-data-pomogut-strakhovym-kompaniyam-i-kak-oni-povliyayut-na-tarify-v-budushchem
14. Simanova, I.: “Real” Recruiting and Internet Recruiting: Pros and Cons. [Electronic resource] (2015). http://aviconn.com/press/realnyiy-rekruting-i-internet-rekruting-za-i-protiv.html
15. “Smart Driving Laboratory” for the year increased the number of customers 10 times. [Electronic resource]. http://www.cnews.ru/news/line/2019-02-13_laboratoriya_umnogo_vozhdeniya_za_god_uvelichila
16. Staff recruitment service - what is it? [Electronic resource]. http://znaydelo.ru/personal/trudoustroystvo/rekruting.html
17. What is a business model. Review definitions. [Electronic resource]. https://daily10.ru/chto-takoe-biznes-model/#hcq=4bkgsqr
18. What is recruiting. [Electronic resource]. http://hr-portal.ru/article/chto-takoe-rekruting

Formation and Development of Regional Innovation Systems

Liubov V. Plakhova(✉), Natalia V. Zakharkina, Ivan V. Ilin, Viktor P. Bardovskii, and Nikolay V. Pokrovskiy

Orel State University of Economics and Trade, 12 Oktjabr'skaja St., Orel, Russia
plahova.l.05@yandex.ru, 1278orel@mail.ru,
super-ya-57@mail.ru, osuet@mail.ru, my-orel-57@mail.ru

Abstract. The gradual transformation of the regional economy has led to the creation of innovative systems. Innovative systems are an important link in the modification of the regional economy. Firstly, innovative systems form the technological economy. Secondly, innovative systems provide the necessary level of development of the region. These circumstances shape the relevance of the topic of a scientific article. The purpose of the scientific article is to consider the ways of formation and development of regional innovation systems. The objectives of a scientific article are to examine the methodological level of research of regional innovation systems; to define the definitional gradation of the notion "regional innovation system"; to identify the features of the regional innovation system; to suggest ways of forming regional innovation systems; to form the ways of implementation of regional innovation systems. The methodical research tools based on basic and additional scientific methods. The basic scientific methods include the definition and bibliographic method, the axiomatic method, the method of species structurization, the method of implication of circumstances. The additional scientific methods include the method of conceptual gradation, the method of systematic exposure, and the method of complex energy.

Keywords: Innovation · Innovation processes · Innovation system · The institute · The agent · The region · Paths of development · Paths of formation

1 Introduction

The region is developing in accordance with the concepts and strategies of the innovative formation of society. The innovation is an important aspect. New tools and technologies form an innovative economy. Technologies determine the efficiency and effectiveness of the development of the economy of the future. Innovative systems are the place of accumulation of processes and methods. The processes and methods of innovation systems provide a breakthrough in territorial development. Innovation systems are important for the development of the regional level. Innovative systems are changing the economy of the subjects of the Russian Federation. This thesis is based on the following circumstances.

E. G. Popkova and B. S. Sergi (Eds.): ISC 2019, LNNS 91, pp. 510–520, 2020.
https://doi.org/10.1007/978-3-030-32015-7_58

Firstly, the digitization of the socio – economic paradigm can't be achieved without creating an innovation system. This thesis is relevant for the regional level of the economic development. The innovation system is the basic element of the digitalization of society. The introduction of innovations is a paramount factor in the economy of the region. Innovations regulate the technological basis of society. The regional innovation system allows you to simulate the processes of socio – economic policy in the region. The regional innovation system determines the digital factors of the development of a constituent entity of the Russian Federation. The regional innovation policy includes the digitization of society. Digitalization is based on the generation of knowledge of the regional innovation policy. The regional innovation system allows you to move to a digital paradigm of social development.

Secondly, the regional innovation system has economic agents. Economic agents are business, regional authorities and science. Economic agents of the regional innovation system are developing in innovative areas. Economic agents of the regional innovation system are divided into:

- to business is an agent engaged in project activities in the innovation sphere of the region on the basis of innovation financing;
- to regional authorities are the agent forming and planning the development of innovative processes through legislative and project initiatives;
- to science is an agent formed at the expense of scientific and educational institutions of developing innovations in science.

Economic agents are interrelated and complementary. The principle of interconnectedness of economic agents demonstrates the usefulness of the regional innovation system. The regional innovation system consists of parts. The principle of complementarity reflects the denial of one economic agent from the activities of another economic agent. This circumstance will not allow the formation of a regional innovation system. Resources and economic agents must be complete. This condition is necessary for the formation of a regional economic system.

Thirdly, the diffusion of knowledge is an important regulation in the formation of the innovation system in the region. Knowledge is the source of innovation. Knowledge isn't included in the analysis of the economic development of the region at the present time. Knowledge is an indicator of the educational and scientific process of the socio – economic development of the region. This approach isn't correct. Knowledge is the means and direction of innovation. Knowledge forms the innovation process. Knowledge develops an innovative system. Knowledge is a parity formalization of the regional innovation system.

The highlighted theses determine the significance of the chosen topic of the scientific article. The purpose of the scientific article is to consider the ways of formation and development of regional innovation systems. The tasks of the scientific article are:

- to consider the methodological level of research of regional innovation systems;
- to determine the definitional gradation of the notion of "the regional innovation system";
- to identify the features of the regional innovation system;
- to propose ways of forming regional innovation systems;
- to form ways to implement regional innovation systems.

The highlighted tasks confirm the scientific and practical validity of the topic of a scientific article. Scientific validity lies in the definition of the term “the regional innovation system”. To address this issue, the concept of “the regional innovation system” is graded according to factors and conditions. Practical significance is seen in the ways of formation and implementation of regional innovation systems.

2 Methods

Methodical research tools based on the features of the theoretical perception of regional innovation systems. Methodical tools are divided into basic and additional scientific methods. The main methods characterize the basic components of the study. Basic methods allow you to implement the objectives of scientific research. The main methods include:

– the definition-bibliographic method is a scientific tool that allows you to explore the conceptual apparatus based on the analysis of bibliography;
– the axiomatic method is a scientific tool that sets the initial theoretical positions of the research to identify additional scientific forms and qualities of the subject of scientific research;
– the method of species structuring is a scientific tool that groups the species structures of the subject of scientific research;
– the method of implication of circumstances is a scientific tool reflecting the author’s proposals on the topic of research.

Additional scientific methods reveal the essence of the main research methods. Additional methods consider the specific features of the subject matter. Additional methods are:

– the method of conceptual gradation is a scientific method allowing to distinguish the characteristics of the definition in time;
– the method of systematic exposure is a scientific method that defines the group characteristics of the subject of scientific research;
– the method of complex energy is a scientific method revealing the properties of the proposed measures to improve the subject of scientific research.

Methods of scientific research reflect a multitasking system of the question of the formation and implementation of regional innovation systems. The definition–bibliographic method and the conceptual gradation method form an idea of the concept of “the regional innovation system”. The axiomatic method, the method of species structuring, the method of systematic impact distinguish features of the regional innovation system. The method of implication of circumstances, the method of complex energy form and improve the foundations of regional innovation systems. Dedicated methods form the development of regional innovation systems.

The concept of “the regional innovation system” is systemic and periodic. The concept of “the regional innovation system” was rejected at the initial stage (Fig. 1). The concept of “the technological system of the territory” reflected the essence of the topic under study. This concept was common. The concept of “the technological system of the territory" changed in the 1960s. This period formed a new technological structure of the economy.

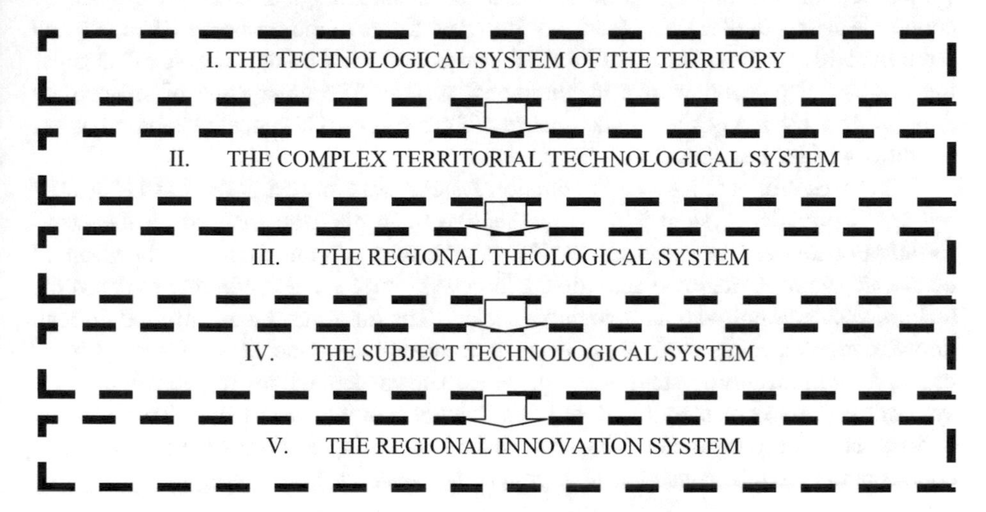

Fig. 1. Gradation of the notion "the regional innovation system"

The technological system of the territory is a collection of scientific and technical objects formed in a single technological policy (Ruban 2016). The peculiarity of this concept is the priority of creating a new territorial infrastructure. This concept reflects the peculiarities of the definition "the territorial technological policy". The system is a set of scientific and technical objects. The innovation factor is manifested in the creation of a new infrastructure.

The concept of "the territorial technological policy" is supplemented by the principle of complexity in the 80s of the 20th century. A complex territorial technological system is an interconnected set of industries with common functions and processes in a specific territory (Lundval 1999). The principle of complexity reflects the properties and phenomena of the technological system. The technological system does not form the aspect of innovation. The regional factor is absent in the technological system. The principle of territoriality represents the place of accumulation of technological systems.

The regional factor is reflected in the concept of "the regional technological system". This concept was formed under the influence of the development of technological systems in accordance with the strategies and economic policies (Galeeva 2008). The regional technological system is a hierarchy of principles, processes and factors of technological development aimed at the implementation of strategic goals in the field of development of economic agents (Dedov 2018). This definition has some peculiarities. Firstly, the regional technological system is shaped by the development aspect. Development defines the process of developing a regional technological system. This role confirms the dynamism of the regional technological system. Secondly, the regional technological system realizes strategic goals. The regional technological system is characterized on the basis of the functions of territorial development. The regional technological system is an object.

The principle of regionality goes into the background in the development of the concept of "the regional technological system". Subjectivity is a new principle. The subjectivity of a technological system is a set of technological actions regulated by economic agents to meet the needs and form new forms of the economy (Filippetti and Archibugi 2011). The concept of "the subjectivity of a technological system" changes the process of perception of a technological system. The emergence of subjectivity expresses the totality of the satisfied needs of the territory. This concept is based on the conditions of technologization.

The concept of "the regional innovation system" was formed in the late 1990s. The regional innovation system is a comprehensive open platform focused on the manifestation of innovation processes and the organization of communication between all economic agents of regional activities (Ciborowski 2016). This definition replaced the technological system with an innovative system. The innovation is manifested through knowledge, science, institutions, subjects. A feature of this concept is the perception of the regional innovation system as an open area. The concept of "the regional innovation system" is complemented by a different number of characteristics. Transparency defines the scale component. The scale is limited to the territory of the region. For consideration of this concept, we highlight the features of the regional innovation system.

3 Results

The regional innovation system has features. The features of a regional innovation system are scaled based on tools. Scalable tools are provided with innovative processes. The basic conditions consider the features of the regional innovation system. The main condition of the regional innovation system is the invariant character. Economic agents don't affect the particular manifestations of the regional innovation system (Fig. 2). The regional innovation system is unchanged.

The regional innovation system is divided into the socio – economic subsystem and the intellectual subsystem. The regional socio – economic subsystem is a collection of objects, processes and other forms that satisfy the social and economic needs of the population. The intellectual subsystem is a set of technical and scientific areas created for the implementation of the technological concept of the development of society.

Subsystems are interconnected, but not complementary. The feature is manifested in the following aspects. Firstly, intellectual systems have various components of social and economic activity. Social – economic subsystems don't regulate the functional side of the intellectual process. Secondly, science and technology develop the socio – economic subsystem. Social – economic subsystems are interconnected with science and technology. Convergence hasn't retroactive effect.

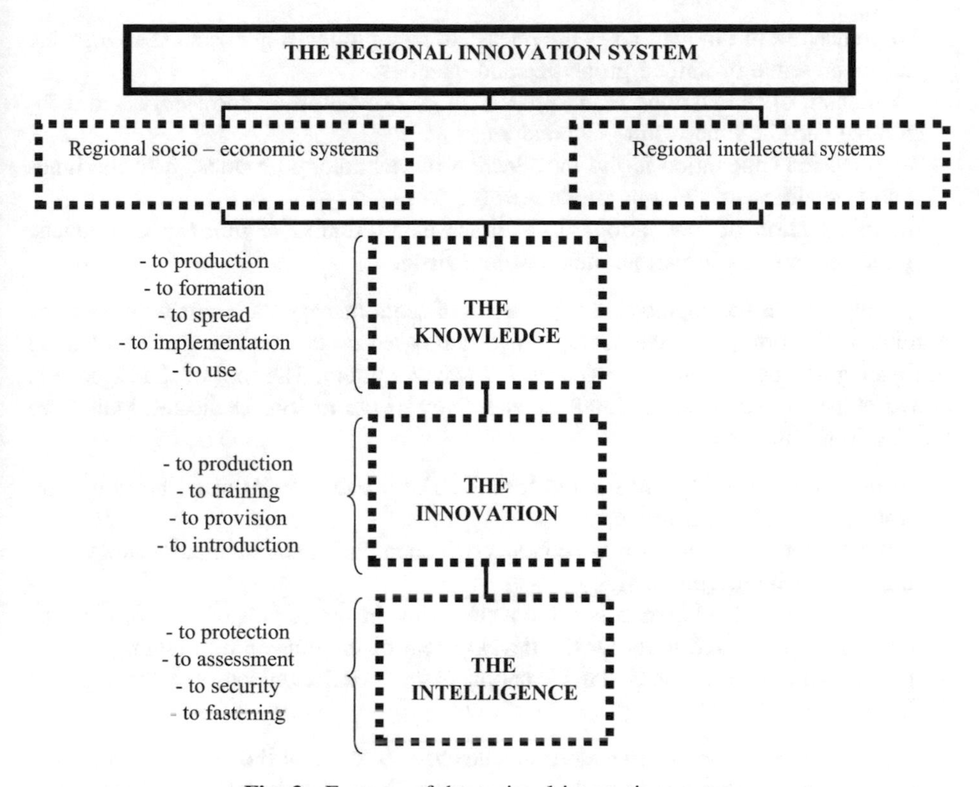

Fig. 2. Features of the regional innovation system

Features of the regional innovation system are associated with knowledge, innovation, intelligence. Knowledge is expressed as a process of creating an innovative environment. The peculiarity of knowledge lies in the identification of the functional apparatus for creating a base of innovative processes. Dedicated tools have the following functionality:

– to production knowledge is the definition of knowledge as a process for creating innovation;
– to formation knowledge is the organization of the process of mastering knowledge and obtaining a result;
– to spread knowledge is the action of the distribution of knowledge and skills among economic agents;
– to implementation knowledge is the introduction of knowledge components into subsystems to implement the result of innovation;
– to use is the cumulative use of knowledge as intended.

The innovation is a product of regional innovation. Innovations are a feature of the regional innovation system development. The regional innovation system can't be formed without innovation in the region. The lack of innovation negates the result of the development of a regional innovation system. Dedicated tools have the following functionality:

- to production of innovations is the release of new products, goods and services due to the presence of unique properties and qualities;
- to training of innovations is the process of consolidation of knowledge and technology into the final result – innovation;
- to provision innovation is the introduction of technology products with the innovative qualities of a unique product or service;
- to introduction of innovations is a diffuse manifestation of innovative products, goods or services in various fields of knowledge.

Intellect forms the cognitive competencies of economic agents. These competencies regulate innovation processes. Competences are based on the implementation of ways of forming and developing the regional innovation system. The tool of intelligence is aimed at the formation of the intellectual activity of the region. Dedicated tools have the following functionality:

- to protection of intelligence is the formation of a base for ensuring the safe functioning of intellectual property;
- to intelligence assessment is the conduct of a quantitative and qualitative analysis of the tools of intellectual activity;
- to provision of intelligence is the concentration in the region of intellectual processes, functions and tools for the development of the innovation system;
- to fastening of intellect is the formation of the intellectual base of the region's innovative development.

The highlighted features predetermine the characteristics of the regional innovation system. Modern experience is based on the use of bureaucratic measures for the formation and development of a regional innovation system. The author of a scientific article has shaped some ways of forming and developing regional innovation systems.

Ways of formation of regional innovation systems are institutional arrangements for the accumulation of innovative processes. Ways of formation reflect the actions of a system-bureaucratic nature. The key economic agent is the state.

The main author's ways of forming regional innovation systems will be discussed below. The way 1 is to create a regulatory framework for regional innovation (Fig. 3).

The way 1 proposes to reform the existing regulatory framework in the field of innovative development of the territory. Regulatory innovation base system operates at the regional level. The existing regional regulatory framework of innovation is reduced to the consolidation of issues of innovation. The existing regional regulatory framework for innovation duplicates federal legislative measures. State measures for the financial support of innovation are different at the federal and regional levels. The federal center provides innovation from the federal budget. Regions provide innovation through limited direct funding. Indirect measures to improve innovation are investment in nature. The legislative problem of innovation is in a vacuum of legislative regulation of regional innovations. This problem doesn't allow to develop the regulatory direction in accordance with the required changes.

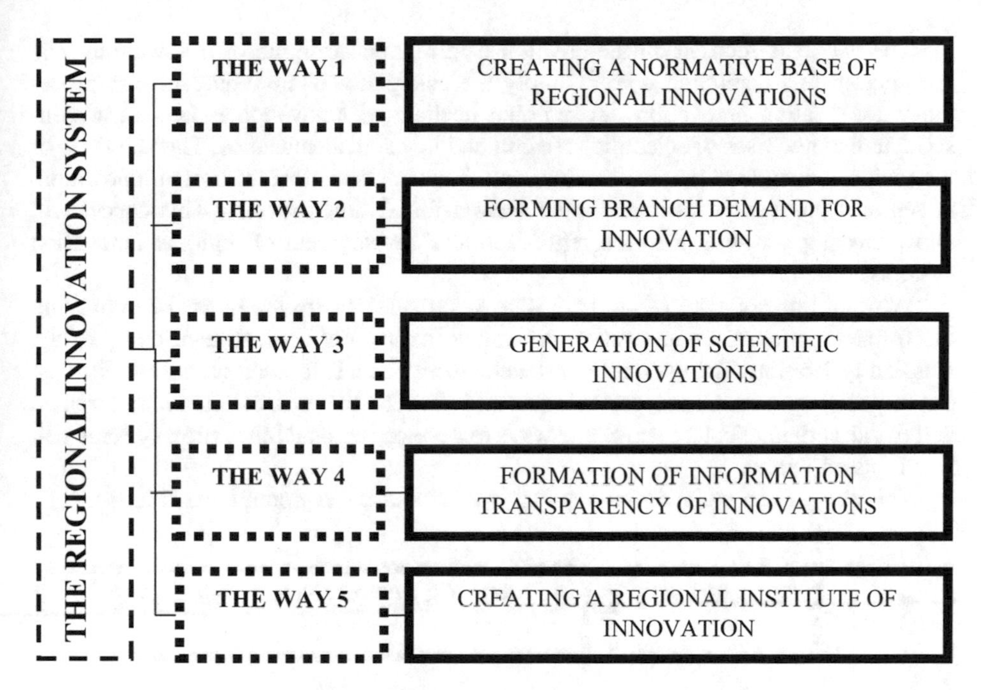

Fig. 3. Ways of forming a regional innovation system

The way 2 of the formation of a regional innovation system has a microeconomic basis. This aspect determines the importance of two economic agents. One agent is business. Another agent is the state. Business creates innovation. Innovations are concentrated in narrow branches of the national economy. Industries limit the use of innovation. This problem is associated with high costs of creating and implementing innovations. The state should become an integral link in shaping the demand for innovation. Firstly, technological renewal of industries should not occur at the expense of import innovations. Secondly, a reformed innovation base should expand the list of industries gradually shifting to an innovative development path. Tough government measures for the transition to an innovative path of development of industries of the national economy should not be applied.

The way 3 is based on the generation of scientific innovations. Innovations are based on research and development. The generation of scientific ideas into practice is inevitable. The lack of centralized scientific institutions slows down the process of the formation of regional innovation systems. To generate scientific innovations, it is necessary to increase the stock of annual innovation developments, form personnel for innovative breakthroughs, regulate the needs of regional economies in innovations.

The way 4 is to form the information transparency of the economy. This event is in a constant information and communication flow. Informational transparency is focused on mass manifestation. For informational transparency, it’s necessary to create a platform for ideological innovations. The platform of ideological innovations consists in combining innovative ideas, innovative products and innovative investors.

The way 5 is focused on the creation of a regional institute of innovations. The presence of this institution doesn't imply the emergence of new bureaucratic procedures for creating innovations. A regional institute of innovation is needed to consolidate the processes of scientific research and investment subsidies. The activities of the regional institute of innovation shouldn't focus on the nationalization of innovation. A regional innovation institute should have a foundation structure. This aspect will allow creating a new link in the regulation and development of regional innovation systems.

Ways of development of regional innovation systems are the basis for improving the formed model of the innovation process at the level of a constituent entity of the Russian Federation. The main role of development paths is focused on finding the best convergence options. Convergence is founded between the regional innovation system and the ideal theoretical model. Some ways to develop regional innovation systems will be discussed below.

The way 1 is to maintain innovation processes at the regional level (Fig. 4).

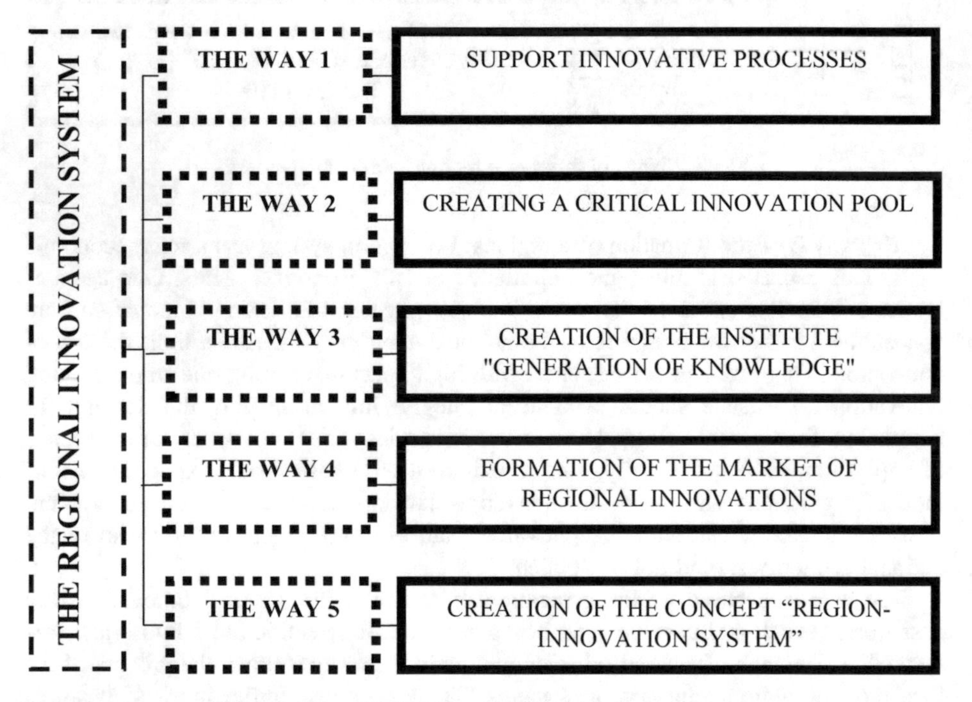

Fig. 4. Ways to develop a regional innovation system

The maintenance process is long and multi-stage. Innovation processes are a key element of the regional innovation system. Innovative processes create a connection between economic agents. Innovation processes form the conditions for the manipulation of various actions of the regional innovation system. Slowing down innovation processes leads to stagnation of the regional innovation system. This action can lead to the fatalism of the regional innovation system. Innovation processes should be supported by economic agents.

The way 2 is governed by the creation of the pool of "critical innovations". The pool of "critical innovation" is a list of innovative products and processes needed to maintain a minimum level of innovative development in the region. The peculiarity of this pool is manifested in multitasking. Firstly, pool of "critical innovation" reveals the level of regional innovation development. Secondly, the pool of "critical innovations" forms the innovative foundation of the regional economy. Thirdly, the pool of "critical innovations" creates a system for evaluating the technological component in the region. The pool of "critical innovation" is a tool for developing a regional innovation system. The introduction of the pool will lead to an increase in the innovation factor in the regions.

The way 3 is to create an institution of "knowledge generation" in the region. The development of the regional innovation system is projected through the improvement of the model of the regional institute of innovation. The process of improvement transforms the role of factors of the regional innovation system. The regional innovation institute maneuvers technology as the only subject of innovation creation. The institution of "knowledge generation" pushes into first place. Knowledge is primary in relation to technology. Knowledge isn't a condition for the formation of the digital economy. Knowledge forms the economy of knowledge. Knowledge as well as technology are sources of innovation creation.

The way 4 requires the creation of the regional innovation market. The way 4 intersects with the way 5 - the creation of the concept "region – innovation system". This concept isn't a strategy or program. The concept involves the representation of the territory as an object. The region favors the innovation system. The region is planning and forecasting innovative processes. The concept allows to revise the existing system of innovative development. The concept excludes components that are not defined in the foresight model for the development of a regional innovation system.

4 Conclusion

The conducted research allows to draw the following conclusions.

1. The methodological level of research of regional innovation systems is manifested through the use of basic and additional scientific methods. The main methods include the definition-bibliographic method, the axiomatic method, the method of species structuring, the method of implication of circumstances. Additional methods include the conceptual grading method, the systematic impact method, the complex energy method.
2. The deformation gradation of regional innovation systems is determined on the basis of five stages. These stages include the following concepts: "the technological system of the territory", "the integrated territorial technological system", "the regional technological system", "the subject technological system", "the regional innovative system".
3. Features of the regional innovation system are reflected in the socio – economic subsystem and the intellectual subsystem. These subsystems are manifested through knowledge, innovation and intelligence. Knowledge, innovation and intelligence

are tools for the development of the regional innovation system. Knowledge, innovation and intelligence are transposed through the principle of functionality.
4. Ways of forming and developing a regional innovation system differ in their essence. Ways of formation accumulate innovative processes. These include: the creation of a regulatory base for regional innovations, the formation of industry-wide demand for innovations, the generation of scientific innovations in practice, the formation of information transparency of innovations, the creation of the regional institute of innovations. Ways of development can improve innovation processes. These include: maintaining innovation processes, creating a pool of "critical innovations", forming the institute of the "knowledge generation", having a market for regional innovations, creating a concept "region - innovation system".

References

Ciborowski, R.: Innovation systems in the terms of Schumpeterian creative destruction. Eureka Soc. Humanit. **4**(4), 29–37 (2016)

Dedov, S.V.: The role of effective management of innovation processes of socio–economic systems in ensuring competitiveness. Finan. Econ. **6**, 1697–1701 (2018)

Filippetti, A., Archibugi, D.: Innovation in times of crisis: national systems of innovation, structure, and demand. Res. Pol. **40**(2), 179–192 (2011)

Galeeva, E.I.: Innovatsionnyye tekhnologii v upravlenii sotsial'no–ekonomicheskimi sistemami. Problemy sovremennoy ekonomiki **3**(27), 68–73 (2008)

Lundvall, B.A.: National business and national systems of innovation. Int. Stud. Manage. Organ. **29**(2), 60 (1999)

Ruban, D.A.: Upravleniye innovatsionnymi sistemami i investitsionnaya sreda regionov. Vestnik NGUEU **2**, 82–95 (2016)

Innovative Import Substitution in Russia and the World: A Comparative Analysis

Alla V. Litvinova(✉), Natalya S. Talalaeva, and Marina V. Ledeneva

Volzhskii Branch of Volgograd State University, Volzhskii, Russian Federation
litvinova_av@mail.ru, talalaeva_ns@mail.ru,
mledenjova@yandex.ru

Abstract. The purpose of the paper is to theoretically prove the role of innovations in import substitution, develop the performance benchmarks reflecting the impact of innovation on the import substitution as well as reveal the trends of innovative import substitution in different countries depending on the features of their external economic balance. The research methodology relies on the fact that the foreign trade turnover of each country where innovative import substitution occurs suffers structural changes, which are manifested in the growing proportion of innovative goods in exports and decreasing proportion of similar goods in imports. Therewith, the augmented export of innovative goods is proportional to the drop in the import thereof. An indicator of the positive impact of innovation on import substitution is simultaneous growth of costs associated with the innovation activity, and increasing export of innovative goods; moreover, the innovation component of exports should outstrip the innovation costs. The condition for innovative import substitution in countries with a very high or very low export-import coverage ratio of suitable values hereof (100–115%) against the background of the growing share of innovative goods' exports. The authors investigated the dynamic changes of innovative import substitution in twelve countries of the world that representing various models of the establishment of foreign trade balance and applied the method of analytical equalization of the dynamic series. Analysis of target indicators of the positive impact of innovations on import substitution shows that all the countries under consideration to some extent perform innovative import substitution. We want particularly note Russia, the United States, Great Britain, Mexico, and South Africa. Regardless of the results achieved by the countries, the innovative import substitution should be developed towards the promotion of technological modernization of domestic production, enhancement of its efficiency and development of new competitive types of goods with high added value.

Keywords: Import substitution · Innovations · Import · Export · Foreign trade balance · Cross-country comparisons

JEL Code: F14 · O 39 · O57

E. G. Popkova and B. S. Sergi (Eds.): ISC 2019, LNNS 91, pp. 521–533, 2020.
https://doi.org/10.1007/978-3-030-32015-7_59

1 Introduction

The role of innovation as a leading factor in economic development is recognized by an absolute majority of researchers and proved by the experience of the global economy. All economically developed countries have significant innovation capacity and high rates of innovation performance, which makes their products competitive in the long term.

From the point of state interests, import substitution secures the growth of exports, which leads to an increase in GDP and employment rate. Therefore, it is reasonable that the countries successfully pursuing the policy of import substitution moved at the appropriate times to foreign markets to attain sustainable economic growth.

If export-oriented and import-substituting strategies have been long considered as conflicting alternatives, then in current studies the export-oriented strategy acts as a follow-up of the import substitution, type and new stage hereof (Vatolkina and Gorbunova 2015; Zudin et al. 2016).

It should be noted that one can theoretically rely on low-tech, including primary commodities, in implementation of the export-oriented strategy. However, the prevailing focus on low-tech export is inexpedient for some reasons. Firstly, the unitary cost of products of one or another sector rises with subsequent processing; besides, this effect enhances the growth of the innovation component of the processed products due to the intellectual rent. Secondly, the export of high-processed end products is relatively independent of the climatic conditions, transport-geographic and geopolitical factors, while the profit margins for processing are little sensitive to commodity cost fluctuations. At the same time, the need for primary commodities is limited, the demand is low-elastic, and therefore their excess on the market can cause a sharp decline in their prices. All these facts are an objective prerequisite for import substitution based on the build-up of high-tech non-resource exports and the growth of its innovative component.

Many writings of domestic and foreign scientists are dedicated to the impact of innovation on import substitution. Most Russian studies theoretically justify the essence and the mechanism of import substitution, assess the effectiveness of measures taken by the government to encourage innovation in favor of import substitution, develop its new directions (Ivanchenko and Selishcheva 2014; Plakhin 2015; Babkin 2015; Ershova and Ershov 2016; Polyakov et al. 2016; Chernova and Klimuk 2016; Ukhanova and Raiskaya 2016; Bondar and Kobzev 2017).

The research of export-oriented import substitution strategy and the role of innovations in the delivery hereof were contributed by foreign scientists, such as Baer (1975), Zhu (2006), Ogujiuba et al. (2011), Neumann (2013) et al. The works that carry out quantitative diagnostic assessment of innovation impact on import substitution with the application of econometric methods are in particular group. So, M.A. Kantemirova, V.A. Kuchieva, and V.T. Balikoev studied 120 economic entities in agricultural production operating in the Republic of North Ossetia-Alania. Using the correlation-regression analysis, the authors proved that there is a positive relation between investments in R&D, innovation and agricultural output (Kantemirova et al. 2016). Using the example of panel data from Turkey and the Republic of Korea, Y. Kilinaslan and I. Temurov empirically proved that import substitution has a positive effect on the competitiveness of manufacturing industry, while the influence of import substitution persists in sectors with both low- and medium-tech development (Kilicaslan and Temurov 2016). A.I. Panov and N.Yu. Fedorov established the point that a regional industrial policy aimed at the growth of new technologies to take a path of innovative development and import substitution should be differentiated depending on the trends hereof revealed in the constituent entities of the Russian Federation. The authors analyzed the distribution of the constituent subjects of the Russian Federation in term of two indicators: the number of researchers per 100 thousand of the working population and the number of technologies developed. We discovered a very high statistical dependence of technology development on the research activity (Panov and Fedorov 2016).

U. Kivikari developed a methodology for quantitative assessment of Russian export efficiency based on the calculation of the absolute efficiency as the ratio of the export price of a product to its domestic price adjusted for the currency exchange rate. The calculations show that in Russia, absolute efficiency generally decreases from the export of commodities to export of end products (Kivikari 1997). N. Fedorenko and G. Chagalov came to similar conclusions, having applied the index of macroeconomic comparative export efficiency that reflects the ratio of production costs to foreign currency earnings from exports. According to the authors' estimates, Russia has the biggest comparative advantages in the sector of fuel and raw materials (natural gas, crude oil, round timber). At the same time, as the products pass a higher stage of processing, the comparative advantages decrease. The exceptions are high-tech products with high added value (Fedorenko and Shagal 2002).

However, we clearly see an insufficient number of studies that assess the impact of innovations on the import substitution through cross-country comparison with regard to the establishment of a system of performance benchmarks displaying the effectiveness of innovative import substitution in different countries. Such diagnostic assessment makes it possible to reveal the position and role of each country in world innovation processes from the standpoint of achieving the generally acknowledged priorities of import substitution. In addition, we should develop the research where cross-country comparisons are made on the example of countries representing different models of foreign trade balance establishment, which is of fundamental importance in understanding the methods used by different countries in solving import substitution problems and their resource provision.

2 Methodology

The positive impact of innovation on import substitution is manifested in observing some trends. Firstly, the foreign trade turnover of each country where innovative import substitution occurs suffers structural changes, which are manifested in the growing proportion of innovative goods in exports ($\uparrow Sh_{ie}$) and decreasing proportion of similar goods in imports ($\downarrow Sh_{ii}$). Therewith, the augmented export of innovative goods is proportional to the drop in the import thereof.

$$+\Delta Sh_{ie} \geq -\Delta Sh_{ii}, \tag{1}$$

or the growth rate of the share of innovative goods in exports is higher than the slowdown rate hereof in imports:

$$R_{growth}Sh_{ie} > R_{slowdown}Sh_{ii}, \tag{2}$$

Where $R_{growth}Sh_{ie}$ is the growth rate of the share of innovative goods in exports; $R_{slowdown}Sh_{ii}$ is the slowdown rate hereof in imports.

Secondly, the innovation processes in import substitution are the effect of innovative activity of the economic entities manifested in R&D, technological, marketing, organizational innovations, participation in the technological exchange, etc. An indicator of the positive impact of innovation on import substitution is simultaneous growth of costs associated with the innovation activity, and increasing export of innovative goods; moreover, the innovation component of exports should outstrip the innovation costs. It mainly concerns R&D costs, which are essential for the implementation of all types of innovations. If R&D costs and export volumes of innovative goods are expressed in absolute units, the following conditions should be simultaneously fulfilled:

$$\uparrow V_{ie}, \uparrow V_{r\&d\,costs}, \tag{3}$$

$$R_{growth}V_{ie} > R_{growth}V_{r\&d\,costs}, \tag{4}$$

Where V_{ie} is the export volume of innovative goods;
$R_{growth}V_{ie}$ is the growth rate of export volume of innovative goods;
$V_{r\&d\ costs}$ is R&D costs;
$R_{growth}V_{r\&d\,costs}$ is the growth rate of R&D costs.

In the event that these variables are expressed in relative units (export of innovative goods is in shares of total exports; R&D costs is in shares of GDP in accordance with the method common to Russian and international statistics), the following conditions should be fulfilled:

$$\uparrow Sh_{ie.}, \uparrow Sh_{r\&d\,costs}, \tag{5}$$

$$R_{growth}Sh_{ie} > R_{growth}Sh_{r\&d\,costs}, \tag{6}$$

Where Sh_{ie} is the share of innovative goods in total exports;
$R_{growth}D_{ie}$ is the growth rate of the share of innovative goods in exports;
$D_{r\&d\ costs}$ is the share of R&D costs in GDP;
$R_{growth}V_{r\&d\,costs}$ is the growth rate of the share of R&D costs in GDP.

Thirdly, in addition to the indicators described above, other ones (for example, the overall level of innovative activity of companies (L_{ia})), the growth hereof with a simultaneous increase in the export of innovative goods Sh_{ie} means positive trends accompanying the innovative import substitution.

Besides, we are certainly interested in revealing the trends of innovative import substitution in different countries depending on the peculiarities of their foreign economic balance historically developing under the influence of many factors: the size of countries, the number of their population, the type of natural resources, specialization in inter-country exchange of goods, features of domestic production and consumption of goods, the level of technological development, the economic policy of the country, foreign economic and geopolitical situation, etc.

The fundamental features of the established foreign trade balance model are the nature of its balance (positive, negative) and associated export-import ratio (higher than 100% – under a surplus, lower than 100% – under a negative balance). The analysis shows that the models of foreign trade balance of different countries greatly vary in terms of these parameters. The surplus, which means the excess of exports over imports and the positive difference between revenues and expenditures on foreign trade transactions respectively, exactly acts a positive trend in the development of the country's economy and shows the high demand for the goods on the international market. However, very high surplus and export-import coverage ratio (Russia, Azerbaijan, Ireland, Norway, Kazakhstan, Brazil) ("Russia and the World", 2018) are mostly the consequence of the prevalence of low-tech primary commodities in the export. On the other hand, some countries with non-resource exports also have a high surplus (Germany, China, Republic of Korea). The foreign trade policy of these countries is based on building up exports of competitive high-demand products, restraining the growth of the population's income, limiting domestic consumption and import respectively. At the same time, the trade surplus of these countries means a forced balance deficit of their trading partners. The foreign trade turnover of most highly developed countries with powerful innovative capacity (the USA, Great Britain, France, Canada, Austria, Turkey, Israel, etc.) ("Russia and the World", 2018) has a negative balance and low values of export-import coverage ratio. The explanation of this phenomenon lies in the conscious restraining of inflation, maintaining a high standard of living of the population due to the outsourcing of labor-intensive production (Chizhikov 2017). However, the deficit of the country's foreign trade balance arising out of it has to be met by the growing volume of foreign borrowing, which generally determines the adverse impact of the negative balance. In addition, international experience shows that the extremely low values of the import-export coverage ratio are common to Ethiopia, Sudan, Armenia, Kyrgyzstan, Pakistan, the Republic of Moldova, Albania, and other countries (Russia and the World, 2018) means a low level of their economic development, non-competitive exports, inability to pay for imports and, as a consequence, persistent depreciation of the national currency. At present, there is no uniform understanding of the suitable value of the export-import coverage ratio in international practice and in numerous scientific studies devoted to the quantitative assessment of trends in the external economic balance of Russia and other countries. Only the regulatory requirements of the European Union limiting the values of the trade surplus of the eurozone countries in the amount of 6% of GDP over three consecutive years can serve as an example, (Dobrov 2017). International practice of foreign trade transactions shows that in most economically developed countries with a surplus (Italy, Singapore, Japan, Belgium, the Netherlands, South Africa, etc., except for Germany, China, and the Republic of Korea), the average values of the export-import coverage ratio ("Russia and the World", 2018) are equal to 100–115%, which can be a benchmark in understanding the most reasonable way to establish foreign trade balance of the country.

The fundamental point of innovative import substitution in countries with a high value of the coverage ratio is the simultaneous fulfillment of two conditions: lowering the coverage ratio ($\downarrow K_{пэи}$) to the suitable value and increasing the share of innovative goods in exports ($\uparrow D_{иэ}$). Accordingly, for countries with a negative balance and low values of the coverage ratio, the latter should grow ($\uparrow K_{пэи}$) simultaneously with the export of innovative goods ($\uparrow D_{иэ}$).

For diagnostic assessment of innovative import substitution in various countries under analysis, we selected 12 developed and developing countries. Six of them are characterized by negative foreign trade balance (Canada, Mexico, Great Britain, USA, Republic of Moldova, Ethiopia), and the other six – by a surplus (Russia, Azerbaijan, Germany, China, South Africa, Japan). Moreover, we see a surplus in Japan and South Africa only in 2016–2017. Due to the lack of complete data on the share of the total volume of imported innovative goods in official statistics, which is required to compare the associated trends of growth (slowdown) of the export of innovative goods and in the import hereof, we employed the data on the share of import and export of goods in section 7 "Machines and Equipment" of the Standard International Trade Classification (SITC) represented in statistic reports of the most countries of the world ("Russia and the World" 2012–2018).

Since the database involved in the study on the dynamic trend of the processes accompanying innovative import substitution in different countries is big, in Table 1 we presented indicators only for Russia. For all other countries, we used the same set of indicators for the same term.

Table 1. Dynamic trend of innovative import substitution performances in Russia

Performances	2010	2011	2012	2013	2014	2015	2016	2017
Exports, million U.S. dollars	397 068	516 718	524 735	525 976	497 359	343 512	285 652	357 767
Imports, million U.S. dollars	228 912	305760	317 263	315 298	287 063	182 902	182 448	227 464
Balance, million U.S. dollars	168 156	210 958	207 472	210 679	210 296	160 610	103204	130 303
The export-import coverage ratio of goods Rc_{ei}	159.8	161.8	157.1	152.9	161.4	176.9	147.1	148.5
The share of innovative goods in the export D_{ie}, percent	9.1	7.9	8.4	10.0	11.4	13.8	10.7	11.5
Growth rates of the share of innovative goods in exports R_{growth} Sh_{ie}, percent		86.8	106.3	119.0	114.0	121.1	77.5	107.5
The volume of innovative export V_{ie}, million U.S. dollars	36 133	40 820	44 078	52 598	56 700	47 405	30 565	41 143
Growth rates of innovative export volumes R_{growth} V_{ie}, percent		113.0	108.0	119.3	107.8	83.6	64.5	134.6
The share of exported goods in section 7 of the Standard International Trade Classification $Sh_{моэ}$, percent	3.2	2.5	2.7	4.1	5.4	5.4	5.1	5.5
The rates of growth (slowdown) of export of goods in section 7 of the		78.1	108.0	151.9	131.7	100.0	94.4	107.8

(*continued*)

Table 1. *(continued)*

Performances	2010	2011	2012	2013	2014	2015	2016	2017
Standard International Trade Classification $R_{growth}D_{МОЭ}$, percent								
The import of goods under section 7 of the Standard International Trade Classification $Sh_{МОИ}$, percent	41.1	45.0	31.5	44.9	41.8	41.8	41.2	45.6
The rates of growth (slowdown) of import of goods in section 7 of the Standard International Trade Classification $R_{growth}D_{МОИ}$, percent		109.5	70.0	142.5	93.1	100.0	98.6	110.7
R&D costs $V_{r\&dcosts}$, million U.S. dollars	33083	35 183	37 913	38 610	40 339	39 734	39 874	…
Growth rates of R&D costs $R_{growth}V_{r\&dcosts}$, percent		106.3	107.7	101.8	104.5	98.5	100.4	
Share of R&D costs in GDP $D_{r\&dcosts}$, percent	1.13	1.01	1.03	1.03	1.07	1.10	1.10	…
The growth rates of the share of R&D costs in GDP R_{growth} $D_{r\&dcosts}$, percent		89.4	102.0	100.0	103.9	102.8	100.0	

Source: developed by the authors based on 20 "Russia and the World" (2018); 23 "World Bank Statistics" (2019); 24 "International Trade Statistics Yearbook. Vol. 1. Trade by country" (2011–2017).

For diagnostic assessment of innovative import substitution in the economies of the countries under study, we used the method of analytical equalization of the dynamic series, which is reduced to the selection by Microsoft Excel tools of the most suitable approximating function to calculate the equalized values of empirical data used. For sequences of observations for which it was not possible to determine the long-term trend of indicators due to the significant impact of random factors, we set the intervals for reaching the target indicators. The target indicator is considered "reached" if it occurred at least in four periods out of seven (as an exception, Table 2 shows the indicators of the suitable values of export-import coverage ration reached by China for 2011–2013).

3 Results

The results of the diagnostic assessment of innovative import substitution in the countries under study are presented in Table 2.

As the data analysis of Table 2 shows, in 2010–2017 none of the countries under consideration, regardless of the model of their trade balance, did not reach simultaneously all target indicators of the positive impact of innovations on import substitution. Particular indicators were reached discretely, i.e. not along the entire period under study.

Diagnostic assessment of import substitution in countries with a positive balance showed the following results.

In Russia, all target indicators, except for the criterion of R&D cost-efficiency were reached only in 2012–2014. However, it should be noted that the criterion associated with R&D costs was not reached in any of the countries with a positive trade balance along the entire period of observation. Decrease in the number of reached target

Table 2. Achievement of target performances of innovative import substitution in different countries

Target performances of innovative import substitution	Countries achieving the target performances		Countries not achieving the target performances	
	Countries with a positive balance	Countries with a negative balance	Countries with a positive balance	Countries with a negative balance
Share of innovative goods in exports $\uparrow D_{ie}$	Russia, Azerbaijan (2010–2013), Germany (2011–2013, 2014–2016), China (2011–2013, 2014–2015), South Africa (2010–2014)	Canada (2011–2013, 2014–2015), Mexico (2010–2013, 2014–2017), USA (2012–2016), Ethiopia	Japan	Great Britain, the Republic of Moldova
$R_{growth}\ Sh_{МОЭ} > R_{slowdown}\ Sh_{МОИ}$, Where $R_{growth}\ Sh_{МОЭ}$ is the growth rate of the share of innovative goods in export according to section 7 of the Standard International Trade Classification; $R_{growth}\ Sh_{МОИ}$ is the rate of the slowdown of the import of innovative goods in import according to section 7 of the Standard International Trade Classification	Russia (2012–2015), Azerbaijan (2012–2013, 2015–2016), South Africa (2012, 2014–2016)	Mexico (2012–2015, 2017), Great Britain (2011–2012, 2014, 2016), Republic of Moldova (2011–2012, 2015, 2017)	Germany, China, Japan,	Canada, USA, Ethiopia
$\uparrow V_{ie}$, $\uparrow V_{r\&dcosts}$, where V_{ie} is the export volume of innovative goods; $V_{r\&dcosts}$ is R&D costs	Russia (2011–2014), Germany, China, South Africa (2010–2011, 2012–2014)	Canada (2010–2013), Mexico, Great Britain (2010–2011, 2012–2013), USA, Ethiopia	Azerbaijan, Japan	Republic of Moldova
$R_{growth}V_{ie} > R_{growth}V_{r\&dcosts}$, where $R_{growth}V_{ie}$ is the growth rates of innovative goods exported; $R_{growth}V_{r\&dcosts}$ is the growth rates of R&D costs. The excess of the growth rate of export volume of innovative goods over the growth rates of R&D costs	–	Canada (2011–2013, 2017), Mexico (2011–2012, 2016–2017)	Russia, Azerbaijan, Germany, China, South Africa, Japan	Great Britain, USA, Republic of Moldova, Ethiopia
$\uparrow D_{ie}$, $\uparrow D_{r\&dcosts}$, Where D_{ie} is the share of innovative goods in the total export volume; $D_{r\&dcosts}$ is the share of R&D costs in GDP	Russia (2011–2015), China (2011–2013, 2014–2015)	Great Britain (2012–2013, 2014–2016), USA (2012–2016)	Azerbaijan, Germany, South Africa, Japan	Canada, Mexico, the Republic of Moldova, Ethiopia
$R_{growth}\ Sh_{ie} > R_{growth}Sh_{r\&dcosts}$, where $R_{growth}\ Sh_{ie}$ is the growth rates of the share of innovation goods in exports; $R_{growth}V_{r\&dcosts}$ is the growth rates of R&D costs in GDP	Russia (2012–2015, 2017), Germany (2012–2013, 2015–2016), South Africa (2011–2014)	Great Britain (2011–2012, 2015–2016), USA (2012, 2014–2016)	Azerbaijan, China, Japan	Canada, Mexico, Republic of Moldova, Ethiopia

(*continued*)

Table 2. (*continued*)

Target performances of innovative import substitution	Countries achieving the target performances		Countries not achieving the target performances	
	Countries with a positive balance	Countries with a negative balance	Countries with a positive balance	Countries with a negative balance
For countries with a positive balance and a relatively high export-import coverage ratio: ↓R_{cei} decreases to the level of 100–115% and ↑Sh_{ie} increases simultaneously, where R_{cei} is export-import coverage ratio of goods; Sh_{ie} is the share of innovative goods in exports	China (2011–2013)	–		
For countries with a negative balance and relatively low export-import coverage ratio: ↑R_{cei} increases to the level of 100–115% and ↑Sh_{ie} increases simultaneously, where R_{cei} is the export-import coverage ratio of goods; Sh_{ie} is the share of innovative goods in exports	–	Canada (2011, 2013, 2014), Mexico (2012)		

indicators from five to two by 2016–2017 (the latter includes the growth of the export of innovative goods and the excess of the growth rates of share of innovative goods in exports over the growth rates of R&D costs in GDP) is associated with difficulties in the Russian economy, slowdown of economic growth due to drop in prices for main export goods, deterioration of the foreign policy situation. Thus, we can conclude that the innovative import substitution in Russia is being implemented in general, but its performance can be assessed as satisfactory.

In another energy-exporting country, Azerbaijan, only two target indicators reached the required values, but over a limited period of time: (1) a growth of export of innovative goods (2010–2013), which slowed down in 2014 and (2) excess of the growth rates of the share of innovative goods in exports over the slowdown rates of innovative goods in imports (only in 2012–2013, 2015–2016). At the same time, the first indicator was reached due to fall in prices for the main export commodity of Azerbaijan – crude oil (70% of the total export volume), which entailed an annual reduction in the value of commodity exports during 2012–2016. Thus, there is a reason to believe that the processes of innovative import substitution in Azerbaijan are weak.

South Africa belonging to the countries with diversified foreign trade reached no more than four indicators out of six during the period under consideration. In general, the country passes through innovative import substitution, however, its effectiveness is insufficient: the growth rate of export volume of innovative goods over more than half of the period under consideration did not exceed the growth rates of R&D costs; export of innovative goods in total exports did not grow simultaneously with R&D costs in GDP.

Germany, one of the leading exporting countries of the world, reached target value only for half of the indicators in 2011–2017. In particular, the growth of export of innovative goods was accompanied not by a slowdown, but the growth of import of corresponding goods. Export volumes of innovative products had been growing, but the growth rates of exports hereof were lower than the growth of R&D costs, which evidences the low effectiveness of innovative import substitution. Therewith, R&D costs in GDP did not increase. The growth rates of the share of innovative goods in exports exceeded the growth rate of R&D costs in GDP. It should also be noted that, although Germany has already made great progress in the implementation of innovative import substitution, the country focuses mainly on the export of medium-tech goods, rather than high-tech ones.

Like Germany, China reached the same indicators as well as an additional one that is extremely important for countries with a high trade balance surplus, namely, a slowdown of the export-import coverage ratio to suitable 100–115% under simultaneous growth of export of innovative goods. Thus, in general, China performs innovative import substitution, however, but reaches only a part of the indicators. It is explained by the increasing volumes of consumption in the domestic market due to the growing income of the high-number population, which causes a decrease in the export orientation of the Chinese economy.

Japan reached none of the indicators during the period under consideration, which can be explained by stagnation, GDP falldown over 2013–2015 and decreased share of the country in global gross product, escalation of competition from China and other countries of Southeast Asia. Nevertheless, despite the stagnation in innovation development in recent years, Japan achieved high values in terms of the export of innovative goods.

As for the countries with a negative trade balance, the countries with the highest scientific and technical capacity (the USA and Great Britain) achieved four indicators out of six. In particular, both countries do not observe the criterion of cost-effective R&D, which can be explained by the high capacity of the domestic market. In Great Britain, the country with the highest scientific and technical capacity in Europe, the share of innovative goods in exports doesn't grow. The United States, on the other hand, demonstrates the rapid growth of innovative goods in imports against exports due to the transfer of production of high-tech goods to developing countries. So, since 2007, the US import of manufactured goods from developing countries first has increased the corresponding imports from developed countries (Fischer et al. 2010).

Canada achieved half of the target indicators. In particular, we can note positive trends of a growing share of innovative goods in exports, increasing the volume of exports of innovative goods and R&D costs. The criterion for cost-effective R&D was also fulfilled. On the other hand, like in the United States, the growth of innovative goods in imports of Canada outstrips the growth hereof in exports. In addition, the share of R&D costs in GDP doesn't increase, and the growth rates of innovative goods in exports do not exceed the growth rates of R&D costs in GDP. The export-import coverage ratio is close to optimum values.

Mexico with its rather high share of innovative goods in exports (42 manufacturers of auto vehicles operate in the country, which makes Mexico's auto market one of the most developed in the world and provides the innovative component of exports at the level of developed countries) achieved four target indicators. Like in Canada, the share of R&D costs in GDP does not increase in Mexico and the growth rate of innovative goods in exports does not exceed the growth rate of R&D costs in GDP. Thereat, the criterion associated with the necessity to achieve the optimum value of the export-import coverage ratio is fulfilled.

As for Ethiopia and the Republic of Moldova, we can roughly talk about innovative import substitution. Ethiopia achieves only two target indicators: Growth of the share of innovative goods in exports as well as a simultaneous growth of export of innovative goods and R&D costs. Moldova reached only one target indicator, namely, the excess of growth rates of the share of innovative goods in exports over the slowdown rates of the share of innovative goods in imports.

The study revealed patterns that distinguish countries with positive and negative trade balance. Almost all countries with a positive trade balance, except for Japan, has a growing share of innovative goods in exports during the period under consideration. On the other hand, almost all countries with a negative balance, except for the Republic of Moldova, are characterized by the simultaneous growth of export costs for innovative goods and R&D costs. The excess of the growth rates of export volume of innovative goods over the growth rates of R&D costs is found out in two countries with a negative trade balance (Canada, Mexico) and in none of the countries with a positive one. However, making more accurate conclusions requires more extensive sample scope.

In general, the largest number of indicators was reached in countries with the highest scientific and technical capacity (USA, UK), as well as in developing countries actively pursuing the strategy of innovative import substitution (Mexico, Russia, South Africa).

4 Conclusions/Recommendations

Thus, innovations play a key role in import substitution, since their introduction allows improving goods and services, manufacturing goods with new consumer properties and lower prime cost, and developing the production and export of high-tech goods.

The outcome of the import substitution policy in the country should be higher competitiveness of domestic goods through the encouragement of process upgrade, enhancement of its efficiency and launch of new competitive goods with high added value. The policy of import substitution should lay the basis for the development of a sustainable export-oriented economy in the long term.

The use of benchmarks of import substitution performance developed by the authors allows a comprehensive assessment of the impact of innovation on the dynamic trend of indicators of the country's foreign trade activity, as well as cost-effectiveness associated with the implementation of innovation activities.

The research of the impact of innovations on import substitution in twelve developed and developing countries of the world revealed that all of them undergo, to some extent, the processes of innovative import substitution, while none simultaneously reached all the indicators. The positive impact of innovation on import substitution was the most considerable in Russia, the United States, Britain, Mexico, and South Africa, and to a lesser extent, in Canada, Germany, and China.

Acknowledgments. The publication was prepared with the assistance of the Autonomous Non-Profit Organization "Institute of Scientific Communications" (Volgograd).

The research was conducted under financial support of the Department of Human and Social Sciences of the Russian Foundation for Basic Research in the framework of the research project "Comprehensive Assessment of the Import Substitution Efficiency and Its Impact on Economic Growth of Russia" (Project No. 19-010-00519 A).

References

Vatolkina, N.Sh., Gorbunova, N.V.: Import substitution: foreign experience, instruments and effects. Sci. Tech. J. St. Petersburg State Pedagogical Univ. Econ. Sci. **6**(233), 29–39 (2015)

Zudin, N.N., Kuzyk, M.G., Simachev, Yu.V.: Foreign experience of pursuing the policy of import substitution: who look up to? Russia Trends Dev. Prospects **11-3**, 267–273 (2016)

Ivanchenko, A.D., Selishcheva, T.A.: Innovations as an engine of import substitution. In: Scientific Community of Students of the 21st Century. Economic Sciences: Collection of Mathematical Papers of the 27th International Student Research and Practice Conference, no. 12(27), pp. 287–291 (2014)

Plakhin, E.S.: Import substitution as the main direction of improving the efficient use of innovations in the agrarian sector. In: Import Substitution of Agricultural Products at the Regional Level: Problems and Prospects: Proceedings of the Research and Practice Conference, Kursk: Publishing House of Non-State Educational Institution of Higher Professional Education "Regional Open Social University", pp. 47–55 (2015)

Babkin, A.V. (ed.): Innovation and Import Substitution in the Manufacturing Industry. Publishing House of Polytechnical University, St. Petersburg (2015)

Ershova, I.G., Ershov, A.Yu.: Assessing the effectiveness of public regulation of the import substitution policy. Fundam. Res. **3**, 375–379 (2016)

Polyakov, R.K., Balyasnikova, E.V., Chumakov, A.S.: Sectoral sanctions: focus on import substitution and the development of innovations in the Russian Federation. Bull. Murmansk State Techn. Univ. **19**(2), 502–511 (2016)

Chernova, O.A., Klimuk, V.V.: Reasonable import substitution as a demand for the implementation of a new model of development of the Russian economy. Bull. Samara State Univ. Econ. **5**(139), 34 (2016)

Ukhanova, R.M., Raiskaya, M.V.: A model providing innovative import substitution in the Russian industry. Bull. Econ. Law Soc. Stud. **3**, 78–81 (2016)

Bondar, A.V., Kobzev, I.I.: Innovative import substitution. Consum. Cooperation **1**(56), 13–18 (2017)

Baer, W.: Import substitution and industrialization in Latin America: experiences and interpretations. Lat. Am. Res. Rev. **7**(1), 95–122 (1975)

Zhu, T.: Rethinking import-substituting industrialization. Development strategies and institutions in Taiwan and China. UNU-WIDER Research Paper, no. 76, pp. 260–279 (2006)

Ogujiuba, K., Nwogwugwu, U., Dike, E.: Import substitution industrialization as learning process: sub Saharan African experience as distortion of the «Good» business model. Bus. Manage. Rev. **1**(6), 8–21 (2011)

Neumann, S.: Import substitution industrialization and its conditionalities for economic development – a comparative analysis of Brazil and South Korea. Central European University, Master thesis (2013)

Kantemirova, M.A., Kuchieva, M.V., Balikoev, V.T.: Innovations as a factor of import substitution in regional agriculture. Fundam. Res. **4–2**, 392–396 (2016)

Kilicaslan, Y., Temurov, I.: Import substitution, productivity, and competitiveness: evidence from the Turkish and Korean manufacturing industry. Optimum J. Econ. Manage. **3**(2), 67–83 (2016)

Panov, A.I., Fedorov, N.Yu.: Import substitution, innovations, human capital in the context of regional economic policy. MIRBIS Res. Rev. **1**, 52, 43–55 (2016)

Kivikari, U.: Foreign trade liberalization during the economic transformation in Russia. Econ. Issues **8**, 57–72 (1997)

Fedorenko, N., Shagal, G.: The efficiency of Russia's involvement in the international division of labor. Econ. Issues **7**, 83–93 (2002)

Russia and the World: Statistical Book. Rosstat (2012–2018)

Chizhikov, Yu.N.: Analysis of the trade balance in the russian federation. Sci. Methodol. Electron. J. "Concept" **39**, 311–315 (2017). http://e-koncept.ru/2017/970389.htm. Accessed 12 Apr 2019

Dobrov, D.: Germany's Trade Surplus Causes Concern in the USA and Europe (2017). https://inosmi.ru/politic/20170317/23889602.html. Accessed 12 Apr 2019

World Bank Statistics (2019). https://data.worldbank.org/indicator/TX.VAL.TECH.MF.ZS?locations=ET&view=chart. Accessed 12 Apr 2019

International Trade Statistics Yearbook, vol. 1. Trade by country (2011–2017). United Nations, New York. https://comtrade.un.org/pb/. Accessed 12 Apr 2019

Fischer, K., Reiner, C., Starlitz, C.: Globale Güterketten. Weltweite Arbeitsteilung und ungleiche Entwicklung. Promedia, Wien (2010)

Innovation and Investment Potential of Regions as a Vector for Their Development in the 21st Century

Evgenia M. Kolmakova[1(✉)], Ekaterina M. Kolmakova[2], and Irina D. Kolmakova[1]

[1] Chelyabinsk State University, Chelyabinsk, Russia
janenet@mail.ru, kolmirina@mail.ru

[2] South Ural State University (national research university), Chelyabinsk, Russia
katekol_mn@mail.ru

Abstract. The activation of innovation and investment initiatives is a necessary condition for stable functioning and progressive development of the country's economy. The paper discusses a new methodology for assessing the innovation and investment potential of municipalities in the region. This approach uses the analysis of statistical indicators in five groups: production and finance, labor, infrastructure, investment, social sphere. The main stages of methodology: to identify indicators that characterize the innovation and investment potential; determine the indicator's value for each municipality; to assess a complex indicator of the territory's potential; compare the complex indicator with data base, which uses the average indicators of the region.

The paper considers the structure of the complex indicator proposed for assessing innovative and investment potential of the territories. Due to the polarization of social and economic space of the region, it is of vital importance to classify municipal formations according to the type of pursued social and economic policy and the prerequisites for innovative development.

It is necessary to use the analysis of innovative and investment potential in the activities of municipal authorities since it allows: to assess the state and readiness of municipalities for innovative reforms; to identify advantages and disadvantages, to forecast the main development trends; to prepare recommendations for the formation of an innovation and investment development strategy.

Keywords: Comprehensive index · Innovative potential · Investment potential · Municipality · Region

JEL Code: R11 · R58

1 Introduction

The activation of innovation and investment initiatives is a necessary condition for stable functioning and progressive development of the country's economy. But in Russia this problem has a pronounced regional content, due to significant

E. G. Popkova and B. S. Sergi (Eds.): ISC 2019, LNNS 91, pp. 534–543, 2020.
https://doi.org/10.1007/978-3-030-32015-7_60

differentiation of territories in terms of the level of the gross municipal product, the rate of inflation, the balance of regional and municipal budgets, and the standard of living. As a result, the national economy's transition to the stage of stable functioning and sustainable growth should be preceded by institutional support for reducing the risks of economic activity. In other words, only an investment mechanism of expanded reproduction that is effective in the regions of Russia can guarantee the national economy a way out of the recession and form the foundations of an innovative structure that makes it possible to take a leading position in the world and provide the population with a high quality of life.

Hypothesis 1. Innovation and investment potential is determined by a number of factors; it is necessary to determine the most significant of them.

Hypothesis 2. The types of pursued social and economic policy predetermine the possibility of increasing the innovative and investment potential.

2 Methodology

One of the most important elements of the mechanism for financing the innovation and investment processes in municipalities of a region is the methods for assessing the potential of their innovative and investment development. State financial support for the territories in the process of innovative and investment development should be based on the creation and promotion of long-term competitive advantages. In this regard, we propose the application of a model for assessing the potential of innovation and investment development of municipalities in the region using an integrated approach. It is based on a system of indicators, including an analysis of the dynamics of the social and economic development of the region, an assessment of the resource potential for innovation and investment development.

The methodology proposed by the authors for assessing the innovation and investment potential of municipalities is based on indicative analysis. This approach is often used in the economic literature to assess social and economic phenomena in regional economic systems (Gurban and Sudakova (2015), Vasilieva et al. (2014), Kolmakova (2014), Pykhov and Kashin (2015), Shindina et al. (2016)).

The basis of this methodology is the method of identifying a set of indicators that characterize the level of social and economic development of municipalities. These indicators are grouped according to the elements of the potential characterizing its individual components, forming blocks of indicators. For a more complete description of the situation in some areas, synthetic indicators are also used, which are a number of particular indicators. The application of this hierarchical structure and the corresponding set of indicators allows us to analyze more deeply the conditions for the formation of innovative and investment potential of a particular territory.

The assessment of a region's innovation and investment potential includes a number of stages:

1. Determining the score of the municipal entity in the region on the basis of the system of indicators for assessing innovation and investment potential.

2. Defining the potential of innovative and investment development of the region for each group of indicators.
3. Introducing a comprehensive assessment indicator of the innovation and investment potential of municipalities.
4. Comparing the obtained integral indicator with a certain base.
5. Drawing conclusions, making decisions.

At the first stage, it is necessary to determine the sum of scores for each municipality in the region assessing the indicators which characterize its innovative and investment potential, which are presented in Table 1.

Table 1. System of indicators for assessing the innovation and investment potential of municipalities

Indicators of the blocks	Indicators of the elements of block 1
Production and finance	A_1 - shipped goods of own production, performed works and services on their own (without small businesses) per capita, one thousand rubles; A_2 - fixed assets for urban districts and municipal areas (at the end of the year; at full cost; million rubles per 1 employee of organizations); A_3 - surplus (+), deficit (−) of the municipal budget (local budget), executed, thousand rubles, value of the indicator for 2016
Investments	C_1 - volume of investment in fixed capital (except for budgetary funds) per 1 person, rubles, value of the indicator for the year C_2 - number of sites open to the investor (according to the investment passports of the municipality on 22.12.2016) C_3 - share of innovative goods, works, services in the total volume of goods shipped, works performed, services of industrial production C_4 - number of patents for inventions per population of municipality
Infrastructure	E_1 - length of motor roads per unit area of the territory under consideration (km/km^2) E_2 - area of municipality per capita as of January 1, 2015, km^2 E_3 - share of the territory influenced by the megapolis from the total area of the territory under consideration E_4 - volume of ore mines (thousand tons) E_5 - volume of nonmetallic minerals (thousand m^3) E_6 - volume of rare-earth minerals (kg)
Labor resources	G_1 - dynamics of the population in municipalities (the analysis period of 3 years) G_2 - average annual number of employees of organizations in % of the total population G_3 - level of registered unemployment in % of the economically active population G_4 - characteristics of the territory by the number of scientific personnel per capita
Social block	K_1 - ratio between fertility and mortality (coefficient of vitality) K_2 - level of average wages of the population (thousand rubles per month) K_3 - total area of residential premises, an average for one inhabitant (m^2/person) K_4 - consumer price index

The method was tested in 12 urban districts within the Chelyabinsk region. The study provides a comparative analysis of the innovative and investment potential of urban districts in the Chelyabinsk region for 2011–2013 (I period) and for 2014–2016 (II period). Criteria for evaluating the innovation and investment potential of municipalities are given in Table 2.

Then we determine the indicator of the innovative and investment potential of municipality for each block of indicators according to the formula:

$$PMO_i = Mf_i / Mn_i, \quad (1)$$

where Mf_i – actual sum of points for each block of indicators;

Mn_i – minimum set sum of points for each block of indicators.

Table 2. Criteria for assessing the innovation and investment potential of municipalities (fragment).

Indicator	Criterion for assessing
Block "Production and Finance"	
A_1 - shipped goods of own production, performed works and services on their own (without small businesses) per capita, one thousand rubles;	Above average regional level - 2 Points At the level of the regional average - 1 Point Below the regional average - 0 Points
A_2 - fixed assets for urban districts and municipal areas (at the end of the year; at full cost; million rubles per 1 employee of organizations);	Above average regional level - 2 Points At the level of the regional average - 1 Point Below the regional average - 0 Points
A_3 - surplus (+), deficit (−) of the municipal budget (local budget), executed, thousand rubles, value of the indicator for 2016	Budget surplus of municipality - 2 Points Budget deficit of municipality - 0 Points
Block "Human Resources"	
G_1 - dynamics of the population in municipalities (the analysis period of 3 years)	Positive dynamics of the indicator - 2 Points Unstable dynamics - 1 Point Negative dynamics - 0 Points
G_2 - average annual number of employees of organizations in % of the total population	Above average regional level - 2 Points At the level of the regional average - 1 Point Below the regional average - 0 Points
G_3 - level of registered unemployment in % of the economically active population	Above average regional level - 2 Points At the level of the regional average - 1 Point Below the regional average - 0 Points
G_4 - characteristics of the territory by the number of scientific personnel per capita	Above average regional level - 2 Points At the level of the regional average - 1 Point Below the regional average - 0 Points

At the next stage we define a complex indicator of the innovative and investment potential of the territory by the formula:

$$KAP = \sum_{i=1}^{n} PMO_i \Big/ n \quad (2)$$

where n is a number of indicators of innovation and investment potential of municipality.

The fourth stage involves comparing the obtained integral indicator (KAP) with the base, which is used as integral indicators, the average for the region.

The final stage involves drawing conclusions and making decisions.

3 Results

To illustrate the application of the proposed methodology, we tested it on the example of municipalities within the Chelyabinsk region of the Russian Federation.

Table 4 presents a fragment of calculating the score for each municipality in the context of the blocks of indicators proposed by us before (Table 1). The scoring for each block of indicators is based on the analysis and evaluation of statistical data according to Table 2.

Table 3 presents the score for each municipality in the context of the groups of indicators presented in Table 1 for 2011–2013 (I period) and for 2014–2016 (II period).

The scoring for each group of indicators is based on the analysis and evaluation of statistical data (economic, social and other indicators) according to Table 2.

For each group of indicators, the value of the smallest sum of points gained by any municipality becomes the base (Mn_i) for calculating the indicator of the innovation and investment potential of the municipality for each group of the indicator system (PMOi). Next, it is necessary to define a comprehensive indicator of the innovative and investment potential of each municipality. For example, for the first municipality:

$$\sum_{i=1}^{n} PMO_1 = PMO_1 + PMO_2 + PMO_3 + PMO_4 + PMO_5 \quad (3)$$

Similarly, we determined the comprehensive indicator of innovation and investment potential of other municipalities. The results of the calculation of comprehensive indicators of evaluating innovation and investment potential of municipalities of the Chelyabinsk region (fragment) are presented in Table 4.

As it can be seen from the table, the greatest innovation and investment potential is that of Chelyabinsk City, which is followed by Miass District. Kopeysk District and Troitsk district had in the first period rating B, in the second period fell to a position lower in the ranking, Kyshtym and Ust-Katav Districts are in group D. Out of the 12 studied city districts in the Chelyabinsk region in the second rating period 3 municipalities are rated as A, 4 – as B, 3 are rated as C, and 2 municipalities are rated as D. At the same time, three municipalities (Miass, Kopeysk and Troitsk) have worsened their

Table 3. Score of municipalities on the basis of indicators of the innovation and investment potential (fragment)

Indicators		Miass District		Kyshtym District		Ust-Katav District		Kopeysk District		Troitsk District		Chelyabinsk City	
		I period	II period	I period	II period	I period	II period	I period	II period	I period	II period	I period	II period
Block "Production and Finances"	Mf_1	5	4	4	3	2	2	4	2	4	2	5	4
	Mn_1	2	2	2	2	2	2	2	2	2	2	2	2
	PMO_1	**13**	**2,0**	**2.0**	**1,5**	**1,0**	**1,0**	**2,0**	**1,0**	**2,0**	**1,0**	**2,5**	**2,0**
Block "Investments"	Mf_2	14	12	10	10	3	3	16	11	18	18	21	26
	Mn_2	3	3	3	3	3	3	3	3	3	3	3	3
	PMO_2	**4.7**	**4,0**	**3,3**	**3,3**	**1,0**	**1,0**	**5,3**	**3,7**	**6,0**	**6,0**	**7,0**	**8.7**
Block "Infrastructure"	Mf_3	6	4	2	7	5	6	8	8	7	5	7	8
	Mn_3	2	4	2	4	2	4	2	4	2	4	2	4
	PMO_3	**3,0**	**1,0**	**1,0**	**1,8**	**2,5**	**1,5**	**4,0**	**2,0**	**3,5**	**1,2**	**3,5**	**1,3**
Block "Labor Resources"	Mf_4	6	5	2	2	2	2	4	4	4	3	7	7
	Mn_4	2	2	2	2	2	2	2	2	2	2	2	2
	PMO_4	**3,0**	**2,5**	**1,0**	**1,0**	**1,0**	**1,0**	**2,0**	**2,0**	**2,0**	**1,5**	**3,5**	**3,5**
Social block	Mf_5	5	4	4	3	2	3	4	2	2	1	6	4
	Mn_5	2	1	2	1	2	1	2	1	2	1	2	1
	PMO_5	**2,5**	**4,0**	**2,0**	**3,0**	**1,0**	**3,0**	**2.0**	**2,0**	**1,0**	**1,0**	**3.0**	**4,0**
$\sum_{i=1}^{n} PMO_i$		15,7	13,5	9,3	9,6	6,5	7,5	15,3	10,7	14,5	10,7	19,5	19.5

Resources: 1. Regions of Russia. The main social and economic and urban indicators: Statistic data 2016.
2. Municipal statistics. Chelyabinskstat. [Electronic resource] http://chelstat.gks.ru/wps/wcm/connect/rosstat_ts/chelstat/ru/municipal_statistics/. (Date of review 10.03.2018).

Table 4. Rating of innovation and investment potential of municipalities on the basis of integrated indicators (fragment)

Indicators	Miass District	Kyshtym District	Ust-Katav District	Kopeysk District	Troitsk District	Chelyabinsk City
I period (2011–2013)						
$KAP = \sum_{i=1}^{n} PMO_i \Big/ n$	3,14	1,86	1,30	3,06	2,90	3,90
Rating	A	D	D	B	B	A
II period (2014–2016)						
$KAP = \sum_{i=1}^{n} PMO_i / n$	2,70	1,92	1,50	2,14	2,14	3,90
Rating	B	D	D	C	C	A

positions in the ranking. The reason for this situation was that in these municipalities there was no process of active updating of fixed assets of production; the level of investment was insufficient for innovation. Thus, the social and economic policy implemented in municipalities directly affects the level of innovation and investment potential of municipalities. The characteristics of municipalities in terms of innovation and investment potential are given in Table 5.

Table 5. Characteristics of municipalities in the level of innovation and investment potential

Group of municipalities	Characteristics of municipalities in the group
A Municipalities with the most favorable conditions for innovation and investment activity	When active renovation of the material and technical base of production and investment is concentrated in the real sector of the economy, we can talk about a high level of innovative filling of investments in this municipality, creating favorable conditions for the formation of innovation environment in municipalities
B Municipalities primarily operating the existing economic complex	The situation in which the existing permissible level of the material and technical base of production is not supported by new capital investments, testifies to the exploitation of the existing economic complex and the absence of long-term development plans for business representatives
C Municipalities mainly operating financial or natural resources	If most of the investments are directed to short-term and rapidly recouping sectors of the economy, compared to investments in the real sector of the economy, this leads to moral and physical deterioration and aging of equipment in the industrial sectors of the national economy and a gradual lag in economic development. In this situation, the financial or raw material resources of municipalities are being exploited
D Crisis “depressed” municipalities with worn out material and technical base and low level of investment activity	Low investment activity of the municipal formation in combination with a heavily worn out material and technical base indicates a deep investment and economic crisis, that is, the lack of conditions for the implementation of innovations

Municipal formations belonging to group A, thus, have a fairly stable innovation and investment potential, municipalities belonging to groups B and C are medium-stable, and those in group D have innovative investment potential with low sustainability.

From the above, it can be concluded that sustainable municipal development can be interpreted as a dynamic, complex state of the system based on a balanced set of social and economic, ecological, political and other interrelated processes implemented on the basis of rational use of all resources of the territory, not exceeding the maximum permissible loads for environment, and enabling to consistently increase the potential of the municipality to improve the quality of life and the needs of citizens residing on its territory.

Thus, we can conclude that the administrative and territorial units of the Chelyabinsk region are quite differentiated, despite the fact that on the whole the region has the status of an industrial region specializing in metallurgy.

4 Conclusion

Determining the level of development of innovation and investment potential allows us to take measures to ensure the growth of new technologies, innovative goods, works, services, increasing the investment attractiveness of the municipal formation, not only in the foreseeable future, but also in the strategic perspective.

The results of territory ranking by innovative and investment potential allow us to determine the most problem areas in levels and criteria necessary for taking measures of ensuring innovation and investment development of a given territory. Besides, the results obtained on the basis of potential opportunities of the territory (fossils, energy, recreational zones, etc.) make it possible to develop program measures to increase the investment attractiveness of a particular territory.

Municipalities in Group C and D (for example, Kyshtym and Troitsk Districts) demonstrate a decline in the population for the analyzed period, respectively, the potential for labor resources is low.

The financial basis for the development of the regional economy and its constituent municipalities in terms of financing the public sector are territorial budgets. The municipalities from group A demonstrate high positive dynamics in the formation of the financial result of the activity. They demonstrate the highest indicators of budgetary security.

The obvious lag of most municipalities is observed in terms of indicators of the "Infrastructure" block, which is a confirmation of hypothesis 1. This confirms the conclusions of Sylvie (2001), Kumar et al. (1997), Kelejian and Robinson (1997). Munnell (1992), Doloreux (2002) on the importance of this factor in the development of the territory.

Assessing the development level of the innovation and investment potential of the municipality results in the identification of potential investment objects. This greatly facilitates the work of the investor in the search for sources of investment, as well as reduces the risk of non-return of the funds invested by them.

Prospective directions of investing in the economy of territories belonging to the first group can be:

- development of innovative activities. For the development of municipal education as an economic center, it is necessary to pursue a policy of advancing development of science and technology in relation to other branches and spheres of activity of the municipality;
- construction of international logistics centers.

Prospective directions of investing in the economy of territories belonging to the second group can be:

- construction of large and small electric power facilities.

Prospective directions for investing in the economy of territories belonging to the third group can be:

- development of the tourism and recreation industry;
- organization of economic activities that reduce the negative impact on the environment and the health of the population, preserve the biological and landscape diversity of the territory;
- diversification of economy in single cities.

Prospective directions for investing in the economy of the territories belonging to the fourth group can be:

- realization of "breakthrough" investment projects and import substituting technologies;
- introduction of energy and resource saving;
- maximum use of natural, infrastructure and human capital opportunities of small towns and rural areas within the framework of comprehensive programs of social and economic development.

In the conditions of uneven development of municipalities within the social and economic space of the region, it is possible to single out, in our opinion, the following types of social and economic policy conducted by municipalities:

- stimulating municipal policy. The municipal authorities actively search for investors and use all the means at their disposal to accelerate economic development by stimulating the introduction of innovations, the development of modern industries (as well as the curtailment of old ones), through infrastructure and information preparation of the territory (Chelyabinsk City, Magnitogorsk City).
- adapting municipal policy. The municipal authorities use their available resources to mitigate the negative consequences associated with the lack of funds in the local development budget. Such a policy is also oriented at receiving subsidies, subventions, benefits from higher-level budgets for the performance of assigned powers (for example, Kyshtym District, Karabash District, Troitsk District).
- compensatory municipal policy. The municipal authorities promote the adaptation of more mobile and manageable components of municipal development to inertia, less manageable ones; they use the resources at their disposal to mitigate the

negative consequences associated with the lack of resources in the local development budget. This policy is also associated with obtaining subsidies, subventions, benefits from budgets of a higher level, but for the purpose of transition to a new type of production related to development (Ust-Katavsk District, Chebarkul District).

One of the ways to increase the innovation and investment potential of municipalities is their entry into the composition of agglomerations and territories of advanced social and economic development.

Acknowledgments. The work was supported by Act 211 Government of the Russian Federation, contract No. 02.A03.21.0011.

References

Sylvie, D.: Infrastructure development and economic growth: an explanation for regional disparities in China. J. Comp. Econ. **29**, 95–117 (2001)

Doloreux, D.: What we should know about regional systems of innovation. Technol. Soc. **24**(3), 243–263 (2002)

Gurban, I., Sudakova, A.: An assessment methodology for the development of higher education in Russia. Mediterr. J. Soc. Sci. **6**(5), 197–210 (2015)

Kelejian, H., Robinson, D.: Infrastructure productivity estimations and its underlying economic specifications: a sensitivity analysis. Pap. Reg. Sci. **76**, 115–131 (1997)

Kolmakova, Ek.M.: Potential and limitations of the spatial-investment-innovation model of the region's growth. Bull. Chelyabinsk State Univ. **21**(350) (2014). Economics (47), 82–86

Kumar, A., Gray, D., Hoskote, M., von Klaudy, S., Ruster, J.: Mobilizing Domestic Capital Markets for Infrastructure Financing: International Experience and Lessons for China. World Bank Discussion. World Bank, 377, Washington, DC (1997)

Munnell, A.: Policy watch: infrastructure investment and economic growth. J. Econ. Perspect. **6**, 189–198 (1992)

Pykhov, P.A., Kashina, T.O.: Infrastructural security of the UrFD regions: assessment methodology and diagnostic results. Econ. Reg. **3**, 66–77 (2015)

Shindina, T.A., Kolmakova, Ek.M., Vlasova, G.A., Orlova, N.A., Kolmakova, I.D.: Labour and social relations as the economical category and the efficiency of the regional system of labour and social relations (on the example of the constituent parts of ural federal okrug of Russian Federation). J. Appl. Econ. Sci. **XI**(4(42)), 781–784 (2016)

Vasilieva, E.V., Kuklin, A.A., Lykov, I.A.: Program of complex diagnostics of the quality of life in the region. Functional characteristics and possibilities of its application. In: Living Standards of the Population of Russian Regions, no. 1, pp. 118–123 (2014)

Scenario of Hi-Tech Growth of Innovative Economy in Modern Russia

Anna I. Pakhomova[1(✉)], Rustam A. Yalmaev[2], Elena V. Belokurova[3], and Larisa V. Shabaltina[4]

[1] Institute of Service Sphere and Entrepreneurship (branch) of Don State Technical University, Shakhty, Russia
paxomoval202@mail.ru

[2] Chechen State University, Grozny, Russia
r.yalmaev@chesu.ru

[3] Tyumen Industrial University (Nizhnevartovsk branch), Nizhnevartovsk, Russia
e.belokurowa@yandex.ru

[4] Plekhanov Russian University of Economics, Moscow, Russia
lvs-28@mail.ru

Abstract. Purpose: The purpose of the paper is to determine the current scenario according to which innovative economy is developing in modern Russia and to develop recommendations for transition (or improvement) to scenario of hi-tech growth.

Design/methodology/approach: The methodological research base includes structural analysis, correlation and regression analysis, scenario development, and graphical methods.

Findings: The results of the performed analysis showed that in modern Russia (2019) innovative economy is implemented according to scenario of low-tech and medium-tech growth, which does not allow Russian entrepreneurship to conquer the leading positions in the world markets of hi-tech products and envisages delayed development as compared to developed countries. Implementation of innovations in entrepreneurship envisages its temporary competitive advantages, due to which effectiveness of innovative activities is low.

Originality/value: For practical implementation of the Strategy of innovative development of the Russian Federation until 2020, adopted by the Decree of the Government of the Russian Federation dated December 8, 2011, No. 2227-p, it is offered to correct the development of innovative economy in Russia and ensure its implementation according to scenario of hi-tech growth. As the key factor of development of hi-tech production in Russia is expenditures for R&D it is recommended to increase them.

Keywords: Hi-tech growth · Innovative economy · Sustainable advantages · Innovations · Modern Russia

JEL Code: C41 · O31 · O32 · O33 · O38 · O47

E. G. Popkova and B. S. Sergi (Eds.): ISC 2019, LNNS 91, pp. 544–551, 2020.
https://doi.org/10.1007/978-3-030-32015-7_61

1 Introduction

In the modern global economic practice, innovative economy of national economic systems develops according to two scenarios. Scenario of hi-tech growth is implemented by developed countries (e.g., the USA and Germany) and envisages rapid development: creation of new markets of hi-tech and created products. The advantage of this scenario is a possibility of obtaining monopolistic profit in the first years of existence of the created new markets and formation of sustainable (which are not recreated by rivals) advantages in them due to usage of unique patented technologies.

Scenario of medium-tech and low-tech is implemented by developing countries (e.g., China, India, Brazil, and South Africa) and envisages delayed development: moving domestic companies to the existing markets of hi-tech or strengthening of their positions in other existing markets through implementation of innovations. The advantage of this scenario is mass accessibility, as its implementation requires lower volume of financial resources; if there are no resources of other types (e.g., personnel and technologies), it is possible to attract them (labor migration and import).

A large drawback of this scenario is instability (high risk of recreation by rivals, loss of uniqueness) of advantages due to foundation on relatively new (only for the studied countries, but widespread in the world) and/or unpatented technologies. In the Strategy of innovative development of the Russian Federation until 2020, adopted by the Decree of the Government of the Russian Federation dated December 8, 2011, No. 2227-r, the most preferable variant (scenario) of innovative development of economy, "which conforms to long-term goals and tasks of the Strategy, is achieving leadership in the leading scientific and technical sectors and fundamental research".

Though the period of activity of the Strategy ends, its basic principles will probably be preserved in the next strategy, which is to be adopted in late 2019 – early 2020. This means that the strategic priority of innovative economy in modern Russia is its development according to scenario of hi-tech growth. The purpose of the paper is to determine the current scenario, according to which innovative economy in modern Russia is developing, and to develop recommendations for transition (or improvement) to scenario of hi-tech growth.

2 Materials and Method

The fundamental and applied works on the topic of growth and development of innovative economy, which are the basis for this article, include Bezrukova et al. (2017), Bogoviz (2019), Mateut (2018), Nuruzzaman et al. (2018), Pritvorova et al. (2018), Popkova and Sukhodolov (2017), Popkova (2018), Richter et al. (2017), Sergi et al. (2019), Sibirskaya et al. (2019), Sitanggang (2018), and Vanchukhina et al. (2018). The methodological base of the research includes structural analysis, correlation and regression analysis, development of scenarios, and graphical method.

For determining the role of medium-tech and hi-tech production in the structure of GDP of Russia in 2018, let us use statistical data of the World Bank (Fig. 1).

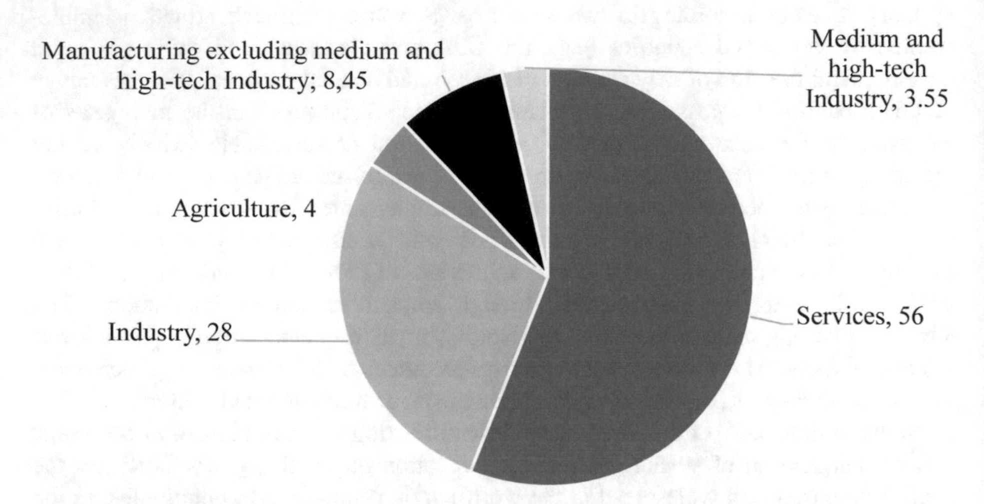

Fig. 1. Sectorial structure of Russian GDP in 2018. Source: compiled by the authors based on World Bank (2019).

Figure 1 shows that the share of medium-tech and hi-tech production in the structure of Russian GDP in 2018 constitutes 3.55%. At the same time, according to the Federal State Statistics Service of the Russian Federation (2019), the share of innovative products in the structure of the processing industry (manufacturing) of Russia constitutes 10.9%. Therefore, the share of medium-tech and hi-tech products in the structure of gross added value, created in the Russian innovative economy, constitutes 32.57% (3.55 * 100/10.9), giving way to low-tech products. In order to determine the reasons for this, let us consider the structure of the developed and used leading production technologies in 2018.

Out of 1,402 developed leading production technologies in Russia in 2018 only 13.55% are completely new – i.e., they could be tools of creating new markets. Out of 240,054 used leading production technologies in 2018 almost half (45.25%) were purchased abroad. 34.59% of developed and 3.80% of used leading production technologies are patented – i.e., capable of forming sustainable competitive advantages.

Thus, at present (2019), innovative economy of Russia is developing according to scenario of medium-tech and low-tech growth, as it actively (almost in 50% cases) used relatively new technologies – borrowed, unprotected by patent law, and possessing novelty only for Russia. Transition to scenario of hi-tech growth of innovative economy in modern Russia requires significant correction of the existing economic practice.

3 Results

In order to determine future perspectives of development of innovative economy of Russia and to offer recommendations for managing this process in the interests of transition to scenario of hi-tech growth let us determine the factors of development of the Russian innovative economy. The potential factors are export of hi-tech products, internal expenditures for R&D, and foreign investments; and the resulting indicator is the share of medium-tech and hi-tech production in the structure of the processing industry (Table 1).

Table 1. Dynamics of the result and the factors of development of innovative economy of Russia in 2010–2019.

Year	Medium and high-tech Industry (including construction), % manufacturing value added	High-technology exports, % of manufactured exports	Research and development expenditure (% of GDP)	Foreign direct investment, net inflows (BoP, current US$)
	y	x_1	x_2	x_3
2010	25,037	9,066	1,130	1,864
2011	24.709	7.972	1.013	2.290
2012	24.838	8.375	1.027	2.119
2013	23.140	10.006	1.025	2.137
2014	24.131	11.452	1.070	1.861
2015	25.598	13.760	1.097	2.411
2016	27.520 (forecast)	10.719	1.096	2.458
2017	29.587 (forecast)	11.524	1.178 (forecast)	1.950
2018	31.809 (forecast)	12.389 (forecast)	1.267 (forecast)	2.096 (forecast)
2019	34.198 (forecast)	13.320 (forecast)	1.362 (forecast)	2.254 (forecast)

Source: compiled by the authors based on Federal State Statistics Service of the Russian Federation (2019).

The calculated coefficients of autocorrelation of the selected factors with the result of development of the Russian innovative economy in 2010–2019 are shown in Fig. 2.

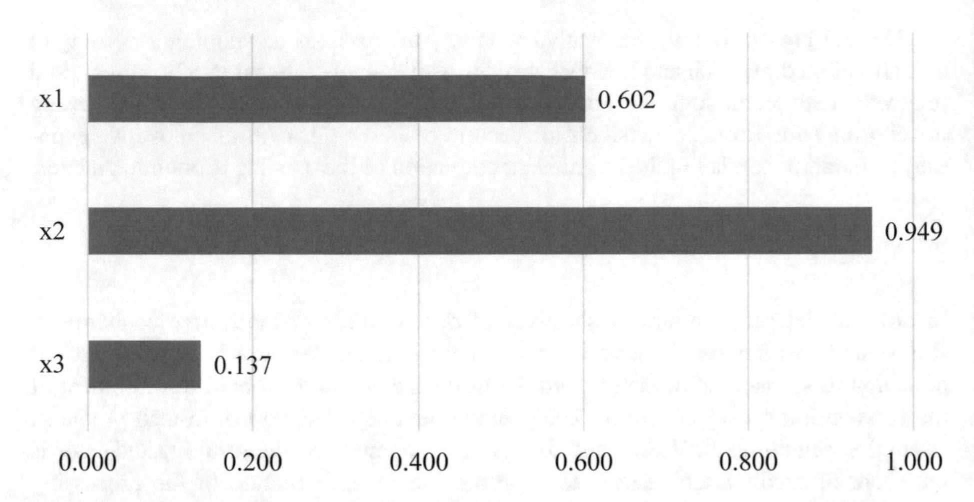

Fig. 2. Coefficients of autocorrelation of the selected factors with the result of development of the Russian innovative economy in 2010–2019. Source: calculated and built by the authors.

Figure 2 shows that the only significant (R > 0.90) factor of development of the Russian innovative economy (R = 0.949) is expenditures for R&D (x_2). Regression analysis of dependence of the share of medium-tech and hi-tech production in the structure of the processing industry on expenditures for R&D is performed in Table 2.

Table 2. Regression analysis of dependence of the share of medium-tech and hi-tech production in the structure of the processing industry of Russia on expenditures for R&D.

Regression statistics	
Multiple R	0.9493
R-square	0.9012
Adjusted R-square	0.8888
Standard error	1.2212
Observations	10

Dispersion analysis

	df	*SS*	*MS*	*F*	*Significance F*
Regression	1	108.8155	108.8155	72.9658	2.71608E-05
Residue	8	11.9306	1.4913		
Total	9	120.7461			

	Coefficients	*Standard error*	*t-Stat*	*P-Value*	*Lower 95%*	*Upper 95%*
Y-intercept	-7.4620	4.0595	-1.8382	0.1033	-16.8232	1.8992
x_2	30.6424	3.5873	8.5420	2.71608E-05	22.3702	38.9146

Source: calculated by the authors

Table 2 shows that coefficient of correlation $R^2 = 0.9012$ – therefore, the change of the share of medium-tech and hi-tech production in the structure of the Russian processing industry in 2010–2019 is by 90.12% explained by the change of expenditures for R&D. The obtained values of coefficients allow compiling an equation of paired linear regression: $y = -7.4620 + 30.6424 * x_2$. Significance F does not exceed 0.05 – which shows statistical significance of regression equation with $\alpha = 0.05$.

Therefore, increase of expenditures for R&D by 1% GDP leads to increase of the share of medium-tech and hi-tech production in the structure of the Russian processing industry by 30.6424%. This shows strong direct dependence. The structure of expenditures for R&D in Russia is shown in Fig. 3.

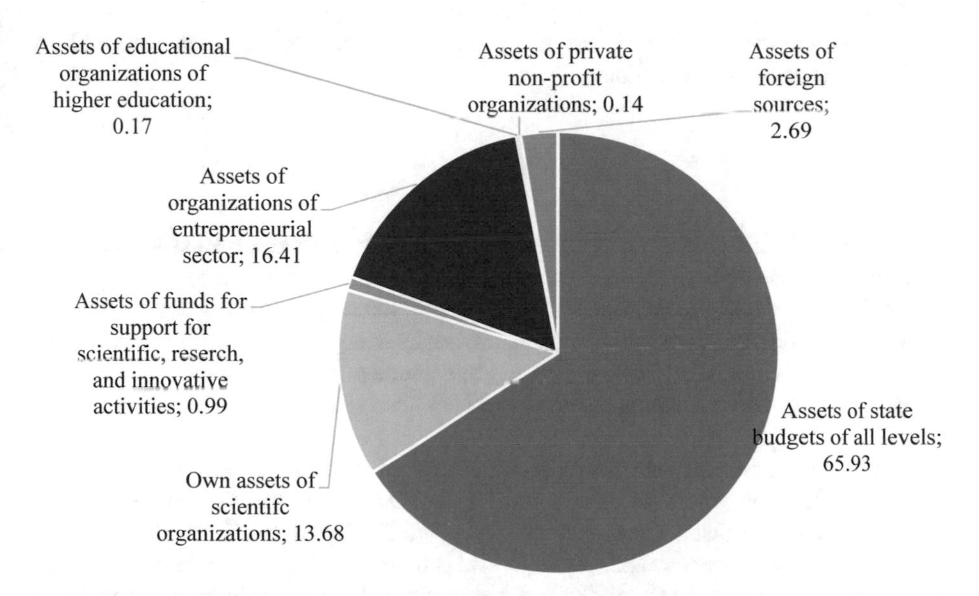

Fig. 3. Structure of expenditures for R&D in Russia in 2018. Source: compiled by the authors based on Federal State Statistics Service of the Russian Federation (2019).

Figure 3 shows that the key source (65.93%) of expenditures for R&D in Russia in 2018 was assets of state budgets of all levels (federal, regional, and local). The most significant sources also include assets of organizations of the entrepreneurial sector (16.41%) and own assets of scientific organizations (13.68%).

Scenario of hi-tech growth of innovative economy in modern Russia envisages increase of the share of medium-tech and hi-tech production in the structure of the processing industry to at least 75%, increase of the share of developed completely new technologies to 80% (all have to be patented), and increase of the used leading production technologies, which are purchased in Russia, to 90% (all have to be patented).

For achieving the set tasks, the share of medium-tech and hi-tech production in the structure of the Russian processing industry should be increased to 78.75% (34.198 * 75/32.57). Setting the calculated value y into the compiled regression equation, we have: 75.75 = − 7.4620 + 30.6424 * x_2. Solving this equation, we have: x_2 = 2.7156((75.75 + 7.4620)/30.6424). Aggregate expenditures for R&D in Russia should be increased from 1.362% of GDP in 2019 to 2.7156% of GDP (by 1.99 times). The following recommendations are offered:

- increase of the volume of assets of state budgets of all levels that are used for financing (including co-financing) of R&D by 2.5 times;
- increase of the volume of assets of organizations of the entrepreneurial sector that are used for financing of R&D by 2 times;
- increase of the volume of assets of scientific organizations that are used for financing of R&D by 2 times.

4 Conclusion

The results of the performed analysis showed that in modern Russia innovative economy is implemented according to scenario of medium-tech and low-tech growth, which does not allow the Russian entrepreneurship to conquer the leading positions in the world markets of hi-tech products and envisages delayed development as compared to developed countries. Implementation of innovations into entrepreneurship ensures its temperate competitive advantages, due to which effectiveness of innovative activities is low.

For practical implementation of the Strategy of innovative development of the Russian Federation until 2020, adopted by the Decree of the Government of the Russian Federation dated December 8, 2011, No. 2227-p, it is offered to correct the development of innovative economy in Russia and ensure its implementation according to scenario of hi-tech growth. As the key factor of development of hi-tech production in Russia is expenditures for R&D it is recommended to increase them.

References

Bezrukova, T.L., Popova, E.V., Korda, N.I., Kuznetsova, T.E., Bezrukov, B.A.: Institutional traps of innovative and investment activities as an obstacle on the path to the well-balanced development of regions. Contrib. Econ. (9783319606958), 235–240 (2017)

Bogoviz, A.V.: Industry 4.0 as a new vector of growth and development of knowledge economy. Stud. Syst. Decis. Control **169**, 85–91 (2019)

Mateut, S.: Subsidies, financial constraints and firm innovative activities in emerging economies. Small Bus. Econ. **50**(1), 131–162 (2018)

Nuruzzaman, N., Singh, D., Pattnaik, C.: Competing to be innovative: foreign competition and imitative innovation of emerging economy firms. Int. Bus. Rev. **2**(1), 46–54 (2018)

Popkova, E.G.: Contradiction of economic growth in today's global economy: economic systems competition and mutual support. Espacios **39**(1), 20 (2018)

Popkova, E.G., Sukhodolov, Y.A.: Theoretical aspects of economic growth in the globalizing world. Contrib. Econ. 5–24 (2017)

Pritvorova, T., Tasbulatova, B., Petrenko, E.: Possibilities of Blitz-psychograms as a tool for human resource management in the supporting system of hardiness of company. Entrep. Sustain. Issues **6**(2), 840–853 (2018). https://doi.org/10.9770/jesi.2018.6.2(25)

Richter, C., Kraus, S., Brem, A., Durst, S., Giselbrecht, C.: Digital entrepreneurship: Innovative business models for the sharing economy. Creativity Innov. Manage. **26**(3), 300–310 (2017)

Sergi, B.S., Popkova, E.G., Bogoviz, A.V., Ragulina J.V.: Entrepreneurship and economic growth: the experience of developed and developing countries. In: Entrepreneurship and Development in the 21st century, pp. 3–32. Emerald Publishing Limited (2019)

Sibirskaya, E., Popkova, E., Oveshnikova, L., Tarasova, I.: Remote education vs traditional education based on effectiveness at the micro level and its connection to the level of development of macro-economic systems. Int. J. Educ. Manage. **33**(3), 533–543 (2019). https://doi.org/10.1108/IJEM-08-2018-0248

Sitanggang, A.S.: Information systems interest talent in developing system (Independent and innovative creative economy) on child with special needs disabled in Bandung City. IOP Conf. Ser. Mater. Sci. Eng. **407**(1), 012133 (2018)

Vanchukhina, L.I., Leybert, T.B., Khalikova, E.A., Khalmetov, A.R.: New approaches to formation of innovational human capital as an element of institutional environment. Espacios **39**, 22–32 (2018)

World Bank: Data catalog: indicators (2019). https://data.worldbank.org/indicator. Accessed 26 May 2019

Government of the Russian Federation: Decree dated December 8, 2011, No. 2227-r, "Strategy of innovative development of the Russian Federation until 2020" (2019). https://www.garant.ru/products/ipo/prime/doc/70006124/. Accessed 26 May 2019

Federal State Statistics Service of the Russian Federation: Russia in numbers – 2018: statistical collection (2019). http://www.gks.ru/bgd/regl/b18_11/Main.htm. Accessed 26 May 2019

Innovative Economy in the 21st Century: Contradiction and Opposition of Developed and Developing Countries

Vladislav A. Shalaev[1(✉)], Elena A. Vechkinzova[2], Anna L. Shevyakova[3], and Oksana Y. Vatyukova[4]

[1] Tyumen Industrial University (Nizhnevartovsk branch), Nizhnevartovsk, Russia
shhel77.77@mail.ru

[2] V.A. Trapeznikov Institute of Problems of Management of the Russian Academy of Sciences, Moscow, Russia
kvin07@list.ru

[3] Rational Solutions LLP, Karaganda, Kazakhstan
shevyakova.anna@gmail.com

[4] Volgograd State University, Volgograd, Russia
vatukova_o_u@mail.ru

Abstract. Purpose: The purpose of the research is to perform a comparative analysis of scientific and practical approaches to formation and development of innovative economy in modern developed and developing countries and to determine scenarios of opposition of developed and developing countries in the global innovative economy until 2030.

Design/methodology/approach: The authors use the dialectical method of scientific cognition and use the Law of unity and fighting of oppositions for describing the contradiction (difference in the approaches of formation and development of innovative economy) and opposition (competition in common world markets of innovations and hi-tech products with commonness of the internal mission) of the modern developed and developing countries in the global innovative economy of the 21st century. The research objects are 6 developed and 6 developing countries, which are leaders (in their categories) in the global rating of countries as to the level of development of innovative economy of Cornell University, INSEAD, and WIPO for 2018.

Findings: It is substantiated that innovative economy in the 21st century is a source of contradiction and the field for opposition of developed and developing countries. They use different scientific and practical approaches to formation and development of innovative economy. The approach that is used by developed countries stimulates their quick development and supports their competitiveness. The approach that is used by developing countries does not fully conform to their interests and requires correction.

Originality/value: It is recommended to increase all indicators of innovative economy be developing countries: share of medium-tech and hi-tech industry in the structure of added value of the real sector, share of expenditures for R&D in GDP, publication and patent activity, share of the number of researchers, and share of export of hi-tech products in the structure of industrial export in 2018. It is shown that this will created an opportunity for implementing the most optimal

E. G. Popkova and B. S. Sergi (Eds.): ISC 2019, LNNS 91, pp. 552–560, 2020.
https://doi.org/10.1007/978-3-030-32015-7_62

scenario of long-term (until 2030) development of the global economy, within which its balance and sustainability are achieved.

Keywords: Innovative economy · Developed countries · Developing countries · Well-balanced · Sustainable development

JEL Code: C62 · F12 · Q01 · O31 · O32 · O33 · O38

1 Introduction

Innovative economy of the 21st century is a result of the evolution of the Theory of economic development, which foundations were set by J. Schumpeter in early 20th century. The initial main idea of innovative economy was innovative activity of companies, which source was acceleration of entrepreneurial capabilities as a new production factor.

In the mid-20th century, during determination and management of innovative economy the main attention was paid to the source of creation of companies' profit and gross product of the national economy – it had to be not only capital (financial resources) and the real (material) sector, but also non-material resources: knowledge and information. This explains multitude of scientific notions, which denote innovative economy, including information economy and "knowledge economy".

In the 21st century, under the influence of the Fourth industrial revolution and restoration of interest to the real sector of economy as a tool of provision of its sustainable development and mass distribution of digital technologies, innovative economy acquired a new treatment – as a progressive socio-economic system, open for innovations and generating new technologies, conducting hi-tech production and export of products, and attractive for venture investing due to high competitiveness (including digital). The source of growth of the modern innovative economy is intellectual resources.

The new concept of innovative economy formed in the 21st century and aggravated the contradiction and opposition of developed and developing countries. This is an important scientific and practical problem of modern times, as the results of this opposition determine the future of the global innovative economy – in particular, possibility of implementation of the global goals in the sphere of sustainable development, which sense is brought down to reduction of disproportions of developed and developing countries and leveling of their socio-economic position in the global economic system.

The purpose of the research is to perform a comparative analysis of scientific and practical approaches to formation and development of innovative economy in modern developed and developing countries and to determine scenarios of opposition of developed and developing countries in the global innovative economy until 2030.

2 Materials and Method

The new concept of innovative economy, which envisages combination of post-industrialization and transition to Industry 4.0, human and artificial intellectual resources, is developed in the works Bezrukova et al. (2017), Bogoviz (2019), Pritvorova et al. (2018), Sibirskaya et al. (2019), and Vanchukhina et al. (2018).

The issue of contradiction of the interests and opposition of developed and developing countries in the modern global economy, as well as disproportions in their development, which lead to imbalance and cause unsustainability of the global economic system, is studied in the works Popkova and Sukhodolov (2017), Popkova (2018), and Sergi et al. (2019).

For systemic research of the above issues, which are studied separately in most of the existing publications, here the dialectical method of scientific cognition and the Law of unity and fighting of oppositions are used. the contradiction (difference in the approaches of formation and development of innovative economy) and opposition (competition in common world markets of innovations and hi-tech products with commonness of the internal mission) of the modern developed and developing countries in the global innovative economy of the 21st century are described.

Analysis of statistical data is performed for detailed research of the opposition of developed and developing countries during formation and development of innovative economy in the modern economic practice: share of medium-tech and hi-tech industry in the structure of added value of the real sector, share of expenditures for R&D in GDP, publication and patent activity, share of the number of researchers, and share of export of hi-tech products in the structure of industrial export in 2018.

In their categories) in the global rating of countries as to the level of development of innovative economy of Cornell University, INSEAD, and WIPO for 2018. The category of developed countries includes Switzerland (1st position), the Netherlands (2nd position), Sweden (3rd position), the UK (4th position), Singapore (5th position), and the USA (6th position). The category of developing countries includes Russia (46th position), Chile (47th position), the Republic of Moldova (48th position), Romania (49th position), Turkey (50th position), and Qatar (51st position). The selected statistical data are shown in Figs. 1, 2 and 3.

Figure 1 shows that the share of medium-tech and hi-tech industry in the structure of added value of the real sector in developed countries (54.7% on average) is by 1.69 times higher than in developing countries (32.3% on average). The share of expenditures for R&D in developed countries (2.5% of GDP on average) is by 4.2 times higher than in developing countries (0.6% on average).

Figure 2 shows that publication activity in developed countries (98,100 publications on average) is by 1.45 times higher than in developing countries (67,700 publications on average). The patent activity in developed countries (52,397.3 patents on average) is by 9.65 times higher than in developing countries (5,427.8 patents on average).

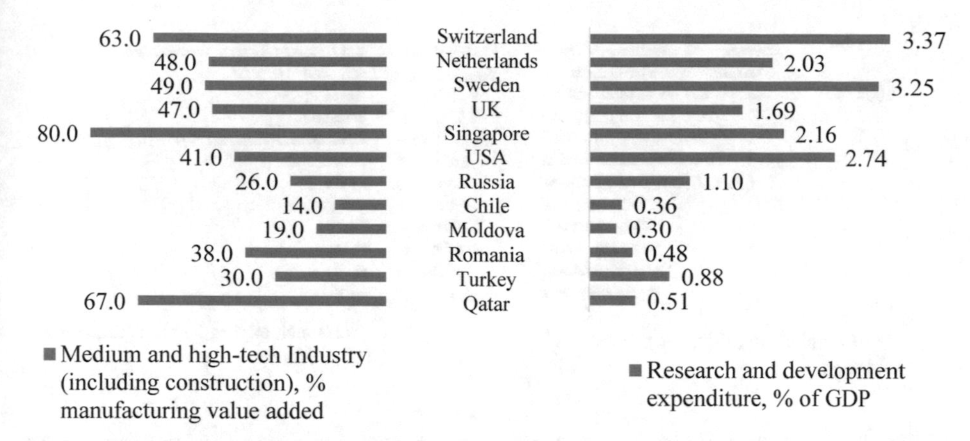

Fig. 1. Share of medium-tech and hi-tech industry in the structure of added value of the real sector and share of expenditures for R&D in GDP in the selection of developed and developing countries in 2018. Source: built by the authors based on World Bank (2019).

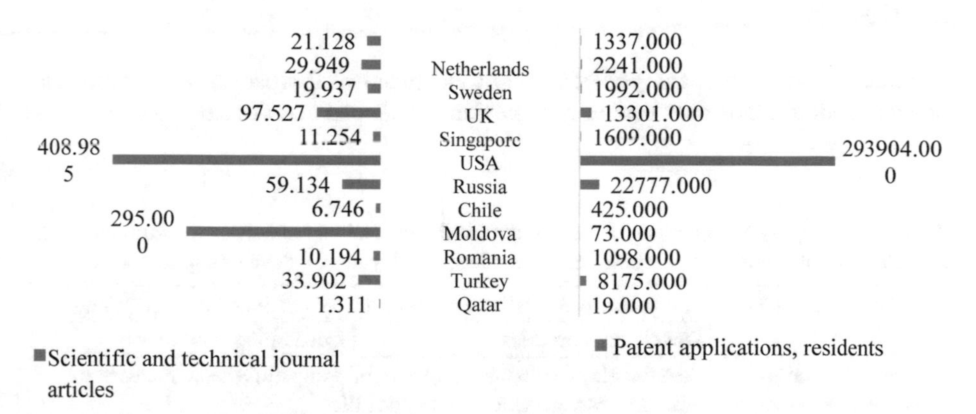

Fig. 2. Publication and patent activity in the selection of developed and developing countries in 2018. Source: built by the authors based on World Bank (2019).

Figure 3 shows that the number of researchers in developed countries (3,780.3 per 1 million people on average) is by 3.27 higher than in developing countries (1,156.2 per 1 million people on average). The share of export of hi-tech products in the structure of industrial export in developed countries (32.2% on average) is by 5.22 times higher than in developing countries (6.2% on average).

Thus, the performed analysis of statistical data showed that developed countries – as to all quantitative indicators – show a much higher level of development of innovative economy.

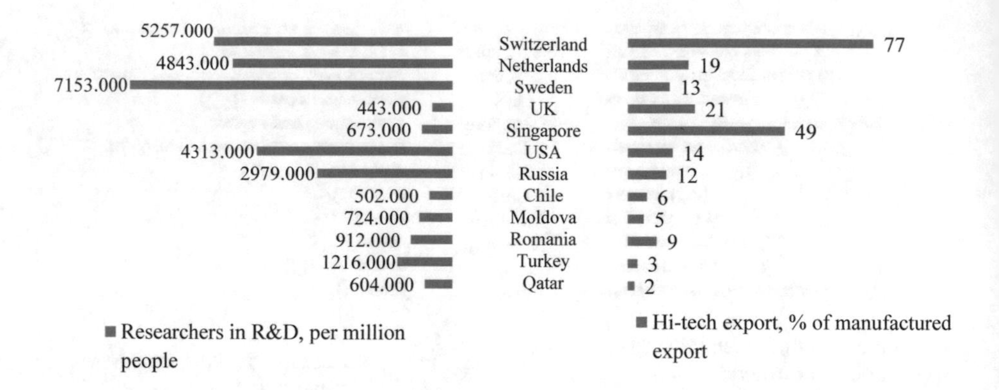

Fig. 3. Share of the researchers and share of export of hi-tech products in the structure of industrial export in the selection of developed and developing countries in 2018. Source: built by the authors based on World Bank (2019).

3 Results

Qualitative comparative analysis of scientific and practical approaches to formation and development of innovative economy in modern developed and developing countries is performed in Table 1.

Table 1. Comparative analysis of scientific and practical approaches to formation and development of innovative economy in the modern developed and developing countries.

Characteristics of thc approach	Approach to formation and development of innovative economy	
	Developed countries	Developing countries
Internal mission of innovative economy	Sustainable development, growth of competitiveness, increase of effectiveness, growth of quality of life	
Target markets of innovative economy	World markets of innovations and hi-tech products with priority	
	of markets of hi-tech products	of markets of innovative products
External goal of innovative economy	Preservation of leadership in the global markets of innovations and hi-tech products	Attraction of foreign investments, increase of export of innovations and hi-tech products
Source financing of innovative economy	Private investments, public-private partnership	State investments
Growth vector of innovative economy	Hi-tech entrepreneurship	Innovations-active entrepreneurship
Tool of formation and development of innovative economy	Development and implementation of own breakthrough technologies by separate companies	Mass implementation of the leading technologies that are purchased by companies

(*continued*)

Table 1. (*continued*)

Characteristics of the approach	Approach to formation and development of innovative economy	
	Developed countries	Developing countries
Technological mode and used technologies	Fourth: robototronics, technologies of virtual and alternate reality, "smart" technologies	Third: hi-speed broadband Internet, RFID-technologies, cloud technologies

Source: compiled by the authors.

Table 1 shows that developed and developing countries seek the common internal mission of innovative economy, connected to provision of sustainable development, growth of competitiveness, increase of effectiveness of economic activities, and growth of quality of population's life. The target markets of innovative economy are world markets of innovations and hi-tech products, though developed countries prefer the markets of hi-tech products, and developing countries prefer the markets of innovative products.

The external goal of innovative economy of developed countries is to preserve leadership in the world markets of innovations and hi-tech products, and the goal of developing countries is to attract foreign investments and increase export of innovations and hi-tech products. The source of financing of innovative economy in developed countries is private investments and public-private partnership, and in developing countries – state investments.

The growth vector of innovative economy in developed countries is hi-tech entrepreneurship, and in developing countries – innovations-active entrepreneurship. The tool for building and developing innovative economy in developed countries is development and implementation of own breakthrough technologies by separate companies, and in developing countries – mass implementation of the leading technologies that were purchased by companies.

The fourth technological mode is already formed or is in the process of formation in developed countries; within this mode, robototronics, technologies of virtual and alternate reality, and "smart" technologies are used. Developing countries still stick to the third technological mode and use the corresponding technologies: hi-speed broadband Internet, RFID-technologies, and cloud technologies.

Depending on the long-term results of opposition of developed and developing countries, we determined the scenarios of development of global innovative economy until 2030 (Table 2).

Table 2 shows that three possible scenarios of long-term development of the global innovative economy are distinguished. The most probable scenario is domination of developed countries. This scenario envisages preservation of the current rate and approach to development of innovative economy in developed and developing countries. The result of implementing this scenario will be increase of disproportions between developed and developing countries and increase of imbalance of the global innovative economy, which makes the goals of sustainable development unobtainable.

Table 2. Scenarios of opposition of developed and developing countries in the global innovative economy until 2030.

Characteristics of scenarios		Initial data for 2019		Scenarios of development of innovative economy					
				Domination of developed countries		Well-balanced development		Progress of developing countries	
		P-d*	P-ing**	P-d*	P-ing**	P-d*	P-ing**	P-d*	P-ing**
Conditions	Research and development expenditure, % of GDP	2.5	0.6	3.75	0.9	3.75	3.6	3.75	5.4
	Researchers in R&D, per million people	3,780.3	1,156.2	5,670.45	1,734.3	5,670.45	5,202.9	5,670.45	8,671.5
	Qualitative condition	–		Preservation of the current rate and approach to development of (const)			Accelerated development (*1,2;*9)	Const	Breakthrough development (*2;*15)
Results	Scientific and technical journal articles	98.1	67.7	147.15	101.55	147.15	121.86	147.15	203.1
	Patent applications, residents	52,397.3	5427.8	78,595.95	8,141.7	78,596	73,275.3	78,596	122,126
	Medium and high-tech Industry (including construction), % manufacturing value added	54.7	32.3	82.05	48.45	82.05	72.675	82.05	96.9
	High-technology exports, % of manufactured exports	32.2	6.2	48.3	9.3	48.3	46.5	48.3	93

*P-d – developed countries;
**P-ing – developing countries.
Source: compiled by the authors.

The most optimal and rather probable scenario is scenario of well-balanced development of the global innovative economy. It envisages preservation of the current rate and approach to development of innovative economy in developed countries and its accelerated (by 1.2–9 times as to different indicators) development in developing countries. Due to this, underrun of developing countries from developed countries will be almost overcome and will be insignificant.

The least probable scenario is scenario of progress of developing countries. It envisages preservation of the current rate and approach to development of innovative economy in developed countries and its breakthrough (by 2-15 times as to different indicators) development in developing countries. As a result, developing countries will reduce the underrun from developed countries, which will change the balance of powers in the global innovative economy, but the problem of its imbalance will be preserved.

4 Conclusion

It is possible to conclude that innovative economy in the 21st century is a source of contradiction and a field for opposition of developed and developing countries. They use different scientific and practical approaches to formation and development of innovative economy. The approach used by developed countries stimulates their quick development and supporting their competitiveness.

The approach used by developing countries does not fully conform to their interests and requires certain correction. In particular, it is necessary to increase all indicators of innovative economy: share of medium-tech and hi-tech industry in the structure of added value of the real sector, share of expenditures for R&D in GDP, publication and patent activity, share of the number of researchers, and share of export of hi-tech products in the structure of industrial export in 2018. This will allow for implementation of the most optimal scenario of long-term (until 2030) development of the global economy, within which its balance and sustainability are achieved.

References

Bezrukova, T.L., Popova, E.V., Korda, N.I., Kuznetsova, T.E., Bezrukov, B.A.: Institutional traps of innovative and investment activities as an obstacle on the path to the well-balanced development of regions. Contrib. Econ. (9783319606958), 235–240 (2017)

Bogoviz, A.V.: Industry 4.0 as a new vector of growth and development of knowledge economy. Stud. Syst. Decis. Control **169**, 85–91 (2019)

Cornell University, INSEAD, WIPO: Global Innovation Index 2018 rankings (2019). https://www.globalinnovationindex.org/gii-2018-report#. Accessed 25 May 2019

Popkova, E.G.: Contradiction of economic growth in today's global economy: economic systems competition and mutual support. Espacios **39**(1), 20 (2018)

Popkova, E.G., Sukhodolov, Y.A.: Theoretical aspects of economic growth in the globalizing world. Contrib. Econ. 5–24 (2017)

Pritvorova, T., Tasbulatova, B., Petrenko, E.: Possibilities of Blitz-psychograms as a tool for human resource management in the supporting system of hardiness of company. Entrep. Sustain. Issues **6**(2), 840–853 (2018). https://doi.org/10.9770/jesi.2018.6.2(25)

Sergi, B.S., Popkova, E.G., Bogoviz, A.V., Ragulina, J.V.: Entrepreneurship and economic growth: the experience of developed and developing countries. In: Entrepreneurship and Development in the 21st Century, pp. 3–32. Emerald Publishing Limited (2019)

Sibirskaya, E., Popkova, E., Oveshnikova, L., Tarasova, I.: Remote education vs traditional education based on effectiveness at the micro level and its connection to the level of development of macro-economic systems. Int. J. Educ. Manage. **33**(3), 533–543 (2019). https://doi.org/10.1108/IJEM-08-2018-0248

Vanchukhina, L.I., Leybert, T.B., Khalikova, E.A., Khalmetov, A.R.: New approaches to formation of innovational human capital as an element of institutional environment. Espacios **39**, 22–32 (2018)

World Bank: Data catalog: indicators (2019). https://data.worldbank.org/indicator. Accessed 25 May 2019

Structuring the Added Value of Biomedicine Products in the Innovative Economy

Natalia G. Varaksa(✉), Sergey A. Alimov, Maria S. Alimova, and Victor A. Konstantinov

Orel State University named after I.S. Turgenev, Orel, Russia
natalia.varaksa@yandex.ru, alimov_sergei@mail.ru, mashasmsl@gmail.com, neovitek@mail.ru

Abstract. In modern conditions, biomedicine, as a priority sector of the innovation economy, is developing dynamically, due to the desire of the world community to increase the quality of life, duration, financial security, and social well-being. The basis of the structuring of biomedical products is a step-by-step process of transformation of the scientific idea and innovative project into the final product of the industry, having a value assessment and high practical relevance to all stakeholders. One of the criteria of success of an innovative product is the indicator of added value. The aim of the research is to determine the multi-element composition of added value of innovative products in biomedicine and develop its gradual structuring on the basis of a generalization of modern domestic and world level research. The article analyzes the specific structure of biomedicine products, reveals the essence of added value as an indicator integrating assessment and analytical procedures that determine the importance and relevance of a particular biomedical product for the population, government, business community, as well as the degree of need for its early development, approbation, and implementation. A step-by-step structuring of the value added of biomedicine products has been developed and its element-wise composition has been evaluated. Practical significance of which lies in the possibility of making alternative choices at the government level for investment purposes and at the business community level in order to introduce the most promising option from the point of view of mass production and obtaining the maximum socio-economic and financial effect.

Keywords: Innovative product · Value added · Innovations · Biomedicine · Economy

JEL Code: D24 · L65

1 Introduction

In the modern conditions of the functioning of the innovation economy and the expansion of its influence on all spheres of activity of the population, the branch of medicine has strategic importance. Improvement of methods, techniques, tools of medicine has led to the transformation in a variety of areas in the industry, based on the

E. G. Popkova and B. S. Sergi (Eds.): ISC 2019, LNNS 91, pp. 561–569, 2020.
https://doi.org/10.1007/978-3-030-32015-7_63

use of exclusively advanced and relevant technologies - biomedicine. Biomedicine products today can save a huge number of people: timely diagnosis of complex and dangerous diseases; improved treatment of serious diseases such as cancer, HIV infection, tuberculosis, diabetes, heart diseases; by the implementation of highly effective preventive measures and the introduction of personalized medicine technologies.

The annual growth rate of the global biotechnology market is about 10%, and by 2020 its volume is projected at 600 billion dollars (Frost & Sullivan 2014). The volume of the Russian market of biomedicine is about 2% of the world market and indicates a significant lag behind foreign countries in the volume of production of biomedical products, despite the presence of high demand.

Each innovative product in the field of biomedicine has its own value and competitive advantages, which are reflected in its value. The cost of an innovative product in the field of biomedicine is formed at each stage of creation and commercialization, which requires the need to analyze the element-by-element composition of added value and evaluation of the value of each element in the process of creating the final product (Potashnik et al. 2018). To date, there are practically no studies of the formation of the structure of added value taking into account the specifics of innovative products in a particular branch of the economy (Yashin et al. 2018), in particular, biomedicine, which actualizes this study.

2 Methodology

Methodical tools for the study of the formation and structuring of the added value of innovative biomedical products include methods of analysis, synthesis, comparison, analogy, specification, graphical interpretation tools, and a system - structural approach. These methodical tools allowed:

- to analyze the specific structure of biomedicine products and to identify the indicator characterizing the success of an innovative product—added value;
- formulate the author's definition of added value and offer its gradual structuring for innovative products in biomedicine.

The theoretical and methodological base of the research consists of scientific developments in the field of:

- theories of value and value added, its element-by-element evaluation. So, V. Petty, A. Smith, D. Ricardo considered added value as part of the surplus value theory. K. Marx, F. Engels introduced the term surplus value. Z.B. Sei defined the essence of added value in terms of product utility, production costs, demand, and supply. Modern Researchers E.V. Peshina, P.A. Avdeev differentiated the concept of added value and surplus value. Maslova (2012) developed a methodology for creating and updating the added value in accounting and analytical space. The study of value flows, the establishment of communication links and the element-wise estimation of the value added of products, including innovative ones, were carried out by many scientists, in particular, Malkina (2016), Maslov (2009), Alimova (2014), Dedkova (2017), Garina et al. (2017).

– the specific structure of biomedicine products and the step-by-step process of its creation and commercialization. Fundamentals of classification and role of biomedicine products were considered in researches of a number of scientists, in particular, Pankrushina et al. (2017), Krutikov et al. (2013), Creation stages of innovative products were analyzed in works of Vasiliev (2012), Starodubov and Kurakova (2014), Honle (2013), etc.

Despite the available research of scientists, the process of gradual structuring of added value and its element-by-element evaluation in relation to innovative products of biomedicine is not sufficiently covered.

3 Results

3.1 Specific Structure of Biomedicine Products

The structure of biomedicine products is conditioned by a high degree of innovation in the conditions of application of modern technologies for the creation of pharmaceutical and diagnostic tools, methods, instruments, which have the highest performance of competitiveness, many times higher than the effectiveness of obsolete and irrelevant medical products. The specific structure of biomedicine products in modern innovative socio-economic conditions is presented in Fig. 1.

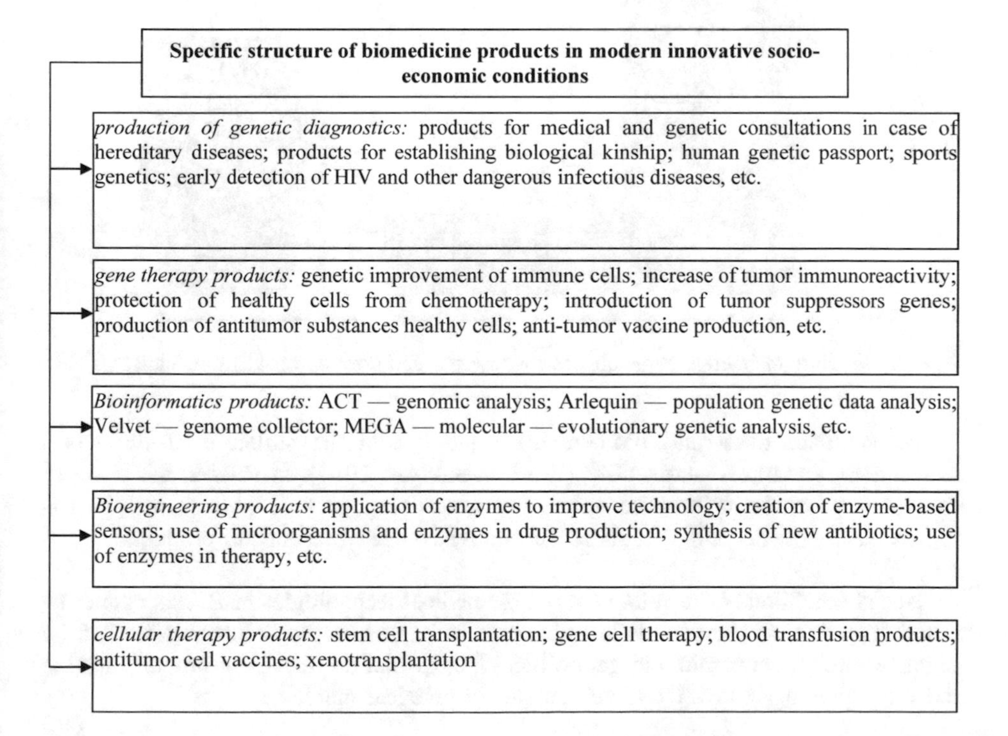

Fig. 1. Specific structure of biomedicine products in modern innovative socio-economic conditions

The formation of this structure is influenced by many factors that interact with each other and organize a single international biomedical space: social and legal, financial and economic, business-oriented, factors of state impact on the development of the industry (stimulating or constraining nature), etc.

In addition, the basis of the structuring of biomedical products is a step-by-step process of transformation of the scientific idea and innovative project into the final product of the industry, which has a high practical significance for all participants of the world reproduction process in the form of equipment, medicine, method of treatment, forecasting, diagnostics etc.

The structure of biomedicine products is also determined by the set of state priorities in this area and independent expertise, as well as by the formation of a single innovative environment based on medical scientific platforms and mechanisms of translational medicine.

Every year in Russia the size of budgetary expenses in the structure of expenses for research and development in the medical industry increases. Thus, in 2017, according to Rosstat, 25.7 billion rubles were allocated from budgets of different levels for scientific research and development in medicine (Fig. 2).

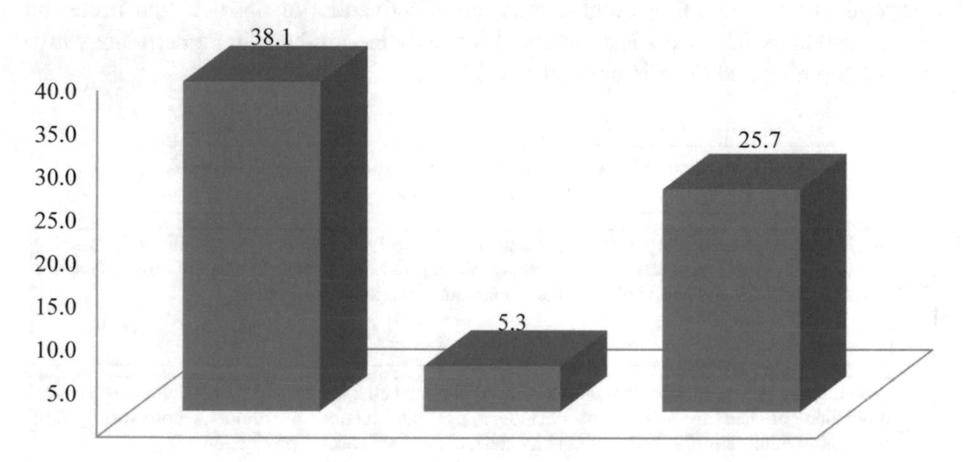

Fig. 2. Structure of internal expenditures for research and development in medicine in 2017.

Biopharmaceuticals and biomedicine stand out in the structure of the world biotechnology market—about 60% (Frost & Sullivan 2014). In Russia, the share of innovative products in the total volume of goods shipped by high-tech activities is approximately 18%, and the share of innovative pharmaceutical products is 9% (Fig. 3).

Actual conditions of development of biomedical technologies in Russia cannot be called favorable, as the stimulating government impact on the industry is leveled by negative foreign economic and geopolitical factors that do not allow properly testing and implementing innovative developments in practical medicine.

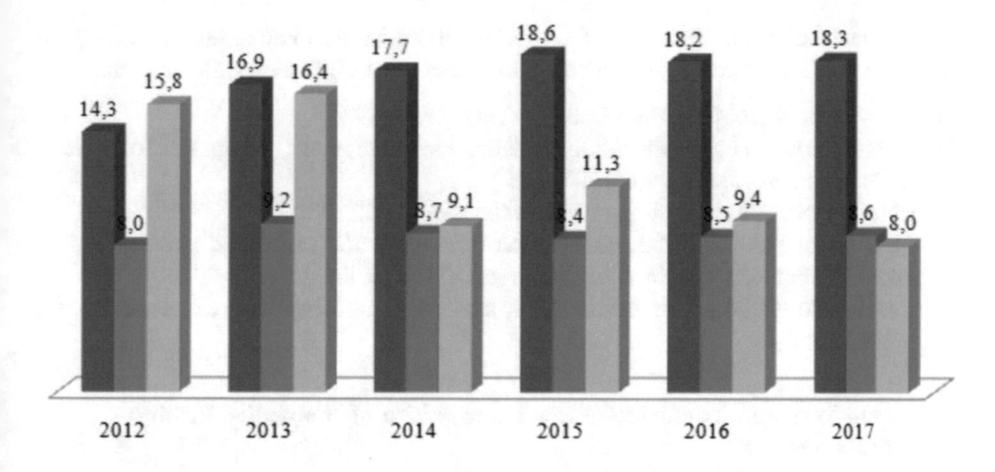

Fig. 3. Proportion of innovative pharmaceutical products in the total volume of shipped products

In this regard, the issue of assessing the effectiveness of innovative biomedical products in terms of achieving the necessary and required social effect, compliance with the standards economic feasibility indicators, as well as investment and financial parameters. Added value is a category, which allows characterizing the future success of an innovative biomedicine product from different perspectives.

3.2 The Essence and Element Composition of Added Value of Biomedicine Products

Analysis of the terminological apparatus available in the economic literature in relation to the category "value added" allowed to formulate the author's definition of the value of products in Biomedicine.

The value added of biomedical products is an indicator that integrates assessment and analytical procedures that determine the importance and relevance of a particular biomedical product for the population, government, business community, as well as the degree of need for its early development, approbation, and implementation in the practice of medical centers.

Production of biomedical products is accompanied by a long and multi-stage process of formation of value and added value of the final product, which according to experts on average is from 2 to 12 years. Complexity of development procedures, preclinical and clinical trials multiplies the size of added value created in this industry, which causes multi-component element-by-element composition indicator for biomedical products.

The main elements that have the greatest share in the total value added of biomedicine products, taking into account the influence of determining factors, are:

(1) Wages and social payments from the wage fund;
(2) Depreciation charges, including experimental equipment, equipment on which prototypes and samples are produced;
(3) Rent payments for rented premises, expensive equipment;
(4) Costs associated with the examination of project documentation and innovative product, obtaining approval for the practical use of the drug;
(5) Costs of scientific and industrial trips, seminars, multichannel conferences, round tables, etc.

3.3 Step-by-Step Structuring of the Value Added of Innovative Products in Biomedicine

The value added of innovative products in biomedicine is formed and evaluated taking into account the stages of its creation, as well as the specifics of the industry. In practical activity of development of innovative products of biomedicine, the following scheme of gradual evaluation of the structural composition of value added in conditions of functioning of innovation economy (Fig. 4).

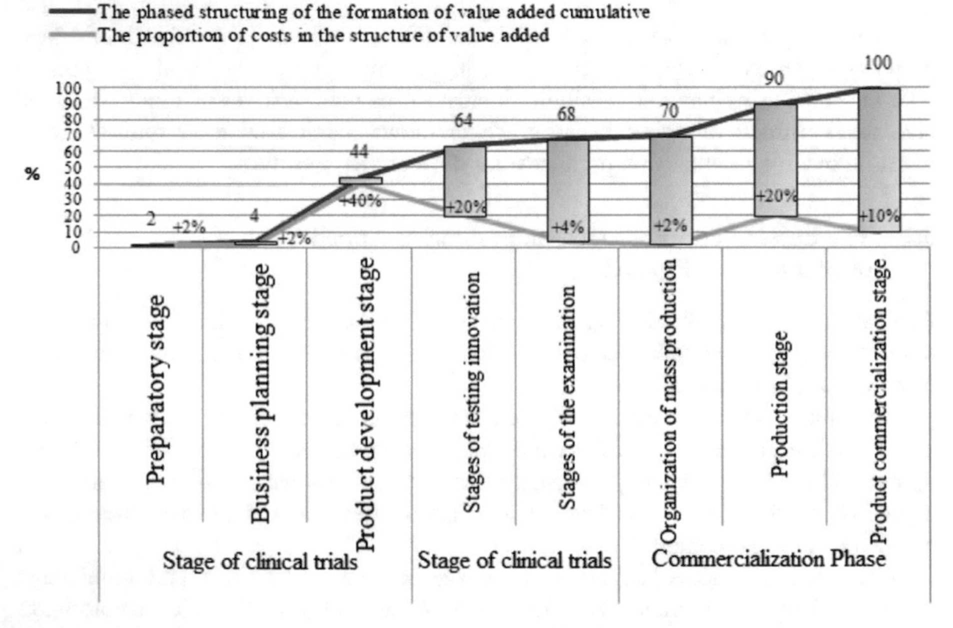

Fig. 4. Structuring of formation of value added of biomedicine products in conditions of the innovative economy

Thus, the process of formation of value added is continuous and includes a number of steps presented in Fig. 4, each of which increments the total value of the indicator, in turn, taking into account its element composition and specific structure of biomedical products.

On the basis of the developed step-by-step structuring of formation of value added of innovative biomedicine products, the cost of creation of a drug for treating cardiovascular diseases was calculated (Table 1).

Table 1. Elemental assessment of the value added structure of innovative biomedical medicines for the treatment of cardiovascular diseases

Composition of costs of gradual creation of an innovative product of biomedicine	Total costs for the development of medicines for the treatment of cardiovascular diseases, mln. Rub.
Costs of the preparatory phase: salaries of employees engaged in market research, insurance premiums, payments for the use of paid information resources	21.2
Cost of the business planning stage: salaries of employees involved in the preparation of the business plan, insurance premiums, payments for the use of paid information resources	22.8
Costs of the stage of development of an innovative product: salaries of employees associated with key developments, employees of the pilot center, insurance premiums, depreciation charges, including experimental equipment, equipment on which prototypes and samples are produced, taxes included in the cost, rent payments, costs of research and production trips	456
Costs of the innovation testing phase: salaries of employees associated with the conduct of clinical trials, insurance premiums, depreciation charges, including experimental equipment, equipment on which prototype models and samples are produced, taxes included in the cost, rent payments	228
Costs of project expertise and innovation: costs associated with the examination of project documentation and innovative product, obtaining approval for the practical use of the drug	45.6
The costs of the stage of financing the launch of innovation into mass production: costs associated with financing R&D to address national scientific and technical problems	22.8

(continued)

Table 1. (*continued*)

Composition of costs of gradual creation of an innovative product of biomedicine	Total costs for the development of medicines for the treatment of cardiovascular diseases, mln. Rub.
The costs of the stage of production of an innovative product: wages of production workers, management personnel, insurance premiums, depreciation deductions, taxes included in the cost, rent payments, travel expenses	228
Costs of the stage of commercialization of the innovative product: potential taxes, rate of return	114
Subtotal	1138.4

4 Conclusions and Recommendations

Thus, the production of innovative products in the field of biomedicine should be accompanied not only by constant financial support of federal and regional authorities but also by the provision of a certain share of independence in making certain managerial decisions to the subjects of innovation sphere on a high level of responsibility. This will make the process of formation of value added of innovative products more transparent both for users of information support from the state and from the side of developers and innovators. The success of the innovative biomedicine product should be evaluated on the basis of the value added index and its structure. The practical significance of the proposed structuring of biomedicine products in modern innovative socio-economic conditions, as well as the evaluation of the element-by-element composition of the added value of a biomedical product, is the possibility of making alternative choices both at the state level for investment purposes and at the level of the business community for mass production. innovations of the most promising available options for the development of methods, tools, methods of treatment, prevention and agnostic diseases in terms of obtaining the maximum socio-economic and financial effect.

References

Petty, V., Smith, A., Ricardo, D.: Anthology of economic classics, 199 p. (1993)

Alimova, M.S.: Essence and elemental structure of added value. Polythematic Netw. Electron. Sci. J. Kuban State Agrarian Univ. **100**, 1423–1433 (2014)

Vasilyev, A.N.: Qualitative preclinical research - the necessary stage of development and introduction into clinical practice of new drugs. Antibiot. Chemother. **57**, 41–46 (2012)

Dedkova, E.G.: Communication interaction of elements in the information system of assessment, accounting, analysis and control of cost flows. Manag. Acc. **10**, 52–57 (2017)

Krutikov, V.E., Kostina, O.I., Shakhmetova, E.A.: Innovations in the development of the region: cluster of pharmaceuticals, biomedicine and biotechnologies. Bull. Bryansk State Univ. **3**, 25–32 (2013)

Marx, K., Engels, F.: Capital. Criticism of political economy. Volume one, 2nd edn., vol. 23, 632 p. (1960)

Malkina, E.L.: Conceptual bases of the theory of cost flows in the framework of the modern understanding of the category "cost". Manage. Econ. Syst. **10**, 25–36 (2016)

Maslov, B.G.: Construction of a model of management accounting of formation of value in the conditions of integrated data processing. Econ. Hum. Sci. **6/212**(580), 28–34 (2009)

Maslova, I.A.: The methodology of formation of branch added value and value of products in fair valuation. Manage. Acc. **4**, 92–99.4 (2012)

Pankrushina, A.N., Dementieva, S.M., Ivanova, S.A.: The role of biotechnology in the formation of knowledge-based economy. Fundam. Res. **6**, 26–35 (2017)

Peshina, E.V., Avdeev, P.A.: Formation of gross added value of high-tech and knowledge-intensive products (goods, services). News UrGEU **6**(50), 46–56 (2013)

Starodubov, V.I., Kurakova, N.G.: Financing of medical science: new principles and financial instruments. Health Manager **3**, 48–60 (2014)

Honle, T.A.: Expenses for development of innovative drug. Problems of accounting and finance, no. 2(10) (2013). https://cyberleninka.ru/article/n/zatraty-na-razrabotku-innovatsionnogo-lekarstvennogo-preparata

Garina, E.P., Kuznetsova, S.N., Garin, A.P., Romanovskaya, E.V., Andryashina, N.S., Suchodoeva, L.F.: Increasing productivity of complex product of mechanic engineering using modern quality management methods. Acad. Strateg. Manage. J. **16**(4), 8 p. (2017)

Frost & Sullivan Review of the market of biotechnology in Russia and assessment of prospects of its development (2014). https://www.rvc.ru/upload/iblock/e21/20141020_Russia_Biotechnology_Market_fin.pdf

Potashnik, Y.S., Garina, E.P., Romanovskaya, E.V., Garin, A.P., Tsymbalov, S.D.: Determining the value of own investment capital of industrial enterprises. Adv. Intell. Syst. Comput. **622**, 170–178 (2018)

Yashin, S.N., Trifonov, Y.V., Koshelev, E.V., Garina, E.P., Kuznetsov, V.P.: Evaluation of the effect from organizational innovations of a company with the use of differential cash flow. Adv. Intell. Syst. Comput. **622**, 208–216 (2018). https://doi.org/10.1007/978-3-319-75383-6_27

Organizational Innovation in Cost Management as a Factor of Increasing the Competitiveness of the Enterprise

Alexander N. Vizgunov[1(✉)], Yuri V. Trifonov[2], Anna A. Abrosimova[2], Tatyana E. Maslova[2], and Pavel S. Shalabaev[2]

[1] National Research University Higher School of Economics, Moscow, Russian Federation
vizgunovhse@yandex.ru

[2] Nizhny Novgorod State University named after N.I. Lobachevsky, Nizhny Novgorod, Russian Federation
kei@ef.unn.ru, ann-serova@yandex.ru, maslova1703@mail.ru, p.shalabaev@mail.ru

Abstract. In the condition of global competition and development of the innovative economy, organizational innovations in the field of cost management play an important role. Introduction of modern methods of cost management should be carried out taking into account specifics of the economic environment of the enterprise. The aim of the article is to develop a method of cost management based on the use of the basic provisions of the ABC/ABM method and taking into account conditions of dynamically changing economic environment. The approach proposed by the authors involves identification of the most important, key types of costs in the structure of the production cost. On the basis of data on key types of costs, rules for allocation of indirect costs between types of output are determined. Also, data on key types serve as information for analysis of cost optimization options and minimization of risks associated with the use of certain types of resources. This work discusses the peculiarities of an organization of cost management procedures, such as analysis and control, based on the use of data on key costs. Practical results of application of the technique are illustrated on the example of the Russian enterprise of the wood processing industry.

Keywords: Cost management · Cost analysis · Activity-Based Costing (ABC) · Woodworking industry · Organizational innovations

JEL Code: D24 · L73

1 Introduction

Efficient cost management is one of the key factors for increasing the competitiveness of the enterprise. Activity-Based Costing (Cokins 2004) is the main management accounting method for accurate allocation of indirect costs. The main characteristics of the ABC method were formulated in the 80s by R. Cooper and R. Kaplan (Cooper and

E. G. Popkova and B. S. Sergi (Eds.): ISC 2019, LNNS 91, pp. 570–575, 2020.
https://doi.org/10.1007/978-3-030-32015-7_64

Kaplan 1988; Cooper 1988). This method is based on the concept of action (activity). Actions are activities carried out in an enterprise with the ultimate goal of producing and marketing products; ordered sets of actions form the business processes of the enterprise (Cokins 2006). The calculation of the cost of production in accordance with the ABC method includes the distribution of the cost of the spent resources between the various actions and, further, the distribution of the calculated cost of actions between different types of products. Accordingly, the introduction of the ABC method involves the creation of a system of rules for the exact allocation of resources between actions and actions between products. The development of ABC was the emergence of the concept of Activity-Based Management (ABM), which formulated principles of optimizing structure and composition of costs based on ABC-model data—a clear providing information on actions and their costs allows you to identify actions that do not add value, as well as actions that are unreasonably high (Armstrong 2002; IMA 1998). The introduction of the ABC/ABM method can be considered as one of the types of organizational innovation, involving both fundamental changes in accounting procedures and reorganization of the cost management system.

Currently, the ABC/ABM method is widely used in the world, but it poses a number of challenges (Stouthuysen et al. 2010). Thus, the most important drawback is the high complexity of maintaining the ABC-model in the context of dynamic changes in the economic environment (Kaplan and Anderson 2004). This problem is particularly relevant for enterprises in emerging economies. Dynamic changes in the conditions of production and economic activity in such countries are due not only to increased competition and the emergence of new technologies (as is the case in developed countries), but also constant changes in legislation, rising energy tariffs, high volatility of the national currency, etc. This is confirmed by the research conducted by the authors at 20 Russian enterprises of various types of activity, which showed that none of these enterprises ABC methodology is implemented. The ABC elements are used locally only in two enterprises—to account for the costs of individual units (Garina 2017), (Yashin and Trifonov 2018). The reasons for this situation identified by employees of enterprises are high labor intensity of construction of the initial ABC-model, insufficient level of qualification of specialists carrying out functions of accounting and analysis costs, and the main reason, according to the respondents, is the difficulty of maintaining the ABC-model in conditions of constant, and poorly predicted changes.

2 Methodology

In the paper (Vizgunov and Trifonov 2017) we proposed a cost management methodology based on accounting, analysis, and control of only the most important key costs. Accordingly, the application of this technique simplifies the construction of the ABC-model and its maintenance up to date, as it is assumed that the rules of the exact distribution are developed only for those indirect costs that have been identified as key ones. For other types of costs, the exact allocation mechanism may not be applied, and their distribution will be done in simpler ways—based on conditional coefficients.

Key types of costs are determined on the basis of two groups of criteria. The first group of criteria consists of the criteria of the share of costs in the total volume and regulation (regulation characterizes the possibility of managers to influence the size of costs). Costs classified as key by these criteria should serve as an information base for finding cost reduction opportunities, as they have the greatest impact on its size. The second group of criteria is the criteria characterizing the level of production and economic risks (risks of disruption of the production process, supply of raw materials, etc.) (Potashnik 2018). These are such qualitative criteria as the level of limited supply of used resources, the level of dependence of the price of resources on changes in the exchange rate of the national currency, etc.

In the framework of this article, we will consider the management aspects of the application of this methodology related to the development of rules for the allocation of key costs and organization of procedures for their analysis and control, focus on practical examples.

Once the key costs are determined, actions in which the resources constituting these costs are expended are determined and the allocation rules that adequately reflect a distribution of costs between the types of products produced. The ABC model, built on the basis of key costs, should provide a visual representation of the types of costs that are most important to the enterprise, accurately distributing such costs between different types of products. It should be noted that the development of rules for the exact distribution of the cost of resources between different types of products is important not only for the costs attributed to the key criteria of availability of reserves reduction cost but also for costs classified as key risk criteria. The importance is due to the need to understand the cost of which products will increase most in the event of adverse events defined in the framework of risk analysis.

3 Results

Let's consider the issues of organization of key cost management procedures on the example of one of the Russian multi-profile woodworking enterprises, the analysis of which was carried out by authors in 2010s.

In the beginning, we will consider the results of determining the key costs for different types of production of this enterprise. For most types of products, the key types of costs according to the criteria of regulation and the share of costs in total cost are assigned to the main materials (this item of expenditure is a cost of lumber) and wages of production workers. So, for example, in the cost price of the board of the subwindow share of "Basic materials" is 56%, and the item "Basic salary of production workers" - 7.2%.

Other significant items of expenditure (transportation costs, maintenance costs, fuel costs for technological needs and storage costs) are composite. In general, the volume of these cost items is allocated to the most significant components, after which precise allocation rules are determined for them. So, for example, more than 80% of the total amount of transportation costs is paid for transportation of lumber by rail. Accordingly, only these costs (that is, associated with the implementation of "delivery of lumber by rail") can be attributed to the key, and as a base, distribution can be used the amount of consumption of sawn timber by each production.

The main type of cost, which is classified as key in terms of risk criteria, is the cost of electricity for technological needs. The constant increase in electricity tariffs implies the need for accurate accounting and constant monitoring of the use of this type of resources. The problem of the exact distribution of electricity costs can be solved by installing additional electricity meters in individual areas.

As an example of costs, which are classified as key on the criterion of dependence on changes in the exchange rate, can be given the cost of hardware used in the production of window blocks. In production, only accessories of foreign manufacturers are used, as Russian analogs of acceptable quality are absent. This fact determines the dependence of the value of costs on changes in the exchange rate of the national currency.

After defining the key costs, we will move on to the organization of procedures for managing such costs. The most important aspect of managing key costs is organization of an effective control system. This applies primarily to costs classified as key in terms of share of costs and level of regulation. Resources constituting key costs should be monitored at all stages of their use. For each type of key resources, it is necessary to establish a list of responsibility centers involved in the acquisition and processing of resources and define cost parameters that can be regulated by each responsibility center. For example, for sawn wood, controlled usage parameters in the section of responsibility centers are defined as follows: see Table 1.

Table 1. Controlled parameters of acquisition and processing of lumber in the context of responsibility centers.

Responsibility center	Controlled parameters
Supply Division	Price, terms of delivery
Warehouse	Discarding of lumber characterized by falling knots
Drying shop	Volume of material not dried to the required humidity
Joinery shop	Volume of flaw in sawing

Supply Division is responsible for the price at which sawn wood is purchased, terms of delivery (in particular, the supplier or recipient pays the shipping costs). In the warehouse of finished products, sorting of lumber is performed on the basis of the presence of falling knots. This type of timber cannot be used for the production of floor boards, but the presence of dropping knots is not a problem for those industries where the surface of sawn timber is covered with plastic. Careful sorting of lumber at this stage will avoid the use of lumber in those industries where it should not be used. Also, reducing unproductive consumption of lumber is facilitated by minimization of defect during drying, sawing, and gouging.

Next, we will focus on the approach proposed by authors to optimize costs, including the cost of gas for drying chambers and maintenance of warehouses. These costs are the key to most types of products; their minimization will significantly reduce the cost of production. The analysis showed that the company cannot do its own warehouse premises, which causes the need to lease additional warehouse areas. The need for additional storage space arises from the fact that lumber enters the plant in large quantities at low intervals. Authors propose to use the method of daily procurement, and the size of the shipment is planned on the basis of the daily demand for sawn timber and minimization of drying costs. Procurement planning in accordance with the proposed model will allow the company to abandon the lease of additional storage areas and minimize the cost of drying sawn timber.

Let's focus on the procedures for managing the costs, which are classified as key in terms of risk criteria. The following organizational arrangements have been proposed to reduce the risks to be taken into account in determining key costs:

- long-term contracts with suppliers of components on the market, is limited,
- creation of insurance stocks of materials and components, supply of which is limited and value of which is determined by changes in the exchange rate of the national currency,
- creation of own production of certain types of window and door fittings (to eliminate suppliers dependence),
- organization of monitoring of new energy saving technologies (to find opportunities to reduce risks caused by uncontrolled increase in electricity tariffs).

4 Conclusion

The main results of the proposed cost management methodology are presented in Table 2.

Table 2. Results of implementation of cost management methodology based on key cost data

Initial situation	Content of organizational innovations	Results
Indirect costs are allocated on the basis of conditional coefficients (proportional to the salary of production workers, etc.)	Development and application of the rules for allocation of key costs within the framework of the ABC-model	Precise allocation of key indirect costs based on rules that reflect the specific use of resources
Common methods of control are applied for all types of costs	Organization in relation to key costs of more accurate control across responsibility centers	Reduction of the size of the defect (for example, in relation to the processing of lumber—by 5–10%), prompt identification of places occurrence of inefficient costs
There is no systematic approach to search for cost reduction reserves	Development of procedures for optimization of key costs, including, with the help of tools of economic-mathematical modeling	Reduction of key costs; in particular, reduction of storage and drying costs of sawn timber by more than 10%
The risk analysis of cost increases is carried out for only a few types of costs, determined on the basis of subjective management assessments	Building a risk management system based on key cost data and developing measures to reduce these risks	Minimize the risks of uncontrolled increase in key costs

In general, organizational innovations associated with the introduction of the proposed cost management methodology provide the enterprise with a stable competitive advantage for the following reasons:

1. Obtaining operational, visual and detailed information about the most important costs of the enterprise; accurate distribution of key costs between different types of products.
2. Easily maintain key indirect cost allocation rules in the context of dynamic changes.
3. Creation of information base for search of reserves of cost reduction and identification of risks associated with a possible increase in cost of used resources.

References

Armstrong, P.: The costs of activity-based management. Acc. Organ. Soc. **27**(1–2), 99–120 (2002)

Cokins, G.: Measuring customer value: how BPM supports better marketing decisions. Business Performance Management, February, pp. 13–18 (2006)

Cokins, G.: Performance Management (Finding the Missing Pieces to Close the Intelligence Gap). Wiley, New York (2004)

Cooper, R., Kaplan, R.S.: Measure Costs Right: Make the Right Decisions. Harvard Business Review, September, pp. 96–103 (1988)

Cooper, R.: The rise of activity-based costing- part one: what is an activity-based cost system? J. Cost Manag. Summer, pp. 45–58 (1988)

Drury, C.: Cost and management accounting: an introduction, South-Western Cengage Learning (2011)

Garina, E.P., Kuznetsova, S.N., Garin, A.P., Romanovskaya, E.V., Andryashina, N.S., Suchodoeva, L.F.: Increasing productivity of complex product of mechanic engineering using modern quality management methods. Acad. Strateg. Manag. J. **16**(4), 8 (2017)

IMA. Statements on Management Accounting. Implementing Activity-Based Management: Avoiding the Pitfalls, Institute of Management Accountants, Montvale, NJ (1998)

Kaplan, R.S., Anderson, S.R.: Time-driven activity-based costing. Harvard Business Review, November, pp. 131–138 (2004)

Potashnik, Y.S., Garina, E.P., Romanovskaya, E.V., Garin, A.P., Tsymbalov, S.D.: Determining the value of own investment capital of industrial enterprises. Adv. Intell. Syst. Comput. **622**, 170–178 (2018)

Samusenko, S.: Methodological approaches to overhead costs accounting in works of american and russian specialists in XX century, Economic analysis: theory and practice (Russian), no. 19, pp. 44–52 (2010)

Stouthuysen, K., Swiggers, M., Reheul, A., Roodhooft, F.: Time-driven activity-based costing for a library acquisition process: a case study in a Belgian University. Library Collections, Acquisitions, and Technical Services, no. 34, pp. 83–91 (2010)

Vizgunov, A., Trifonov, U.: Management of corporate business process cost performance based on key costs data. Int. J. Econ. Finan. Issues **7**(2), 1–6 (2017)

Yashin, S.N., Trifonov, Y.V., Koshelev, E.V., Garina, E.P., Kuznetsov, V.P.: Evaluation of the effect from organizational innovations of a company with the use of differential cash flow. Adv. Intell. Syst. Comput. **622**, 208–216 (2018). https://doi.org/10.1007/978-3-319-75383-6_27

Zeleny, M.: Human systems management: integrating knowledge management and systems. World Scientific, Hackensack (2005)

Industrial Policy: Peculiarities of Understanding and Dependence on the Level of Implementation

Oleg V. Trofimov[1(✉)], Andrey P. Kostyrev[1], Lyudmila V. Strelkova[1], Yuliya A. Makusheva[1], and Tatyana V. Trofimova[2]

[1] National Research Nizhny Novgorod State University N.I. Lobachevsky, Nizhny Novgorod, Russia
oleg_trofimov@mail.ru, ksore09@mail.ru, strelkova412@mail.ru, sjm2@yandex.ru

[2] Institute of Management — branch of RANEPA, Nizhny Novgorod, Russia
oleg_trofimov@mail.ru

Abstract. One of the actual problems in the implementation of industrial policy at present is a multiplicity of approaches to its understanding both within a single country and when comparing views of authors from various countries. This problem is deeper than just the difference in definitions. At different levels of implementation of industrial policy there may be inconsistency of its goals, objectives, and means. The aim of this research is to analyze the multiplicity of interpretations of industrial policy among Russian and foreign authors and to consider the problem of definition of industrial policy in normative documents on federal and regional levels in Russia. Division of goals, objectives, and means of industrial policy into general (set by the higher level) and specific (reflecting the interests of a certain level) is proposed, concept of prioritization of objectives is stated, objectives and means of industrial policy at a certain level of its implementation. It is concluded that it is important to harmonize the content of industrial policy and take into account interests of subjects and stakeholders at different levels. In conclusion, it is noted that industrial policy at the regional level should combine federal and regional goals and objectives, unite interests of sectoral groups and territories. Achieving harmonization of goals, objectives and means of industrial policy will contribute to emergence of a synergistic effect in terms of improvement of quantitative and qualitative indicators of the industry.

Keywords: Industry · Industrial policy · Content of industrial policy · Level of industrial policy · Prioritization

JEL Code: L500 · L 510 · L520 · L600

E. G. Popkova and B. S. Sergi (Eds.): ISC 2019, LNNS 91, pp. 576–582, 2020.
https://doi.org/10.1007/978-3-030-32015-7_65

1 Introduction

In conditions of gradual deployment of the digital economy, problem of transformation of a system of national management economy and its individual components has a fairly broad understanding and field for analysis, as solutions can be considered and proposed, given differences in scale, sample, and detail. In the field of industrial production this problem is concretized into the problem of formation and implementation of industrial policy adequate to the latest trends in development of technologies, economy, and society.

Despite a considerable number of works devoted to industrial policy, there is no common understanding and approach to its analysis in the scientific community.

The purpose of this article is to analyze the multiplicity of interpretations of industrial policy and its reflection in the normative documents of the federal and regional levels in Russia and to formulate the concept of prioritizing objectives, and means of industrial policy at a certain level.

Realization of the goal is achieved by solving the following tasks:

- analysis of approaches to the definition of industrial policy among Russian and foreign authors;
- Consideration of the interrelationship of objectives, levels and means of industrial policy;
- Formulation of recommendations for party's implementing industrial policy at a certain level.

2 Methodology

1. Analysis of approaches to defining industrial policy

By denoting the range of studied issues on the theory of organization of industry (theory of organization of industrial markets), researchers note that in Russian the concept of "industry" has two meanings — wide and narrow V.I. Dahl gives the following interpretation: "Industry is all human activity carried out as a fishery and aimed at creating, transforming or moving economic goods." The narrower meaning of this word in the economy of socialist industry meant "aggregate" branch (subdivision) of the national economy" (Juha et al. 2014, p. 6).

The domestic term "manufacturing" and closest foreign English term "industry" differ. The English-language concept of "industry" into Russian is translated both as "manufacturing" and as "industry". It is broader than the Russian word "manufacturing" and, in general, means industry. Modern understanding of industry in domestic science and economic practice, in our opinion, is most convenient to consider through the interpretation in official statistics, now enshrined in Art. 3 Federal Act No. 488-FL, which defines industrial production as "totality of economic activities related to extraction of minerals, provision of electricity, gas and steam, air conditioning, water supply, water disposal, waste collection, and elimination of pollution."

3 Results

In Table 1 we will give some results of the generalized analysis of definitions of industrial policy contained in scientific papers.

Table 1. Results of the generalized analysis of definitions of industrial policy in the works of domestic authors (based on materials of considered works)

Authors who conducted a summary analysis and enumeration of presented definitions	Main findings
I. G. Idrissov (2016) Foreign definitions given in works of V.C. Price (1981), P.R. Krugman and M. Obstfeld (2006), D. Rodrik (2004), World Bank Documents (1992), UNIDO (2013)	There are many definitions of industrial policy, ranging from the most general to the more specific ones. Some authors deny industrial policy as a separate form of economic policy, "emphasizing that it is more just an industry view than a predetermined plan action" (Idrissov 2016, p. 7). Modern understanding of industrial policy is broad and inextricably linked with a large number of other economic policies (customs, tax, monetary, etc.) and" is characterized by some integral effect of impact on the economy as a whole" (Idrissov 2016, p. 9)
I. Simachev (2014) Definitions given in works of foreign authors are given: P. Krugman (1991), Ha-JoonChang (1984), V. Price (1981), H. Pack and K. Saggi (2006), K. Warwick (2013)	Two types of definitions are reflected: - which focus on specific sectors to achieve the required results (P.Krugman, Ha-JoonChang), - which refer to a set of measures to influence structural changes or improve business environment. (V. Price, H. Pack and K. Saggi, K. Warwick)
Shcheglov E. V. (2015) The analysis of definitions of industrial policy contained in works of domestic and foreign authors, as well as contained in some regional laws devoted to industrial policy (according to as of 2014—2015)	The author makes a number of generalizations, including the following interpretations of industrial policy (Shcheglov 2015, p. 14): - a system of measures aimed at the development of industrial enterprises (A.R. Safiulin (2010), J.Tirole (1996), V. Kondratiev (2003), N.I. Atanov (2012), etc.); - a set of measures for the development of industry (E.Y. Salikova (2006), E. Smirnov (2007)); - set of actions of state authorities on development of industrial enterprises (V. Zavadnikov (2007), O.L.Graham (1994), Yu. Karmanov (2008), etc.)

(*continued*)

Table 1. (*continued*)

Authors who conducted a summary analysis and enumeration of presented definitions	Main findings
Nikitin A. S. (2009) The analysis of approaches to the definition of industrial policy offered by Russian politicians and business communities is given	Author draws attention to allocation of the innovative component and existing emphasis on technological renewal and modernization of the industrial base. At the same time, it is noted that the lack of a unified approach to industrial policy and a fixed single document (as of 2008—2009) made it difficult to carry out industrial policy

An example of the analysis of foreign approaches to definition of industrial policy can be found in the work "New Industrial Policy of the European Union" held at the Collegium of World Economy at the Warsaw School of Economics (Ambroziak (eds.) 2017, pp. 6–7).

Regarding official definition of industrial policy, it should be noted that in the Russian Federation it first appeared at the regional level. The first regional laws on industrial policy were adopted in 1996—1997, while the Federal Law was adopted only at the end of 2014.

According to Art. 3 of the Federal Law of 31.12.2014 No. 488-FL "On industrial policy in the Russian Federation" industrial policy is defined as "a complex of legal, economic, organizational and other measures aimed at developing the industrial potential of the Russian Federation, ensuring the production of competitive industrial products." The adoption of Federal Law No. 488-FL contributed to the strengthening of the trend of unification in terms of regulatory and legal definition of industrial policy.

There are, however, a number of exceptions to this trend. In some regional laws, the definition of industrial policy differs from definition given in the Federal Law in the direction of abstraction, or vice versa, in detail and expansion.

2. Selection of objectives and means of industrial policy depending on the level of implementation

In implementation of industrial policy, different levels can be distinguished depending on the territorial scale and type of economic activity.

Although industrial policy represents a single system with many interconnections, the level of implementation imprints on the formulation of goals and choice of means (Usov et al. 2018). The development of industrial production at territorial levels is considered by researchers on examples of different countries (Bačić and Aralica 2017; Crawley and Munday 2017; Piekkola 2018).

The content of industrial policy of the region will largely depend on the sectoral structure of gross value added. According to official statistics (Regions of Russia. Socio-economic indicators, 2018) mining in gross value added in Russia in 2016 amounted to 10.9%. However, in some regions this value is several times larger and is more than half of the total structure of gross value added: Nenets Autonomous Region —74.5%, Khanty-Mansi Autonomous Region—Ugra—66.1%, Yamal-Nenets

A) Priority is given to specific industrial development objectives of the region

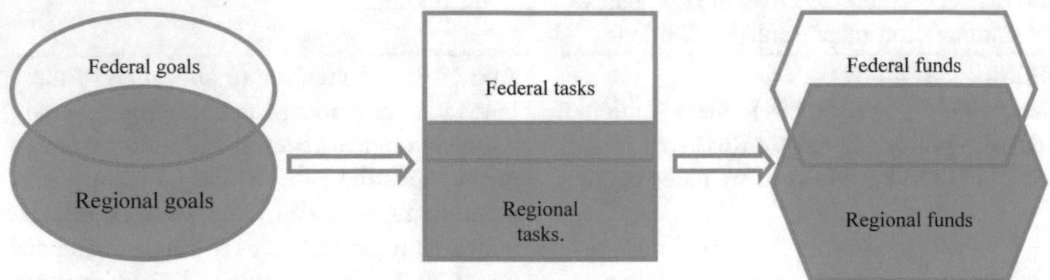

B) Priority is given to general development objectives of the country's industry

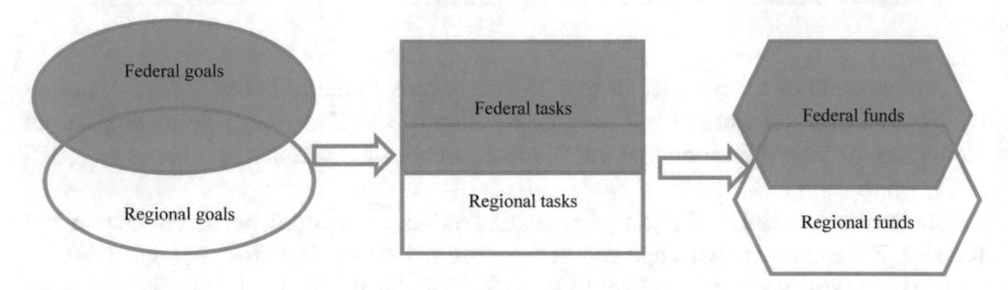

Fig. 1. Prioritization in relation to general and specific objectives and tools of industrial policy

Autonomous Region—54.5%, Sakhalin Region—54.0%, Sakha Republic (Yakutia)—51.6%, Chukotka Autonomous Region—50.1%. At the same time, in 13 Russian regions, the share of extractive industries in a structure of gross value added is 0.1% or less. It is obvious that the content of industrial policy in the first and second groups of regions will differ significantly. For the first group, determining factors will be the world prices for raw materials, as well as availability of opportunities for transporting raw materials. In this group the problem of diversification of economy and development of processing industries is acute (Garina 2017). For the second group, commodity market conditions are of little importance, mainly there is an acceptable level of diversification. Regions of this group are more sensitive to measures to support the manufacturing sector of the economy (Potashnik 2018).

Goals and means of the higher level (for example, federal) will be considered as common goals and means. Under the specific goals and means implied goals of a certain level, different from the general or complementary which are specifying the general once. Figure 1 shows possible priorities in relation to general and specific goals, objectives and means of industrial policy on the example of interaction between federal and regional levels.

Figure 1 illustrates two priority situations. In the first case, priority is given to specifics of the industry of region. In the second case there is an opposite situation - priority are the general goals and objectives of development of the country's industry. In other words, in the first case, the center "adjusts" to the regions, which reflects the formation of industrial policy "bottom-top". In the second case, the regions "adjust" to the center, which illustrates the formation of industrial policy "top-bottom".

In practice, there is a mixture and interaction of these approaches: regions are forced to take into account the federal goals and objectives, and federal subjects of industrial policy formation should take into account the specifics and capabilities of individual regions.

4 Conclusions

The domestic term "manufacturing" and the foreign term "industry" differ in their content. The foreign term has a broader meaning and generally refers to the industry (Müller et al. 2018). The uniform official interpretation of industrial production is enshrined in Article 3 of Federal Law No. 488-FL. The Act also contains an official definition of industrial policy, which has now become a framework and is used in most regional regulations.

However, it was at the regional level that the first formal definitions of industrial policy emerged in the second half of the 1990s. Acceptance in 2014 Of Federal Law No. 488-FL served to strengthen trends of unification in the understanding of industrial policy by implementing agents at different levels.

The content of industrial policy pursued by various actors is revealed through its objectives, objectives, and means, which are proposed to be divided into general (inherent to the higher level, most often—federal) and specific (inherent in lower levels). Taking into account the multilevel nature of industrial policy there is a problem of prioritization of goals, objectives, and means, which should take into account the opinion of subjects and stakeholders at each of considered levels.

The content of industrial policy varies depending on the type and scale of the territory, and specifics of individual economic activities related to industrial production.

Industrial policy at the regional level should combine federal and regional goals and objectives, unite sectoral and territorial interests. The goals and objectives of industrial policy cannot be realized without the competent selection of means and tools adequate to the existing conditions. When selecting tools, it is necessary to take into account the best foreign experience, where indicative and program-target planning plays a special role.

To find an optimal balance of goals, objectives, and means of industrial policy at different levels will contribute to its harmonization, which should have a positive synergistic effect on the quantitative and qualitative indicators of national industry.

Conclusions and Recommendations

Differences between goals, objectives, and means of industrial policy at different levels should be minimized on the basis of consensus between participants of the federal and lower levels (regions or types of economic activities). This is particularly important in the context of increasingly popular cluster approach to industrial policy.

Federal goals and objectives should have an unconditional priority in the field of national security, development of military-industrial complex, space technologies and other strategic directions. However, the placement of production from the above spheres should also largely rely on the capabilities and resources of individual territories.

Acknowledgments. "The research was carried out with the financial support of RFBR and the Government of Nizhny Novgorod region within the scientific project No. 18-410-520009"

References

Federal law of 31.12.2014 No. 488-FL (ed. 27.06.2018) "On industrial policy in the Russian Federation" [Text]. SZ RF, no. 1 (part I), Art. 41 (2015)

Danilov, P., Mikhailova, S.I., Morozova, N.V.: Industrial policy: Russian experience of generalizing the results of scientific research. Bull. Econ. Law Soc. **4**, 46–49 (2014)

Juha, V.M., Kuritsyn, A.V., Stapova, I.C.: Economics of industrial markets [Text]: textbook, 3rd edn., KNORUS, 288 p. (2014). (Bachelor)

Idrissov, G.I.: Industrial policy of Russia in modern conditions [Text]. Publishing House of Gaidar Institute, 160 p. (Scientific works/ Institute of Economic Policy named after E. T. Gaidar; No. 169P) (2016)

Nikitin, A.S.: Approaches to the definition of industrial policy [Text]. ArsAdministrandi, no. 1, pp. 57–61 (2009)

Egorenko, S.N., et al. (ed.): Regions of Russia. Socio-economic indicators. 2018 [Electronic resourcc]: Stats/ Rosstat, 1162 p. (2018). http://www.gks.ru/free_doc/doc_2018/region/reg-pok18.pdf. Accessed 05 March 2019

Simachev, Y.: Industrial policy in Russia: institutional features, groups of interests, lessons for the future [Electronic resource] (2014). http://www.iacenter.ru/publication-files/197/174.pdf. Accessed 18 Jan 2019

Usov, N.V., Trofimov, O.V., Frolov, V.G., Makusheva, Y.A., Kovylkin D.Y.: Assessment of innovative potential of priority industries of Nizhny Novgorod region. Russian entrepreneurship. vol. 19, no. 10, p. 2921–2930 (2018)

Shcheglov, E.V.: Development of organizational-economic mechanism of formation and realization of industrial policy [Text]. Perm, 219 p. (2015)

Müller, J.M., Kiel, D., Voigt, K.I. (ed.): Drives the Implementation of Industry 4.0? The Role of Opportunitiesand Challenges in the Context of Sustainability. [CrossRef], What Sustainability,vol. 10, 247 p. (2018)

Garina, E.P., Kuznetsova, S.N., Garin, A.P., Romanovskaya, E.V., Andryashina, N.S., Suchodoeva, L.F.: Increasing productivity of complex product of mechanic engineering using modern quality management methods. Acad. Strateg. Manag. J. **16**(4), 8 (2017)

Bačić, K., Aralica, Z.: Regional competitiveness in the context of "New industrial policy"—The case of Croatia, ZbornikRadovaEkonomskogFakultet au Rijeci. vol. 35, no. 2, pp. 551–582 (2017)

Crawley, A., Munday, M.: Priority sectors in city regions? Some issues from a study of the Cardiff Capital Region. Local Econ. **32**(6), 576–589 (2017)

Ambroziak, A.A. (ed.): The New Industrial Policy of the European Union [Text], 276 p. Springer, Cham (2017). (Contributions to Economics)

Potashnik, Y.S., Garina, E.P., Romanovskaya, E.V., Garin, A.P., Tsymbalov, S.D.: Determining the value of own investment capital of industrial enterprises. Adv. Intell. Syst. Comput. **622**, 170–178 (2018)

Piekkola, H.: Internationalization via export growth and specialization in Finnish regions. Cogent Econ. Finan. **6**(1), 1–25 (2018)

Methodological Aspects of Assessing the Creditworthiness of Municipalities as an Important Condition for Improving the Efficiency of Bank Lending Operations and Managing Municipal Borrowings

Nadezhda I. Yashina(✉), Svetlana D. Makarova, Natalia N. Pronchatova-Rubtsova, Igor A. Makarov, and Oksana I. Kashina

National Research Nizhny Novgorod State University N.I. Lobachevsky, Nizhny Novgorod, Russia
oksana_kashina@mail.ru, pronat89@mail.ru, makartolk@mail.ru, makarovasd@iee.unn.ru

Abstract. Relevance. At present, authorities of various levels actively use attracted financial resources as income sources. This is due to the transition to predominantly project financing, which involves co-financing from own and borrowed funds. This approach makes it imperative to have a mandatory assessment of the creditworthiness of borrowers, which are authorities at various levels. The purpose of the study is to develop methodical approaches to assessment of creditworthiness of territories taking into account specifics of their functioning.

Methods. This article proposes a methodology for assessing the creditworthiness of constituent entities of the Russian Federation and municipalities, based on the calculation of an integrated standardized indicator that allows you to compare condition of a certain territory with similar indicators by banks and interested authorities in the dynamics and at the current moment.

Results. The ranking of territories of Nizhny Novgorod region by creditworthiness level was carried out in order to assess the effectiveness of government bond issues in 2015–2017.

Discussion. Application of the methodology ensures transparency of assessment of the creditworthiness of borrower territories by banking organizations, which allows to minimize their risks on loans and expand the range of targeted funding sources. This approach is applied for the first time to credit rating of territories on the basis of integral indicators based on grouping on the basis of their minimization or maximization.

Conclusion. The methodology can be used by territorial authorities and banking management to improve the management of territorial borrowing and credit operations of banks.

Keywords: Creditworthiness of territories · Budgetary stability · Integral standardized indicator · Efficiency of bank credit operations · Management of municipal borrowings

E. G. Popkova and B. S. Sergi (Eds.): ISC 2019, LNNS 91, pp. 583–589, 2020.
https://doi.org/10.1007/978-3-030-32015-7_66

JEL Code: H66 · H63 · H11

1 Introduction

Features of the modern market mechanism of management bring the need to develop forms of attraction of funds to all economic entities, including territorial authorities. Currently, regional and municipal securities and bank lending have become particularly relevant sources of investment. To attract investments in the form of loans in territorial securities, the most important for the investor (lender) is the ability of a borrower to timely fulfill both previously assumed and future debt obligations (Yashin and Trifonov 2018).

Integral assessment of the borrower's ability to fulfill its debt obligations on the attracted resources to potential creditors, taking into account a wide range of possible domestic and external risks, this reflects the credit rating of territory. It allows to make the process of raising funds in the financial market as transparent as possible for borrowers and creditors.

The most famous in Russia method of assigning a credit rating and determining the quality of management of territorial finances is the Standard & Poor's method with the Russian scale of ratings. [1] However, it should be noted that the practice of banking organizations uses slightly different approaches to assessing the creditworthiness of potential territories for various purposes and formed the basis of the study.

Lending of subjects of the Russian Federation and municipalities is mainly engaged in large banks, which are interested in carrying out analysis of a whole set of indicators, including possible risks, such as payment, financial, risk of investment projects, etc. Which are arising in the market of territorial borrowing depending on the current economic situation and forming a credit rating of a territorial entity.

Since borrowings are carried out under the guarantees of territorial budgets, the preferred definition of payment and financial risks is due, because it depends directly on the structure of budget revenues and expenditures, revised plan and budget performance indicators (Potashnik 2018).

In the presented article authors propose a method of assessing the creditworthiness of municipalities and subjects of the federation. In contrast to the existing methods, the purpose of which is, first of all, the assessment of the territory from the point of view of managing state and municipal borrowings, the proposed approach solves a complex task and improves the efficiency of the bank's lending operations.

2 Theoretical Basis of Research

The works of Russian and foreign scientists are devoted to the evaluation of various parameters of municipalities and regions in order to improve their functioning. Anikeeva (2005) and Vlasova (2005) analyzed the need to establish territorial ratings in order to improve the investment climate. Grechenyuk A.V., Dmitriev M., and Gimadieva L.S. (Dmitriev et al. 2011) focused their attention on strategic directions of investment policy in regions. Marchenko and Machulskaya in their works (Marchenko 2004), suggested methods of rating territories from the point of view of their investment attractiveness. Vilenchik (2002) assessed possible risk factors in the issuance of securities by local authorities. In the work of Scherer (1997) presents approaches to the rating of the Russian Federation from the point of view of investment attractiveness, etc. It should be noted that at the depth of the study of the issue of peculiarities of formation of favorable investment climate, the above studies touched on the issues of financial stability of the territories, characterizing their investment opportunities mainly from the point of economic indicators. Assessment of financial stability is one of the most important characteristics of the financial condition of an economic entity from the point of view of the potential of long-term sources of financing (Garina 2017). The issues of assessing the financial condition of the territory, as one of the types of an economic entity, from the point of view of assessing the level of financial sustainability in regions, are analyzed in the works of Polyak (1998). In the works of Yashinoy (2015) [Yashin et al., 14], Poyusheva et al (2018), Dogaeva (2012) reviewed methodological approaches to determining the financial sustainability of territories using integral indicators.

However, the problems of multidirectional assessment of creditworthiness of territories as an important condition for increasing the efficiency of long-term credit operations of bank and territorial borrowing management at the same time remain insufficiently explored, although the issue is of particular relevance at present.

3 Research Methodology

The study uses methods of economic, system analysis and economic statistics. Approbation of the methodology was carried out on the official data of the report on the implementation of the consolidated budget of the Nizhny Novgorod region for 2015–2017.

The economic meaning of this method is to calculate the value of the ranks: If the rank value is lower than the higher the creditworthiness of the analyzed territory. The method of determining the integral standardized indicator of the creditworthiness of territories, taking into account the coefficients included in the system of criteria for assessing their creditworthiness. Also makes it possible to increase both the efficiency of the bank's lending operations and managing municipal borrowings in conditions of economic volatility.

Indicators are classified into two groups: an increase in some indicators - contributes to a decrease in the creditworthiness of the territories, while an increase in other indicators causes an increase in their creditworthiness. The first group should include indicators, the increase of which indicates the increase in efficiency of management of budgetary funds of the territory and creditworthiness. The second group includes indicators, the decrease of which is a positive indicator of management of the territory and increases its status as a potential borrower.

The first group includes coefficients that characterize financial independence and sustainability, the ratio of own income to grants, the security of own income financing of social expenditures and productive sectors, investment processes in human capital, etc.

The second group includes coefficients that characterize interest costs, the ratio of arrears to tax income and accounts payable, etc.

By means of expert methods of evaluation of each indicator, the limit values of indicators of conditional municipalities of high (first class or class A) have been determined, satisfactory (second or B) and low (third or C) creditworthiness class.

Thus, each of the thresholds reflects the boundary state of the territory as it moves to the state of the other level. The coefficients system shall be formed taking into account the following requirements:

1. Maximum information content;
2. Reliability;
3. The ability to assess current indicators and indicators in dynamics;
4. The presence of a minimum confidence interval.

For analysis of budgetary stability of territories, all coefficients will be relative indicators, which are calculated as a ratio of absolute values (Table 1).

Table 1. Factors included in the credit rating for bank lending and effective territorial management

No.	Coefficient	Economic content of the coefficient
P1	Coefficient of financial independence	ratio of own budget revenues to total income
P2	Financial sustainability ratio	ratio of tax income to total income
P3	Endowment ratio	ratio of tax and non-tax revenues to non-tax payments
P4	Coefficient of social expenditure	ratio of income to social expenditure (this includes budget items such as health, education, social policy, physical education, culture, and cinematography)
P5	Coefficient of the production sectors	ratio of own income to expenditures on the national economy and housing and communal services
P6	Ratio of investment in human capital	ratio of social expenditure to total expenditure
P7	Interest cost ratio	ratio of public and municipal debt servicing costs to tax revenues
P8	Human capital development financing ratio	Share of expenditure on education, health, culture in total budget expenditure
P9	Financing coefficient of economy	ratio of national economy expenditure to total expenditure
P10	Debt ratio	ratio of arrears to tax revenues
P11	Ratio of receivables and payables	relation to arrears payable on socially significant budget expenditures
P12	Coefficient of underfunding of socially significant expenses	ratio of accounts payable for socially significant budget expenditures to its total expenditures
P13	Debt sustainability ratio	ratio of the volume of loans and loans made by the regional government to the total amount of tax revenues of the budget
P14	Service ratio of socially significant expenses	ratio of expenses on debt servicing of the territory to the value of socially significant expenses

In the course of the study, standardization of the above indicators was carried out and unified integral indicator of the creditworthiness of municipalities was defined, which is the sum of the above coefficients. Its values vary from 0 to 1.

The standardized indicators are calculated according to the following formulas (Table 2).

Table 2. Standardized indicators for groups formed on the basis of the objectives of minimization and maximization of calculated coefficients

Group No.	Direction of change of indicator	Formula
1	minimization of indicators	$P_{st} = \frac{Pi - Pmin}{Pmax - Pmin}$
2	maximizing indicators	$P_{st} = \frac{Pmax - P_i}{Pmax - Pmin}$

where P_{st} is a standardized indicator;
Pi—index of territory;
P_{max}—the maximum value of the indicator from the population of territories;
P_{min}—the minimum value of the indicator from the population of territories.

The obtained value of the i-th indicator is compared with corresponding value of the integral standardized indicator reflecting the normative value for the region belonging to the group with high, medium and low creditworthiness. The smaller the value of the integral standardized indicator of the solvency of municipalities, the higher the investment opportunities of the territory. This allows you to rank the territories by levels of reliability or creditworthiness of municipalities: issuers (borrowers) with high reliability (class A), issuers (borrowers) with satisfactory reliability (class B) and with low reliability (class C).

4 Results

The results of the calculations are presented in Table 3.

Analysis showed that the leader is consolidated and regional budget of the Nizhny Novgorod region. In 2016, the Nizhny Novgorod region placed bonds for 10 billion rubles circulation period of 7 years, in 2017 for 12 billion rubles circulation period of 5 years, in 2018 - 10 billion rubles circulation period of 5.5 years. At the same time, the demand for bonds always exceeds the offer. The main investors are banks, investment companies and pension funds.

The Ministry of Finance of the Nizhny Novgorod region issued a bond loan in the amount of 10 billion rubles. The coupon rate was 8.68%. Circulation term—5.5 years. This once again confirms the reliability of the issuer - the Ministry of Finance of the Nizhny Novgorod region and testifies to the confidence of investors.

Table 3. Creditworthiness rating of municipalities of Nizhny Novgorod region in 2015–2017

Name	Subtotal						Result	
	2015	Rank 2015	2016	Rank 2016	2017	Rank 2017	2015–2017	Rank 2015–2017
Consolidated budget	1.45	1	1.19	1	0.86	1	3.49	1
Regional budget	1.64	2	1.35	2	1.10	2	4.10	2
Class A issuer	2.47	3	2.61	3	2.82	3	7.90	3
Dzerzhinsk	3.02	4	3.29	5	3.43	5	9.75	4
Pavlovo	3.92	13	3.29	4	3.85	8	11.07	5
…								
Bogorodsk	4.04	15	3.61	7	4.14	19	11.78	11
Issuer class B	3.71	8	3.94	22	4.17	21	11.81	12
Balakhna	3.86	11	3.86	18	4.18	23	11.91	13
Arzamas	4.18	18	3.74	13	4.00	13	11.92	14
…								
Pervomayskiy	5.01	34	4.99	52	5.52	55	15.52	53
Sharangskiy	5.76	58	5.11	54	5.39	52	16.26	54
Krasnooktyabrsky	5.55	54	5.25	55	5.56	56	16.36	55
Issuer of Class C	5.47	53	5.46	56	5.67	58	16.60	56
Bolshemurashkinsky	5.56	55	5.52	57	5.57	57	16.65	57
Vadsky	5.73	57	5.79	58	5.41	54	16.93	58

Source [Author's table]

5 Conclusion

Conducted studies showed the validity of the proposed methodology and full compliance of the received rating values of the territories with their real financial reliability. It should be noted that the proposed methodology can be implemented independently by interested parties on the basis of open indicators. This approach will allow banking organizations to independently rank potential customers and avoid additional costs by ordering expensive ratings from rating agencies. Territories will receive a simple and transparent tool to assess the effectiveness of investment management and the basis for the possibility of applying to bank institutions for the resources.

Acknowledgement. The research was carried out with the financial support of the Russian Foundation for Fundamental Research within the framework of the scientific project 18-010-00909 A.

References

Anikeeva, A.: Actual problems of investment ratings of regions in Russia. Investments Russia **5**, 3–7 (2005)

Vlasova, M.A.: Analysis of existing methods of assessment of the investment climate of the region. Ind. Policy Russ. Fed. **9**, 27–30 (2005)

Grechenyuk, A.B.: Analysis of the effectiveness of regional investment policy and directions of its improvement. Reg. Econ. Theory Pract. **3**, 37–50 (2005)

Dmitriev, M.: Investment strategies of the Russian regions: New challenges and opportunities. Econ. Policy **4**, 19–30 (2006)

Gimadieva, L.S.: Strategy and methods of attraction of investment resources at municipal level. Sci. J. KubSAU, 74(10) (2011)

Dogaev, A.V.: Assessment of the dependence of the results of investment policy implementation in the constituent entities of the Russian Federation on their credit ratings. http://www.e-rej.ru/Articles/2012/Dogaev.pdf. (In Russian)

Marchenko, G.: The main results of the rating: rating of investment attractiveness of regions. Expert **45**, 98–108 (2004)

Marchenko, G., Machulskaya, O.: Rating results: rating of investment attractiveness of regions. Expert **44**, 116–125 (2005)

Pushcheva, E.V., Yashina, N.I., Kashina, O.I.: Methods for assessing the effectiveness of public health financing on the basis of financial and non-financial indicators. http://nbpublish.com/library_read_article.php. (In Russian)

Scherer, F., Ross, D.M.: Structure of industrial markets. INFRA M, 698 p. (1997)

Sharp, W., Alexander, G., Bailey, J.: Investments. Translated from English. INFA-M, XII, 1024 p. (1997)

Yashina, N.I., Pronchatova-Rubtsova, N.N.: Estimation of influence of state policy of various countries on efficiency and stability of socio-political and economic processes. Finan. Credit **23**(647), 2–16 (2015)

Garina, E.P., Kuznetsova, S.N., Garin, A.P., Romanovskaya, E.V., Andryashina, N.S., Suchodoeva, L.F.: Increasing productivity of complex product of mechanic engineering using modern quality management methods. Acad. Strateg. Manag. J. **16**(4), 8 (2017)

Potashnik, Y.S., Garina, E.P., Romanovskaya, E.V., Garin, A.P., Tsymbalov, S.D.: Determining the value of own investment capital of industrial enterprises. Adv. Intell. Syst. Comput. **622**, 170–178 (2018)

Yashin, S.N., Trifonov, Y.V., Koshelev, E.V., Garina, E.P., Kuznetsov, V.P.: Evaluation of the effect from organizational innovations of a company with the use of differential cash flow. Adv. Intell. Syst. Comput. **622**, 208–216 (2018). https://doi.org/10.1007/978-3-319-75383-6_27

Stress-Testing at the Bank of Albania: Methodology of Approaches and the Quality of Forecasting

Ela Golemi[1(✉)] and Vasilika Kota[2]

[1] University "Aleksandë Moisiu", Durrës, Albania
golemiela31@yahoo.com
[2] Bank of Albania, Tirana, Albania
ekota@hotmail.com

Abstract. The paper aims to present for the first time the methodology for building the stress-testing at the Bank of Albania, as well as to evaluate the quality of its forecasts. Through the forward- looking stress-testing analysis the financial system stability and capital adequacy in the banking sector are estimated for a period of up to two years. Regardless of the purpose of stress-testing to assess whether in the event of large losses the banking sector has the ability to absorb them and not to accurately predict the indicators of the banking sector, it is still important to assess whether the attitude toward risk is sufficiently conservative.

The results suggest that the forecast of the capital adequacy ratio is quite close to its actual values. However, disintegrating these developments by the contribution that comes from the underestimation of the regulatory capital, in order to preserve the conservative trend of the exercise, has eased the underestimation of the risk-weighted assets, mainly reflecting the changes in the regulatory framework of the Bank of Albania.

Following the results of the analysis, the paper proposes several ways to further improve the quality of forecasting. They relate mainly to the transition towards a dynamic stress test forecasting, the consolidation of the conservative trend for the forecasting of the regulatory capital and a preliminary assessment of the regulatory changes and their inclusion in the stress test.

Keywords: Stress-testing · Capital adequacy · Quality of forecasting

1 Introduction

The stress-testing exercise has obtained a special importance in assessing the risks to which the banking sector is exposed. The beginning of this exercise in the Bank of Albania is around 2005, in a simplified form of it, while the standardization and comprehensive inclusion of the main risks of the sector, was completed around 2010. The history of the stress-testing exercise in its actual form is relatively short. The International Monetary Fund refers to the stress-testing exercise in its FSAP analysis, starting from 2001. However, the global financial crisis showed that this new instrument of risk assessment failed to foresee their accumulation and the exposure or the real vulnerability of the banking sector. Haldane (2009) extensively discusses the

E. G. Popkova and B. S. Sergi (Eds.): ISC 2019, LNNS 91, pp. 590–607, 2020.
https://doi.org/10.1007/978-3-030-32015-7_67

reasons for the failure of stress-testing exercises, ranging from underestimation of the size of the applied shocks to problems in the assessment of the interlinkages between financial institutions. On the other hand, Jorion (2009) argues that, despite the large size of losses after the financial crisis, it should not imply that the risk assessment methodologies were wrong. Updating models with the newest approaches and recent data, adjusting the volatility or the correlations, helps build more accurate estimations of the stress-testing.

The purpose of this paper is to present for the first time the methodology for building the stress-testing at the Bank of Albania, as well as to evaluate the quality of its forecasts. Through the forward-looking stress-testing analysis the financial system stability and capital adequacy in the banking sector are estimated for a period of up to two years. Stress-testing focuses only on risks to the banks' capitalization based on macroeconomic scenarios, while risks associated with funding and liquidity are not part of these mechanisms.

The Bank of Albania has build two methodologies for stress-testing. Starting from 2010, the sustainability of the banking sector and individual banks in terms of severe but plausible economic shocks is estimated on quarterly basis through the top-down approach. In this approach, the Bank of Albania builds macroeconomic scenarios and assesses the risks of the banking sector based on the data of each bank's balance sheet. The results are presented to the Administrators and published in an aggregated way, drawing attention to the exposure of the banking sector and the individual banks to the risks analyzed. In the second approach, the bottom-up approach, the Bank of Albania builds macroeconomic scenarios, while the commercial banks assess the exposure. This exercise is performed once a year, starting from 2012. In this case, the results draw attention to the improvement of the procedures of the exercise and to the risk exposure, presented directly to the participating banks.

This paper evaluates the forecasting quality for the methodology of the top-down approach. Regardless of the purpose of stress-testing to assess whether in the event of large losses the banking sector has the ability to absorb them and not to accurately predict the indicators of the banking sector, it is still important to assess whether the attitude toward risk is sufficiently conservative. The results of stress-testing concisely present the banking sector's vulnerability to adverse circumstances (Blaschke et al. 2001). Consequently, stress-testing does not assess the probability of the materialization of a certain scenario, but answers the question "How large could the loss be?" rather than "How probable is the loss?" (Blaschke et al. 2001). In this context, the stress-testing exercise is expected to be biased towards risk overestimation to provide a solvency buffer (cushion).

The results suggest that the forecast of the capital adequacy ratio is quite close to its actual values. However, disintegrating these developments by the contribution that comes from the underestimation of the regulatory capital, in order to preserve the conservative trend of the exercise, has eased the underestimation of the risk-weighted assets, mainly reflecting the changes in the regulatory framework of the Bank of Albania. The quality comparison of the time-extended stress-test forecasting reflects a better quality for the first year of the exercise, in line with the expectations.

The paper is structured as follows. The second section details the methodology of the stress-testing exercise at the Bank of Albania. Section 3 summarizes the steps of stress-test forecasting and the indicators used to estimate its quality. Section 4 presents the results for the forecasting errors of the capital adequacy ratio, regulatory capital and risk-weighted assets. Section 5 discusses the implications of the results and suggests some ways of improving the quality of the exercise. Lastly, the main findings are summarized.

2 The Methodology of Stress-Testing at the Bank of Albania

The stress-testing exercise at the Bank of Albania took full shape starting from 2010. Previously, starting from 2005, the stress-testing mainly consisted in compiling some special sensitivity analysis.

The process of building macroeconomic scenarios is the first challenge in performing a stress-testing exercise. In general, the stress-test scenarios should be adverse, but plausible, and at the same time coherent, which means that the risk factors need to develop consistently. The consistency of the results for the stress-testing exercise is closely related to the consistency of the adverse scenarios. Constructing the right scenarios, in line with developments in the economic cycle, allows the analysis of the situations in real time and the construction of countercyclical policies for the banking sector. If the model does not include sufficiently "extreme" scenarios, thus "tail events" in the probability distribution, the final results do not represent the necessary risk and the relevant uncertainties.

Generally, stress to banks consists of a set of risks that can or cannot be statistically measurable. For this purpose, to assess and plan the impact of different scenarios to the banking sector, intermediate models have been built to evaluate the banks' exposure to different scenarios.

The baseline scenario of stress-testing is built based on MEAM, the macroeconomic model of the Bank of Albania. The model generates projections mainly for the macroeconomic indicators such as economic growth and other indicators[1]. It gives the possibility to perform simulations and obtain quantitative effects of economic policies and shocks on economic variables.

The baseline scenario is based on the official forecast of the Bank of Albania for the economic growth, interest rates, credit expansion and developments in the exchange rate.

The adverse scenario is based on assumptions of lending deterioration, weakening of the domestic currency and interest rate growth. In order to analyze the banking sector's stability to the stress-test scenarios, the shocks of the adverse scenarios maintain the same structure. Generally, the adverse scenarios in the stress-testing exercise are built based on two approaches:

1. Reduction by 50% of the annual growth of lending used in the baseline scenario, 20% depreciation of the domestic currency and an interest rate increase by

[1] For more details refer to: Zoltan M. J., V. Kota and E. Dushku, 2007, "Macro econometric model of Albania: A follow up", 7th Conference of the Bank of Albania, "Monetary policy strategies for small economies".

2. standard deviations to the baseline scenario. Based on these shocks, MEAM estimates the economic growth for the adverse scenarios through the transmission channels.

Meanwhile, following the implementation of the IMF recommendations, the Bank of Albania builds an adverse scenario based on the cumulative decline of the GDP by 2 standard deviations (2-year cumulative) calculated as follows:

$$GDP^{BASELINE}_{year-year(+2)} = \frac{GDP^{baseline}_{realyear(+2)}}{GDP_{realyear}} \tag{1}$$

while the economic growth after 2 years is estimated as follows:

$$g^{adverse}_{year-year(+2)} = g^{brseline}_{year-year(+2)-0.123} = \frac{GDP^{adverse}_{realyear(+2)}}{GDP_{realyear}} \tag{2}$$

where $GDP^{baseline}$ is the Gross Domestic Product in the baseline scenario,
$GDP^{adverse}$ is the Gross Domestic Product in the adverse scenario
g – is the growth rate of the GDP.

2.1 Credit Risk

Credit risk represents the main exposure in the stress-testing exercise. This risk assesses the possibility that the banking sector borrowers do not repay their debt. In this case, banks are exposed to losses due to lack of inflows on their income on one hand, and increased spending on provisioning on the other hand (Blaschke 2011).

The literature of credit risk assessment is focused on the assessment of losses stemming from credit risk and the relevant impact on capital. Credit risk generally includes expected losses, which must have been provisioned by banks, and unexpected losses, which should be recently provisioned. However, credit risk in the Albanian banking sector is assessed through the non-performing loans ratio, which summarizes the assessment on the share of delinquent loans over 90 days to total loans outstanding. The ratio of non-performing loans is regressed to the main macroeconomic factors that evaluate the sensitivity of banks' borrowers to macroeconomic risk factors. This assessment is separately done for loans in ALL and foreign currency loans, including borrowers who are exposed to exchange rate fluctuations through indirect credit risk. The general form of the equations is as follows:

$$NPL\_ALL_t = \alpha_0 + \alpha_1 NPL\_ALL_t + \alpha_2 \Delta GDP_{i,t} + \alpha_3 i_t + \varepsilon_{i,t} \tag{3}$$

$$GDP\,FX_t = \alpha_0 + \alpha_1 NPL\_FX_t + \alpha_2 \Delta GDP_{i,t} + \alpha_3 i_t + \alpha_4 \Delta FX_{i,t} + \varepsilon_{i,t} \tag{4}$$

where NPL_ALL - is the non-performing loans ratio in ALL
GDP - is the Gross Domestic Product, percentage change
i - is the reference rate for lending, respectively the 12-months rate of Treasury Bills for ALL and *euribor* for foreign currency

FX - is the ALL/EUR exchange rate (percentage change) since the highest share of lending in foreign currency is in this currency.

Assuming a linear exposure to risks, the volatility of non-performing loans can be calculated as:

$$\sigma_{\left(\frac{RKP}{Aktiveve}\right)} = \sqrt{\beta^2\sigma_i^2 + \gamma^2\sigma_p^2 + \cdots + 2\rho_{i,p}\sigma_i\sigma_p + \cdots} \tag{5}$$

According to this approach, a simultaneous integration of credit risk and market risks is performed.

The assessment of additional provisioning expenses is based on the structure of loan categories for each individual bank. Their impact is assessed in terms of capital adequacy as follows:

$$\textit{Capital Adequacy Ratio}_{\textit{stress test}} = \frac{\textit{Regulatory capital} - \textit{additional provisioning expenses}}{\textit{Risk weighted assets} - \textit{additional provisioning expenses}} \tag{6}$$

The interest rate and the exchange rate that determine the non-performing loans ratio are also determinants of market risk (interest rate risk and exchange rate risk are presented below).

2.2 Interest Rate Risk

The banking sector is exposed to interest rate risk if there is a mismatch between assets and liabilities that are sensitive to interest rates. The change in interest rate affects the interest income and interest expenses, the net result of which is the main source of income for the banking sector in Albania. The assessment of interest rate risk can be done through two approaches:

The first approach, the re-pricing model, estimates the gap between assets and liabilities that are sensitive to interest rates. These assets/liabilities are allocated in several specific "baskets" according to their sensitivity to the interest rate and maturity. For each basket a gap between revenue flows in assets and the flow in liabilities is estimated. For any change in the interest rate ΔR_i, the re-pricing gap determines the change in net interest income for the basket i and the total portfolio:

$$\Delta \textit{Net interest income}_i = \textit{Gap}_i \times \Delta R_i \tag{7}$$

The repricing model is simple and can be applied to aggregated data from banks' balance sheets. The main disadvantage is that this model does not include the impact of interest rate changes in the market value of the assets, basically using the value of assets/liabilities according to the balance sheet value and omitting profits/losses in capital.

The second approach, the gap model, includes the remaining maturity of assets and liabilities of the banking sector in the analysis. In this case, the weighted-average maturity of assets/liabilities is estimated as follows:

$$M^A = \sum\nolimits_{i=1}^{N} w_i^A M_i^A \qquad M^L = \sum\nolimits_{i=1}^{N} w_i^L M_i^L \tag{8}$$

ku M^A is the weighted average maturity of assets.
M^L is the weighted average maturity of liabilities.
M_i^A is the maturity of an asset with a given maturity *i*.
M_i^L is the maturity of a liability with a given maturity *i*.
$w_i^L M_i^L$ is the share of the asset in the total portfolio.
w_i^L is the share of the liability in the total portfolio.

The banking sector's exposure to the interest rate risk depends on the size of the maturity gap mismatch.

$$Gap^{Maturities} = M^A - M^L \tag{9}$$

The first approach, the re-pricing model, where the gap between assets and liabilities sensitive to interest rate changes is built, has been used by the Bank of Albania until the end of 2013. Further on, the methodology has improved by using the assessment of the maturity gap.

Regarding the size of shock in terms of interest rate, a simpler approach in accordance with the methodology applied, is the parallel shift of the interest rate curve. This shock is applied to the interest rate in the base period and it does not include changes in volatility or correlation.

2.3 Exchange Rate Risk

The exchange rate risk is also important for the banking sector. This risk assesses the exposure of bank assets and liabilities to exchange rate fluctuations. The exchange rate risk may be direct, in accordance with the open foreign currency position of the banking sector, or indirect, in the event that the banking sector has provided loans to borrowers with income in domestic currency.

For the assessment of the direct risk, the shock means a higher depreciation of the domestic currency and an assessment of the corresponding impact in loss/profit on the long/short FX position

$$\Delta\, loss/profit = Net\, open\, FX\, position * \Delta FX \tag{10}$$

The highest value between the total net open long FX position and the total net open short FX position represents the total net open FX position of the bank, or the bank's overall exposure to the exchange rate risk. Depending on the long or short position, the exchange rate depreciation can generate profit or loss for the banking sector.

The assessment of indirect exchange rate risk is included in the equation of non-performing loans in foreign currency and the corresponding impact on provisioning expenses.

2.4 Profitability, Regulatory Capital and Risk Weighted Assets

The capital adequacy ratio is the comprehensive leading indicator of the stress-testing estimation. One of its determinants, the regulatory capital, is significantly affected by the performance of the banking sector's financial result after relevant shocks. The financial result includes the impact that comes from net interest income through interest rate risk, losses or profit from exchange rate shocks in the net open FX position, change in provisioning expenses through credit risk and the applied haircuts in the securities portfolio through market risk.

$$\textit{\textbf{financial result}} = f\left\{\textit{\textbf{NII}}, \frac{\textit{\textbf{loss}}}{\textit{\textbf{profit}}}\textit{\textbf{FX}}, \textit{\textbf{provisioning expenses}}, \textit{\textbf{haircut}}\right\} \tag{11}$$

Other profitability indicators such as commission income, administrative or other expenses are kept in line with the historical trend. The regulatory capital in the base period is corrected with the financial result of the forecasted periods, precisely at the end of the first year and at the end of the second year of the exercise. The adjustment with the financial result is performed in the same measure for both cases, profit and loss. For the exercise, it is also assumed that the banking sector will not distribute dividends.

$$\textit{\textbf{Regulatory capital}}_{t+1} = \textit{\textbf{Regulatory capital}}_t + \textit{\textbf{RezultatifinanciarFinancial result}}_{t+1} \tag{12}$$

Regarding the performance of risk-weighted assets, their forecast is divided into two periods. For the 2010–2014 period, the banking sector was under Basel 1 approach, implying the need to assess only credit risk through forecasting the growth of outstanding loans. Starting from January 2015, the banking sector switched to the Basel 2 approach, and the capital requirements expanded not only for credit risk but also for market and operational risk. Regarding the operational and market risk, risk-weighted assets reflect the expert judgment on net income and the potential of asset quality deterioration.

At the end of the first year and of the second year of the exercise, the capital adequacy ratio is estimated for each bank and for the whole banking sector.

$$\textit{\textbf{Capital Adequacy Ratio}}_{t+1} = \frac{\textit{\textbf{Regulatory Capital}}_{t+1}}{\textit{\textbf{Risk Weighted Assets}}_{t+1}} \tag{13}$$

3 Assessing the Forecasting Quality of Stress-Testing

This section aims to assess the forecasting quality of the stress-testing exercise and its structure. The stress-test forecasting is not the same as the macroeconomic variables forecasting, mainly because it is based on scenarios with a low plausibility. Consequently, the comparison of the actual values with the forecasted values of the adverse scenario is not valid for its quality assessment. In the baseline scenario, the comparison of the stress-testing forecasted values with the actual values of the banking sector indicators is still difficult. The baseline scenario includes some static assumptions on the development of profitability indicators, capitalization or the interest rate effect. On the other hand, the actual data include a dynamic behavior of banks in search of profit, preliminary budgeting of their activity, a corrective behavior towards meeting their targets etc. However, the purpose of stress-testing is the assessment of the banking sector's stability even in the baseline scenario, which generally aims to maintain a buffer of a higher capital adequacy ratio than the actual one. For this purpose, considering the forecasted estimations in the baseline scenario, the forecasting errors that overestimate the risks are more acceptable than the forecasting errors that underestimate them.

3.1 The Forecasting Structure of Stress-Testing

Another aspect of stress-testing at the Bank of Albania is that exercise is static and not dynamic. Despite the quarterly frequency of the exercise, the forecasting horizon extends up to two years, thus up to 8 quarters, fixing the ending period at the end-year of the exercise. More explicitly, if it is December 2010, the forecasting exercise extends until December 2012, i.e. 8 forecasted quarters. The next exercise will be conducted in March 2011, however, it extends again until the end of 2012, i.e. 7 forecasted quarters, and so on. The indicators of the banking activity will be estimated in each case for the end of the first year (December 2011) and the end of the second year (December 2012). A graphical illustration of the forecasting structure is presented below.

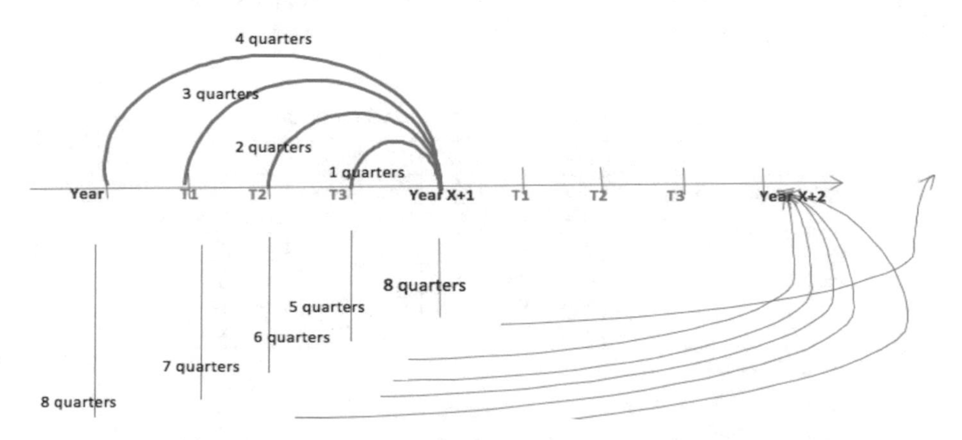

Chart 1. The structure of building the stress-testing exercise. Source: Authors

According to Gersl and Seidler (2010) the forecasting of the baseline scenario should assess slightly increasing risks compared with the actual performance of the banking sector indicators. In this case, it is important that the entire banking sector maintains a higher level of capitalization to face all the uncertainties related with losses assessed in the event of a weaker performance of the economy. Consequently, this means that the forecasting errors of stress- testing should be assessed differently from the forecasting errors of macroeconomic variables, for which the direction of the forecasting error is considered in the same way as "poor quality". In the case of stress-testing, the forecasting errors can also be biased toward overestimation of risks and underestimation of capital adequacy ratio, aiming that the banking sector is hedged against them.

To assess the quality of stress-testing at the Bank of Albania, we rely on quarterly data for the period of March 2010 - December 2015. The indicators for which the forecasting quality is assessed are the regulatory capital, the risk weighted assets and capital adequacy ratio. However, as the NPL ratio has a significant impact on the financial result of the banking sector and on the capital adequacy ratio, we also assess the forecasting quality of this assumption.

The assessment for each indicator of the stress-testing is compared with its actual performance according to the dynamics presented in Chart 1. This means that the forecasting errors are assessed according to the different periods of its estimations. More specifically, if we are at the beginning of the period, December 2010, the stress-testing exercise has forecasted the performance of the banking sector indicators up to the end of 2011 and 2012, thus a forecasting 4 and 8 quarters. In the second period, March 2011, the stress test has forecasted the performance of the banking sector indicators again until the end of 2011 and the end of 2012, updating the financial situation of banks with the relevant period. However, in this case the forecast extends to 3 quarters (end of 2011) and 7 quarters (end of 2012). Moving along, the forecasted periods narrow further down to 2 quarters (end of the first year) and 6 quarters (end of the second year) for the exercise of June, etc. The following table presents precisely the estimation structure of the forecasted periods where the actual values are compared with the forecasted values of the stress-testing exercise (Table 1).

Table 1. Forecasted periods based on quarters

	Q1, YearX		Q2, YearX		Q3, YearX		Q4, YearX	
The forecasted period	Year X + 1	Year X + 2	Year X + 1	Year X + 2	Year X + 1	Year X + 2	Year X + 1	Year X + 2
Forecasted quoters	3	7	2	6	1	5	4	8
Actual Values	Q4, Year X +1	Q4, Year X +2	Q4, Year X +1	Q4, Year X +2	Q4, Year X +1	Q4, Year X +2	Q4, Year X +1	Q4, Year X +2

Source: The authors

As of the above, the forecasted horizon is divided in accordance with the number of periods in advance on which the stress-testing exercise is built and varies from 1 quarter to 8 quarters, which represents the longest forecasting period. Compared to the end of the first year of stress-testing, the forecasting periods are considered as 4 quarters of forecasting, 3 quarters, 2 quarters and lastly, 1 quarter. Compared with forecasting up to the end of the second year of stress-testing, the forecasted period are estimated as 8 quarters, 7 quarters, 6 quarters and lastly, 5 quarters. As a result, the forecasting errors will also be evaluated for 1–8 quarterly periods of forecasting.

3.2 Forecasting Errors

The analysis of forecasting errors focuses on evaluating the size of the error for each certain number of periods (periods in advance i), as well as on analyzing the direction of the error. Regarding the size of the forecasting error, the literature presents several indicators that focus on comparing the forecasted values with the corresponding actual values as follows.

a. **Forecasting Errors**

The average error (ME) helps to evaluate the deviation of the forecasted values to the actual ones by determining if there bias in forecasting. A positive value of the ME means overestimation and a negative value means underestimation. But the ME indicator can also be misleading. A value of zero could mean that the model has perfectly forecasted the actual value (less likely) or that the positive and negative values eliminate each other. In both cases, the ME indicator underestimates the error. The ME indicator is defined as:

$$ME = \frac{1}{n}\sum\nolimits_{t=1}^{n} (F_t - A_t) \tag{14}$$

where n is number of the forecasted periods, F is the forecasted period, A is the actual value and t refers to the period.

The average absolute error (MAE) assesses the extent of the forecasting errors. MAE is a way to address the underestimation errors by the ME indicator. Using absolute values of the errors, the average gives a better indication about the size of forecasting errors. The average absolute error (MAE) is determined as follows:

$$MAE = \frac{1}{n}\sum\nolimits_{t=1}^{n} \left|F_t - A_t\right| \tag{15}$$

The square root of the average error (RMSE) is an indicator that assumes that errors with a higher value are worse than the ones with a small value. The RMSE indicator eliminates the problem of positive/negative values by obtaining the square of the error. This result gives more importance to a bigger error, therefore the assessment of the forecasting errors is more conservative compared to the AAE indicator.

$$RMSE = \left(\frac{1}{n}\sum\nolimits_{t=1}^{n} (F_t - A_t)^2\right)^{1/2} \tag{16}$$

The indicators presented above are "series specific," meaning they provide an assessment of forecasting errors in units of each of the indicators. Meanwhile the two measures presented below enable the comparison between the forecasting quality of various indicators. This process is performed using the error rate regarding the actual values.

The average percentage error (MPE) is a relative measure of forecasting errors. It shows the weight of the forecasting error to the actual value of the indicator, but it also is a subject of "averaging" positive and negative values of the forecasting error (likewise the AE indicator).

$$MPE = \left(\sum_{t=1}^{n} \left(\frac{F_t - A_t}{A_t} \right) \times 100 \right) / n \tag{17}$$

The average absolute percentage error (MAPE) is a benchmark indicator that does not include the problem of averaging the positive and negative errors. This indicator provides the weight of the absolute error to the actual value and it is a simple way to communicate the forecasting quality.

$$MAPE = \left(\sum_{t=1}^{n} \left(\left| \frac{F_t - A_t}{A_t} \right| \right) \times 100 \right) / n \tag{18}$$

b. **The Contribution of Assumptions on the Forecasting Errors**

The forecasting errors can reflect two main factors. Firstly, the impact of the macroeconomic or financial assumptions and the credit quality is very important to the financial result. The assumed interest rates changes affect net interest income, credit quality, etc.; assumptions regarding the exchange rate impact the income statement in accordance with the net open FX position; the impact of the assumption on the performance of non-performing loans is reflected in the performance of provisioning expenses and the financial result. The second element that influences the forecasting error is related to the internal structure of stress- testing through the assumed interlinkages for each indicator.

In order to separate the contribution of the assumptions from the one that comes from the stress-testing structure, the stress-testing exercise is re-estimated for the entire period under consideration by replacing the macroeconomic assumptions with their actual performance. The remainder of the forecasting error reflects the rigidity of its structure and the interlinkages within the banking sector indicators, the implicit risk aversion of the stress- testing and its tendency to underestimate the positive outcome.

4 The Results

4.1 Comparing the Actual Performance with the Stress-Test Forecasting

The main indicator of stress-testing is the Capital Adequacy Ratio. The actual values of this indicator are generally above the forecasted value up to the end of 2013. After this period, it is noted that the stress-testing exercise tends to overestimate the capital adequacy ratio until the end of 2015, where the difference between the two estimations is moderated. In the first part of the period under consideration, until 2013, only a small share of the forecasting error is due to errors in the stress-testing assumptions. However, after the second period, starting from the end of 2013, it is noted that the error correction in the underlying assumptions partially smoothes the difference from the actual values and aims at underestimating the Capital Adequacy Ratio (Chart 2).

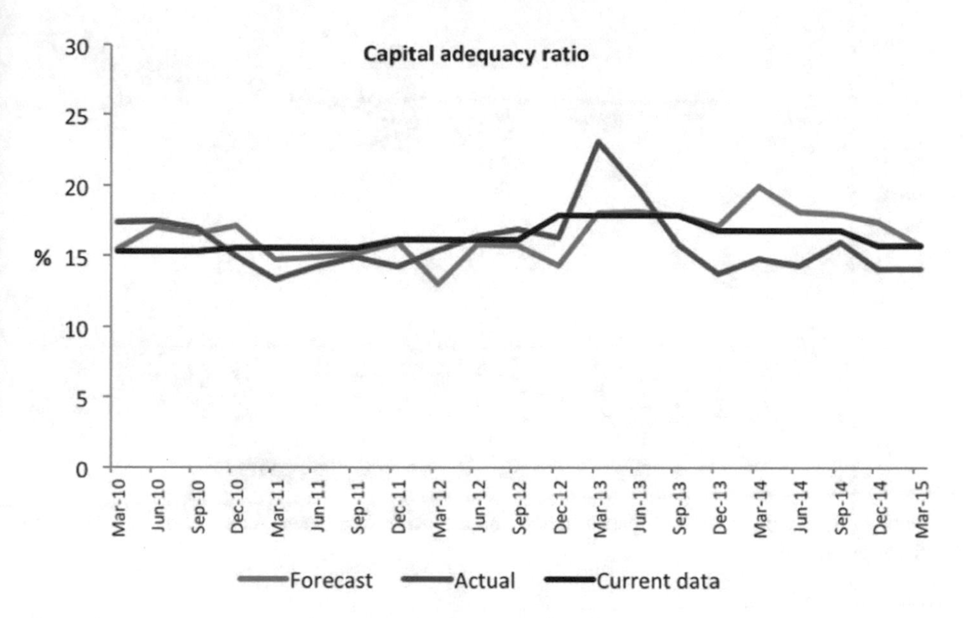

Chart 2. The comparison of the performance for the Capital Adequacy Ratio. Source: Authors' estimations

The difference between the actual performance and the forecasted capital adequacy ratio reflect assessments for both the Risk Weighted Assets and the Regulatory Capital.

Considering the risk-weighted assets, the stress-testing exercise seems to have underestimated their performance persistently. This underestimation becomes even more significant after 2013, where the difference from the actual value is even more profound. The error correction for the assumptions of the stress-testing smoothes this difference significantly. For the period after 2013, the current assumptions values are quite close to the actual value of Risk Weighted Assets, implying that incorrect assumptions have had the main contribution for the underestimation of this indicator (Chart 3).

Focusing further on this assessment, this development reflects two fundamental changes in the estimation of the risk-weighted assets, which are not part of the macroeconomic assumptions.

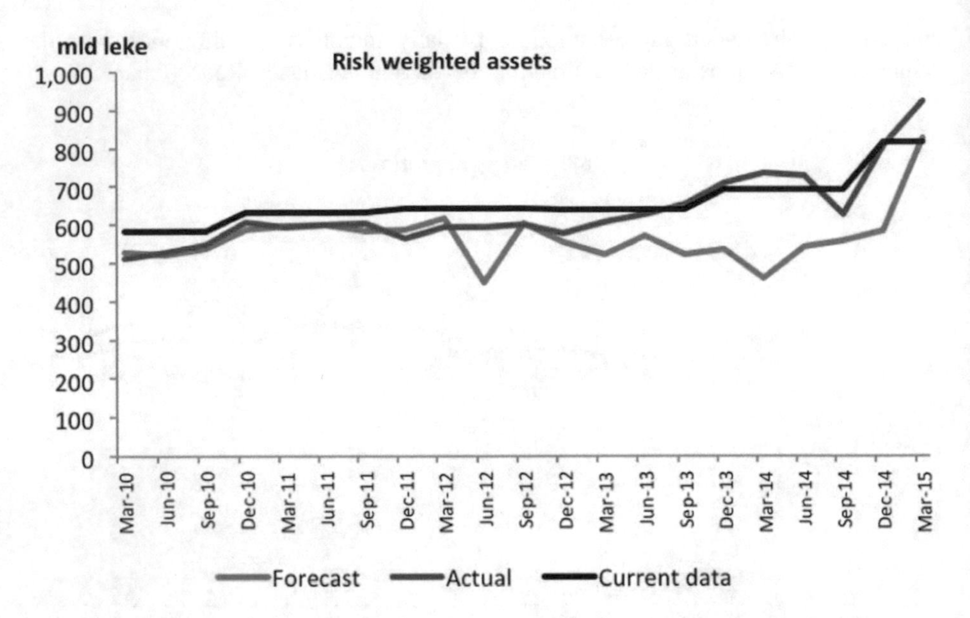

Chart 3. The comparison of performance for the Risk Weighted Assets. Source: Authors' estimations

- Firstly, in May 2013, the Bank of Albania presented a package of countercyclical/macroprudential measures which aimed at supporting and stimulating lending and reducing the credit risk. The package of measures included incentives to channel the excess liquidity of the banking sector towards lending to the economy and supported the prompt restructuring of loans, prior to becoming non-performing loans. Measures related to stimulating lending aims:
  - i. zeroing coefficients for calculating the risk-weighted assets for credit growth within the range of 4-10% and
  - ii. an increase to 100% of the risk coefficients for the additional net position of the investments abroad.

The package of countercyclical measures gave a direct impact on the performance of risk-weighted assets, mainly in terms of their growth for the additional net investment position to other countries, which have not been part of the stress-testing assumptions.

- Secondly, the risk-weighted assets increased after the implementation of the new regulation on capital adequacy in January 2015 under Basel 2. In addition to the credit risk, the Risk Weighted Assets include the market risk and the operational risk. In the condition where the exposure to these additional risks was unknown and there were no estimations, the stress-testing assumptions have not incorporated the new implemented regulation. As a result, for forecasted periods after 2015, the changes in regulation gave a significant impact in changes between actual and forecasted values.

The second element that determines the capital adequacy ratio is the regulatory capital. Unlike the risk-weighted assets estimation, the forecasted regulatory capital is generally underestimated for all the period under consideration. In this case, the main impact comes from the conservative trend of expectations for the increase of the financial result, which is the main contributor in the performance of the regulatory capital in the stress-testing exercise (Chart 4).

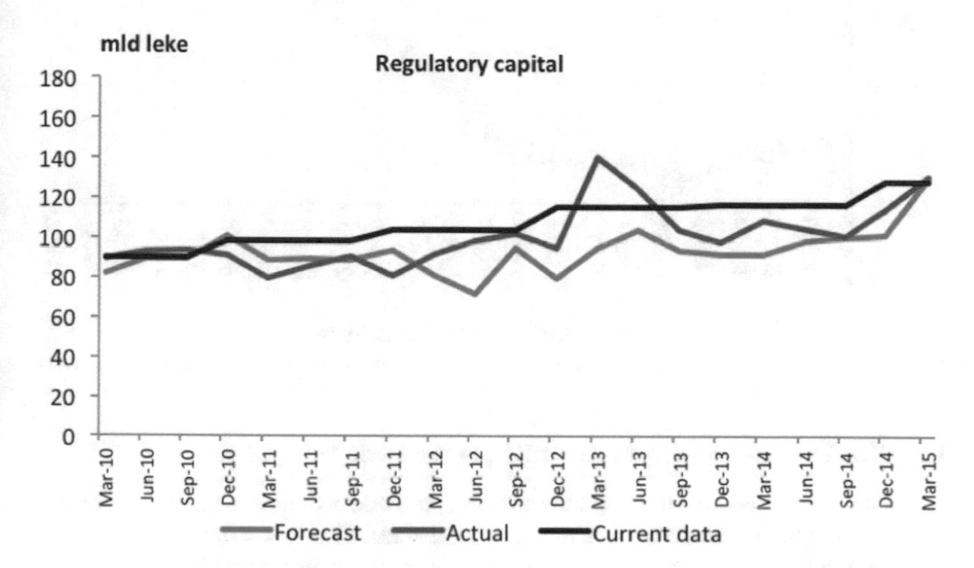

Chart 4. The comparison of the Regulatory Capital performance. Source: Authors' estimations

The disintegration of the errors contribution in estimating the capital adequacy ratio by the inaccurate forecasting of the risk-weighted assets, as well as the inaccurate forecasting of the regulatory capital, shows that these two elements have offset one another. The underestimation of the risk-weighted assets was offset by the underestimation of the regulatory capital, smoothing the forecasting error of the capital adequacy ratio.

Finally, the developments in non-performing loans are important for stress-testing, as the provisioning expenses are the main item that affects the financial result, particularly in the adverse scenarios. Generally, the comparison of actual and forecasted values show a qualitative forecasting for the non-performing loan ratio. There is an exception for the period after 2015, when the Bank of Albania required the write-off of non-performing loans which had been classified as lost for more than 3 years. In this period, the impact of the non- performing loans write-off is high, resulting in an underestimation of credit quality. This trend reflects a deterioration of the credit quality, which has smoothed the positive impact arising from non-performing loans write-off, as well as an overestimation of the outstanding loans that should have been written-off.

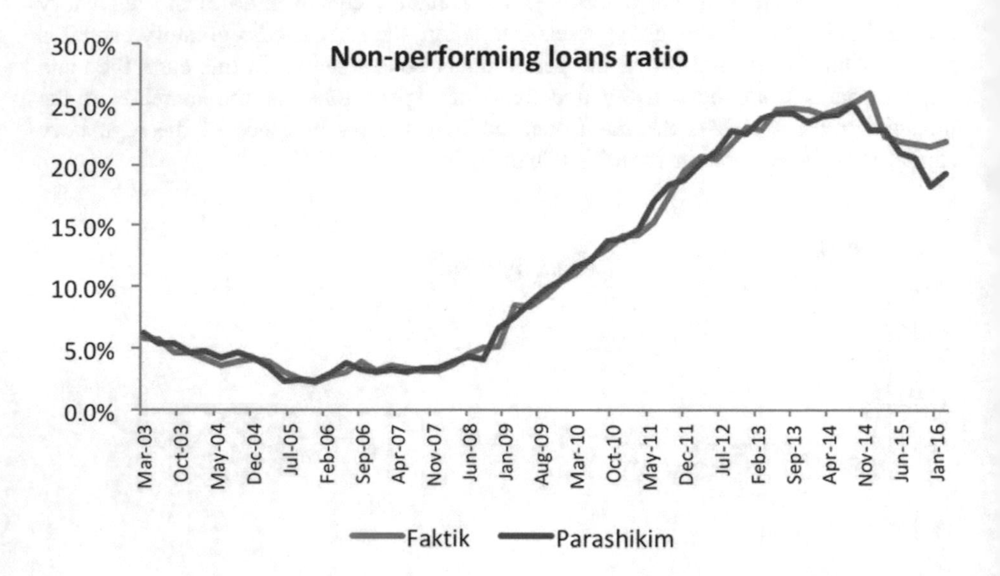

4.2 The Forecasting Errors According to the Periods in Advance

As explained in section two, the stress-testing exercise includes a variable period of forecasting, which extends up to 8 quarters in advance. The following section presents two main indicators of forecasting errors, for its main elements, while the rest of the forecasting errors is presented in the annex.

The absolute forecasting error (ME) of the capital adequacy ratio scores -0.1 percentage points for 4 quarters in advance and 0.2 percentage points for 8 quarters in advance. This means for example that stress-testing forecasts a capital adequacy ratio of respectively 15.9% and 16.2%, instead of the actual value of 16%. The size of the forecasting error increases for the second year of the forecasting, showing a trend of overestimation for the indicators, although the size is quite small. The corrected assumptions of risk-weighted assets have led to higher forecasting errors for the capital adequacy ratio (actual) due to lack of smoothing from the performance of regulatory capital (Chart 5).

In relation to the forecasting errors for the risk-weighted assets, they show significant bias due to underestimation for the entire period. The corrected assumptions for risk-weighted assets through the re-estimated assumptions and the inclusion of two important changes in the regulations of the Bank of Albania, has significantly reduced the forecasting errors. The forecasting errors in the first and second year of the exercise do not change significantly, even though they show an increase during the end-year period, due to adjustments made by banks in their activities with the objective of reaching their targets (Chart 6).

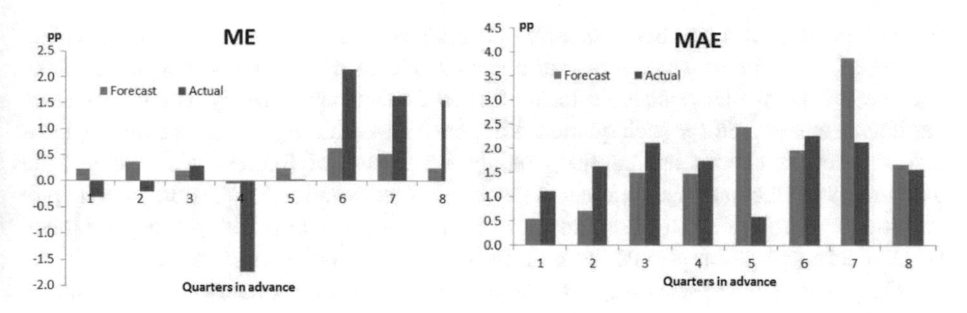

Chart 5. Forecasting errors, Capital Adequacy Ratio. Source: Authors' estimations

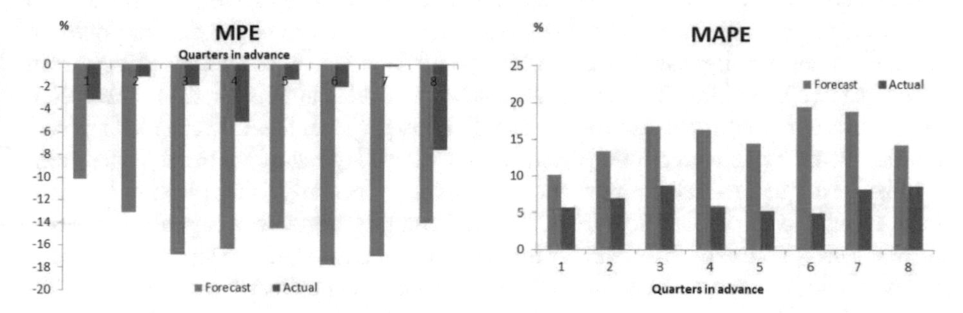

Chart 6. Forecasting errors, Risk Weighted Assets. Source: Author's estimations

The forecasting errors of the regulatory capital reflect the trend of their underestimation. The re-estimated assumptions with their actual performance has slightly smoothed these errors. In relative terms (MAPE), the regulatory capital is underestimated by 15% at the end of the first year of the exercise and by 13% at the end of the second year of exercise (Chart 7).

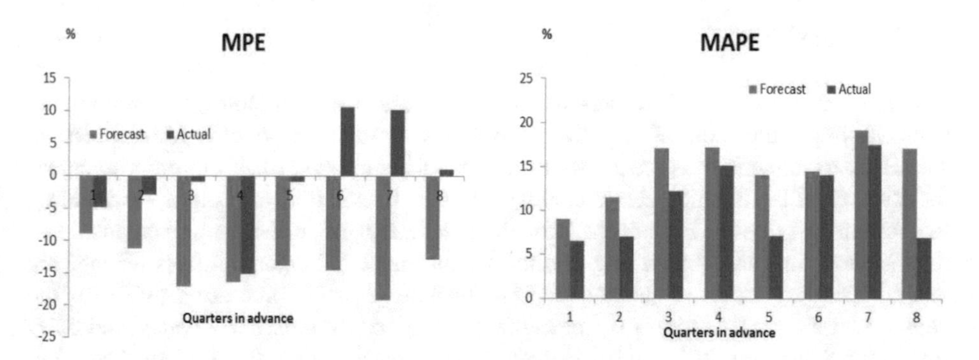

Chart 7. Forecasting errors, Regulatory Capital. Source: The author

The main impact on this forecasting comes from the structure of the stress-testing exercise. The static and not dynamic nature of the exercise means that items of the income statement that contribute to the financial result and directly to the regulatory capital are annualized for each quarter. After the annualization process, the assumptions on the relevant shocks are applied, mainly the impact of interest rate changes, the exchange rate fluctuations and the non-performing loans ratio. As a result, the structure and the formation of the financial result in the baseline scenario does not depend much from shocks and assumptions, but mainly from the annualization process.

Despite that the assessment process of the stress-testing quality generally shows that this exercise is relatively qualitative, with a good forecast of the capital adequacy ratio, it is still necessary to improve it further. The results show that the static nature of the exercise restricts the dynamic estimation of the banking sector activity and the lower difference with its actual values. Also, the static nature of the exercise does not maintain a uniform forecast horizon, which would enable a more consistent assessment of its quality. Therefore, the first recommendation of this paper is related to the need to switch to a dynamic stress-test forecasting. For this purpose, the financial result of each quarter would serve as the base period, which varies depending on the financial result for the next quarter and the relevant assumptions. Secondly, stress-testing seems to have maintained a conservative trend on forecasting the regulatory capital. However, a further consolidation of this trend for the baseline scenario would mean that the change in this indicator reflects the financial loss but not the positive financial result. The second recommendation is precisely the restriction of changes in the regulatory capital only for the financial losses. Thirdly, it is noted that the lack of information on the expected impact of the regulatory changes of the Bank of Albania on the Risk Weighted Assets has had a significant impact on the stress-testing results. It is necessary that such changes also include a preliminary estimation of their impact, with an eligible marginal error, to correctly assess the risk exposure of the banking sector. Finally, the assessment of the stress-testing forecasting quality is important and should be carried out on frequent basis for its continuous improvement.

5 Conclusions

The purpose of this paper is to present for the first time the methodology of building the stress-testing at the Bank of Albania, as well as to evaluate the quality of its forecasts. The Bank of Albania uses the stress-testing exercise on regular basis in order to assess the stability of the banking sector against extreme shocks. For this purpose, evaluating the forecasting quality is important for the assessment of further improvements. The results show that the forecasting quality for the main indicator of stress-testing, the capital adequacy ratio, is qualitative. The absolute error of forecasting (ME) for the capital adequacy ratio scores - 0.1 percentage points for 4 quarters in advance and 0.2% points for 8 quarters in advance. However, these results are influenced by two main elements: the persistent conservative trend to changes in the regulatory capital, which has smoothed the underestimating trend of the Risk Weighted Assets. In relation to the latter ones, the regulatory changes of the Bank of Albania have given the main impact in the underestimation of the banking sector risks. Following these results, the paper proposes several ways to further improve the quality of forecasting. They relate mainly to the transition towards a dynamic stress test forecasting, the consolidation of the conservative trend for the forecasting of the regulatory capital and a preliminary assessment of the regulatory changes and their inclusion in the stress test.

References

Blaschke, W., Jones, M., Majnoni, G., Martines Peria, S.: Stress testing of financial systems: an overview of issues, methodologies and FSAP experiences. IMF Working Paper (2001)

Geršl, A., Seidler, J.: Conservative stress testing: the role of regular verification. Working Paper, Czech National Bank (2010)

Haldane, A.G.: Why Banks Failed the Stress Test, Speech given at the Marcus Evans Conference on Stress-Testing, 9–10 February 2009

Jorion, Ph.: Risk management lessons from the credit crisis. In: European Financial Management (2009)

Zoltan, M.J., Kota, V., Dushku, E.: Macro econometric model of Albania: a follow up. In: 7th Conference of the Bank of Albania, Monetary Policy Strategies for Small Economies (2007)

Macro Determinants of Real Exchange Rates: Albanian Case

Ermira Kalaj[1] and Ela Golemi[2(✉)]

[1] Department of Finance and Accounting,
University of Shkodra "Luigj Gurakuqi", Shkodër, Albania
ekalaj@unishk.edu.al

[2] Economics Department, University "Aleksandër Moisiu" Durrës University,
Durrës, Albania
golemiela31@yahoo.com

Abstract. This paper presents an effort to empirically analyze the factors affecting the real exchange rate in Albania, using data for the period 1995–2015. Real exchange rate behavior is at the centre of policy debates because exchange rates play a fundamental role in global trading and portfolio investments. This study applies the VAR model. Johansen Cointegration technique is applied to find long run relationship among the variables.

The data are obtained from the Bank of Albania, World Development Indicators (WDI) published by the World Bank and the International Financial Statistics (IFS) published by the International Monetary Fund. In our analysis we make use of macroeconomic variables such as; budget deficit, the volume of money flows, the net foreign assets, gross domestic product (GDP), and the oil prices.

According to the results of research, this paper suggests that the central bank can decrease the real exchange rate fluctuations more than volume of money flow and inflation by decreasing monetary policies and increasing fiscal policies. Study findings indicate significant and long run relationship between real exchange rate and trade.

Keywords: Real exchange rate · VAR model · Monetary policy

JEL Classification: C2 · C8 · E52

1 Introduction

Over the last two years, in Albanian economy has been observed a domestic currency appreciation in relation to the currencies of the main partner countries. The nominal effective exchange rate (NEER) was overvalued in annual terms, on average by 3.2% in 2016 and by 4.2% in the first six months of 2017. The appreciation of the lek against the euro has been the main contributor to this development by its very heavy weight (about 80%) that this currency occupies in the basket of major currencies.

The rapid exchange rate appreciation is an uncommon event in the history of Albanian financial markets and may potentially have implications for monetary policy. For this purpose, this paper aims to contextualise the real exchange rate behaviour in

E. G. Popkova and B. S. Sergi (Eds.): ISC 2019, LNNS 91, pp. 608–614, 2020.
https://doi.org/10.1007/978-3-030-32015-7_68

Albania, explain the possible factors that have influenced it, and evaluate its implications in Albanian economy.

The paper is structured in four sections: (i) in the first section, a background is presented; (ii) the second section analyzes the literature review; (iii) in the third section presents the methodology and data; and (iv) the last section presents the conclusions.

2 Background

The exchange rate is an important variable of an economy, its levels and fluctuations affect real and nominal macroeconomic indicators. The effect of exchange rate developments on inflation is significantly reduced when inflation stays low and shows low volatility – conclude Tanku et al. (2009). This result is confirmed by the updating of Istrefi and Semi (2007), which results that exchange rate developments are fully transmitted to inflation with a lag of up to 5 month and the coefficient of delivery is reduced from 70% to 23%. Based on these studies, short and medium-term forecasts, the exchange rate appreciation has had a direct downward impact from –0.2 to –0.3% points in January 2016 to May 2017. Indirect impact, through negative impact in the competitiveness of products and the lowest economic growth is estimated to be low, ranging from –0.04 to –0.05% points.

Economic theories on trade show that domestic currency appreciation increases the relative value of relevant exports by making them more expensive compared to competitors. Consequently, exports of goods and services would be adversely affected by currency appreciation. In the case of Albania, statistics do not show a reduction in export flows. On the contrary, both goods and services exports have grown at high rates both during 2016 and in the first quarter of 2017. This implies that any impact that may come from price appreciation is quite low compared with the effects of other factors - foreign demand, foreign prices and local supply factors. In Fig. 1 we can notice the fluctuations of real GDP and CPI for the period 2008 to 2017

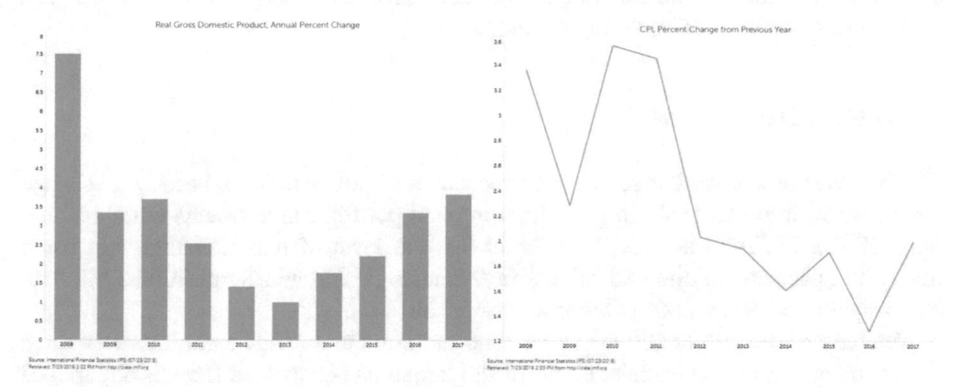

Fig. 1. Fluctuations of Real GDP and CPI

There are some practical arguments that support the low exchange rate impact on exports. The first argument relates to the structure of export of goods when such branches dominate such as “Fuel”, “Minerals”, “Metals” and “Textiles”. In all of these branches, the dynamics of international demand and foreign prices offered are the main influencing axis of exports. Moreover, major providers enjoy dominant positions. Albania’s exports are concentrated in a few destinations and in no case would our country pose a threat to the dominant position of large exporters. Secondly, in order that competitively channel to work, buyers need to increase the demand for domestic currency to buy albanian exports. But trade in these segments is dominated by currency transactions (Euros or Dollars). This also reflects on the high growth of business deposits, particularly in dollars, over the last quarters at a time when mineral trade with China expanded considerably. So in the absence of the need to convert and in the conditions when domestic exporters are generally “receiving price”, the impact of the export rate would be negligible. In the case of imports, the exchange rate channel makes foreign goods relatively cheaper, thus increasing demand for the latter. Even in this case, the impact of the rate on demand would be considered negligible, given that the domestic economy’s ability to replace imports is estimated to be relatively low. On the other hand, domestic currency appreciation for importing businesses is a positive element given that they already need a lower domestic currency to buy foreign currency. Thus, positive aspects would be related to those importing businesses operating in the domestic market with transactions in Lek.

Exchange rate developments play an important role in explaining the behaviour of demand for money. Generally, studies based on the methods of the co-integrating relationship find that exchange rate depreciation is perceived to lower the demand for money in the economy (Shijaku 2016; Sergi 2003).

We can conclude that, exchange rate appreciation is estimated to have had a declining impact on inflation, but, in line with empirical estimates, this impact is low in low inflation conditions. On the other hand, Albanian exports have marked an increase in the pursuit of price increases for core and improved foreign demand. However, counter factual analysis would help us to estimate how much exports would have been in the event of a lack of such appreciation.

3 Literature Review

The behavior of real exchange has been the center of policy debates because exchange rates play an important role in global trading and portfolio investments (Adekola and Sergi 2016). Different studies have found that the level of real exchange rate has a relevant impact on exports and private investments (Hsing and Sergi 2009a, b, 2010; Naghshpour and Sergi 2009; Onour and Sergi 2011).

Moore and Pentecost (2006) using long-run structural VAR technique examined the factors affecting the exchange rate of the Indian Rupee against the US dollar. The paper results showed that the real exchange rate is non-stationary and the real shocks have permanent effects on the exchange rate, making their management risky and difficult and possibly harmful to the economy.

The long-run behaviour equilibrium real exchange rate in Nigeria was investigated by using a vector error correction model (ECM) by Rano (2009). The econometric investigation starts by analyzing the stochastic properties of the data and subsequently estimates a vector error correction model. Regression results show that most of the long-run behavior of the real exchange rate could be explained by terms of trade, index of crude oil volatility, index of monetary policy performance and government fiscal stance. The results further suggest that deviations from the equilibrium path are eliminated within one to two years. Large inflow of oil revenues into the country and stable macroeconomic performance, for instance, were discovered to account for undervaluation of the real exchange rate.

Structural VAR model was used by Khan et al. (2010) to investigate the sources of real exchange rate fluctuation in Pakistan. Structural decomposition shows that about sixty percent of the variance in forecasting the real foreign exchange rate at a sphere of four quarter is due to nominal shocks in macroeconomics variables. The finding gives some empirical support to sticky price model (Dornbusch 1976) of exchange rate.

Chan et al. (2011) explore in their study the mean reversion behaviour of three Japanese real exchange rates during January 1980 to January 2010. The CPI- and PPI-based real yen/USD rates and real effective yen rates are examined using newly improved unit root tests allowing for endogenous breaks in the linear and non-linear manner. They identify structural breaks in 1985 and 1997/98, but the results were mostly against the PPP hypothesis. The exchange rate misalignment is somewhat less evident after the Plaza Accord 1985, and stronger evidence for PPP is found in the post-1999 period.

4 Methodology and Data

In our analysis we use the following macroeconomic variables to estimate the real exchange rate equation:

$$LnRE = \beta_! + \beta_! LnBD + \beta_! LnM + \beta_! LnNFA + \beta_! LnY + \beta_! LnOP + \varepsilon$$

Where: LnRE is the natural logarithm of real exchange rate,

BD is the budget deficit, M is the money flows, NFA represent the net foreign assets, Y is the Gross Domestic Product, and OP is the oil price level. Moreover, in this study we use annual data for the period of 1995 to 2015. The data are obtained from the Bank of Albania, World Development Indicators (WDI), and the International Financial Statistics (IFS) published by the IMF.

To investigate the response of macroeconomic variables we use an unrestricted vector autoregressive model (VAR). The Vector Autoregression (VAR) model is one of the most flexible and easy to use models for the analysis of multivariate time series. It is a natural extension of the uni-variate autoregressive model to dynamic multivariate time series. The VAR model has proven to be especially useful for describing the dynamic behavior of economic and financial time series and for forecasting. It often provides superior forecasts to those from uni-variate time series models and elaborate theory-based simultaneous equations models. Forecasting from VAR models are quite flexible because they can be made conditional on the potential future paths of specified variables in the model. Johansen Cointegration technique is applied to find long run relationship among the variables.

In order to specify the VAR, test for unit roots and co-integration are conducted, we check the unit root by using Augmented Dickey-Fuller test (ADF). The test indicated in Table 1 shows that the null hypothesis of the unit root cannot be rejected for all the variables. The first differences are confirmed to be stationary.

Table 1. Results of ADF unit root test

Variables	Level	First difference
LnRE	5.72***	–2.36
LnBD	–3.40**	–2.41
LnM	–2.64*	–2.02
LnY	2.57*	–1.88
LnNFA	3.52**	–2.06
LnOP	–3.95***	–1.87

Note *, ** and *** denotes 10%, 5%, and 1% significance level respectively.

After this we check for the co-integration by using the Johansen test to specify long-run equilibrium relationship between proper variables. Also the variance decomposition is conducted. Long-run relationship between variables was estimated and presented in the form of normalized co-integration coefficients as shown in Table 2.

Table 2. Johansen co-integration test of the real exchange rate

Variables	Coefficient	Standard error
LnRE	1	–
LnBD	0.0021	0.0001
LnM	0.84	0.081
LnY	–0.88	0.162
LnNFA	0.59	0.03
LnOP	–0.64	0.073
Coefficient	7.5	–

Note *, ** and *** denotes 10%, 5%, and 1% significance level respectively.

The results of Table 2 show that there is a positive relationship among budget deficit, volume of money flow, and the real exchange rate but a negative relationship between GDP, net foreign assets, and the oil price. All the findings are in line with the theoretical principles and similar empirical analyses in other countries.

5 Conclusions

In this study we analyzed the factors affecting the real exchange rate in Albania for the period during 1995 to 2015 using a VAR model. We also conducted a Johansen co-integration test in order to confirm the convergence between variables. The test confirms the long-run equilibrium relationship between the chosen variables.

According to the empirical results, this paper suggests that the Central Bank can decrease the real exchange rate fluctuations more than the volume of money flow and inflation by decreasing monetary policies and increasing fiscal policies when oil price increase. The fiscal policy plays an important role in the fluctuations of exchange rate by paying attention to the amount of revenues and reducing redundant costs is necessary to reduce or prevent constant budget deficits. The government can decrease the real exchange rate by adopting taxes on import of goods. Consequently, this will direct to increase in domestic production and gross domestic product.

References

Adekola, A., Sergi, B.S.: Global Business Management: A Cross-Cultural Perspective. Routledge, New York (2016)

Chan, C., Lee-Lee, C., Wooi, C.: Japan-US real exchange rate behaviour. Evidence from linear and non-linear endogenous break tests. Asian Acad. Manag. J. Account. Financ. 95–109 (2011)

Dornbusch, R.: The theory of flexible exchange rate regimes and macroeconomic policy. Scand. J. Econ. **78**, 255–275 (1976)

Haw, C., Lee, C., Wooi, C.: Japan-US real exchange rate behaviour. Evidence from linear and non-linear endogenous break tests. Asian Acad. Manag. J. Account. Financ. **7**, 95–109 (2011)

Hsing, Y., Sergi, B.S.: The dollar/euro exchange rate and a comparison of major models. J. Bus. Econ. Manag. **10**(3), 199–205 (2009a)

Hsing, Yu., Sergi, B.S.: Analysis of the CZK/USD exchange rate: a comparison of four major models. Int. J. Monetary Econ. Financ. **2**(2), 91–102 (2009b)

Hsing, Yu., Sergi, B.S.: Test of the bilateral trade J-curve between the USA and Australia, Canada, New Zealand and the UK. Int. J. Trade Glob. Markets **3**(2), 189–198 (2010)

Istrefi, K., Semi, V.: Exchange rate pass-through in Albania. Bank of Albania, Working Paper (2010)

Khan, M.L., Alamgir, Q., Sulaiman, M.: The source of real exchange rate fluctuation in Pakistan. Eur. J. Soc. Sci. I **14**, 32–43 (2007)

Moore, T., Pentecost, E.: The source of real exchange rate fluctuation in India. J. Indian Econ. Rev. I **41**, 9–23 (2006)

Naghshpour, S., Sergi, B.S.: World trade indicators and a new approach to measure globalisation and countries' openness. Int. J. Trade Glob. Mark. **2**(1), 1–24 (2009)

Onour, I.A., Sergi, B.S.: Modeling and forecasting volatility in the global food commodity prices. Agric. Econ. **57**(3), 132–139 (2011)

Rano, S.: Real Exchange Rate Misalignment: An Application of Behavioral Equilibrium Exchange Rate in Nigeria. Social Science Research Network (2009)

Sergi, B.S.: Economic Dynamics in Transitional Economies: The Four-P Governments, the EU Enlargement, and the Bruxelles Consensus. Routledge, New York (2003)

Shijaku, G.: The Role of Money as an Important Pillar of Monetary Policy - Albanian Case. Bank of Albania, Working Paper (2016)

Tanku, A., Gjermeni, M., Vika, I.: The role of exchange rate in an IT framework, what do we do? Bank of Albania (2009)

Analysis of the Effects of Macroprudential Measures on GDP's Trend – Simulation Using a Macro Financial Model for Albania

Ela Golemi(✉)

Department of Economic Sciences, University "Aleksandër Moisiu", 2000 Durrës, Albania
golemiela31@yahoo.com

Abstract. This study provides an assessment of the impact of macroprudential policy measures taken from the Bank of Albania, on the main financial indicators and real economy's dynamics, as well as their impact in raising the resilience of financial system and its stability. Based in Albania's financial system composition, the level of market development and the quality of data, this study finds appropriate to make use of a Macro Financial Model for Albania to assess the effects of countercyclical macro prudential measures taken by the Bank of Albania, as a toolkit to address credit revival.

Our analyses support that all measures implemented individually improve the main financial variables and affect positively Albania's GDP growth, although the impact of the simultaneous implementation of these three measures is higher. The implementation of macroprudential policy measures can help contribute to a stable financial intermediation by raising the resilience of the financial system against risks.

Keywords: Macroprudential policy · Systemic risk · Financial stability

JEL Classification: C81 · E5 · G38

1 Introduction

Experience from the last global crisis shows that an internal shock can deepen through the procyclical behavior of institutions and individuals and spread to the real economy and across borders. So, the debate among researchers has been focused on identifying systemic risks and developing an appropriate response known as "macroprudential policy" - a framework of high-end and intermediate objectives and of relevant tools (mainly with prudential nature) to address the risks that threaten the stability of the entire financial system. It is understood as the ability to adopt prudential measures for addressing systemic risk.

Given the significant correlation among the financial system and the real economy, whereby the destabilization of the financial system leads to the stagnation of the real economy and, in turn, to further destabilization of the financial system (Mishkin 2008); and given the important role of macroprudential policy to prevent further financial crisis and/or to reduce the impacts of a crisis, meaning at the same time to prevent

E. G. Popkova and B. S. Sergi (Eds.): ISC 2019, LNNS 91, pp. 615–623, 2020.
https://doi.org/10.1007/978-3-030-32015-7_69

and/or reduce the large costs on the public budget; this paper sought to focus on and explore this issue, by raising *the research question*: *To what extend Macroprudential policy measures affect the real economy dynamics?*

Conventional macro stress testing fails to fully capture the interaction between the financial system and the real economy, assessing only the impact of a slowdown in the real economy on the financial system without taking into account the negative feedback loop. This research emphasizes the importance of the feedback effects and in order to evaluate the impact of financial regulations, such as macroprudential measures in Albania, it uses a macrofinancial model that incorporates the interrelation between the financial sector and the macroeconomic sector.

2 Literature Review

Regarding the macroprudential tools, the discussion is also on-going. Galati and Moessner (2011) point out that there have been investigated a range of possible macroprudential measures, without identifying a primary instrument or a standard taxonomy of instruments. Weistroffer (2012) state that macroprudential tools (measures) are mainly derivations of microprudential tools that incorporate a system-wide perspective. In addition, one has to consider other macroeconomic tools that support financial stability. In fact, Borio and Shim (2007), and Caruana (2010), argue that prudential policies are not enough to achieve financial stability and that fiscal and monetary policies can help to mitigate the build-up of financial imbalances (Fig 1).

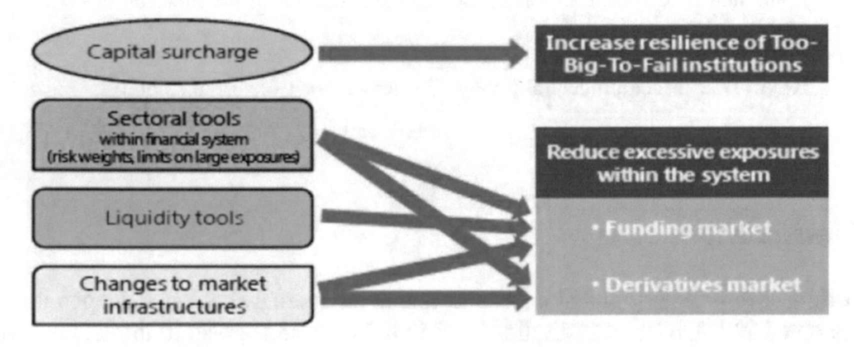

Fig. 1. Mapping tools to objective: structural dimension. Source: IMF

Macroprudential measures can be classified in various ways, which can also be overlapping (Galati and Moessner 2011). One important distinction among them is linked with the two dimensions of the systemic risk, that is its time dimension and cross-sectional dimension. Some of the macroprudential tools linked with the time dimension feature, capturing the evolution of risk over time and targeting its procyclicality, include the countercyclical capital requirements, forward-looking statistical provisioning, practices related with valuation of collateral and maximum loan-to-value (LTV) ratios. Shin (2009) finds an important contribution of countercyclical capital requirements for banks, in moderating the fluctuations in their leverage and size of balance sheet. Discussing the loan- loss provisioning, various authors have noticed its pro-cyclical behavior, being lower at times of credit booms and rising at times of distress (Borio et al. 2001). Hence, referring to the case of Spain, Shin (2009) finds that forward-looking statistical provisioning, through its direct impact on capital, can reduce the lending ability of the bank during the capital buoyancy. Some of the macroprudential tools linked with the cross-sectional dimension focus on systemic risk arising by similar or common exposures arising from banks' balance sheet interlinkages. Galati and Moessner (2011) find out that those measures target the bank's capital and/or the amount of short-term debt in relation to bank's total liabilities. These vulnerabilities spillover to the rest of the system through credit chains, payment and settlement systems or bank runs which are triggered also by the asymmetric information and the inability to distinguish solvent from insolvent institutions (Galati and Moessner 2011). More specific macroprudential tools in this case, are those known as *net stable funding ratio and liquidity coverage ratio* (BIS, BCBS 2010), targeting the maturity structure of banks' balance sheets.

Another distinction of macroprudential tools is whether they are applied based on rules or discretion (Borio and Shim 2007). By making an analogy to monetary policymaking, rule-based macroprudential tools can offer accountability, transparency and efficacy (Galati and Moessner 2011). On the other hand, discretion-based tools can prove to be time- inconsistent. Referring to the work of Goodhart (2004), Galati and Moessner (2011) find that loan loss provisions, capital requirements and surcharges, or loan-to-value ratios can be designed in a rule-based way. As examples of discretionary tools Galati and Moessner (2011) mention supervisory reviews or warnings, in the form of speeches or reports targeting the build-up of risk in the system.

Another distinction between macroprudential tools is whether they represent quantity or price restrictions. Examples of price restrictive tools are measures that act as a "tax" on variable margins, i.e. on the difference between liquid assets and short-term liabilities. Examples of quantity restrictive tools include the net funding ratio of a bank (BIS, BCBS 2009). Perotti and Suarez (2011) find that such tools may be used to target different incentives for risk creation. Their analysis suggests that combining "price" and "quantity" macroprudential tools may be desirable to better manage systemic risk externalities and control risk's appetite of banks. Galati and Moessner (2011) confirm that some studies make another classification of macroprudential tools, in the context of industrial or emerging market countries. Interestingly, they find that some emerging market countries have been using macroprudential tools, without calling them by this name (McCauley, 2009, as referred by Galati and Moessner 2011).

Khandokar.I; Apostolos.S argue that the leverage of financial intermediaries not only helps to price financial assets but is an important macroeconomic variable with a potential to affect macroeconomic conditions. This suggests that monetary policymakers should be considering the procyclical nature of leverage in order to have an accurate assessment of the effect of their policies on the macroeconomy. In particular, they should be monitoring the balance sheets of broker-dealers and other financial intermediaries and avoid contractionary monetary policy in times of low leverage and expansionary monetary policy in times of high leverage.

3 Data and Methodology

To analyze the impact on the macro economy landscape, of using macro prudential policy measures to directly affect the financial system, it is necessary to use a model that incorporates the feedback loop between the financial sector and real economy. The Macro Financial Model (MFM) from Dushku and Kota (2012) that it is used in this study is a small and medium-sized structural model, comprising two sectors, a financial sector introducing mainly by banking sector in Albania, and some macroeconomic variables. The Model focuses the banks' soundness in the Albanian financial system. To these banks, the Model provides a quantitative framework for assessing the transition mechanism of different shocks to banks' balance sheets, taking in consideration the macro-credit risk, the interaction between banks and feedback loop displayed in two sides of balance sheet (assets & liabilities). *The MFM is a model that explicitly incorporates the feedback loop between the financial sector and the real economy* in Albania. Through this mechanism, it allows to know how the banks act to macro-prudential measures and how this shock is transmitted in real economy through GDP trends. The Macro Financial Model (MFM) has in total 49 financial and macroeconomic variables. The MFM emphasizes the importance of financial activities, where 40 variables being included in the financial sector, and 9 variables are included in macroeconomic sector.

Among total 35 equations of the model, eight are behavioral equations and the rest are identities equations. Estimation of the equations is based on regressions with fix effects, to account for the dynamic relationship at individual bank level, using the quarterly annualized growth rates as the main variables and we have paid attention to enter all variables as stationary variables in all behavioral equations. All the dates are quarterly from 2002T1–2014T3. The estimated equations are:

- household and corporate lending volume equations,
- lending interest rate equation,
- net interest income equation,
- credit cost equation,
- credit risk equation,
- portfolio risk (or non-performing loans) equations for households and business

Using statistical software E-views 7.2, with panel data, observing banks several time it is analyzed the linear relationship between endogenous variables and explanatory variables, or exogenous variables. A general approximation of a multiple linear regression for banks i = 1, 2, 3 …, N, who is observed at several time periods t = 1, 2, 3…, N is given as below:

$$\mathbf{Y_{it}} = \alpha + \mathbf{x}'_{it}\beta + c_i + u_{it}$$

Where:

Y_{it} is the dependent variable,

x'_{it} is a K-dimensional row vector of explanatory variables excluding the constant;

α is the intercept;

β is a K-dimensional column vector of parameters;

c_i is an individual-specific effect and u_{it} is an idiosyncratic error term.

The linear regression is estimated based on the so-called balanced bank **i**, in all times period **t**. The **T** observations for individual **i** can be summarized as follows:

$$y_i = \begin{bmatrix} y_{i1} \\ y_{it} \\ y_{iT} \end{bmatrix}_{T\times 1} \quad X_i = \begin{bmatrix} X'_{i1} \\ X'_{it} \\ X'_{iT} \end{bmatrix}_{T\times K} \quad u_i = \begin{bmatrix} u_{i1} \\ u_{it} \\ u_{iT} \end{bmatrix}_{T\times 1}$$

NT observations for all banks and time periods are presented as:

$$Y = \begin{bmatrix} y_i \\ y_i \\ y_N \end{bmatrix}_{NT\times 1} \quad X = \begin{bmatrix} X_1 \\ X_i \\ X_N \end{bmatrix}_{T\times K} \quad u_i = \begin{bmatrix} u_1 \\ u_i \\ u_N \end{bmatrix}_{NT\times 1}$$

Data generation process (DGP) is described by linearity and independence, while idiosyncratic error term $\mathbf{u_{it}}$ is assumed uncorrelated with the explanatory variables of the same individual. There are chosen to estimate fixed versus random effect equations, to see how the main relationships variables vary across individuals at the same point in time, and possibly over time for all banks all together. Due to the lower (cross-section) banks number than the number of the period we use in the model, we have been oriented towards fixed- effects regressions, by not considering GMM models.

4 Simulations and Results

Regarding the macroprudential measures taken from the Bank of Albania, through the Macro Financial Model for Albania, this paper analyses the impact that three instruments from the package of macroprudential measures have on the main financial and real indicators, and the impact of the three measures used jointly. The simulations analysis is the way to evaluate the performance of these measures, by observing the reactions of all endogenous variables and feedback loop between financial and macroeconomic sectors. Generation of the baseline[1] - current level of endogenous variables, determined according to the assessed equations and the connections provided in the model, with assumption that exogenous variables have a determined value and behavior and all other exogenous shocks are equal to zero. The simulations results for the entire banking system are given as the differences between the simulation results and the baseline, expressed as a percentage or in base points. As shocks results are taken within the sample, the deviations of scenarios from the model baseline bears also their current behavior during the period of assessing the equations in the model (Fig. 2).

[1] The baseline is the behavior of variables when no policy measures are used.

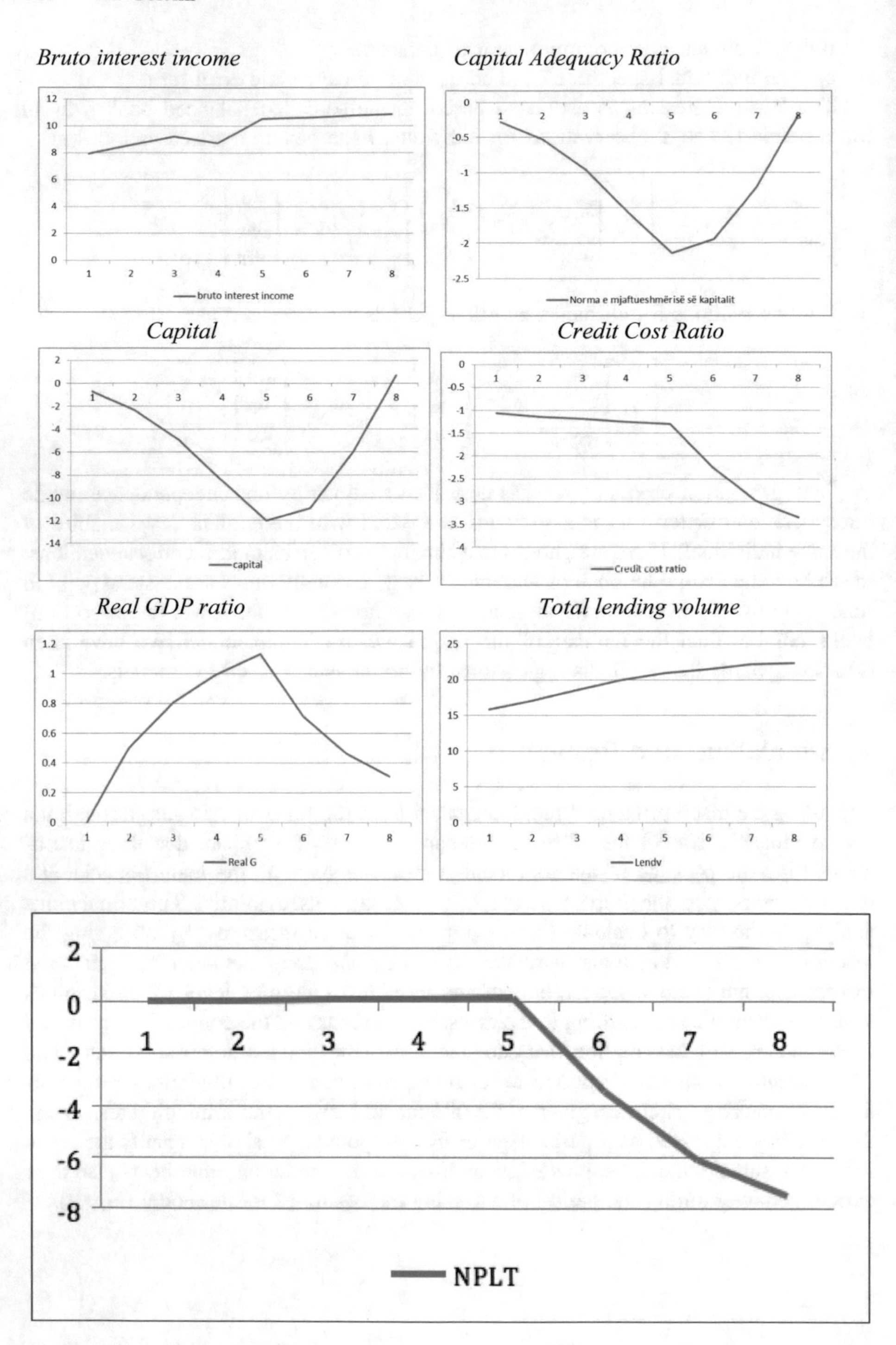

Fig. 2. Results of the Scenario of measures' combination

Assumptions:
Combination of:

- increase of total credit stock by 10% for a period of two years;
- general reduce of regulatory liquidity indicator by 5%;
- increase of provision by 10%, from credit restructuring in regular categories

The impact of a combination of all three measures causes a significant rise of lending volume by an average of 17.84% during first year and 21.73% during the second year. Credit cost ratio is reduced by an average of 1.8 percentage points during eight quarters and capital adequacy ratio is shrunk by an average of 0.84 pp and 1.57 pp during first and second year respectively. Impact on the GDP is on average 0.7 pp during second year.

5 Discussion of the Results

The increase of total lending by 4% and 10% causes the improvement of NPL rate by an average of 0.81 and 1.54 percentage points; the increase of the Real GDP rate respectively by an average of 0.22 and 0.42 percentage points during eight quarters; and slight increase of capital adequacy ratio by an average of 0.049 percentage points in case of credit growth by 10% for two years.

The general decrease by 5% of the regulatory liquidity indicator, as per banks' risk profile will be associated with improvement of the NPL ratio by an average of −0.79 pp during two years; slight decrease of Capital Adequacy ratio by an average of −0.24 pp; and improvement of real GDP rate of 0.3 pp during eight quarters.

The increase of 10% of provisions for credit restructuring when it is considered as a good credit – does not have any impact on real GDP ratio. This measure serves mainly to prevent further deterioration of credit quality.

The combination of all three measures impact positively Real GDP growth by an average of 0.62 pp during eight quarters; Lending volume and Total assets by an average of 19.7% and 9.64% respectively. CAR decreases on average by 1.10 pp during eight quarters.

6 Conclusion

As most of the theories conclude, a primary objective of macroprudential policy measures is to increase the resistance of the financial system to shocks and ensure financial stability. Given the significant interaction between the financial sector and the macroeconomic one, this study emphasizes the high economic costs of financial instability. The package of macroprudential measures undertaken by the Bank of Albania, beyond addressing the considerable slowdown in crediting and the worsening quality of credits given, considered increasing the resilience of the financial system to shocks from different grounds, that is eliminating risk from financial crisis.

As a general conclusion, a single macroprudential policy measure has a slight positive affect on financial and economic variables, but their role is not ideal. While using multiple macroprudential policy measures is a better alternative, because of their significant positive impact on main financial and economic variables and the ability to maintain the efficiency of policy measures by responding to the multiple source of risks.

The performance of NPL and CAR indicators shows that the package of measures addressed also the resistance of financial system to shocks, the parameters of NPL and CAR stand in good levels compared to their current respectively regulatory thresholds.

References

Adekola, A., Sergi, B.S.: Global Business Management: A Cross-Cultural Perspective. Routledge, New York (2016)

Apostolos, S., Dennis, N.: International Monetary Policy Spillovers. International Monetary Policy Spillovers, Working Papers 2018-06

BIS, BCBS.: Basel III: International Framework for Liquidity Risk Measurement, Standards and Monitoring, Bank for International Settlements, pp. 1–53, December 2010

Bogoev, J., Petrevski, G., Sergi, B.S.: Reducing inflation in ex-communist economies independent central banks versus financial sector development. Probl. Post-Communism **59** (4), 38–55 (2012)

Borio, C., Furfine, C., Lowe, P.: Procyclicality of the financial system and financial stability: Issues and Policy Operations, BIS Working Papers, no.1, pp. 1–57, March 2001

Borio, C., Shim, I.: What Can (Macro-) Prudential Policy Do To Support Monetary Policy, BIS Working Paper no. 242, pp. 1–44, December 2007

Caruana, J.: Macroprudential Policy: Working Toward A New Consensus, Remarks at the high level meeting on "The Emerging Framework for Financial Regulation and Monetary Policy" Washington D.C., pp. 1–6, 23 April 2010

Dushku, E., Kota, V.: Macro Financial Model in Albania: Approach towards panel data. Materials for Discussion, Bank of Albania (2012)

Goodhart, C.A.E.: Some New Directions for Financial Stability? The Per Jacobsson Lecture, Zürich, Switzerland, 27 June 2004

Khandokar, I., Apostolos, S.: Monetary Policy and Leverage Shocks. Int. J. Finan. Econ. 23 November 2016

Masood, O., Sergi, B.S.: China's banking system, market structure, and competitive conditions. Front. Econ. China **6**(1), 22–35 (2011)

Matousek, R., Sergi, B.S.: Management of non-performing loans in eastern Europe. J. East-West Bus. **11**(1/2), 141–166 (2005)

Matousek, R., Dasci, S., Sergi, B.S.: The efficiency of the Turkish banking system during 2000–2005. Int. J. Econ. Pol. Emerg. Econ. **1**(4), 341–355 (2008)

McCauley, R.: Macroprudential policy in emerging markets. Paper presented at the Central Bank of Nigeria's 50th Anniversary International Conference on "Central banking, financial system stability and growth", Abuja, 4–9 May 2009

Mishkin, F.S.: Monetary Policy Flexibility, Risk Management, and Financial Disruptions Speech at the Federal Reserve Bank of New York, NY (2008)

Perotti, E., Suarez, J.: A Pegovian Approach to Liquidity Regulation, Duisenberg School of Finance - Tinbergen Institute, Discussion Paper, pp. 1–32, February 2011

Petrevski, G., Bogoev, J., Sergi, B.S.: The link between central bank independence and inflation in central and eastern Europe: are the results sensitive to endogeneity issue omitted dynamics and subjectivity bias? J. Post Keynesian Econ. **34**(4), 611–651 (2012)

Sergi, B.S., Qerimi, Q.: The Political Economy of Southeast Europe from 1990 to the Present: Challenges and Opportunities. Continuum, New York (2008)

Sergi, B.S.: A new index of independence of 12 European national central banks: the 1980s and early 1990s. J. Transnatl. Manag. Dev. **5**(2), 41–57 (2000)

Sergi, B.S.: Economic Dynamics in Transitional Economies: The Four-P Governments, the EU Enlargement, and the Bruxelles Consensus. Routledge, New York (2003)

Sergi, B.S., Bagatelas, W.A., Kubicova, J. (eds.) Industries and Markets in Central and Eastern Europe. Ashgate Publishing (2012)

Shin, H.S.: Financial Intermediation and the Post-Crisis Financial System, Prepared for 8th BIS Annual Conference, pp. 1–32, 25–26 June 2009

Weistroffer, C.: Macroprudential supervision. In: Search of an Appropriate Response to Systemic Risk, Deutche Bank Research, Current Issues-Global Financial Markets, pp. 1–20, May 2012

Agricultural Lease as a Perspective Mechanism of Development of Infrastructure of Entrepreneurship in the Agricultural Machinery Market

Tatiana N. Litvinova(✉)

Volgograd State Technical University, Volgograd, Russia
litvinova1358@yandex.ru

Abstract. Purpose: The article seeks the goal of determining the current role and importance of agricultural lease in the structure of financial mechanisms of development of infrastructure of entrepreneurship in the agricultural machinery market in modern Russia and developing the perspective directions of improving the practice of application of the mechanism of agricultural lease for development of infrastructure of entrepreneurship in the agricultural machinery market.

Design/methodology/approach: For determining the role and importance of agricultural lease in the structure of financial mechanisms of development of infrastructure of entrepreneurship in the agricultural machinery market in modern Russia the authors use the method of determined factor analysis. The factor model is the function of the volume of new fixed capital formation in agriculture, presented in the form of the sum of volume of financing of this process with the help of various mechanisms that are available in modern Russia.

Findings: It is determined that in modern Russia the agricultural lease is the least actively used mechanism of financing of new fixed capital formation (agricultural machinery) in agriculture, as the conditions of its provisions are least favorable as compared to other accessible mechanisms (agricultural crediting, state subsidizing, and own investments of agricultural companies). The perspective directions of improving the practice of application of the mechanism of agricultural lease for development of infrastructure of entrepreneurship in the agricultural machinery market is partnership with leasing companies, cooperation with leasing companies within clusters on production and distribution of agricultural machinery, and independent provision of agricultural lease services to agricultural companies by manufacturers and suppliers of agricultural machinery.

Originality/value: The improved mechanism of agricultural lease will allow for significant increase of domestic demand for agricultural machinery and better usage of production capacities of companies in the Russian market of agricultural machinery and will stimulate their further innovative development, thus stimulating practical implementation of the strategy of development of agricultural machine-building of Russia until 2030.

E. G. Popkova and B. S. Sergi (Eds.): ISC 2019, LNNS 91, pp. 624–630, 2020.
https://doi.org/10.1007/978-3-030-32015-7_70

Keywords: Agricultural lease · Financial mechanism · Development · Agricultural companies · Infrastructure of entrepreneurship · Market of agricultural machinery · Russia

JEL Code: D53 · D92 · H54 · Q12 · Q13 · O13

1 Introduction

The market of agricultural machinery in modern Russia has a contradiction that hinders the development of entrepreneurship and slows down the rate of strategic growth of this market. The essence of the contradiction is that the government sets to entrepreneurship in the market of agricultural machinery the requirements in the sphere of increase of the volume of production and manifestation of high innovative activity in the interests of import substitution and supply to the domestic market of the modern agricultural machinery, which is peculiar for high level of the global competitiveness and has innovative features that are valuable for consumers (agricultural companies). These requirements are established in the Strategy of development of agricultural machine building of Russia until 2030, adopted by the Decree dated July 7, 2017, No.1455-r. (Government of the Russian Federation 2019).

At the same time, consumers show low volume of effective demand – for innovative (including digital) and new agricultural machinery. According to the data of the Federal State Statistics Service of the Russian Federation (2019), in 2017 the level of ageing of the fixed funds in agriculture constituted 38.2% - i.e., it was rather high. Therefore, low demand for new agricultural machinery is explained not by absence of the need for it but by deficit of financial resources for its purchase – i.e., low payment capacity of agricultural companies in Russia.

As a result, the signs of crisis of entrepreneurship in the Russian market of agricultural machinery are observed. One of them is critically low level of usage of production capacities for a lot of types of the issued products – i.e., for agriculture (11.76%), machines for inter-row and row soil cultivation (21.18%), and self-moving harvesters (23.29%) (National Research Institute "Higher School of Economics", 2019). However, dependence on import of agricultural machinery in Russia reduces annually – its share in 2016 constituted 60%, and in 2017 – 46%.

Another sign of the crisis of entrepreneurship in the Russian market of agricultural machinery is the negative value of the production index in 2013 (−18%), in 2014 (−6.7%), and 2015 (−14.3). As a response to this tendency, entrepreneurship in the Russian market of agricultural machinery had to reduce prices for export of the products in 2013 by 7.3% and in 2015 by 1.87%. However, ratio of export of the Russian agricultural machinery to sales in the internal market in 2016 reduced to 5.3%, as compared to 5.6% in 2015 (National Research Institute "Higher School of Economics" 2019).

This shows that the possibilities of the export activities of the studied market's companies are limited and the domestic market is the key market for them. For solving this contradiction and stimulating the development of entrepreneurship in the Russian market of agricultural machinery it is necessary to improve the financial mechanisms of development of its infrastructure for expanding the possibilities of products; sales in the Russian economy. One of the most perspective of these mechanisms is agricultural lease. Its advantages are reduction of the tax load onto agricultural companies (due to its shift onto leasing companies), possibility to purchase the leased agricultural machinery, etc.

That's why the purpose of the paper is to determine the current role and importance of agricultural lease in the structure of financial mechanisms of development of infrastructure of entrepreneurship in the agricultural machinery market in modern Russia and to develop the perspective directions of improving the practice of application of the mechanism of agricultural lease for development of infrastructure of entrepreneurship in the agricultural machinery market.

2 Materials and Method

Fundamental and applied issues of development of infrastructure of entrepreneurship in the agricultural machinery market are studied in the works Bogoviz et al. (2019), Litvinova et al. (2016, 2017, 2019), Morozova et al. (2018), Popkova (2019), and Troyanskaya et al. (2017). The theoretical aspects and the existing practice of application of agricultural lease are discussed in the works De Castro (2017), Perthen-Palmisano and Jakl (2005), and Saqib and Zafar (2017).

However, the drawback of the existing works and publications on the topic is studying agricultural lease from the positions of agricultural companies. The beneficiaries of agricultural lease are companies that specialize in production of agricultural machinery, which interests are usually neglected or considered indirectly. Due to this, agricultural lease - as a mechanism of development of infrastructure of entrepreneurship in the agricultural machinery market – is studied insufficiently and requires independent research.

For determining the role and importance of agricultural lease in the structure of financial mechanisms of development of infrastructure of entrepreneurship in the agricultural machinery market in modern Russia, this work uses the method of determined factor analysis. The factor is model is the function of the volume of fixed capital formation in agriculture, which is presented in the form of the sum of the volume of financing of this process with the help of various mechanisms that are accessible in modern Russia:

$$TFfc_t = AL_t + AC_t + SS_t + OI_t \quad (1)$$

where TFfc – total cost volume of fixed capital formation (agricultural machinery) in agriculture;

AL – volume of agricultural lease;

AC – volume of agricultural crediting;

SS – volume of state subsidizing of a part of direct expenditures for creation and modernization of the objects of the agro-industrial complex;

OI – own investments of agricultural companies into fixed capital;

t – time period (calendar year).

Dynamics of the values of the indicators for formula (1) in Russia in 2016–2017 are shown in Table 1.

Table 1. Data for factor analysis.

Year	Structural mechanisms of financing of fixed capital formation (agricultural machinery) in agriculture				Total fixed capital formation (agricultural machinery) in agriculture, RUB million
	Agricultural lease, RUB million	Agricultural crediting, RUB million	State subsidizing, RUB million	Own investments of agricultural companies, RUB million	
	AL	AC	SS	OI	TFfc
2016	23,744.00	139,450.00	116,000.00	327,149.00	606,343.00
2017	37,230.00	231,010.00	155,000.00	234,552.00	657,792.00

Source: Compiled by the author based on Bank of Russia (2019), Ministry of Agriculture of the Russian Federation (2019), Expert Rating Agency (2019), Federal State Statistics Service (2019).

Let us use the factor analysis for determining the contribution of agricultural lease into growth of total fixed capital formation in agriculture (ΔTFfc(AL)) in Russia in 2017, as compared to 2016. For this we use the following formula:

$$\Delta TFfc(AL) = (AL_{2017} + AC_{2016} + SS_{2016} + OI_{2016}) - TFfc_{2016} \quad (2)$$

Let us determine growth rate of total fixed capital formation in agriculture depending on agricultural lease ($TP(TFfc_{AL})$) with the following formula:

$$TP(TFfc_{AL}) = \Delta TFfc(AL) * 100\%/(TFfc_{2017} - TFfc_{2016}) \quad (3)$$

The offered formulas (2) and (3) are standard for determined factor analysis and here only their variables are determined.

3 Results

Calculations by formulas (2) and (3) led to the following results:

ΔTFfc(AL) = (37,230 + 139,450 + 116,000 + 327,149)−606,343 = 13,486 RUB million

$TP(TFfc_{AL})$ = 619,829 * 100%/(657,792 − 606,343) = 26.21%.

This means that increase of the volume of agricultural lease by RUB 13,486 million in 2017, as compared to 2016, leads to increase of total fixed capital formation (agricultural machinery) in agriculture in Russia by 26.21%. This emphasizes the important role of agricultural lease in formation of financial infrastructure of entrepreneurship in the Russian market of agricultural machinery.

The share of agricultural lease in the structure of the mechanisms of financing of fixed capital formation (agricultural machinery) in agriculture in Russia is very small, constituting 5.66% in 2017 – though it grew by 1.5 times, as compared to 2016 (3.92%). The key barrier on the path of expansion of usage of agricultural lease in modern Russia is high complexity of obtaining it (requirements to leaseholder, financial guarantee, service maintenances, insurance of agricultural machinery which is the object of lease, etc.).

The barrier of agricultural crediting is lower due to state support, which agricultural lease does not receive. That's why the perspectives of improving the practice of application of the mechanism of agricultural lease for development of infrastructure of entrepreneurship in the agricultural machinery market are connected to implementation of own initiatives of the companies in this market in the three following (not necessarily alternative) directions.

1st direction: partnership with leasing companies for providing the agricultural lease services by more profitable terms. This envisages vertical integration with preservation of full economic independence of companies in the market of agricultural machinery and agricultural lease companies. Their partnership is implemented on the basis of an agreement, according to which purchase of agricultural machinery of the manufacturer/supplier is possible with the help of services of this agricultural lease company. Agricultural company does not select a supplier of agricultural lease services but uses the services of a partner of the manufacturer/supplier of agricultural machinery. This allows obtaining the "scale effect" by the agricultural lease company and reducing the cost and prices.

2nd direction: cooperation with leasing companies within clusters on production and distribution of agricultural machinery, which ensures provision of agricultural lease services by more profitable terms. In this case, an agricultural company applied for agricultural machinery to the cluster. There's no necessity for purchasing certain types of agricultural machinery from different manufacturers/suppliers or concluding a lot of separate agreements of agricultural lease (as within the first direction); it suffices to place or order in the cluster, receiving agricultural leasing services for all require agricultural machinery by profitable terms.

3rd direction: independent provision of agricultural leasing services by manufacturers and suppliers of agricultural machinery. Agricultural lease is an accompanying service, which is provided by the manufacturer/supplier of agricultural machinery. This envisages diversification of the companies' activities in the market of agricultural machinery and thus is the most complex direction – though in case of successful implementation it might increase the volume of sales, profit, and profitability of these companies.

4 Conclusion

As a result of the research, it is determined that in modern Russia agricultural lease is the least used mechanism of financing of fixed capital formation (agricultural machinery) in agriculture, as the conditions of its provision are least favorable as compared to other accessible mechanisms (agricultural crediting, state subsidizing, and own investments of agricultural companies).

The perspective directions of improving the practice of application of the mechanism of agricultural lease for development of infrastructure of entrepreneurship in the agricultural machinery market are partnership with leasing companies, cooperation with leasing companies within the clusters on production and distribution of agricultural machinery, and independent provision of agricultural leasing services by manufacturers and suppliers of agricultural machinery. Though the offered directions have their specifics, they provide the following advantages for companies in the market of agricultural machinery:

– reduction of cost of agricultural leasing services and increase of their accessibility for agricultural companies, which stimulates the increase of the volume of effective demand for agricultural machinery;
– expanded opportunities for usage of agricultural lease for selling innovative (including digital) agricultural machinery for increasing the consumers' awareness of its advantages and formation of sustainable demand.

Due to the above advantages, the improved mechanism of agricultural lease will allow for significant increase of domestic demand for agricultural machinery, ensure better usage of production capacities of companies in the Russian market of agricultural machinery, and stimulate their further innovative development, thus supporting the practical implementation of the Strategy of development of agricultural machine-building of Russia until 2030.

References

Bogoviz, A.V., Sandu, I.S., Demishkevich, G.M., Ryzhenkova, N.E.: Economic aspects of formation of organizational and economic mechanism of the innovational infrastructure of the EAEU countries' agro-industrial complex. In: Advances in Intelligent Systems and Computing, vol. 726, pp. 108–117 (2019)

De Castro, L.F.P.: Dimensions and logic of rural leasing in family agriculture: A case study | [Dimensões e lógicas do arrendamento rural na agricultura familiar: Um estudo de caso]. Revista em Agronegocio e Meio Ambiente, vol. 10(2), pp. 437–457 (2017)

Litvinova, T.N., Khmeleva, G.A., Ermolina, L.V., Alferova, T.V., Cheryomushkina, I.V.: Scenarios of business development in the agricultural machinery market under conditions of international trade integration. Contemp. Econ. **10**(4), 323–332 (2016)

Litvinova, T.N., Kulikova, E.S., Kuznetsov, V.P., Taranov, P.M.: Marketing as a determinant of the agricultural machinery market development. Contributions to Economics, (9783319606958), p. 465–47 (2017)

Litvinova, T.N., Tolmachev, A.V., Saenko, I.I., Iskandaryan, G.O.: Role and meaning of the ICT infrastructure for development of entrepreneurial activities in the russian agricultural machinery market. In: Advances in Intelligent Systems and Computing, vol. 726, pp. 793–799 (2019)

Morozova, I.A., Popkova, E.G., Litvinova, T.N.: Sustainable development of global entrepreneurship: infrastructure and perspectives. Int. Entrepreneurship Manag. J. 1–9 (2018)

Perthen-Palmisano, B., Jakl, T.: Chemical leasing: cooperative business models for sustainable chemicals management - summary of research projects commissioned by the austrian federal ministry of agriculture, forestry, environment and water management. Environ. Sci. Pollut. Res. **12**(1), 49–53 (2005)

Popkova, E.G.: Preconditions of formation and development of industry 4.0 in the conditions of knowledge economy. Studies in Systems, Decision and Control, vol. 169, pp. 65–72 (2019)
Saqib, L., Zafar, M.A.: A realistic approach to the concept of Ijārah (leasing) in the Sharī'ah and its possible role in the development of agricultural sector in Pakistan (development of ijārah based model for financing agriculture). Hamdard Islamicus, vol. 40(1), pp. 71–99 (2017)
Troyanskaya, M.A., Ostrovskiy, V.I., Litvinova, T.N., Matkovskaya, Y.S., Bogoviz, A.V.: Possibilities and perspectives for activation of sales in the agricultural machinery market within sectorial development of Russian and European economies. Contributions to Economics, (9783319606958), pp. 473–480 (2017)
Bank of Russia.: Statistical bulletin of the Bank of Russia, no. 9 (304) 2018 (2019). http://www.cbr.ru/Collection/Collection/File/7510/Bbs1809r.pdf. Accessed 11 May 2019
Ministry of Agriculture of the Russian Federation.: The main results of the work of the Ministry of Agriculture of the Russian Federation for 2017 (2019). http://government.ru/dep_news/32261/. Accessed 11 May 2019
National Research Institute "Higher School of Economics".: The agricultural machinery market of Russia – 2017 (2019). https://dcenter.hse.ru/data/2018/02/03/1163430452/Рынок%20agricultural%20машин%202017.pdf. Accessed 11 May 2019
Government of the Russian Federation.: Strategy of development of agricultural machine-building of Russia until 2030, adopted by the Decreee dated July 7, 2017, no.1455-p. (2019). http://government.ru/docs/28393/. Accessed 11 May 2019
Expert Rating Agency.: The leasing market as a result of 2017: upward movement – the volume and dynamics of the market of agricultural lease in Russia (2019). https://raexpert.ru/researches/leasing/2017/att1. Accessed 11 May 2019
Federal State Statistics Service.: Fixed capital: the level of wear of fixed capital and fixed capital formation (2019). http://www.gks.ru/wps/wcm/connect/rosstat_main/rosstat/ru/statistics/enterprise/fund/. Accessed 11 May 2019

State Support for Digital Logistics

Vera V. Borisova[1(✉)], Tamila S. Tasueva[2], and Bella K. Rakhimova[3]

[1] St. Petersburg State University of Economics, St. Petersburg, Russia
verabrsv@yandex.ru
[2] Grozny State Oil Technical University named after M.D. Millionshikov, Grozny, Russia
[3] Complex Research Institute named after K.I. Ibragimov of the RAS, Grozny, Russia

Abstract. The article is devoted to the issues of state support for digital transformation in logistics. The authors focus on the issues of strategic and tactical actions of the state that are aimed at increasing the efficiency of digital logistics. It is shown that preconditions for digital transformation of logistics are created at the sectorial and state levels. The purposes of state support for digital logistics are predetermined by the measures that are envisaged by the programs of strategic development of the country and regions, necessity for execution of state functions and responsibilities, and execution of international obligations by the government bodies. The task of the state is to take into account favorable preconditions of digitization and to create conditions for their implementation in the spheres and regions of the country. Thus, the problem of transformation of institutional environment digital logistics appeared. The authors substantiate the necessity for formation of digital logistics as an institutional structure, which performed the integrating and regulating role in the hybrid (virtual and real) world. Emergence of new institutes and transformation of the existing ones, predetermined by digital transformation of logistical activities, predetermined the topicality of this research.

Keywords: State support · Institutional environment · Cyber security · Strategies of digitization · Supply chain management · Digital logistics

JEL Code: M110

1 Introduction

Creation of a digital eco-system in Russia is implemented within the Federal projects of the national program "Digital economy of the Russian Federation". The Russian Federation is the initiator of digital projects that are aimed at provision of normative regulation of digital environment; creation of global competitive information infrastructure; training of skilled personnel for the digital economy; provision of digital security with the usage of domestic developments during transfer, processing, and storing of data; creation and implementation of digital "breakthrough" technologies and digitization of the system of state management (National program "Digital economy of the Russian Federation" 2019). The sense of the strategic national projects consists not in increasing

E. G. Popkova and B. S. Sergi (Eds.): ISC 2019, LNNS 91, pp. 631–638, 2020.
https://doi.org/10.1007/978-3-030-32015-7_71

the budget expenditures for digitization but reformation of the structure of a range of the spheres of economy, regulation, and implementation of new technologies.

Digitization becomes the key trend of transformations in the logistical sector of economy. Technological revolution leads to essential changes in the structure and contents of the logistical systems (Afanasenko and Borisova 2019). State, business, and society adapt to the conditions of a quickly changing digital environment. The digital platform is the basis and the core of transformations of the logistical sector of economy. The digital transport and logistics environment of the Russian Federation ensures creation of unique digital services for consumers of logistical services with application of innovative Russian technologies and software. Within the digital transport platform, previously separate information flows of the elements of the logistics systems are integrated, and transition to a completely new level of multi-modal logistics takes place. Russian leaders of the transport sphere and logistical operators united unto the association "Digital transport and logistics". The association's participants initiate the pilot projects that are aimed at Russia's integration into the international transport & logistics digital environment. Expert evaluations show the efficiency of this work in the form "state-business-society". A breakthrough in formation of the main transit space between Asia and Europe is based on international integration of supply chains and adoption of joint – together with European and Asian leaders – logistical decisions in the sphere of digitization of logistics. Stimulation of the processes of digital transformation of the transport and logistics segment by the state will allow increasing the transit capabilities of Russia at the global arena.

2 Methodology

The principles and methods of state support for digital logistics receive new scientific interpretations. This is caused by the changes of the market competitive environment.

New methods of competition include fighting for user data and market share. These methods are predetermined by increase of the dominating positions in the market of large technological platforms, which built business empires based on user data (Amazon, Apple, Facebook, Google, Microsoft). The national states are to assess the correctness of using the parsers – programs on processing and analysis of the required information according to a certain algorithm for access to the data of other companies and to develop the regulator of stopping the data by other users. New methods of competitive struggle include prioritization of search results during placement of information, limitations during dissemination of competing services, coordination of prices, and fix-ups during trading with the usage of software. It is necessary to develop standards and influence the behavior of participants of the digital environment.

Germany's experience in this direction is connected to limitation of the volume of data on German citizens that is collected by Facebook. To which level the regulation of digital environment between countries should be unified? The answer to this and other questions could be found in the national program "Digital economy of the Russian Federation", which includes six federal projects: "Normative regulation of digital environment", "Information infrastructure", "Personnel for digital economy", "Information security", "Digital technologies", and "Digital state management". The program is to be implemented in 2018–2024. The planned volume of financing of this national program in 2019–2024 will constitute RUB 1.8 trillion. Most of the finances – RUB 1 trillion – will be allocated from the federal budget.

The federal projects combine application of the methodology of strategic planning and project and platform management of digital transformations. Federal projects aim at regulation of the digital environment, creation of a flexible legal basis, unified regulations of digital document turnover, and digital operations. More than RUB 1.5 billion will be spent for these purposes by 2024. The strategic task is creation of the global information infrastructure that can transfer, store, and process large arrays of data on the basis of digital platforms with the usage of innovative equipment. The project "Information infrastructure" will require RUB 772 billion by 2024. Expenditures for training of personnel for the digital economy until 2024 will constitute RUB 143 billion.

The state pays special attention to information security of the national digital environment. The logistical infrastructure is directly connected to usage of the information and communication technologies. The current requirement is development of the Russian software, services, and apps that could oppose external threats and guarantee protection of cargo transportations, personal data, and payment systems from cyber criminals (Volkova and Titova 2018).

Thus, digital technologies should be developed based on national projects. For this, the Russian law in the sphere of digital technologies should be nationally-oriented, up to direct limitations of competition for foreigners in the sphere of IT. As of now, foreign manufacturers of IT in Russia are in a better position than Russian ones. At the same time, most Western countries have legally support the national developers of IT.

The law "Regarding information, information technologies, and protection of information" was adopted in Russia in 2015. There is a register of Russian programs for computers and data bases. There's a ban for purchasing foreign software during state purchases – if there's a Russian analog. When purchasing foreign software, governmental customers will have to report and prove the absence of analog of these products in the Russian market. For example, the Chinese market is fully closed for foreign manufacturers of information products (Kasperskaya 2019).

Creation of Russian breakthrough digital technologies of the international level was the basis of export-oriented development of the national sector of information and computer technologies. There's a task of increasing the volumes of export of telecommunication services, software products, information and communication equipment, and other information and communication products. For this, a digital platform, which stimulates the export-oriented development of the Russian IT sector, will be created. Also, creation of the intellectual transport and telecommunication systems and formation and keeping of the leading positions in the international transport and logistical corridors implements the national strategic task of "connection of the territory of the Russian Federation" (National program "Digital economy of the Russian Federation" 2019).

Interaction between the state and business in solving the tasks of digitization of logistics allows balancing the resources and stimulates the increase of the speed of the logistical cycle at the set level of reliability and quality of the supply system. New forms of cooperation, balance of interests, responsibility and constructive regulatory environment are the key conditions of development of digital logistics. Such approach to digital transformation of logistics allows involving the representatives of small and medium business in solving new tasks. By 2024, their number in the entrepreneurial activities of Russia should reach six million people. More than RUB 400 billion is allocated for this task (National program "Digital economy of the Russian Federation" 2019).

3 Results

Strategic settings of digital transformation of logistics are connected to growth of labor efficiency in the sphere. At present, this indicator in the Russian economy is by three times lower than in Germany and the USA (USD 26 per hour). The Russian logistical sector, being the "blood" system of commerce, allows implementing the reserves of growth of labor efficiency in the country's economy on the whole. According to the expert evaluations, by 2022 the digital technologies will replace 80 million human workers and will created more than 130 new jobs.

Increase of labor efficiency is the goal of a new project of the Russian railway sphere "Digital railroad". Transport platforms, storage robotized services, etc. are developing quickly. It should be noted that research reports and studies focus on the foreign experience of digitization of logistics. One of the examples of the Russian digital inventions is startup SYNGERA and its technological platform SIMPLATE. Working with large arrays of data (Big Data) raises flexibility of coordination of the supply chain by redirecting managerial decisions to software. At present, implementation and support for such digital systems are performed by large logistical operators, and very soon these systems will become accessible for most participants of the market (Borisova & Gordei).

Robotization of storage operations is conducted by the Russian company RoboCV, which focuses on production of fully automatized trucks and forklifts. This equipment could work autonomously with different categories of goods in the storage (Borisova et al. 2019).

New reality of digital logistics changes the nature of labor and becomes an impulse for emergence of completely new processes with application of unmanned cars, chat bots, and services that are aimed at increase of labor efficiency.

Intensity of product flows predetermines the necessity for expansion and reconfiguration of the main transport corridors. For example, the global logistics corridor Belt and Road, which was offered several years ago, is developing quickly. The sales volume within this commercial initiative increased by 14% in 2018. This tendency preserves, due to delayed demand from European and Asian consumers. On the path of product glows from China to Europe, Russia has a strategically important geographic position. Development of the Russian logistical potential is connected to increase of the export flows of the Russian goods in the international markets. At the same time, increase of transit attractiveness of Russia envisages regulatory measures in the part of interaction between participants of the logistical chains of goods' supply. It is necessary to reconsider the existing business models and to distribute tasks between people and digital technologies. This sets the necessity for preparing the economic subjects for the fundamental changes before the state and business.

Analysis of foreign experience of support for digital transformations in economy showed that for overcoming the modern challenges the governments' efforts are focused on complex implementation of digital technologies – the Internet of Things, system of mobile communications of the 5th generation (5G), Big Data, AI, robototronics, and blockchain. During development of the national programs of digitization, governments use a wide range of initiatives: supporting free Internet access; development of the telecommunication infrastructure; achievement of more even Internet coverage; improvement of qualification of digital personnel; stimulating the elimination of normative and trade barriers on the path of distribution of digital technologies, etc.

The leaders in this process are the USA and China. The volume of U.S. investments into digitization of economy grows very quickly and will reach USD 500 billion by 2025. The main goal is reduction of costs and increase of effectiveness of managing the state and private sectors of economy. The experts note that the share of the digital economy in the U.S. GDP constitutes 11%, and the rates of its annual growth since 2006–2016 were at the level of 6% (with growth level of other sectors of economy of 1.5% per year). Growth of value that is created by digital technologies in the USA will reach USD 1.6–2.2 by 2025. The information and communication technologies and services on their basis provide about 9% of export of goods and more than 24% of export of services (Borisova 2019).

Being the leader in export of information technologies and contents, the USA supports maximum openness of the Internet and limitation of requirements of certain countries as to localization of data storing, noting the fact that this could reduce competitiveness of American companies and commercial companies from any countries. On the one hand, this policy of the American government aims at democratization of the Internet space; on the other hand, local companies in countries with lower development of digital economy experience expansion from digital giants.

Another informal leader in the world of manufacturers of digital technologies with large potential of growth is China. In 2015, China adopted the strategy "Internet Plus", which in 2016 became a part of the five-year plan of China's socio-economic development. This strategy should integrate mobile and cloud technologies, IoT, and Big Data into the modern production and help Internet companies of China to enter the global market and achieve further development of e-commerce and online banking. China has selected the sectorial landmark of the digital strategy "Internet Plus". This strategy is aimed at digitization of the spheres of industry, agro-industrial production, finances, medicine, and state services. Development of China's digital economy is actively stimulated by the state in the part of motivation of local companies for usage of digital services and program products. According to the 13-year plan of development of the national science and technology of China, average expenditures for R&D per person should grow by 35% - up to USD 78,000 in 2020, which reduces the gap between China and the USA as to this indicator. As to the number of supercomputers, China has already exceeded the USA (Borisova 2019).

It should be noted that in 1990–2016, information & telecommunication equipment, technologies, and software were mainly imported in Russia. This was caused by the country's socio-economic course at becoming a part of the global economic system and destroying the previous economic relations.

Revival of the market of information and telecommunication technologies, program products, and information & technological solutions started in Russia in 2014 – after introduction of the US and European sanctions. Russia adopted the Program of liquidation of digital inequality, aimed at development of the Internet and provision of settlements of the country (14,000 with 5 million people) with broadband Internet.

The logistical sector of economy is presented as the key part of the national digital eco-system. Its competencies include "information support and current data bases, new technologies of management and modeling of business processes, digital document turnover, usage of the systems of scanning and bar-coding, radio frequency automatic identification of cargoes, and satellite communications and navigation systems which allow tracking cargo flows in real time" (Afanasenko and Borisova 2019).

Digital logistical projects of development of transport corridors and sea and air navigation with the help of AI are implemented with the tools of public-private partnership. For state institutes, these tools are a means that make the research and innovative policy more susceptible to the changing character of innovations and to social and global challenges. For business, partnership with the state will allow developing new markets and creating value through cooperation and joint production (Borisova et al. 2019).

Partnership relations between state and private business acquire the form of concessions, contract agreements, rent, and agreements on usage of the results of joint activities. Each of these forms is applicable for conducting joint digital projects, including within implementation of the policy of import substitution and proactive import substitution of information technologies.

If the Russian economy starts using foreign digital technologies in the economic practice, it will become dependent and vulnerable. "Russia depends on the Western information technologies, and if the Russian production and transport is controlled by AI that is developed by Google or Microsoft, if they will receive Big Data on the economy, nuclear power plants, citizens, and government establishments – we will become a digital colony of the USA" (Borisova 2019).

Solving the task of import substitution and proactive import substitution of information technologies guarantee full independence from foreign software by 2025. At present, "the list of the Russian software has more than 4,000 Russian software products that cover the necessary technological line: operational systems for services, PC's, and smartphones, office apps, graphic editors, systems of automatic design, information security software, games, search systems, etc." (Kasperskaya 2019).

Another aspect of development of digital technologies in logistics is connected to regulatory and legislative limitations. Experts agree that there's a need for preventing laws that would prevent the emergence of problems and risks. In the practical aspect, this could be implemented by starting experimental platforms, projects, and "legislative sandpits" for spheres and territories. These digital training grounds will have the conditions for implementing the procedure of quick feedback, when problems and risks that emerge in the sphere of new technologies lead to quick change of the law and constant setting of its regulation (Kasperskaya 2019).

The work group on normative regulation with the autonomous non-profit organization "Digital economy" is planning to approve the draft law on experimental legal regimes – regulatory sandpits. This document is aimed at creation of special regulatory regimes, in which the effect of the Russian laws will be limited for tests of practical application of digital innovations, verification of their usefulness in the conditions of refusal from limitations that are set by current laws without a risk of their violation. For example, "sandpits" could become training grounds for a breakthrough in application of unmanned transport and blockchain projects" (Information materials on the national program "Digital economy of the Russian Federation" 2019).

4 Conclusions

State support for digitization of logistics is implemented in the conditions of ongoing reforms. A lot of experts note that the system of state management is constantly transformed, being at a new stage of transformations (Volkova and Titova 2018).

It is possible to conclude that state support of essentially new products and services that are based on digital technologies provides a new impulse for development of intellectual logistical activities. The efforts of the Russian government are aimed at "formation of the national digital eco-system in which the data in the digital form are the key factor of production in all spheres of socio-economic activities and which provides effective interaction, including trans-border, business and scientific & educational community, government, and citizens; creation of necessary and sufficient efforts of the institutional and infrastructural character. There's a task of eliminating the existing obstacles and limitations for development of hi-tech businesses and preventing the obstacles and limitations in traditional and new spheres of economy and hi-tech markets" (National program "Digital economy of the Russian Federation" 2019).

The important condition of development of digital transformations in logistics is provision of confidence of all economic subjects that the collected, stored, and used data are protected. Such protection could be provided only by the government, by creating legal norms for fighting cybercrimes and motivating training of skilled cybercops, and developing technological solutions for data protection.

References

Afanasenko, I.D., Borisova, V.V.: Digital Logistics: Study Guide for Universities, p. 272. Piter, SPb (2019)

Borisova, V.V., Gordei, K.G.: Digital technologies warehouse logistics. In: Science of the XXI Century: Problems and Prospects of Researches, Warsaw, vol. 2, pp. 3–8 (2017)

Borisova, V., Taymashanov, K., Tasueva, T.: Digital warehousing as a leading logistics potential. In: International Conference (Sustainable Leadership for Entrepreneurs and Academics), Prague. Springer Proceedings in Business and Economics AG (2019)

Borisova, V.V.: State support for digital transformation in logistics. In: Modern Management: Problems and Perspectives: Collection of Articles as a Result of the 14th International Scientific and Practical Conference, 801 p. SPbSUE, SPb (2019)

Borisova, V.V., Bataev, D.K.-S., Tasueva, T.S.: Logistic agro-industrial cluster as a strategic tool for regional development. In: The International Scientific and Practical Conference "Contemporary Issues of Economic Development of Russia: Challenges and Opportunities", 12–13 December 2018, No: 53, pp. 492–499. Published by the Future Academy (2019)

Volkova, O.A., Titova, M.V.: The Russian Federation as a cyber state. Digital transformation with the method of platform management. Innovative economy: perspectives of development and improvement, No. 6 (32), pp. 66–73 (2018)
Information materials on the national program "Digital economy of the Russian Federation" (2019). http://static.government.ru/media/files/3b1AsVA1v3VziZip5VzAY8RTcLEbdCct.pdf. Accessed 12 Apr 2019
Kasperskaya, N.: Digital economy and risks of digital colonization. Theses of a speech at the Parliamentary hearings in the State Duma (2019). https://ivan4.ru/news/traditsionnye_semeynye_tsennosti/the_digital_economy_and_the_risks_of_digital_colonization_n_kasperskaya_developed_theses_of_the_spee/. Accessed 12 Apr 2019
National program "Digital economy of the Russian Federation" (2019). http://government.ru/rugovclassifier/614/events/. Accessed 08 Apr 2019
Rakhimova, B.K.: Regional logistical system of the agro-industrial complex, 198 p. KNII of RAS, Grozny (2017)
Tasueva, T.S., Rakhimova, B.K., Dagaeva, K.K.: Digital technologies of logistics in the agro-industrial complex. In: Improving the Methodology of Cognition for the Purpose of Development of Science: Collection of Articles of the International Scientific and Practical Conference, 28 October 2017, p. 1, 257 p. AETERNA, UFA (2017)

Basic Approaches to the Understanding of Cooperation and Corporation: Historical and Philosophical Aspect

Olga Bezgina(✉), Olga Evchenko, and Tatiana Ivanova

Togliatti State University, Togliatti, Russia
{bezgina, IvanovaT2005}@tltsu.ru, evchenko75@mail.ru

1 Introduction

In contemporary science, much prominence is given to the problems of the formation of organizations and their effective functioning. The processes and methods of organization administration are addressed in both foreign and Russian studies. This paper analyzes the main studies that contain historical, social philosophical and economic philosophical approaches to such concepts as cooperation and corporation. In this context, the works of such foreign authors as Marx, K., Tönnies, F., Durkheim, E. (Durkheim 1996; Tönnies 1994) are of great importance. From among Russian researchers, we should distinguish Mikhail Ivanovich Tugan-Baranovsky, Chayanov, A.V., Prokopovich, S.N. et al. (Tugan-Baranovsky 1989; Prokopovich 1919; Chayanov 2006).

For historical reasons, human activity requires unification. People used to unite not only within the framework of labor relations, but, for example, in order to stand against the forces of nature, to settle their ground, to organize their leisure. Thus, people who were united by the solidarity of goals, interests, and other factors entered into corporate, in a general sense, relations throughout the history of human development (corporate derives from Latin "corpore" - to relate, to interrelate, to unite). A family can also be treated as a kind of corporation, since it formally has all the characteristic features of the latter. For example, a corporation, like a family, "is theoretically formed on the principles of close internal unity and solidarity" (Apresian 2001). In addition, family members and corporations are "bound together by moral and legal responsibility", they have "collective values and standards of behavior" (Concise Philosophical Dictionary).

Having started in the XIX century, the active development of public relations led to the emergence of various forms of associations and mechanisms of their activity. At the same time, theoretical justifications of these processes appeared.

More often than not, corporation is only defined as an organization which unites individuals for the protection of their common interests, whereas cooperation is understood to mean partnership. However, in order to have a more complete grasp of phenomena determining these definitions, the analysis of the understanding of cooperation and corporation should be started from etymological and historical excursus.

E. G. Popkova and B. S. Sergi (Eds.): ISC 2019, LNNS 91, pp. 639–653, 2020.
https://doi.org/10.1007/978-3-030-32015-7_72

The analysis of the concept of cooperation has an important methodologic value in the study of history and theory of cooperative movement. The word "cooperation" derives from Latin (cooperation – partnership).

In a general sense, the word "cooperation" means a partnership of individuals, organizations or states in a particular activity for the achievement of common goals, implementation of common interests (cooperation of labor, cooperation of industrial enterprises, cooperation of science and production, interstate cooperation). In a more restricted sense, cooperation means public and business relations occurring in the course of joint activity and based on equality of rights, collectivism, mutual aid, and mutual benefit.

On the other hand, cooperation is a set of cooperatives functioning in different sectors of economy. A distinction is made between consumer, loan, industrial, agricultural and other cooperative organizations.

Theoretical apprehension of cooperation began in the 1st half of the XIX century in the countries of Western Europe. The concepts of cooperation and cooperative were introduced by the English economist, public figure and theorist of cooperation of socialism and industrialist Robert Owen at the beginning of the 1820s. The most generally accepted studies in the field of cooperation include the papers of B. Laverney in France, Marx, K. in Germany, Mikhail Ivanovich Tugan-Baranovsky, Chayanov, A. V., Prokopovich, S.N. et al. in Russia.

Marx, K. defined cooperation as a systematic partnership with others (Maslov 1928). He defined cooperation more extensively as "a form of labor in which many people systematically work side by side and interact with each other in the same process production or different unrelated production processes" (Marx et al. 1988).

In his paper "Capital", Marx studied the growing role of cooperation in the historical progress of society. The Marxist general methodological statement on cooperation comes down to the statement that "cooperation as a form of economic management and mutual relations between people in the process of activitics and voluntary association forms the basis of all organizations of society, only acquiring further specification in each of them" (Martynov, V.D.). This German theorist placed exceptional importance on cooperation as the most important basis of production and social life in general. Marx, K. positively assessed the role of cooperatives; he believed that cooperation, while not eliminating the causes of exploitation of workers, contributes to its certain reduction, creates the best labor conditions.

The domestic theoretic cooperative idea has started its development at the late XIX - early XX century with the development of the institute of cooperation in Russia. This process continued until the 30s of the XX century, when theoretical disagreements between scientists on this issue turned into a political struggle.

The concept of cooperation has been interpreted in different ways throughout almost two-hundred-year history of development of the cooperative movement. From the original meaning of this word, meaning a simple partnership, the concept of cooperation has been repeatedly modified, acquiring new characteristics and details. The formation of the theory and ideology of the Russian cooperation was influenced by various, sometimes exactly opposite ideas – conservatism and liberalism, populism and Marxism, Russian socialism and cooperativism. At the same time, the entire variety of theories can be conveniently classified into those essential categories which are borrowed by various authors for the definition of cooperation.

In Russia, the principle of cooperation has been interpreted as the subject of social philosophical analysis in papers of, for example, Ziber, N.I., who relied on Marx works in his interpretation of socialism as a cooperative regime. According to the American sociologist of Russian descent P. Sorokin, a cooperative is an organism that develops naturally, without compulsory collectivism. The abovementioned principle is reflected in papers of the sociologist, primarily in his idea of the priority of superorganic systems of values. Philosophical and sociological and scientific rationale for cooperation was formed by Kropotkin, P.A. According to Kropotkin, efficient cooperation is “a result of joint efforts exerted all social classes of the country” (Martynov, V.D.). Kropotkin believed that cooperation is the source of innovations and breakthroughs in technological development at the same time, provided that the rights of people in it are thoroughly protected, if they have been enforced from top to bottom since the beginning of self-management.

Consideration of cooperation as one of the levers in the process of modernization of socioeconomic relations provides us with the opportunity to present various points of view with regard to the definitions of concepts of cooperation and cooperative movement. It is apparent that since the beginning of its extensive use in Russia, cooperation acted not only as a business or commercial enterprise, but as a broad social movement with its inherent theory and ideology, based on classical principles – common availability, democracy and equality for all participants regardless of their property and social origin. While analyzing the nature of cooperation, the researchers found social functions, in addition to economical functions, in it.

Thus, one of the most prominent theorists of Russian conservatism Tikhomirov, L. A. considered the idea of cooperation, which had been stifled by individualism prior to that, as almost the only advantage of socialism. He believed that “society should constantly take care of the development of the internal popular organization and independent action, because it is impossible to help people who do not care about themselves” (Tikhomirov 1997).

A social ideal which underlies cooperative theories, is reflective of the aspiration of various theorists and practicians of the cooperative movement towards the common good, that is also expressed in the definition of cooperation. Petrashevskiy, M.V. defined cooperation, relying on Owen’s theory as a new form of organization of society (Petrashevskiy 1998). He pointed out that association built on the grounds he proposed, corresponds to the nature of a person that has developed in a civilized society, and it can organically develop, “not in the least being detrimental to the already established public relations” (Petrashevskiy, M.V.).

Many definitions of cooperation concur in their dedication to identify such essential principles serving as a basis for cooperative activity, as voluntariness, consciousness, openness, equality of rights, material interest, responsibility. Thus, already in the early XX century, Nikolayev, A.A. defined cooperation as “voluntary and self-managing union of individuals established with a view to achieving their common business goals and based on the principles of democracy and labor” (Prokopovich 1919).

By the early XX century, Western European and Russian theorists have conducted extensive historical and socioeconomic research of cooperation. Despite the fact that the approaches of scholars differed from each other, in general, they all noted that, although any cooperative organization amounts to association, in other words - a union, not every association amounts to cooperation. Thus, according to Laverney, B, not every voluntary association of individuals for the achievement of common goals may be referred to as a cooperative. Only such association may be regarded as cooperative which is aimed at "eliminating personal unearned revenues and hence chooses not to distribute its profits according to capitalistic principles" (Martynov, V.D.). However, in our opinion, such an approach to the understanding of cooperation and corporation is quite restricted.

Mikhail Ivanovich Tugan-Baranovsky was a renowned theorist of cooperation at the early XX century. Co-operation built on voluntary self-organization seemed to him a prototype of society, in the creation of which he saw the meaning of human history. The researcher generalized his theoretical views on cooperation in his book "Social foundations of cooperation". In it, he distinguished between the concepts of cooperative movement and cooperation as a business enterprise. The ideals of cooperative movement include socialistic commune, a new individual. In contrast, a real cooperative is established for economic benefit. The main determinant attribute of a cooperative, its key difference from a privately-owned factory consists in the absence of such goal of activities as realization of profit. In other words, a cooperative enterprise is established not for realization of profit on invested capital; on the contrary, "cooperative means such a business enterprise of several voluntarily united individuals which is intended not to obtain the maximum possible profit on invested capital, but delivery of some other benefits to its members by means of joint business activities" (Tugan-Baranovsky 1989).

In the conception of Mikhail Ivanovich Tugan-Baranovsky, the moral and ethical aspects of cooperation prevail. He believed that cooperation is the environment where collectivist principles prevail. "He matched the production collectivism based on individualism (which was aimed at realizing profit) against the organization of a consumer society in which collectivism expressed the principles of solidarism. He thought of cooperation as the basis of socialism, its organizational source and ideal, the prototype of the highest value of the future" (Chedurova 2007).

Tugan-Baranovsky, M.I. used to emphasize that cooperation is an association of workers. Drawing on the generalization of historical experience of development of cooperation in the countries of Western Europe and in Russia, he arrived at conclusion that the main features of true cooperation are self-sufficiency and independence. If these principles of organization of a cooperative enterprise are violated for some reason, cooperatives lose their essence.

Antsiferov, A.A. saw the essence of cooperation primarily in its democratic principles. He believed that a cooperative association is a "voluntary association of a group of individuals established with a view to achieving their common business goals and based on the principles of complete equality of rights of participants and self-management, in which every member takes direct personal part and is financially liable, while the profit made from business transactions is not intended for the return on invested capital" (Prokopovich 1919).

Prokopovich, S.N. laid emphasis on the social component of cooperation: "The main social peculiarity of a cooperative enterprise lies in the fact that it is a union of not capitals, but individuals having equal rights and obligations" (Prokopovich 1919). He formulated his own definition of "socioeconomic nature of cooperative association":

"This is a special form of collective enterprise established by a voluntary association of an unlimited number of individuals having equal rights and obligations in order to increase the productivity and profitability of their work, as well as to cheapen their private household; a capital that is needed for cooperative activity is paid by the market lending rate, while the entire net profit from entrepreneurial activities is distributed between the members of the association in proportion to their participation in the joint activity" (Prokopovich 1919). At the same time, the author added that this is about the "desirable, ideal nature of the cooperative movement".

A prominent practician and theorist of cooperation Chayanov, A.V. wrote: "If we strive to derive a single definitory formula for all types of cooperation, then we must include in it what is common to all industries and what can be taken out of the brackets" (Chayanov 2006).

Chayanov, A.V., paying particular attention to economic functions of cooperation, gave a very concise definition of it: "Cooperative is a part of economic activity of a particular group of individuals that is organized upon terms of collectivity and intended to serve the interests of this group alone" (Vakhitov 2005). Litoshenko, L.N. in his definition of cooperative pointed out that cooperation is the most efficient for small, small-scale enterprises: "Cooperation is a means of self-help for small-scale enterprises" (Litoshenko 1995).

Maslov, S.L. differentiated the concept of a simple partnership, which has existed since olden times, from the concept of cooperation. He believed that "cooperation as a public form of the union of labor efforts was born in the XIX century in countries that have progressed far away from the primitive forms of the economy" (Maslov 1922).

According to Brutskus, B., cooperation cannot be developed spontaneously. "It requires conscious activities for the unification of its members around specific goals. These goals appear from certain social relations; they color cooperative literature and make it biased" (Brutskus 1995). It appears that Brutskus, B. assumed that it was quite natural for the nature of cooperation to change in accordance with "certain public relations". Brutskus, B. wrote that "the interest of researchers in cooperation was related to their social appetences which left their mark on theoretical work" (Brutskus 1995).

Hence the definition of cooperation which was given by Brutskus, B.: "Cooperation is not an intrinsically originated community. This is a voluntary association established with a view to achieving certain business goals; they do not grow into such an association, but voluntarily enter it, and voluntarily leave it" (Brutskus 1995).

Literally from the outset, Soviet historiography began to draw a line between pre-revolutionary and Soviet cooperation. For example, the member of the editorial board of the newspaper Pravda and the member of the board of the Central Union of Consumer Societies (Tsentrosoyuz) Meshcheriakov, N.L. defined cooperation as follows: "If we mean by cooperation, not socialist communities, but only those consumer, productive, loan, etc. associations that could be commonly found towards the end of the XIX century and at the beginning of the XX century, it must be recognized that the cooperation is an exclusive phenomenon of capitalist regime. Such organizations did not exist before capitalism" (Meshcheriakov 1922).

An original view on the definition of this concept was developed in the Soviet era. "According to Lenin's theory, cooperation under the proletarian dictatorship should become some kind of a bridge between the proletariat and the peasantry, some kind of a path in which the proletariat can exercise its influence on the village" (Krotov and Plakitin 1924). Therefore, cooperation was no longer perceived as a means to improve the lives of its members, but became a means for solving the socio-economic and political problems of the country's leadership. The concept of cooperation has transformed, having divided into cooperation "under capitalism… and in the era of the development and evolution of the socialist society" (Dmitrienko et al. 1978). Socialist cooperation appears as "one of the forms of the class struggle of the proletariat" (Dmitrienko et al. 1978).

Besides, the prominent practicians and theorists of cooperation were forced to adjust their own views while defining this concept. In 1925, Chayanov, A.V. pointed out that certain authors, introducing cooperation into new sociopolitical conditions, "go even further and directly define cooperation as a transitional form towards the establishment of the socialist regime" (Chayanov 2006). In this regard, the definition of cooperation by the Soviet economist Pazhitnov, K.A., in which sociopolitical sense is clearly dominant, can be considered typical of the Soviet era: "The cooperative is such a voluntary union of several individuals which is aimed at exerting joint efforts for combating exploitation on the part of capital and gaining welfare of its members during production, exchange or distribution of economic benefits, that is, as producers, consumers or sellers of labor power" (Chayanov 2006).

Maslov, S.L. emphasized the transformation of the very concept of cooperation in accordance with the changing sociopolitical environment. Defining cooperation, he noted that cooperation in its "initial sense" has been existing long ago and in the literal translation of the term "cooperation" as such means partnership. In this regard, cooperation shall mean the technical partnership of the two or more individuals for the performance of any work" (Maslov 1922). "However, nowadays, when people talk about cooperation, it should be treated as the system of completely new relations, a new form of community-based economic organization. Contemporary cooperative organization constitutes an association of workers; secondly, this association and partnership is based on the free agreement, devoid of compulsory nature; the purpose of such an association consists in the satisfaction of economic interests of the working class which faces the economic conditions of the contemporary capitalist society" (Maslov 1922).

In point of fact, Soviet Encyclopedic Dictionary recognized dual interpretation of the concept of cooperation: "Under capitalism, cooperation is a collective privately-owned factory; under socialism, it is a form of association and involvement of common workers, primarily the peasantry, in socialist construction" (Prokhorov 1982).

Some changes in the assessments of the essence of cooperation are recorded in the Concise Political Dictionary, which was released in the midst of "perestroika". There is also a dual understanding of the nature of cooperation – in a capitalistic and socialistic environment, but the assessments are less categorical now: "In a capitalistic environment, K. develops in accordance with its economic laws, but confronts big business and is more democratic in nature. Under socialism, K. is an important link in the economic and political systems of society, and in the period of transition from capitalism to socialism it also acts as the main means of socialist transformation of small commodity production" (Onikov and Shishlin 1989). The bare fact that the assessment of the market law "On Cooperation" (1988) which, was actually the first law of its kind in the USSR, is still in the tideway of Lenin's ideas, is also indicative of inertness in the understanding of the essence of cooperation: "The Law on K. in the USSR (1988) determines the economic, social, organizational and legal conditions for the activities of cooperatives on the basis of Lenin's ideas about K. with regard to the modern stage of building of socialism in the USSR: voluntariness, economic self-sufficiency, self-management, inviolability of cooperative property, complete self-accounting, self-financing, balance between expenditures and income, consolidation of the role of economic interests" (Onikov and Shishlin 1989).

Having considered the formation of a theoretical study of cooperation, we shall turn to the contemporary interpretation of this concept. The interpretations of the term "cooperation" are usually compatible with each other in philosophical and sociological encyclopedias and dictionaries; quite the contrary - they complement each other, filling this concept with sense. We shall take the definition given in the dictionary "Russian Philosophy" as a basis.

"Cooperation - (lat. cooperatio - partnership) is a "form of organization of labor and, when broader defined - public life, which implies joint participation of individuals in the implementation of a particular activity" (Sapov 1995). Cooperation is sometimes interpreted as additional division of human labor; to be more exact, division and cooperation of activities constitute the aspects of the same process. Cooperation can be treated as any useful joint activity. It is a universal form which penetrates the whole economy with its division of labor, and, consequently, the need for cooperation. In this regard, any association can be called a cooperative - joint-stock companies, financial industrial groups, etc. "Its features are as follows: voluntary membership in a cooperative, electivity of governing bodies and gratuitousness of their work, democratic management (one member - one vote), the distribution of income in accordance with equity share in the work of the cooperative, localized area of activity, etc." (Martynov, V.D.).

Contemporary researchers, interpreting the concept of cooperation in different ways, kind of continue the controversy that was started more than 100 years ago. A well-known contemporary researcher Rogalina, N.L. defines cooperative movement as a "spontaneous economic movement stimulated by the real economic interest", hence the fact that "it responded sensitively and directly to changes" (Rogalina 2010). At the same time, Petrova, V.P. denies the very possibility of cooperation to change depending on sociopolitical factors: "The nature of cooperation cannot be capitalistic, socialistic or state capitalistic, as it was claimed in the literature. Cooperative organizations can be either cooperative or not" (Petrova 2004). Modern Economic Dictionary defines cooperation as an enterprise, an organization that was established by means of voluntary association of individuals on a share basis for the carrying out of entrepreneurial activities. Cooperatives are legal entities and operate on a footing of self-financing and self-management (Raizberg et al. 1996).

Saratov historian Konovalov, I.N. associates functioning of cooperation exclusively with the market-driven economy. "Cooperation is a part of the market system which has its own principles and functioning mechanism", – claims the author (Konovalov 1999).

Modern historiography interprets cooperation as a comprehensive phenomenon. The definitions vary from the broadest to those which identify only one of directions of cooperative activity as an essential attribute. Thus, Korelin, A.P., reflecting on fundamental principles of cooperation, defines it as "the efficient means of mobilization of internal forces for the revival and modernization of the economy, and… the means of mitigation of and powerful promotion of activity of the wider population…" (Korelin 2009). At the same time, the modern researcher of the history of cooperative movement Dianova, E.V. pays special attention to its cultural and educational activities in her definition of cooperation: "…cooperation means a partnership of individuals and community-based organizations in cooperative agitation and propaganda, joint activity in cultural and educational establishments for the achievement of socially important goals" (Dianova 2017).

The transformation of the definition of cooperation is also related to the subjects of research interest and to the fact what role was assigned to it in a particular phase of history. Researchers, identifying both stable and situationally specific features of cooperation, affirm the multifaceted nature of this phenomenon every time.

At the same time, the understanding of cooperation as a dualistic phenomenon remains unaltered. It's hard not to agree with Lubkov, A.V., who emphasizes the dual nature of the cooperative ideal: "On the one hand – it is the focus on the future without a clearly defined framework (hence the commonsensical utopianism); on the other hand – clear pragmatism, focus on solving specific problems of life (hence the fully justified conservatism). It is no wonder that the theory of Russian cooperation is very diversified and complex: in the meantime, the synthesis of ideas and concepts was deeply creative in nature (Lubkov 1998). As if to continue this thought, Korelin, A.P. pointed out that "the study of the Russian cooperative best practices is still one of the most important tasks both in scientific academic and practically political terms" (Korelin 2009).

The last statement is quite fair, since modern politicians do not remain uninvolved in the apprehension of the essence of cooperation. Thus, the leader of political party "Spravedlivaya Rossiya" Mironov, S., reflecting on the role of cooperation in modern economic life of the country, emphasizes its social function: "The most important function of cooperation is that this socially oriented form of economic management provides an opportunity to find the balance between economic interests of all its participants, increase efficiency of agricultural production. These processes contribute to solving the food problem, increasing the production of fast-moving consumer goods and expanding the services sector, improving labor and living conditions in the village" (Mironov 2005).

Let us turn to one more definition of the concept of cooperation given by Encyclopaedia Britannica: "cooperation in a special and technical sense means the association of a number of persons or societies for the mutual benefit or for the purposes of purchasing and distributing the items of consumption" (Encyclopedia Britannica). It should be noted that the concept of cooperation in this definition is very close to the concept of corporation.

The emergence of the term "corporation" is dated Late Antiquity (approximately 160 A.D.). According to Bandurin, A.V., this term derives from Latin "corpus habere", which denoted the rights of a legal personality; such rights were first recognized for private unions of Ancient Rome (Bandurin, A.V.). The corporations as such appeared as a phenomenon in the Middle Ages. At that time, a corporation was regarded as one of the types of numerous estate-professional organizations of the guild type. In the XIV-XV centuries, in many European cities, even the system of corporate political power was taking shape, when municipal public authorities were formed by the guilds.

The fundamentals that are universal for a corporate organization as such have emerged and developed in the medieval corporation: it is a union of individuals who follow a common interest; delegation of authority to a small group of their representatives; strict hierarchy of power; turning the general interest into a special interest of the highest ranks; value-conscious and rational type of behavior of the members of corporation. Certain systems of corporate views began to form as early as in the Middle Ages. Their most characteristic features were as follows: solidarity of individuals of the same profession, self-organization, strict regulation of rights and obligations of professional activity, etc.

Two traditions are most vividly distinguished in the contemporary understanding of the term "corporation": economic and social philosophical.

In economic terms, corporation is an organizational form of association. Ansoff, I. emphasizes this feature in his own definition: "corporation is a form of organization of entrepreneurial activities that is widely used in developed market economies, provides for equity ownership, legal status and concentration of management functions in holdfast of the top echelon of professional managers working as employed persons" (Ansoff 1999). In Russia, corporation usually means a large enterprise.

In philosophy and sociology, the concept of corporation is used to denote not a complex organization with an impersonal system of relations and rules of procedure, as it is seen by economy, but a certain stable system of interpersonal relations. Such interpretation can be found in papers of, for instance, F. Tönnies, E. Durkheim, and several other prominent sociologists. It should be noted however, that the concept of corporation is fairly rarely used in philosophy.

The concept of corporation can be repeatedly found in sociological papers of Ferdinand Tönnies. Sometimes in his works the term "corporation" appears in a historical and cultural context to designate real-life objects (for example, he regards top church hierarchs as corporations; he calls them "everlasting corporations") (Tönnies 1994). However, the concept of corporation acquires conceptual meaning more frequently. According to Tönnies, corporations are "such unions of individuals which act as a certain unity for the achievement of a goal (no matter what goal), able to express their will and to act. A form is referred to as corporation if it has an internal organization, which means that its members perform certain functions and their actions are actions of the corporation" (Osipov 1999). Therefore, a corporation, as interpreted by the sociologist, means ideal types of forms of social life, and not actual organizations similar to contemporary economic corporations.

The theme of the corporation is presented in the doctoral thesis, the first big book "The Division of Labor in Society" ("De la division du travail social") by Emile Durkheim. This sociologist shows us the relation between the division of labor and social solidarity, concluding that "the ideal brotherhood of humans can only be achieved to the extent to which the division of labor progresses" (Durkheim 1996). As a result, "the division of labor matches not individuals, but social functions against each other. However, the society is interested in the activities of the latter (individuals): it will be healthy or sick depending on whether they collaborate correctly or not. Therefore, its existence depends on them, and the more divided they are, the more closely it depends on them, which is why it cannot leave them in a state of uncertainty; and yet, they are determined by themselves. This is how these rules are formed, the number of which increases as labor is divided, and the absence of which makes organic solidarity either impossible or imperfect" (Durkheim 1996). Durkheim matches organic solidarity, which is inherent in society with advanced division of labor functions, against mechanical solidarity, which, according to him, can be found in primitive nations, where the division of labor is in its infancy ("If we use Durkheim's definitions, mechanical solidarity is a solidarity in consequence of similarity. When this form of solidarity is dominant in the society, individuals have little differences from each other. Being the members of the same team, they resemble each other because they have the same feelings, are committed to the same values, they recognize the same thing as sacred. Society is united, because individuals are not yet differentiated." An organic form of solidarity is born as a result of differentiation or is attributable to it. "The individuals here do not resemble each other; they are different, and to a certain extent a consensus is reached due to this difference. Durkheim refers to solidarity that is based on the differentiation of individuals as organic solidarity by analogy with the organs of a living being, each of which exercises its functions and is not similar to any other organ; however, all of them are equally important for life") (Aron, R.).

Durkheim most fundamentally addressed the issue of corporation in his book "Le Suicide". In this work, the sociologist seeks to understand the social origins of suicides; he argues that "there is the highest spiritual reality, namely the collective, above an individual" (Durkheim, E.). This statement provides sociologist an opportunity to primarily see social conditions at the heart of a suicide. It also brings Durkheim to the theoretical concept of overcoming the negative development of society, which actually causes the growth in the number of suicides. This concept is actually his concept of corporation.

Durkheim states that corporation consists of individuals who are engaged in the same work, who have common interests. Thus, corporation can inherently act as a collective body which is jealous about its autonomy and power over its members, "therefore there is no doubt that it can be a moral environment for them. There is no reason for corporate interest not to acquire that higher nature which social interest always has compared to private interests in any well-organized society" (Durkheim, E.).

According to the theory of the sociologist, a professional group, which is a corporation, has an advantage over all of others in the strengthening of social solidarity; the power of corporation constantly manifests itself: "since professional life is almost the whole life, the influence of the corporation makes itself felt in every detail of our classes which are thus directed towards a collective goal" (Durkheim, E.). Durkheim's conclusion lies in the fact that corporation has everything it needs to embrace an individual and withdraw it from the state of loneliness.

He emphasizes that a corporation will have influence only if it will be organized on completely different grounds. A corporation must cease to be a private group ignored by the state; it must become a recognized body of public life. "The researcher moves from the sphere of theory to the sphere of design of social reality, taking up the position that it is corporations that should be in charge of insurance offices, loan societies, and pension funds; it is corporations that would have to resolve frequently occurring professional conflicts, provide various categories of enterprises with appropriate conditions that contracts should meet to be valid, prevent the strong from exploiting the weak for the sake of common interests" (Lukov, VA). (Lukov, V.A.). It follows as a logical consequence that corporation means both the form of organization and various relations in it – economic (profit), coordination of various interests (professional and personal self-fulfillment, satisfaction of the need for communication, support of the team, etc.). "In order to apply and maintain each specialist discipline, a special body is needed. Who should constitute its team if not workers collaborating in the same function?" - asks sociologist rhetorically and claims that it is corporation that provides an opportunity to particularize moral standards. As the division of labor develops, the law and the morals take various forms in each particular function. Apart from rights and obligations that are common for all people, there are such rights and obligations that depend on the specifics of profession, and they grow in numbers, their significance grows too along with the development and variability of professional activity.

In the preface to the second edition of his book "The Division of Labor in Society" ("De la division du travail social"), Durkheim conceives a corporate regime where corporation acts as a single organization for the whole country, which differs from the old corporations, strictly communal by nature, the traditionalism of which constituted only one of aspects of communal traditionalism (Durkheim 1996). "The corporation of the future will have even more complex functions exactly due to its larger size. Other functions, that are now under the supervision of communes or private societies, will be grouped around its own professional functions... We can even put forward an assumption that a corporation is intended to become the foundation or one of the foundations of our political organization ... it can be foreseen that if the development continues in the same direction, it will have to take a central, dominant place in society" (Durkheim 1996). This is how the prospects for further development and the value of corporation for the society are perceived by Durkheim, E.

Thus, it may be said that a corporation, according to the sociologist, must exercise a wide range of public functions - from production to moral and cultural, develop and implement new forms that will regulate relations between people and promote personal enhancement.

Now we shall turn to modern interpretations of the term "corporation". The word "corporation" translated into Russian (English Corporation from Latin Corporiu – union, community) means a community, a union, a group of individuals united by a community of professional or class interests. The contemporary understanding of the term "corporation" is constantly evolving, acquiring new meanings. In this respect, a comparative analysis of two terms that can often be found in Roman law is interesting: "corpus" (body) and "universitas" (totality, community). Both terms were used in reference not only to the unions of individuals, but also a collective concept of items, things, united into a single whole, for example, a herd of draft cattle, a herd of horses, a house, a ship. However, the word "corpus" is more often used as a synonym for "collegium" (i.e. the union of individuals united by a common profession, cult et al.), and the word "universitas" is a common name for all corporations, i.e. both for communities and for boards.

In the broad, philosophical sense, "corporation is a special social institution, relatively closed association, which on certain conditions expresses the interests of its members and protects them" (Bakshtanovskiy and Sogomonov 2005). Besides, one may come across definitions in which any work collective is referred to as corporation (for example, Soviet labor collectives were corporations in point of fact, although they were not referred to as corporations). "According to the extensive version, corporation is characterized by a rather complex organization, focused on the achievement of a particular predetermined goal that requires coordination of the actions of the members of this organization, implementation of the management function and, hence, trained personnel. However, although a certain organizational effect is clearly expressed in corporations, not every organization and not any autonomous group can be referred to as corporation" (Bakshtanovskiy and Sogomonov 2005).

In the sociological sense, the concept of corporation has a more specific meaning – "it is a socioeconomic organization, an association of individuals who have provided their money or their labor for its establishment and existence and, therefore, having common goals and interests with regard to its development. That said, the presence of common interests does not mean that the interests of these groups are identical" (Beslikoeva 2004). In classical sociology, the concept of corporation is sometimes used to denote "not a complex organization with an impersonal system of relations and rules of procedure, as it is seen by a contemporary economist, but a certain stable system of interpersonal relations" (Lukov et al.). The concept of corporation has a high level of sociality, since it shows certain forms of relations between people arising in the process of public production and ensuring the integrity and stability of this process.

In contemporary legal literature, the word "corporation" is usually used to denote a legal entity, an organization. The term "corporation" is used in cases when it is necessary to emphasize that the organization is treated as a single whole and can act as the participant in civil transactions. In the legislation of the Russian Federation, the term "corporation" is only used as a component of the names of governmental commercial organizations (Great Law Dictionary).

All definitions of corporation given above are quite correct, but they do not fully express the essence of corporation. Therefore, we shall turn to the economic meaning of this term. In the work of Kochetkov, G.B. and Supian, V.B., specifically dealing with the analysis of economic corporations, the following definition is given: "corporation is a system of business organization in which the owners are isolated from the operational administration of economic functions which is entrusted to professional managers. The latter constitutes a fundamental feature which distinguishes the corporate form of business organization and management from any other form, existing in the modern economic paradigm" (Kochetkov and Supian 2005). We can quote one more definition, being a refinement definition: "Modern corporations are large enterprises and their associations, in which economic entities voluntarily joined their resources (finance, real property, information, labor, capacity for innovative risky activity et al.) in a consolidated capital and established a system of management on the basis of free division and cooperation of labor, a common strategy of entrepreneurship in order to realize a sustainable entrepreneurial profit" (Andronov 2003).

Having considered the various meanings of the term “corporation” presented above, we would like to emphasize some specific features of this term. In the strict sense, corporation means associations in which ethical regulation is established with all relevant infrastructure, codes, committees, etc. that is determined by the need to direct the behavior of a large number of individuals working in the same company, often absolutely unacquainted, in a single direction, to create a unified organizational culture. Such regulation is carried out by means of norms, codes that are coordinated and obligatory for all. In small teams, codes are non-mandatory, since everyone here knows each other personally, they can agree, the rules of personal morality and interpersonal relations apply here.

2 Conclusion

Hence, having analyzed historical, sociological and philosophical sources which present the basic approaches to the understanding of cooperation and corporation, we may conclude that, being a particular case of cooperation, corporation inherits the main features and peculiarities of cooperation. However, an important distinction should be noted as well. The primary goal of corporation is the realization of profit, while cooperation, on the contrary, is only aimed at gaining welfare of its members, while solving both economic and sociocultural problems at the same time.

References

Durkheim, E.: De la division du travail social. Kanon Publishing House (1996). Le suicide [electronic resource]. http://trushkovvv.narod.ru/library/samoubiistvo_str330.html

Tönnies, F.: Die Entwicklung der sozialen Frage. In: Dobrenkov, V.I., Belenkova, L.P. (eds.) Texts in the History of Sociology of XIX—XX Centuries: A Reading Book. Nauka Publishing House (1994)

Tugan-Baranovsky, M.I.: Social foundations of cooperation [reprint from the 3rd enlarged edition]. Ekonomika (1989)

Prokopovich, S.N.: Cooperative associations and their classification (1919)

Chayanov, A.V.: Main ideas and forms of organization of farming cooperation. Economical Heritage. TONCHU Publishing House (2006)

Apresian, R.G.: Corporatism. In: Apresian, R.G., Guseinov, A.A. (eds.) Ethics. Encyclopedic Dictionary, pp. 225–226. Gardariki Publishing House (2001)

Concise Philosophical Dictionary [electronic resource]. http://phenomen.ru/public/dictionary.php?article=892

Maslov, S.L.: Economic foundations of agricultural cooperation, p. 5 (1928)

Marx, K., Engels, F., Lenin, V.I.: About cooperation, pp. 35–36 (1988)

Martynov, V.D.: Кооперация. Great Soviet Encyclopedia [electronic resource]. http://slovari.yandex.ru/~PєPSPёPiPё/P‘PЎP/PљPsPsPïPµCЂP°C†PёCЏ/

Tikhomirov, L.A.: Socialism in terms of state and public life. A criticism of democracy, pp. 328, 345–347 (1997)

Petrashevskiy, M.V.: Owenism: Cooperation. Chapters of history: In 3 volumes, vol. 1. Selectas of Russian Economists, public figures, and cooperators-practicians: In 3 Volumes, vol. 1, 30–40 years of the XIX century - beginning of the XX century, p. 120. Nauka Publishing House (1998)
Petrashevskiy, M.V.: "Complete" and "absolute" reform of the social life. Cooperation. Chapters of history: In 3 volumes. Selectas of Russian Economists…, vol. 1, p. 120
Chedurova, E.M.: An outstanding theorist of Russian cooperation (2007)
Tugan-Baranovsky, M.I.: Personality in the History of Siberia of XVIII-XX centuries. A collection of biographical sketches, pp. 210–215. Sova Publishing House, Novosibirsk (2007) [electronic resource]. http://sibistorik.narod.ru/project/person/20.html
Vakhitov, K.I.: Cooperation: Theory, history, practice: Selected notabilia, facts, materials, and comments, p. 27. Publishing and Trading Corporation "Dashkov i K" (2005)
Litoshenko, L.N.: Cooperation, socialism and capitalism. Voprosy Ekonomiki, no. 10, p. 142 (1995)
Maslov, S.L.: Cooperation in a farm household, pp. 5–6, 10 (1922)
Brutskus, B.: Revisiting a theory of cooperation. Voprosy Ekonomiki, no. 10, p. 125 (1995)
Meshcheriakov, N.L.: The Problems of Modern Cooperation. Publishing House of the All-Russian Central Union of Consumer Societies, p. 14 (1922)
Krotov, P.G., Plakitin, M.P.: A thesis summary of cooperation. To the 30th anniversary of cooperation on the Samara-Zlatoust railway, Samara, p. 20 (1924)
Dmitrienko, V.P., Morozov, L.F., Pogudin, V.I.: Party and cooperation, p. 3 (1978)
Prokhorov, A.M. (ed.): Soviet Encyclopedic Dictionary. "Soviet Encyclopedia", p. 634 (1982)
Onikov, L.A., Shishlin, N.V. (eds.): Concise Political Dictionary, p. 254. Politizdat Publishing House (1989)
Sapov, V.V.: Cooperation. In: Maslin, M.A. (ed.) Russian Philosophy: a Dictionary. Respublika Publishing House (1995) [electronic resource]. http://terme.ru/dictionary/183/word/
Rogalina, N.L.: Power and agrarian reforms in Russia in the XX century. A learning guide. Encyclopedia of Russian villages, p. 76 (2010)
Petrova, V.P.: The history of agricultural cooperation in Ural (1917–1930). Synopsis of the thesis … of the Doctor of Historical Sciences, Tyumen, p. 30 (2004)
Raizberg, B.A., Lozovskiy, L.S., Starodubtseva, E.B.: Modern Economic Dictionary, p. 163 (1996)
Konovalov, I.N.: Agricultural cooperation in Russia at the end of XIX century – beginning of XX century (based on materials collected in governorates of Northern, Priuralsky and Volga regions): Thesis research … of the Doctor of Historical Sciences, Saratov, p. 40 (1999)
Korelin, A.P.: Cooperation and cooperative movement in Russia, 1860–1917. "Rossiyskaya Politicheskaya Entsiklopediya" (ROSSPEN), p. 5, 14 (2009)
Dianova, E.V.: Cultural and educational activities of cooperation of the European North in the first third of the XX century): Synopsis of the thesis … of the Doctor of Historical Sciences, St. Petersburg, p. 4 (2017)
Lubkov, A.V.: Cooperative movement in central regions of Russia, 1907–1918. Thesis research … of the Doctor of Historical Sciences, p. 6 (1998)
Mironov, S.: Promoting the development of cooperation. Ekonomika Selskogo Khoziaystva Rossii, no. 2, pp. 3–4 (2005)
Cooperative: Encyclopedia Britannica [electronic resource]. http://www.britannica.com/EBchecked/topic/136330/cooperative
Bandurin, A.V.: Historical and legislative analysis of the origin of corporations [electronic resource]. http://cfin.ru/bandurin/article/sbrn02/23.shtml
Ansoff, I.: A New Corporate Strategy, p. 145. Piter Publishing House, St. Petersburg (1999)

Osipov, G.V. (ed.): The history of sociology in Western Europe and the United States, p. 109. NORMA-INFRA Publishing Group (1999)

Aron, R.: Stages of development of sociological thought [electronic resource]. http://www.gumer.info/bibliotek_Buks/Sociolog/aron/07.php

Lukov, S.V.: Sociological interpretation of the corporation by Emile Durkheim [electronic resource]. http://www.mosgu.ru/nauchnaya/publications/SCIENTIFICARTICLES/2006/Lukov_S_V/

Bakshtanovskiy, V.I., Sogomonov, Y.V.: The ethics of the profession: Mission, Code, Act. A Monograph. Tyumen: Research Institute of Applied Ethics at Tyumen State Oil and Gas University, p. 80 (2005)

Beslikoeva, E.V.: Modern corporation: a sociological analysis of ownership, power and management: Thesis research … of the candidate of social sciences: 22.00.08, St. Petersburg (2004) [electronic resource]. http://www.lib.ua-ru.net/diss/cont/124855.html

Lukov, V.A., Lukov, S.V.: Interpretations of corporation in classical sociology [electronic resource]. http://www.mosgu.ru/nauchnaya/publications/SCIENTIFICARTICLES/2006/Lukov%20V.A.&Lukov%20S.V/

Refer to: Great Law Dictionary [electronic resource]. http://slovari.yandex.ru/~книги/Юридический%20словарь/Корпорация

Kochetkov, G.B., Supian, V.B.: Corporation: The American Model, pp. 9–10. Piter Publishing House, St. Petersburg (2005)

Andronov, V.V.: Corporate management in modern economic relations. In: Balabanov, V.S. (ed.) Science, p. 116. Russian Entrepreneurship Academy, ZAO “Izdatelstvo Ekonomika” (2003)

Analytical Procedures for Assessing the Risks of Introducing Innovative Technologies into the Organization's Activities

Maxim M. Kharlamov[1(✉)], Tatyana S. Kolmykova[2], Tatiana O. Tolstykh[3], Evgenia S. Nesenyuk[2], and Ekaterina P. Garina[4]

[1] Scientific-Research Test Centre of Cosmonauts Training named after Yuri Gagarin, Star City, Russia
kgtu_fk@list.r

[2] Southwestern State University, Kursk, Russia
t_kolmykova@mail.ru, kgtu_fk@list.ru

[3] National Research Technological University «MISiS», Moscow, Russia
tt400@mail.ru

[4] Minin Nizhny Novgorod State Pedagogical University, Nizhny Novgorod, Russia
e.p.garina@mail.ru

Abstract. Nowadays risks invariably accompany all socio-economic phenomena and processes. Their influence can be considered both from the point of view of the influence of external, environmental factors, and internal, connected with the specifics of the organization itself. Achieving the necessary effect of the risk management process involves timely and correct identification of risk factors and their evaluation. Evaluation and management of risk factors are important, as their untimely accounting and elimination can have a negative impact on the performance of research and developmental tests. Forecasting risk factors at early stages of the project implementation allow estimating the financial losses that can occur during the implementation of projects. Thus, the task of analyzing and assessing risks is relevant, allowing to compensate risks for the project cost and the terms of implementation.

The article explores modern approaches to risk management in the process of developing and implementing innovative technologies in the organization's activities. Forecasting risk factors in the early stages of design allow estimating the financial losses that may occur during the execution of projects. Thus, the task of analyzing and assessing risks is relevant, allowing to compensate risks for the project cost and the terms of implementation.

The process of risk analysis is proposed, which involves a complex of external and internal factors with the help of special economic and mathematical methods, which makes it possible to obtain an assessment of the level of risk and calculate the effectiveness of anti-risk measures. Application of the proposed approach makes it possible to reduce the likelihood of occurrence of a risk situation, and also provides an opportunity to find an administrative solution in the event of unfavorable factors.

E. G. Popkova and B. S. Sergi (Eds.): ISC 2019, LNNS 91, pp. 654–662, 2020.
https://doi.org/10.1007/978-3-030-32015-7_73

Keywords: Innovation management · Innovative technologies · Risk analysis · Organization risk management

1 Introduction

Crisis phenomena in the economy, the unstable political situation in the world, and sanctions pressure negatively affect the innovation activity of Russian organizations. In this case, we should talk about external risk factors, reducing the negative effect of these factors may be due to the organization's unique competencies that allow it to remain competitive for a long period of time. Internal risk factors, arising from the activities of the enterprise itself, on the contrary, are well predictable and are identified in the early stages of their formation. Let's join the opinion of scientists that the management of internal risk factors at the enterprises of the organization can be reduced to the timely removal of pain points in the internal activity of the enterprise [6, 15]. Accordingly, from the point of view of the effectiveness of the risk management process, it is precisely the analysis of risk factors that should be given special attention.

As part of this work, approaches to risk management are built in the process of developing and implementing innovative technologies, that is, the assessment and management of risk factors that may have a negative impact in the research and development of experimental designs. Forecasting risk factors at early stages of the project implementation allow estimating the financial losses that can occur during the implementation of projects. Thus, the task of analyzing and assessing risks is relevant, allowing to compensate risks for the project cost and the terms of implementation.

2 Theoretically the Basis of the Study

In order to define approaches to analysis and risk management during the development and implementation of innovative technologies, we will consider the approaches that exist in risk management. According to the widespread approach, the risk is understood as a certain phenomenon, leading with some probability to the onset of an event associated with losses. Specialists note that these are, as a rule, material, temporary or financial losses [3, 7, 11–13]. On an intuitive level, the concept of the risk of implementing innovative projects is due to the inability to fully, 100% predict the conditions in which the project will develop in the future. Thus, the risk is considered in relation to the future and is based on the construction of various kinds of forecasts and plans. The risk is influenced by those decisions that are made by the management of enterprises and integrated structures in general regarding the development process.

Mathematically, the concept of risk can be formalized as follows. The standard approach to probabilistic risk assessment is that two random variables are considered. The first random variable describes the probability of occurrence of the risk, and the second random variable describes the damage from the implementation of the risk situation. Thus, we will consider the following pair of random variables, which we call a probabilistic description of the risk

$$R = < \xi_R, \xi_L > \quad (1)$$

where ξ_R - a random variable describing the likelihood of risk situation at the considered moment, ξ_L - the random amount of economic damage in case of realization of this risk.

It is accepted to distinguish three concepts that are closely related to the formation of the notion of risk: uncertainty of the event; losses; indifference.

Let's consider each of these factors separately.

3 Research Methodology

The uncertainty factor of the event can be characterized as follows. Risk can exist only when there are several options for the development of an event. The loss factor entails, as a rule, an unintended reduction in value as a result of the realization of a negative event. Finally, the factor of non-indifferencc is characterized by the fact that the risk must cause damage to a certain subject of activity that opposes the possibility of an unfavorable event for him.

By origin in economic theory, the following types of risks stand out: net risk and speculative risks.

Net risk refers to unpredictable or unplanned losses without an alternative to a possible win. Such factors the organization can not restrict or change, they are external in nature. Net risks can be understood as risks of a natural and geographical nature (for example, the destruction of enterprise facilities as a result of a natural disaster), of a military nature (for example, entry of an enterprise into a military operation zone), of a political nature (for example, due to a sharp change in tax legislation) majeure circumstances. The consequences of such factors can not be avoided.

Net risk is characterized by the following concepts:

- objectivity;
- the possibility of undesirable consequences;
- the difficulty of measurement.

To assess the net risks and their characteristics, the methods of mathematical statistics and probability theory are usually used. A description of possible net risks is made using a risk map. It is a two-dimensional diagram reflecting the ranking of risks according to the degree of their dangerous impact on the basis of information about the probability of occurrence of losses and their magnitude.

There are general methods for managing net risks. According to the theory of risk management, the management of pure risks can occur in two ways:

- hedging (limiting) risks in order to reduce vulnerability to risks before occurrence of adverse events;
- adaptation as an effective response to new conditions and new information to mitigate the consequences.

Another class of risks is the speculative risk. In general, such a risk provides an opportunity not only to incur losses but also to derive some benefits from certain variants of the development of events. This risk depends largely on the management decisions of the organization's management and is explained by the current state of the economy. In order to manage such risks, a detailed analysis of economic activities is carried out and forecasts are made for various economic and production performance indicators.

The risk is closely related to another economic category - uncertainty. The state of uncertainty characterizes the economic environment and is its integral part. Uncertainty, according to a number of analysts, is a state of the environment in which it is completely unknown which event is most likely from a number of possible [4, 5, 14]. Thus, uncertainty is described by estimating the likelihood of possible outcomes. The formation of uncertainty occurs under the influence of a number of factors. Among them there are three key ones.

The first factor is temporary uncertainty. It is caused by the impossibility of absolutely reliably predicting the value of a parameter in the future. It is also impossible to establish reliably the effect of a parameter on the activities of an organization. Thus, forecasts always occur in a certain confidence interval.

Another factor is the uncertainty of the exact values of the parameters of the economic environment at the present time. At present, a huge number of economic and economic-mathematical methods and models for estimating various parameters have been developed. Different methods can lead to even opposite results. Much also depends on the interpretation of the results. In this way, estimates of the values of the parameters at the present moment of time are subjective, which makes a fairly large contribution to the formation of a state of uncertainty.

Finally, the third key factor is the unpredictability of behavior of participants in the economic activity. For his research, there is a developed mathematical apparatus of game theory and adaptation to this or that behavior of participants in economic activity can be performed on the basis of the corresponding game-theoretical models.

The risk arising from uncertainty is a probabilistic concept. Therefore, for its description, the concepts of probability theory and mathematical statistics are often used. Probabilistic methods also estimate the likelihood of adverse events. Thus, the risk is a characteristic of uncertainty that can be assessed.

Uncertainty, which can not be eliminated, is called residual in the economic literature.

There are three types of residual uncertainty:

Interval uncertainty. It is known that the uncertainty can take any value from a certain interval, but the distribution of the random variable within the interval is unknown. An example of interval uncertainty is shown in the Fig. 1, where the uncertainty takes on a value from the interval $[A;\ B]$.

Fig. 1. Interval uncertainty

– distribution of uncertainty according to a discrete law. In this case, the uncertainty is represented by a set of possible values of a random variable indicating the probability of their occurrence. An example of such a distribution is shown in Table 1. It shows the law of distribution of values of a a_i random variable X with an indication of the corresponding occurrence probability;

Table 1. Uncertainty distribution by discrete law

X	a_1	a_2	$\cdots$	a_n
p	p_1	p_2	$\cdots$	p_n

– some probability distribution of uncertainty. In this case, the distribution of the random variable is known, and the specific value that it takes is unknown. An example of interval uncertainty can serve as a forecast value of any economic indicator, for example *NPV*. At the same time, the boundaries are defined, in which the value of this indicator can vary. In the second form of residual uncertainty, for example, a discrete law of the probability distribution of the acceptance of *NPV* values from a particular set can be known.

The mean square deviation (scattering of a quantity relative to its mathematical expectation) is an important numerical characteristic of the probability distribution. This value characterizes the degree of dispersion around predicted value. Any economic or production indicator can serve as a random variable. The higher the value of the standard deviation, the greater the variance of the random variable relative to the central value (mathematical expectation).

The following classification of risks is based on their different origins. Depending on the origin of risks, different methods of their evaluation and management are applied. Depending on the origin due to uncertainties of various kinds, the risks can be divided into the groups described in Table 2.

Table 2. Classification of risks by origin

Risk level	Name of risk	Risk assessment	Risk management
0	Global risks that cannot be predicted	Cannot measure	Investing in flexibility and resource security
I	Strategic risks are the risks to which the organization is consciously implementing the strategy	Map of strategic risks	Identification of the main risks inherent in the strategy. Constant monitoring of strategic risks. Definition and execution of risk management strategies

(*continued*)

Table 2. (*continued*)

Risk level	Name of risk	Risk assessment	Risk management
II	Traditional risks, which are typical for all organizations (commercial, credit, operational, and liquidity risks)	Risk map, quantitative vulnerability assessments	Quality development of policies and procedures. Internal control system for following policies and procedures
III	Industry risks unique to enterprises of a particular industry	Risk map, quantitative vulnerability assessments	Quality development of policies and procedures. Internal control system for following policies and procedures

A source: compiled by the authors.

At the zero level, so-called global risks are located. They are also called "unknown known". These risks consist of the appearance of unique events that can not be predicted and which lead to great losses. Global risks are almost impossible to assess. An effective way to manage such risks is to invest in flexibility and resource security. Such management, while related to the expenditure of the budget of the organization, but leads to a reduction in losses in the event of risk. An example of global risk is the operation of international sanctions, which can be directed both against the entire economy of Russia and against specific enterprises or individuals.

At the first level, there are strategic risks. These risks include the rule known to the leadership of the organization factors, which it can not always quickly and effectively affect. Examples of such risks are the emergence of new competing enterprises (for example, the possible emergence of a competing science-intensive enterprise in China), technological change. Identification and measurement of such risks occur on the basis of a chart of strategic risks. For all its subjectivity, the map is a fairly effective means of formalizing the management of strategic risks. In addition to mapping, there is also a need for continuous monitoring of strategic risks, as well as the definition and implementation of risk management strategies.

At the second level are traditional risks. These risks are typical for all enterprises and organizations operating in the same economic conditions. These are, first of all, market risks, credit risks, and liquidity risks. The methods for assessing and managing such risks are similar for all enterprises. Differences can be due only to specific features of the industry or type of activity.

Risks of the third level are sectoral. Enterprises of different industries have different sectoral risks, which in general will not intersect in the methodological tools of their management (for example, specific risks of helicopter construction).

4 Analysis of Research Results

To make strategic decisions in terms of risk management, it is useful to consider several levels of residual uncertainty. Experts distinguish four levels of residual uncertainty in terms of practical significance:

– quite accurately predictable future;
– alternative options for the future;
– blurry future alternatives;
– full unpredictability.

With a fairly accurately predicted future, one forecast is usually developed. This forecast can be used as the basis for the strategic development of the enterprise (organization). This forecast is usually quite accurate since the residual uncertainty is very small. Forecasting, in this case, is carried out by classical methods and consists in researching prospective segments, analyzing the activities of competing enterprises, determining demand, and so on. The final forecast of the company's performance results will be quite in line with reality.

At the next level of residual uncertainty, alternative options for the future are located. In conditions of a significant level of residual uncertainty, it is not enough to have one single forecast. At the beginning of the forecast creation, the parameters influencing the achievement of the desired result are determined. Further on the basis of this analysis, several alternative models for the development of the situation are being developed. At the initial stage, it is impossible to establish which of the models for the development of the situation is reliable. However, it is possible to develop methods that allow you to calculate with a certain accuracy the probability of implementing a particular scenario. Having several alternative forecasts, it is possible in the process of economic activity to determine which of the development models is being realized.

The next option of residual uncertainty is blurred alternatives of the future. Unlike the simple case of alternative future variants, here it is necessary to set interval estimates of the values of variables in the implementation of each of the possible scenarios. Such a situation arises when the forecast tools can give a parameter value only in a certain interval. In such circumstances, it is much more difficult to formalize the process of making strategic decisions.

The most difficult is the situation of complete unpredictability. This situation characterizes the last level of residual uncertainty. In this case, a large scale of uncertainty does not allow us to determine the parameters on which the future depends to the greatest extent, to develop development models for different scenarios and to predict the ranges of possible outcomes. In place of a quantitative method of evaluation in this situation, qualitative methods come. To obtain estimates it is necessary to systematize all the available information, on the basis of which it will be possible to form some idea of the future. It is important in this situation to monitor economic signals that indicate positive or negative changes. The residual uncertainty of this type is quite rare and in the course of time shifts towards uncertainty of the first three types. Such a situation can be typical for pioneering projects of public-private partnership in science-intensive industries.

5 Conclusion

Hereby, the systematized traditional approaches to the identification of risks, their classification and identification of characteristics have made it possible to show the possibility of their application in assessing the risks of introducing innovative technologies. The main compensation parameters that allow neutralizing the negative consequences of risk situations include existing risk analysis technologies, for example, decision tree analysis, sensitivity, scenario method. Since the task of introducing innovative technologies is topical [1, 2, 8–10], it is worth noting a number of indicators characterizing the probability of occurrence of a risk situation, such as the standard deviation, the probability of an undesired event, the concept and indicator of VAR, the economic added value EVA, and risk map.

References

1. Edler, J., Fagerberg, J.: Innovation policy: what, why, and how. Oxford Rev. Econ. Policy **33**(1), 2–23 (2017)
2. Kolmykova, T., Merzlyakova, E., Bredikhin, V., Tolstykh, T., Ovchinnikova, O.: Problems of formation of perspective growth points of high-tech productions. In: Advances in Intelligent Systems and Computing (2018)
3. Kolmykova, T.S., Emelyanov, S.G., Merzlyakova, E.A.: Research of innovative potential of the region. J. Appl. Eng. Sci. **15**(3), 276–279 (2017)
4. Kuznetsov, V., Garina, E., Garin, A., Kornilov, D., Kolmykova, T.: A creative model of modern company management on the basis of semantic technologies. Commun. Comput. Inf. Sci. **754**, 163–176 (2017)
5. Kuznetsov, V., Kornilov, D., Kolmykova, T., Garina, E., Garin, A.: A creative model of modern company management on the basis of semantic technologies. Commun. Comput. Inf. Sci. **754**, 163–176. https://doi.org/10.1007/978-3-319-65551-2_12
6. Kuznetsov, V.P., Garina, E.P., Andriashina, N.S., Kozlova, E.P., Yashin, S.N.: Methodological solutions for the production of a new product. In: Managing Service, Education and Knowledge Management in the Knowledge Economic Era - Proceedings of the Annual International Conference on Management and Technology in Knowledge, Service, Tourism and Hospitality, SERVE 2016, pp. 59–64 (2017)
7. Machová, R., Tóth, Z., Mura, L., Haviernikovà, K.: The entrepreneur's network as a cooperation form of entrepreneurship: case of Slovakia. J. Appl. Econ. Sci. **12**(1(47)), 160–169 (2017)
8. Serebryakova, N., Ovchinnikova, T., Bulgakova, I., Sviridova, S., Tolstykh, T.: Innovational methods of development of intellectual labor for economy's security. Eur. Res. Stud. J. (2017)
9. Tsepelev, O.A., Serikov, S.G.: Peculiarities of regional development and industrial specialization of the far east of Russia. J. Appl. Econ. Sci. **12**(5(51)), 1422–1432 (2017)
10. Weber, K.M., Truffer, B.: Moving innovation systems research to the next level: towards an integrative agenda. Oxford Rev. Econ. Policy **33**(1), 101–121 (2017)
11. Goncharov, A.Yu., Sirotkina, N.V.: The mechanism of management of the balanced development of regions with dominant types of economic activity. News of Higher Educational Institutions. Technology of the Textile Industry **4**(358), 35–43 (2015)

12. Klevtsov, S.M., Kharchenko, E.V.: Reproduction of the regions material assets: theoretical and applied aspects. Sci. Bull. Belgorod State University Ser. Econ. Comput. Sci. **19**(16–1), 48–55 (2010)
13. Ovchinnikova, O.P., Kharlamov, M.M.: Digital transformation of the organization in the implementation of innovative development projects. In: Actual Problems of International Relations in the Conditions of the Formation of a Multi-Polar World: A Collection of Scientific Articles of the 6th International Scientific and Practical Conference, pp. 138–141 (2017)
14. Preobrazhensky, B.G., Tolstykh, T.O., Shkarupta, E.V.: Analysis of the development of the human potential of the region in the conditions of digital transformation. Region Syst. Econ. Manag. **1**(36), 59–66 (2017)
15. Khmeleva, G.A., Semenychev, V.K., Koroleva, E.N., Agaeva, L.K., Korobetskaya, A.A., Zvereva, I.V., Zenina, K.S.: Innovative development of Russian Regions Under Sanctions: Monograph, Samara (2017)

Transformational Period of Russian Development in the Digital Economy

Svetlana N. Kuznetsova(✉), Victor P. Kuznetsov, Elena P. Kozlova, Yaroslav S. Potashnik, and Sergey D. Tsymbalov

Minin Nizhny Novgorod State Pedagogical University, Nizhny Novgorod, Russia
dens@52.ru, kuzneczov-vp@mail.ru, elka-a89@mail.ru, yaroslav.sandy@mail.ru, sergey.cymbalov@mail.ru

Abstract. *Relevance:* This article deals with the question of a transformation period of Russia, which is significant in the context of the development of a market effective innovation economy. Reforms in Russia are accompanied by a deep polysystemic crisis. In order to get Russia out of a polysystemic crisis on the trajectory of stable economic growth, it is necessary to implement a systematic and integrated approach: entrance to the trajectory of steady growth in the level and quality of national human capital; economic diversification and creation of an effective national innovation system and an innovation economy or a knowledge economy; decriminalization of the country.

Modernization of traditional production and service industries as a result of the penetration of information technologies and digitalization of economic processes creates the basis for the formation of new markets and new conditions for the functioning of the Russian market, as well as new approaches to analytics, forecasting, and management decisions.

Objective: formulate development directions for the formation and maintenance of the most favorable organizational, infrastructural and regulatory characteristics of the Russian digital jurisdiction for business development in the new economic order, as well as advanced development of national digital economy institutions.

The methodology for studying a content of discussions consists of: development of conceptual apparatus, statistical (quantitative) methods, forecasting methods, estimation methods of state policy in the field of digital economy.

The results of the study: General points for the conceptual apparatus were identified, trends in digitalization were generalized, remaining differences were revealed, and it was shown that the government is a powerful driver of digitalization.

Keywords: Transformation · Development · Diversification · Efficiency · Competitiveness · Digitalization

E. G. Popkova and B. S. Sergi (Eds.): ISC 2019, LNNS 91, pp. 663–669, 2020.
https://doi.org/10.1007/978-3-030-32015-7_74

1 Introduction

The goal is to develop an effective trajectory of stable sustainable growth of the Russian economy; as well to create favorable organizational and regulatory conditions in Russia for the effective development of institutions digital economy with the participation of the government, the national business community, and civil society. At the same time ensuring rapid growth of the national economy through qualitative changes in the structure and system management of national economic assets, and achievement of the effect of "Russian economic miracle" in conditions of formation of the global digital ecosystem.

The main tasks are:

– Providing the country's technological leadership in the context of the formation of a global digital space;
– Formation of a qualitatively new structure of economic assets that meet the economic priorities of the digital economy;
– Formation of approaches to organization of production industries, trade, and services sectors, taking into account the achievements of the digital economy and effective in conditions of formation and development of global digital space;
– Creation of conditions for improving the quality of life of the population by changing a structure and quality of social services and creating new opportunities for business and labor activity;
– Ensuring effective participation of the country in processes of formation of the global ecosystem of the digital economy and global digital space.

The level of scientific development of the problem. Questions about the transformation period of Russia have been developed in the works of scientists such as Pavroz A.V., Fedotova V.G., Ponomarev M.V., and others.

The main part.

The main obstacles to realization of the trajectory of sustainable growth and negative results (Fedotova and Kolpakov 2008):

– Slow modernization and diversification of the economy; Dutch trouble of the Russian economy;
– Low growth rates of the level and quality of Russian human capital; high inflation rates; weak banking system; undeveloped securities market.

According to results of the first quarter of 2019, the federal budget of the Russian Federation was executed with a surplus of 2.2% of GDP, or 546 billion rubles (Pavroz 2008).

In the first quarter of 2019, the net outflow of capital from Russia amounted to 25.2 billion dollars; the money supply in the first quarter of 2019 amounted to 46.2 trillion rubles. Export: +1.7% ($66.5 billion against $65.4 in the first quarter of 2018), imports: −1.7% ($33.3 billion against 33.9).

The authors consider results of macroeconomic indicators within the framework of adaptation changes (Held 2004):

- GDP growth in 2018 accelerated from 1.6–2.3%, which is provided by sectors such as mining, transportation and storage, construction, financial and insurance, in the result is expected slow down of economic growth to 1.3%;
- Growth rate of retail trade turnover increased in 2018 from 1.3–2.6% due to increased demand for durable goods (passenger car sales increased from 11.9–12.8% in 2018);
- High rates of salary growth +6.8% in 2018 in real terms, real disposable incomes of the population decreased by 0.2%;
- Current account surplus in 2018 reached a maximum value of 114.9 billion dollars;
- Inflation in January 2019 amounted to 1.0%, at the end of the year if the ruble remains stable, inflation will fall below 5%;
- Economic growth by the end of 2018 accelerated from 1.6–2.3%, industry and construction were key drivers of economic growth (contribution to economic growth 0.6 p.p. and 0.3 p.p.). Trade (0.3 p.p.) and transport (0.2 p.p.) also contributed positively to GDP dynamics, while agriculture was slightly negative.

At the end of the year, the positive dynamics was demonstrated by all the enlarged branches of industry. Mining and manufacturing continued to grow, electricity and water supply growth rates returned to a positive area after the recession in 2017. At the same time, the structure of growth of industrial production changed during the year. If in the first half of the year the expansion of industrial output was based on manufacturing industries, in the second half of the year the growth driver of steel, and extractive industries (Ponomarev 2004).

- Main contribution to GDP growth in 2018 was made by consumer demand, which increased by 2.2%.

Retail turnover in 2018 increased from 1.3–2.6%, due to increased demand for durable goods. Automotive market for the second year in a row showed double-digit growth (12.8% in 2018). In addition, the growth of public catering turnover accelerated to 3.6% in 2018.

- Growth of gross fixed capital formation slowed from 5.5 to 2.3% in 2018, while maintaining investment activity around the same level as last year (4.1%).

With the weak ruble, the growth rate of imports of goods and services declined from 17.4 to 3.8% in 2018, while exports of goods and services continued to grow steadily (by 6.3% in real terms after 5.0% a year earlier). As a result, the contribution of net exports to GDP growth in 2018, according to estimates, amounted to 0.8 p.p. (compared to −2.3 p.p. a year earlier). Therefore, net exports more than offset the slowdown in domestic demand and became the main driver of accelerating GDP growth in 2018 (Artemieva 2017).

- Active growth of salaries in both social and non-budgetary sectors, in 2018 the growth of real wages amounted to 6.8% (in 2017—2.9%).

- Unemployment rate in 2018 fell to 4.8%, the total number of labor force decreased by 95.3 thousand people (−0.1%), the number of employed population increased by 215.7 thousand people (+0.3%) for a reduction in the total number of unemployed persons by 311,0,000 (−7.8%);
- Inflation in January 2019 was 1.0%, inflation in January accelerated to 5.0% after 4.3% in December 2018. A significant contribution (+0.24 p.p.) to inflation in January was made by indexation of regulated prices and tariffs, while the cost of utilities increased by 2.6%, including as a result of higher VAT and changes in payment procedure for removal of solid municipal waste;
- In 2019 inflation will be 0.5–0.6% (5.2–5.4%), with peak values (5.5–5.7%) inflation will take place in March—May of this year. At the end of the year, if the ruble remains stable, inflation will fall below 5%. In 2018, real disposable incomes of the population decreased by 0.2% compared to the previous year. At the same time, the dynamics of the main components of income were mixed;
- In 2018, the current account surplus reached the highest value in the history of observations. At the end of the year, the current account surplus increased to 114.9 billion dollars. USA (6.9% of GDP) after 33.3 billion dollars. In 2017, USA 2.1% of GDP (Garina et al. 2017).

The factors that predetermined the negative dynamics of the consolidated indicator were:

- Reduction of income from property, including from savings on deposits in the banking system. The latter was a consequence of lower nominal interest rates on deposits amid lower inflation. The total contribution of this factor reached minus 0.5 p.p.;
- Growth of interest payments on bank loans. Their volume is recorded in disposable income statistics with a negative sign. The two-digit growth rate of lending to the population while maintaining interest rates on unsecured consumer loans at a high level led to an increase in debt servicing costs by 10.5% to the level of last year. The negative contribution to the dynamics of real disposable income amounted to minus 0.2 p.p.;
- Growth of other mandatory payments, including tax payments. Increase in real estate tax payments, and in income tax revenues led to an increase in other mandatory payments by 14.8%. Contribution to the total indicator of real disposable income amounted to minus 1.0 p.p. (Kuznetsov 2018a, b).

The factors that have had a positive or neutral impact were (Kuznetsov 2018a, b):

- Growth of the average salary level. Remuneration of employees increased by 6.7% in real terms. Along with positive situation in the labor market, an increase in wages in real terms was facilitated by an increase in the minimum wage and indexation of wages for employees of the public sector;
- Dynamics of social payments close to inflation. Pensions and other social payments made a slightly negative contribution (less than 0.1 p.p.) to dynamics of real disposable incomes of the population, which was connected with the excess of the actual the average annual inflation rate in 2018 (2.9%) above the base inflation rate.

In order to substantiate the main elements of the transformation period of Russia's development in the digital economy, a survey was conducted. This study was conducted on the basis of an expert survey in Nizhny Novgorod, Vladimir and Ivanovo regions. The quota sample was 250 respondents.

The main findings of the survey are:

– About half of companies from the number of respondents considered themselves to the mature stage of digital development. Projects in the field of electronic document management were mainly implemented. While some business sectors tend to dominate other digital applications—such as big data storage, processing, and analytics (in the banking sector), management of production equipment and monitoring of its work (exporting enterprises).
– Main initiators of projects using digital technologies were general directors of enterprises or other persons on the highest administrative position in the company. Such factors as the size of the company, availability of export products and stage of digital development of the company determine the role of an initiator of implementation of projects for corporate needs in the implementation of digital solutions. The larger the company and the higher stage of digital development, the more likely it is not only the CEO but also the Director of information technology.
– Today the impact of digital technologies on the business of companies is estimated as quite high: for 7 points out of 10 maximum points. In most cases, companies estimate the impact of projects as corresponding to, or even exceeding, their expectations. The main effect is obtained in simplification and acceleration of processes, as well as in improving accuracy and quality of work. In the long term after 5 years, an assessment of the impact of digital technologies on business is even higher: 8 points.
– Mainly business today is affected by technologies such as Internet of Things and production automation, digital design and modeling, virtualization technologies, remote access, mobile technology, and cross - channel communications. In the future, the impact of these technologies will continue, but two more areas will be added to them: social networks and supercomputer systems. There is also a noticeable increase in the influence of three directions: virtual, augmented and mixed reality systems, additive technologies, cloud technologies.
– At least 2/3 of companies give fairly high estimates of the level of awareness and competence of their employees in assessing the impact of digital technologies on the business of companies. The higher the level of digital development of companies, the higher the level of awareness of their employees about the development of digital technologies and higher their understanding of the extent to which technology can impact their business of the company. Large companies and companies exporting products often state that their specialists have all the necessary information about the development of digital technologies and are well aware of their possible impact on the company's business.
– With a fairly high level of awareness about digital technologies, most companies do not have clear digital strategies. At the same time, a significant proportion of companies (40%) plan to implement certain specific significant projects on the use of digital technologies. Key motives for the implementation of digital projects in the

company are compliance with the industry-wide level of development and having some advantage over competitors.

Results of the Study. Consider the dynamics of the main indicators of the base forecast for 2019–2024 (Table 1)

Table 1. Dynamics of indicators of the basic development forecast for 2019–2024*

Indicators/Year	2018	2019	2020	2021	2022	2023	2024
The ruble exchange rate for 1 dollar.	62.5	65.1	64.9	65.4	66.2	67	68.6
Price per barrel of oil	70	63.4	59.7	57.9	56.3	55	53.5
Share of investments in GDP,%	20.6	21.2	22.5	23.3	23.9	24.6	25.2
Investments in fixed assets,% y/y	4.3	3.1	7	6.3	5.8	5.6	5.3
Inflation,%	4.3	4.3	3.8	4	4	4	4
GDP,% y/y	2.3	1.3	2	3.1	3.2	3.3	3.3
Real wage,% y/y	6.8	1.1	2	2.7	2.7	2.7	2.7
Real disposable income,% y/y	–	1	1.5	2.2	2.3	2.3	2.4

*https://www.rbc.ru/economics/12/04/2019/5caf07de9a7947c3f64cdc2f

There are risks of not achieving indicators on real incomes and salaries, pension reform will delay a large number of people before retirement age in the labor market, and competition for jobs will lead to a slowdown of wage growth. The growth of real income will be held back by the high level of credit of the population, and the income from savings will fall due to the decrease in deposit rates (Kuznetsova 2018).

Due to the surge in inflation, the dynamics of real incomes of the population in 2019 will be near zero, inflation is expected to be only 4.3% at the end of the year.

According to the results of 2019, the growth of investments in fixed assets is expected at 3.1% after an increase of 4.3% by the end of 2018. Overall, the forecast for investment growth throughout the forecast period:

in 2020—from 7.6 to 7%;

in 2021—from 6.9 to 6.3%;

The forecast for subsequent years is also reduced: now instead of an increase of 6.1–6.6% is expected to accelerate to only 5.3–5.8%.

At the same time, the estimate of the share of fixed investment in GDP decreased. By 2024, the government should increase it to 25% (in 2018 it was 20.6%). According to forecasts, the share of investments in 2019 will be 21.4% and by 2024 reached 26.4% in the base scenario and 25.3% in the conservative scenario. In the updated forecast, the share of investments in GDP will reach 25.2%, but in the conservative—not (24.1%) (Potashnik 2018).

Digital innovation is thus an important lever of economic development, offering progressive solutions to global problems, increasing the efficiency of management decisions and stimulating active participation of business and civil society in the formation of the economic welfare of the country.

With the generally recognized role of the digital economy as a driver of growth and a tool for qualitative change of indicators of welfare of states, analytical forecasting tools should be taken into account social and ethical aspects of the digital economy. As a result of implementation of adaptation changes, the following positive results will be achieved (Shkunova 2018): macroeconomic stability of the economy; financial stabilization, except for the reduction of inflation; living standards of the population (income) steadily growing; implementation of large national projects.

References

Pavroz, A.V.: Groups of interests and transformation of the political regime in Russia. St. Petersburg University Publishing House, 360 p. (2008)

Fedotova, V.G., Kolpakov, V.A., Fedotova, N.N.: Global capitalism. Three great transformations. Cultural Revolution, 608 p. (2008)

Held, D., et al.: Global transformations. Politics, Economics and Culture: monograph. Praxis, 576 p. (2004)

Artemieva, M., Kuznetsova, S., Bakhtiarov, Y., et al.: Peculiarities of Innovative Activities in the Low-Tech Sector. Overcoming uncertainty of institutional environment as a tool of global crisis management: Conference on Overcoming Uncertainty of Institutional Environment as a Tool of Global Crisis Management Location: Athens, GREECE Date: Apr, 2017. Book Series: Contributions to Economics, pp. 289–294 (2017)

Garina, E., Kuznetsov, V., Yashin, S., et al.: Management of Industrial Enterprise in Crisis with the Use of In company Reserves. Overcoming uncertainty of institutional environment as a tool of global crisis management: Conference on Overcoming Uncertainty of Institutional Environment as a Tool of Global Crisis Management Location. Contributions to Economics, Athens, Greece, pp. 549–555, April 2017

Kuznetsov, V.P., Garina, E.P., Romanovskaya, E.V., Kuznetsova, S.N., Andryashina, N.S.: Organizational design and rationalization of production systems of a machine-building enterprise (by the example of the contract assembly workshop), Espacios, 39 article # 25 (2018a)

Kuznetsov, V.P., Romanovskaya, E.V., Egorova, A.O., Andryashina, N.S., Kozlova, E.P.: Approaches to developing a new product in the car building industry. AISC, vol. 622, pp. 494–501 (2018b). https://doi.org/10.1007/978-3-319-75383-6_63

Kuznetsova, S.N., Romanovskaya, E.V., Artemyeva, M.V., Andryashina, N.S., Egorova, A.O.: Advantages of residents of industrial parks (by the example of AVTOVAZ). AISC, vol. 622, pp. 502–509 (2018c). https://doi.org/10.1007/978-3-319-75383-6_64

Potashnik, Y.S., Garina, E.P., Romanovskaya, E.V., Garin, A.P., Tsymbalov, S.D.: Determining the value of own investment capital of industrial enterprises. AISC, vol. 622, pp. 170–178 (2018). https://doi.org/10.1007/978-3-319-75383-6_22

Shkunova, A.A., Yashkova, E.V., Sineva, N.L., Egorova, A.O., Kuznetsova, S.N.: General trends in the development of the organizational culture of Russian companies. J. Appl. Econ. Sci. **12**(8), 2472–2480 (2018)

Methodological Bases of the Assessment of Sustainable Development of Industrial Enterprises (Technological Approach)

Elena P. Kozlova(✉), Victor P. Kuznetsov, Ekaterina P. Garina, Elena V. Romanovskaya, and Natalia S. Andryashina

Minin Nizhny Novgorod State Pedagogical University, Nizhny Novgorod, Russia
elka-a89@mail.ru, kuzneczov-vp@mail.ru, e.p.garina@mail.ru, alenarom@list.ru, natali_andr@bk.ru

Abstract. In the article, the authors address the issue of methodological approaches to an assessment of the level of sustainable development of industrial enterprises. Authors analyzed already existing approaches and found that most often researchers use the generalized integral indicator in the areas of economic, environmental and social sustainability. At the present stage, the importance of technologies and technological transformations for industrial enterprises increases so much that makes it possible to speak about the need to include this approach when assessing the level of sustainable development. Therefore, the authors in the framework of the study proposed to allocate technological stability and evaluate it with the help of a separate integral indicator. The effectiveness of an assessment of sustainable development depends on clearly defined goals and objectives of sustainable development, and it is proposed that a targeted approach should be used, making it possible to assess the extent to which the qualitative and quantitative objectives have been achieved.

It was found that the application of the generalized integral indicator requires an objective, deeply grounded and logically built methodology. Working on this methodology for industrial enterprises, it is necessary to take into account the current trends. The proposed methodology includes five stages and is a structured model with data, has a flexible support structure, which, if necessary, can be supplemented and increased taking into account features of primary and boundary conditions of business activity of the enterprise and specifics of established goals.

Keywords: Sustainable development · Integral indicator · Technological stability

E. G. Popkova and B. S. Sergi (Eds.): ISC 2019, LNNS 91, pp. 670–679, 2020.
https://doi.org/10.1007/978-3-030-32015-7_75

1 Introduction

Modern economic trends allow to emphasize the theoretical and practical significance of a problem of increasing efficiency and sustainability of industrial enterprises on the basis of technological approach. At the same time, sustainable industrial development will entail technical and technological progress of the economy as a whole. In order to determine how the enterprise develops, what character this development has, it is necessary to conduct its analysis, assess the degree of sustainable development, which in turn will allow the enterprise on the basis of obtained results, to make the right management decisions in conditions of a dynamically changing environment. Such an assessment requires an objective, well-reasoned and logically structured methodology.

2 Methods of Research

The theoretical and methodological basis of the work was the fundamental provision reflected in works of researchers in the field of assessment of sustainable development. The methodology of research is based on a technological approach that allows to assess the level of sustainable development of an industrial enterprise in conditions of scientific and technological development.

3 Results of the Study

Various economic indicators and criteria play a key role in the analysis and diagnosis of the state of sustainable development of an organization. In order to assess the level of quality work of any organization, it is necessary to conduct a complex integrated assessment characterized by expansion of indicators of economic growth, compliance with performance of its activities tactical objectives and requirements.

Classic variant of the model of sustainable growth, which is often noted in scientific works, is the formula proposed in 1977 by the American scientist R. Higgins. This model has been modernized many times, improved, and the simplest version of it has the form [3]:

$$\Delta M = [Sp\,Kp\,Ad\,/\,(Vp)] * 100\% = Rh\,Kp\,(A/E)\,d * 100\%$$

where ΔM - the possible growth rate of sales volumes of products that does not disturb the financial equilibrium of an organization,%;

Sn.p - amount of net profit of the organization;
Kn.p - net profit capitalization ratio;
A - value of assets of an organization;
d = the turnover rate of assets;
Vp - volume of sales of products (revenue);
E - amount of equity capital of an organization;
Rn - net profitability.

The basic idea of this formula is: sustainability increases directly in proportion to the share of equity capital, and no matter how changed and rewritten it, this idea will remain to this day.

L.A. Bazarova considers disadvantage of above model in a weak focus on the analysis of trends of the external environment of an organization. According to this reasoning, if the economy is recovering, the asset turnover ratio (d) will be evidence of effective management action. In the event of economic decline, this indicator will contribute to the decline in the sustainability of an organization and building up stocks. The author to reflect the efficiency of the management process suggests taking into account such parameters as: innovative potential of the enterprise—γ, as well as the level of organization of the team—α, which will help or slow down the process of joint actions. The modified formula is as follows [3]:

$$\Delta B = Rn\,Kn.p\,(A/E)d * \gamma * \alpha * 100\%$$

O.A. Singer offers a system of indicators that take into account benefits of both internal and external users in order to assess the sustainability of an enterprise. She emphasizes sustainability indicators such as [6]:

- Financial sustainability;
- Marketing sustainability;
- Production sustainability;
- Innovative sustainability.

A.V. In order to determine the essence of the mechanism of sustainable development, A.V. Shankin identifies some constituent elements involved in activities of an enterprise, expressing them in three subsystems: managing, operating, and supporting. He proposes to classify the key indicators in terms of these three subsystems [11].

At the present stage, both domestic and foreign authors often allocate 3 complex indicators, with help of which it is possible to assess the degree of sustainable development of an organization. These include indicators of economic, environmental and social sustainability.

For example, O.I. Averina identifies three groups, adding to them indicators characterizing the stability of the external environment. It includes ratios of average increase in energy prices, GDP growth, unemployment, inflation, reduction in fixed capital investments, ratios of scientific and technical, political and natural risk, as well as consumer prices [1].

N.A. Khomyachenkov in addition to three above complex indicators of sustainability offers also risk stability. By it, she understands ensuring sustainable growth of the market value of an industrial enterprise due to its ability to combine risk and performance management processes [10].

In general, we share this point of view, developed and tested methodology for assessing the sustainability of industrial enterprises allows to rank weight values by individual kinds (economic, environmental, social and risk) sustainability and thus take into account the sectoral and individual specificities of each enterprise.

Identification of main improvement directions of methodical tools of assessment of the level of sustainable development of enterprises is difficult by presence of many different points of view on this problem that requires analysis and systematization (Table 1).

Table 1. Analysis of approaches to assessing the level of sustainable development

Author	Features of method	Number of stages
Solomka A.V. [11]	"calculation of the integral index of sustainability on the basis of indicators of production, economic, social and environmental sustainability, which are formed by the use of piecewise linear conversion functions to normalize them"	5
Bogomolova I.P., Gusev I.S. [4]	"assessment of the level of sustainable development of a branch enterprise is based on the study of indicators of economic, social, environmental and risk aspects of its activity"	4
Borodin M.A. [5]	"methodology is based on a comparative assessment of the level of economic security of an enterprise in the industry, allowing to determine its strategic positions and outline the trajectory of further development"	3
Ilyicheva A.V. [7]	"application of fuzzy sets approach in the economy, which allows to quantify local and integral indicators of sustainable development"	5
Tarasova N.P., Kruchina E. B. [9]	"two approaches to building sustainable development indicators: (1) construction of a system of indicators, each of which reflects separate aspects of sustainable development (environmental, economic, social, institutional); (2) construction of the integral indicator"	3

Speaking about the sustainable development of industrial enterprises, we consider the development and sustainability of technologies as the most important. Therefore, from our point of view, when assessing the sustainable development of an enterprise, it is necessary to apply a technological approach. We propose to allocate technological stability, which should be evaluated through a separate integral indicator.

4 Conclusion

In accordance with the idea of allocating technology ability and dividing goals into qualitative and quantitative, the following methodology for assessing sustainable development of the enterprise has been proposed (Fig. 1).

This methodology is a data model, has a flexible support structure, which, if necessary, can be supplemented and (or) expanded taking into account the specifics of initial and boundary conditions of operation and nature of objectives. It includes five stages.

At the first stage, an analytical base of the system of assessment indicators of sustainable development of the enterprise is compiled. Within this phase, experts from various types of staff are selected.

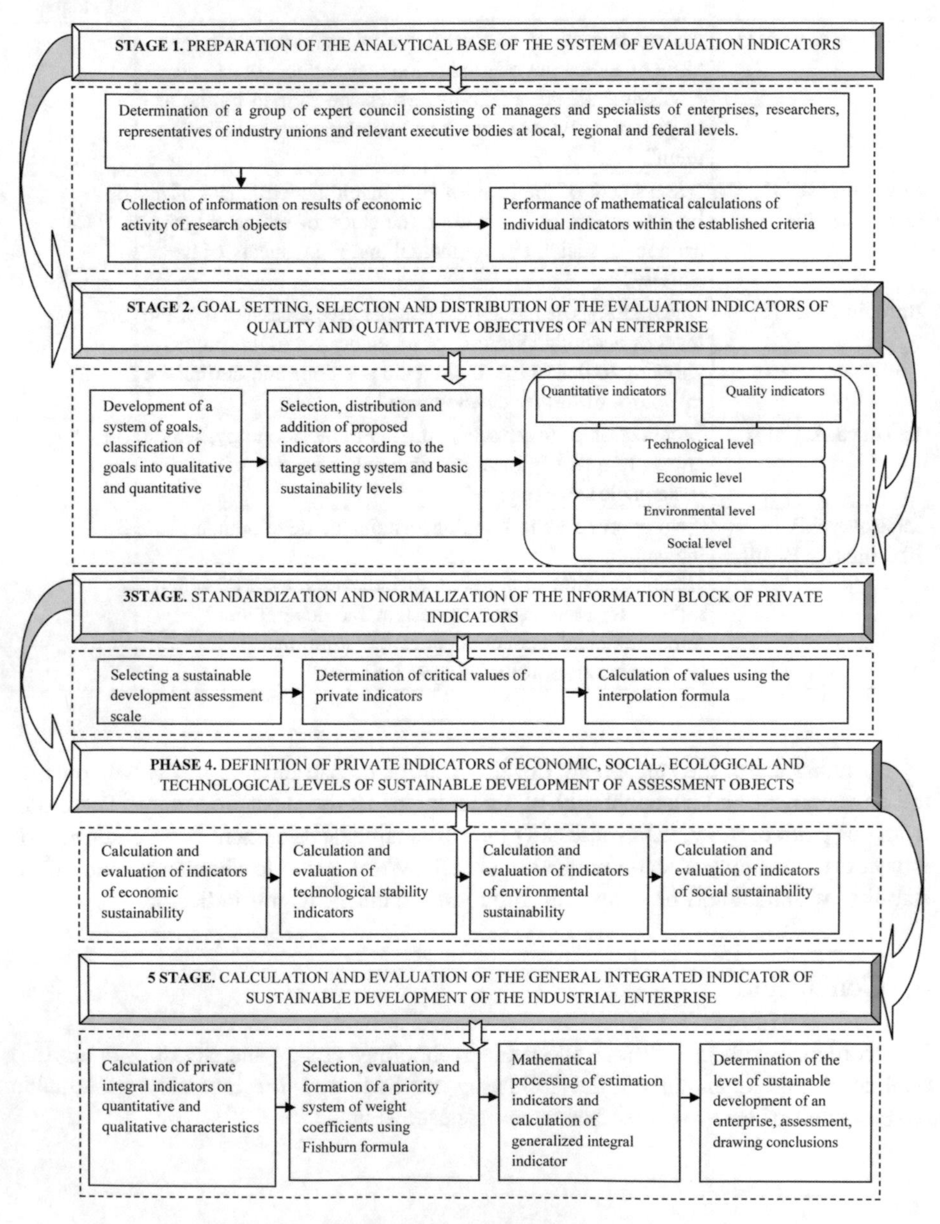

Fig. 1. Stages of assessment of the level of sustainable development of industrial enterprises

At the same time, information material on the results of functioning of the investigated objects is selected, and calculations of selected coefficients are carried out on the basis of established criteria.

Within the framework of the study a set of indicators, which can be selected by experts to assess the level of sustainable development of the enterprise (Fig. 2).

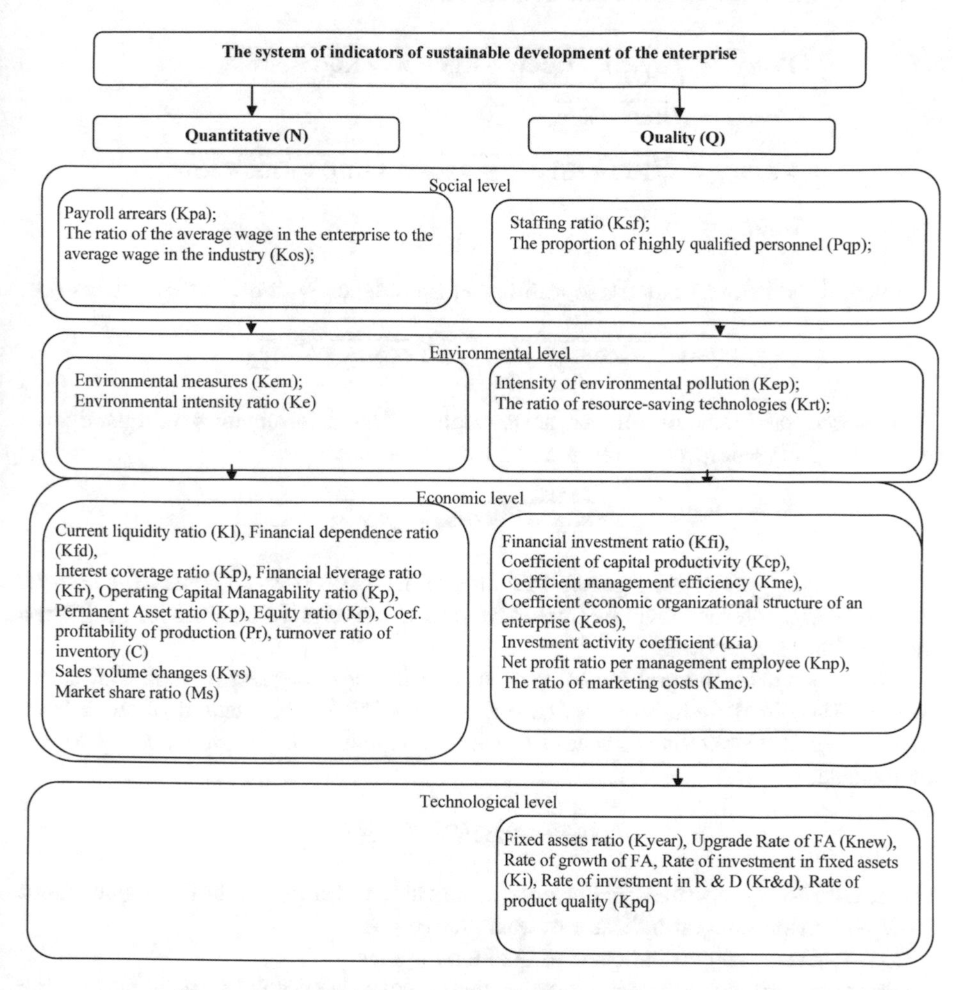

Fig. 2. Systematization of sustainable development indicators.

In the second stage, objectives of an enterprise are set and coefficients are grouped to measure the achievement of qualitative and quantitative objectives, taking into account the levels of economic, environmental, social and technological sustainability.

The third stage consists of the process of standardization and normalization of information module of coefficients and their integration into a set of computed data. It is allowing to calculate normalized values of zero and negative coefficients.

The fourth stage is to determine the private integral coefficients of economic, technological, social and environmental levels of sustainable development of enterprises under study, including indicators of various kinds of sustainability are calculated and evaluated. In accordance with the proposed division of coefficients, depending on the qualitative (Q) and quantitative (N) goals of an enterprise, private integral indicators have been formed, which look like as follows:

$$Ytec_Q = \sqrt[6]{\text{Kyear} * \text{Knew} * \text{Kgr} * \text{Ki} * \text{Kqp} * \text{Kr}\&d}$$

$$Yeco_Q = \sqrt{\text{Kep} * \text{Krt}},$$

$$\text{Yecon}_\text{Q} = \sqrt[7]{\text{Kia} * \text{Kfi} * \text{Kcp} * \text{Keos} * \text{Knp} * \text{Kmc} * \text{Kme}}$$

$$Ysoc_Q = \sqrt{\text{Ksf} * \text{Pgp}}$$

Integral indicator of enterprise stability in accordance with quality objectives (Q):

$$Q = \sqrt[4]{* \text{Ytec}_\text{Q} * \text{Yeco}_\text{Q} * \text{Yecon}_\text{Q} * \text{Ysoc}_\text{Q}},$$

The integral indicator formed as a result of the determination of quantitative (N) goals of the enterprise is represented by the formula:

$$N = \sqrt[3]{\text{Yeco}_N * \text{Yecon}_\text{N} * \text{Ysoc}_\text{N}}, \text{ where}$$

At the final, *fifth stage*, a generalized integral indicator of sustainable development of an industrial enterprise will be formed by calculation of private values for qualitative and quantitative purposes.

The generalized integral indicator of sustainable development of industrial enterprises (GISD) in this study is carried out on the basis of the method of sums from private integral quantitative indicators (N) and of a qualitative (Q) group, calculated by the formula:

$$GISP = 0.67N + 0.33Q$$

where: GISD - a generalized indicator of sustainable development of the organization;

N — private integral indicator of quantitative goals;

Q — private integral indicator of qualitative goals;

Practical application of developed methodology to assess the level of sustainable development of industrial enterprises was carried out on the basis of enterprises of Nizhny Novgorod region (PJSC "GAZ", PJSC "NMZ", and "Hydromash") (Table 2).

Table 2. Calculation of the generalized integral indicator of sustainable development

Indicator name	PJSC "GAS"			PJSC "NMZ"			NOJSC "Hydromash"		
	2016	2017	2018	2016	2017	2018	2016	2017	2018
GISD	0.394	0.385	0.375	0.373	0.375	0.357	0.486	0.500	0.547
Private integral stability indicator for quality objectives (Q)	0.339	0.362	0.334	0.289	0.289	0.282	0.452	0.490	0.579
$Yecon_Q$	0.330	0.336	0.354	0.350	0.340	0.356	0.492	0.465	0.524
$Yeco_Q$	0.336	0.428	0.337	0.159	0.153	0.202	0.392	0.544	0.753
$Ysoc_Q$	0.538	0.524	0.490	0.509	0.528	0.440	0.745	0.775	0.788
$Ytec_Q$	0.151	0.158	0.155	0.136	0.134	0.131	0.179	0.176	0.252
Private integral stability index by quantitative objectives (N)	0.422	0.396	0.395	0.414	0.418	0.394	0.502	0.505	0.531
$Yecon_N$	0.330	0.336	0.354	0.350	0.340	0.356	0.492	0.465	0.524
$Yeco_N$	0.397	0.328	0.396	0.383	0.386	0.396	0.370	0.376	0.381
$Ysoc_N$	0.538	0.524	0.435	0.509	0.528	0.430	0.645	0.675	0.688

Graphically obtained results can be displayed using a pie (petal) chart (Fig. 3).

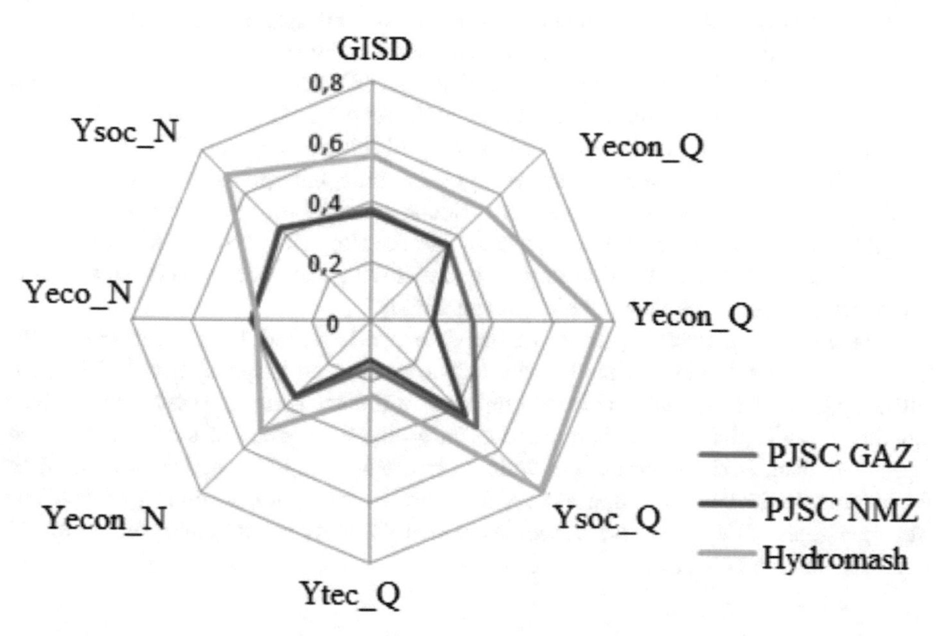

Fig. 3. Generalizing and private integral indicators of enterprises achieved by 2018

We propose to interpret the results of the evaluation of an integral indicator of sustainable development by means of Table 3. This table was developed by N.A. Khomachenkova, it allows to determine the kind of stability depending on the obtained generalized integral index [10].

Table 3. Decoding the obtained results of the sustainability of enterprise development

Kind of sustainability	Value of the indicator
Absolute sustainable development	$0{,}9 < y \leq 1$
High sustainable development	$0{,}8 < y \leq 0{,}9$
Normal sustainable development	$0{,}7 < y \leq 0{,}8$
Medium sustainable development	$0{,}6 < y \leq 0{,}7$
Weak sustainable development	$0{,}5 < y \leq 0{,}6$
Unsustainable development	$0{,}4 < y \leq 0{,}5$
Critical condition	$0{,}3 < y \leq 0{,}4$
Crisis condition	$y \leq 0{,}3$

Therefore, based on calculations and table of decoding the level of sustainable development, we can conclude that the highest level of sustainable development is Hydromash - 0.547, which allows it to implement sustainable development policies. On the scale of sustainable development of the industrial enterprise “Hydromash” is at the level of “unsustainable development”. When measured separately, quantitative and qualitative indicators are at the level of unsustainable development.

The lower level has PJSC “GAZ” and PJSC “NMZ”, according to the results of this study on the generalized integral index in 2015. At the level of unsustainable development. PJSC “GAZ” is on the lower boundary of this degree, and 2013–2014 was at all in a critical stage of development. At the same time, enterprises achieve qualitative goals much better than quantitative goals. Social sustainability has the highest score.

In order to achieve the goals of sustainable development PJSC “GAZ” and PJSC “NMZ” we need to pay attention to their quantitative goals, namely to make efforts to increase the level of its technological sustainability.

Therefore, a distinctive feature of the developed methodology for assessing the level of sustainable development is the technological and target approach. Which is allowing to assess the sustainable development of the enterprise in accordance with its qualitative and quantitative characteristics in each of the areas of sustainable development (social, environmental, economic, technological), and taking into account their interaction in levels. This approach will allow to take into account specifics and strategic goals of the enterprise, as well as to identify the cause of failures on the way to achieve the common goal.

References

1. Averina, O.I., Gudkova, D.D.: Analysis and assessment of sustainable development of enterprise [Text]. Scientific Information Publishing Center and Editorial Office of the Journal "Actual Problems of Humanities and Natural Sciences", no. 1–3, pp. 10–19 (2016)
2. Andrianov, V.D.: Competitiveness of Russia in the world economy. World Economy and International Relations, no. 3, pp. 53–57 (2013)
3. Bazarova, L.A.: Management of sustainable development of industrial enterprise: methodology and technology [Text]: dissertation doctor of economic sciences, St. Petersburg, p. 32 (2014)
4. Bogomolova, I.P., Gusev, I.S.: Sustainable development of enterprises of meat industry in the conditions of market economy [Text]. Modern Economy: Problems and Solutions. Economics of Industry, no. 3(63), pp. 29–37 (2015)
5. Borodin, M.A., Korshunov, L.A., Markina, T.V., Belov, V.V.: Development of a methodology for calculating the integral index of tax security of the region and assessment of its value in Altai region [Text]. 4th Interregional NPK "Effective Tax System - The Basis of Economic Development": Collection of Scientific Materials, pp. 167–170. Publisher AltSTU, Barnaul (2003)
6. Singer, O.A.: Formation of strategy of sustainable development of industrial enterprise [Text]: abstract of dissertation: 08.00.05, Penza, 22 p. (2010)
7. Ilyicheva, A.V.: Methodology of integral assessment of sustainability of enterprises of territorial-industrial meat complex, on the example of Krasnodar region [Text]. Science and Business: Ways of Development, no. 9(27), pp. 168–172 (2013)
8. Kucherova, E.N.: Formation of the mechanism of sustainable development of machine-building enterprises in modern conditions [Text]: abstract of dissertation: 08.00.05, 31 p. (2011)
9. Tarasova, N.P., Kruchina, E.B.: Indexes and indicators of sustainable development. Sustainable development: nature-social-human: materials from the Intern. Conf., vol. 1 (2006)
10. Khomyachenkova, N.A.: Mechanism of integral assessment of sustainability of industrial enterprises [Text]: abstract dissertation: 08.00.05, 21 p. Moscow State Institute of Electronic Technology (2011)
11. Shankin, A.V.: Improvement of the mechanism of sustainable development of food industry enterprises [Text]: abstract of dissertation: 08.00.05, Moscow, 25 p. (2012)
12. Garina, E., Kuznetsov, V., Yashin, S., et al.: Management of Industrial Enterprise in Crisis with the Use of Incompany Reserves. Overcoming uncertainty of institutional environment as a tool of global crisis management: Conference on Overcoming Uncertainty of Institutional Environment as a Tool of Global Crisis Management Location: Athens, GREECE Date: APR. Contributions to Economics, pp. 549–555 (2017)
13. Kuznetsov, V.P., Romanovskaya, E.V., Egorova, A.O., Andryashina, N.S., Kozlova, E.P.: Approaches to developing a new product in the car building industry. Advances in Intelligent Systems and Computing, vol. 622, pp. 494–501 (2018)
14. Kuznetsov, V.P., Garina, E.P., Romanovskaya, E.V., Kuznetsova, S.N., Andryashina, N.S.: Organizational design and rationalization of production systems of a machine-building enterprise (by the example of the contract assembly workshop). Espacios **39**(1), 25 (2018)

Understanding the Challenges in the Research and Innovation Ecosystem in India

Sandeep Goyal[1,3(✉)], Sumedha Chauhan[2], and Amit Kapoor[3]

[1] L M Thapar School of Management, Derabassi, Punjab 140507, India
sandy2u@gmail.com
[2] O. P. Jindal Global Business School, Sonipat, India
sumedha.chauhan@gmail.com
[3] The Institute for Competitiveness, Gurugram, Haryana, India
amit.kapoor@competitiveness.in

Abstract. The purpose of this article is to understand the research and innovation ecosystem in India. How India has made progress over the last five years in terms of enhancing its performance with respect to competitiveness, innovation and intellectual property indices? What kind of contribution has been made by the Government in transforming the research and innovation landscape? What are the challenges faced by the leading research institutions in India? What kind of orientation is needed in the coming years to enhance the competitiveness, innovation and intellectual property rankings of India? The qualitative multi-case based research methodology has been applied for data collection and analysis. The key challenges faced by the research institutions in India involve complexities in the patenting process; lack of willingness among the talented resources to take up research as a career option; outdated research infrastructure; issues in commercializing the patented innovations; limited availability of research funding and incentives.

Keywords: Patents · Innovations · Research · India · Commercialization

1 Introduction

Technology, Research and Innovation are emerging as strategic levers for competitive advantage of nations and companies globally. The global socio-economic landscape is undergoing a significant transition due to the rapid advancement and high penetration of internet and communication technologies. Fast pace of digitization has impacted all sort of social and economic activities at country, organizational and individual levels. Globally, countries are trying to enhance their innovation capacity in order to attract more and more investments. New start-ups are coming up every day with disruptive business models and innovative products and services offerings. These start-ups are posing a huge challenge for the traditional organizations across industries having obsolete business models.

Developing economies like India and China are making considerable efforts to enhance their research and innovation capabilities at par with developed economies. No longer, research and innovation are confined to the boundaries of developed economies

E. G. Popkova and B. S. Sergi (Eds.): ISC 2019, LNNS 91, pp. 680–690, 2020.
https://doi.org/10.1007/978-3-030-32015-7_76

like USA and European countries. Globally, developing economies are making conscious attempts to simplify their research and innovation ecosystem so that indigenous talent can be attracted and motivated in making a contribution to enhance the research and innovation capacity rather than migrating to developed economies.

According to GII 2019[1] (Global Innovation Index) rankings, global landscape of science, innovation and technology has undergone a significant shift over the last decade. Middle-income economies especially, Asian economies are raising their stakes and contribution in global R&D (Research & Development) as well as registering global patents via WIPO's (World Intellectual property Organization) international patenting system. Four among the top five innovation clusters are based in Asia. Innovation clusters have been ranked on the basis of patents and publishing. At other end, R&D investment in high-income economies is getting slow or no change over the last few years. This reversal of trends in the rate of growth for R&D investments signifies a role reversal for the middle income versus high income economies in the future of world.

To understand the increasing role of Middle-income economies especially Asian countries, let's evaluate India's performance on different global rankings during the last few years. During 2015–2019, India has made a consistent progress in enhancing its global competitiveness and innovation rankings.

According to GII 2019 rankings, India has gained twenty nine positions to move from a global ranking of 81st in 2015 to 52nd in 2019. This is considered to be the biggest jump by any economy over the last 5 years. This gain in performance is mainly attributed to increase in scientific publications and universities, productivity growth, increasing export of services related to digital technologies, growing R&D investments and emergence of world's top science and technology clusters.

According to GCI 2018[2] (Global Competitiveness Index) rankings, India has gained thirteen positions from a global ranking of 71st in 2015 to 58th in 2018. India has jumped five places from 63rd in 2017 to 58th in 2018, which is considered to be the largest gain by any G20 economy. India's performance on GCI 2018 is attributed to relative growth of market size, innovation capability, transport infrastructure, and business dynamism.

Despite the positive growth story in last five years and global jump in GCI and GII rankings, India still needs to make significant enhancements in its innovation and R&D ecosystem to come at par with R&D leaders like USA, UK, European countries, and Asian countries like China etc.

[1] GII 2019 has ranked 129 economies on the basis of 80 indicators including R&D investments, international patent and trademark applications, mobile-phone app creation, and high-tech exports etc. https://www.wipo.int/pressroom/en/articles/2019/article_0008.html (last accessed 31 July 2019).

[2] GCI 2018, also known as GCI 4.0 by World Economic Forum, has ranked 140 economies globally. GCI4.0 framework involves 12 main drivers of productivity including institutions, infrastructure, information and communication technology adoption, macroeconomic context, health, education and skills, product market, labor market, financial system, market size, business dynamism, and innovation capability. http://www3.weforum.org/docs/GCR2018/05FullReport/TheGlobalCompetitivenessReport2018.pdf (last accessed 31 July 2019).

China has been ranked 28th on GCI, 2018 and 14th on GII, 2019 respectively. This requires systemic focus by the government, research institutions, and companies on increasing the research contribution in global publications, motivating more and more young graduates in taking up research as a career option, enhancing the filing and registration of global patents and spending more on R&D activities. Table 1 highlights India's performance vis-à-vis other leading economies on key parameters like **R&D spending, number of research publications, number of researchers, and patents.**

Table 1. R&D performance – country level analysis (Year 2015)

Country	Researchers (millions)	Research Publications (%age share)	Patents (per million people)	R&D Investments (%age of GDP)
South Korea	0.322	4	4,451	4.15
Japan	0.661	5.8	3,716	3.47
China	1.484	20.2	541	1.9
USA	1.265	25.3	910	2.8
UK	0.259	6.9	–	1.8
Brazil	0.139	2.9	34	1.2
India	0.193	4.2	17	0.88

Source: Sivaram (2016)

The objective of this study is to gain a better understanding of the patent-driven innovation ecosystem across research institutions in India. Research institutions are considered to be one of the major drivers for stimulating the research and innovation ecosystem in a country. The similar expectation lies from premier research and innovation institutions in India. The study focuses understanding the key challenges faced by the research institutions in driving the innovation ecosystem in India?

The chapter is divided into the following sections. Section 2 highlights the research methodology including the sample of research institutions being selected for this study. This section also provides a detailed overview of patented innovations undertaken by the selected research institutions. Section 3 focuses on the key research findings and recommendations. **Finally,** Sect. 4 summarizes and concludes the research.

2 Research Methodology

The objective of this research involves understanding the key challenges and optimal choices adopted by the research institutions in India.

A detailed qualitative field study has been undertaken to understand the complexities, challenges faced and strategic actions adopted by the research institutions in India. There are many historical pieces of evidence in research and practitioner based studies, which highlight the significant role of interactive paradigm and field studies in understanding a particular phenomenon, especially where contextual factors play an important role and understanding requires interactive mode of enquiry with diverse stakeholders (Dana and Dumez 2015; Yin 2009; Dana and Dana 2005; Eisenhardt 1989). Moreover, qualitative case-based research holds relevance here because any study involving evolutionary paradigm requires multi-stakeholder interaction (Siggelkow 2007; Eisenhardt 1989).

2.1 Sampling and Data Collection

Sampling involved making a choice of government research institutions in India on the basis of the following criteria. One, research institution was India based and had done the innovation in the context of India. Second, research institution had filed a patent for the innovation in India as well as globally. Third, patent filed by the research institution involved names of researchers based out of India. Fourth, patented innovation had been under commercialization or had already been commercialized. Fifth, innovation was scalable and had potential to create significant socio-economic or environmental impact in India.

Table 2. Sampling - Research institutions in India

Research Institution	Year of Setup	Innovator	Innovator Patents	Innovation Year	Innovation	Commercialization
IIT*, Chennai	1959	Prof. T. Pradeep	>70	2016	Low-cost nano-particle water filtration system	**Research Lab:** Thematic Unit of Excellence at IIT, Chennai **Commercialization Start-up:** InnoNano Research Pvt. Ltd.
IIT*, Kharagpur	1951	Prof. Rabibrata Mukherjee & Nandini Bhandaru	>4	2014–15	Nano fabrication and Nano patterning of soft polymeric blends	**Research Lab:** Instability and Soft Patterning Laboratory (ISPL) **Commercialization:** Industry wide application in areas like electronic devices, data storage media, optoelectronic devices and cancer therapeutics
CSIR-IMTECH**	1984	–	–	2009	Low-cost thrombolytic drug to dissolve blood-clots during heart attack	Commercialization: SK-based protein licensing to Nostrum Pharma

***IIT: Indian Institute of Technology; ^JNTBGRI: Jawaharlal Nehru Tropical Botanical Garden and Research Institute;**
****CSIR-IMTECH: Council of Scientific and Industrial Research – Institute of Microbial Technology**
Source: Research Institution Websites and Discussion with Innovators at IITs

Table 2 provides a quick overview of the three leading research institutions, which were selected as a part of the research study via convenience sampling. Multiple sources of inputs were used to ensure the richness of information and to deepen the findings (Creswell 1998).

2.2 Data Analysis

The data analysis stage involved four major steps linked to data conceptualization, categorization, mapping and linking during iterative analysis (Corbin and Strauss 1990). The first step involved transcribing the interviews and focus group discussions followed by organizing all the transcripts, field notes and data collected from secondary sources (company and online) into project database (Van de Ven and Poole 1990). The second step involved finding the key themes from the project database related to challenges, actions and strategic choices adopted by the sampled research organizations during the evolution and commercialization phases (Eisenhardt 1989). These themes were developed by doing the open and axial coding (conceptualization and categorization) of the field transcripts and notes (Corbin and Strauss 1990). The concepts and categories were compared in an exploratory manner for data analysis and reduction (Miles and Huberman 1994). The third step involved enfolding the themes on the basis of comparison with the extant literature (Eisenhardt 1989). During data analysis, the focus had been on interpretive orientation. This implies focusing on understanding the phenomenon under investigation on the basis of undertaking the iterative review of the inputs independently (actors and key activities) as well as synthesizing those inputs as part of the whole (inter-relationships) (Klein and Myers 1999).

Appendix A highlights the comparative overview of the sampled research organizations from India.

3 Research Findings

A detailed level of interaction with the innovators at government research institutions in India has brought significant insights into the research and innovation ecosystem in India. Following are the research findings from the detailed analysis of the sampled research institutions in India:

3.1 Patenting Involves Significant Duration, Efforts and Costs

Research Institutions and Innovators find patenting to be a tedious process, which needs proper guidance as well as significant investment of efforts and capital. To compete globally, patenting process involves filing patents both in India and other global markets thereby increasing the lead time, costs and efforts without any assurance of tangible outcomes or economic returns. Moreover, this enhances the risk of competitive replication or substitution by better technologies. Also, research institutions in services industries especially information and communication technology believe that time and money spent on filing and getting patents across global markets for the majority of technology innovations may not yield the desired competitive advantage due to decreasing technology life-cycle and advancements. By the time, patent gets awarded, new and better technology replaces the one which has been patented thereby shifting the competitive dynamics.

Prof. Rabibrata from IIT, Kharagpur emphasized upon the need for a better patenting ecosystem in terms of stronger IP laws, availability of patent advisors for the research institutions, individual innovators and small businesses, as well as increasing number of skilled resources at patent offices to enable quick decisions and grants.

3.2 Talented Resources Are not Willing to Take up Research as a Career

Premier research institutions in India like IITs, CSIR etc. find it challenging to attract the talented resources for taking up research as a career option. Majority of the talented resources in India prefer to take up a highly paid corporate job rather than investing their time and efforts as a researcher in research-based institutions. This preference and mindset is driven by multiple factors like future career prospects, limited recognition for a researcher, uncertainty of outcomes in terms of earning potential and recognition, and lack of job stability due to risks of failure after spending long time in specific research area. Research as a career in India is driven more by patience, persistence and self-motivation of an individual.

Prof. Pradeep and Prof. Rabibrata highlighted lack of skilled and motivated researchers as a major challenge in pacing up the high quality research outcomes.

Government of India need to review and upgrade the higher education and research ecosystem by offering better incentives, rewards and recognition, and easy grants, as well as dedicated support for filing the patents, incubation, funding and commercializing the technology invention. The real value-addition will happen when researchers and Corporates will find the incentive to collaborate and commercialize the indigenous research into a sustainable business model.

3.3 Need for Advanced Research Ecosystem with Latest Technologies

Research and innovation ecosystem in India lacks quick and continuous advancement to remain at par with global institutions in terms of technology infrastructure, skill-building options, and research opportunities on latest technologies. Higher education and research institutions like IITs, Indian Institute of Sciences (IISc), All India Institute of Medical Sciences (AIIMS), CSIR organization, and public and private universities in India are relatively slow in adoption of modern technologies. The research labs are not at par with global standards. Learning and skill-building opportunities are not readily available as and when new technologies are introduced globally. Research options on latest breakthrough technologies are also not readily available in the premier research institutions in India.

This has led to a persistent challenge regarding the lack of availability of talent pool having the research aptitude, motivation, and right skills regarding clinical or non-clinical understanding, artificial intelligence and product engineering. The talented resources prefer to move into global universities for high end research opportunities rather than building their career in Indian universities or research institutions. Indian universities are picking pace in modernizing their research and innovation ecosystem but still there is a long way to go before coming at par with global research universities and institutions. Prof. Rabibrata and Dr. Pradeep echoed the presence of an invaluable research and innovation ecosystem at IITs. However, they also emphasized upon the need for rapid changeover and incorporation of futurist research and innovation culture in academic institutions.

The latest trends in India indicate the primary focus of these universities in churning out the graduates, the majority of whom are unemployable in any industry (Pratap 2018).

3.4 Commercialization of Patents Need Collaborative Interface Between Industry and Research Institutions

There are many challenges preventing the commercialization of inventions in India. Research institutions in India have been developing and patenting increasing number of inventions but the majority of those inventions fail to get converted into a commercial business model. One of the key reasons attributed to this is the lack of strong interface between industries, academic institutions and research institutions. All these entities tend to operate in silos and once any invention is made, the effective utilization and value-addition from the invention remains unknown due to lack of strong collaboration between these three entities. Dr. Pradeep and Prof. Rabibrata attributed the commercialization gap to lack of deep connect and interface between industry and research institutions right during the development or ideation stage. Once idea gets developed into an invention, then industry wants to see a prototype and lack of prototyping limits the commercialization of an invention. Dr. Pradeep became successful after years of learning and hard work, when he started focusing on developing the prototypes of his inventions. This required significant investment of time and money but helped Dr. Pradeep in licensing the technologies to industries at high valuation.

There is a need for strong industry-academia-research interface right from ideation to discovery to patenting stage. Also, there is a need for push to develop an ecosystem where there is access to funding and resources for converting the technologies/ inventions into working prototypes in shorter time frame. It is likely that a growing network of incubators and accelerators are going to play an important role in streamlining the transformation of an invention into an innovation.

3.5 Need for Early Stage Access to Funding and Incentives

The majority of researchers take up research as a career option in India purely from self-motivational perspective. Research career in India is perceived as an uncertain and risky proposition where time and efforts are not commensurate with the rewards and outcomes. Due to lack of liberal funds and incentives in a research career, most of the talented resources in India either take up professional career in high-growth industries or move to universities abroad for a research career. Research institutions in India rely on government funding for research. Limited availability of research grants and funding from the government has an adverse impact on the research ecosystem. The morale of the researchers gets impacted and they are less likely to work on advanced research areas due to high risks of failure and limited funding options.

There is a need for government to review the research ecosystem and enable a steady stream of funding and incentives to attract high-end talent pool into research career as well as to motivate the researchers in taking up complex issues as potential research areas.

4 Conclusion

There is no denying the fact that India has made significant progress on global innovation and competitiveness rankings. However GCI 2018 ranking of India (58th) versus China (28th) and GII 2019 ranking of India (52nd) versus China (14th) brings forth the fact that India still has a long way to go before gaining a significant position in terms of innovation and competitiveness. Research institutions in India are showing an increasing orientation towards enhancing their innovation capacity and capability. The same is evident from the progress made by India in the annual US Chambers of Commerce Intellectual Property (IP) Index, 2019[3]. India has climbed by eight positions from 44th rank in 2018 to 36th rank in 2019 on Intellectual Property Index. This index involves selection of 50 countries, representing 90% of the global GDP and ranking of these economies on the basis of 45 indicators related to patent, trademark, copyright and trade secrets protection. The improvement in global IP Index 2019 ranking is attributed to large-scale adoption of digitization and wide-spread policy reforms instituted by the Indian government across academic and research institutions in India (HBL 2019). Key reforms implemented by government of India involves accession to the WIPO Internet Treaties, announcing a dedicated set of IP incentives for small businesses and academic institutions, and administrative reforms for simplifying the global patenting process. However, India still lies in the bottom 30% of the top 50 countries in terms of IP ecosystem. The key issues, which need to be tackled for making a significant improvement in global IP rankings, include licensing barriers, technology transfer complexities, inflexible registration requirements, lengthy pre-grant opposition proceedings, non-availability of funding, weak collaboration or interface between industry and research institutions, and lack of access to patenting advisory for the small-scale businesses, research institutions and individuals.

[3] https://www.uschamber.com/press-release/us-chamber-releases-2019-international-ip-index (last accessed 2-August-2019).

Appendix A

	IIT Chennai	IIT Kharagpur	CSIR-IMTECH
Overview	Prof. T. Pradeep identified arsenic contaminated water as a huge challenge for the society and decided to develop a low cost, easily deliverable, serviceable, and environment friendly technology Thematic Unit of Excellence (TUE) set up at IIT Madras in 2008, designed a system which filtered out Arsenic in real-time InnoNano Research Pvt. Ltd. signed an agreement in 2016 for USD 18 million with US-based energy and water investment firm to scale the research globally	Prof. R. Mukherjee and his research student's leveraged nano patterning expertise to develop nanostructure based surfaces having self-cleaning and cancer remedial properties Nanotechnology innovation provided a low-cost, water and oil repellant alternative having self-cleaning properties and real-world application in areas like optoelectronic devices, plastic solar cells, biological scaffolds, mobile phones and cancer therapeutics etc.	IMTECH recognized cardio-vascular diseases as one of the key focus areas in 1992 and launched a research program to develop a novel drug for dissolving the blood clot with maximum efficacy and minimal side-effects IMTECH launched a novel hybrid SK-based protein, filed global patents for the same and licensed the technology to Nostrum Pharma during 2009
Need Addressed	Non-Availability of Clean Drinking Water	Non-availability of low-cost technology having self-cleaning and cancer remedial properties	Lack of access to affordable and efficient treatment for dissolving life-threatening blood clots during heart attack
Innovator	Prof. T. Pradeep	Prof. R. Mukherjee & Dr. Nandini Bhadaru	–
Innovation	Design of affordable Nano-particle filtration systems for water Purification. Filters pesticides and minerals like arsenic, fluoride, iron, mercury etc.	Developed Nano patterning technique to manufacture Nano materials on a nanometer scale	Affordable and highly effective thrombolytic drugs for dissolving the blood clots with-in six hours of heart attack
Value Offering	Affordable Nano filtration solution for clean drinking water in semi-urban and rural areas	Cost-effective Nano-technology to fabricate Nano-scale patterned surfaces having self-cleaning properties	Affordable and highly effective thrombolytic drugs for cardio-vascular issues

(*continued*)

(*continued*)

	IIT Chennai	IIT Kharagpur	CSIR-IMTECH
Challenges	Lack of Innovation culture in majority of academic institutions and universities Non-encouraging socio-cultural mindset towards research as a career option in India Lack of institutional support for faster commercialization of patented technologies in academic Institutions	Infrastructure limitations for doing cutting-edge research in academic institutions Limited institutional support and guidance for patenting and commercializing research outcomes in academic institutions Lack of willingness among youth for taking up research as a career	Delays in availability of government funding for research Limited opportunities for collaboration with private enterprises for research based innovations Time and cost investments in filing global patents
Enablers for Success	Focused approach towards water related issues Leveraged the IIT Ecosystem for research and funding Strong research background Prototype first, full solution later Self motivation, persistence and doer attitude	Focused approach towards Nano-patterning Leveraged the IIT Ecosystem for getting top talent for research Strong research background Self motivation	Integrated research outcomes with global collaborations and licensing opportunities R&D Institutions of national importance play an important role

Source: Research Institution Websites and Discussion with Innovators at IITs.
IMTECH: https://www.wipo.int/ipadvantage/en/details.jsp?id=2916; https://www.who.int/en/news-room/fact-sheets/detail/cardiovascular-diseases-(cvds); https://www.imtech.res.in/achievements/patents; https://www.imtech.res.in/about/about-IMTECH
IIT, Chennai: Prasad (2016); http://www.dstuns.iitm.ac.in/t-pradeep.php#researchprojects; http://www.dstuns.iitm.ac.in/filesdec2015/1.%20Amrit%20Brochure.pdf; http://www.dstuns.iitm.ac.in/AMRIT%20Report.pdf
IIT, Kharagpur: https://sites.google.com/site/rmresearchgroup/; IANS (2016); Subramanian (2017)

References

Corbin, J., Strauss, A.: Grounded theory research: procedures, canons, and evaluative criteria. Qual. Sociol. **13**(1), 3–21 (1990)

Creswell, J.W.: Qualitative Inquiry and Research Design: Choosing Among Five Traditions. Sage Publications, Thousand Oaks (1998)

Dana, L.P., Dana, T.E.: Expanding the scope of methodologies used in entrepreneurship research. Int. J. Entrep. Small Bus. **2**(1), 79–88 (2005)

Dana, L.P., Dumez, H.: Qualitative research revisited: epistemology of a comprehensive approach. Int. J. Entrep. Small Bus. **26**(2), 154–170 (2015)

Eisenhardt, K.: Building theories from case study research. Acad. Manag. Rev. **14**(4), 532–550 (1989)

HBL: India vaults eight slots to 36th rank in IP index. The Hindu Business Line, 7 February 2019. https://www.thehindubusinessline.com/economy/indias-ip-ranking-goes-up-eight-notches-to-36-in-us-chambers-index/article26203662.ece. Accessed 2 Aug 2019

IANS: IIT-Kharagpur student Nandini Bhandaru bags European Materials Research Society award. Indian Today, 17 May 2016. https://www.indiatoday.in/education-today/news/story/nandini-bhandaru-323958-2016-05-17. Accessed 3 Aug 2019

Klein, H., Myers, M.: A set of principles for conducting and evaluating interpretive field studies in information systems. MIS Q. **23**(1), 67–93 (1999)

Miles, M.B., Huberman, A.M.: Qualitative Data Analysis: An Expanded Sourcebook. Sage Publications, London (1994)

Prasad, R.: With $18 million funding, IIT Madras professor breaks the glass ceiling. The Hindu Business Line, 17 July 2016. https://www.thehindu.com/sci-tech/science/With-18-million-funding-IIT-Madras-professor-breaks-the-glass-ceiling/article14492695.ece. Accessed 3 Aug 2019

Pratap, R.: Worthless degrees and jobless graduates. The Hindu Business Line, 27 January 2018. https://www.thehindubusinessline.com/news/education/worthless-degrees-and-jobless-graduates/article9660619.ece. Accessed 3 Aug 2019

Siggelkow, N.: Persuasion with case studies. Acad. Manag. J. **50**(1), 20–24 (2007)

Sivaram, S.: India's Spend on Science, Technology and Innovation: The Budget 2016-17 (2016). http://www.swaminathansivaram.in/media/data/INDIA%E2%80%99S%20SPEND%20ON%20SCIENCE,%20TECHNOLOGY-May-2016.pdf. Accessed 1 Aug 2019

Subramanian, A.: A Window That Cleans Itself? A Material That Doesn't Need Dusting? IIT Students Have Made It Possible! The Better India, 28 January 2017. https://www.thebetterindia.com/84133/iit-students-nanotechonology-window/. Accessed 3 Aug 2019

Van de Ven, A.H., Poole, M.S.: Methods for studying innovation development in the Minnesota Innovation Research Program. Organ. Sci. **1**(3), 313–334 (1990)

Yin, R.K.: Case Study Research: Design and Methods, 4th edn. Sage Publications Inc., California (2009)

Conclusions

A systemic view on the issues of the influence of technological progress on the socio-economic problems in the 21st century, presented in this volume, allowed determining a contradiction of this influence, which consists in the fact that, on the one hand, the tendencies of intellectualization, digitization, and innovative development of the modern global economy are to reduce the inequality of developed and developing countries and to overcome the disproportions in the level and rate of their technological progress.

On the other hand, developed and developing countries are characterized by different level of involvement into the Fourth industrial revolution and have various capabilities in the sphere of transition to Industry 4.0. Due to the differences in the conditions (social environment, investment and business climate, and global competitiveness), digital modernization of economies of developed and developing countries is developing according to different scenarios. As a result, differentiation of the participants of the global economic system increases, and disproportions in development of the global economy deepen.

The contradiction is also connected to the fact that acceleration of technological progress in early 21st century, which led to the start of the Fourth industrial revolution, was initiated by the interests of increasing the sustainability of the modern global economy. Global goals in the sphere of sustainable development, adopted by the UN in 2015 and supported by the participants of the global economic relations, include overcoming of famine in the world and provision of food security, protection of environment, and creation of favorable (inclusive) social environment with equal opportunities for everyone.

Technologies of Industry 4.0 do stimulate the realization of global goals in the sphere of sustainable development, allowing developing hi-tech and agricultural production, which is peculiar for high efficiency regardless of the natural and climate conditions. Secondly, Industry 4.0 offers new "green" technologies, which allow reducing the volume of consumption of natural resources and minimizing production waste, thus protection the environment. Thirdly, creation of AI allows overcoming the inequality in human society and putting humans into equal

E. G. Popkova and B. S. Sergi (Eds.): ISC 2019, LNNS 91, pp. 691–692, 2020.
https://doi.org/10.1007/978-3-030-32015-7

conditions, opposing them to intelligent machines. “Smart” homes and “smart” cities ensure full inclusion and favorable environment for life and self-realization of each human.

However, apart from the above advantages, the socio-economic systems of the Fourth technological mode have new problems, risks, and threats. One of them is provision of digital security for protection of digital data and information. Another is preservation of uniqueness of innovative technologies, knowledge, and information and protection of intellectual property and measuring the level of innovativeness and stimulating the creation and implementation of true innovations that stimulate new organizations of production and consumption and manufacture of new goods and services.

Thus, this volume raised new questions and actualized new problems of modern economics, social sciences, and law, and showed the gaps in the system of the existing scientific knowledge and perspectives for further research. These perspectives are connected to the search for the means of harmonizing the interests of various participants of domestic and international economic relations and developing the scientific and methodological provision of systemic management of the processes of intellectualization, digitization, and innovative development of socio-economic systems in the 21st century.

Elena G. Popkova
Leading Researcher of the Chair of Management Theories
and Business Technologies
Plekhanov Russian University of Economics
Moscow
Russian Federation

Author Index

A
Aboimova, Irina S., 142
Abrosimova, Anna A., 570
Ageeva, Elena L., 446
Akopova, Elena S., 353
Alimbekova, Anastasia S., 96
Alimov, Sergey A., 561
Alimova, Gulbar B., 484
Alimova, Maria S., 561
Ambartsoumyan, Arthur E., 316
Andreeva, Elena V., 77
Andryashina, Natalia S., 163, 170, 298, 670
Artashina, Irina A., 156
Astafyev, Igor V., 254
Atabekova, Nurgul K., 415

B
Bakhtiyarova, Lyudmila N., 460
Bardovskii, Viktor P., 510
Batsyna, Yana V., 316
Baturina, Nataljya A., 282
Belik, Elena B., 239
Belinova, Natalia V., 422
Belokurova, Elena V., 544
Belyaeva, Tatyana K., 430
Belyakov, Vladimir I., 30
Beskorovaynaya, Natalia S., 208
Bezgina, Olga, 639
Bicheva, Irina B., 422
Borisov, Sergey A., 316
Borisova, Vera V., 631
Borzenko, Ksenia V., 353
Buevich, Angelika P., 397

C
Chaikina, Zhanna V., 275
Chauhan, Sumedha, 680
Chelnokova, Elena A., 275
Cherney, Olga T., 275

D
Demeneva, Nadezhda N., 439
Deryugina, Tatyana V., 372
Digilina, Olga B., 503
Dotdueva, Zukhra S., 208
Dudina, Elena V., 282
Dzyuba, Elena M., 477

E
Egorov, Evgeniy E., 430
Egorova, Anastasia O., 149
Egorova, Elena N., 346
Elsukova, Yuliana Yu., 179
Epinina, Veronica S., 104, 191
Ermakov, Ilya V., 208
Ermolina, Lilia V., 346
Evchenko, Olga, 639

F
Filatova, Olga N., 452

G
Garin, Alexander P., 156, 298
Garina, Ekaterina P., 163, 170, 298, 654, 670
Gashenko, Irina V., 247
Giyazov, Aydarbek, 415
Golemi, Ela, 590, 608, 615

E. G. Popkova and B. S. Sergi (Eds.): ISC 2019, LNNS 91, pp. 693–696, 2020.
https://doi.org/10.1007/978-3-030-32015-7

Goncharov, Alexander I., 3
Goncharuk, Aleksey G., 119
Gordeeva, Irina A., 446
Goyal, Sandeep, 680
Grazhdankina, Lilianna Y., 219
Grigoryeva, Mariya O., 264
Grinevich, Julia A., 134
Gritsai, Olga V., 24
Gruzdeva, Marina L., 275
Gryaznova, Elena V., 119
Gulzat, Kantoro, 339
Gutsu, Elena G., 439

H
Hizhnaya, Anna V., 422

I
Ilchenko, Natalia M., 477
Ilin, Ivan V., 510
Inshakova, Agnessa O., 3, 372
Inyushkin, Alexey N., 30
Inyushkin, Andrey A., 30
Ismanaliev, Kurmanbek I., 229
Ivanova, Tatiana, 639
Ivashchenko, Tatyana N., 14

K
Kalaj, Ermira, 608
Kamerilova, Galina S., 446
Kapoor, Amit, 680
Karnaukhov, Igor A., 387
Karpova, Anna A., 239
Karpushova, Svetlana E., 68
Kartavykh, Marina A., 446
Kashina, Oksana I., 583
Katkova, Marina A., 96
Kayl, Iakow I., 104, 179, 191
Khalikova, Elvira A., 83
Khanova, Tatyana G., 422
Kharlamov, Maxim M., 654
Khartanovich, Konstantin V., 291
Khizhnaya, Anna V., 452
Khokhlova, Elena V., 208
Kitsay, Yuliana A., 346
Klimovets, Olga V., 54
Kochetova, Elena V., 439
Kolesova, Oksana V., 439
Kolmakova, Ekaterina M., 534
Kolmakova, Evgenia M., 534
Kolmakova, Irina D., 534
Kolmykova, Tatyana S., 654
Konina, Olga V., 495
Konstantinov, Victor A., 561
Kostyrev, Andrey P., 576
Kota, Vasilika, 590
Kotlyarova, Vera V., 24
Kozlova, Elena P., 663, 670
Krasilnikova, Larisa V., 422
Krupoderova, Elena P., 460
Krupoderova, Klimentina R., 460
Kryukova, Elena S., 30
Kulagina, Alexandra A., 142
Kulueva, Chinara R., 229, 484
Kuryleva, Olga I., 149
Kuznetsov, Victor P., 163, 170, 229, 484, 663, 670
Kuznetsova, Natalya P., 264
Kuznetsova, Svetlana N., 663
Kvintyuk, Yurij M., 104

L
Lagunova, Marina V., 468
Lamzin, Roman M., 104, 179, 191
Lang, Vitaly V., 208
Lapygin, Denis Yu., 306
Lapygin, Yuri N., 306
Latuhina, Anna L., 477
Ledeneva, Marina V., 521
Leibert, Tatyana B., 83
Letiagina, Elena N., 134
Lezer, Viktoria A., 387
Litvinova, Alla V., 521
Litvinova, Tatiana N., 624
Lvova, Maria V., 142

M
Makarenko, Elena N., 247
Makarov, Igor A., 583
Makarov, Pavel Yu., 306
Makarova, Svetlana D., 583
Makusheva, Yuliya A., 576
Malikov, Evgeny Y., 372
Maltceva, Svetlana M., 119
Markova, Svetlana M., 452
Maslova, Tatyana E., 570
Mayasova, Tatyana V., 439
Menshchikova, Vera I., 38
Merkulova, Elena Yu., 38
Mikhailova, Ekaterina V., 24
Milenky, Alexander V., 291
Mironenko, Nadezhda V., 14
Mosina, Ekaterina I., 282
Mukhomorova, Irina V., 415
Muraveva, Nataliia N., 219

Muzayev, Mahomed Z., 170
Myrzaibraimova, Inabarkan R., 484

N
Neizvestnykh, Anna A., 77
Neprokina, Irina V., 407
Nesenyuk, Evgenia S., 654
Novikova, Irina V., 219

O
Onour, Ibrahim A., 325
Oveshnikova, Lyudmila V., 264

P
Pakhomova, Anna I., 544
Pastushenko, Elena N., 96
Patsyuk, Elena V., 68
Petrenko, Elena S., 239
Petrushkina, Anna V., 47
Pisarev, Georgy A., 239
Plakhova, Liubov V., 510
Pokrovskiy, Nikolay V., 510
Ponachugin, Aleksandr V., 460
Popkova, Elena G., 339
Popov, Vasily V., 96
Popovicheva, Natalya E., 14
Potapova, Tatyana K., 430
Potashnik, Yaroslav S., 156, 663
Pronchatova-Rubtsova, Natalia N., 583
Przhedetskaya, Natalia V., 353

Q
Qerimi, Qerim, 360

R
Rakhimova, Bella K., 631
Romanovskaya, Elena V., 163, 170, 229, 484, 670
Ruzanova, Valentina D., 30
Rybakova, Svetlana V., 96

S
Safargaliev, Mansur F., 346
Salikov, Daniil A., 3
Samerkhanova, Elvira K., 460
Samsonova, Irina V., 77
Samsonova, Marina V., 191
Sapfirova, Apollinariya A., 47
Sedykh, Catherine P., 452
Seitov, Bolotbek M., 298
Semenov, Sergey V., 149
Semenova, Elena E., 282
Shabaltina, Larisa V., 544
Shabanova, Tatyana L., 430
Shabatura, Liubov N., 387
Shalabaev, Pavel S., 570
Shalaev, Vladislav A., 552
Sharipov, Bahrom K., 125
Shcherbak, Natalia A., 316
Shchukina, Natalia V., 68
Sherbakova, Elena E., 142
Sheveleva, Tatyana N., 477
Shevyakova, Anna L., 552
Shlyakhov, Mikhail Y., 430
Shokhnekh, Anna V., 179, 191
Shpilevskaya, Elena V., 163
Shulga, Andrey V., 61
Shulus, Aleksei A., 353
Sibirskaya, Elena V., 264
Sineva, Nadezhda L., 149
Skriabin, Vasilii V., 77
Smirnova, Zhanna V., 156, 275
Sokolov, Dmitry P., 254
Spiridonov, Sergey P., 38
Starikova, Tatyana V., 291
Stepanova, Maria A., 282
Strelkova, Lyudmila V., 576
Suanov, Vladimir M., 125
Surkova, Olga A., 68
Syrbu, Anzhelika N., 104, 179

T
Talalaeva, Natalya S., 521
Tashkulova, Gulzat K., 415
Tashmurzaeva, Gulzat T., 298
Tasueva, Tamila S., 631
Tenetova, Evgeniya P., 264
Terskaya, Galina A., 397
Teslenko, Irina B., 503
Tikhonyuk, Natalya E., 291
Tolsteneva, Alexandra A., 468
Tolstykh, Tatiana O., 654
Trifonov, Yuri V., 570
Trifonov, Yury V., 134
Trifonova, Elena Y., 134
Trofimov, Oleg V., 576
Trofimova, Tatyana V., 576
Tronina, Larisa A., 219
Tsymbalov, Sergey D., 663
Tsyplakova, Svetlana A., 452

U
Ubaidullayev, Mirlanbek B., 229

V
Valinurova, Liliya S., 83
Varaksa, Natalia G., 561
Varvus, Svetlana A., 397

Vatyukova, Oksana Y., 552
Vechkinzova, Elena A., 552
Veryaskina, Marina A., 446
Vilkina, Marina V., 54
Vinnik, Valeria K., 468
Vizgunov, Alexander N., 134, 570
Vladimirov, Alexander A., 119
Volkova, Victoria V., 47
Voronkova, Anna A., 468

Y
Yakushin, Vladimir A., 61
Yalmaev, Rustam A., 544
Yashin, Sergey N., 316
Yashina, Nadezhda I., 583
Yashkova, Elena V., 149
Yudin, Andrei V., 24
Yusupov, Ramil' Z., 24

Z
Zabaznova, Tatiana A., 68
Zaitseva, Svetlana A., 156
Zakharkina, Natalia V., 510
Zakharova, Victoria T., 477
Zanozin, Nikolai V., 119
Zhdankin, Nikolay A., 125
Zhelnakova, Nina Y., 219
Zhilina, Natalia D., 468
Zima, Yulia S., 247
Zimina, Evgenia K., 142
Zviagintceva, Yuliia A., 14

Printed in the United States
By Bookmasters